G. Matthess F. Frimmel P. Hirsch
H.D. Schulz H.-E. Usdowski (Eds.)

Progress in Hydrogeochemistry

Organics – Carbonate Systems – Silicate Systems – Microbiology – Models

Foreword by D. Maronde

With 285 Figures
and 108 Tables

Springer-Verlag
Berlin Heidelberg New York
London Paris Tokyo
Hong Kong Barcelona
Budapest

Professor Dr. G. Matthess
Geologisch-Paläontologisches Institut
Universität Kiel
Olshausenstraße 40–60
W-2300 Kiel 1, FRG

Professor Dr. F. Frimmel
Engler-Bunte-Institut
Technische Universität Karlsruhe
Richard-Willstätter-Allee 5
W-7500 Karlsruhe, FRG

Professor Dr. P. Hirsch
Institut für Allgemeine Mikrobiologie
Universität Kiel Biozentrum
Am Botanischen Garten 19
W-2300 Kiel, FRG

Professor Dr. H.D. Schulz
FB 5 – Geowissenschaften
Universität Bremen
Klagenfurter Straße
W-2800 Bremen 33, FRG

Professor Dr. H.-E. Usdowski
Sedimentopetrographisches
Institut
Universität Göttingen
Goldschmidtstraße 1
W-3400 Göttingen, FRG

ISBN-13:978-3-642-76631-2 e-ISBN-13:978-3-642-76629-9
DOI: 10.1007/978-3-642-76629-9

Library of Congress Cataloging-in-Publication Data

Progress in hydrogeochemistry: organics, carbonate systems, silicate systems, microbiology, models/G. Matthess ... [et al.], (eds.) p. cm. "This volume summarizes the main results of a priority programme of the Deutsche Forschungsgemeinschaft (DFG), Bonn-Bad Godesberg" – Pref. Includes bibliographical references and index
ISBN-13:978-3-642-76631-2 (U.S.)
1. Hydrogeology – Congresses. 2. Water chemistry – Congresses. I. Matthess, Georg. II. Deutsche Forschungsgemeinschaft. GB1001.2.P77 1991 551.48–dc20 91–22482 CIP

Softcover reprint of the hardcover 1st edition 1992

Typesetting: International Typesetters, Inc., Manila, Philippines
32/3145-5 4 3 2 1 0 – Printed on acid-free paper

1 Foreword

Earth scientists, who have worked together for 6 years in the priority program "Hydrogeochemical Processes in the Hydrological Cycle Within the Unsaturated and Saturated Zones", have summarized the results of their research in this volume. This is the occasion to take stock and then look ahead. The priority program was set up by the Senate of The Deutsche Forschungsgemeinschaft in October 1981. This was preceded by lengthy and careful preparation by a Program Committee, and was finally recommended by the Senate Commission for Joint Research in Earth Sciences. The main aim was the interdisciplinary research of geochemical processes in natural systems in the total underground water cycle, whereby water pollution was not to be considered.

Officially started on 1 September 1982, the program has received a grant of DM 10.9 Mio. from the DFG, and this has enabled it to support a total of 50 projects. Although at the beginning, practice-oriented projects, which were funded elsewhere, were not to be included, many of the results are applicable. The four categories presented were investigated with varying intensity. An early concentration on certain representative fields of measurement and research areas proved to be the right approach – this avoided a waste of effort in other fields. During the period of the priority program, new and topical questions arose, e.g. in context with the research on the influence of microbacteria, the interaction of humic substances, as well as with the development and application of hydrogeochemical and geochemical-hydraulic models. This was one of the main reasons why the program could be extended into the 6th year.

Cooperation in the priority program was strengthened by annual workshops, in which all the scientists involved in the program participate. In addition, there were also working group meetings, where results and future plans could be discussed, as well as many meetings in the field. It is but natural that at the end of the program, not all the questions originally raised can be answered. Many new problems arose and have to be worked on in the future. For example, a number of working groups from this priority program have been involved for quite a while in another priority program "Pollutants in Ground Water" and several scientists working in the priority program are keen on participating in other interdisciplinary projects in the future.

The major part of the grant – a DFG average of 75% – was utilised for the financing of personnel. In contrast, there was a marked decrease in expenses for equipment, working material and travel. This enabled many young scientists to complete their doctorates within the projects and some were later even able to work there in a postdoctorate capacity.

The published results speak for the success of the program. While many lectures given at congresses and workshops, as well as various publications informed in detail about progress made in the projects, this volume summarizes most of the research undertaken by the participants and includes some special contributions. It is due to the intensive efforts and active support from the group of expert reviewers and the editing committee, to whom we owe our heartfelt thanks. Special thanks go to the coordinator, Prof. Georg Mattheß and his helpers, without whose untiring efforts the program could not have been successfully completed. It is to be hoped that even in the future, cooperation will be a basis for further interdisciplinary projects between earth scientists and scientists from related fields like chemistry, hydrology, microbiology, physics and soil science.

Bonn-Bad Godesberg *Dietrich Maronde*

Preface

This volume summarizes the main results of a priority programme of the Deutsche Forschungsgemeinschaft (DFG), Bonn-Bad Godesberg, on *Hydrogeochemical Processes in The Hydrological Cycle Within the Unsaturated and Saturated Zones*, in which interdisciplinary projects of scientific institutions and single scientists were involved and financially supported. Between 1982 and 1989, annual workshops were held at Bonn-Bad Godesberg or Neustadt/Weinstraße which offered opportunities for intensive discussions of preliminary results and for exchange of methodological experiences.

During the last workshop in February 23/24, 1989, at Kiel, the results were publicly presented in lectures and posters to an international scientific forum. Invited speakers contributed overviews of the state of the art as bases for critical discussion of the new achievements. The participants of the programme have to thank Professor M. Schnitzer (Land Resources Center, Agriculture Canada, Ottawa), Professor P. Deines (Dept. of Geochemistry, Pennsylvania State University, State College, Pa.), Dr. W.M. Edmunds (Institute of Hydrology, Wallingford, UK), Professor D.L. Balkwill (Dept. of Biological Sciences, Florida State University, Tallahassee, Fla.), Professor W. Kinzelbach (Technical Hydraulics and Engineering Hydrology, Department of Civil Engineering, Gesamthochschule – University Kassel), and Dr. Tomas Paces (Geological Survey of Czechoslovakia, Prague) for their valuable contributions to this event.

It is quite normal that a research programme which lasts for up to 10 years should lead to a number of results which have to be published immediately. Therefore this volume is a mixture of review-type chapters with references to already published experimental data and research publications on work which has been done recently.

The contributions presented here were reviewed by international experts for the respective chapters. We would like to express our very sincere thanks to Professor P. Fritz, Professor P. Deines, Dr. W.M. Edmunds, Professor W.C. Ghiorse (Cornell University, Ithaca, N.Y.), Dr. T. Paces, and Professor E.M. Perdue (Georgia Institute of Technology, Atlanta, Ga.) for their help in the preparation and improvement of chapters of this book.

The participants of the priority programme thank the DFG (Deutsche Forschungsgemeinschaft), Bonn-Bad Godesberg, for the generous funding of our programme, which enabled the interdisciplinary approach. Stimulating scientific discussion and team work became possible by the DFG providing adequate platforms, such as specific working groups, or the annual workshops mentioned above. We especially thanks Drs. D. Maronde and W. Eder, who have accompanied our efforts by helpful advice and active help.

Our work could not have succeeded without the willing help of the board of scientific programme reviewers Professors E. Althaus (Karlsruhe), G. Einsele (Tübingen), E. Jackwerth (Bochum), H. Kobus (Stuttgart), B. Meier (Göttingen), K.O. Münnich (Heidelberg), E. Thofern (Bonn), and G. Tölg (Bochum). This board supported the programme by criticism and many helpful suggestions.

A highlight during the 6 years' programme was the team work and friendships which became established between different scientific institutions and single researchers. As an obvious effect of this, cross references and information on help from co-participants can be found in different sections of this volume. All these authors are particularly grateful for the various help provided by the respective laboratories and scientists.

At July 1, 1990 Professor Reinhard Schweisfurth passed away unexpectedly after a heart attack. We all feel sorrow for the loss of an honest, creative and enthusiastic partner of scientific research in the subsurface ecosystems. We have to thank him for all his engagement to this work.

On behalf of the participants
of the priority programme

G. Matthess, coordinator

Contents

3 Carbonate Systems

6 Hydrogeochemical and Geochemical-Hydraulic Models and Model Concepts

Authors

Abbt-Braun, Dr. Gudrun, Engler-Bunte-Institut Universität Karlsruhe, Richard-Willstätter-Allee 5, Postfach 6380, W-7500 Karlsruhe, FRG

Alexander, Dr. Irmgard, Waldfriedhofstr. 10, W-8000 München 70, FRG

Amberger, Annette, Lehrstuhl für Allgemeine Chemie und Biochemie, Technische Universität München, W-8050 Freising-Weihenstephan, FRG

Bauhus, Dr. Wilhelm, Forschungstransfer, Dezernat 5.4, Westfälische Wilhelms-Universität Münster, Königstraße 22/23, W-4400 Münster, FRG

Baumann, Dr. Joachim, Bundesanstalt für Geowissenschaften und Rohstoffe, Stilleweg 2, Postfach 510153, W-3000 Hannover 51, FRG

Becker, Dr. Renate, Badische Anilin- und Sodafabrik, W-6703 Limburgerhof, FRG

Böttcher, Dr. Jürgen, Niedersächsisches Landesamt für Bodenforschung, Stilleweg 2, W-3000 Hannover 51, FRG

Brühl, Dr. Hanskarl, Prof., FU Berlin, FB Geowissenschaften, Institut für Angewandte Geologie, Wichernstr 16, W-1000 Berlin 33, FRG

Buddrus, Dr. Joachim, Institut für Spektrochemie und angewandte Spektroskopie, Postfach 101 352, W-4600 Dortmund 1, FRG

Bugner, Carola, Dipl.Ing., Institut für Wasserwirtschaft und Hydrologie, Callinstr. 32, W-3000 Hannover, FRG

Burba, Dr. Peter, Institut für Spektrochemie und angewandte Spektroskopie, Postfach 101 352, W-4600 Dortmund 1, FRG

Cevrim, Dr. Mustafa, Zweckverband Verbunddeponie Bielefeld-Herford, Amtshausstr. 1b, W-4900 Herford, FRG

Cordt, Dr. Thomas, Institut Fresenius, Im Maisel 14, W-6204 Taunusstein, FRG

Dahmke, Dr. Andreas, FB 5 Geowissenschaften, Universität Bremen, Postfach 330 440, W-2800 Bremen 33, FRG

Dietze, Dr. Gabriele, Geologisches Landesamt Baden-Württemberg, Albertstr. 5, W-7800 Freiburg, FRG

Dörr, Dr. Helmut, Institut für Umweltphysik, Universität Heidelberg, Im Neuenheimer Feld 366, W-6900 Heidelberg, FRG

Dreybrodt, Dr. Wolfgang, Prof., Universität Bremen, FB 1-Physik/ Electrotechnik, Bibliothekstr. Postfach 330440, W-2800 Bremen 33, FRG

Duynisveld, Dr. Wilhelmus H.M., Bundesanstalt für Geowissenschaften und Rohstoffe, Stilleweg 2, W-3000 Hannover 51, FRG

Ebhardt, Dr. Götz, Prof., Geologisch-Paläontologisches Institut der TH Darmstadt, Schnittspahnstr. 9, W-6100 Darmstadt, FRG

Echle, Dr. Wolfram, Prof., Institut für Mineralogie und Lagerstättenlehre, Rheinisch-Westfälische TH Aachen, Wüllnerstr. 2, W-5100 Aachen, FRG

Eichinger, Dr. Lorenz, Fa. Hydroisotop, Birkenstr. 7a W-8053 Attenkirchen, FRG

Finger, Dr. Wilfried, Fachbereich Chemie, Universität Marburg, Lahnberge, W-3550 Marburg 7, FRG

Forster, Dr. Markus, Institut für Hydrologie, Ingolstädter Landstraße 1, W-8042 Neuherberg, Post Oberschleißheim, FRG

Freitag, Dr. Gertrud, Technische Werke Stadt Stuttgart AG, Lautenschlagerstr. 21, W-7000 Stuttgart 1, FRG

Frimmel, Dr. Fritz H., Prof., Engler-Bunte-Institut, Universität Karlsruhe, Richard-Willstätter-Allee 5, Postfach 6380, W-7500 Karlsruhe, FRG

Fritsch, Dr. Peter, Geologisch-Paläontologisches Institut, TH Darmstadt, Schnittspahnstr. 9, W-6100 Darmstadt, FRG

Goebelbecker, Jürgen, Dipl.-Mineraloge, Kernforschungszentrum, Postfach, W-7500 Karlsruhe, FRG

Gottfreund, Dr. Joachim, Dipl.-Biol., Bachstr.2,
W-7526 Zentern, FRG

Grossmann, Dr. Jörg, Ingenieurbüro WATEC, Marktplatz 5,
W-8015 Markt Schwaben, FRG

Hack, Dr. Alfons, Institut für Hygiene, Ruhr-Universität Bochum,
Universitätsstraße 150, W-4630 Bochum 1, FRG

Heumann, Dr. Klaus Gustav, Prof., Institut für Anorganische Chemie,
Universität Regensburg, Universitätsstr. 31, W-8400 Regensburg,
FRG

Hirsch, Dr. Peter, Prof., Institut für Allgemeine Mikrobiologie,
Universität Kiel, Olshausenstr. 40, W-2300 Kiel, FRG

Hötzl, Dr. Heinz, Prof., Institut für Mineralogie und Geologie
Universität Karlsruhe, Kaiserstr. 12, W-7500 Karlsruhe, FRG

Hoffmann, Dr. Marius, Stettiner Str. 6, W-5010 Bergheim, FRG

Jacob, Dr. Helmut, Institut für Umweltphysik, Universität Heidelberg,
Im Neuenheimer Feld 366, W-6900 Heidelberg, FRG

Kölbel-Boelke, Dr. Jutta, Bruker-Franzen Analytik GmbH,
Kattenturmertleer Str. 122, W-2800 Bremen, FRG

Klamberg, Dr. Horst, Prof., Fachbereich Chemie, Universität Marburg,
W-3550 Marburg/Lahn 7, FRG

Kölle, Dr. Walter, Institut für Wasserforschung/Stadtwerke Hannover AG,
Postfach 5747, W-3000 Hannover 1, FRG

Kölling, Dr. Martin, FB 5 Geowissenschaften, Universität Bremen,
Postfach 330 440, W-2800 Bremen 33, FRG

Kössl, Dr. Hartwig, Institut für Straßenbau und Eisenbahnwesen,
Universität Karlsruhe, Kaiserstr. 12, W-7500 Karlsruhe, FRG

Kußmaul, Dr. Horst, Prof., Institut Fresenius, Im Maisel,
W-6204 Taunusstein 4, FRG

Langer, Bettina, Dipl.-Biologin, Hygiene-Institut Universität Bonn,
Sigmund Freud-Str. 25, W-5300 Bonn 1, FRG

Langguth, Dr. Horst-Robert, Prof., Lehrgebiet für Ingenieurgeologie
und Hydrogeologie, Rheinisch-Westfälische TH Aachen,
Brunnenstr. 5, W-5100 Aachen, FRG

Larue, Jürgen, Dipl.-Geol., Pastor-Lüpschen-Str. 10, W-5160 Düren, FRG

Leuchs, Dr. Wolfgang, Landesamt für Wasser und Abfall NW, Auf dem Draap 25, W-4000 Düsseldorf 1, FRG

Loehnert, Dr. Ekkenhard P., Prof., Geologisch-Paläontologisches Institut, Westfälische Wilhelms-Universität Münster, Gievenbecker Weg 61, W-4400 Münster, FRG

Loosli, Dr. Heinz H., Physikalisches Institut Bern, Sidlerstr. 5, 3012 Bern, Switzerland

Matthess, Dr. Georg, Prof., Geologisch-Paläontologisches Institut, Universität Kiel, Olshausenstr. 40/60, W-2300 Kiel 1, FRG

Merkel, Dr. Broder, Institut für Wasserchemie und Chemische Balneologie, Marchioninistr. 17, W-8000 München 70, FRG

Michaelis, Dr. Jochen, Sedimentpetrographisches Institut Universtät Göttingen, Goldschmidtstr. 1, W-3400 Göttingen, FRG

Moschick, Dr. Anna, Institut für Angewandte Geologie, FU Berlin, Wichernstr. 16, W-1000 Berlin 33, FRG

Mull, Dr.-Ing. Rolf, Prof., Institut für Wasserwirtschaft und Hydrologie, Callinstr. 32, W-3000 Hannover, FRG

Nehrkorn, Dr. Alexander, Prof., Fachbereich 2, Mikrobiologie, Universität Bremen, Leobener Str., W-2800 Bremen 33, FRG

Obermann, Dr. Peter, Prof., Ruhr-Universität Bochum, Institut für Geologie, Universitätsstr. 150, W-4630 Bochum 1, FRG

Otremba, Heike, Dipl.-Biologin, Hygiene-Institut, Universität Bonn, Sigmund Freud-Str. 25, W-5300 Bonn 1, FRG

Petersen, Dr. Axel, IMS – Ingenieurgesellschaft mbH, Bothfelder Str. 35, W-3000 Hannover 1, FRG

Post, Dr. Bernhard, Stadtwerke Frankfurt/Main, Börneplatz 3, Postfach 4269, W-6000 Frankfurt/Main 1, FRG

Rades-Rohkohl, Dr. Ellen, Colegio Alemain, Casillo, 4442 La Paz, Bolivia

Regenberg, Dr. Werner, Institut für Umweltphysik, Universität Heidelberg, Im Neuenheimer Feld 366, W-6900 Heidelberg, FRG

Reifenhäuser, Dr. Christine, Institut für Anorganische Chemie, Universität Regensburg, Universitätsstraße 31, W-8400 Regensburg, FRG

Schaefer, Dr. Karl Wolfram, Radiocarbon Research, Australian National University, GOT Box 4, Canberra 2601, Australia

Schenk, Dr. Dietmar, Geologisch-Paläontologisches Institut Universität Kiel, Olshausenstr. 40/60, W-2300 Kiel 1, FRG

Schlosser, Dr. Peter, Institut für Umweltphysik, Universität Heidelberg, Im Neuenheimer Feld 366, W-6900 Heidelberg, FRG

Schmidt, Dr. Hans-Ludwig, Prof., Lehrstuhl für Allgemeine Chemie und Biochemie, Technische Universität München, W-8050 Freising-Weihenstephan, FRG

Schulten, Dr. Hans-Rolf, Prof., Fachhochschule Fresenius, Abt. für Spurenanalytik, Dambachtal 20, W-6200 Wiesbaden, FRG

Schulz, Dr. Horst Dieter, Prof., FB 5 Geowissenschaften, Universität Bremen, Postfach 330 440, W-2800 Bremen 33, FRG

Schulz, Reiner, Dipl.-Geologe, Analytisches Labor Aachen, Lothringer Str. 17, W-5100 Aachen 1, FRG

Selenka, Dr.med. Fidelis, Prof., Institut für Hygiene, Ruhr-Universität Bochum, Universitätsstr. 150, W-4630 Bochum 1, FRG

Smykatz-Kloss, Dr. Werner, Prof., Institut für Mineralogie und Geologie, Universität Karlsruhe, Kaiserstr. 12, W-7500 Karlsruhe, FRG

Sonntag, Dr. Christian, Institut für Umweltphysik, Universität Heidelberg, Im Neuenheimer Feld 366, W-6900 Heidelberg, FRG

Strebel, Dr. Otto, Bundesantalt für Geowissenschaften und Rohstoffe, Stilleweg 2, Postfach 51 01 53, W-3000 Hannover 51, FRG

Stute, Dr. Martin, Lamont Doherty Geological Observatory, Columbia, University, Palisades N.Y. 10964, USA

Sztuka, Dr. Frank, Prätoriusweg 10, W-2000 Hamburg, FRG

Tuschewitzki, Dr. Georg-Joachim, Hygiene-Institut, Universität Bonn, Sigmund Freud-Str. 25, W-5300 Bonn 1, FRG

Udluft, Dr. Peter, Institut für Geologie und Paläontologie, Pleicherwall 1, W-8700 Würzburg, FRG

Ullsperger, Dr. Irmgard, Ingenieurbüro WATEC, Marktplatz 5, W-8015 Markt Schwaben, FRG

Ulrich, Dr. Bernhard, Prof., Institut für Bodenkunde und Waldernährung, Büsgenweg 2, W-3400 Göttingen, Goldschimdtstr. 1, W-3400 Göttingen, FRG

Usdowski, Dr. Eberhard, Prof., Sedimentpetrographisches Institut, Universität Göttingen, Goldschmidtstr. 1, W-3400 Göttingen, FRG

Valeton, Dr. Ida, Prof., Geologish-Paläontologisches Institut und Museum, Universität Hamburg, Bundesstr. 55, W-2000 Hamburg 13, FRG

Verleger, Dr. Hartmut, Plantagenstr. 19, 1000 Berlin 41, FRG

Voerkelius, Dr. Susanne, Lehrstuhl für Allgemeine Chemie und Biochemie der Technischen Universität München, W-8050 Freising-Weihenstephan, FRG

Weise, Dr. Stephan, Institut für Hydrologie, Ingolstädter Landstr. 1, W-8042 Neuherberg, Post Oberschleißheim, FRG

Weyandt, Dr. Ralph, Institut Fresenius, Im Maisel 14, W-6204 Taunusstein, FRG

Units and Abbreviations

Prefixes for Multiples and Fractions

G	giga	10^9	m	milli	10^{-3}
M	mega	10^6	µ	micro	10^{-6}
k	kilo	10^3	n	nano	10^{-9}
c	centi	10^{-2}	p	pico	10^{-12}

In general, chemical nomenclature follows Nomenclature of Inorganic Chemistry, American version, IUPAC Inorganic Rules, as Published in J. Am. Chem. Soc. 82:5525 (1960)

Units of Measurement

Length	m	metre
Area	m^2	square metre
	ha	hectare = 10^4 m^2
Volume	m^3	cubic metre
	l	litre = 10^{-3} m^3
Mass	g	gram
	kg	kilogram
	D	dalton = mass unit = 1.6608×10^{-24} g
Time	s	second
	min	minute
	h	hour
	d	day
	a	year
Temperature	K	Kelvin (SI)
	°C	Celsius ($T = t + T_0$; T in °C $T_0 = 273.15$ K)

Standard temperature		25 °C = 298.15 K
Pressure	Pa	Pascal (SI) = Newton per metre^2 (n.m^{-2})
	bar	bar = 1000 mbar = 10^5 Pa
Standard atmospheric pressure	atm	atmosphere = 1013.24 mbar = 760 torr
Radioactivity	Bq	becquerel (=s^{-1})
	dpm	decay per minute
	TU	tritium unit (= 0.119 Bq/l = 10^{-18} ^{3}H/H)
	TR	tritium ratio (=^{3}H/H; for 1 TU TR = 10^{-18})
	pmc	percent modern carbon (100%-,modern ^{14}C)
Voltage	V	volt

Abbreviations

a, a$_o$	fracture aperture, at time t and initially
a	specific extinction (absorption) at a certain wavelength
A	extinction (absorption) measured
A	area
A_{min}	surface area of mineral grain
α_f	relation between fluorescence intensity *I* after and I_o before metal addition (=I/I_o)
α_k	kinetic constant
α_m	moisten angle
α_T	partition coefficient
α_x	dispersitivity
AAC	Seffent springs sampling sites
AAS	atomic absorption spectrometry
AC	acetate
AFA	aquatic fulvic acid
AHA	aquatic humic acid
Ah	Ah soil horizon
AIR	AIR standard
AODC	acridine orange direct count
ATP	adenosine triphosphate
b, b$_o$	fracture breadth at time t, and initially
b.g.l.	below ground level
BM	Brunnenseemoor
BOC	Bocholt
BOC	biochemical oxygen consumption
BS	base saturation
B(s)h	B(s)h soil horizon

C_{org}	organic carbon
c	concentration
CC	complexation capacity
CD	Canon Diablo troilite standard (sulfur isotopes)
CDN	Dehlhi/Ontario, Canada
CEC	cation exchange capacity
CFU	colony forming unit
CH_{HS}	carbohydrate associated to HS
CI	chemical ionization
CPMAS	cross-polarization magic angle spinning
CS	closed system
δ	thickness of water layer
D	deuterium
D	coefficient of hydrodynamic dispersion
D'	coefficient of hydromechanic dispersion
D_{eff}	effective coefficient of diffusion
D_M	coefficient of molecular diffusion
D_t	coefficient of transversal hydrodynamic dispersion
d	average diameter of grains
d_v	average diameter of pores
d.a.wt.	dry ash free weight
ddW	double distilled water
DEPT	distortionless enhancement by polarization transfer
DFG	Deutsche Forschungsgemeinschaft
DIC	total dissolved inorganic carbon
DNA	desoxyribonucleic acid
DOC	dissolved organic carbon
DSM	bacterial standard (Deutsche Sammlung für Mikroorganismen)
d.wt.	dry weight
ε	thickness of diffusion boundary layer
EA	alkaline earth/alkaline ion equivalent ratio
EDAX	energy dispersive X-ray analysis
EDTA	ethylene diamine tetra acetate
EI	electron impact
epm	equivalent per million
EPS	extracellular polymeric substance
F	flux (dissolution rate)
FA	fulvic acid
FBR	fluidized bed reactor
FBR	fixed bed reactor
FD	field desorption
FDA	esterase activity
FI	field ionization

FID	flame ionization detector
FOR	Förste/Harz
FRG	Federal Republic of Germany
FT-IR	Fourier transformed infrared spectrometry
γ	conversion factor
G	gravitational constant (= 6.6720×10^{-11} $N.m^2/kg$) (Sec. 2.1.3.6)
g	gravitational acceleration; m/s^2
¢G_r	free reaction enthalpy
GC	gas chromatography
GP	gel permeation
GPC	gel permeation chromatography
GW	ground water
h	hydraulic head
HA	humic acid
HPLC	high pressure liquid chromatography
HS	humic substance
hs	smaller molecular subunit
I	hydraulic gradient
I_{BA}	index of base exchange
I_g	solute input into GW
I_n	solute input into GW
I_t	solute input per total recharge area
IAP	ion activity product
IC	ion chromatography
ICP-AES	inductive coupled plasma atom emission spectrometer
ID	ion desorption
IDMS	isotope dilution mass spectrometry
IHS	International Humic Substances standard
IHSS	International Humic Substances Society
IR	infrared spectrometry
k_1	maximum sorption capacity
k_2	binding constant
k_u	hydraulic conductivity of unsaturated zone
k	rate constant [kinetic constant (factor)]
K	equilibrium constant
K'_C	conditional stability constant
K_C	stability constant
K_H	Henry's law constant
K_S	solubility product
K_W	ion product of water
K_{CO_2}	solubility constant of CO_2

L	length
L_c	characteristic dimension
L_f	length of flow path
λ	distribution coefficient in presence of complexing ligands
λ_1	reaction length (distance)
λ_0	distribution coefficient
LPM	light phase contrast microscopy
M	molarity (mol/l)
M_0	geometrical factor
Me	metal
MM	mineral medium
MPN	most probable number (microbiology)
MRT	mean residence time
MS	mass spectrometry
MSP	Mackenroder Spitze/Göttinger Wald
MT	morphotype
MUC	Munich
MV	measured values
MW	molecular weight
m/z	mass charge ratio
N	normal solution (gram equivalent/l)
ν	kinematic viscosity
$NADH_2$	reduced nicotine amide dinucleotide
NMR	nuclear magnetic resonance
NTA	nitrilotriacetic acid
O_g	solute output with pumped GW
OD	optical density
OS	open system
P	perimeter
p	partial pressure
p_c	capillary potential
PDB	Peedee Belemnite standard
PE	polyethylene
PP	polypropylene
PVC	polyvinylchloride
PW	pore water
P_y	pyrolysis
PYGV	peptone-yeast extract-glucose-vitamin medium
Q	deposition factor
Q	flow rate
QUAT	quaternary-only carbon spectra

R	gas constant
R	flow resistance
r	correlation coefficient
R	rate
R_C	rate of carbonate dissolution
R_n	ground water recharge
R_t	ground water recharge in total catchment
ρ_w	water density
Re	Reynolds' number
RI	refraction index
RP	reversed phase
rpm	rotation per minute
RRR	relative release ration
S	total reactive grain surface
S_g	storage change of solute in the aquifer system
S_{SM}	Sokal & Michener matching coefficient
σ_w	surface tension of water
SAW	Sarchinger Weiher/Regensburg
SDH	Sandhausen/Heidelberg
SDW	Schwetzendorfer Weiher/Regensburg
SEG	Segeberger Forst
SEM	scanning electron microscopy
SFC	sulfate formation capacity
SF	soil fulvic acid
SHA	soil humic acid
SI	saturation index [$SI = \log (IAP/K_s)$]
SMOW	standard mean ocean water (isotope standard)
SL	Schwarze Laaber/Regensburg
SSA	soil extract agar
SS	sediment samples
SW	surface water
SYC	synthetic organic compounds
T	absolute temperature
T	breakthrough time
$T_{1/2}$	half-life
τ	time constant
τ_{diff}	diffusion relaxation time
τ_t	turnover time
TDS	total dissolved solids
TEM	transmission electron microscopy
TM	total mineralization
TN	Tunesia
TOC	total organic carbon
TON	total organic nitrogen
TR	Turkey

UV	ultraviolet
v	flow velocity
V_g	gas volume
V_l	solution (liquid) volume
V_p	pore volume of medium
VIS	visible range
vpm	siehe: Sect. 3.2.2:3 (volume per million)
w	soil moisture (pore water content)
x	mole fraction
XAD	polystyrene resin
x_S	saturation length
Z	boundary layer
z	charge
z	depth
$\bar{z}$	relaxation depth

1 Scope

Georg Matthess

Ground water originates from the infiltration of atmospheric precipitation and, where a landward gradient of the water table occurs, of surface water, too. After flowing underground for varying distances it feeds the surface water systems, rivers and lakes. Only a small proportion flows directly into the sea or evaporates where the water table comes close to the ground surface. During its movement from the surface of the ground through the soil, the water-unsaturated and the water-saturated zone to the points where it re-emerges, water is subject to numerous interactions between the aqueous and the solid phases through physical, chemical and microbial processes such as dissolution, precipitation, oxidation and reduction, complexation, ad-and desorption, filtration, gas exchange, evaporation, biological metabolism, isotopic redistribution and anthropogenic influences, which lend it its typical properties.

From 1982 to 1988, the Deutsche Forschungsgemeinschaft (DFG) supported a priority programme entitled *Hydrogeochemical Processes in the Hydrological Cycle within the Unsaturated and Saturated Zones*, in which scientists from various disciplines such as analytical chemistry, geochemistry, geophysics, ground water hydraulics, hydrogeology, isotopic hydrology, microbiology, mineralogy/petrography, pedology and physics, investigated the complex processes which control the quality changes taking place in the subterranean part of the hydrological cycle.

The programme is of special current interest because the natural regenerative capacity of this system of seepage and ground water (hydrosphere), soil and rock (pedo- and lithosphere), gas phase (atmosphere) and microbial biosphere is based on material transformation and migration. It may be surmised that the regulatory capacity of this system, especially in individual sections of its subsurface passage, is exhaustible. Consequently, both the changes in the water during its passage as well as the geochemical changes occurring in the soil and rock, through which the water flows, were studied.

The research was focussed on natural or near-natural seepage and ground water systems. As the water proceeds along its subterranean path (unsaturated zone, ground water fluctuation zone, ground water transport zone, and ground water discharge zone) its composition was examined to elucidate the interactions between water, rock and biota. Chemical analysis was performed to examine individual or combined processes of material change, to isolate zones of material transformation and to calculate material and energy balances. This also involved consideration of the hydrological boundaries and the hydrodynamics of the catchment area and its water budget, with special reference to its geological structure. In the course of this work, existing field facilities such as those in Bocholt, the Fuhrberger Feld near Hanover, the Munich alluvial plain and the Segeberger Forst were used. Furthermore, results of a site with similar environmental conditions in Dehli, Ontario/ Canada, and investigations on karst phenomena in a gypsum karst area in Foum Tatahouine, South Tunesia, and a paleokarst in the Aladag Region, Southeast Turkey, were included. The sites are listed and generally characterized in Fig. 1.1 and Table 1.1

The subsurface is hydrologically separated to unsaturated and saturated zones (Fig. 1.2). The unsaturated zone includes the soil sensu strictu and, in most cases, the subsoil (intermediate layers) lying under it down to the ground water table. In this zone solid materials coexist with ground air and water in the form of adhesion, capillary, and percolating water, which are referred to here as pore water (Matthess 1982). The boundary between unsaturated and saturated zones is approximately

Fig. 1.1. Investigation sites in West Germany

Table 1.1. Location and general characteristics of investigation sites

Investigation site	Code	Coordinates	Geology	Soil	Precipitation mm/a	Vegetation	Sections
Segeberger Forst 40 km N' Hamburg	SEG	53°50'N 10°10'E	Pleistocene	Podsol	600–800	Mixed forest	2.1.2; 2.1.4; 2.1.5 2.1.6; 2.1.8; 2.1.9 2.2.1–2.2.3; 2.2.5 3.1.2; 4.3.2; 5.1.1 5.2.1; 5.3.3; 6.1.3 6.2.4
Wingst W' Hamburg	WIN	53°43'N 9°03'E	Pleistocene sands covered with quick-sand	Podsol	600–800	Spruce	4.2.1
Hamburg	HAM	53°33'N 9°59'E	Miocene and Pleistocene sands		600–800		4.1.7
			Morain of Saalian age covered by melt water sands	Luvisol Podsol			4.2.2
Fuhrberger Feld NE' Hanover	HAN	52°34'N 9°50'E	Cretaceous age, Quaternary cover	Podsol to gley-podsol	600–800	Pine, birch, black alder, spruce	4.1.2; 4.1.3; 4.2.1 5.2.3; 6.2.5
Berlin	B	52°31'N 13°24'E	Quaternary sediments	Peat, lake sediments			4.1.5
Bocholt (Liedern)	BOC	51°50'N 6°36'E	Pleistocene to Holocene age	Brown earth	800–1000	Agriculture (pasture ground, corn)	2.1.3; 2.1.5; 2.1.6 2.1.8; 2.2.1; 3.2.3 4.1.1; 5.1.1; 5.1.3 5.2.1; 5.3.2; 6.2.2 6.2.3; 6.2.4; 6.2.6 6.3
Dorsten	DOR	51°39'N 6°58'E	Upper Cretaceous age	Sand	600–800		4.1.6
Hils NW' Göttingen	HIL	51°55'N 9°48'E	Mud of Hilssandstein with loess cover	Podsolized brown earth	600–800	Beech	4.2.1

Table 1.1. (*Continued*)

Investigation site	Code	Coordinates	Geology	Soil	Precipitation mm/a	Vegetation	Sections
Solling NW' Göttingen	SOL	51°40'N 9°41'E	Buntsandstein with loess cover	Podzolized Brown earth	600-800	Spruce, beech	4.2.1
Osterode/Förste	FOR	51°44'N 10°11'E	Zechstein-2 partly covered with loess	Podsol	1000–1200	Agriculture, partly forest	3.2.4
Mackenröder Spitze 7 km E' Göttingen	MSP	51°31'N 9°55'E	Middle Triassic	Mull rendzina terra fusca rendzina	700	Dense forest	3.2.6
Seffent Springs 4 km N' Aachen	AAC	50°47'N 6°05'E	Upper Cretaceous		800–1000		3.2.5; 3.4.2 6.3
Bensheim	BEN	49°41'N 8°00'E	Quaternary sands, gravel	Cambisol	700	Agriculture, forest	4.1.4
Sandhausen 7 km SW' Heidelberg'	SDH	49°25'N 8°43'E	River sediments, quicksand dunes	Podsol	600–800	Pine, beech	2.3; 3.2.3; 4.2.3 5.2.1; 5.3.1
Munich Munich-Gravel-Plain	MUC	48°05'N 11°20'E	Pleistocene carbonate gravel over Tertiary silt layers		800–1000	mixed forest, grass land	2.1.1; 2.1.4 2.1.5; 2.1.7; 2.2.1 3.2.1; 3.2.2; 5.3.4 6.1.2
Aladag-Region (SE'-Turkey)	TR	37°30'N 35°10'E	Devonian to Cretaceous age				3.3.3; 3.5
Delhi, Ontario 7 km SE' Trout Creek (Canada)	CDN	42°51'N 80°30'E	Pleistocene Norfolk Sand underlying Devonian limestones				3.1.2
Foum Tatahouine Jabal Dahar (S. Dahar)	TN	33°00'N 10°30'E	Upper Triassic to Middle Jurassic age		200–280		3.3.1; 3.3.2; 6.3

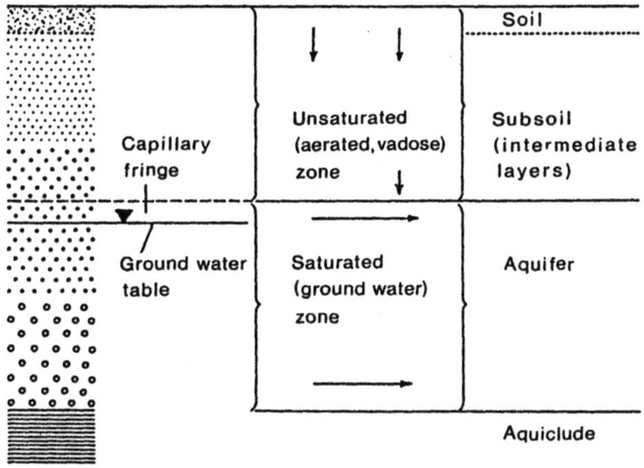

Fig. 1.2. Hydrological subsurface categories

described by the ground water table, which in a strict sense is defined as the surface in unconfined material along which the hydrostatic pressure is equal to the atmospheric pressure. The zone below the water table is generally called the saturated or ground water zone, neglecting the fact that the lower part of the capillary fringe is also saturated and participates in the ground water flow with almost the same velocity as the ground water just below the water table (Davies and De Wiest 1967, Matthess and Ubell 1983). This simplification is acceptable in this context in as far as no special research dealt with the processes in the capillary fringe.

The material for examination was acquired from field samples obtained from wells, springs, lysimeters and suction heads (measuring and observation devices). Bore holes and exploratory excavations produced depth-specific soil samples from which the pore solutions could be extracted and the solid matter subjected to further analysis. These field observations facilitated the derivation of individual processes which could subsequently be simulated in the laboratory. The results permitted the study of the kinetics of the carbonate dissolution under various environmental conditions and of the silicate dissolution under the conditions of pressure and temperature found in regions of moist-temperate climate.

The practical aspects of this research have demanded the development of special hydrochemical methods, such as methods for the analysis of silicic acid derivatives, carbon and nitrogen compounds and organic metal complexes, and methods for the characterization of thin sediment layers (coatings) on mineral surfaces. In addition, methods had to be developed and tested for recording the number and activity of the various groups of geochemically significant microorganisms. The aim of this priority programme was to coordinate the work of scientists from different disciplines in order to produce a combined effort to develop material transport and material change models based on hydraulic, physico-chemical and microbiological principles which will allow the study of the flow of materials in the subsurface sections of the water cycle. This quantitative treatment of the material cycle provides a geoscientific contribution to ecological research. Models developed now form a basis for initial prognostic studies involving the functional capacity of ecosystems.

Little is at present known about the structure, composition and concentrations of dissolved and solid subsurface organic substances, although they evidently play an important role as complexing and sorptive agents, and also as nutrients for microorganisms. A quantitative determination of vertical distribution, nature and concentration of these substances gives important insights into their origin, stability functions of their complexes and their transport behaviour. Data could be derived to assess the contribution made by the organic substances washed out of the humus layer or the incidence of these substances occurring at deeper levels below the surface. Such information is of great importance for the forecasting of potential effects of nitrates from agricultural fertilizers on the ground water and for other environmentally important features of ground water systems.

Specialized microorganisms are involved in the processes described above. Geochemists, hydrogeologists and microbiologists, working in close cooperation, conducted the first-ever systematic research into biotopes in the unsaturated and saturated zones. Collecting representative soil and water samples was especially demanding for the scientists. The microbiologists recorded the microflora resident in the biotopes and studied their potential activities. They defined the conditions under which these various groups of organisms can live together in the different hydrogeologically and geochemically described environments. Direct cooperation between scientists in the various subjects has been directed towards investigating the influence of these microorganisms on the kinetics of important geochemical processes. One of their endeavours was aimed at the processes of the sulfur, iron, manganese, phosphorus and nitrogen cycles, in the nutrient-deficient parts of the water-unsaturated and -saturated zones.

2 Polar Organic Substances and Their Role in the Water-Saturated and -Unsaturated Zones

2.0 Introduction

F.H. Frimmel

Humic substances (HS) are products formed by the microbial degradation and biochemical rearrangement of plant material and dead organisms in soil (Ziechmann 1980; Stevenson 1982) and water (Christman and Gjessing 1983). HS derive their heteropolycondensed structure from their parent compounds such as lignins, tannins, proteins, fatty acids, carbohydrates and numerous other organic and inorganic substances (Gjessing 1976). The variety of starting materials and the huge number of possible reaction pathways (Flaig 1964) are the reason for the difficulties in suggesting a precise structure, which usually is no problem in the field of pure chemical compounds.

Because it seems impossible to characterize HS by their molecular identity (Hayes et al. 1989), we have to rely on operational definitions. This includes the necessity of a detailed description of the origin and treatment of the samples we are looking at and of the methods applied. This is no doubt troublesome but at the moment the only reasonable way to obtain meaningful information on the complicated nature of HS (Frimmel and Christman 1988).

A widely used differentiation of HS uses the fulvic acid fraction (FA) and the humic acid fraction (HA). FA are soluble in acids and bases, whereas HA are only soluble in alkaline solutions. The dissolved state is mostly defined according to the application of 0.45 μm membrane filtration.

There has been a strong interest in terrestrial HS for many years, owing to the important role they play in plant productivity (Schnitzer and Khan 1972). More recently there has been an awareness of the ubiquitous presence of HS in aquatic systems, and a number of reactions have been studied which are of environmental significance (Fig. 2.0.1). Unquestionably, it is the FA fraction which plays the dominant role in the reactions that take place in solution.

HS can interact with a number of inorganic and organic water constituents. Synthetic organic compounds (SYOC) as environmental pollutants can be complexed. As a consequence, there is a change in their physical and chemical properties which leads to a changed behaviour in the environment. For example, the water solubility of micropollutants can be enhanced by HS (Chiou et al. 1986) and transport mechanisms in the aquifer are influenced (Abdul et al. 1990). Even though there are practically no data available, it is reasonable to assume that the toxicity of organics to microorganisms will also be influenced by HS. For metals (Me) and

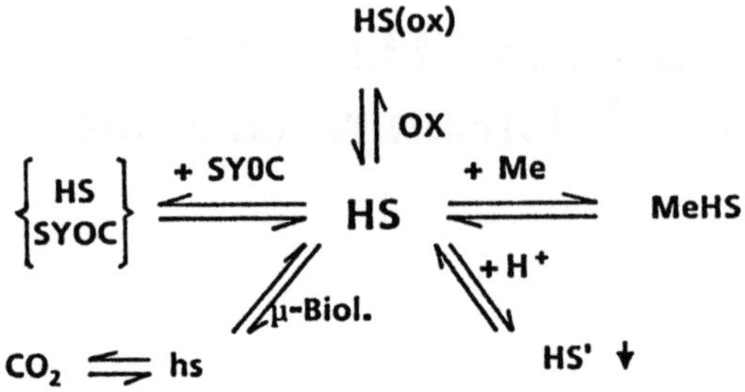

Fig. 2.0.1. Environmental reactions including humic substances (HS)

especially for the so called heavy metals, similar effects can be discussed. The influence of HS on the physical and chemical characteristics of metals in the underground will be the focus of the following chapters. In addition, the distribution of HS between the solid and liquid phase of the aquifer system plays an important role for the distribution of organic carbon. These equilibria are strongly dependent on the pH value. Microbial degradation of HS is a further pathway which has to be kept in mind when dealing with the genesis and fate of HS and their role in the environment (Perdue and Gjessing 1990). Smaller molecular subunits (hs) and finally CO_2 have to be expected as products. Technologically important oxidation reactions (e.g. influence of O_3, UV-radiation and disinfectants like chlorine products) have not been the subject of this book.

This chapter deals with three main topics, the isolation and characterization methods of humic substances, the interaction of inorganic substances with humic substances and the characterization of other organic acids in subsurface systems. The first topic begins with the methods used for the isolation and general characterization of organic acids from pore solutions in the unsaturated zone (Frimmel, Sect. 2.1.1), of soil humic matter (Finger, Post and Klamberg, Sect. 2.1.2) and of organic substances in ground water (Selenka and Hack, Sect. 2.1.3). Specific methods are presented with the chromatographic characterization of acid soluble humic substances (Frimmel, Sect. 2.1.4), the use of spectroscopic techniques in the ultra violet and visible region and of infrared spectroscopy (Abbt-Braun, Sect. 2.1.5), by temperature-programmed/time-resolved pyrolysis field ionization mass spectrometry (Schulten, Sect. 2.1.6) including their interpretation (Abbt-Braun, Sect. 2.1.7), and by ^{13}C nuclear magnetic resonance spectroscopy (Buddrus and Burba, Sect. 2.1.8). The use of organic solvents for the extraction of humic substances is discussed (Post and Klamberg, Sect. 2.1.9).

The interaction of humic substances includes their complexation ability, investigated by polarography and fluorescence spectroscopy (Frimmel, Sect. 2.2.1) and by the ion-exchange method (Finger and Klamberg, Sect. 2.2.2), their sorption effects on metals (Becker and Klamberg, Sect. 2.2.3) and on iodine (Heumann and Reifenhäuser, Sect. 2.2.4), and with the influence of organic ligands upon the kinetics of feldspar weathering (Petersen, Matthess and Schenk, Sect. 2.2.5).

Little is known on the nature and behaviour of simple organic acids in the subsurface aquatic systems, but there are indications for their importance for the water/rock interaction in the weathering zone. A contribution to their characterization is presented in Cordt and Kussmaul (Sect. 2.3).

2.1 Isolation Procedures and Characterization Methods

2.1.1 Isolation and General Characterization of Organic Acids from Pore Water

F.H. Frimmel

2.1.1.1 Origin and Collection of Samples

For transport mechanisms in the subsurface, dissolved substances and their reactions are of fundamental importance. The understanding of these reactions is severely limited by the small number of data which have been available for the water-unsaturated zone, where many experimental problems are connected to in situ measurements. An accessible research shaft (Fig. 2.1.1.1) offers the possibility of measuring and sampling at different levels under relatively undisturbed natural conditions (Frimmel 1985). The research shaft is situated close to the Institut für Wasserchemie at the Technische Universität of München (Großhadern, Fig. 1.1, Table 1.1). The water-unsaturated zone has a depth of more than 10 m and is typical for the so-called Munich Gravel Plain formed by Quaternary calcium carbonate and dolomite gravel. The ground water (height of a few meters) flows on a Tertiary layer. The soil on top (ca. 40 cm deep) belongs to the type of cultosol (cultivated rendzina). The ranges of dissolved organic carbon (DOC) as determined

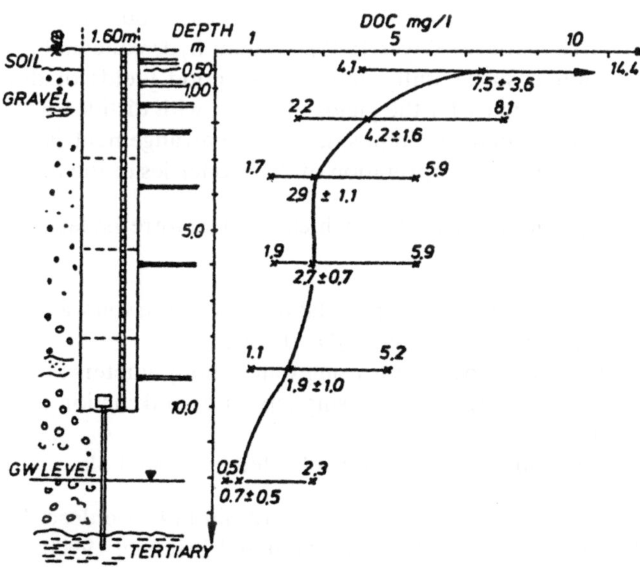

Fig. 2.1.1.1. Accessible research shaft MUC3 in the Munich Gravel Plain and DOC of seepage water sampled at different levels

in more than 30 individual sample sets taken over a period of 1 year are given for the different levels.

Mean values and standard deviations for each data set are given. It is obvious that the meteorological influence, which is strongest at the soil surface, decreases with depth. Microbiological degradation and sorption lead to the final concentration of the DOC in the ground water of about 0.7 mg/l.

The DOC can be quantified reliably (standard deviation < 0.1 mg/l) in small sample volumes. This parameter is well suited for balances of the organic matter, and a high portion of it reflects the dissolved humic substances as relatively stable and therefore long-lived organics. However, we should keep in mind that DOC and dissolved humic matter are normally not identical on the basis of the above given operational definition of HS.

The isolation of higher amounts of water-soluble HS from the soil needs aquatic extracts of the soil. They can be gained by different experimental approaches:

- Based on the natural conditions, seepage water can be collected by means of suction tubes (Merkel and Prömper 1984). The method is without question the closest to reality, but it is limited by the randomness of meteorological situations. The technical difficulties in sampling in deeper levels and in transferring pore water in tubing systems are severe. In addition, there are uncontrollable degradation reactions caused by microbial action during the long collection periods necessary to obtain enough sample volume.
- Elution experiments with soil samples and double distilled water (ddW) under defined conditions are well suited to combine the advantage of a wide range of laboratory conditions with the aquatic matrix close to rain water.
- Soil extractions with acids, bases and aqueous salt solutions are performed in a broad variety (Hayes 1985). Sodium hydroxide and pyrophosphate have been applied successfully and lead to good yields of dissolved organic matter, exceeding by far the yields obtained with distilled water. However, the degree of molecular decomposition and rearrangement is still unknown and consequently the interpretation of the properties of the reaction products is limited.

The aquatic samples from which the HS were isolated in this project were the result of

1. elution experiments in the laboratory where soil was treated with NaOH (0.1 M) or pyrophosphate ($Na_4P_2O_7$; 0.1M);
2. irrigation experiments using water and lysimeters in nature;
3. irrigation experiments using tap water and suction tubes installed in the upper layer of the soil;
4. the natural ground water collected in a bore hole.

The details of the procedures (1), (2) and (4) are described elsewhere (Frimmel 1985; Weis et al. 1989; Abbt-Braun and Frimmel 1990b, Sects. 2.1.2 and 2.1.3). In experiment (3) a set of suction tubes (Al_2O_3 ceramics with an effective pore size of 0.45 μm) were installed at a distance of 50 and 20 cm below the soil surface in the field of the Institut für Wasserchemie in München. The suction pressure was between

–500 and –700 kPa and the area was irrigated by tap water [10 l/(m^2d)]. The DOC of the resulting pore water ranged between 6 and 12 mg/l. It was desalted by ultrafiltration (Amicon cell; membrane 500) with a loss of 11 to 18% of the DOC. The FA from the desalted solution was isolated by the XAD method.

2.1.1.2 Isolation Procedure from Aqueous Solutions

The isolation procedure with the widest application uses the principle of pH-dependent adsorption/desorption on macroporous resins. The method, based on the work published by Mantoura and Riley (1975), was altered in details (Thurman and Malcolm 1981, Frimmel and Niedermann 1980) and is recommended by the International Humic Substances Society (IHSS). The general principle used in this work is outlined in Fig. 2.1.1.2.

The initial filtration and application of low pH leads to the fulvic acid fraction as dissolved part of the humic matter. Because the filtration through 0.45-μm membranes can be very time-consuming for some samples, wider pore sizes can be used at the beginning of this procedure. In these cases the fractionation into FA and HA has to be done at the end of the protocol by applying once more a low pH value followed by 0.45 μm filtration. The danger of hydrolysis and oxidation at high pH values can be minimized by working under exclusion of air and by decreasing the time in which the humic material is exposed to base. To avoid photochemical denaturing reactions the isolation procedure has also to exclude light. This is important because most of the HS in soil have not seen sunlight and therefore are expected to be photoreactive (Frimmel and Bauer 1987).

The samples isolated by the procedure described are given in Table 2.1.1.1.

Even though the method of isolation is internationally widely accepted, it is necessary to keep in mind that the material obtained can only partly reflect nature due to the operationally defined fractionation and the undefined degree of denaturing reactions. Alternative procedures like membrane separation and analytical methods which are suited to measure natural samples in situ are most important to solve these problems.

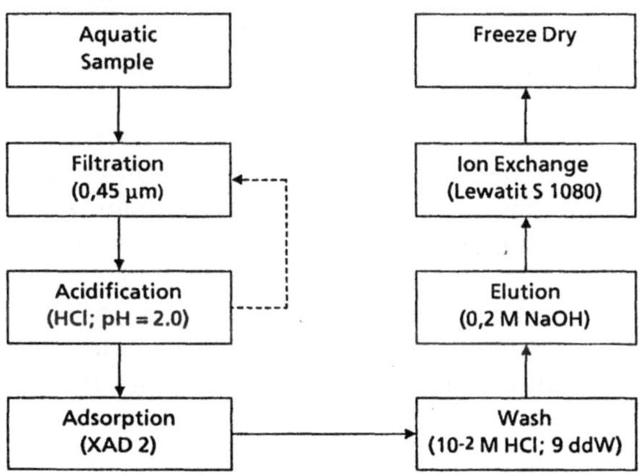

Fig. 2.1.1.2. Isolation of dissolved aquatic humic material on a column filled with polystyrene resin. *ddW* double distilled water

Table 2.1.1.1. Humic samples isolated from different origin

Code	Origin[a]	Extraction	Fraction	Remark
MUC1	a)	NaOH (0.1 M)	FA, HA	Lab. extr.
MUC2–1	a)	ddW	FA	Lab. extr.
MUC2–2	a)	Pyrophosphate (0.1 M)	FA, HA	Lab. extr.
MUC2–2 GRA	a)	AS MUC2–2	FA	After gravel equilibration
MUC3	a)	Tap water	FA	Field
SEG-TA	b)	Seepage water	FA, HA	Lysimeter
SEG-KA	c)	Seepage water	FA, HA	Lysimeter
BOCl/n[b]	d)	Ground water	HS, FA	Field
BOC3/n[b]	e)	Ground water	HS, FA	Field

[a]a) Cultosol, Institut für Wasserchemie, München; b), c) podsol, Segeberger Forst; d), e) brown earth, Bocholt aquifer system.
[b]n = depth in m.

2.1.1.3 Elemental Composition of Humic Substances

Elemental analysis is one of the powerful tools for characterization of HS (Huffman and Stuber 1985). Care must be taken to avoid pitfalls from uncontrolled water content. The availability of experimental O_2 determination has brought an increase in the exactness of the elemental composition compared with the classical method where oxygen was calculated from the difference of the mass balance. The ash content is commonly used as an indicator for the quality of the sample. The smaller the amount of ash in a sample the better the isolation procedure is thought to be. From the atomic ratios interesting information on the origin, genesis and functionality of the humic material can be obtained (Steelink 1985).

Table 2.1.1.2 provides the elemental analysis for the fulvic acids isolated after different extractions from a cultosol sample of the Munich Gravel Plain.

The values are in good agreement with the results known from literature (e.g. Stevenson 1982). It is interesting to note that the extractions with pure water lead to fairly ash-rich samples. To gain some insight into the possible ash components, metal analysis was performed for the isolated samples and the results were compared with the concentrations in the soil extract on a matched DOC basis (Table 2.1.1.3).

As could be expected, the amounts of the trivalent metals Al and Fe in the FA samples were fairly high. The metals in the isolated samples account for up to 17% of the original dissolved concentrations (Abbt-Braun and Frimmel 1990b). This is surprising, as the isolation procedure was done at a pH of 2, which should have led to dissociation of most of the electron donor-acceptor complexes. The results raise the question on the type of bonding of the metals in the humic substances. The presence of metal traces, especially of paramagnetic ones, has also to be taken into account when interpreting the results of structural investigations by spectrometric methods.

Table 2.1.1.2. Elemental composition of some isolated fulvic acids (numbers are % of dry weight)

Sample	C	H	N	O	S	Ash
MUC 1	45.2	5.0	4.8	40.1	0.9	4.0
MUC 2–1	47.9	3.5	2.9	38.5	1.0	6.2
MUC 2–2	49.4	3.8	3.9	40.5	0.9	1.5
MUC 2–2 GRA	48.5	3.7	5.0	39.5	0.9	2.4
MUC 3	47.3	4.0	4.3	37.0	1.3	6.1

Table 2.1.1.3. Metal concentrations (μg/g DOC) in the original aquatic samples (OA) and in the isolated products (FA)

Sample	Ca	Mg	Cu	Zn	Al	Fe
MUC 2–1						
FA	6.4×10^3	39	95	280	2355	760
OA	1.3×10^6	3.7×10^5	897	6250	1.4×10^5	4690
%	0.4	0.01	10.5	4.5	1.7	16.3
MUC 2–2						
FA	65	31	89	20	760	800
OA	2.5×10^4	5.8×10^4	840	8330	1.9×10^5	1.6×10^4
%	0.3	0.05	10.6	0.2	0.4	4.9
MUC 2–2 GRA						
FA	0.7	182	45	49	1500	400
OA	9.10^5	5.6×10^4	424	2380	8650	2780
%	0.001	0.3	10.7	2.1	17	14.4

2.1.2 Isolation and Characterization of Soil Humic Matter

W. Finger, B. Post and H. Klamberg

2.1.2.1 Origin and Collection of Samples

All investigated soil humic matter was isolated from different soil samples from the research site in the Segeberger Forst (Fig. 1.1, Table 1.1; Pekdeger 1977; Holthusen 1981; Ohse 1983; Ohse et al. 1984). The samples in April 1984 (1) and May 1985 (2) were taken by G. Matthess and co-workers, those of October 1985 (3) and February 1987 (4) by H. Klamberg and co-workers. Samples were taken from the four horizons Ahe, Ae, B(s)h and B(h)s and will be abbreviated in the following by SEG-Ah, SEG-Ae, SEG-Bh and SEG-Bs. The two horizons Ah and Bh had the biggest amount of humic matter. Therefore several samples from these horizons were taken in the time period 1984 to 1987, and the soil humic material was isolated for the investigation of the interaction of humic material with metal ions (Post 1987; Post et al. 1988; Finger 1989). The soil humic acids (HA) were isolated using the traditional way of NaOH extraction and precipitation with HCl. The soil fulvic acids (FA) were extracted with water and concentrated on a resin. The abbreviation SEG stands for Segeberger Forst and the numbers 1, 2, 3 and 4 stand for the sequence of the samples taken.

2.1.2.2 Isolation of Soil Humic Acids

Based on the protocol of other working groups (e.g. Pierce and Felbeck 1975; Senesi et al. 1977; Stevenson 1982) and on our results (Kneißel 1985; Post 1987), the standard extraction method with diluted NaOH was used. Despite the warning of Ziechmann (1980) we relied on the findings of Schnitzer and Skinner (1968) and Mehta et al. (1963), who did not observe any changes in the structure of humic substances by applying diluted cold NaOH under exclusion of air. Figure 2.1.2.1 gives the steps of the procedure used.

According to the origin of the samples and the solubility, soil HAs, FAs and humin fractions were obtained. The main fraction was the HA fraction. The other humic fractions and the fractions of lipids and lipid acids were also used for chromatographic and spectroscopic characterization.

The soil samples were dried at 60 °C and extracted with ethanol/toluene (1:1). After treatment with diluted HCl, the HA were extracted with diluted NaOH at ambient temperature under exclusion of air. The humic acids were finally precipitated by addition of HCl at a pH value of 1. The material was washed with water and reprecipitated for purification.

The humin fraction, which is the non-extractable organic residue, was demineralized with a mixture of HCl/HF. From the HCl-extract, from the solution of the HA centrifugation and from the washing of the HAs, fulvic acid fractions (FA1 to FA3) were obtained by means of adsorption on macroporous resin of the XAD-2 type according to Thurman and Malcolm (1981). From the NaOH-extract of the lipid fraction the lipid acids were precipitated by acid.

A detailed description of the isolation procedure and cleaning steps and a discussion of their influence on the properties of isolated soil humic material is given in Post (1987).

2.1.2.3 Isolation of Soil Fulvic Acids

The FAs are the water-soluble fraction of the humic material. For the amounts needed for the experiments, the investigations were focussed mainly on the Ah- and Bh-horizons and developed a new method aiming at the most gentle reaction conditions, to minimize denaturing. In contrast to the NaOH extraction and the method according to Thurman and Malcom (1981), which also uses diluted NaOH for the desorption of the humic substances from the macroporous resin, conditions did not exceed 60 °C and pH values > 7. Figure 2.1.2.2 shows the schematics of the whole procedure.

The original soil material was extracted in a Soxhlet-apparatus for 120 h using ddW and vacuum at a temperature below 60 °C. The dark brown solution was cation-exchanged on the strongly acid DOWEX 50W X-8. Filtration through a glass

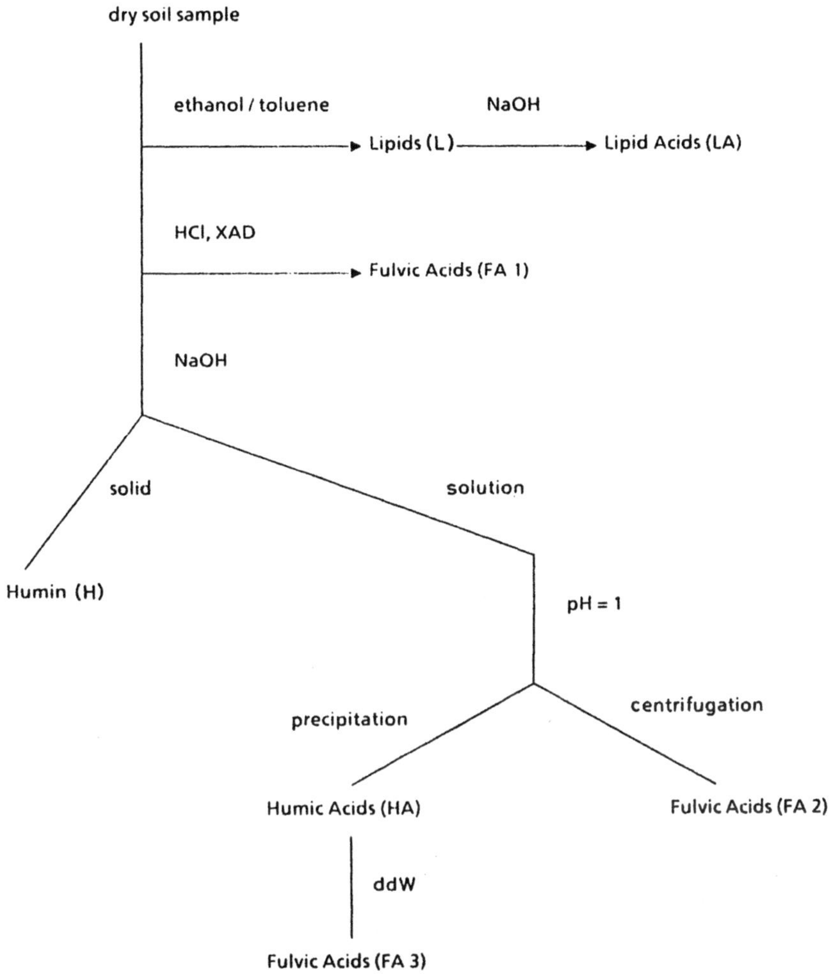

Fig. 2.1.2.1 Isolation of soil humic substances

filter (D3) and under pressure through a membrane (pore size 0.22 μm) removed the ion exchanger and other particulate matter. The brown filtrate was acidified with HCl (c = 2 mol/l) to a pH value of 1. According to the International Humic Substances Society (1985), changes in the structure of fulvic acids can be minimized under these conditions. To concentrate the FAs and to separate them from low molecular organic substances, the humic matter was adsorbed on XAD-8 and desorbed with water. The solution was passed through an ion-exchange column (DOWEX 50W X-8) and isolated by a rotavapor under vacuum at a maximum temperature of 60 °C.

A detailed description of the procedure and the results are given by Finger (1989).

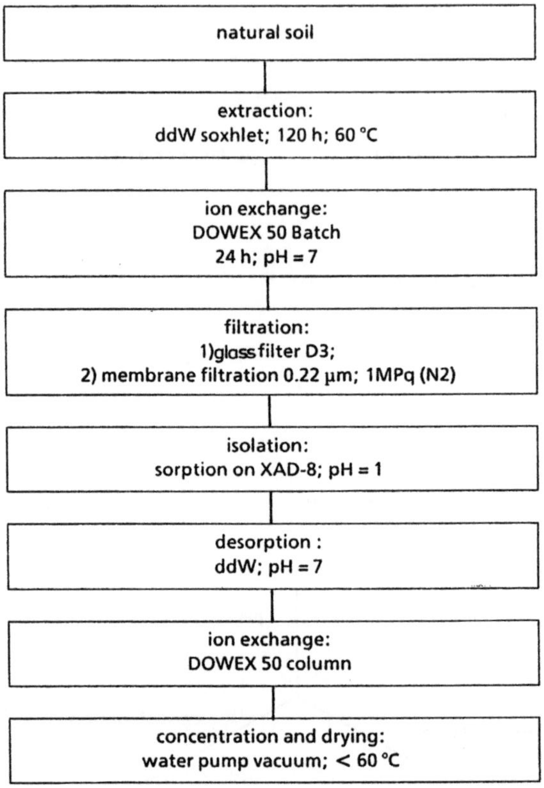

Fig. 2.1.2.2 Isolation of soil fulvic acids

2.1.2.4 Characterization of Soil Humic Substances

The elemental composition for the two humic acids SEG-Ah(3)-HA and SEG-Bh(3)-HA and the two fulvic acids SEG-Ah(4)-FA and SEG-Bh(4)-FA, the concentration of functional groups, the average molecular weight and the moisture and ash content of the samples are given in Tables 2.1.2.1 and 2.1.2.2. The values identified as d.wt. and d.a.wt. refer to dry weight and dry ash-free weight respectively.

The CHN contents were determined by combustion (Elemental analyzer CHN-Rapid, Heraeus, Hanau). The initial amount of material was used according to the ash content (3–15 mg). For the determination of S and Cl, the HS were oxidized in O_2 according to Schöniger; the H_2SO_4 produced was titrated photometrically by $BaClO_4$ using Thorin as indicator; the chloride was titrated potentiometrically by $AgNO_3$ (Photometer 662, Titroprocessor 636 and Dosimat E 635, Metrohm, Herisau). All other elements were determined under standardized conditions by means of flame atomic adsorption spectrometry (Spectrometer Type 5000, Perkin-Elmer, Überlingen). The HS were digested by $HF/HClO_4$; for the determination of Si the sample was melted with Li-metaborate.

The functional groups were determined according to Schnitzer (1975) and Perdue (1985). The average molecular weight was determined by the vapour pressure osmometer method (vapour pressure-osmometer, Knauer, Berlin) in dimethylformamide and calibrated with a polystyrene standard (M_n = 2100, Pressure Chemical Company, Pittsburgh). In addition, gel permeation chromatography, using Fractogel TSK HW-50 (s) (Merck, Darmstadt), a phosphate buffer (pH = 7), and calibration with peptide standards (Serva, Heidelberg) was used.

The water content was determined by drying the samples at 110 °C for 24 h and calculating the weight differences. For gravimetric determination of the ash content, the dried humic samples were treated at 850 °C for 5 h.

A detailed description of the experimental procedure and the results were published by Post (1987), Post et al. (1988) and Finger (1989).

2.1.2.5 Discussion

The CHN content is in good agreement with the humic material described in the literature. There are no significant differences between the samples Ah and Bh. The

Table 2.1.2.1. Elemental composition and average molar masses of HA of SEG-Ah(3)-HA and SEG-Bh(3)-HA

	SEG-Ah(3)-HA	SEG-Bh(3)-HA
Elemental composition (% d. wt.)		
C	57.93	58.15
H	4.24	4.31
N	2.72	2.47
Si	0.44	1.73
Ti	0.0956	0.485
Al	0.183	0.696
Fe	0.194	2.19
Cr	0.0036	0.0084
Mn	0.0003	0.0022
Co	0.0078	0.0199
Ni	0.0190	0.0295
Cu	0.0043	0.0019
Zn	0.0059	0.0070
Mg	0.0057	0.0348
Ca	0.0004	0.0037
Functional groups (mmol/g d.a.wt)		
Total acidity	8.00	8.20
Carboxyl groups	3.20	3.30
Average molar masses (g/mol)		
Vapour pressure osmometry in dimethylformamide	2230	1700
Gel permeation chromatography TSK HW-50 (s)	7940	6960
Water content (%)	3.25	5.10
Ash content (% d. wt)	2.10	10.80

Table 2.1.2.2. Elemental composition and average molar masses of the SEG-Ah(4)-FA and SEG-Bh(4)-FA

	SEG-Ah(4)-FA	SEG-Bh(4)-FA
Elemental composition (% d. wt.)		
C	54.00	53.92
H	2.20	2.93
N	0.93	0.81
S	1.03	0.72
Cl	0.63	0.50
Si	0.04	0.03
Ti	0.187	0.196
Al	0.157	0.160
Fe	0.0659	0.0277
Cr	0.0002	0.0154
Pb	<0.0001	<0.0001
Mn	0.0011	0.0033
Co	<0.0001	0.0002
Ni	0.0102	0.0150
Cu	0.0016	0.0029
Zn	0.0031	0.0014
Mg	0.0488	0.0378
Ca	0.0410	0.0050
Na	0.0131	0.108
K	0.0446	0.0367
Functional groups (mmol/g d.a.wt.)		
Total acidity	9.09	9.16
Carboxyl groups	5.19	5.29
Hydroxyl groups	7.38	9.16
Phenolic hydroxyl groups	3.90	3.87
Aliphatic hydroxyl groups	3.48	5.29
Carbonyl groups	3.30	3.86
Average molar masses (g/mol)		
Vapour pressure osmometry in dimethylformamide	4993	5049
Gel permeation chromatography TSK HW-50 (s)	8732	9016
Water content (%)	14.47	10.42
Ash content (% d. wt.)	2.11	1.80

relatively low amounts of C, H and N in FAs compared with HA suggest a relatively high O content of FAs as a result of oxidation reactions (Stevenson 1982). This interpretation is also supported by the higher values of the total acidity and the carboxyl groups in FAs, which also explains their better water solubility.

All HS show a relatively high ash content, especially the humic acid SEG-Bh(3)-HA. This is surprsing, because all samples have passed at least once the totally dissolved state during the isolation procedure. The elemental composition of the ash shows that there is a small amount of heavy metals which are possibly complexed in the original humic material, and a fairly high amount of Ti, Al, Fe and Si, which can be derived from the silt and metal oxide content of the humics. The average molecular weights can only be seen as estimates according to the bias of the method used. Gel permeation chromatography leads to values which are nearly twice the values of vapour pressure osmometry.

It is attractive to assume that the different solvents used and the aggregation of HS and aggregated solvents have a significant impact on the results. It is striking that in both methods the isolated soil FAs show higher molecular weights than the corresponding soil HA. It is also interesting to note that the interaction between HS and water is so high that drying only proceeds slowly and is mostly incomplete. The application of hard drying conditions includes the risk of structural degradation. In some experiments the results from the water determination, according to Karl Fischer, leads to values 25% higher than the values derived from the dry weight, which cannot be explained yet.

2.1.3 Isolation and Characterization of Organic Substances in Ground Water and Sediments

F. Selenka and A. Hack

2.1.3.1 Background

In the Bocholt aquifer system (Fig. 1.1, Table 1.1) organic substances in ground water corresponding sediments were estimated as total organic carbon (TOC), total organic nitrogen (TON), extractable humic substances (HS), and carbohydrates associated to humic substances (CH_{HS}). In addition, isolated HS were characterized by gel permeation chromatography.

The low concentrations of organic compounds in sandy aquifers demand special sampling techniques and methods for isolation and analytical detection to guarantee reliable results.

Sediment and water samples were taken at seven Bocholt sampling sites, named BOC1 to BOC7 (Sect. 4.1.1). Sediment material was taken from bore holes at intervals, ranging from 1 m in the upper parts to 2 m in depths between 7 and 20 m and 4 m below 20 m depth. Ground water samples were taken from multi-level wells, constructed in these bore holes with plastics or stainless steel, and bearing filter devices at the same depth where sediment samples had been previously taken. The sampling techniques of sediments and ground water, as well as the results of this investigation, are presented in Section 5.1.3 (this Vol.).

If direct examination of organic material in sediment samples is not possible, this material has to be removed from the particles. This was done by extraction with water or other solvents. Shaking with glass beads (ø 4 mm) and water is the least destructive method, but it proved to be of poor efficiency. Better results were obtained by means of NaOH, Na-pyrophosphate, NaF, organic acid salts, organic chelates, formic acid or other extractants. With NaOH at least two to three times more organic carbon can be isolated than with other methods (Stevenson 1982). Extraction with alkaline solutions, however, can cause chemical alteration of the mobilized humic material, such as modification of esters and acceleration of oxidative decay. Working under oxygen-free conditions can minimize these effects (Stevenson 1982). Aquatic suspensions of mechanically or chemically desorbed organic material are analytically treated like water samples.

2.1.3.2 Analytical Detection of Organic Compounds

A survey of kind and quantity of organic substances in sediment and water can be achieved by general parameters on the one hand and analysis of special organic compounds on the other. Of great value is the knowledge of the concentrations of total organic carbon (TOC) and nitrogen (TON) in original samples and in extracts of samples.

Total Organic Carbon (TOC)

In sediment samples the first step was the removal of inorganic carbon by 1 M HCl. Then the total organic carbon was determined by pyrolysis of the sample (800 °C) in a constant stream of nitrogen and oxygen. The reaction gases were dried and quantitatively transferred to a gas analyzer or coulomat, where the CO_2 content was quantified. About 10 μg/kg d.wt. organic carbon could be detected by this method. The following procedure was most efficient: lyophilized sediment fractions smaller than 2 mm were pulverized in a mortar or a mill and 0.05 to 1 g of this material was weighed into previously annealed (900 °C) ceramic receptacles. Then they were mixed with 1 ml 1M HCl and after 30 min incubation at 30 °C they were dried under vacuum. Finally the sample was pyrolyzed in a pyroreactor (e.g. ANTEK, Model 772C) in a continuous stream of argon and oxygen (1:1; 12 l/h) using a temperature gradient from 70 to 800 °C with an increase of 1.5 °C/s.

The gas was dried with silica and after elimination of chlorine with zinc granules finally brought into an IR-gas analyzer (Siemens, Ultramat 3.1). It was used alternatively within the range of 0–20 or 0–50 mmol/l. This procedure allowed detection of less than 10 μg C/kg d.wt. with a standard deviation of ± 1%. The same method was used to analyze organic material removed from sediment and subsequently dried.

The organic carbon in aqueous solution, however, was subjected to oxidation by ultraviolet radiation directly in water. This method has a very low detection limit (< 2 μg C/l), is very fast, convenient to operate and requires only a minimum of sample handling, thus minimizing contamination.

The photooxidation reactor developed by Graentzel (1980), used in the present project, consists of two vertically placed concentric quartz glass cylinders with a "low pressure" mercury lamp placed in the centre of the inner cylinder. The lamp has a power input of 100 W and emits UV-radiation at wave-lengths 254 nm (80%) and 185 nm (15%). Around the lamp a continuous flow of sample water (1 ml/min) automatically acidified to pH 2.8 is distributed over the inner surface of the outer glass cylinder by means of Teflon pins placed in the inner rotating quartz cylinder.

A thin film of water (50 to 100 μm) is generated in this way. The upper zone of the reactor is shielded from radiation. Here water of pH 2.8 is mixed with the sample and inorganically bound carbon is thus released and stripped completely. The CO_2 formed in the oxidation zone below is also continuously removed from the reactor with a continuous nitrogen flow. After drying the reaction gas in a cold chamber and eliminating ozone, the CO_2-concentration is determined in a gas analyzer. It is proportional to the carbon content of the sample. The high sensitivity of this device allows its use as an online detector in high resolution liquid chromatography (Fuchs and Raue 1981; Fuchs 1982; Mueller and Bandaranayake 1982).

Total Organic Nitrogen (TON)

Organically bound nitrogen in sediment samples or extracts from sediment samples was measured after the exchangeable NH_4^+ has been substituted by Na^+-ions (0.2 M NaOH) and emerging NH_3 is stripped with inert gas. Then the total amount of bound nitrogen in these samples was determined (Hernandez 1981; Stedman 1981) and the amount of nitrogen present as NO_3^- or NO_2^- – determined by ion chromatography (Rossner and Schwedt 1984) – was subtracted.

The total amount of bound nitrogen was detected by means of the ANTEK Model 707 Chemiluminescent Nitrogen Analyzer (ANTEK, Houston, Texas). Organic compounds are pyrolyzed in the front part of a pyroreactor in a stream of gas containing argon and oxygen in the ratio of 1:1, using a temperature gradient from 70–850 °C with an increase of 1.5 °C/s. The temperature of 850 °C is held constant until no further nitrogen is detected. Downstream, the temperature in the reactor is held at 1050 °C, thus converting the chemical equilibrium of formed nitrogen oxides nearly completely to nitric oxide (NO): the gas is finally dried in a NAFION filter and then contacted with ozone. NO reacts with ozone to metastable nitrogen dioxide (NO_2^*), which chemiluminesces as it relaxes. The quantity of light emitted is detected photometrically. It is proportional to the concentration of total nitrogen in the sample (Hernandez 1981; Stedman 1981). Up to 1 g of solid sample is pyrolyzed in quartz or ceramic receptacles. The detection limit is about 20 μg N/kg d.wt., and the standard deviation is around ± 1%.

2.1.3.3 Isolation of Humic Substances

The method of isolation of humic substances depends on their origin from aqueous or solid samples. The isolation of HS from sediments yields humic acids (HA) and fulvic acids (FA); the water- and alkali-insoluble humins stay undissolved in the sediment material.

Isolation from Sediments

One hundred grams lyophilized and pulverized sediment fraction (< 2 mm) were weighed into 250-ml glass bottles filled with nitrogen or argon. Then 100 ml 0.2 N NaOH were added and the samples were rotated at 100 rpm in a revolving shaker for 24 h.

The extracts of three successive extraction procedures were united after sedimentation of suspended particles. The pH value of this raw extract was adjusted to pH 2.2 with 2 N HCl. Then the solution was incubated for 30 min at a temperature of 4 °C and finally centrifuged with a relative centrifugal force of 1500-*g* for 10 min. The supernatant (fulvic acid fraction) was treated as a water sample with XAD-2 for cleaning and desalting purposes.

Removal of adsorbed humic substances from the concentrating column was achieved with organic or alkali extractants.

Although fulvic acids were isolated nearly completely, a small part of the humic acids in the sample was eliminated during desalting of the extract with ion exchanger.

Isolation from Water

For isolation purposes, the organic substances in water are concentrated. This can be done by extraction, by sorption to macroreticular adsorber resins, ion exchange, vacuum drying at a temperature of 35 °C or lyophilization (Aiken 1985).

An adsorption chromatographic method, based on the macroreticular polystyrene resin Amberlite XAD-2 (Mantoura and Riley 1975; Frimmel and Niedermann 1980; Aiken 1985, 1988), was employed to isolate humic substances from water in these investigations. Extraction yielded a yellow or brown powder that could be homogenized with a glass rod in a glass flask before further use. The extracts were stored in the dark in airtight bottles or ampoules under nitrogen.

For isolation of small amounts (some milligrams) of organic matter, glass extraction columns of 10 mm inner diameter with a filling of 7 g d.wt. XAD-2 were used and charged with 1 l acidified sample.

Preparative isolation was performed with glass columns of 1 m length and a capacity of 3 kg XAD-2. About 3000 l online acidified water is pumped through the columns at a rate of some 30 l/h and column. Up to 10 g organic material at a time per sample were isolated from ground water this way.

2.1.3.4 Characterization of Humic Substances

Gel Permeation Chromatography

Gel permeation chromatography was performed using a TSK G3000 SW gel, a flow rate of 1 ml/min, a phosphate buffer (0.7 g KH_2PO_4 + 1.1 g $Na_2HPO_4 \cdot 2H_2O$ + 0.002g NaN_3) as mobile phase, and a UV detector (256 nm) and a TOC detector.

Gel chromatographic examination of HS extracts that came from water samples (2–45 m) showed no significant differences in the distribution of molecular weight of the organic substances that were analyzed, despite differences in the total org. C content of nearly 10 mg/l extractable material. Only HS extracted from water originating from the ground water table were different. A small amount of organic substances with higher molecular weight was found.

All samples gave chromatograms with five or six incompletely separated fractions by TOC detection as well as by means of photometric detection (UV_{256}, UV_{436};

Fig. 5.1.3.5). Nearly identical patterns were also found by Vartiainen et al. (1987), who analyzed organic material extracted from ground water of 28 localities in Finland. This means that the processes responsible for sorption, decomposition and alteration of organic substances in aquifers are due to extractable humic substances as a whole, and are not bound to distinct fractions of it.

The same results were obtained when organic substances of the FA fraction were removed from sediment by shaking with 0.2 N NaOH followed by a cleaning and concentration step with XAD-2. The retention times of HS fraction maxima were the same when HS came from water or sediment extracts. The chromatographic separation of substances derived from sediment, however, was less effective than that of substances from water. In addition, HS separated from sediments showed a small portion with apparent molecular weights that were higher than those of substances extracted from water. Despite this, there was great similarity in patterns, indicating an interchange of organic material between sediment and water.

Detection of Carbohydrates Associated with HS

HS-associated carbohydrates are either covalently bound carbohydrate residues derived from HS precursors from plant material or covalently bound or adsorbed carbohydrates of other origin, e.g. microbial products. As yet, affiliation of the HS-associated carbohydrates to these sources has not been demonstrated. Gas chromatographic analysis of the products of hydrolysis with methanolic hydrochloric acid according to Preuss and Thier (1982) reveals the composition of HS-associated carbohydrates.

A lyophilyzed HS sample containing about 1 to 5 mg org. C was hydrolyzed with 1 ml methanolic HCl (2 M) at 100 °C in a tightly closed test tube for 4 h. Then the sample was dried under vacuum at 35 °C. The monosaccharides in the sample were silylated with 25 µl N, O-bis-(trimethylsilyl)-trifluoroacetamide containing 1% trimethylchlorosilane (FLUKA, Buchs) within 15 min at 60 °C. They became volatile and could be separated by gas chromatographic methods (Model 1400; Varian). They were detected with a flame ionization detector (FID). For separation, fused silica capillary columns DB 13.1 or 3.1 or alternatively glass capillary columns with DE-30 phase could be used.

Because carbohydrates were present in form of several isomerics with different chromatographic retention times, most of the monosaccharides yielded several peaks that had to be summed up for quantification. The detection limit was 0.1 µg/l for every monosaccharide. Sorbit or myo-inositol were used as internal standards. They were added to the sample after hydrolysis. The standard deviation of this detection method was ± 10% if sorbit was used and ± 5% if inositol was used as internal standard.

2.1.4 Chromatographic Characterization of the Acid-Soluble Part of Humic Substances

F.H. Frimmel

2.1.4.1 Background

The chromatographic behaviour of HS, HA and FA does not normally lead to fractions with base line separation. Even though the chromatograms are poorly resolved, they can add significant information to the characterization of HS. Furthermore, chromatography is preparative scale leads to fractions which can be examined by a number of physical and physico-chemical methods. The polar structure of HS, which is the reason for their water solubility, and their non-volatility, exclude gas chromatography as an analytical tool to a great extent. Only sample pretreatment (e.g. pyrolysis, derivatization by methylation etc.) can lead to suitable volatile reaction products (Schnitzer and Khan 1972; Greenland and Hayes 1978; Christman et al. 1989). Liquid chromatographic investigations are more favourable. They are mostly done by high pressure liquid chromatography (HPLC), whereas preparative work is normally performed at medium or even atmospheric pressure. From the wide selection of different fractionation principles gel permeation (GP) and reversed phase (RP) techniques are most common in humic chemistry (Gardener and Landrum 1983; Swift 1985; Leenheer 1985).

2.1.4.2 Gel Permeation Chromatography

The determination of sizes and shapes of humic substances by gel permeation chromatography (GPC) was reviewed in the context of chromatographic behaviour of HS (Gardener and Landrum 1983; Swift 1985; Leenheer 1985) and was critically discussed recently. The results depend on the type of gel, the ionic strength and kind of buffer. Severe problems can arise from partly irreversible absorption effects. Due to the lack of authentic substances, calibration has to be done by more or less humic like substances, which introduces some uncertainty. Comparison with the results from other methods for molecular weight determination is also limited because of the polydispersity of HS and the operationally defined values. However, GPC is a promising method which adds to the knowledge of HS, especially when standardized isolation procedures and chromatographic conditions are used. A rewarding approach is the combination of different detection systems with the aim of obtaining multidimensional chromatograms. In addition to UV, fluorescence, and RI (refraction index) (Gjessing 1976; Frimmel and Trampisch 1984), DOC detection is well suited as a basis for mass balances and therefore it can add significantly to the understanding of the chromatographic behaviour of HS and consequently their properties.

2.1.4.3 Experimental Procedures

Samples were isolated and treated as described in Section 2.1.1. The FA samples of MUC 1, MUC 2–1, MUC 2–2 and SEG-TA were chromatographed. Furthermore MUC 2–2 FA was dissolved in water (21 mg/l) and equilibrated with $CaCO_3$-gravel (120 g gravel per 5 l solution). From the equilibrated solution FA were isolated according to the XAD-2 method. The isolated FA sample was assigned as MUC 2–2 GRA. The sample MUC3 was isolated by the XAD-2 method from water gained by suction tubes (Al_2O_3 sinter, –500 to –700 kPa) which were installed 20 cm below surface in the cultivated rendzina of the research field in Munich/Großhadern. The soil had been irrigated with ground water at an intensity of ca. 10 $l/(m^2 \cdot d)$.

Chromatography was done by high and medium pressure using TSK-HW-40 and TSK-HW-50 gels. As detection systems a DOC analyzer and a spectrometer (λ = 254 nm) were used simultaneously as described earlier (Frimmel and Trampisch 1984, Fuchs 1985, Weis et al. 1989). The DOC detector was a continuously working photochemical reactor (UV, O_2) digesting the organic substances to CO_2 which was quantified by IR measurement. The chromatograms were integrated.

2.1.4.4 Results

A typical gel chromatogram is shown in Fig. 2.1.4.1. It is fairly featureless. The fraction with nominal MW around 200 is dominant. About 60% of the DOC has a

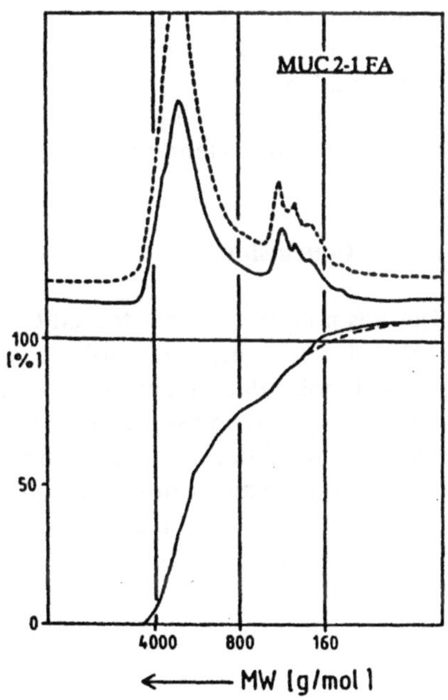

Fig. 2.1.4.1. Typical gel chromatogram with DOC-(bold line) and A(254 nm)-(dashed line) detection molecular weight distribution (sample: MUC 2–1; Gel: TSK 40; mobile phase: phosphate buffer pH 7.0)

Table 2.1.4.1. Characteristic data for the chromatographic separation of FAs detected by DOC and UV absorption (254 nm). Gel: TSK 50; TSK 40; mobile phase: phosphate buffer pH 7; MW: nominal molecular weight according to calibration with polyethylene glycol and carbohydrates; yield: mass balance between influent and effluent

		DOC			A(254 nm)		
FA	TSK	MW max	MW > 800 %	Yield %	MW max	MW > 800 %	Yield %
MUC 1	50	2000	72	95	1950	63	95
MUC 2–1	40	3000	75	100	2500	75	100
MUC 2–1	50	1900	58	90	1900	54	83
MUC 2–2	40	4200	82	95	4200	82	98
MUC 2–2 GRA	40	3500	87	100	3600	78	95
MUC 2–2 GRA	50	1800	77	100	1500	68	98
MUC 3	40	2500	80	98	2500	87	98
SEG-TA	40	4000	33	100	4000	96	100

nominal MW >800. DOC and UV detection show similar chromatograms. Nearly 10% of the FAs is adsorbed irreversibly into the column.

The influence of the origin and sample treatment on the character of the FAs is reflected in Table 2.1.4.1.

It is obvious that the yield is normally better than 90%. Different extractions lead to somewhat different chromatograms. Double distilled water and ground water extract substances (MUC 2–1; MUC3) with nominally lower MW than aqueous pyrophosphate and NaOH. The maximum of the MW distribution shifts to lower values when FAs are equilibrated with carbonate gravel prior to a second isolation procedure (XAD-method; MUC 2–2 FA and MUC 2–2 GRA FA). Differences in DOC and UV detection are most obvious in the case of the pyrophosphate extraction and after gravel equilibration, which can be interpreted to mean that UV-absorbing substances tend to be more readily adsorbed.

2.1.4.5 Conclusions

There is no question that the GPC is well suited for the characterization of humic material even though the resolution of the chromatograms is relatively poor, and the original molecular weight numbers must be highly speculative. The use of a DOC detection system and standardized chromatographic conditions lead to comparable results for the different samples and their special kind of treatment. Keeping in mind all the pitfalls which can result from calibration with non-humic model compounds and from unknown separation mechanisms including adsorption, size exclusion, ionic effects etc., the interpretation of the results is limited to their operational definition.

2.1.4.6 Reversed Phase Chromatography

Reversed phase chromatography (RPC) uses alkylsilanes with non-polar alkyl chains commonly from 4C to 18C as stationary phases, e.g. the code RP C-18 means that there are octadecyl groups.

As mobile phase (eluent), polar solvents like water, methanol and acetonitrile are common. This makes the method most attractive for the investigation of substances like FAs which do not dissolve in non-polar organic solvents. Surprisingly, RPC has not found a broad application in humics research (Leenheer 1985). Nonpolar substances will interact more strongly with the stationary phase than polar compounds, which leads to higher selection times for the non-polar ones. Therefore, the fractions of unknown mixtures can be assigned as more or less polar according to their retention time. A stepwise fractionation was used by Saleh et al. (1989) to differentiate between hydrophilic and hydrophobic constituents. According to the UV-VIS spectra, aliphatic acids, aliphatic ketones and phenols could be assigned. There was evidence for the presence of vanillic acid structures. In the work of Susic and Boto (1989) the use of fluorescence detection leads to a detection limit of 15 ng humic matter. However, samples from different origin are all eluted as a single, relatively sharp peak.

Typical experimental conditions used in our chromatographic work are given in Table 2.1.4.2 and a RP chromatogram of a FA sample is shown in Fig. 2.1.4.2. In addition, gradient elution, the use of buffers or ion pair formation are well suited to improve the chromatograms. The fractionation of HS by RPC shows some typical results.

- There is an irreversible adsorption of humic matter (mostly about 20% of the initial DOC concentration) on the stationary phase when water is used as mobile phase. Methanol can elute most of the deposited material.
- The concentrations of an aquatic buffer used as mobile phase strongly influences the chromatograms. Good results have been gained by 0.01 M phosphate buffer (pH 7.0).
- Different additions of salt to the injection sample also alters the chromatogram, e.g. an increased amount of salt leads to a decrease of the more polar fraction(s).
- Fluorescence intensity and UV-absorption of the fractions show no linear correlation.

Table 2.1.4.2. Experimental conditions for reversed phase chromatographic fractions of fulvic acids

Column:	C 18	250 × 4 mm
Detection:	UV (254 nm)	
Flow rate:	1 ml/min.	
Mobile phases:	• ddW • 0.011 M Phosphate (pH 7.0) • 0.011 M Phosphate Water: methanol = 95:5	

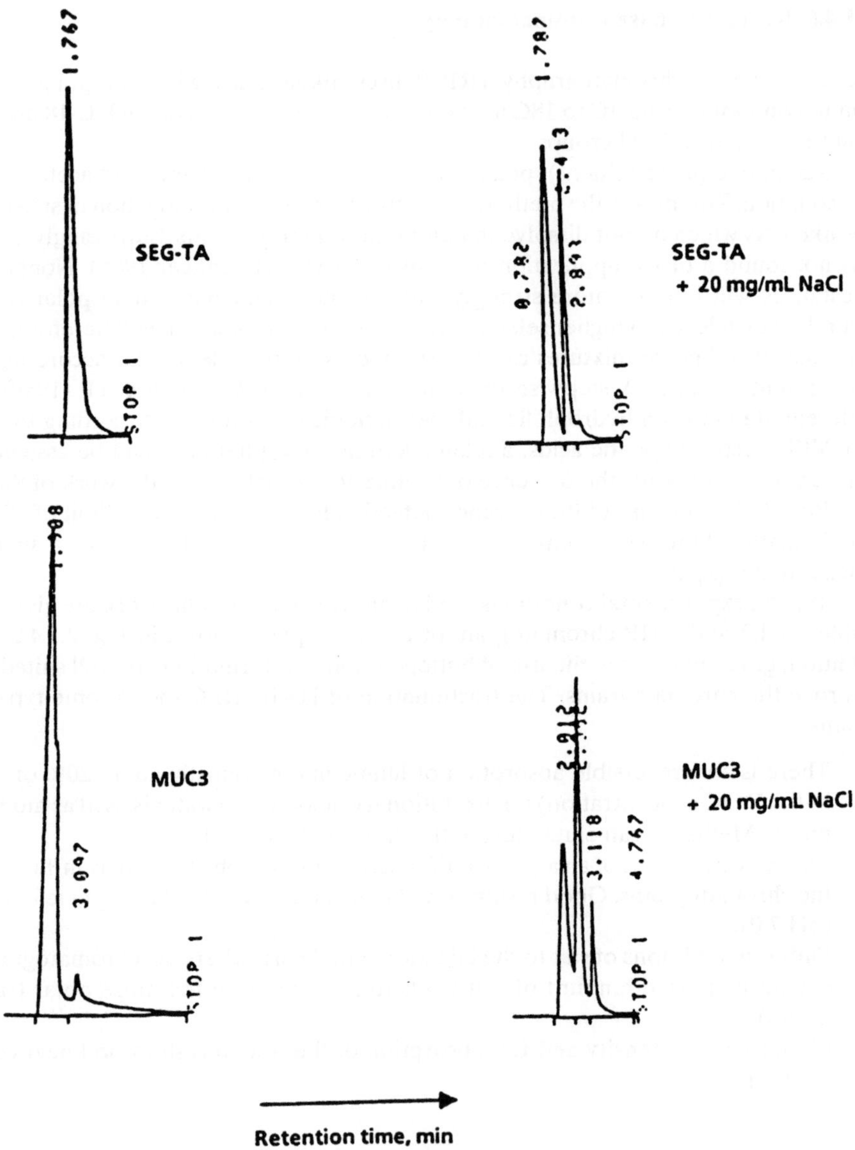

Fig. 2.1.4.2. Typical reversed phase chromatograms for isolated fulvic acids

2.1.5 Spectroscopic Characterization of Humic Substances in the Ultraviolet and Visible Region and by Infrared Spectroscopy

G. Abbt-Braun

Spectroscopic measurements in the different regions of the electromagnetic spectrum provide valuable information for the characterization of HS. Most frequently used are the ultraviolet (UV = 200 to 380 nm), visible (VIS = 380 to 800 nm) and the infrared regions (IR = 2.5 to 25 μm, 4000 to 400 cm^{-1}).

2.1.5.1 Ultraviolet and Visible Region

The absorption in the UV region is mainly caused by the excitation of lone pair electrons of e.g. oxygen ($n \rightarrow \Pi^*$) and by conjugated C = C double bonds ($\Pi \rightarrow \Pi^*$). These systems are thought to be located in the core of HS (Ziechmann 1980). The absorption in the visible range, responsible for the yellow brown colour of HS, is due to lone pair electrons and to charge-transfer systems. Some typical chromophores have quinone structures or conjugated and unsaturated C = C, C = O, and N = O groups (Stevenson 1982). The origin and code name of the HS, FA and HA samples investigated are shown in Table 2.1.5.1.

HS, HA and FA show relatively featureless spectra in the UV- and VIS-range (Fig. 2.1.5.1a-c). The optical density usually decreases as the wavelength increases. However, a slight maximum is indicated around 270 to 280 nm in case of the MUC 1 HA (Fig. 2.1.5.1a), SEG-TA FA and SEG-TA HA (Fig. 2.1.5.1b). The spectra of

Table 2.1.5.1. Origin and code name of the samples

Code name	Origin[a]
MUC 1	Soil extracted with NaOH-solution (0.1 M)
MUC 2–1	Soil extracted with ddW
MUC 2–2	Soil extracted with $Na_2P_2O_7$-solution (0.1 M)
MUC 2–2 GRA	MUC 2–2 FA equilibrated with gravel
SEG-TA	Pore water, irrigated with synthetic rain (pH = 3.8, metal concentration of 10 mg/l)
SEG-KA	Pore water, irrigated with synthetic rain (pH = 5.8)
BOC 1/2	Ground water, 2 m depth
BOC 1/33	Ground water, 33 m depth
BOC 3/3.5	Ground water, 3.5 m depth
BOC 3/5.5	Ground water, 5.5 m depth
BOC 3/9.5	Ground water, 9.5 m depth
BOC 3/13.5	Ground water, 13.5 m depth
BOC 3/17.5	Ground water, 17.5 m depth
BOC 3/20.5	Ground water, 20.5 m depth
IHS	Suwannee River, Standard of the International Humic Substances Society (IHSS)

[a]HS, FA and HA were isolated by the XAD method (see Sect. 2.1.1).

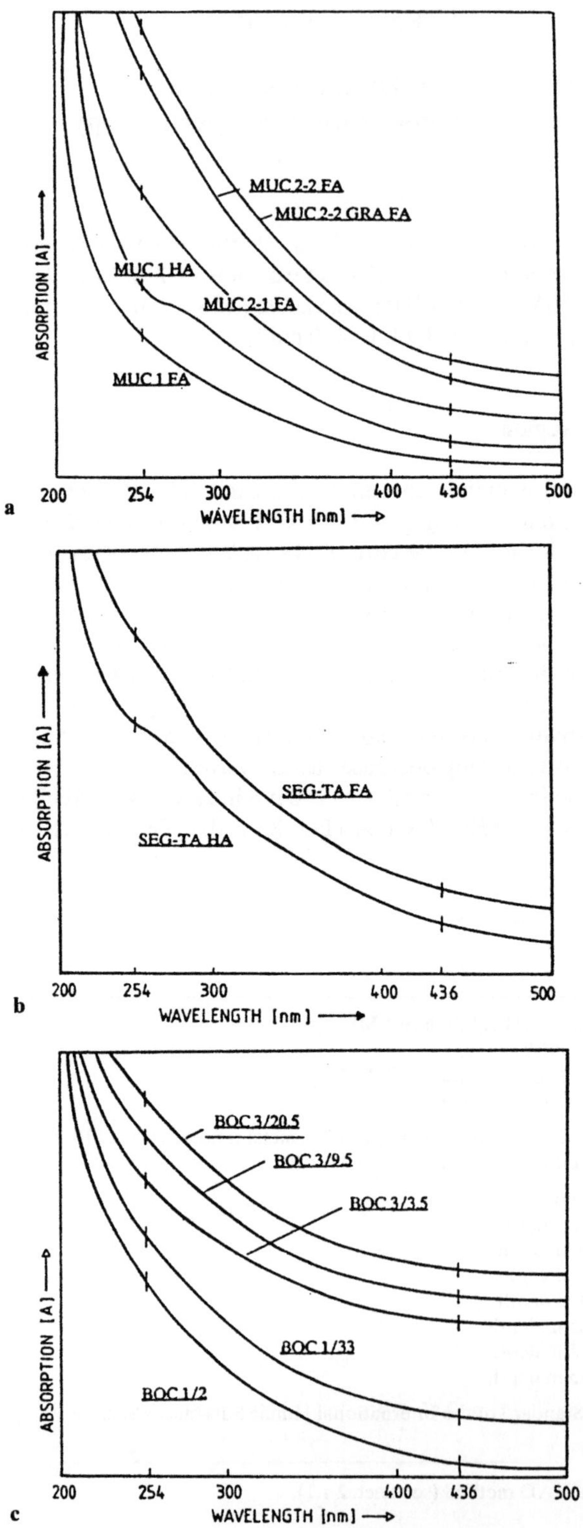

Fig. 2.1.5.1. UV-VIS spectra of **a** MUC1 FA, MUC1 HA, MUC2–1 FA, MUC2–2 FA and MUC2–2 GRA FA; **b** SEG-TA FA and SEG-TA HA; and **c** BOC1 and BOC3 (1 cm quartz cells (ddW); pH = 11; DOC 10 mg/l)

isolated MUC 1 FA and MUC 1 HA show differences in the UV region: the spectral absorbance from 300 to 200 nm is steeper for FA compared with HA. Differences are also found for soil samples which were extracted with different procedures. So the spectral absorbance in the range of 300 to 200 nm is steeper for MUC 1 FA, MUC 2–1 FA and MUC 2–2 FA. The shapes of the UV spectra of MUC 2–2 FA and MUC 2–2 GRA FA are fairly similar. In addition, the spectra of BOC 1 are steeper in UV region compared with BOC 3. The extraction procedures obviously mobilize different substances responsible for different UV absorption because all spectra were obtained for samples with similar DOC concentrations. In addition, irreversible changes of the molecular structure have to be considered especially in the case of high pH values. The specific UV-absorption (normalized to DOC) is higher for HA compared to FA. This can be explained by a core with higher aromaticity in the case of the HA which is also supported by the findings of functional group analysis and elemental composition of the FA. No obvious changes in the UV/VIS spectra appear after equilibration with gravel (MUC 2–2 FA and MUC 2–2 GRA FA). It can be concluded that the optical properties are the same for the fraction which is adsorbed on the gravel surface and the rest which remains in solution.

For comparison, specific spectral absorption coefficients are useful. The ratios of *a*(254 nm)/DOC, *a*(436 nm)/DOC are given for some HS in Fig. 2.1.5.2a. In addition, the (254 nm)/(436 nm) ratio is a parameter well suited for characterization of HS solutions (Fig. 2.1.5.2b).

It is striking that there are great differences concerning the samples from different origins and isolation procedures. The differences are most obvious in the specific UV absorption [*a*(254)/DOC]. It is higher in the case of FA and HA from the rendzina and pore water from the soil than in the case of HS from ground water, which reflects the variation in functional groups. In addition, the values for FAs and HAs from MUC and SEG, with the exception of MUC 1 FA and SEG-KA HA, show a narrow range from 4.5 to 5.5 l/(mg.m), whereas the range for HS from the ground water (BOC) is very broad [2.7 to 4.2 l/(mg.m)]. This leads to the conclusion that the depth from which the samples were taken may have a stronger influence on the optical properties than the geographical location. The values for MUC 1 FA and SEG-KA HA do not fit into the general scheme [2.2 or 25 l/(mg.m)]. It is obvious that in the case of MUC 1 FA the isolation procedure and the extraction by NaOH influences the chemical structure.

The differences concerning the specific yellow colour [*a*(436)/DOC] are very low. Again, the HS from ground water (BOC) show the lowest ratio, which is consistent with the relatively low concentration in chromophoric groups.

The ratio of *a*(254 nm)/*a*(436 nm) is in general lower for HAs than FAs. Therefore it can be said that chromophoric groups responsible for the yellow colour show higher amounts in case of the HAs compared to the FAs. In addition, these functional groups show relatively low abundance in the deeper ground water. It is important to note that the spectral absorbance in the visible range is highly dependent on the pH of the HS solution. The absorbance doubles by increasing the pH value from 6.0 to 8.9 (Frimmel 1977). Therefore the chromophoric groups responsible for the colour must include functional groups with protonatable hydrogen.

This assignment goes along with the observed increase in absorbance of these bands by methylation with diazomethane of the FA (Abbt-Braun and Frimmel 1990a).

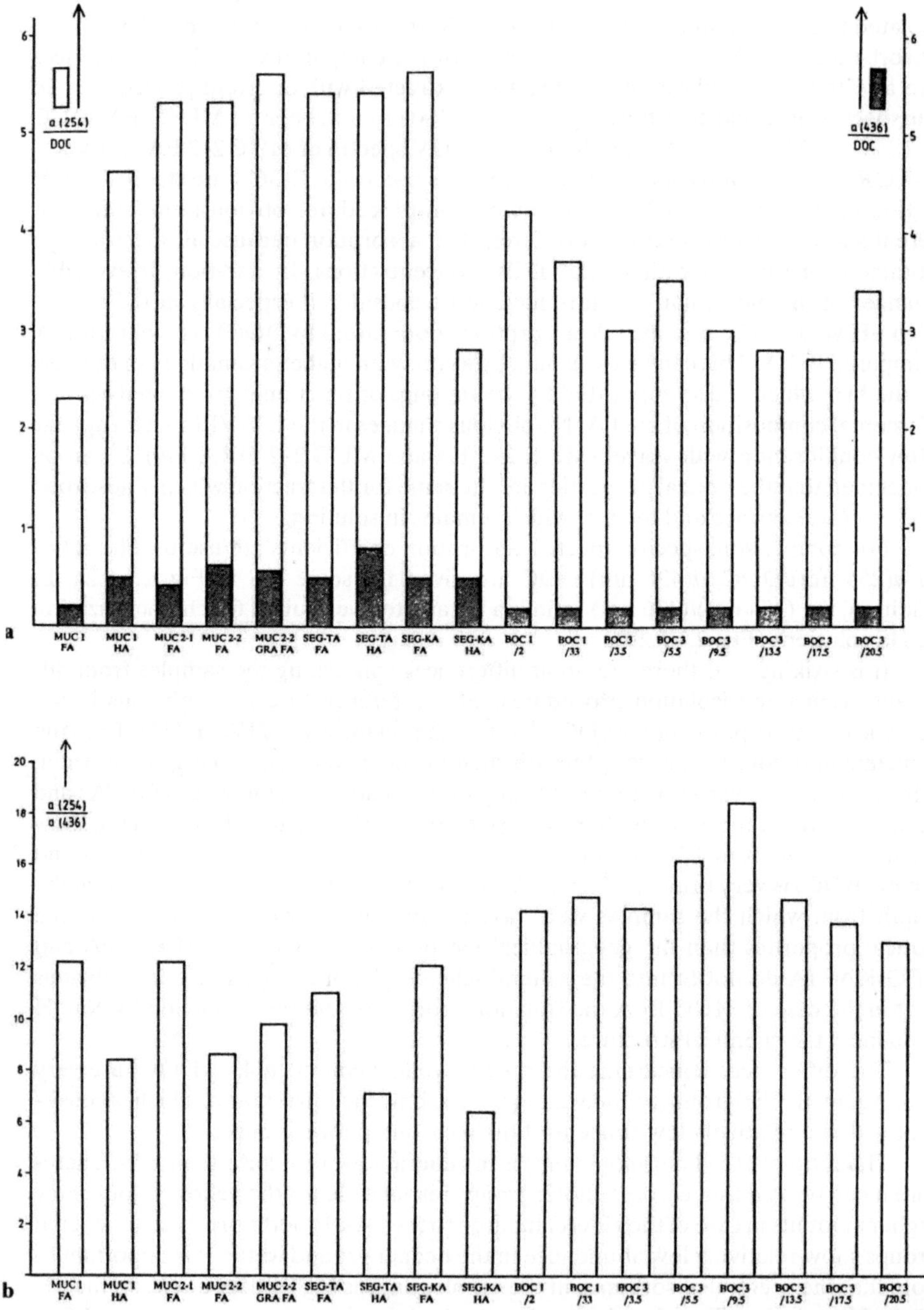

Fig. 2.1.5.2. Specific spectroscopic data of isolated HS. **a** a(254)/DOC: 1/(mg.m); a(436)/DOC: 1(mg.m); pH = 11, **b** a(254)/a(436); pH = 11

2.1.5.2 Infrared Spectrophotometry

In contrast to the featureless UV-VIS spectra, the solid-state IR spectra of HS contain a variety of bands which are characteristic for specific molecular bonds and for inter- and intra-molecular interactions. The assignments to the main absorption bands are given in Table 2.1.5.2 and typical IR spectra of FA and HA are shown in Fig. 2.1.5.3.

The region around 3400 cm^{-1} reflects the OH groups of alcohol, phenols and carboxylic acids. The high shift of the absorption band suggests that the OC-containing functional groups are substituted at the same subunit (e.g. hydroxy benzene carboxylic acid). Due to the low N content in HS, amines and amides do not appear in the spectra. The absorption bands between 2970 and 2936 cm^{-1} are caused by C-H stretching mode of methyl or methylene groups. It is unlikely that amines and amides contribute significantly to the absorption bands in that region. Two of the main bands appear at 1720 cm^{-1} and 1630 cm^{-1}. They are caused by the stretching frequences of C = O bonds of COOH and ketones as well as to quinones, esters and amides. Additionally, the band at 1630 cm^{-1} is caused by the C = C vibration of aromatic and aliphatic molecular units conjugated with C = O. The absorption band at 1720 cm^{-1} is shifted to higher wavenumbers at 1738 cm^{-1} by methylation. Therefore it can be concluded that it is caused mainly by the stretching frequency of the C = O bond of COOH. The intensity of the band at 1720 cm^{-1} is considerably stronger in the FA compared with the HA. This fits well with the observation that the content of COOH groups, measured by chemical and spectroscopic methods, is higher in the case of the FAs (Schnitzer and Khan 1972; Wilson et al. 1981) compared to HAs. By methylation of FA the absorption band at 1630 cm^{-1} is shifted to 1615 cm^{-1}. Moreover, the relatively strong absorption band at 1630 cm^{-1} in HA

Table 2.1.5.2. Main IR Absorption bands of HS and their assignments. (Bellamy 1975, 1980; MacCarthy and Rice 1985; Schnitzer and Khan 1972; Stevenson and Goh (1971)

Frequency (cm^{-1})	Assignment to bonds
3400–3300	O-H, N-H
2940–2900	C-H (aliphatic)
1725–1720	C = O (acid, ketone)
1660–1630	C = O (amide, quinone, conjugated ketone)
1630	C = C (aliphatic)
1620–1600	C = C (aromatic, conjugated ketone)
1590	C = C (aromatic)
1590–1517	COO^-, N-H (amide)
1517	C = C (aromatic)
1460–1450	C-H (aliphatic)
1420	C = C (aromatic)
1400–1390	O-H, C-O (phenol) C-H (aliphatic)
1280–1200	C-O, O-H (carboxylic acid) C-O (aryl ethers) C-N (amide)
1170–950	C-O (polysaccharide) Si-O (silicate impurities)

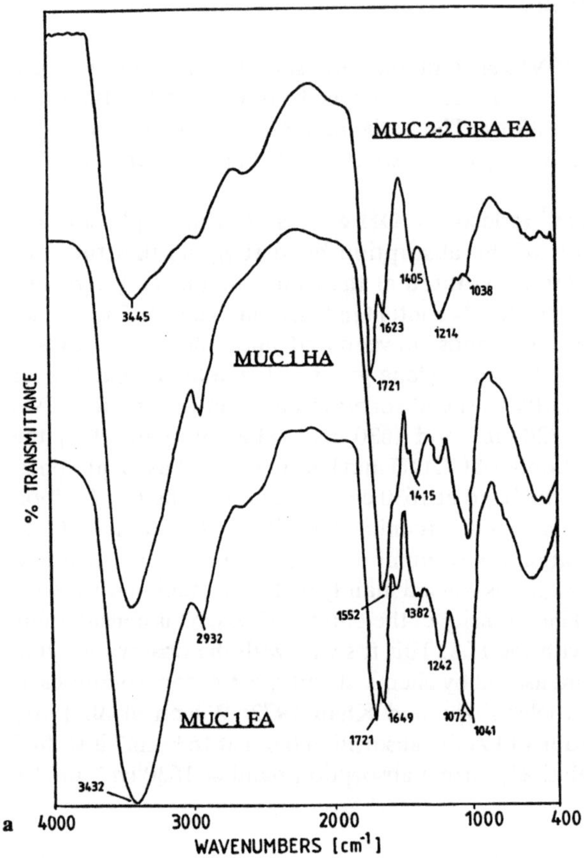

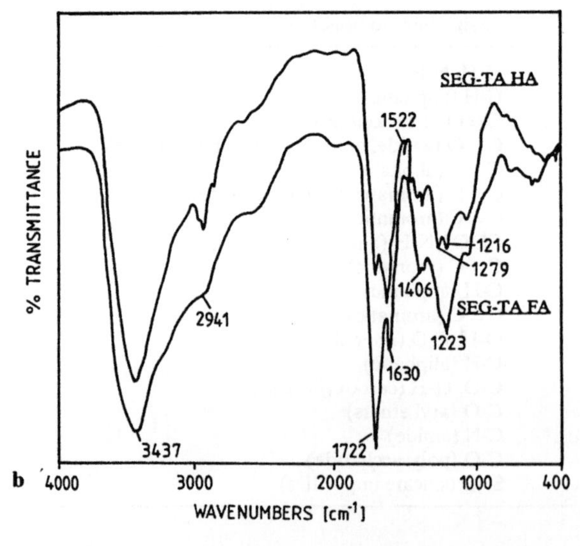

Fig. 2.1.5.3. Typical FT-IR spectra of isolated MUC (**a**) and SEG (**b**) samples (KBr disk, resolution = 2 cm^{-1}, 256 scans)

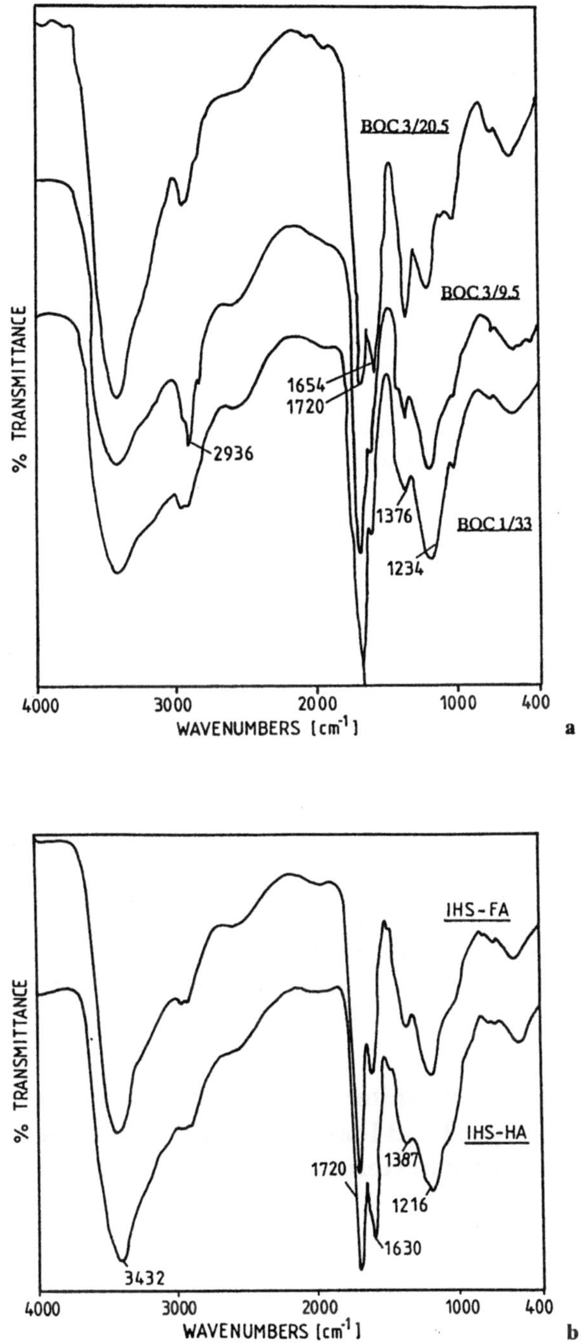

Fig. 2.1.5.4. Typical FT-IR spectra of isolated BOC **a** and IHS **b** samples (KBr disk, resolution = 2 cm^{-1}, 256 scans)

samples is in good agreement with the lower O content of HA compared to FA. The fingerprint region (below 1500 cm^{-1}) is difficult to assign because of a wide variety of combination bands.

The IR spectra of the HS isolated from soil, pore water and ground water shows some differences concerning origin, isolation procedure and the HA and FA fraction. MUC 2–1, MUC 2–2 and MUC 2–2 GRA, extracted by double-distilled water or aqueous pyrophosphate solution, show IR spectra similar to the IR spectra of the IHS-FA from the International Humic Substances Society (Fig. 2.1.5.4b).

The IR spectra of MUC 1 FA and MUC 1 HA (Fig. 2.1.5.3a) extracted from soil by aqueous NaOH are quite different from the IR spectra of the IHS samples (shown in Fig. 2.1.5.4b). It is obvious that the aliphatic portion is high. From the absorption bands at 2932 cm^{-1} and at 1382 cm^{-1} it is obvious that the contribution of aliphatic structures is fairly high in the case of the MUC 1 samples. The strong absorption bands at 1721 cm^{-1} and 1649 cm^{-1} show the same intensities in the case of the FA, whereas the band at 1721 cm^{-1} is not present in the IR spectra of MUC-HA. It is reasonable to assume that the band at lower frequency is caused by the C = O and N-H stretching of amides and amines. The band at 1532 cm^{-1} may be due to the N-H stretching of secondary amides. Looking at the elemental composition of FA and HA extracted from soil, it is interesting to note that soil samples have a higher N content than HS isolated from water. The broad absorption bands at 1041 cm^{-1} and 1072 cm^{-1} can be interpreted by the C-O vibration of polysaccharides. Concerning the great differences in the IR spectra of samples gained by alkali extraction and by soft dissolutions like double distilled water and pyrophosphate, it is reasonable to assume that significant denaturation occured in case of the alkali treatment. No differences were observed between the FA equilibrated with gravel and the FA before gravel contact.

The IR spectra of FA from pore water are similar to those of FA isolated from rivers. HA, however, have a different intensity pattern. It is striking that the absorption bands for HA in the region of 1225 cm^{-1} are split, which can be explained by the C-O and OH vibrations of carboxyl groups and the C-O vibrations of arylethers (Fig. 2.1.5.3a,b). The IR spectra of HS from groundwater, taken at different depths, show also significant differences (Fig. 2.1.5.4a), e.g. the amount of aliphatic structural units seems to be highest at 9.5 m below the water table.

2.1.6 Temperature-Programmed/Time-Resolved Pyrolysis Field Ionization Mass Spectrometry – a New Method for the Characterization of Humic Substances

H.-R. Schulten

2.1.6.1 Background

Rapid progress has been made by mass spectrometric methods mainly in two directions. Firstly, in the soft ionization of thermally labile and/or polar compounds of low volatility and secondly, in the expansion of the available mass range which presently extends well above 20000 mass units. Despite these development, direct molecular-specific characterization of biomacromolecules such as humic substances and organic matter in soils, appears feasible only after chemical or physical degradation.

In recent years, pyrolysis (Py) mass spectrometry (MS) has played an increasingly important role in the analyses of a wide range of biomaterials (Meuzelaar et al. 1982). Most of this work has been carried out using electron impact (E1) MS, mainly with low electron energy. First studies of soft ionization methods were reported for field ionization (FI) MS of bacterial Py products (Schulten et al. 1973b) and Py field desorption (FD) MS of nucleic acids (Schulten et al. 1973a). Essential features of these techniques are the extension of the recorded mass range and the almost exclusive formation of molecular ions of the pyrolyzates (Schulten 1984).

In view of the fact that the importance of humic substances and their role in the environment cannot be overestimated (Schnitzer and Khan 1972; Schnitzer 1973; Frimmel and Christman 1988), it appeared promising and necessary to study these complex materials by the newly developed methods. Therefore, recently mass spectral fingerprinting and time-resolved Py-FIMS of aquatic and terrestrial humic substances have been reported (Schulten 1987).

In the following, a survey of the methodology of thermally programmed, time-resolved Py-FIMS for humic substances (HS) and their fractions such as fulvic acid (FA), humic acid (HA) and humin will be given. The aim is to highlight the progress achieved in analytical mass spectrometry of these complex biomaterials in surface, pore and ground water.

2.1.6.2 Experimental Procedure

Humic substances were mainly isolated, prepared and fractionated according the technique described by Mantoura and Riley (1975). For pyrolysis-field ionization mass spectrometry of HS, approximately 50 µg of dried sample was placed in the quartz micro-furnace of the modified direct introduction system (Schulten et al. 1987b) of a double-focussing Finnigan MAT 731 mass spectrometer. A schematic diagram of the high-temperature, high-sensitivity setup for Py-FIMS is shown in Fig. 2.1.6.1. With this modification of the commercial device (Schulten et al. 1987), an augmentation of the total ion count by a factor of approximately 10 is achieved, thus allowing direct Py-FIMS of whole soils and freeze-dried water. In contrast to previous experiments, the opening in the housing of the ion source opposite to the direct probe was closed by a metal plug in order to enhance the vapour pressure of the pyrolysis products near the FI-emitter. For controlled thermal degradation, the samples were heated linearly from 50 to 750 °C at a heating rate of 1 °C/s. The ion source was maintained at a pressure below 10^{-3} Pa and at a temperature of 200 °C. For electric detection, magnetic scans were recorded in 10 °C intervals in the mass range from m/z 50 to 1000 at + 8 kV emitter potential. Between the repetitive scans the emitter was flash-heated to ca. 1500 °C in order to avoid condensation and to regenerate its surface. About 50 spectra were registered in the whole temperature range. All these spectra were then integrated by using the Finnigan MAT SS200

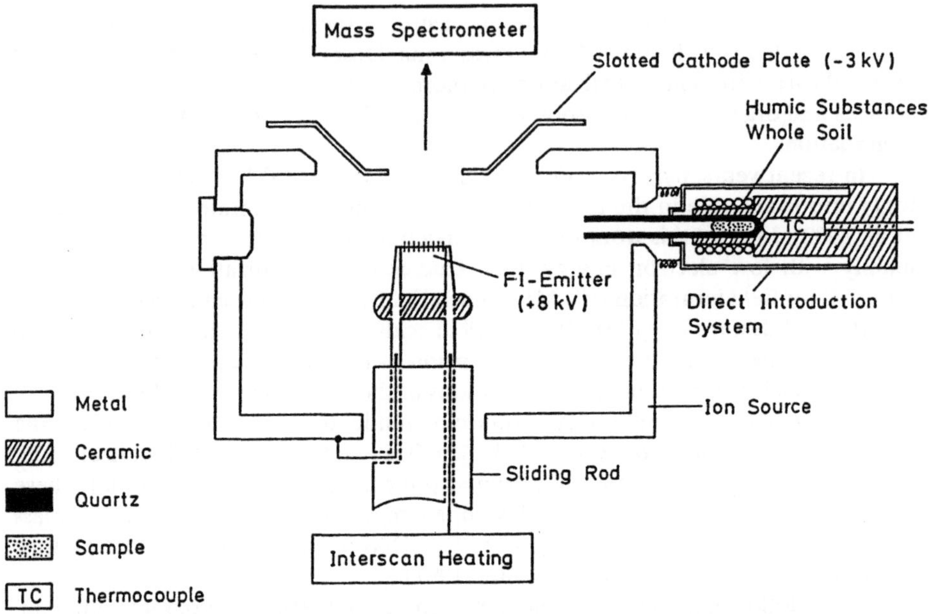

Fig. 2.1.6.1. Schematic drawing of the novel experimental setup for Py-FIMS. The sample of humic substance or soil is deposited inside a quartz micro-furnace which is mounted in the FI ion source of the high performance mass spectrometer

data system. The summed spectra of three separate runs were normalized and averaged to yield a survey spectrum (Schulten et al. 1987b). The integrated intensity of the detected FI signals in a spectrum is expressed in arbitrary units of the total ion count. The latter may be plotted against time, temperature, and scan number to give a thermogram (Schulten 1984). Previously, details of electric and photographic detection of FI mass spectra of standards of the International Humic Substances Society (IHSS) such as Suwannee river FA and HA have been reported (Schulten et al. 1987a; Schulten 1987).

Pattern recognition was done by uni- and multivariate statistical methods to discriminate forest humus horizons (Schulten et al. 1988). Similarly, for the present statistical evaluation of aquatic FA and HA preparations from the Munich (MUC), Bocholt (BOC) and Segeberger Forst (SEG) sites, the modified ARTHUR program (Harper et al. 1977) was employed.

2.1.6.3 Results and Interpretation

First, the thermal characterization of HS and second, the correlation between the recorded FI signals of direct, thermally produced subunits of the humic macromolecules and their structure is attempted in the following distinct analytical steps.

Low Mass Resolution Py-FIMS

It took years to optimize the sensitivity, specificity and reproducibility of the Py-FIMS to obtain mass spectra with characteristic patterns of the HS which can be distinguished by visual inspection, or, if the differences are not discerned in this way, by the application of pattern recognition techniques. As illustrated in Figs. 2.1.6.2 and 2.1.6.3, the HS isolated by Selenka and Hack (Sects. 2.1.3 and 5.1.3) from ground water in Bocholt (BOC1) in the depths from 2 to 33 m give intense and characteristic fingerprints. The weight-average molecular weight ($\overline{MW}$) of the pyrolysate of HS in 2 m depth is 262, whereas in 33 m a $\overline{MW}$ value of 306 can be calculated. In this first step of fingerprinting to characterize and differentiate HS preparations from different depths in the ground water levels, chemical information is not needed.

Time-/Temperature Resolved Py-FIMS

Due to the linear heating rate of 1 °C/s, the total ion count can easily be plotted versus the analysis time, scan number and micro-furnace temperature. In this manner a third dimension, i.e. pyrolysis time or temperature, is added to the ordinate (relative abundance) and abscissa (m/z = mass/charge ratio) of conventional mass spectra. From the resulting thermogram a first indication about the thermal behaviour of the HS samples can be derived. For IHSS Suwannee river FA and HA these differences in volatilization and thermal stability have been discussed (Abbt-Braun 1987; Schulten 1987; Schulten et al. 1987a). In addition, the evaluation of the

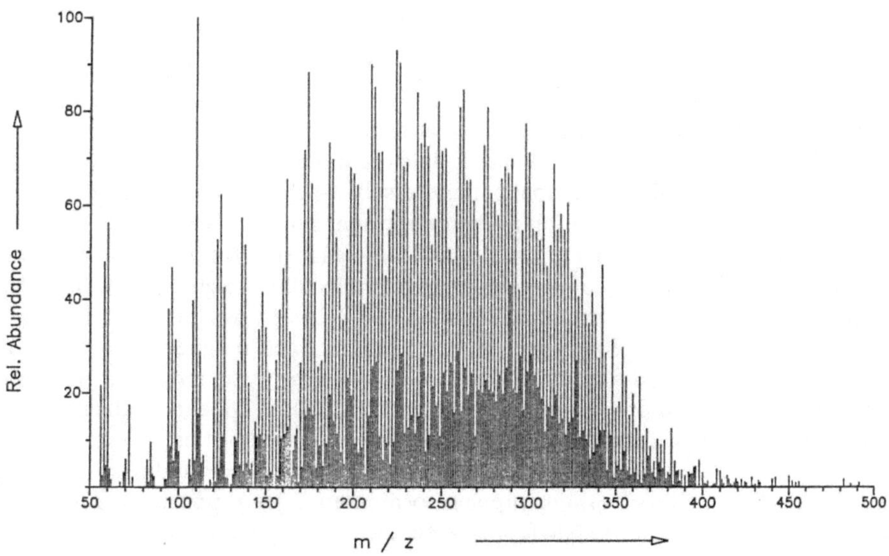

Fig. 2.1.6.2. Py-FI mass spectrum of humic substance isolated from ground water at Bocholt (BOC1; Sect. 5.1.3) at 2 m depth

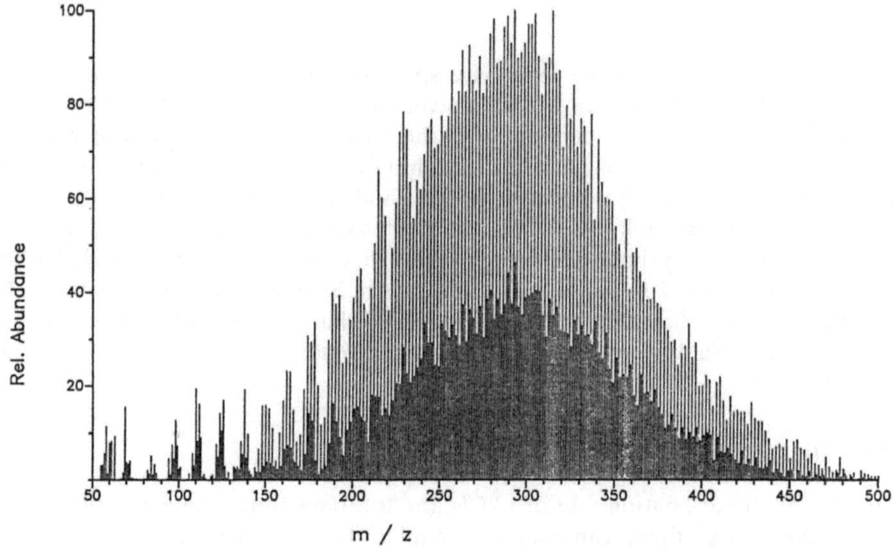

Fig. 2.1.6.3. Py-FI mass spectrum of humic substance isolated from ground water at Bocholt (BOC1; Sect. 5.1.3) in 33 depth

thermograms makes it possible to follow the release of individual volatile products and pyrolysates of the macromolecules HS.

Recently, weight loss curves for IHSS FA and HA were introduced using Py-FIMS as a novel variant to evaluate properties of HS such as moisture, char, ash contents and the evolution of organic Py products (Schulten 1987). Generally, FA decomposes over a wider temperature range (200–650 °C) and gives less residue (8.5%) compared to HA (200–550 °C, 19.2%). This technique represents a kind of thermogravimetry with a highly sensitive and specific detector. However, owing to the higher heating rates, thermodynamic data are more difficult to achieve.

In addition to studies of global parameters by weight loss curves, the characterization of HS in the configuration of their molecular subunits is a decisive task. The first link between the thermal properties and molecular structures of building blocks could be the recognition of weight-average molecular weights of Py products monitored during the Py process (Schulten 1987). As shown in Fig. 2.1.6.4, it is possible to produce a plot of the $\overline{MW}$ values for each magnetic scan and the corresponding Py-FI mass spectrum (ordinate) versus the Py temperature (abscissa). For the HA preparations of four soil horizons (SEG), it appears that during the processes of podzol formation the size and thermal stabilities of the subunits increases from Ahe to B(h)s. In an integrated approach, these findings were compared with the results of vapour pressure osmometry, gel permeation chromatography, UV/VIS-, FTIR-, ^{13}C-NMR spectroscopy and elemental analyses (Post et al. 1988).

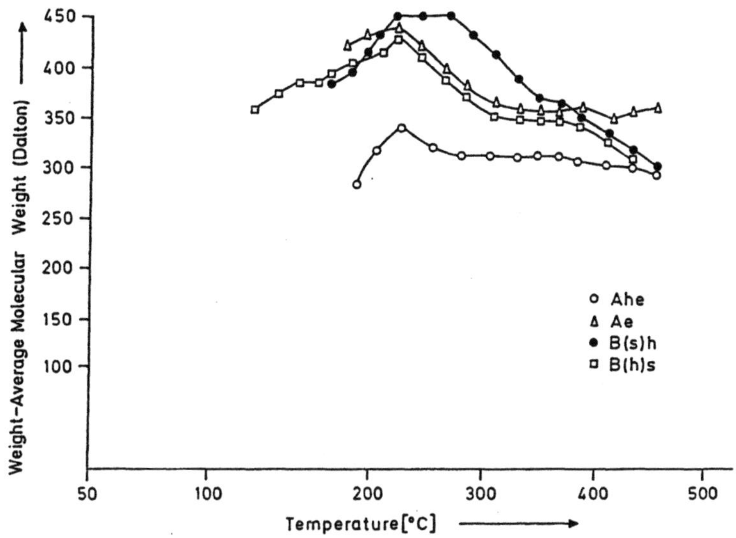

Fig. 2.1.6.4. Plot of the sample temperature versus the weight-average molecular weights of the pyrolysis products of HA (SEG) isolated from four horizons of a podsol (Post et al. 1988; Sect. 2.1.2). The data points represent single FI mass spectra above a threshold of 3% total ion intensity

High Mass Resolution Py-FIMS

In general, the problem of identification of the Py products is very difficult. Despite the fact that mass spectrometric fragmentation is strongly reduced in comparison to other methods of ionization and mainly molecular ions of chemical species generated by thermal fragmentation are observed, the unambiguous identification of HS building blocks is difficult. The high complexity of the spectra is not only reflected in the plurality of nominal masses as demonstrated, for instance in Figs. 2.1.6.2 and 2.1.6.3 with more than 400 peaks. Even at a mass resolution of 5000 to 10000 (10% valley definition), it became clear that most signals are multiplets (Schulten 1987). Thus, high mass resolution and accurate mass measurements within an error of a few millimass units can already reveal the presence of up to six maxima, which are all determined by their elemental composition. Considering the number of possible isomers which all follow these proposals for elemental formulae makes the exhaustive identification an "endless story"!

However, using high resolution by electric or photographic detection, it is possible to follow the evolution of individual chemical species from HS in time, spectrum number or temperature as illustrated in Fig. 2.1.6.5. The thermograms, i.e. the plots of intensity versus time, spectrum number and sample temperature, show **a** the molecular ion current of dimethylfuran, $(C_6H_8O)^+$, calc. 96.0575 Dalton; **b** the molecular ion current of furfural, $(C_5H_4O_2)^+$, calc. 96.0211 Dalton. The ion currents were recorded with a mass window of 20 millimass units. This analytical approach is valid only in favourable cases where the intensity of the investigated signal is high, one elemental composition dominates and the temperature resolution during

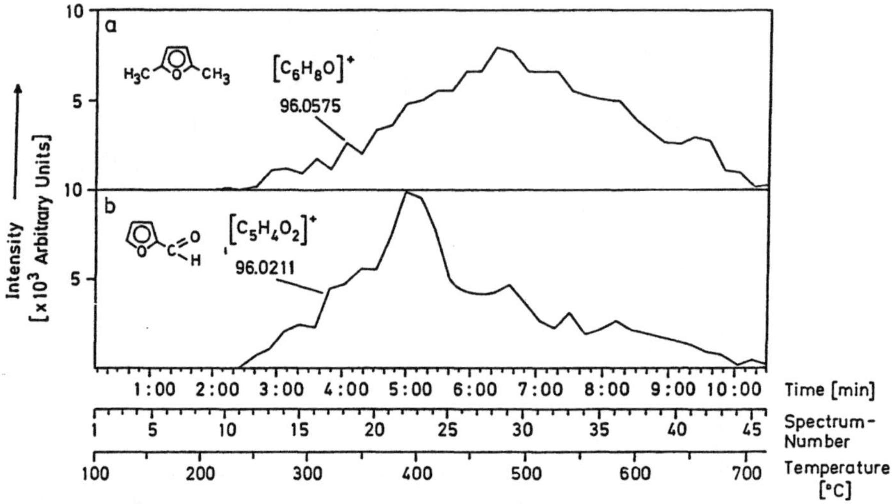

Fig. 2.1.6.5. Time-resolved and electrically recorded FI ions with the nominal mass at m/z 96. The recorded FI ions are derived from the Py products of FA (MUC)

Py-FIMS is sufficient to resolve the thermal evolution of the individual pyrolysis products. For reliable assignment, the standard compound, library data and confirmation by GC/MS are absolutely necessary.

Flash Pyrolysis-GC-MS

Structural characterization of aquatic humic materials after derivatization (e.g. permethylation) by Py-GC/MS has been reported for conventional electron ionization (EI) and chemical ionization (CI) MS (Liao et al. 1982). Recently, Curie-point Py in combination with GC/EI and GC/FIMS has been employed to compare the identified subunits of HS with the investigations by direct Py-FIMS (Abbt-Braun 1987; Schulten et al. 1987a). The results of this comparison are very encouraging because the previously found structural subunits could be confirmed and novel Py products of higher polarity and mass could be proposed.

2.1.6.4 Pattern Recognition

Integrated Py-FI mass spectra were produced of humic and fulvic acid fractions of a large number of samples of aquatic humic substances isolated from the Brunnseemoor (BM), the Segeberger Forst (SEG) and Bocholt (BOC). The BM specimens were taken at different times in the years 1983–1986 and their physical and chemical properties have been described in detail by Abbt-Braun (1987). Each sample was analyzed several times, yielding 30 integrated mass spectra of HA and 38 mass spectra of FA fractions. For these fractions all mass spectra are summed and the resulting representative Py-FI mass spectra are shown for FA in Fig. 2.1.6.6 and for HA in Fig. 2.1.6.7. The comparison of both mass spectra shows that the

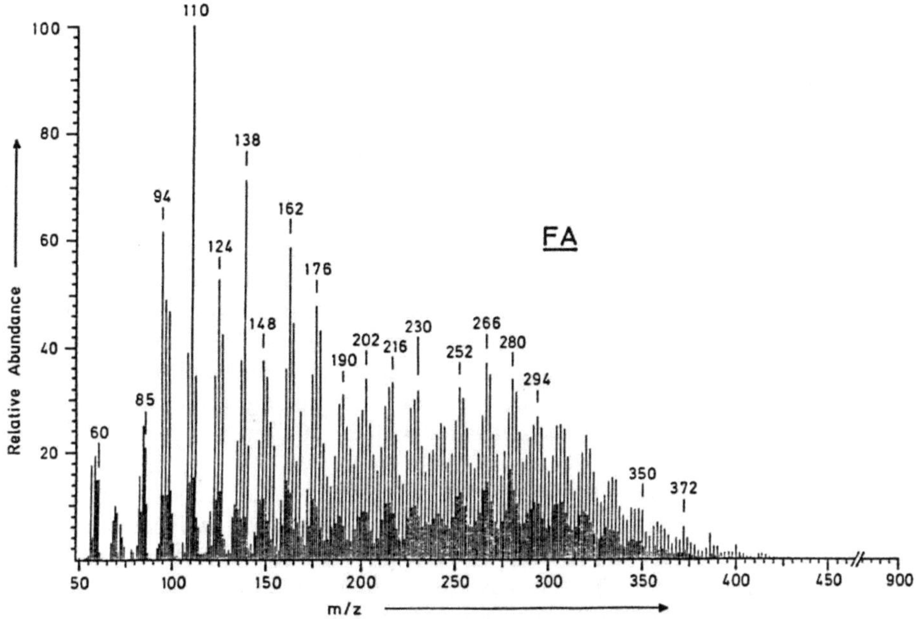

Fig. 2.1.6.6. Representative Py-FI mass spectrum of the normalized and summed ion intensities from measurements of 38 aquatic FA fractions isolated from BM (Brunnseemoor, Bansee), Bocholt (BOC1 and BOC2) and SEG (podsol)

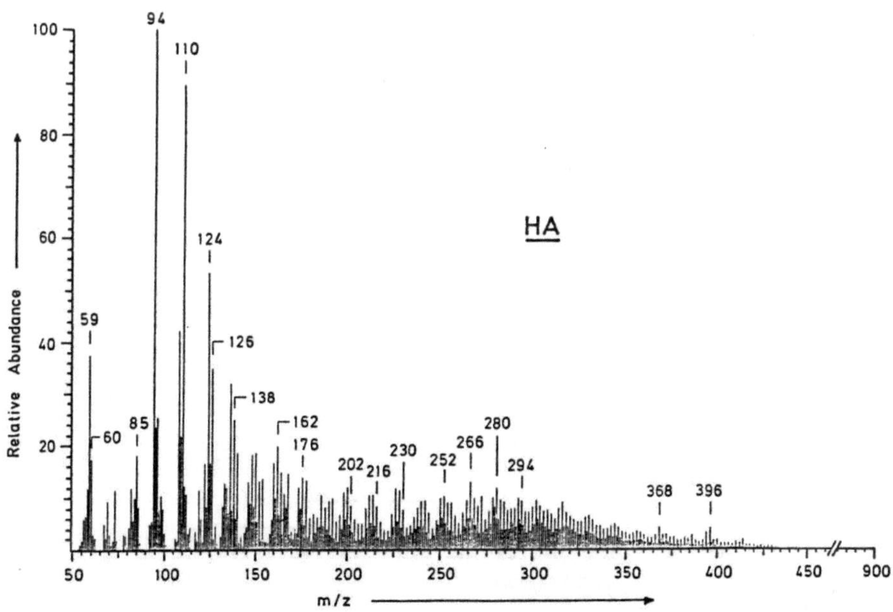

Fig. 2.1.6.7. Representative Py-FI mass spectrum of 30 HA fractions integrated as for FA in Fig. 2.1.6.6 (same origins)

thermal degradation of FA results in much more abundant FI signals in the range m/z 200–300, whereas more intense mass signals in the mass range m/z 300–400 are observed in the Py mass spectrum of HA.

In order to find a significant FI mass signal pattern which is characteristic for the differences between FA and HA, principal component analysis was carried out with pre-selected FI mass signals. Pattern recognition studies using Py-EI mass spectra have been described by Meuzelaar et al. (1982). In the present study, the adopted ARTHUR statistical software package was used for the chemometric evaluation of the mass spectral data set consisting of 68 HA and FA Py-FI mass spectra. The relative intensities of all FI mass signals were normalized. The mass spectral data sets for FA and HA samples were divided into two groups. For selecting the most characteristic mass signals for the distinction of FA and HA, the Fisher weight was calculated for each FI signal. This is the ratio between the differences of the two groups (outer variance) and the variation within one group (inner variance). Principal component analysis was performed with the 50 FI mass signals with Fisher weights above 0.93.

Figure 2.1.6.8 presents the score plot of the first versus the second principal component. Both principal components describe 78.3% of the variance of the data set. As indicated, almost all FA samples are separated from the HA samples by the first principal component. There is only one exception, the four mass spectra of the HA fraction of sample BM-9 which are grouped to the mass spectra of the FA spectra. Looking on the recorded Py-FI mass spectra, it is obvious that the HA mass spectra of BM-9 are very similar to the FA spectra. At present it cannot be ex-

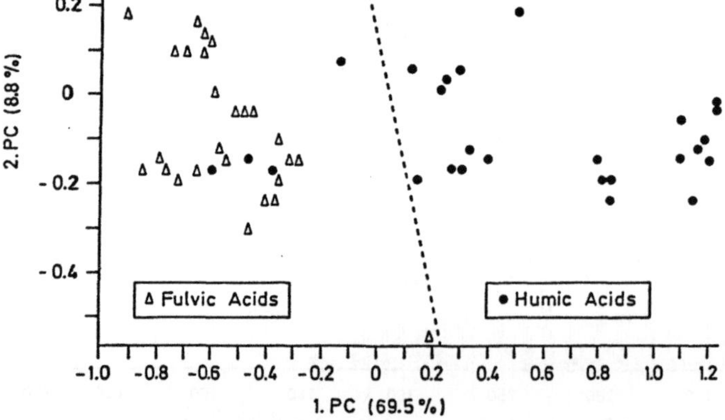

Fig. 2.1.6.8. Score plot of the first versus the second principal component of the 38 FA and 30 HA used for Figs. 2.1.6.6 and 2.1.6.7

plained, but presumably this is due to the specific character of the brown water lake. All results indicate that Py-FI mass spectrometry can be used for quality control of the HS fractionation and allows the distinction of several HA and FA fractions.

Besides principal component scores, the loadings are another result of principal component analysis. Mass signals which significantly load a principal component are very important and intercorrelated and are the basis for chemical interpretation. Hence the mass signals which load the first principal component will indicate the signals which are responsible for the distinction of FA and HA in Fig. 2.1.6.8 and in Fig. 2.1.6.9, indicate the importance of the corresponding FI mass signal for the distinction of FA and HA. According to Abbt-Braun (1987), these FI signals can be assigned to following thermal degradation products of HS: methyl-methoxyphenol at m/z 138, dimeric furan subunits in a homologous CH_2 series at m/z 162, 176, 190, dimeric methoxyphenol-alkylphenol-units in a homologous CH_2 series at m/z 188, 202, 216, 230, 244, 258 and dimeric alkylphenol-alkylphenol-units in a homologous CH_2 series at m/z 238, 252, 266, 280 and 294. Obviously, the monomeric thermal degradation products at m/z 138 and the dimeric thermal degradation products in the mass range m/z 200–300 are significantly more intense in the FA fraction, whereas the trimeric thermal degradation products in the mass range m/z 300–400 are more intense in the HA fractions.

These preliminary results demonstrate that Py-FIMS in combination with pattern recognition techniques is a powerful tool for the universal, objective and quantitative characterization of HS. It can be expected that the consequent application of these novel techniques will open up new insight the nature of HS.

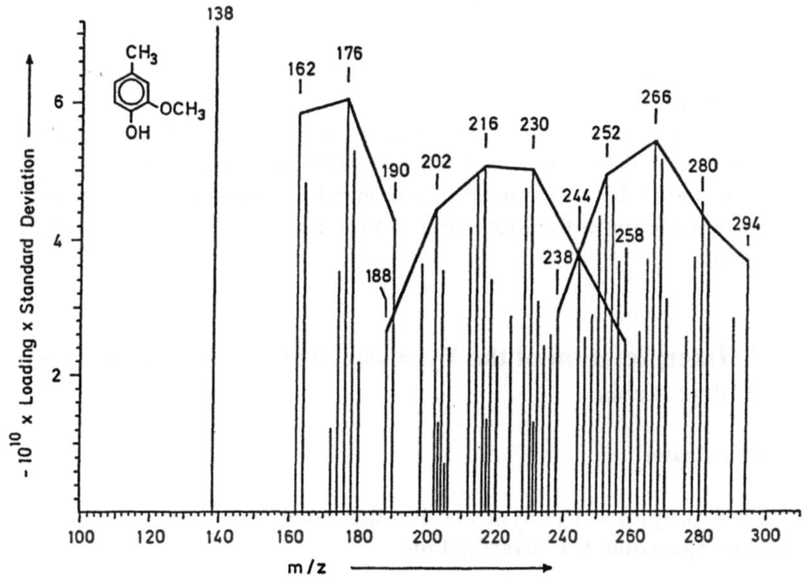

Fig. 2.1.6.9. Principal component spectrum of the first principal component calculated from Py-FI mass spectra of the FA and HA fractions used in Figs. 2.1.6.6–2.1.6.8

2.1.6.5 Conclusions

Critical points are firstly, that the knowledge about the relationship between thermally generated products and the original structure of complex biomaterials is still very limited. Secondly, for statistical work the available data sets on biopolymers and HS are too small and incomplete. Thirdly, it is evident that the term "structure" as used in the conventional way for a chemical compound is not valid for humic materials and has still to be defined in an operational manner. As any isolation and preparation of aquatic or terrestrial humic materials or fractions will inevitably change the original three-dimensional network, one possibility to reduce isolation artefacts may be the direct analysis of whole soils (dried, milled, sieved) and water (filtered, freeze-dried). Investigations using Py-FIMS for direct characterization of whole soils (Haider and Schulten 1985; Schulten 1987; Schulten et al. 1990), of forest soil profiles such as mull, moder and moor (Hempfling et al. 1987; Schulten et al. 1988; Hempfling and Schulten 1988, 1990a), and water with DOC contents between 4 and 160 mg/l (Simmleit and Schulten 1989; Hempfling and Schulten 1990b) have been encouraging.

Summarizing the results obtained by Py-FIMS and pattern recognition of aquatic HS, FA and HA, the following facts can be outlined.

- It is clear that HS can be distinguished from each other reproducibly and quantitatively by using Py-FI mass spectra as fingerprints. The analytical potential of the obtained mass spectral pattern is their relationship to typical, chemical building blocks. Thus, quality control of HS and its fractions is achieved which can serve as basis for differentiation, classification and standardization.
- It appears that the combination of temperature-resolved thermal degradation, low- and high-mass resolutions MS, Py-GC/MS, MS/MS, library search and the utilization of complementary data of other, preferably non-destructive, analytical methods in an integrated approach can supply structural information on large HS subunits.
- It may be possible in future work to correlate quantitative results of the environmental interaction of HS with the highly specific data of Py-FIMS. The aim would be to establish a canonical correlation matrix which allows for selected features the modelling and prediction of the role of HS in the environment.

2.1.7 Interpretation of the Pyrolysis Products of Isolated Humic and Fulvic Acids

G. Abbt-Braun

2.1.7.1 Principles of Pyrolysis and Product Analysis for Mass Spectrometric Investigation

In contrast to the fundamental significance of humic substances (HS), little is known about their chemical structure. One of the reasons is the complicated macro-

Table 2.1.7.1. Summary of the methods applied

Method	Result	Example
Py-GC/MS	Structure of a component	Phenol
Low mass resolution Py-FIMS	m/z Values, m/z = 94 m/z series	 m/z = 94, 108, 120, . . .
High mass resolution Py-FIMS	Exact masses, 94.0419 elemental formula	 C_6H_6O

molecular composition and the great variety of molecular building blocks and their arrangement. Structural information, however, is the basis for the understanding of the reaction mechanisms in which HS are involved. Due to the low volatility of HS, mass spectrometry (MS) and gas chromatography/mass spectrometry (GC/MS) can only be applied after degradation and derivatization of the samples. Pyrolysis (Py) with a high yield of degradation products of the humic material was used as a promising method for structural studies.

By the absorption of thermal energy, vibrational bonds become excited until relaxation takes place by heterolytic and homolytic dissociation of the weaker bonds. Pyrolysis results in some fundamental reactions: for example, decarboxylation of the carboxyl groups, scissions of C-O bonds followed by OH elimination as H_2O, scissions of C-C bonds leading to a double bond. The pyrolysis of biopolymers, such as polysaccharides, lignins or lipids has shown that the thermally produced fragments are highly characteristic and give information on the original structure of the molecule (Irwin 1982; Bracewell et al. 1989).

A detailed description of the methods applied and the experimental procedures are given by Schulten (Sect. 2.1.6, this Vol.).

The different methods and the information obtained are listed in Table 2.1.7.1. In putting these results together and in comparing them with the results of other methods, valid structural information on HS can be gained.

2.1.7.2 Interpretation of the Low Mass Resolution Py-FIMS

The integrated Py-FIMS spectra directly reflect the total amount of all molecular ions produced by the pyrolytic conditions (Figs. 2.1.6.6 and 2.1.6.7 for aquatic FA and HA and Figs. 2.1.6.2 and 2.1.6.3 for HS of ground water in Sect. 2.1.6).

The m/z-patterns seem to be fairly similar. Differences are mainly observed in the intensities. Typical pyrolysis products are reflected by their ion sequences. It is likely that the signals are caused by substances like furans, phenols, methoxyphenols, dimethoxyphenols, alkylbenzenes, dihydroxybenzenes, alkenes and alkadienes. The corresponding m/z values are listed in Table 2.1.7.2. The ion sequences of HS can be compared with typical sequences of structurally related substances with known molecular formula. In addition, the mass spectra can be used to fit the suggested constituents of humic matter known from literature.

Table 2.1.7.2. Dominant sequences of characteristic pyrolysis products

Class of compounds	m/z Series
Furans	68, 82, 84, 96, 98, 110
Phenols	94, 108, 120, 122, 134, 136, 148
Methoxyphenols	124, 138, 150, 152, 164, 166, 178
Dimethoxyphenols	154, 168, 180, 182, 194, 196, 210
Alkylbenzenes	78, 92, 106, 120, 134, 148, 162
Dihydroxybenzenes	110, 124
Alkenes	56, 70, 84, 98, 112
Alkadienes	54, 68, 82, 96, 110

R = alkyl chain partly oxygen substituted

Some characteristic building blocks of the HS structure are polysaccharides, phenol precursors, aromatic hydrocarbon precursors, lipids and lignins (Irwin 1982; Meuzelaar et al. 1982; Haider and Schulten 1985; Saiz-Jimenez and De Leeuw 1986). The pyrolysis products in the higher mass range (> m/z 170) cannot be matched with the components described in Table 2.1.7.2. This is due to the different pyrolysis techniques applied. Most of the data shown in the literature are derived from Curie-point pyrolysis, whereas the method applied in this work was time-resolved pyrolysis combined with soft ionization, which is a less energetic procedure and extends the mass range of intact pyrolysis to higher m/z values.

The m/z values of the higher mass range reflect larger building blocks of the macromolecules. The identification of these products is very difficult by the common GC/MS methods because of their low volatility. By the Py-FIMS method it has been shown for lignins that dimeric guaiacyl fragments occur as higher aggregated structural subunits of soil polymers (Haider and Schulten 1985).

It is attractive to assume that some of the more abundant mass numbers can be attributed to the molecules shown in Table 2.1.7.2 which are linked such as: alkylphenol-alkylphenol units, e.g. at m/z 214, 228, 238, 252, 266, 268, 280, 282 and 294 or methoxyphenol-alkylphenol units, e.g. at m/z 216, 230, 242, 256, 270, 284, 296 and 310.

To obtain additional information about the chemical nature of the m/z values, the influence of oxidative degradation prior to pyrolysis can be examined. The oxidative degradation products of HS can also be identified by GC/MS. Therefore the molecular weight of the identified degradated molecules can be compared with the m/z values of the Py-FIMS spectra. This has been already discussed for FA oxidized after methylation and for FA methylated after oxidation (Schulten et al. 1987a). In the Py mass spectrum of the methylated-oxidized FA the m/z values of aliphatic and aromatic acids show high intensities. Benzene carboxylic acids, methoxybenzene carboxylic acids and monobasic aliphatic acids dominate the spectrum. In the case of the mass spectrum of the oxidized-methylated FA methyl esters of benzene, carboxylic acids and aliphatic dibasic acids are dominant.

2.1.7.3 Interpretation of the High Mass Resolution Py-FIMS

A more detailed interpretation of the m/z values can be achieved by high mass resolution Py-FIMS. The method described by Schulten (Sect. 2.1.6, this Vol.) shows that most of the signals are multiplets for which the elemental composition can be determined. For these elemental formulae chemical structures may be suggested. In addition, results from Curie-point pyrolysis GC/MS investigations can be used for interpretation.

Table 2.1.7.3 shows some selected m/z values representing the furan and phenol series. A detailed description of other masses is given elsewhere (Abbt-Braun et al. 1989). Although it is obvious that each m/z value can stand for one or more chemical structures, they become more meaningful when integrated into the m/z series representing molecular building blocks. In summarizing the results of low and high resolution mass Py-FIMS and Curie-point Py-GC/MS investigation, some characteristic classes of compounds arise: aliphatic and aromatic aldehydes and ketones, aromatic and aliphatic acids, furans, benzenes, phenols and N-containing heterocyclic ring structures.

2.1.7.4 Discussion of Structural Units

One of the main questions using pyrolytic methods is whether the pyrolysates represented by their m/z values, elemental compositions and chemical structures were already present in the original sample or whether they are secondary products of the pyrolytic processes. In the following discussion a critic evaluation of the results of pyrolysis is made, and structural units (building blocks) of HS are derived (Fig. 2.1.7.1).

Carbohydrates (No. 1)

It is still unanswered whether the furans were originally present as structural units or whether the furans were thermally generated. This is because furans can be formed pyrolytically from carbohydrates (Irwin 1982; Meuzelaar et al. 1982) and from aliphatic chains with hydroxy groups (No. 2) (Bracewell et al. 1989). For the HS investigated, it is likely that the amount of polysaccharides containing hexose and pentose units is very low because of the signal pattern in the Py-FIMS spectra. Therefore aliphatic C-chains (C-4 and higher) which carry one or more OH-, COOH-, CH_2OH groups seem to be an important building block in the HS-structure. In the original sample, aromatic rings can be connected by aliphatic chains with oxygenated functional groups. Thermal decomposition in this case leads to aromatic units and to aliphatic subunits which pyrolyze to furans. This is in good agreement with structural assignments, showing aromatic rings connected by two aliphatic chains up to C-5 as a result of oxidative degradation of HS (Christman et al. 1989).

A large fraction of the pyrolysates consist of benzenes substituted with hydroxy-, methoxy-, carboxylic- and carbonyl-groups. Phenol-, methoxyphenol- and dimethoxyphenol derivatives are known to be characteristic pyrolysis products of

Table 2.1.7.3. Comparison of high-resolution MS data of pyrolysis products obtained from aquatic FA (isolated from a bog lake) by photoplate recording with the identified and proposed compounds obtained by Curie-point Py-GC/MS showing the same elemental composition

Class of Compound	Nominal mass	Accurate mass measurement			Curie-point-Py-GC/MS identified compound[a] proposed compound[b]	Reference[c]
		Measured mass	Calculated mass	Probable elemental composition		
Furans	82	82.0424	82.0419	C_5H_6O	2-Methylfuran, or 3-	1–6, 8, 9
					2 Cyclopenten-1-one	1, 5
	84	84.0218	84.0211	$C_4H_4O_2$	Hydroxyfuran	4
					2 H-furan-3-one, or 5H	9
					2-Keto-3-butenal	9
		84.0574	84.0575	C_5H_8O	Dihydropyran	2, 6, 7, 8
					Methyldihydrofuran	2
					Methylcyclobutanone	2
					2-Methyl-1 buten-3-one	9
	96	96.0220	96.0211	$C_5H_4O_2$	2-Furaldehyde	[a]1–10
					3-Furaldehyde	[a]1–10
					Cyclopent-1-ene-3, 5-dione	7, 9
					2 Hydroxyclopentanone	7
		96.0587	96.0575	C_5H_8O	2,5-Dimethylfuran	[a]2–5, 7, 8, 10
					2,4-Dimethylfuran, or 2,5	2, 4, 5, 7, 8
					Ethylfuran	2
					3-Methylcyclopent-2-n-1-one	1, 2, 7, 9, 10
	110	110.0403	110.0368	$C_6H_6O_2$	5-Methyl-2-furaldehyde	[a]1, 3–10
					1,2-Dihydroxybenzene	[a]4, 7, 9
					Acetylfuran	9
		110.0403	110.0480	$C_5H_6N_2O$	1(1H-pyrazol-4-yl)-ethanone	[a]
		110.0765	110.0732	$C_7H_{10}O$	Propylfuran	2, 7, 9
					2-methyl-5-ethylfuran	7, 9
					2, 5-dimethyl-cyclopentenone	9
Phenols	94	94.0432	94.0419	C_6H_6O	Phenol	[a], 1–10
					Vinylfuran	2, 7, 9
	108	108.0586	108.0575	C_7H_8O	2-methylphenol, or 3-, 4-	[a]1, 2–10
					2-methyl-5-vinylfuran	9
					2-methyl-1,3-cyclo-pentanedione	9
	120	120.0601	120.0575	C_8H_8O	Acetophenone	[a]1, 6, 7, 10
					Benzoacetaldehyde	[a]
					2,3-dihydrobenzofuran	[a]10
					Vinylphenol	2, 6, 7
					4-methylbenzaldehyde	9
	122	122.0381	122.0367	$C_7H_6O_2$	Benzoic acid	[a]10
					3-hydroxybenzaldehyde	7, 9
		122.0755	122.0731	$C_8H_{10}O$	2-ethylphenol, or 4-	[a]1, 2, 9
					3,5,-dimethylphenol, or 2,5-,2,4-,2,6-	[a]9
					Methylbenzylalcohole	6

Table 2.1.7.3. (*Continued*)

Class of Compound	Nominal mass	Accurate mass measurement			Curie-point-Py-GC/MS identified compound[a] proposed compound[b]	Reference[c]
		Measured mass	Calculated mass	Probable elemental composition		
	134	134.0750	134.0732	$C_9H_{10}O$	Phenyl-2-propene	7
					Methyldihydrobenzofuran	7
	136	136.0575	136.0524	$C_8H_8O_2$	Hydroxyacetophenone	9
					Hydroxybenzaldehyde	9
	150	150.0711	150.0681	$C_9H_{10}O_2$	Methoxyvinylphenol	5, 6, 9

[a]Identified by comparing the El mass spectra with mass spectral libraries.
[b]Well-known pyrolysis products of biologically produced substances such as lignins, polysaccharides and proteins.
[c]1 = Gadel and Bruchet 1987; 2 = Saiz-Jimenez and de Leeuw 1986; 3 = Martin et al. 1977; 4 = Bracewell et al. 1989; 5 = Irwin 1982; 6 = Saiz-Jimenez et al. 1987; 7 = Van Smeerdijk and Boon 1987; 8 = Wershaw and Bohner 1969; 9 = Powels et al. 1987; 10 = Wilson et al. 1983.

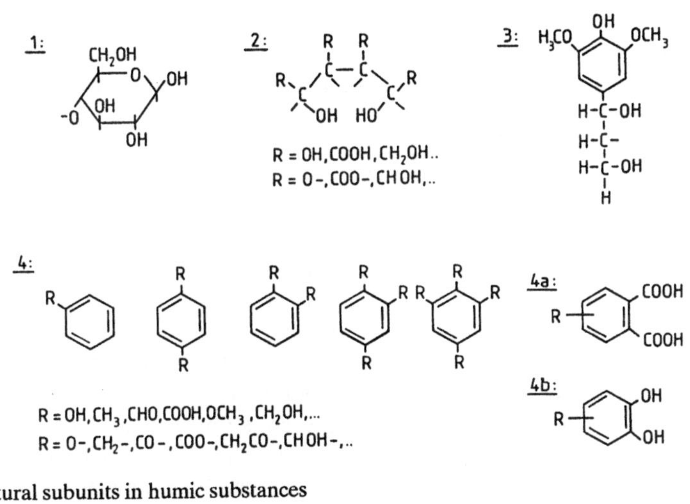

Fig. 2.1.7.1. Some structural subunits in humic substances

lignins (No. 3). In addition, benzenes are obvious (No. 4) which are mono-, 1,2-, 1,4-, 1,2,4- and/or 1,2,4,6-substituted. They show different functional groups, such as -OH, -COOH, $-CH_3$, $-OCH_3$ etc. or they are linked with other subunits (-OR, -COOR, -R). Benzenes substituted with two -O-functions or with two -O- and one -C-function are quiet abundant (No. 4a). In addition, 1,2-benzenedicarboxylic acid and 1,2-dihydroxybenzene are present (No. 4b).

The structural models shown in Fig. 2.1.7.1 represent some characteristic molecular units of HS based on Py-FIMS investigations which are in agreement with the results from other methods like IR, NMR spectroscopy, elemental analysis and titration of the acidic functional groups.

2.1.8 Characterization of Isolated Humic Material by ^{13}C Nuclear Magnetic Resonance Spectroscopy

J. Buddrus and P. Burba

2.1.8.1 Background

The nuclei of certain atoms (e.g. ^{1}H, ^{13}C) possess a magnetic momentum. After their excitation in a magnetic field energy transitions occur. The resulting nuclear magnetic resonance spectra (NMR) provide valuable information on the direct chemical environment of the atomic nucleus under study. Thus, NMR is widely used for the characterization of hydrogen and carbon-containing molecules, including humic substances (HS).

In a series of studies NMR has been applied in order to elucidate the partial structures in HS (Wershaw 1985). Insights into HS were possible by ^{1}H-NMR, and later by ^{13}C-NMR. Other nuclei (e.g. ^{31}P, ^{17}O) are less important for NMR of HS. The application of NMR spectroscopy of HS has been reviewed by several authors (Wilson 1981, 1989; Steelink et al. 1983, 1989; Wershaw 1985; Malcolm 1989).

In general, H-NMR spectra of HS show signals of aliphatic, alkenic, and aromatic protons. Unfortunately, the signals of other relevant protons, particularly bound to functional groups (e.g. alcohols, phenols, carboxylic acids) interfere by producing H-O-D signals from reactions with the solvent D_2O. Thus, the quantitative evaluation of these protons in HS is difficult. Accordingly, ^{1}H-NMR can rarely be used for research in the HS field.

Compared to ^{1}H-NMR, ^{13}C-NMR is of greater importance for the structural investigation of HS: (1) interferences by solvents can be ignored. (2) The ^{13}C signals cover a considerable frequency range (^{13}C 200 ppm, ^{1}H: 10 ppm); however, ^{13}C-NMR spectra of HS show broad and poorly resolved resonances caused in many cases by paramagnetic metal ions [e.g. Cu(II), Fe(III)], free radicals, and the viscosity of the HS solution (Schnitzer and Preston 1986). Insoluble HS samples can be analyzed by solid-state cross-polarization magic angle spinning (CPMAS) ^{13}C-NMR (Wilson et al. 1987). A time-consuming disadvantage of ^{13}C-NMR used for HS is its relatively poor sensitivity which is lower by a factor of 5000 than that of ^{1}H-NMR. However, the ^{13}C sensitivity can be enhanced by using a high-field NMR spectrometer.

The advanced NMR techniques (e.g. spin-echo, j-resolved 2-D) (Buddrus et al. 1989) as well as a chemical derivatization of HS samples (e.g. methylation with C) (Mikita et al. 1981) could improve the accuracy and precision of ^{13}C-NMR spectra. Thus, in the future ^{13}C-NMR can be expected to be a powerful tool for the quantitative determination of partial structures in HS.

2.1.8.2 ^{13}C-NMR of Bocholt HS (BOC HS) and Segeberg HS [SEG HA Bh(3)]

Method

Isolation. The BOC HS were isolated from the sources BOC1 and BOC3 according to the XAD 2 procedure (Sect. 2.1.3, this Vol.) and the SEG HA Bh(3) was separated from soil according to Post (1987) (Sect. 2.1.2, this Vol.).

Complexation with Fe(III). The interfering influence of paramagnetic ions [e.g. Fe(III)] on the ^{13}C-NMR of HS was studied by Fe(III) complexation of BOC HS in the range 0.08 (original Fe content) to 0.8% Fe(III) (with respect to the HS mass). For this purpose BOC HS (e.g. BOC1–2) in solution (pH 4–6) was successively loaded with appropriate amounts of Fe(III) dissolved as iron(III) chloride.

Separation of Fe(III) from HS by Ion Exchange. Fe(III) bound in mobile form to HS was separated using a chelating ion exchanger (Cellulose HYPHAN). For this purpose HS solutions (100 ml, 1–5 mg/ml DOC, pH 5–7) were magnetically stirred with 2 g Cellulose HYPHAN (Riedel-de Haen AG, FRG) for 96 h.

NMR Measurements. NMR spectrometer: JEOL GX 400. The pulse width of a 90° ^{13}C pulse was determined with a solution of 2-butanol in D_2O instead of HS due to the broad and small signals of the latter. The δ-values are relative to sodium 2,2-dimethyl-2-silapentanesulfonate, the largest signal of which was set to zero. This scaling was applied to the subsequent HS measurement. Additionally, 0.03% dioxane was used as internal standard. A frequency range of 48 000 Hz (16 k data points) was used for all 1-D spectra. The conventional technique (pulse acquisition) was applied without proton decoupling. The pulse length was 45°, the acquisition time 0.17 s, and the relaxation delay 0.4 s. Generally, the number of scans was 360 000.

2.1.8.3 Results and Discussion

In Figs. 2.1.8.1–2.1.8.3 conventional ^{13}C spectra of BOC HS (BOC3–3,5, BOC1–2) and SEG HA [HA Bh(3)] measured in the pulse-acquisition mode are shown.

The different signal ranges typical for HS are caused by aliphatic carbons (10–60 ppm), carbohydrates (60–90 ppm), aromatic carbons (110–160 ppm), carboxylates (160–180 ppm), and presumably quinones and ketones (200–220 ppm). As can be seen from the spectra of BOC HS, the carbon in this HS can be attributed to a great extent to aliphatic and aromatic sub-structures and partly to carbohydrates. Additionally, in Fig. 2.1.7.1 the differently substituted aliphatic and aromatic carbons are marked. Using only the one-pulse technique, a quantitative evaluation of the HS spectra is hindered by the rolling baseline caused by signal broadening due to the presence of paramagnetic ions or free radicals. Figure 2.1.8.1b,c [BOC HS, BOC1–2, containing 0.2 and 0.8% Fe(III), respectively] clearly reveals the influence of Fe(III) on the ^{13}C signals, particularly as broadening at <0 and > 220 ppm. This effect can be eliminated by separating the interfering Fe(III) ions from the HS

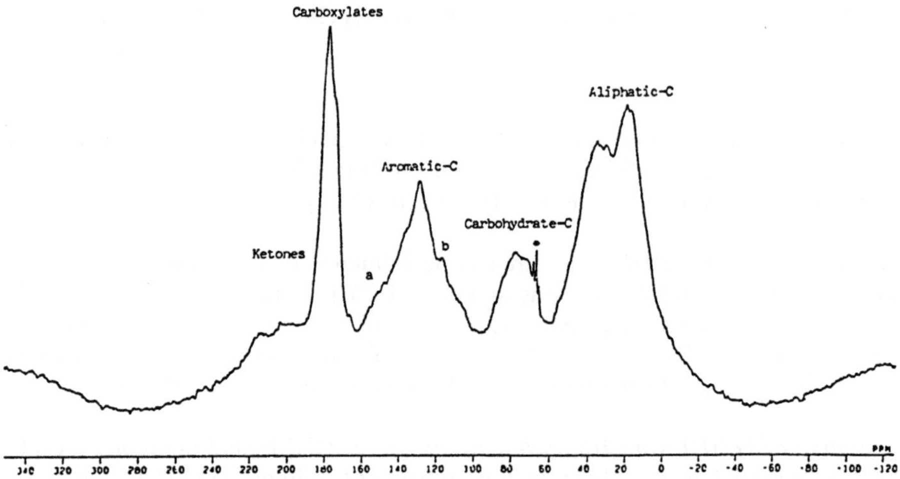

Fig. 2.1.8.1. 100 MHz ^{13}C-NMR spectrum of BOC HS. 450 mg BOC3–3.5 dissolved in 4 ml D_2O, pH 6, 0.03% Fe (referred to as HS mass), 400 000 scans, 0.03% dioxane as internal standard. *a* Phenol-C (ipso); *b* phenol-c (ortho); * dioxane

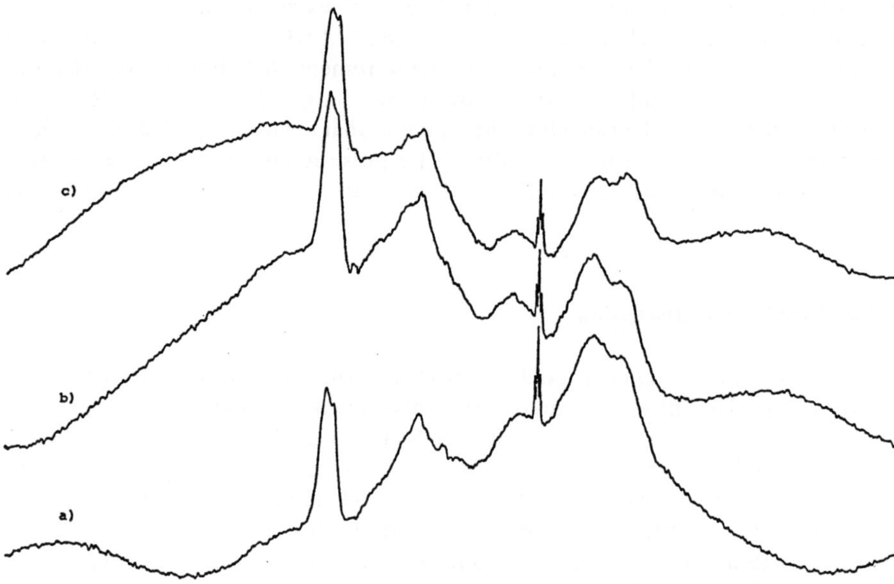

Fig. 2.1.8.2. 100 MHz ^{13}C-NMR spectrum of BOC HS. 400 mg BOC1–2 dissolved in 4 ml D_2O, pH 6.5, 360 000 scans, 0.03% dioxane as internal standard. *a* 0.08% Fe; *b* 0.2% Fe; *c* 0.8% Fe (referred to as HS mass)

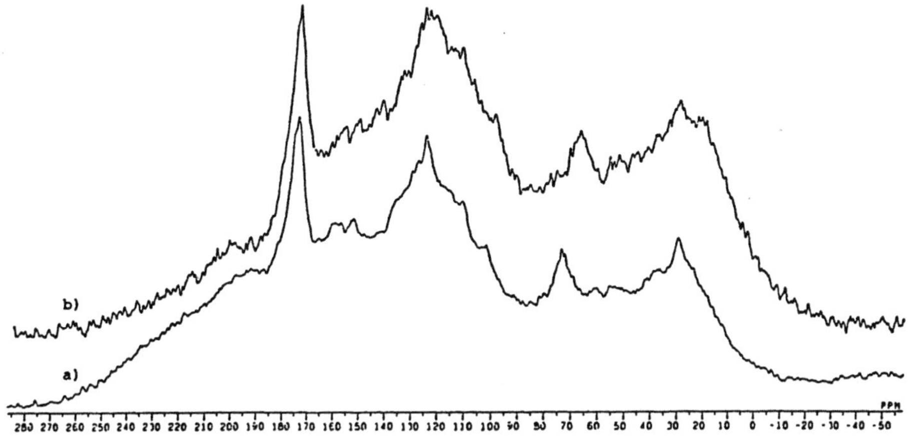

Fig. 2.1.8.3. 100 MHz ^{13}C-NMR spectrum of SEG HA [400 mg SEG Bh(3) dissolved in 4 ml D_2O, pH 7]. *a* SEG Bh(3) containing 0.24% Fe (after purification by ion exchange), 360 000 scans. *b* SEG Bh(3) containing 0.07% Fe (purified according to XAD 8 and subsequently to the ion-exchange procedure), 1000 scans

sample, for example by a chelating ion-exchanger (Cellulose HYPHAN, Burba and Willmer 1988). Using an ion exchange procedure the Fe content in SEG HA Bh(3) [which remained even after isolation of HA by acid precipitation according to Post (1987)] was lowered from 2.10 to 0.24 and 0.07%, respectively. The resulting ^{13}C-NMR spectra free from strong Fe interferences are shown in Fig. 2.1.8.3a,b. Consequently, reliable quantitation of ^{13}C-NMR spectra of HS in solution depends strongly on the absence of paramagnetic ions [e.g., <0.1% Fe(III)]. On the other hand, some ^{13}C-NMR spectra of SEG HA obtained by solid-state CPMAS ^{13}C-NMR (Post et al. 1988) reveal less influence of Fe(III).

2.1.8.4 Conclusions

Further advances of ^{13}C-NMR of HS can be expected from refined techniques, e.g., spin-echo technique, pulse sequences DEPT (distortionless enhancement by polarization transfer), QUAT (quaternary-only carbon spectra), and two-dimensional J-resolved spectra), as described in Buddrus et al. (1989).

2.1.9 Characterization of Humic Substances Extracted by Organic Solvents

B. Post and H. Klamberg

2.1.9.1 Background

The effective and soft isolation of humic substances (HS) from different matrices is most important for their analytical characterization. In general, isolation includes fractionation. From the analytical point of view, the classical separation into humic acids (HA), fulvic acids (FA) and humins proves to be increasingly inefficient. On the other hand, a sophisticated fractionation of soil humic substances along with the isolation process is severely limited by the complex composition of the soil matrix. Fractionation by the extraction of soil samples with organic solvents is known to be a gentle procedure. However, small yields and products with high ash content have to be accepted (Scheffer et al. 1960; Whitehead and Tinsley 1964; Hayes et al. 1975; Ziechmann 1980; Senesi et al. 1983). An interesting approach seems to be the application of organic solvents for the extraction of isolated humic material. Many of the data in the literature are not comparable with one another because of the differences in the origin of the samples, the isolation procedures and characterization methods (Mehta et al. 1963; Salfeld 1964; Kyuma 1966; Otsuki and Hanya 1966; Domke 1972; Korotkov and Frolov 1982; Kneißel 1985; Senesi and Sipos 1985; Kumada and Kawamura 1986). Therefore the aim of the work was to use detailed elemental analysis, spectroscopy and chromatography to characterize the organic solvent extracts, by this contributing to the search of the relation between the chemical structure and the macromolecular functions of humic substances. For the first time quantitative ^{13}C-NMR spectroscopy was used to characterize the organic extracts.

2.1.9.2 Material and Methods

HS were isolated from the A_h and B_h horizons of the mineral soil of Segeberger Forst (Table 1.1, Fig. 1.1). A detailed description of the isolation and characterization procedure of the FA and HA and of the abbreviations applied is given in Section 2.1.2 and in Post (1987), Post et al. (1988) and Finger et al. (1990). Some HS were isolated by the NaOH method and separated into a HA and FA fraction (Fig. 2.1.2.2). FA were concentrated on XAD-2 resin, desorbed with aqueous NaOH, and passed through an ion-exchange column (DOWEX 50 WX-8). The code is SEG-Ah (3.2)-FA and SEG-Bh (3.2)-FA.

The extracts obtained with different solvents were characterized by the following methods: elemental analysis, ash content, UV/VIS, FTIR, ^{1}H-NMR and ^{13}C-NMR, pyrolysis-FIMS, gel permeation chromatography and vapour pressure osmometry (Post 1987; Post et al. 1988).

Successive Extraction by Organic Solvents

The isolated HS were extracted successively by different solvents with different polarities. The extraction was done by the Soxhlet procedure. Extracts were filtrated, concentrated by rotavapor and dried. To avoid the inclusion of N-containing solvents, some dried extracts were cleaned up with acetone.

The choice and the sequence of the solvents applied were mostly done by empirical criteria. The criteria of solubility parameters can only be applied to low molecular weight substances or unpolar homopolymers. For HS the use is limited (Hayes 1985). This is due to the ionic, dispersive and inductive interaction of polar and heterogenic substances. In a multicomponent mixture the individual character of each substance is influenced by phenomena like association, solvation and local orientation (Fuchs 1980). Therefore the solubility of the HS must be discussed in detail. Concerning the acidity of the substances, the alkalinity of the solvents is a dominant parameter. The solvents used are shown in Table 2.1.9.1.

A detailed description of the results is given for SEG-Bh (3)-HA and for SEG-Bh (3.2)-FA. Further results are reported in Post (1987) and Post et al. (1988).

Humic Acids Extracted from Soil

The yields of the extraction procedure and the analytical data of the extracts and of the original sample are listed in Table 2.1.9.1. The extracts show different colour and consistence. By successive extraction the colour of the extracts, changes from yellow (ether) to brown (acetone) and finally to black. In addition, the viscosity of the substances increased and the acetone extract finally resulted in a hard material. This can also be shown by a decreasing E4/E6 ratio (absorption at 465 nm and 665 nm). The atomic ratio C/N decreases from 112 for the acetone extract to 18 for the last fraction while the ash content increases. The elements aluminium, silicon and titanium are most abundant in the last fraction whereas the highest contents of iron and copper are in the extracts with dimethylformamide. It is obvious that metals can be found even in fractions with non-polar character.

Table 2.1.9.1. Characteristic data for the extracts and the original SEG-Bh (3)-HA[a] (n.d. = not determined)

	Yield in%	Ash in%	C in%	H in%	N in%	C/N	E4/E6
SEG-Bh (3)-HA		9.9	57.59	4.31	2.44	27.5	5.99
Extracts							
Ether	0.05	n.d.	68.18	10.82	<0.2	–	n.d.
Acetone	3.78	0.43	65.47	7.73	0.68	112.3	13.69
Tetrahydrofuran	18.58	0.69	61.62	5.55	0.95	75.7	9.98
Methanol	16.23	2.75	56.61	4.81	1.40	47.2	8.40
Dimethylformamide	59.50	6.59	57.55	4.51	2.75	24.4	5.4
Pyridine/water	2.95	23.52	51.52	5.19	3.33	18.1	5.04

[a]For the extracts with known ash content the CHN values are calculated for ash-free material.

The extracts were examined by HPLC with C-18 reversed phases but it did not work, whereas the gel permeation chromatography on fractogel-TSK phase is a good method to characterize the different fractions. The chromatograms are shown in Fig. 2.1.9.1. The TSK HW-50 column was calibrated in the range of 100 to 20 000 molecular weights by the volume of exclusion and total permeation of the gel (Post 1987; Post et al. 1988). The extract obtained with acetone shows a maximum at 4000 mass units. The molecular weight increases with rising polarity of the solvent. The last fractions shows a maximum nominal mass of 10 000. The shape of the chromatogram is dependent on the pH value and the ionic strength of the eluent. In addition, the substances can be adsorbed on the gel. The ash content of the samples is of minor importance. This is shown by the comparison of a sample treated with HCl/HF. The hydrophobicity of the sample has the main influence on the irreversible adsorption. Due to a decreasing polarity of the fractions the irreversible absorbed amount of the fraction doubled from 14 to 28%. The same increase in molecular weight with rising polarity of the fractions is obtained by vapour pressure osmometry. However, the molecular weights found were lower (factor 2 to 3).

The concentration of aliphatic and aromatic compounds is different in the extracts. This can be shown by FTIR and NMR spectroscopy. Methylene groups, belonging to long chains which are not branched, are only present in the non-polar fractions. With increasing polarity of the solvent, the intensity for the aromatic signals is stronger. Different functional groups can be shown by ^{13}C-NMR spec-

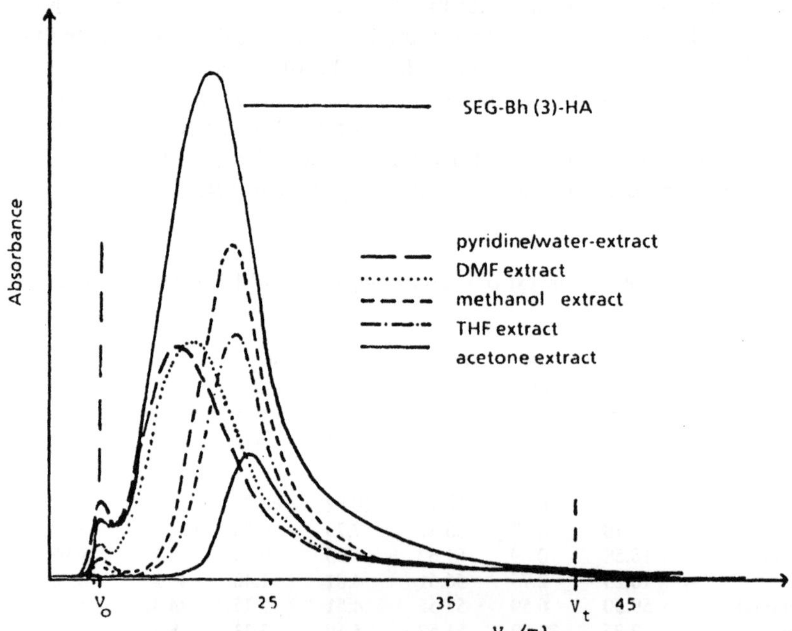

Fig. 2.1.9.1. Gel permeation chromatograms (GPC) of SEG-Bh(3)-HA and the solvent extracts. Fractogel TSK HW-50; pH = 8.25; Tris buffer; c(KCl) = 0.1 mol/l

Table 2.1.9.2. Relative abundance of carbon in the solvent fractions of SEG-Bh (3)-HA derived from solvent ^{13}C-NMR-spectra (in%)

	Range			
	A 0–50 ppm	B 50–100 ppm	C 100–160 ppm	D 160–200 ppm
SEG-Bh (3)-HA	23	8	50	19
Extracts				
Acetone	45	6	32	19
Tetrahydrofuran	25	1	47	27
Methanol	21	5	46	28
Dimethylformamide	11	9	50	30
Residue after extraction with methanol	10	13	52	25

troscopy (inversed gated decoupling) (Post et al. 1988). The results are summarized in Table 2.1.9.2.

By pyrolysis field ionization mass spectrometry (Py-FIMS), some subunits could be identified (Schulten 1987; Post et al. 1988). Highly aliphatic character is typical for the non-polar fractions. The spectrum of the original HA and of the extracts gained by polar solvents show dominant signals related to polysaccharide and aromatic fragments. On the contrary, high contents of lipid-like compounds are shown in the spectrum of the acetone extracted (Figs. 2.1.9.2 and 2.1.9.3). Signals with high intensity appear between 200 and 500 m/z. They can be related to n-fatty acids with long chains. The same substance was hydrolyzed and analyzed by GC/MS. The results confirm the interpretation given above.

Soil Fulvic Acids

The FA fractions show the same tendency concerning colour, consistency and elemental analysis as the HA fractions. However, the differences are not so striking. Table 2.1.9.3 shows the elemental analysis and the E4/E6 ratio for the SEG-Bh (3.2)-FA. The chromatograms for the molecular weight distribution are not shown, because the results are similar to the HAs. According to infrared spectra the FAs show a high content of polysaccharides in the last fraction. The FT-IR spectra are very similar to the spectra of different FA fractions gained by diafiltration, sorption on activated carbon and subsequent extraction with acetone and NaOH (Stevenson and Goh 1971).

2.1.9.3 Discussion

The fractionation of HS can be done by successive extraction with organic solvents. For the HAs distinct fractions with a high content of lipids and aliphatic components are obtained. A fraction with a high content of polysaccharides is typical for

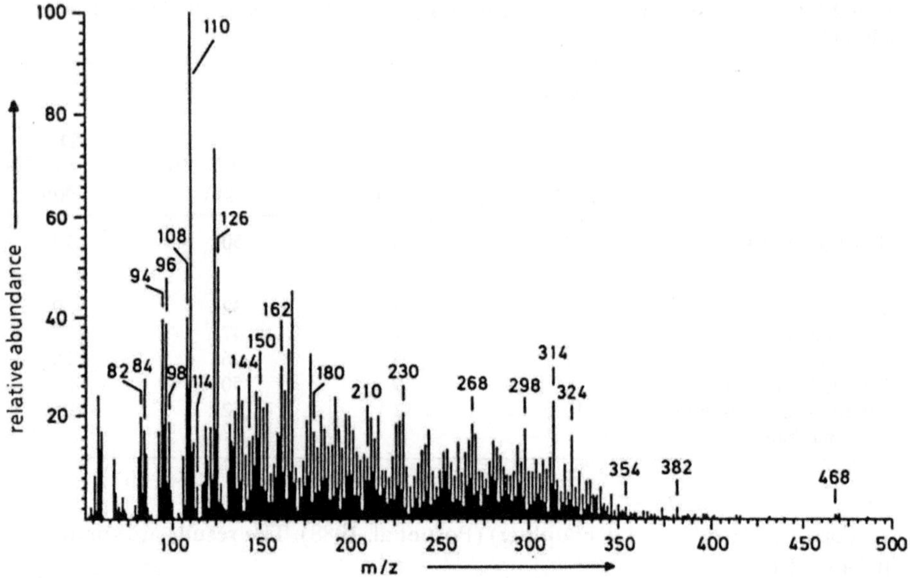

Fig. 2.1.9.2. Pyrolysis field ionization mass spectra (Py-FIMS) of SEG-Ah (3)-HA (recorded by Prof. H.-R. Schulten, Institute Fresenius, Taunusstein)

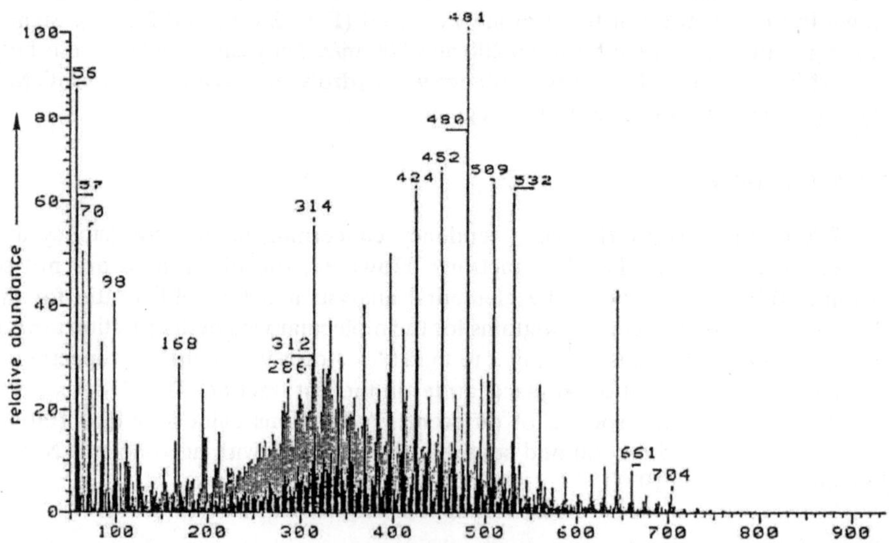

Fig. 2.1.9.3. Pyrolysis field ionization mass spectra (Py-FIMS) of the acetone extract of SEG-Ah(3)-HA (recorded by Prof. H.-R. Schulten, Institute Fresenius, Taunusstein)

Table 2.1.9.3. Characteristic data for the extracts and the original SEG-Bh (3.2)-FA[a] (n.d. = not determined)

	Yield in%	Ash in%	C in%	H in%	N in%	C/N	E4/E6
SEG-Bh (3.2)-FA		0.42	52.14	3.73	1.93	31.2	7.00
Extracts							
Ether	1.5	n.d.	n.d.	n.d.	n.d.	–	n.d.
Acetone	65.5	n.d.	52.54	4.12	1.38	44.4	8.25
Tetrahydrofuran	5.3	n.d.	50.72	5.52	1.75	33.8	n.d.
Methanol	33.7	n.d.	47.06	4.56	2.24	24.5	7.02
Water	2.7	n.d.	41.95	5.14	1.75	27.9	5.02

[a]For the extracts with known ash content the CHN values are calculated for ash-free material.

FA. The binding between organic and inorganic colloids is partly destroyed. Similar to many other isolation and fractionation procedures, the method described cannot eliminate the operational definitions. The successive extraction (Flaig et al. 1975, Khairy and Ziechmann 1981; Fuchsman 1984) can be applied in analytical and preparative scale as well. In combination with well-defined degradation reactions and advanced characterization method, it can add significantly to the identification of the bonding and by this to the understanding of the structure and function of the humic material.

2.2 Interaction of Inorganics with Humic Substances

2.2.1 Investigation of Metal Complexation by Polarography and Fluorescence Spectroscopy

F.H. Frimmel

2.2.2.1 Background

Metal complexation is one of the typical properties of HS. For example, they contribute to a great extent to the dissolved iron concentration especially in oxygenated waters. Of the divalent metals, copper forms the most stable complexes. Even though details of bond formation and kinetics are still unknown, the significance of metal complexation by HS cannot be overestimated. Transport phenomena and ecological functions including biochemical reactions are strongly influenced by the coordination of metals by organic ligands of the type of HS.

Experimental methods and model calculations for the investigation of metal complexation have been reviewed recently (Weber 1988; Buffle 1988). Besides methods which are based on the principle of separating metals which are bound in stable complexes from those which are uncomplexed or involved in labile complexes, electrochemical and spectroscopic procedures are used to differentiate between the various types of bonding. Polarographic and voltammetric methods focus on the metals and their changed response to complex formation whereas fluorescence spectroscopy uses the change of optical properties of the organic ligand to deduce the degree of metal complexation. Both experimental approaches do not necessarily lead to the same results and thus can add independently to a more detailed understanding of metal speciation.

2.2.2.2 Polarography

Polarographic and voltammetric methods have been widely used for quantification of metal-ligand interactions (Raspor et al. 1981). Especially the anodic stripping voltammetric mode using a hanging mercury drop electrode or mercury film electrode has been applied successfully to measure complexation reactions at the low concentration levels of natural aquatic systems (Shuman and Cromer 1979; Figura and MacDuffle 1980; Sterrit and Lester 1984). However, the method has some major pitfalls, especially in the presence of humic matter (Tuschall and Brezonik 1981; Weber 1988).

Problems can arise from:

- uncontrolled adsorption of humic substances on the mercury electrode
- change of the surface tension, resulting in irreproducible diffusion and electrode reactions
- dissociation of metals from the complexes during the deposition step.

Differential pulse polarography with a dropping mercury electrode is less sensitive than inverse voltammetry, but minimizes some of the disturbances by cutting down the drop life and applying a smaller time scale for the measurement.

The complexation capacity (CC) of HS can be measured by titration of a standard Cu (II) solution under defined conditions. Cu(II)-ions form relatively stable complexes with donor ligands and the isolated humic substances applied are practically metal-free. Be-cause of this, the determined CC is close to the maximum complexation ability of the humic substances towards divalent cations. The detailed method is described in Frimmel and Geywitz (1983) and outlined in Fig. 2.2.1.1.

After subtraction of a blank, the CC in μmol/l can be calculated for the DOC concentration of 1 mg/l. This value is operationally defined and can be compared with values for other HS determined under the same standardized conditions (Table 2.2.1.1).

The polarographically determined values range from 1.5 to 3.3 μmol/mg DOC. There seems to be little difference between the samples from different locations. The isolation procedure, however, has a significant influence on the samples from the same origin. It is interesting to note that equilibration with gravel leads to partial adsorption of the FA and thus to a dissolved FA fraction with less complexation ability.

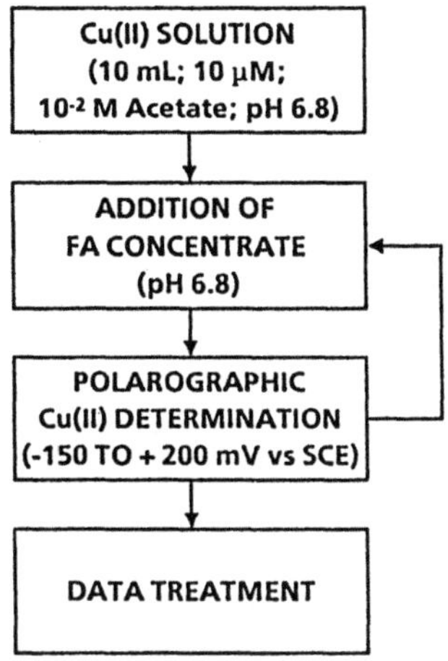

Fig. 2.2.1.1. Procedure for the polarographic determination of copper complexation capacities of humic substances

Table 2.2.1.1. Copper complexation capacity (Cu-CC) for isolated humic substances (polarographic determination, µmol/mg DOC)

Sample	Cu-CC	Sample	Cu-CC
MUC 1 FA	1.55	SEG-TA FA	1.99
MUC 2–1 FA	2.80	SEG-KA FA	2.01
MUC 2–2 FA	3.32	BOC 1–2 FA	1.19
MUC 2–2 GRA FA	2.46	BOC 1–33 FA	1.30
MUC 3 FA	1.72	BOC 3–3.5 FA	1.14
		BOC 3–20.5 FA	1.34

2.2.2.3 Fluorescence Quenching Method

The excitation of aqueous solutions of HS at about 340 nm leads to fluorescence spectra with a maximum of emission around 450 nm. This was first documented by Black and Christman (1963) and used for metal complexation studies by Saar and Weber (1980). Paramagnetic metal ions quench the fluorescence (Fig. 2.2.1.2).

This causes the fluorescence intensity I after metal addition to be less than the fluorescence intensity I_o in the metal-free solution. On the basis of a hypothetical 1:1 complex-formation (2.2.1.1) where Me stands for a paramagnetic metal ion and L for a ligand functional group of the fluoresceing HS, a conditional stability function (K_c') can be calculated.

$$Me + L \rightleftharpoons MeL \qquad (2.2.1.1)$$

$$K_c' = \frac{c(MeL)}{c(Me)c(L)} \qquad (2.2.1.2)$$

It can be shown (Frimmel and Hopp 1986) that the experimental data lead to an equilibrium function which describes the decreasing stability of the complexes formed as the metal concentration in the system increases (Fig. 2.2.1.3).

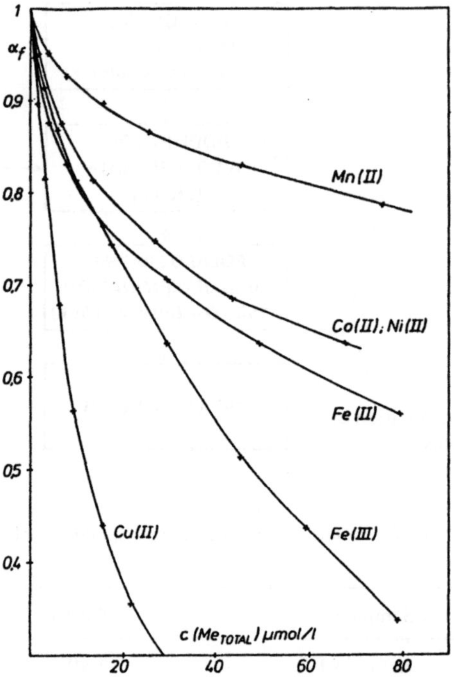

Fig. 2.2.1.2. Quenching of the fluorescence of humic substances by paramagnetic metal ions

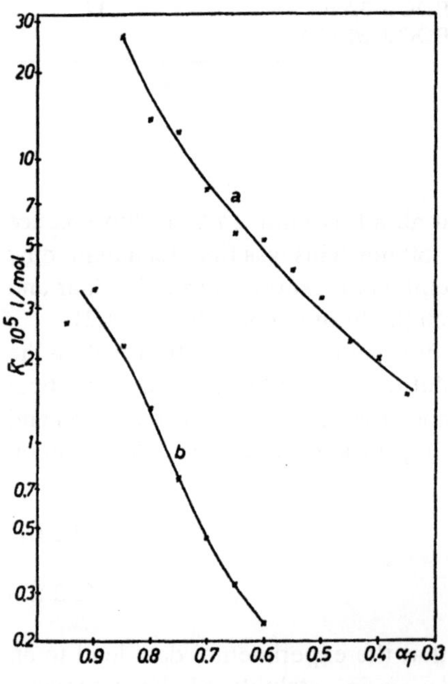

Fig. 2.2.1.3. Stability spectra for the Cu(II)- complexes of soil fulvic acids before (*a*) and after equilibration with carbonate gravel (*b*)

Table 2.2.1.2. Copper complexation capacity for isolated humic substances determined polarographically (DPP) and with the fluorometric method (Flu), µmol/mg DOC

Sample	CuCC-DPP	CuCC-Flu
MUC 1 FA	1.55	0.9
MUC 2–1 FA	2.80	1.62
MUC 2–2 FA	3.32	2.58
MUC 2–2 GRA FA	2.46	1.48
SEG-TA FA	1.99	2.09
SEG-KA FA	2.01	0.79

The method can also be applied under the reaction conditions used for the polarographic determination of the Cu(II)-CC. Thus, the polarographically determined data can be compared to the fluorometric ones (Table 2.2.1.2).

It is obvious that not all values are within the accuracy of the two methods (which is mostly ≤ 10%). This can be due to the different response of some samples to the two methods. Further fractionation leads to a solution of these problem, at least for some of the fractions.

Detailed examination of the reaction conditions and of the influence of major water constituents leads to some general conclusions:

- The quenching effect, which can be correlated with the stability of the complexes formed, decreases for divalent metal ions in the order
  Cu > Fe > Co ≈ Ni > Mn
- quenching increases with increasing pH
- increasing amounts of alkaline earth metal ions (hardness) cause the quenching effect to decrease
- increasing concentrations of alkaline metal salts decrease the quenching effect.

All results are in good agreement with the principles of coordination chemistry known from the literature (Stumm and Morgan 1981).

2.2.2 Determination of Complexation Equilibria by the Ion-Exchange Method

W. Finger and H. Klamberg

2.2.2.1 Material and Methods

Humic substances were isolated from the Ah- and Bh-horizons of the podsol at the Segeberger Forst site (Fig. 1.1, Table 1.1; Sect. 4.3.2) and their interaction with metal ions was investigated. Some other models for metal-humate interaction, such as the polyelectrolyte gel model of Marinsky or discrete ligand and continuous distribution models, are reviewed by several authors (Perdue 1983; Shuman 1983; Dzombak 1986; Ephraim 1986; Fish 1986; Marinsky 1986; Brassard 1990). It is obvious that not all approaches are equivalent. However, they all should obtain similar results. Depending on the scientific background of the respective author, different aspects will be dominant. The fairly simple method described here is based on the ion-exchange model (Schubert 1948). There is no question about the limitation of this approach, but it leads to useful information based on experimental results.

Conditional stability constants and stability functions were determined for the complexes of some selected metal ions. In addition, data on the stoichiometry of the complexes were obtained. The stability constants were determined at a temperature of 25 °C, an ionic strength of 0.1 mol/l (KCl) and at three different pH values, 3, 5, and 7. The stability functions were determined for the pH value of 5.0 and for the temperature and ionic strength mentioned above.

HA fractions were extracted by NaOH (c = 0.1 mol/l) and precipitated with HCl at pH 1. Larger quantities of FA were isolated under gentle conditions by extraction with water and adsorption/desorption using XAD-8 resin (Amberlit, p.a., Serva, Heidelberg). A detailed description of the isolation and characterization of the soil humic-derived substances and the code system for the investigated samples are given in Section 2.1.2.

The molar concentration of HA and FA in aquatic solutions was determined by means of vapour pressure osmometry and gel permeation chromatography. A vapour pressure osmometer (Dampfdruck-Osmometer, Knauer, Berlin) was applied using dimethylformamide (Uvasol, Merck, Darmstadt) as solvent and polystyrene (M_n = 2100, Pressure Chemical Company, Pittsburgh) as standard. The average molar masses for SEG Ah(3)-HA and for SEG-Bh(4)-HA were determined to be 2230 and 1700, respectively. The more gently isolated SEG-Ah(4)-FA and SEG-Bh(4)-FA samples showed the values of 4990 and 5050, respectively. Gel permeation chromatography calibrated by peptides and using a phosphate buffer (pH 7) as mobile phase led to values which were two to three times higher than the osmometric values (Finger 1989). All calculations are based on the vapour pressure osmometric values.

The ion-exchange method used for the determination of the complexation equilibria is based on the following principles.

The distribution of metal ions between an aqueous solution and a suitable cation exchanger is constant in the concentration range investigated. The equilibrium can be given as a distribution coefficient λ_o (Ardakani and Stevenson 1972). In the presence of a complexing ligand, e.g., HS, the distribution is characterized by a coefficient λ. Using the two coefficients λ_o and λ and the mass balance, the rearrangement of the equation leads to the linear function (2.2.2.1) for the complexation of metals by HS.

$$\log\left(\frac{\lambda_o}{\lambda} - 1\right) = \log K_c + n \log c(L)_t \qquad (2.2.2.1)$$

The two distribution coefficients can be determined by batch experiments. By using the total concentration of HS in the system $c(L)_t$, the stability constant K_c can be derived from the intersect with the ordinate; the ligand number n is given by the slope. For the two HAs the conditional stability constants were calculated for the 1:1 complexes; for the two FAs the stability constants were determined for variable metal concentrations as a conditional stability function.

The batch experiments for the determination of the distribution coefficients were performed in 100-ml flasks at constant temperature (25 °C) in an agitation bath (Kottermann, Uetze-Hänigsen). A precleaned and dried, strongly acidic cation exchanger DOWEX 50 WX 8 (p.a., sodium form, Serva, Heidelberg) was used. The pH was controlled by addition of HCl and NaOH (c = 0.001 mol/l). The ionic strength of 0.1 mol/l was maintained by KCl (p.a., Merck, Darmstadt). The metal ions were selected according to their coordination chemical properties following the sequence of Schwarzenbach (1956) (Al^{3+}, Fe^{3+}, Ca^{2+}, Cu^{2+}, Mn^{2+}). In addition, aspects of geochemistry and soil science were included (e.g. concentration range), and the solutions were prepared from standard solutions (Titrisole, Merck, Darmstadt).

Metal concentrations were determined by flame atomic absorption spectrometry (spectrometer 5000, Perkin Elmer, Überlingen) under standardized conditions. Calcium and copper were additionally checked by ion-selective electrodes.

2.2.2.2 Results

In Table 2.2.2.1 the conditional stability constants are given in log values for the approximated 1:1 complex formation for the HA samples. To reach the specific 1:1 complexes different amounts of metal ions were used.

In Table 2.2.2.2 the conditional stability constants for the FA samples are given for pH 5. The log-K_c values can be given as a stability function as shown in Fig. 2.2.2.1.

The stability function can be approximated theoretically by two straight lines which represent the low energy binding sites (Sect. 1 in Fig. 2.2.2.2) and the high energy binding sites (Sect. 2 in Fig. 2.2.2.2).

Extrapolation of the two straight lines leads to maximum values for the stability constants of the two different binding sites. They are given as log-K_c values in Table 2.2.2.3. In addition to the 1:1 complex formation, which is guaranteed by the stoichiometric metal concentration added, complexes with different metal/ligand ratios can be investigated by the addition of different amounts of metal. From the graph of these functions maximum values of conditional stability constants and

Table 2.2.2.1. Conditional stability constants of the HA

HA	Metal	log k		
		pH = 3.0	pH = 5.0	pH = 7.0
SEG-Ah(3)-HA	Ca	3.85	4.25	5.02
	Mn	4.24	5.39	6.09
	Cu	5.10	6.66	7.44
	Zn	4.46	6.20	7.21
SEG-Bh(3)-HA	Ca	3.11	3.71	4.22
	Mn	3.47	4.18	5.48
	Cu	5.10	6.75	7.60
	Zn	4.22	4.63	7.00

Table 2.2.2.2. Conditional stability constants of FA by different metal concentrations (log K-values)

$(M)_t$ (μg/ml)	SEG-Ah(4)-FA					
	Al	Fe	Cu	Zn	Mn	Ca
0.8	–	13.47	6.50	5.98	5.38	4.52
1.0	–	13.30	6.16	5.95	4.46	4.27
1.5	10.37	12.88	6.15	5.79	4.30	3.96
2.0	8.09	10.84	6.05	5.12	4.23	3.92
2.5	6.88	10.84	5.80	5.08	4.06	3.72
5.0	5.21	7.94	5.19	4.63	3.81	3.41
10.0	5.04	6.52	4.70	4.39	3.30	2.56
20.0	4.70	6.16	4.24	3.66	2.67	2.39
40.0	3.95	5.82	3.68	2.98	2.61	2.22
60.0	3.63	5.54	3.40	2.71	2.17	(1.29)
80.0	3.04	5.38	3.10	2.64	1.77	(0.58)
100.0	2.52	5.24	3.05	2.04	1.49	(0.46)
$(M)_t$ (μg/ml)	SEG-Bh(4)-FA					
	Al	Fe	Cu	Zn	Mn	Ca
0.8	–	12.99	7.77	6.23	–	4.12
1.0	–	12.42	7.52	4.66	–	4.07
1.5	10.12	11.23	5.97	4.40	4.15	3.80
2.0	7.60	9.63	5.90	4.23	4.15	3.76
2.5	6.16	9.27	5.15	4.19	3.93	3.51
5.0	4.69	7.63	4.68	3.83	3.69	3.12
10.0	4.48	6.37	4.03	3.67	2.83	2.48
20.0	3.89	6.02	3.81	3.11	2.63	2.26
40.0	3.83	5.79	3.81	2.95	2.00	1.92
60.0	3.70	5.50	3.62	2.41	1.74	(0.51)
80.0	3.10	5.14	2.68	1.80	1.71	(0.51)
100.0	2.84	5.14	2.55	1.50	1.45	(0.28)

ligand numbers can be calculated for the low and high energy binding sites. Values are given in Table 2.2.2.4.

2.2.2.3 Discussion

The conditional stability constants show the expected sequence, i.e., the stability of the three-valent metal ions (Al^{3+}; Fe^{3+}) are higher than the ones of the divalent cations which follow the stability sequence from Irving and Williams (1948). The stability constants for the HS isolated from the two podsol horizons show no significant differences.

The stability constants for the two FAs have a typically strong dependence on the total metal concentration. Figures 2.2.2.1 and 2.2.2.2 show clearly the two linear regions (1 and 2) which can be interpreted as different types of binding sites. According to Schnitzer and Skinner (1968), the low energy binding sites are due to the phthalic acid; structure and high energy sites are of the salicylic acid type. The

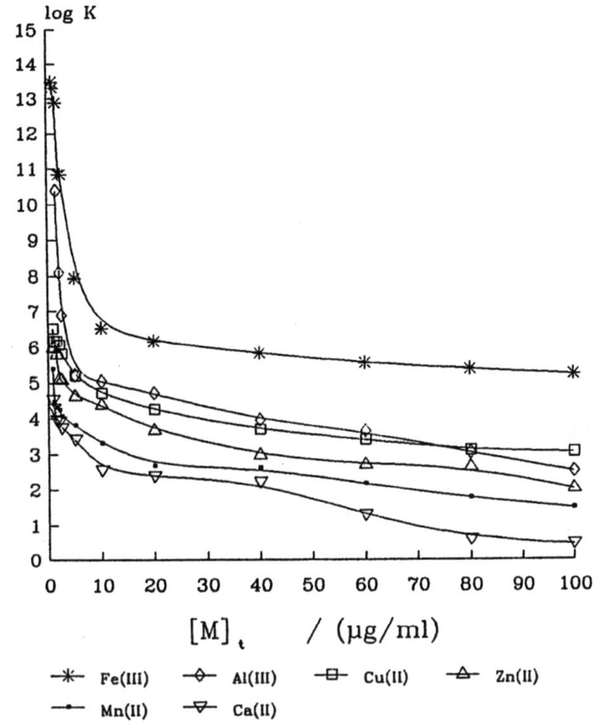

Fig. 2.2.2.1. Conditional stability constants of SEG-Ah(4)-FA

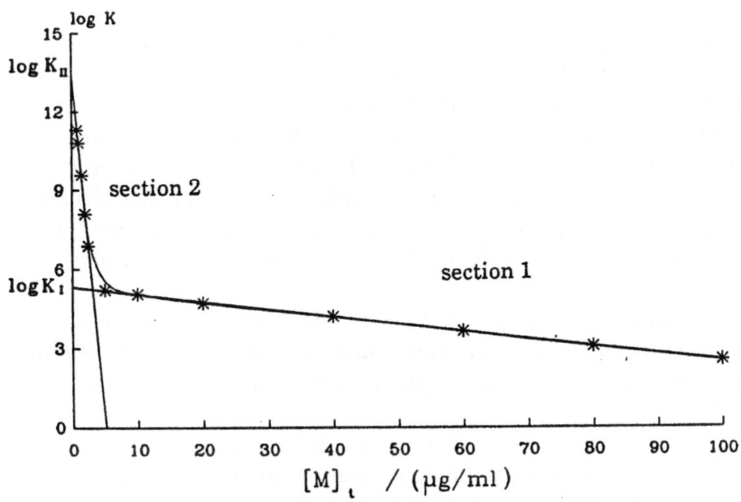

Fig. 2.2.2.2. Conditional stability function. *Section 1* low energy binding sites; *section 2* high energy binding sites

Table 2.2.2.3. Conditional stability constants (log K-values) of the FA. Section 1: low energy binding sites; section 2: high energy binding sites

Metal	SEG-Ah(4)-FA		SEG-Bh(4)-FA	
	Section 1	Section 2	Section 1	Section 2
Al	5.22	15.43	4.48	15.88
Fe	6.31	14.41	6.37	14.71
Cu	4.38	6.58	4.30	8.86
Zn	3.88	6.53	3.78	6.65
Mn	3.28	5.41	2.83	4.46
Ca	2.93	4.74	2.72	4.39

Table 2.2.2.4. Maximum values of the conditional stability constants for the low (section 1) and high (section 2) energy

Metal	SEG-Ah(4)-FA		SEG-Bh(4)-FA	
	Section 1	Section 2	Section 1	Section 2
Al	1.28	3.10	1.11	3.25
Fe	1.13	2.66	1.10	2.36
Cu	0.98	1.25	0.82	1.71
Zn	1.10	1.49	0.96	1.49
Mn	0.85	1.31	0.68	1.04
Ca	0.88	1.15	0.79	1.04

maximum stability constants derived from extrapolation differ by several orders of magnitude and are in good agreement with the interpretation given above. Striking are the great differences in the constants of trivalent and divalent cations in the high energy range and the inversion of the sequence for Al and Fe in the two regions. The extrapolated log-K values for region 2, however, include a strong uncertainty due to the few results available in the low concentration range of the experiments. Similar problems arise for the extrapolated maximum ligand numbers of the two FAs (Table 2.2.2.4) which suggest the formation of 1:1 complexes. The exceptions are the cations Al^{3+} and Fe^{3+}, which form 1:3 complexes in the high energy binding site region.

From these results it can be concluded that most high energy binding sites of FAs in soil are occupied by trivalent metal ions. The low energy binding sites are mainly responsible for the transport phenomena of metal ions, for mobilization, and immobilization.

It has been reported that water-soluble FAs can be precipitated at higher metal concentrations. Also, the batch experiments of this work were severely limited by precipitation at higher calcium concentrations.

2.2.3 Sorption of Metals on Humic Material

R. Becker and H. Klamberg

2.2.3.1 Material and Methods

The interaction between soil HS and some selected metal ions was investigated by different methods. The HS were isolated from Ah and Bh horizons of podsol in the Segeberger Forst site (Fig. 1.1, Table 1.1; Sect. 4.3.2). Besides the formation of metal-humic complexes in solution, the sorption effects of dissolved metal ions on two insoluble HA were investigated. For a detailed understanding of the interactions, batch experiments (static method) and a column method (dynamic method according to Becker (1987)) were used. The sorption process was defined as the reversible phase transfer of a substance between the dissolved and the solid phase, resulting in an enrichment of the substance on the solid surface.

The sorption reactions in the batch experiment were described with a sorption model according to Langmuir (1916). From the linearized form of the sorption isotherm the maximum sorption capacities and the binding constants were calculated. From the column experiments breakthrough curves were derived and consequently the maximum sorption capacities. From the digital simulation of the experimental curves binding constants were calculated.

The soil HA were isolated from the soil horizons by extraction with NaOH (c = 0.1 mol/l) and precipitation with HCl (at pH 1). A detailed description of the isolation and characterization of HA and their code numbers are given in Section 2.1.2 and Finger et al. (1990). The HA were fractionated by means of a certified sieve into sizes (in µm) <32, 32–80, 80–125, 125–200.

The choice of the four metal ions Ca^{2+}, Mn^{2+}, Cu^{2+} and Zn^{2+} was made in respect to their coordination chemical properties. According to Schwarzenbach (1956), the metal ions are grouped in A, B and C metal ions. Ca^{2+} is a A cation, Mn^{2+} is a C/A cation, Cu^{2+} is a C cation and Zn^{2+} is a C/B cation. The metal solutions were made from fixanal (Riedel de Haen, Seelze).

Batch experiments: 30 ml solution; 10 mg HA of defined particle size; 25 °C; pH 5; 24 h agitation for equilibration (constant temperature agitation bath, Kötter-Mann, Uetze-Hänigsen). Metal concentrations at equilibrium were determined by flame atomic absorption spectrometry (type 5000, Perkin-Elmer, Überlingen) under standardized conditions using a calibration curve. After calculation of the adsorbed amount of metals, the data were used in a linear regression analysis with a linearized Langmuir isotherm. From the slope and the intercept with the Y-axis the maximum sorption capacities k_1 and the binding constant k_2 were obtained.

Column experiments: Glass column 2 and 3 mm inner diameter; 10 mg HA of defined particle size. Height of the column material 3 to 5 mm; conditioning of the column with ddW; collection of 60 fractions (1 ml each). The metal concentrations were determined by flame AAS. From the breakthrough curves the maximum sorption capacities k_1 were calculated as the sum of the retained amount of metals.

Simulation of breakthrough curves: The binding constants k_2 cannot be derived directly from the breakthrough curves. They were determined by digital simulation.

To simulate the process of sorption/desorption in the column, a "black box" model was used (Feldberg 1969), which separates the column into n segments equal to the theoretical plate number (Bard and Faulkner 1980). The theoretical plate number was determined according to an approximation described by Glueckauf (1955). Between 1 and 2.4 theoretical plates were found. The simulation was started with chosen binding constants and with a known initial concentration of the metals and the maximum sorption capacities of the HA. It was then repeated iteratively, i.e. the binding constant was varied in order to minimize the approximation error of the simulated curve (Nash 1979). A detailed description of the simulation and of the programme written in Pascal is given in Becker (1987).

2.2.3.2 Results of the Batch Experiments

Table 2.2.3.1 shows the typical results of the batch experiments performed with four selected metal ions and size fraction <80 μm of two HA samples [SEG-Ah(2) and SEG-Bh(2)]. In addition to the maximum sorption capacities k_1 and the binding constants k_2 calculated from the linearized Langmuir isotherms, the correlation coefficient r of the linear regression and the range of the total metal concentration c_t (M) are given. The results for copper and the size fraction 32 to 80 μm of two HA samples [SEG-Ah(3) and SEG-Bh(3)] are given in Table 2.2.3.2.

Table 2.2.3.1. Maximum sorption capacity k_1, binding constant k_2 and correlation coefficient of the linear regression for Cu, Zn, Mn and Ca of SEG-Ah(2)-HA and SEG-Bh(2)-HA (particle size < 80 μm, 3-mm column)

HA	Metal ion	c_t(M) (10^{-4} mol/l)	k_1 (mol/100 g)	k_2 (10^4 l/mol)	r
SEG-Ah (2)-HA	Cu^{2+}	1.31 – 5.24	0.079	5.5	0.9970
	Zn^{2+}	0.77 – 5.10	0.035	7.2	0.9985
	Mn^{2+}	0.61 – 5.46	0.039	4.3	0.9971
	Ca^{2+}	1.52 – 8.32	0.041	2.7	0.9984
SEG-Bh (2)-HA	Cu^{2+}	1.31 – 5.24	0.082	5.1	0.9971
	Zn^{2+}	0.77 – 6.33	0.052	2.7	0.9982
	Mn^{2+}	0.61 – 5.46	0.032	2.6	0.9948
	Ca^{2+}	1.25 – 8.32	0.071	2.0	0.9876

Table 2.2.3.2. Maximum sorption capacity k_1, binding constant k_2 and correlation coefficient of the linear regression for copper of SEG-Ah(3)-HA and SEG-Bh(3)-HA (particle size 32–80 μm, 3-mm column)

HA	c_t(M) (10^{-4} mol/l)	k_1 (mol/100 g)	K_2 (10^4 l/mol)	r
SEG-Ah(3)-HA	1.31 – 5.24	0.079	9.1	0.9984
SEG-Bh(3)-HA	1.31 – 5.24	0.079	7.9	0.9993

The maximum sorption capacity which reflects the available binding sites is highest for copper. The values for the three other metals are roughly the same and about half the value of copper. In general, the constants have the sequence Cu > Zn ~ Mn > Ca, which is in good agreement with the Irving and Williams row (1948) for stability constants of divalent metal ions. The maximum sorption capacities for HA from the two different horizons show no significant differences, whereas the binding constants for the HA of the Ah horizon are slightly higher.

Table 2.2.3.3. shows that the values k_1 and k_2 are also quite similar for the interaction of copper with the three different particle size fractions of SEG-Ah(3)-HA and SEG-Bh(3)-HA. In the case of a wide variation of the total metal concentrations [c_t (M)], a strong concentration dependence of the maximum sorption capacities and the binding constants can be seen. The examples of Table 2.2.3.4 show that for calcium and copper the maximum sorption capacities increase with the concentration and binding constants decrease.

2.2.3.3 Results of the Column Experiments

Figure 2.2.3.1 shows the experimental breakthrough curves for copper and manganese in a 3-mm column packed with SEG-Bh(3)-HA of particle size 80–125 μm.

Table 2.2.3.3. Maximum sorption capacity k_1 and binding constant k_2 for copper [c_t (Cu): 8.3–333.3 mg/l] of SEG-Ah(3)-HA and SEG-Bh(3)-HA from different particle sizes

HA	Particle size (μm)	k_1 (mol/100 g)	k_2 (10^4 l/mol)
SEG-Ah(3)-HA	32 – 80	0.079	9.1
	80 – 125	0.080	8.9
	125 – 200	0.080	8.9
SEG-Bh(3)-HA	32 – 80	0.079	7.9
	80 – 125	0.081	7.3
	125 – 200	0.080	8.9

Table 2.2.3.4. Maximum sorption capacity k_1 and binding constant k_2 of Ca and Cu by different metal concentrations c_t (M) for SEG-Bh(2)-HA (32–180 μm)

	c_t(M) (mg/l)	k_1 (mol/100 g)	k_2 (10^4 l/mol)
Ca	2.7 – 16.7	0.050	18.1
	2.7 – 43.3	0.073	2.0
	2.7 – 66.7	0.092	0.8
Cu	8.3 – 13.3	0.059	19.09
	8.3 – 16.7	0.064	14.2
	8.3 – 33.3	0.083	5.0
	8.3 – 166.7	0.101	1.4
	8.3 – 333.3	0.125	0.6

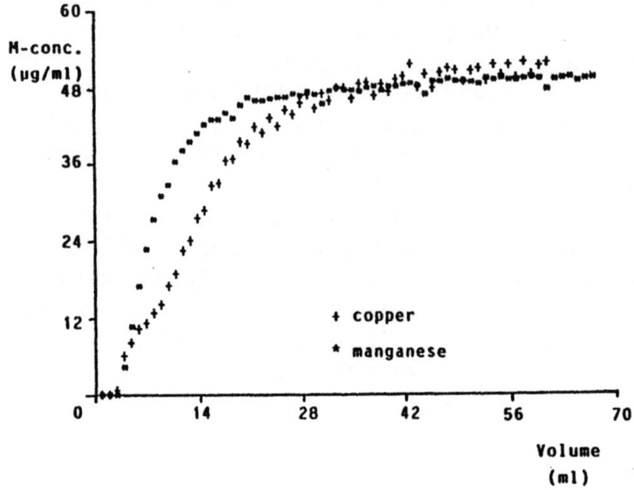

Fig. 2.2.3.1. Experimental breakthrough curves for copper and manganese in a 3-mm column of particle size 80–125 μm of SEG-Bh(3)-HA

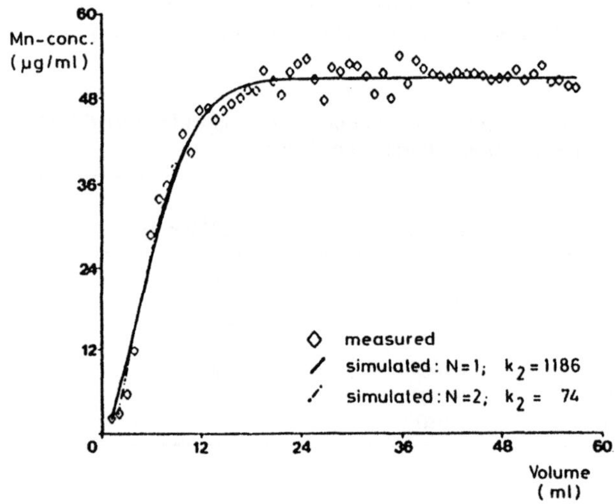

Fig. 2.2.3.2. Experimental and simulated breaktrough curve for manganese in a 3-mm column of particle size 32–80 μm of SEG-Bh(3)-HA

The two curves are quite different, especially in the maximum sorption capacities. From the breakthrough curve k_1-values of 0.09–0.12 mol/100 g and 0.05–0.07 mol/100 g can be derived for copper and manganese, respectively.

Figures 2.2.3.2 and 2.2.3.3 show the results of the simulation for measured breakthrough curves of manganese and copper in 3-mm columns packed with SEG-Bh(3)-HA (size fraction 32–80 μm). For Mn binding constants of 1186 l/mol for $n = 1$ and 74 l/mol for $n = 2$ were calculated. The values for Cu are 1830 and 228 l/mol, respectively. The simulation results show that it is possible to simulate the reaction for both plate numbers using for $n = 1$ higher and for $n = 2$ lower values of the binding constants. In general, a better fit could be achieved for Mn in comparison to Cu (Klamberg et al. 1989).

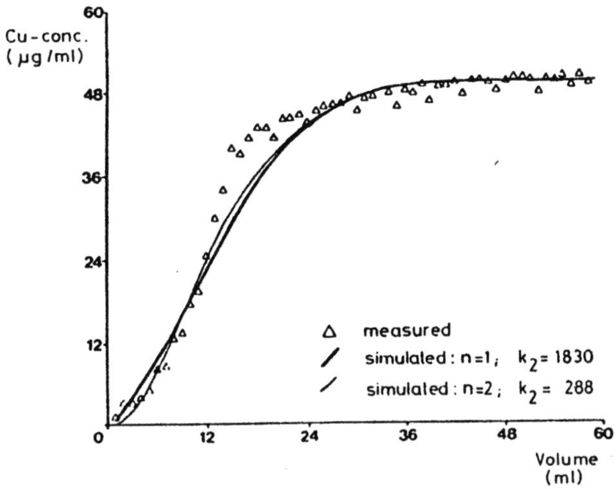

Fig. 2.2.3.3. Experimental and simulated breakthrough curve for copper in a 3-mm column of particle size 32–80 µm of SEG-Bh(3)-HA

It is suggested that Mn, showing the smaller sorption constants, is bound at the most attractive sites of the HA, whereas Cu is bound at the less attractive sites as well. The difference in the two metals is reflected at the beginning of the breakthrough curve, which is a result of the affinity to the binding sites. A good agreement between experiment and simulation at the beginning of the curve can be achieved on the assumption of a lower sorption capacity for Cu, representing only the strong binding sites. Figure 2.2.3.4 shows the experimental curve from Fig. 2.2.3.3 combined with a simulation according to $n = 1$ and a sorption capacity of 0.07 mol/100 g instead of the determined value for k_2 of 0.09 mol/100 g (Table 2.2.3.5). From the early part of the curve a binding constant of 5282 l/mol can be derived, which is of course much higher than the value calculated from the whole function.

2.2.3.4 Discussion

The investigation of metal sorption reactions on isolated humic material has been widely investigated by batch or static methods (e.g. Kerndorff and Schnitzer 1980; Slavek et al. 1982; Thielen et al. 1984; Bodes-Fischer 1985). Recently, column experiments or dynamic methods have been applied for the investigation of the interaction between metal ions and sediments, soils and minerals (Christophersen et al. 1980; Baumann and Schulz 1984; Spies 1984; Schulz and Baumann 1985) and they were used for investigation of isolated humic substances by Becker (1987) for the first time.

Using a definite total metal concentration of a selected metal, maximum sorption capacities from the batch experiments are fairly similar for the different soil horizons and the different size fractions of the HA. It is interesting to note that they

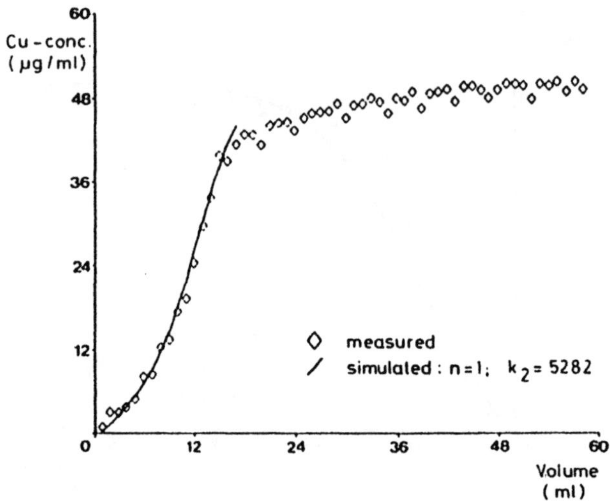

Fig. 2.2.3.4. Experimental and simulated (early part) breakthrough curve for copper in a 3-mm column of particle size 32–80 μm of SEG-Bh(3)-HA

Table 2.2.3.5. Comparison of the batch and column experiments for the maximum sorption capacity of copper

Particle size (μm)	Batch method	Column method
	SEG-Ah(2)-HA	SEG-Ah(3)-HA
32 – 80	0.079	0.092
80 – 125	0.080	0.10
125 – 200	0.080	0.093
	SEG-Bh(2)-HA	SEG-Bh(3)-HA32 – 80
32 – 80	0.079	0.090
80 – 125	0.081	0.106
125 – 200	0.080	0.084

are also in good agreement with the values derived from the breakthrough experiments (Table 2.2.3.5). Different, however, are the binding constants. They are quite smaller than those from the batch experiments. The significance of this observation and the consequence for the model calculation of the interaction between HA and metal ions in natural systems are yet unclear and need further investigations.

The sequence of the maximum sorption capacities proves that the ion exchange due to the chelate formation is a significant mechanism for the sorption effect. The hypothesis is also supported by the change of proton concentration during the sorption reaction. Becker (1987) showed for the column experiments curves for the proton concentration which were reversed to the breakthrough curves.

The observed strong dependence of the determined sorption parameters in a chosen range of the total metal concentration can be explained by the assumption that for low metal concentrations the binding sites that lead to the most stable complexes are first occupied; as the amount of metal increases, also less attractive positions will react leading to an increase in the maximum sorption capacity and to a decrease in the binding constants. Possibly an improved description of the experimental results can be obtained when using stability functions for the concentration dependent change in binding constants as introduced by Gamble et al. (1980) for the description of metal humic complexation (Sect. 2.2.2, this Vol.; Finger 1989).

2.2.4 Interactions of Humic Substances with Iodine

K.G. Heumann and C. Reifenhäuser

2.2.4.1 Iodine in the Environment

With a mass fraction of about 6×10^{-6}% in the Earth's crust, iodine is one of the least abundant elements. The concentration of iodine in aquatic systems therefore lies at a very low level, normally in the range of 100 ng/l to 10 μg/l. Up to now, in addition to total iodine concentrations in the environment also inorganic and organic iodine species have been analyzed.

The most abundant inorganic iodine species are iodide and iodate, but hypoiodous acid has also been detected in the atmosphere. The calculated ranges of stability for the different inorganic iodine species in aquatic systems dependent on the pH value and the redox potential are given e.g. by Bowen (1979). In addition, biologically produced methyl iodide is an organic compound with a substantial transfer rate from the surface of the ocean into the atmosphere (Cicerone 1981; Korzh 1984).

The cycle of iodine in the environment has been studied especially because of the influence of an iodine deficiency on the appearance of goitre and because nuclear fission processes release radioactive iodine isotopes (Bulman 1985; Keller 1986). ^{131}I with a relatively short half-life of about 8 d and the long-lived ^{129}I with a half-life of approximately 1.6×10^7 a are the two most important radioisotopes of this element. From investigations in surface and pore water with a radioiodine tracer it has been concluded that the radioiodine is associated with HS (Behrens 1984, 1985; Albertsen and Matthess 1980). Tikhomirov et al. (1978) also discussed an interaction of $^{131}I^-$ with HA isolated from soil.

One major task of these investigations was the characterization of different iodine species in aquatic systems and their quantification at low concentration levels with a reliable analytical method.

2.2.4.2 Conversion of Radioiodide in Aquatic Systems Containing Humic Substances

The investigations of Behrens (1985) with a radioiodide tracer have shown that the iodide ion undergoes chemical reactions with dissolved organic compounds in aerated surface and pore water samples. These reactions seem to be instigated by microbiological activity, which was indicated by the fact that preheated water samples lost the ability to convert iodide into an organic species. The conversion of iodide resulted in an increase of the amount of iodine which was not longer precipitable with a Ag^+ solution. With carrier-free $^{131}I^-$ tracer concentrations of about 10^{-8} M an equilibrium was always reached after approximately 60 h. In this case about 80% of the initial $^{131}I^-$ was converted.

The results of these radioiodine studies suggested that aquatic systems with HS also contain natural organic iodine compounds similar to those yielded by the radiotracer experiments. Therefore in these investigations the conversion of $^{131}I^-$ into an organic compound and the state of the equilibrium was determined in different water samples with various concentrations of HS. To prove the interaction of iodide with well-defined HS, solutions were also used which were prepared in the laboratory with HS isolated from aquatic systems or from brown coal. Additionally, an analytical procedure was developed which allows the characterization and isolation of the organic iodine compounds by their elution behaviour at a strongly basic anion exchanger resin.

The experiments were carried out with samples of different origin including surface waters from a small tributary of the Danube near Regensburg [Schwarze Laaber (SL)], and from lake waters [the small ponds Sarchinger Weiher (SAW) and Schwetzendorfer Weiher (SDW) in the vicinity of Regensburg, and of the Brunnenseemoor (BM), located in a marshy land in South Bavaria], as well as solutions from brown coal. It can be presumed that the humic substances in all these samples originate from soils or peat deposits.

Figure 2.2.4.1 represents the kinetics obtained when a carrier-free $^{131}I^-$ solution with an activity of about 3.7×10^6 Bq was added to 500 ml of SL water (the concentration in the water sample after this addition was approximately 10^{-11} M). The non-precipitable fraction of the initially added radioiodide is given on the y-axis of Fig. 2.2.4.1. In sample aliquots, taken after different reaction times, a co-precipitation of $Ag^{131}I$ with AgCl was carried out using an excess of chloride, which permits a successful filtration of the radioiodide precipitation (Behrens 1985). A conversion of only 3% was reached after a reaction time of 4 d. After this initial time with a low reaction rate, the portion of non-precipitable iodide increased rapidly until the equilibrium state of about 56% was reached within 14 d. After 20 d the total amount of iodide, which could be precipitated as AgI, was removed from the system. About 2 d after this removal the equilibrium was re-established which means that the non-precipitable iodine compound was partially re-converted into precipitable iodide. This result agrees well with investigations of Behrens (1985).

Behrens (1985) found that the equilibrium state in the systems he investigated contained 80% of the non-precipitable iodine compound. In the experiment with a SL water sample (from November, 1985) a much smaller equilibrium fraction of 56% of this non-precipitable iodine compound was found (Fig. 2.2.4.1). On the

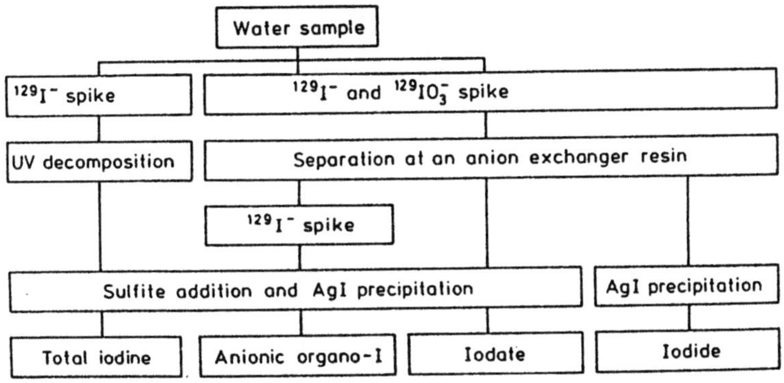

Fig. 2.2.4.1. Conversion of $^{131}I^-$ into a non-precipitable organic iodine compound in an SL water sample depending on the reaction time

other hand, equilibrium was reached after about 2.5 d in Behrens' experiments whereas in the investigation approximately 14 d were used. The fraction of the non-precipitable radioiodine at equilibrium varied in different natural surface water samples. A positive correlation between the fraction of the non-precipitable iodine compound and the concentration of dissolved organic carbon (DOC) could be observed (Table 2.2.4.1). This holds not only for samples of different origin but also for samples from the same source (SL) taken in different seasons.

Further experiments with SL water samples, collected at the same time as the sample which was used for the investigation represented in Fig. 2.2.4.1 have shown the following tendencies:

1. Heating the sample to the boiling point destroys the capability of the system to convert radioiodide into a non-precipitable species.
2. Filtering the water sample through a cellulose nitrate filter with a pore size of 0.8 μm reduces the conversion rate.
3. The addition of a sulfite solution to the system at equilibrium causes an increase of the precipitable radioiodide.
4. An increase of the pH value of the natural sample from 7.6 to 8.5 increases the initial time of conversion with a very low reaction rate from 4 d to approxi-

Table 2.2.4.1. Fractions of the non-precipitable radioiodine compound at equilibrium and DOC concentrations in different natural surface water samples

Sample	Sampling date	Fraction of non-precipitable iodine at equilibrium (%)	DOC concentration (mg/l)
River Schwarze Laaber (SL)	Nov. 1985	56	1.6
	Jan. 1986	61	2.6
	May 1986	68	2.9
Sarchinger Weiher (SAW)	May 1986	70	4.2
Schwetzendorfer Weiher (SDW)	May 1986	77	5.4

mately 12 d. The curve then runs nearly parallel to that given in Fig. 2.2.4.1. By changing the pH value down to 2 or up to 11, a radioiodide conversion is prevented.

5. A sample stored 4 months in a bottle at room temperature in the dark shows the same conversion behaviour for radioiodide as was found for a fresh sample.
6. The addition of $^{131}IO_3^-$ does not show any conversion of the radioiodate with time.

The results of (1)–(3) agree well with those obtained by Behrens (1985) for other surface and pore water samples. The tendencies found in (1) and (2) confirm the assumption that microbiological activity influences radioiodide conversion. Because of the positive correlation between the DOC concentration and the fraction of converted radioiodide (Table 2.2.4.1) and because of the fact that no conversion of radioiodide into another inorganic iodine species, e.g. iodate, was observed it must be assumed that the radioiodine is bound organically, e.g. in HS.

The attempt was made to characterize the non-precipitable organic radioiodine compound by elution in a column filled with the strongly basic anion exchanger resin AG1-X8 (Dowex 200–400 mesh; column height 10 cm, diameter 0.5 cm). For the elution of the anionic compounds $NaNO_3$ solutions of different concentrations were used.

In Fig. 2.2.4.2a the results are shown for $^{131}I^-$ and $31IO_3^-$ added to a SL water sample. The chromatographic separation took place within 1 h after this addition. In this experiment only the iodate and the iodide elution band and no converted compound could be observed.

All other investigations with the same water sample (Fig. 2.2.4.2b), with a solution of aquatic HS (Fig. 2.2.4.2c) which had been isolated by a method described by Frimmel and Geywitz (1983), and with a solution of lignite HS (Fig. 2.2.4.2d) which had been prepared from a commercially available HS substance (Roth, Nr. 2–7824), were carried out with $^{131}I^-$. In this case the chromatographic separation was started after a reaction time sufficient to reach equilibrium (> 14 d). In Fig. 2.2.4.2b-d only one elution band could be observed at an elution volume which is different from the elution volumes of iodate and iodide. The new anionic organoiodine compound is not as strongly bound at the anion exchanger resin as iodide but more strongly than iodate. The elution curves for river water, for an aquatic HS solution, and for a lignite HS solution are nearly identical and suggest that similar anionic iodine HS compounds were formed. The fraction of the initially added ^{131}I activity (corrected for radioactive decay), which is included in the elution band of the anionic organoiodine compounds, lies between 38.0% and 42.7%. The rest of the ^{131}I activity remained at the top of the column and could not be eluted under the experimental conditions used.

A membrane filtration through a cellulose nitrate filter with a pore size of 0.015 μm was carried out with a spiked $^{131}I^-$ sample of the SL and of the lignite HS solution after equilibration (15 d). This membrane filter excludes molecules with masses > 60 000 dalton. Of the total ^{131}I activity, 26% was retained by the filter in the case of the SL sample, 31% in the case of the brown coal HS solution. In both experiments the total amount of the anionic organoiodine compound which could be separated by the anion exchanger resin was in the filtrate. From this it follows that

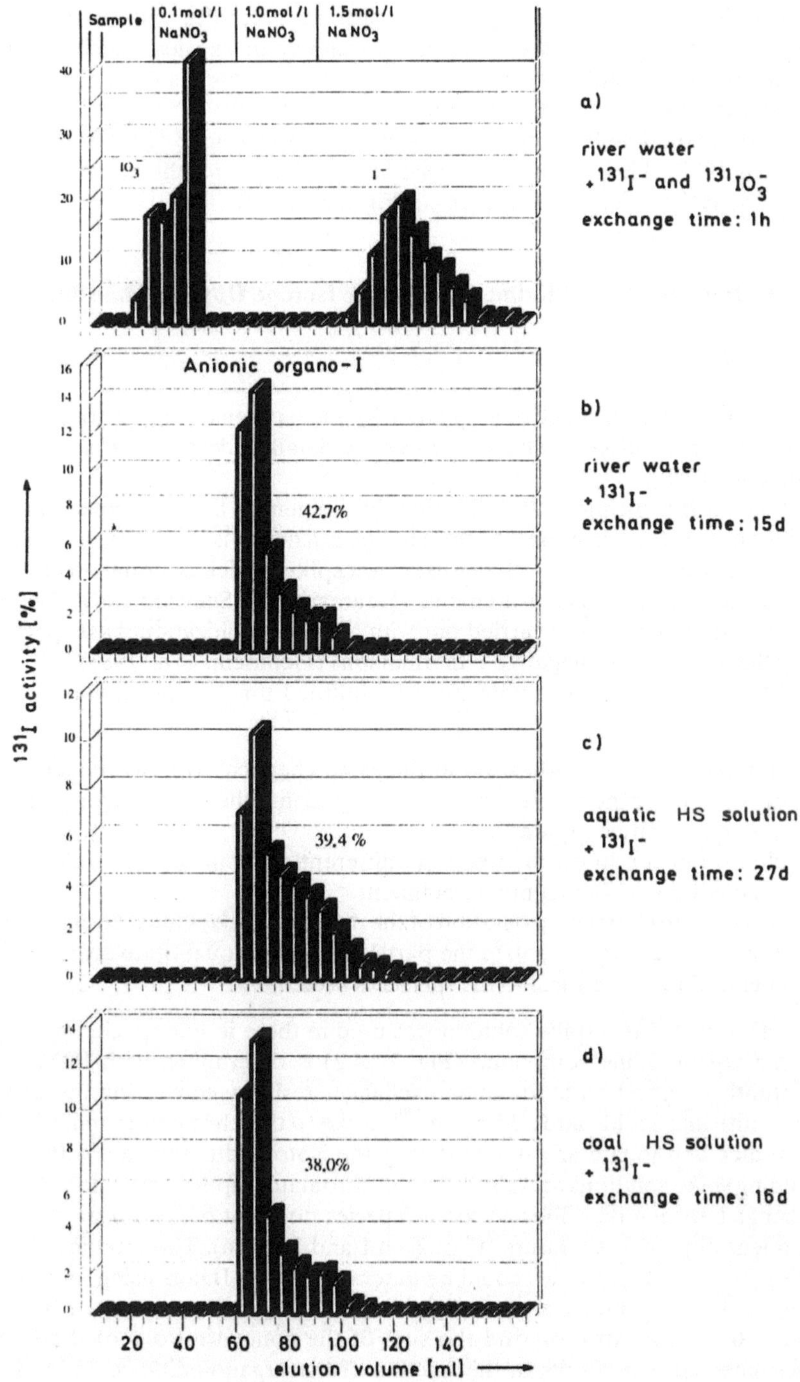

Fig. 2.2.4.2. Chromatographic separation of ^{131}I species after equilibration with HUS-containing solutions

the anionic organoiodine compound consists of molecules with masses < 60 000 dalton and that approximately half of the radioiodine species, which could not be eluted, consists of molecules with > 60 000 dalton. This result completes size exclusion chromatographic investigations of surface and pore water samples which specified radioiodine HS with molecular weights in the range of 1000–10 000 dalton (Behrens 1985). These substances must be associated with the anionic compounds separated at the anion exchanger resin (Fig. 2.2.4.2b-d).

2.2.4.3 Determination of Iodine Species with Isotope Dilution Mass Spectrometry

Fundamentals

Isotope dilution mass spectrometry (IDMS) is a very sensitive and accurate method for the determination of element traces. A detailed description is given e.g. by Heumann (1988).

One basic principle of IDMS is that the element to be determined must consist of at least two stable isotopes. Natural iodine is a mono-isotopic element. Therefore the long-lived radioactive ^{129}I was used as a spike, which is a quasi-stable isotope for mass spectrometric measurements (Heumann and Seewald 1985). The isotope ratio determinations were carried out with a thermal ionization mass spectrometer using the formation of negative I^- thermal ions (Heumann et al. 1985).

Three main presuppositions must be fulfilled for the application of IDMS in species analysis.

1. A spike must be synthesized in the same chemical form as the species to be determined or the species must be converted into the chemical form of the spike after quantitative separation.
2. No isotopic exchange between the different element species is allowed under the conditions of the sample treatment.
3. A complete chemical separation of the different species must be carried out; but it is not necessary to isolate the particular species quantitatively if a spike in a chemical form identical to the species is used.

An $^{129}I^-$ and an $^{129}IO_3^-$ spike solution was used in these iodine species analyses. The anionic organoiodine compound (Fig. 2.2.4.2) is determined with IDMS after its quantitative separation at an anion exchanger resin by converting this compound with sulfite into iodide and adding an $^{129}I^-$ spike to this chromatographic fraction.

Iodide and iodate are analyzed after the isotope dilution step with the corresponding spike solutions and a subsequent chromatographic separation. An isotopic exchange between the different iodine species does not occur during short sample treatment time of a few hours (Figs. 2.2.4.1 and 2.2.4.2a). The total iodine concentration in the water samples could be determined with IDMS using an UV decomposition of the organic substances dissolved in the water. The difference between the total iodine concentration and the sum of the concentrations of the individual determined species results in the fraction of the organoiodine compounds, which could not be eluted at an anion exchanger resin. This calculation of the non-elutable organoiodine species presupposes that no other anionic organoiodine compound exists in natural waters, which could not be observed by the radiotracer experiments with ^{131}I describe above. The complete analytical procedure for the iodine species analysis in water samples with IDMS is represented in Fig. 2.2.4.3. A more detailed description of this procedure is given by Reifenhäuser (1989) and Reifenhäuser and Heumann (1990).

Analysis of Iodine Species

SL and BM water samples collected in different months during 1987 and 1988 from two fairly different aquatic systems with low concentrations of organic compounds (SL) and with high amounts of HS (BM) were analyzed for the various iodine species. The results are listed in Table 2.2.4.2. The given mean values with standard deviations are the result of three or four parallel analyses. The listed concentrations of the different species are always related to the elementary iodine content and not to the molecular weight of the individual species. The results show that the IDMS method is able to analyze traces of iodine species down to the level of some hundred ng/l in water samples with acceptable precision. Because IDMS is also a definitive method the analyzed concentrations should be accurate as well.

Important quality parameters and their values are summarized in Table 2.2.4.3 for the samples at the time of collection. In the case of SL water samples the pH value, the oxygen concentration, the amount of total salt indicated by the electric conductivity and the inorganic carbon concentration are always higher when compared with the BM samples. On the other hand, the BM samples show higher redox potentials and include much higher concentrations of total organic carbon (TOC) or dissolved organic carbon (DOC).

The different water quality data have to be seen in connection with the distribution of the iodine species. Whereas iodate is the most abundant species in the SL, the non-elutable organoiodine compounds display normally the highest concentrations in the BM samples. Although the exact correlation between the different

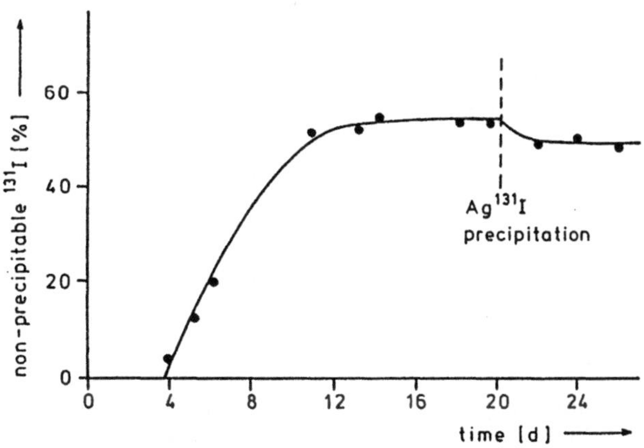

Fig. 2.2.4.3. Analytical procedure for the determination of iodine species in water samples with isotope dilution mass spectrometry

Table 2.2.4.2. Determination of iodine species in water samples of two different origins with isotope dilution mass spectrometry

Sample	Sampling date	Iodine concentration (μg/l)				
		Iodide	Iodate	Anionic organo-I	Total I	Non-elutable organo-I
Schwarze Laaber	Nov. 1987	n.d.[a]	0.9 ± 0.2	0.5 ± 0.1	2.5 ± 0.1	1.1
(SL)	March 1988	n.d.	1.7 ± 0.1	0.7 ± 0.1	4.9 ± 0.1	2.5
	July 1988	1.1 ± 0.1	0.8 ± 0.1	1.1 ± 0.2	3.2 ± 0.2	0.2
Brunnenseemoor	Dec. 1987	n.d.	n.d.	0.7 ± 0.1	5.0 ± 0.3	4.3
(BM)	April 1988	n.d.	n.d.	0.8 ± 0.1	2.2 ± 0.2	1.4
	Aug. 1988	1.6 ± 0.1	0.5 ± 0.1	0.5 ± 0.1	4.2 ± 0.2	1.6
	Nov. 1988	n.d.	n.d.	0.5 ± 0.1	1.9 ± 0.1	1.4

[a]n.d. = not detectable (concentration less than detection limit: I^- = 0.5 μg/l; IO_3^- = 0.1 μg/l; anionic organo-I = 0.2 μg/l; total I = 0.05 μg/l; 2xs definition).

Table 2.2.4.3. Water quality data of the analysed samples

Sample	Sampling date	Water temp. (°C)	pH	O_2 conc. (mg/l)	Conductivity (μS/cm)	Redox pot. (mV)	TOC (mg/l)	DOC (mg/l)	IC (mg/l)
Schwarze Laaber	Nov. 1987	10	7.6	9.6	474	+ 280	2	1	–
(SL)	March 1988	7	7.9	10.7	411	+ 360	2	1	58
	July 1988	11	7.9	7.9	608	+ 375	1	1	68
Brunnenseemoor	Dec. 1987	2	5.0	7.4	80	+ 455	51	46	–
(BM)	April 1988	15	6.3	7.2	145	+ 447	27	26	27
	Aug. 1988	17	6.1	7.8	73	+ 430	45	43	14
	Nov. 1988	10	5.9	7.3	62	+ 413	26	24	11

iodine species and the various parameters is not clearly known to date, the following tendencies can be recognized.

- The iodate concentration increases with increasing oxygen concentration in the aquatic system.
- There is a positive correlation between the redox potential and the concentration of anionic organoiodine compounds, as is represented in Fig. 2.2.4.4 for BM samples. In the other water samples the highest concentrations of these iodine species are also found in samples with a high redox potential. The microbiological activity, which may be responsible for the formation of organoiodine species in water, is possibly increased by relatively high redox potentials in the aquatic system.
- The total iodine concentration substantially depends on the DOC value in aquatic systems with high amounts of HS. In the case of the BM samples the total iodine concentration clearly reflects the DOC content (Fig. 2.2.4.5). This also indicates an association of iodine with the humic substances of the corresponding soil, which is the source for the compounds in the aquatic system.

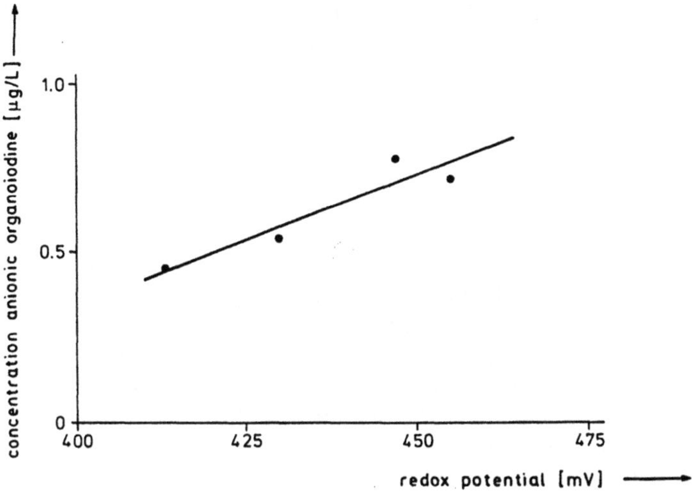

Fig. 2.2.4.4. Correlation between the redox potential and the concentration of anionic organoiodine species in BM samples

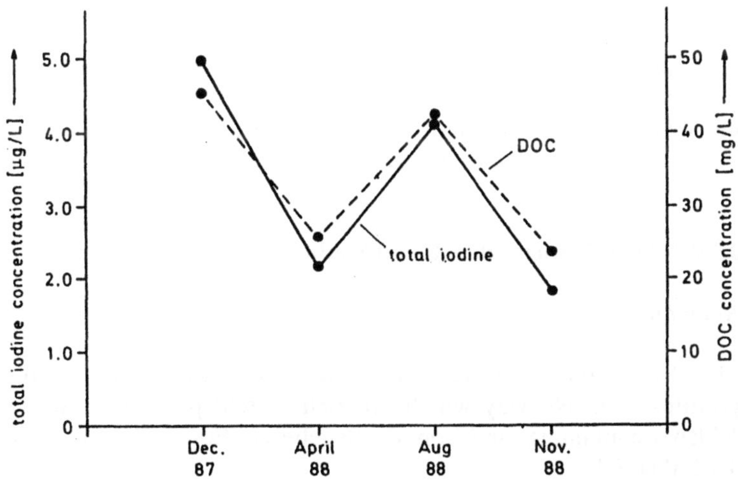

Fig. 2.2.4.5. Correlation between the total iodine concentration and the DOC concentration in different samples of an aquatic system with high HS contents (BM)

2.2.5 Experiments on the Influence of Organic Ligands upon Kinetics of Feldspar Weathering

A. Petersen, G. Matthess and D. Schenk

2.2.5.1 Model Approach

The rate of chemical weathering of silicates, which includes dissolution and precipitation processes, depends on the mineralogical composition and the specific surfaces of the solids, and on the temperature, flow rate, and pH of the water in the unsaturated and saturated zones (Stumm et al. 1985). The presence of organic ligands may influence the kinetics of these reactions and the transport behaviour of dissolved components. The organic ligand-forming substances are metabolites produced by living organisms and complex polymers formed by secondary reactions from the original metabolites (Scheffer and Ulrich 1960; Dubach and Metha 1963; Kononova 1966; Stevenson and Butler 1969).

Recent investigations aim to quantify the influence of the complex-forming behaviour of fulvic and humic acids on the kinetics of feldspar weathering processes, which evidently are important in natural systems. Because humic substances with well known, homogeneous structures and properties are not available in sufficient quantities for the experiments, other complex-forming organic substances with functional groups similar to the humic substances have been used. These reagents include bidentate salicylic and tartaric acids, the tridentate citric acid and the tetradentate phenol pyrocatechol violet. They form chelate complexes with metal ions such as aluminium or calcium, which due to their high stability are important for hydrogeochemical processes in the unsaturated and saturated zone.

2.2.5.2 Investigation Objects and Methods

Rock Material

The dissolution experiments were performed with three samples of pegmatitic granites from Norway which are rich in feldspars and also with glaciofluvial feldspar-containing sand from the Segeberger Forst (SEG, Fig. 1.1, Table 1.1, Sect. 4.3.2, this Vol.):

- an orthoclase-rich rock consisting of 72% vol. orthoclase, 27% vol. albite and 1% vol. anorthite
- an albite-rich rock consisting of 58% vol. albite, 32% vol. anorthite and 10% vol. orthoclase
- an albite- and anorthite-rich strongly weathered rock consisting of 56% vol. albite, 37% vol. anorthite and 7% vol. orthoclase and
- a calcium carbonate-free, medium to coarse, glaciofluvial sand consisting of 81–91% vol. quartz, 3–6% vol. orthoclase, 2–4% vol. plagioclase and 1–3% vol. mica and clay minerals.

Experimental Procedures

The influence of organic complex-forming ligands on the kinetics of feldspar dissolution was investigated with three laboratory experimental arrangements and at different pH values (Schenk et al. 1987). Since the room temperature varied only between 20 and 23 °C and – according to Bailey (1976) – temperature changes in such a small range do not influence the dissolution kinetics of feldspar significantly, no temperature controls were used. The alteration rates of feldspar are referred to the specific surface of the mineral grains [mol/(cm^2.s)] and are derived from the increases in the concentration (dissolution rate) of major components of feldspar (Si or Al) in the fluid before saturation of any secondary alumosilicate has been reached. The preparation procedures for the samples and the performance of the experiments are described in detail in Petersen (1988).

Dissolution Experiments in Solutions with Free and Fixed pH

The experiments were performed with deionized water or aqueous solutions of the organic complex-forming substances and crushed rock material of grain sizes between 100 and 200 µm (concentration: 12.5 g/l). Every day 15 ml fluid were taken from the reaction vessel, and the removed solution volume was replaced by deionized water and the equivalent amount of organic ligands.

In the batch experiments open to the atmospheric CO_2 (Dahmke 1988; Petersen 1988) the pH of the solution/rock system was not buffered. The contact of the fresh feldspar-surface with water and the subsequent hydrolysis of the feldspar leads to an increase in pH (Matthess 1973, 1982) as the experiment progresses, which is not representative of natural systems. These pH changes subsequently cause changes in the interactive system mineral/solution, in the surface properties of the solid material, and in the chemical behaviour of the organic complex-forming ligands.

For a better model approach to the stable natural pH conditions and for a better comparability of the different dissolution experiments, the pH of the solutions was kept constant by titration with solutions of the respective ligand. As a result, the ligand concentration in the reaction vessel is equivalent to the number of protons consumed by hydrolysis of the feldspar at a fixed pH value. The consumption of the organic acids and phenols was measured daily. The pH values in the parallel experiments with deionized water were kept constant by flushing the contact water with CO_2-gas (Fig. 2.2.5.1).

Column Experiments

The column method, adapted from Schultz and Baumann (1985; Fig. 3.1.2.5, this Vol.), is a dissolution experiment in a closed system with respect to atmospheric CO_2. This experiment simulates the situation in an aquifer. The feldspar grains (natural or from crushed rocks) contact each other, protecting a part of the grain surface against attack by the aqueous phase. If solutions of different composition percolate with certain flow rates through the porous grain matrix, the components dissolved from rock are removed from the system. The rock/water ratio is close to the natural conditions in the aquifer.

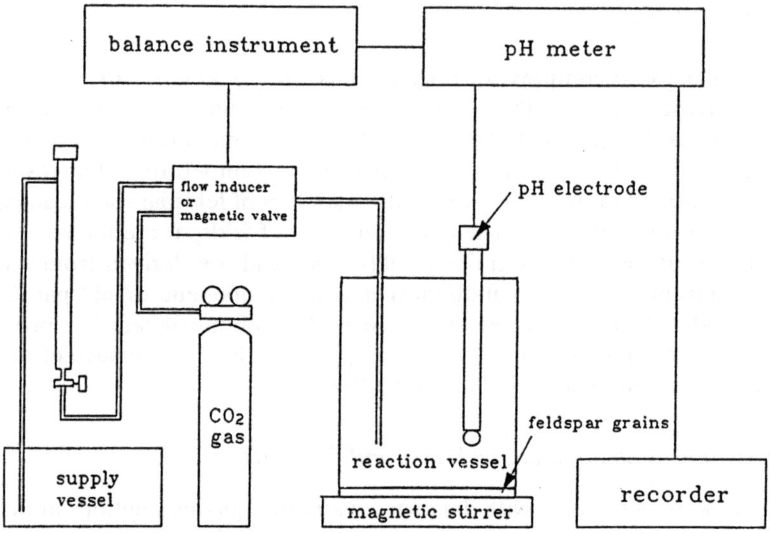

Fig. 2.2.5.1. Setup of the pH-fixed dissolution experiments

2.2.5.3 Results

At pH 4.5 and pH 6.0 only traces of dissolved Al were detected in experiments with deionized water (detection limit 2×10^{-3} mmol/l) in contrast to experiments with organic ligands. On the other hand, relatively high concentrations of Si were measured at both pH values (orthoclase-rich rocks 4.1×10^{-2}, and 2.3×10^{-2} mmol/l respectively; albite-rich rocks 6.8×10^{-2}, and 3.3×10^{-2} mmol/l respectively). Experiments with organic ligands lead to higher concentrations of dissolved Al. During the experiments the cumulative concentrations of dissolved Si and Al versus reaction time follow parabolic curves, converging to plateau values, which depend on the pH of the solution and on the concentration of organic ligands, respectively. The results of the batch experiments are summarized in Table 2.2.5.1.

During the dissolution process, the dissolution rates of feldspars in deionized water and in the solutions containing organic matter depend significantly on the pH value between pH 2.3 and pH 8 (Fig. 2.2.5.2), with a minimum dissolution rate at pH 6. The scattering of the rates between pH 2.3 and pH 2.5 is evidently due to the type of solid species (with highest rates for albite and lowest for orthoclase). Whereas at the low pH values mentioned above, no significant effect of the type of organic ligands is observed, at pH 4.5 and pH 6 the mineral species has a lower effect on the dissolution rates than the type of organic ligands present in the solution. The relationship between dissolution rates and pH are the same for the initial and the final stages of the process, albeit of their absolute levels of dissolution rates, indicating a much higher influence of pH than of the solid species and of organic ligands.

In pH-fixed dissolution experiments with pyrocatechol violet the dissolution process ends at pH 4.5 after 500 h and at pH 6 after 400 h reaction time. It is not

Table 2.2.5.1. Dissolution rates at different pH and different stages of experiments, and with different solvents[a]

	0 to 25 h		0 to 450 h	
	pH	dissol. rate [mol/(s•cm²)]	pH	dissol. rate [mol/(s•cm²)]
Albite				
Tart	2.57	1.9×10^{-13}*	2.44	1.5×10^{-14}*
Cit	2.56	1.6×10^{-13}*	2.67	1.3×10^{-14}*
Sal	2.32	1.6×10^{-13}*	2.64	1.1×10^{-14}*
Vpy	2.53	1.0×10^{-13}*	2.82	9.0×10^{-15}*
Tart	7.65	4.6×10^{-15}	8.42	7.6×10^{-16}
Cit	6.98	8.4×10^{-15}	8.08	7.1×10^{-16}
Sal	7.36	4.5×10^{-15}	7.99	6.5×10^{-16}
Vpy	7.58	$42.\times10^{-15}$	7.97	6.6×10^{-16}
Orthoclase				
Tart	2.41	6.4×10^{-14}*	2.39	5.0×10^{-15}*
Cit	2.54	6.3×10^{-14}*	2.61	3.9×10^{-15}*
Sal	2.41	6.3×10^{-14}*	2.46	3.9×10^{-15}*
Vpy	2.53	5.3×10^{-14}*	2.56	3.1×10^{-15}*
Tart	7.34	2.7×10^{-15}	7.59	7.1×10^{-16}
Cit	6.82	3.9×10^{-15}	7.50	5.8×10^{-16}
Sal	7.07	2.5×10^{-15}	7.26	6.7×10^{-16}
Vpy	7.60	4.1×10^{-15}	7.50	7.8×10^{-16}
Anorthite				
Tart	2.38	1.4×10^{-14}*	2.39	3.7×10^{-15}
Cit	2.51	3.6×10^{-14}*	2.57	2.1×10^{-15}
Sal	2.40	1.6×10^{-14}*	2.46	2.1×10^{-15}
Vpy	2.55	1.8×10^{-14}*	2.60	1.6×10^{-15}
Tart	7.88	5.8×10^{-15}	8.52	8.3×10^{-16}
Cit	7.50	6.0×10^{-15}	7.89	8.8×10^{-16}
Sal	7.67	5.7×10^{-15}	7.88	8.9×10^{-16}
Vpy	8.09	6.1×10^{-15}	7.99	8.9×10^{-16}

	0 to 50 h Init.dissol.rate [mol/(s•cm²)]	50 to 250 h lin. regression [mol/(s•cm²)]	250 to 800 h lin. regression [mol/(s•cm²)]
Albite			
Aqua dest.			
pH 4,5	4.1×10^{-15}	$7.0\times10^{-16}\pm 2\times10^{-17}$	$4.5\times10^{-16}\pm4\times10^{-17}$
pH 6,0	1.5×10^{-15}	$4.9\times10^{-16}\pm3\times10^{-17}$	$1.5\times10^{-16}\pm1\times10^{-17}$
Cit			
pH 4,5	7.3×10^{-15}*	$1.4\times10^{-15}\pm7\times10^{-17}$	$6.6\times10^{-16}\pm5\times10^{-17}$
pH 6,0	3.7×10^{-15}	$2.3\times10^{-15}\pm6\times10^{-17}$	$3.9\times10^{-16}\pm1\times10^{-16}$
Vpy			
pH 4,5	5.8×10^{-15}	$2.3\times10^{-15}\pm2\times10^{-16}$	$5.2\times10^{-16}\pm2\times10^{-17}$
pH 6,0	3.4×10^{-15}	$2.9\times10^{-15}\pm3\times10^{-17}$	$1.7\times10^{-16}\pm3\times10^{-17}$
Orthoclase			
Aqua dest.			
pH 4,5	1.8×10^{-15}	$9.3\times10^{-16}\pm3\times10^{-17}$	$2.7\times10^{-16}\pm3\times10^{-17}$
pH 6,0	8.1×10^{-16}	$3.5\times10^{-16}\pm2\times10^{-17}$	$1.2\times10^{-16}\pm3\times10^{-17}$
Tart			
pH 4,5	2.2×10^{-15}*	$9.3\times10^{-16}\pm6\times10^{-17}$	$2.7\times10^{-16}\pm3\times10^{-17}$
pH 6,0	1.3×10^{-15}	$3.8\times10^{-16}\pm4\times10^{-18}$	$2.7\times10^{-16}\pm3\times10^{-17}$
Cit			
pH 4,5	4.7×10^{-15}*	$1.3\times10^{-15}\pm3\times10^{-17}$	$6.9\times10^{-16}\pm3\times10^{-17}$
pH 6,0	4.7×10^{-16}	$6.8\times10^{-16}\pm1\times10^{-17}$	n.d.
Vpy			
pH 4,5	4.0×10^{-15}*	$1.2\times10^{-15}\pm6\times10^{-18}$	$5.2\times10^{-16}\pm1\times10^{-17}$
pH 6,0	1.8×10^{-15}	$7.2\times10^{-16}\pm3\times10^{-17}$	$1.4\times10^{-16}\pm3\times10^{-17}$

[a*] Dissolution rates derived from the concentration of dissolved aluminium.

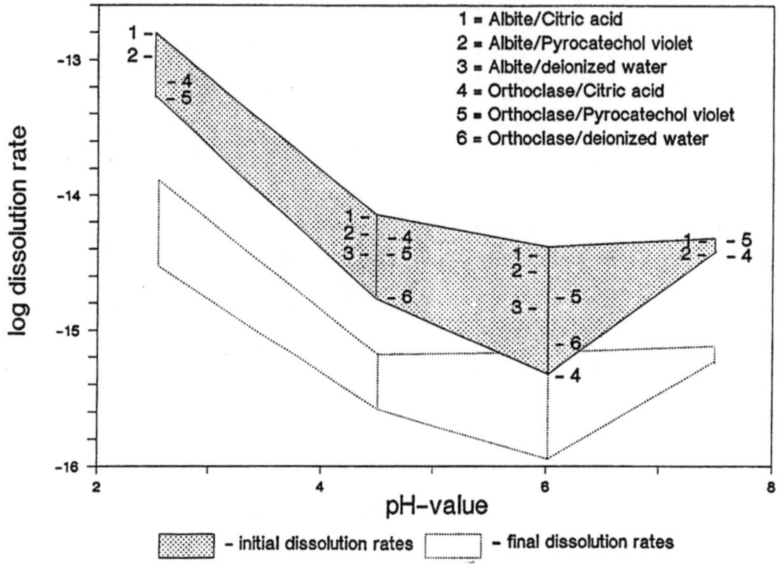

Fig. 2.2.5.2. Dissolution rates of albite- and orthoclase-rich rocks in deionized water and in pyrocatechol violet, and citric acid solutions during initial and final stages of dissolution processes depending on pH

possible to restart the dissolution process by cleaning the feldspar surface with HCl, but heating of rock material up to 500 °C apparently destroys the mineral-organic complex so that the dissolution process may continue.

The tendency of pyrocatechol violet to form stable mineral/organic surface complexes can be confirmed by comparison of the release of Si in two successive experimental runs (Fig. 2.2.5.3). The first run indicates a "normal" dissolution behaviour of Si concentration vs. time and a corresponding decrease of pyrocatechol violet consumption vs. time. The second run shows a very small dissolution rate of Si, which indicates the blocking of reactive surfaces by the organic ligand.

2.2.5.4 Discussion

The dissolution process of the feldspar-rich rocks is obviously affected by the organic ligands after an initial adsorption at the mineral surface forming surface complexes. After saturation of all free valencies, the surplus free organic ligands are able to form complexes with the dissolved cations which result in an increase in the dissolution rates (Petersen 1988).

The observation of a pH dependency of dissolution rates in the pH range 2.3–8 is not consistent with the theoretical predictions of Aagaard and Helgeson (1982), Helgeson et al. (1984), Knauss and Wolery (1986) and Mast and Drever (1987). But the experimental results may be explained according to Blum and Lasaga (1987) by a change of the rate-limiting activated surface complex from an H-complex to an OH complex. Such a change would reasonably explain higher initial dissolution

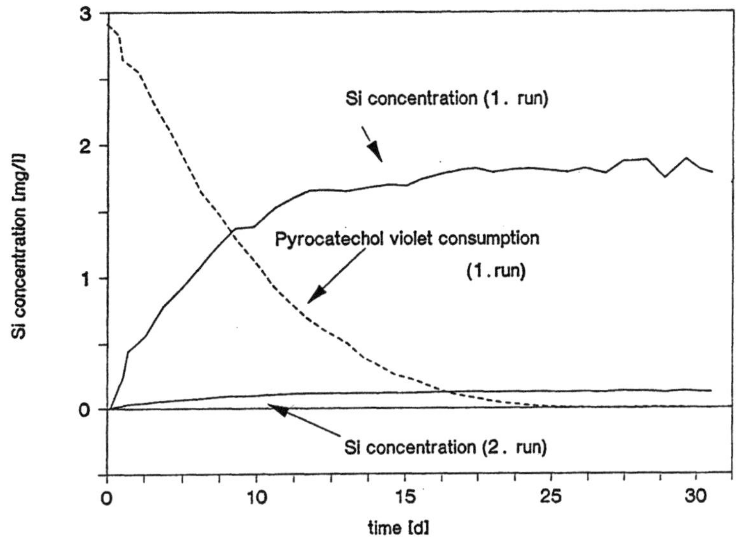

Fig. 2.2.5.3. Si and pyrocatechol violet concentrations vs. time, in the pH-fixed dissolution experiments (pH 4.5) with albite-rich rocks (*first run* original material; *second run* after blocking the minerals surfaces by pyrocatechol violet before heat treatment)

rates in experiments with pH 4.5 and pH 8 compared with the pH-fixed dissolution experiments at pH 6. This may be due to the predominance of OH surface complexes at higher pH of solution in contrast to the mineral-organic surface complexes as a result of the low concentration of complex-forming ligands. At pH values below 6 the concentration of active rate-controlling mineral-organic surface complexes decreases because the complex-forming capacity decreases and the concentration of active rate-controlling H surface complexes increases.

The observed parabolic dissolution of feldspars is explained by Aagaard and Helgeson (1982), Helgeson et al. (1984) and Knauss Wolery (1986) as an artefact of grounding of feldspars which causes an artificial increase of reactive surface sites. Decomposition reduces the concentration of activated surface complexes, which results in successively decreasing dissolution rates. But the same macroscopic effect can be explained by the surface-coordination model (Furrer and Stumm 1986). The possibility of blocking reactive surface sites with stable mineral-organic complexes (chemisorption) is supported by the experimental results with pyrocatechol violet. During the pH-fixed experiment the concentration of passive mineral-organic surface complexes increases with increasing concentration of the organic ligand, finally bringing the dissolution to a total stop. A restart of dissolution requires thermal breakdown of those complexes. These results are consistent with the theoretical approach and the experimental results of Furrer and Stumm (1986).

The application of complex-forming organic ligands leads to significantly higher concentrations of dissolved Al, as mentioned above. The congruency of a dissolution process can be described by the relative release ratio (RRR), suggested by Holdren and Speyer (1985) [Eq. (2.2.5.1)]:

$$RRR = \log \frac{(Al/Si)_{aq}}{(Al/Si)_{solid}} \tag{2.2.5.1}$$

The log RRR indicates a significant dependency of the Al/Si-ratio on the concentration of the organic ligands and the pH values (Fig. 2.2.5.4) and furthermore that the ratio of the dissolved Al/Si is only a function of the specific solubilities of their major components. In the range between pH 4.5 and pH 6, which pH values are fixed by maintaining specific concentrations of the respective organic ligands, the precipitation of aluminium species is prevented by complexation so that the Al/Si ratio in the solution is equal to that in the feldspar. The apparent incongruency of the dissolution process in deionized water may therefore be explained by a specific precipitation of the leached Al. This process may be effective in the dissolution experiments with deionized water and also with citric acid because saturation indices, calculated with the thermodynamic computer program PHREEQE (Parkhurst et al. 1980) with a modified data base, show that the observed lack of dissolved Al is probably caused by the precipitation of Al-hydroxides.

The concentration of dissolved Al in the columns reflects the transport velocity of free organic ligands. The results indicate that the dissolution processes in a natural weathering profile are also controlled by the sorption characteristics of the organic ligand and consequently by their retardation.

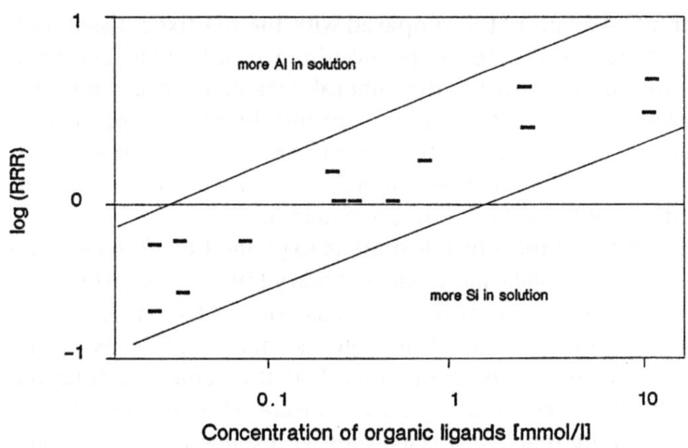

Fig. 2.2.5.4. Al/Si ratio vs maximum concentration of organic complex forming ligands in experiments with albite- and orthoclase-rich rocks and citric acid and pyrocatechol solutions

2.3 Characterization of Some Organic Acids in the Subsurface of the Sandhausen Ecosystem

T. Cordt and H. Kußmaul

2.3.1 Background

The occurrence of simple organic acids in soils and their possible interactions with mineral weathering, metal complexation/chelatization, sorption/desorption, precipitation and biological processes have often been reported, e.g. Stevenson (1982); Matthess (1984) and Thurman (1986). These investigations aimed to achieve more information about the amounts of these substances in natural environments controlling ground water quality.

2.3.2 Sampling Site and Methods

Investigations were carried out in the vicinity of the Sandhausen experimental site (SDH) described by Jacob et al. (Sect. 4.2.3 and 6.2.3, this Vol.). Soil samples were collected in small temporary survey pits. They were kept cool with dry ice until arrival and preparation in the laboratory. Samples of the unsaturated zone down to the ground water were obtained by continuous coring with percussion drilling. Core tubes were sealed at the sampling site, samples were retrieved under a nitrogen atmosphere in the laboratory in order to prevent exposure to atmospheric oxygen thus minimizing alterations of the redox state. Percolating water was separated by centrifugation of the samples immediately after retrieval. Ground water samples were obtained from a multilevel well.

Analysis of inorganic cations and anions was performed by inductive coupled plasma atom emission spectrometry (IC-AES), AAS and ion chromatography (IC), respectively. DOC was determined by UV oxidation followed by IR detection of the produced CO_2.

2.3.3 Geochemical Situation of the Observed Ecosystem

Using the pH value as an acidification intensity parameter, the profile from the surface of the soil down to the ground water reveals a buffer system zonation (Fig. 2.3.1).

The limits of the defined individual zones were chosen according to Ulrich (1981a). Throughout the period of observation a season dependent oscillation between different buffer systems could be observed. The amplitude is approximated

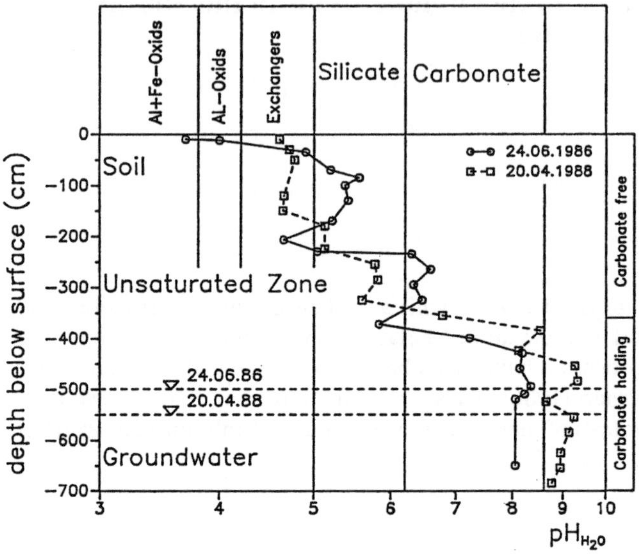

Fig. 2.3.1. Buffer system zonation in the unsaturated zone above the unconfined sandy aquifer of the Sandhausen ecosystem. (Type of buffers according to Ulrich 1981a)

by the two curves of Figure 2.3.1. In the upper part of the profile up to 1 meq/l Al^{3+} could be detected in percolating water. As a consequence of 1 toxicity for bacteria, the major microbial activity can be attributed to acidophile fungi; the system is believed to be in a state of humus disintegration and decomposition as described by Ulrich (1981b). An initial podsolization indicates that translocation processes of matter are controlled by organic substances. Monitoring of the DOC concentrations in percolating water via samples from porous cups as well as centrifugates of core samples exhibit a downward transport of organic substances.

2.3.4 Organic Acids

2.3.4.1 Analytical Methods

The analysis of organic acids was accomplished by GC and GC-MS (Table 2.3.1). A method developed by Littman (1985) for the detection of specific organic acids in foodstuffs was modified to be applied to soil and water samples. In general, soil samples were leached with 0.1 M NaOH, deionized water or canopy throughfall. Extracts and water samples were concentrated by vacuum evaporation at pH 12–13 (NaOH addition). The further procedure included a "column cleanup" (Na_2SO_4/ Cellit mixture, eluens: ethyl acetate) at pH 2 (HCl addition) and derivation of organic acids to methyl esters with methanolic HCl. Glutaric acid was chosen as internal standard for the quantitative analysis because it is believed not to occur in

Table 2.3.1. Apparatus and conditions

GC-MS analysis, qualitative identification of major components. Gas chromatograph HP 5890 with detector MSD 5970 Injector: 270 °C Detector: 280 °C Fused silica column Supelcowax (J&W Scientific Inc.), inner diameter 0.25 mm, film thickness 0.25 μm, length 30 m Temperature programme: 50 to 250 °C, 6 °C/min Carrier gas: helium Interpretation: HP work station 9816 with NBS Mass Spectral Library from 1984
GC analysis of acetic, propionic and butyric acids. Gas chromatograph Varian 3700 with FID detector Injector: 200 °C Detector: 220 °C Column 0V351 (J&W Scientific Inc.), inner diameter 0.322 mm, film thickness 0.25 μm, length 30 m Temperature programme: 120 °C, isothermal Carrier gas: helium Interpretation: internal standard and reference substances
GC analysis of other organic acids. Gas chromatograph Varian 3700 with FID detector Column and conditions like GC-MS Carrier gas: helium Interpretation: internal standard and reference substances

natural environments and could not be detected in the raw samples. However, caution should be taken with respect to the analysis of other soil samples or preparation procedures, as Spiteller (1981) detected glutaric acid after degradation of isolated humic acids with permanganate.

Acetic acid, propionic acid and butyric acid were analyzed in aliquots of the concentrated and acidified samples not subjected to derivatisation. Isovaleric acid was used as internal standard for this analysis.

2.3.4.2 Results and Discussion

In order to obtain insight into the general composition of the "low weight" organic acid pool of the ecosystem under observation, a soil sample with a high content of organic matter was extracted with 0.1 M NaOH for 24 h. Having undergone the procedure described above, the sample was subjected to a qualitative GC-MS analysis. The resulting chromatogram and a list of identified peaks are given in Fig. 2.3.2 and Table 2.3.2.

For further quantitative analysis and organic acid screening, a selection of organic acids identified by GC-MS was employed as reference substances. A comparison of the influence of different aquatic extractants and of centrifugation on the gainable amount of organic acids was carried out using 0.1 M NaOH (pH 12.87), deionized water (pH 6.6) and canopy throughfall (pH 3.79). Soil pH was 4.2.

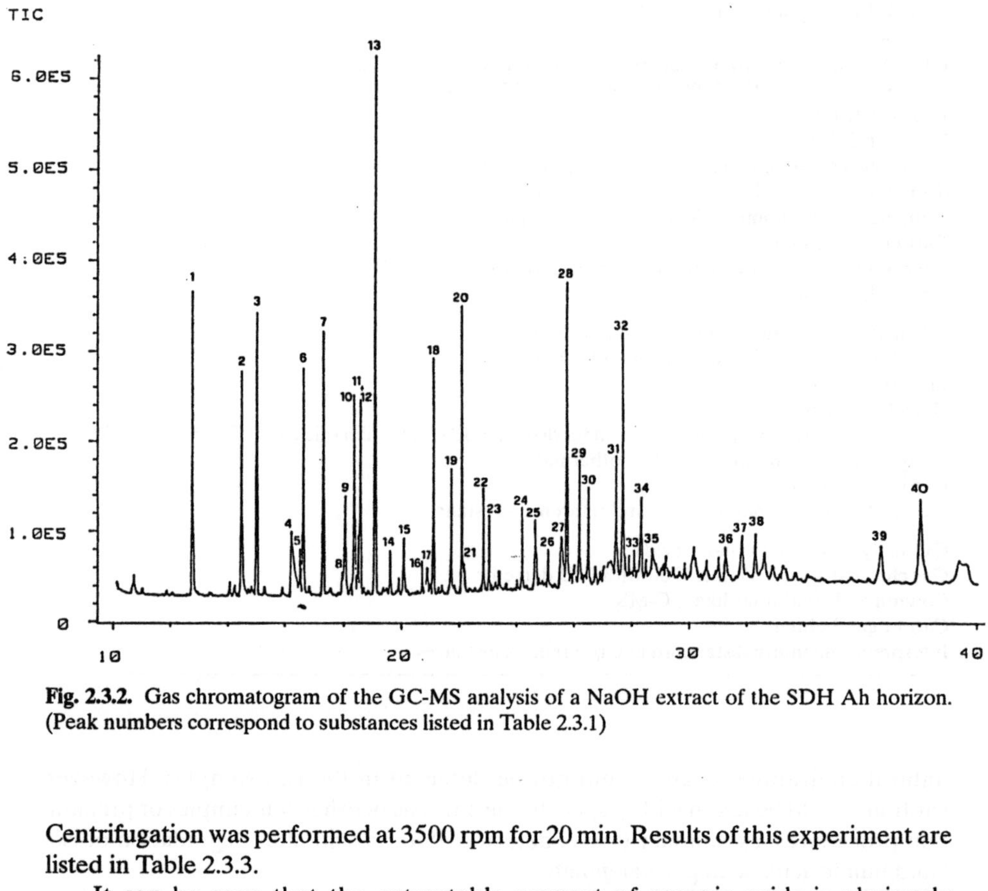

Fig. 2.3.2. Gas chromatogram of the GC-MS analysis of a NaOH extract of the SDH Ah horizon. (Peak numbers correspond to substances listed in Table 2.3.1)

Centrifugation was performed at 3500 rpm for 20 min. Results of this experiment are listed in Table 2.3.3.

It can be seen that the extractable amount of organic acids is obviously pH-dependent and that the solubilization ratios are not the same for different acids. There is also a shift in the substances dominating the gross concentration. Centrifugate concentrations, when transformed to the same units as the extract values via soil water content, lead to the lowest values. This may be due to the fact that only part of the water can be retrieved by this method and that possibly sorption and temperature effects lead to a loss of recoverable amounts. The very high quantities gained by NaOH extraction may have resulted from desorption of loosely bound organic acids from humic substances or even from alterations of the outer structure of these macromolecules, whereas the concentrations obtained by extraction with deionized water and canopy throughfall may be attributed to an increased solubility of corresponding organic salts caused by excess amounts of water during extraction (crystals of Ca-oxalate were found in soil samples via electron microscopy). NaOH extracts seem to represent an overestimation of the total organic acid pool, whereas centrifugates may be an underestimation of the portion solubilized in percolating water.

A comparison of the detected amounts found in centrifugates of upper soil horizons and ground water samples is given in Table 2.3.4.

Table 2.3.2. Organic acids identified as methyl esters by GC-MS analysis of a soil sample extracted with 0.1 M NaOH. The listed peak numbers correspond to the numbers in Fig. 2.3.1.

Peak no.	IUPAC-name of identified esters	Match quality %	Common name and Formula of the corresponding acid	
1	Propanoic acid, 2-hydroxy-, methyl ester	(98)	Lactic acid	$C_3H_6O_3$
2	Acetic acid, hydroxy-, methyl ester	(98)	Glycollic acid	$C_2H_4O_3$
3	Ethanedioic acid, dimethyl ester	(94)	Oxalic acid	$C_2H_2O_4$
4	N.D.			
5	N.D.			
6	Hexanoic acid, hydroxy-, methyl ester	(97)	Hydroxycaproic acid	$C_6H_{12}O_3$
7	Propanedioic acid, dimethyl ester	(89)	Malonic acid	$C_3H_4O_4$
8	Si compound			
9	2-Butenedioic acid (E)-, dimethyl ester	(98)	Fumaric acid	$C_4H_4O_4$
10	Hexanoic acid, hydroxy-, methyl ester	(84)	Hydroxycaproic acid	$C_6H_{12}O_3$
11	Butanedioic acid, methyl-, dimethyl ester	(92)	Methylsuccinic acid	$C_5H_8O_4$
12	Pentanoic acid, -oxo-, methyl ester	(98)	Levulinic acid	$C_5H_8O_3$
13	Butanedioic acid, dimethyl ester	(97)	Succinic acid	$C_4H_6O_4$
14	Benzoic acid, methyl ester	(95)	Benzoic acid	$C_7H_6O_2$
15	Propanoic acid, 2-hydroxy-, methyl ester	(97)	Lactic acid	$C_3H_6O_3$
16	2-Butenedioic acid (E)-, dimethyl ester	(90)	Maleic acid	$C_4H_4O_4$
17	Butanoic acid, hydroxy-methyl-, methyl ester	(84)	Hydroxymethylbutyric acid	$C_5H_{10}O_3$
18	Pentanedioic acid, dimethyl ester	(99)	Glutaric acid (i.S.)[a]	$C_5H_8O_4$
19	N.D.			
20	Benzeneacetic acid, methyl ester	(98)	Phenylacetic acid	$C_8H_8O_2$
21	Hexanoic acid, hydroxy-, methyl ester	(80)	Hydroxycaproic acid	$C_6H_{12}O_3$
22	N.D.			
23	Hexanedioic acid, dimethyl ester	(99)	Adipic acid	$C_6H_{10}O_4$
24	Si compound			
25	Heptanedioic acid, dimethyl ester	(93)	Pimelic acid	$C_7H_{12}O_4$
26	N.D.			
27	N.D.			
28	Butanedioic acid, hydroxy-, dimethyl ester	(99)	Malic acid	$C_4H_6O_5$
29	Heptanedioic acid, methyl-, dimethyl ester	(84)	Methylpimelic acid	$C_8H_{14}O_4$
30	N.D.			
31	N.D.			
32	Nonanedioic acid, dimethyl ester	(85)	Azelaic acid	$C_9H_{16}O_4$
33	N.D.			
34	N.D.			
35	Hexanoic acid, hydroxy-, methyl ester (etc.)	(81)	Hydroxycaproic acid	$C_6H_{12}O_3$
36	Butanedioic acid, monomethyl ester	(90)	Succinic acid	$C_4H_6O_4$
37	(2,4-Pentanediol, 3-methyl-)	(92)		
38	1,2,3-Propanetricarboxylic acid, 2-hydroxy	(96)	Citric acid	$C_6H_8O_7$
39	(Benzaldehyde, 4-hydroxy-3-methoxy-)	(97)		
40	Benzoic acid, hydroxy-methoxy-, methyl ester	(94)	Anisic acid	$C_8H_8O_4$

[a](i.S.) = added internal standard.
N.D. = Not determined.

Table 2.2.3. Comparison of organic acid concentrations obtained with different extractants and by centrifugation of a sample from the SDH Ah horizon

Acid	NaOH (pH 12.87) μmol/kg	H_2O-d (pH 6.60) μmol/kg	H_2O-c (pH 3.79) μmol/kg	Cen-a (pH 4.2) μmol/l	Cen-b (pH 4.2) μmol/kg
Formic	277	69	38	n.d.	
Acetic	836	152	40	129	2.7
Propionic	8.5	2.7	n.d.	n.d.	
Butyric	5.9	4.1	n.d.	n.d.	
Pyruvic	n.d.	20	n.d.	n.d.	
Lactic	368	26	14	147	3.1
Glycolic	251	n.d.	n.d.	n.d.	
Oxalic	203	121	51	528	11
Malonic	181	29	19	68	1.4
Fumaric	32	3.5	n.d.	n.d.	
Hydroxycap.+levulin.	34	5.3	2.1	n.d.	
Succinic	213	19	7.7	n.d.	
Benzoic	24	2.4	n.d.	n.d.	
Maleinic	4.0	n.d.	n.d.	n.d.	
Phenylacetic	3.7	1.6	n.d.	n.d.	
Salicylic	n.d.	n.d.	n.d.	n.d.	
Adipic	6.9	0.7	n.d.	n.d.	
Pimelic	7.6	0.9	n.d.	n.d.	
Malic	22	5.7	n.d.	n.d.	
Azelaic	5.3	1.4	n.d.	n.d.	
Citric	34	n.d.	n.d.	n.d.	
Anisic	10	2.5	n.d.	n.d.	
Σ	2526	467	171	873	18.25

n.d. = not detectable.
NaOH = 0.1 M NaOH. H_2O-d = deionized water, H_2O-c = canopy throughfall.
Cen-a = centrifugate. Cen-b = transformed value using soil water content.

Table 2.3.4. Organic acid concentrations in percolating water of the soil obtained by centrifugation and in ground water samples obtained from a multi-level well. (Acids not listed here could not be detected)

Sample Depth below surface (cm)	Ah-Zen 0–10 μmol/l	Bv-Zen 40–60 μmol/l	ML 500 μmol/l	ML 600 μmol/l
Formic acid	n.d.	n.d.	181	55
Acetic acid	129	85	3.5	2.7
Propionic acid	n.d.	63	n.d.	n.d.
Butyric acid	n.d.	38	n.d.	n.d.
Lactic acid	147	64	2.6	3.0
Oxalic acid	528	208	n.d.	n.d.
Malonic acid	68	n.d.	n.d.	n.d.
Σ	872	458	187	61

Ah-Zen, Bv-Zen = centrifugates of the SDH-Ah and Bv horizons.
ML = ground water samples from multi-level well.
n.d. = not detectable.

In general, water in soil has higher concentrations and a larger spectrum of organic acids than ground water, but there are also indications of microbiological and root-induced alterations of the organic acid concentrations in both water types. This is obvious for the short chained simple fatty acids from C_1 to C_4. An increase of microbiological activity below the ground water table was found by Weyandt and Schweisfurth (Sect. 5.3.1, this Vol.); our own investigations revealed FeS granules in samples from the capillary zone under the electron microscope. Their formation is probably caused by H_2S-producing microorganisms.

Because of very low contents of total organic carbon throughout the unsaturated zone, a screening of organic acids in this compartment was only attempted with NaOH extracts, which were found to yield the highest amounts. Results are listed in Table 2.3.5.

The overall trend of decreasing concentrations with depth is confirmed except for sample HB-12. A look at Fig. 2.3.3 reveals that the assumption of microbial influence on organic acid contents seems to be responsible for the relative concentration peak in the sample mentioned above.

Table 2.3.5. Organic acid concentrations in core samples extracted with 0.1 M NaOH (Sampling points are indicated in Fig. 2.3.3)

Sample Depth below surface (cm)	HB-1 0–10 µmol/kg	HB-2 40–60 µmol/kg	HB-21 210–240 µmol/kg	HB-12 510–540 µmol/kg	HB-9 610–640 µmol/kg	HB-8 650–670 µmol/kg
Formic acid	277	304	148	55	143	158
Acetic acid	836	596	23	283	5.0	13
Propionic acid	8.5	6.8	0.5	3.7	0.5	n.d.
Butyric acid	5.9	8.5	0.5	n.d.	n.d.	n.d.
Pyruvic acid	n.d.	42	n.d.	n.d.	n.d.	n.d.
Lactic acid	368	131	11	34	10	6.0
Glycolic acid	251	n.d.	n.d.	48	n.d.	n.d.
Oxalic acid	203	207	23	44	10	n.d.
Malonic acid	181	192	9.0	53	5.8	4.5
Fumaric acid	32	21	0.8	0.6	0.7	n.d.
Hydroxycap.+Levulin.	34	30	1.7	8.9	0.8	n.d.
Succinic acid	213	153	7.7	44	5.1	3.9
Benzoic acid	24	11	0.7	3.6	0.4	n.d.
Maleinic acid	4.0	4.3	n.d.	n.d.	n.d.	n.d.
Phenylacetic acid	3.7	6.1	0.3	2.7	0.3	n.d.
Salicylic acid	n.d.	n.d.	n.d.	n.d.	n.d.	n.d.
Adipic acid	6.9	5.2	n.d.	2.3	n.d.	n.d.
Pimelic acid	7.6	6.3	n.d.	3.3	n.d.	n.d.
Malic acid	22	26	n.d.	8.5	n.d.	n.d.
Azelaic acid	5.3	14	0.7	6.9	0.3	n.d.
Citric acid	34	n.d.	n.d.	n.d.	n.d.	n.d.
Anisic acid	10	4.1	n.d.	2.7	n.d.	n.d.
Σ	2527	1768	227	604	182	185

n.d. = not detected

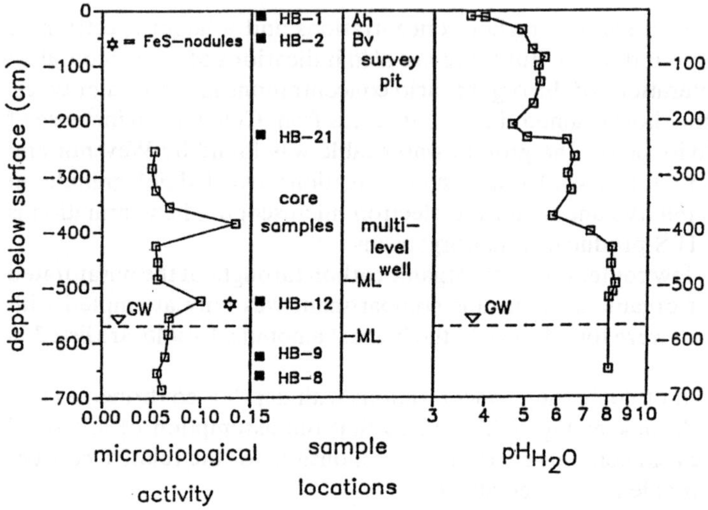

Fig. 2.3.3. Microbial activity, plotted as extinction values of esterase activity (FDA); location of organic acid samples and FeS occurrence

2.3.5 Conclusions

Although there are still unaccounted uncertainties in the definition of organic acid pools in underground environments, partitioning of these substances (in conjunction with humic substances) in geochemical and biological processes may to some degree play a significant role. This may especially be true for mineral dissolution kinetics as was shown in laboratory experiments (Sect. 2.2.5; Schenk et al. 1987; Petersen 1988). Inclusion of concentration levels may broaden the scope of thermodynamic modelling of ecosystem-relevant processes.

3 Carbonate Systems

3.0 Introduction

E. Usdowski

Carbonate rocks constitute about 11% of total sediments of the Earth's crust and may form important aquifers, especially in karstic terrains. The importance of carbonate weathering for substance fluxes is indicated by the total dissolved masses annually delivered by rivers to the ocean which are approximately 5.0×10^{14} g/a Ca^{2+}, 1.4×10^{14} g/a Mg^{2+}, and 19.0×10^{14} g/a HCO_3^-, corresponding to fluxes of Ca^{2+}: 4.5 g/(m^2•a) = 0.11 mmol/(m^2•a), of Mg^{2+}: 0.84 g/(m^2•a) = 0.03 mmol/(m^2•a), and of HCO_3^-: 16.2 g/(m^2•a) = 0.265 mmol/(m^2•a) (Garrels and MacKenzie 1971).

In a standard situation limestones and dolomite rocks are dissolved by infiltrating rain water which has taken up biogenic CO_2 in the soil zone. As this solution percolates downwards, continuously dissolving solid carbonates, the hydrogeochemical system will change from "open" with respect to a CO_2 gas phase in the unsaturated zone to "closed" with respect to that phase in the saturated zone. Thus, the solution passes through various stages of a system which are intermediate between the limiting cases, in general. In modern times the dissolution of carbonate rocks is enhanced by the action of sulfuric acid from atmospheric pollution.

In order to quantify the above processes the chemical development of a carbonate ground water with respect to the major components and the stable isotopes of carbon and sulfur has to be traced along the flowpath of the water. The major components provide information about the saturation of the water, i.e. its capability to dissolve or to precipitate solids. From the stable isotopes the open, closed, or intermediate nature of a hydrogeochemical system, and the extent of carbonate dissolution by acid atmospheric pollution may be deduced. Studies in that respect were carried out in the field on hydrogeologically defined areas as well as by supplementary laboratory studies of reaction kinetics. The combined results may contribute to a further step towards a more comprehensive understanding of surface and subsurface rock dissolution.

The kinetics of carbonate dissolution are a major control for the composition of subsurface waters. In the presented programme they have been studied by theoretical considerations, experimental investigations and field observations. A theoretical model which is based on the fluxes of the various aqueous species at laminar and turbulant flow conditions (Dreybroth, Sect. 3.3.1) compares well to the measurements of dissolution rates which have been carried out at different boundary conditions. Experiments under "closed" system conditions have shown that the "saturation length" is a useful parameter in order to describe the overall effect of dissolution kinetics, hydrodynamic transport and diffusion. These experiments have been extended to show the effect of dissolved inhibitors on the rates of carbonate dissolution (Baumann and Schulz, Sect. 3.1.2).

Case studies included carbonate rocks, but also sandy systems in which carbonates are minor components. Projects dealt with sampling methods in the unsaturated zone of carbonate gravels (Merkel and Grossmann, Sect. 3.2.1) and with the variations of inorganic carbon in this profile (Eichinger and Merkel, Sect. 3.2.2). Measurements of dissolved major components and the carbon isotopes of CO_2 gas, the dissolved carbonate and the solids have been applied to show the "open" and "closed" nature of a carbonate containing sandy aquifer (Dörr, Leuchs, Obermann, Regenberg and Sonntag, Sect. 3.2.3). Considering the influence of acid rain, measurements of stable carbon and sulfur isotopes have been used in order to quantify and to model carbonate dissolution at different depths in a limestone-dolomite-anhydrite-gypsum area (Schaefer and Usdowski, Sect. 3.2.4). The development of a dissolution front at the contact of a sand to a marly limestone aquifer has been investigated by Langguth and Schulz (Sect. 3.2.5). Further studies of stable carbon and sulfur isotopes, and the influence of acid depositions dealt with carbonate dissolution under conditions intermediate between the "open" and "closed" system (Michaelis, Sect. 3.2.6).

On the global scale, rock dissolution in semi-arid areas is not very important, however, it is most significant locally. The rock material, dissolved in short periods of rains, is reprecipitated briefly afterwards by the evaporation of water, mostly as a variety of numerous alkaline and alkaline earth sulfates. As the evaporation occurs rapidly, many of the observed mineral assemblages are thermodynamically metastable. As an example for such conditions the hydrology and mineralogy of a gypsum karst area in Tunisia were studies (Smykatz-Kloss, Hötzl and Kössl, Sect. 3.3.1). Further studies of alteration processes in a semi-arid environment indicate that dedolomitization is the most striking feature with respect to mineral transformations. It appears to proceed via the formation of a calcium-rich protodolomite in a first step (Smykatz-Kloss and Goebelbecker, Sect. 3.3.2).

The fillings of subsurface voids of paleokarst areas act as a trap for a variety of minerals of different origin, thus reflecting geological and mineralogical processes which have occurred near the final stage of a hydrogeochemical system. Fillings range from detrital rocks and minerals to authigenic minerals. The nature of the detritus depends upon the nature of the rocks surrounding the ancient karst area. Authigenic minerals are formed from pore solutions as a function of available cations and anions, and pH-E_H conditions. The transformation processes in paleokarst sediments and the chemistry of modern waters were investigated in the Aladagh Region, Turkey (Cevrim and Echle, Sect. 3.3.3).

3.1 Dissolution Kinetics in the Generation of Carbonate Ground Water

3.1.1 Theoretical and Experimental Results on the Kinetics of Calcite Dissolution and Precipitation

W. Dreybrodt

3.1.1.1 Background

Dissolution and precipitation of carbonates play an important role in nature, such as in the chemistry of sea water, the sedimentation of carbonates at the sea floor, in the geochemical evolution of fresh water aquifers and in the evolution of karst landscapes. To provide a deeper understanding to all these processes, knowledge is required on the physical and chemical parameters which determine dissolution or precipitation rates. Dissolution and precipitation kinetics of calcium carbonate are determined by the chemistry of the system H_2O-CO_2-$CaCO_3$. The chemical reaction is usually given by the overall reaction

$$H_2O + CO_2 + CaCO_3 \rightleftharpoons H_2CO_3 + CaCO_3 \rightleftharpoons Ca^{2+} + 2HCO_3^-, \qquad (3.1.1.1)$$

which summarizes three independent processes:

- CO_2 present in the aqueous solution is converted into the aggressive agent H_2CO_3. This is a slow process and its kinetics have been recently reviewed by Usdowski (1982).
- Ca^{2+}, HCO_3^- and CO_3^{2-} are released from the mineral surface by the following heterogeneous reactions

$$CaCO_3 + H^+ \rightleftharpoons Ca^{2+} + HCO_3^- \qquad (3.1.1.2)$$

$$CaCO_3 + H_2CO_3 \rightleftharpoons Ca^{2+} + 2HCO_3^- \qquad (3.1.1.3)$$

$$CaCO_3 + H_2O \rightleftharpoons Ca^{2+} + CO_3^{2-} + H_2O \qquad (3.1.1.4)$$

The reaction rates depend on the chemical composition of the solution at the surface of the solid (Plummer et al. 1978, 1979).

- Mass transport of the dissolved species Ca^{2+}, HCO_3^- and CO_3^{2-} from the surface into the bulk of the solution and also transport of the aggressive species H_2CO_3 and H^+ from the bulk of the liquid to the mineral surface.

A rigorous attempt to model dissolution and precipitation rates therefore has to consider the above processes.

3.1.1.2 Basic Concepts of a General Model

Figure 3.1.1.1 shows a geometrical model of a calcite surface in contact with the aqueous solution. For simplicity and to render the problem mathematically tractable

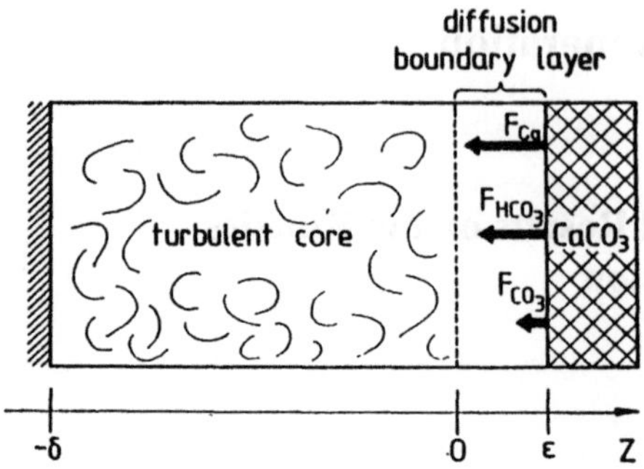

Fig. 3.1.1.1. Geometrical model of the $CaCO_3$-H_2O-CO_2 system with turbulent core and a diffusion boundary layer

a planar calcite surface is assumed. The solution extends from this surface and is bounded at $z = -\delta$ by (a) a CO_2-containing atmosphere, (b) an insoluble solid or (c) a surface of soluble calcite. In the case (a) CO_2 can enter from the atmosphere into the solution. Therefore the CO_2 consumed by the dissolution of calcite is replaced such that the CO_2 concentration in the solution is kept constant (open system). In cases (b) and (c) no replacement of CO_2 is possible (closed system) and CO_2 in the solution is consumed during dissolution of calcite. Next to the calcite surface a diffusion boundary layer of thickness ε exists. This layer is of importance for turbulent flow of the bulk solution. At $z = \varepsilon$, i.e. at the calcite surface, there are fluxes of ions. $F_{Ca^{2+}}$ denotes the flux of Ca^{2+} ions, and $F_{HCO_3^-}$ and $F_{CO_3^{2-}}$ are the fluxes of hydrogen carbonate and carbonate associated with the release of Ca^{2+} (Plummer et al. 1978). Since CO_2 does not react at the surface and one C atom is released with each Ca^{2+} in any case Eq. (3.1.1.5) holds

$$F_{Ca^{2+}} = F_{HCO_3^-} + F_{CO_3^{2-}} . \tag{3.1.1.5}$$

Note that in the case of precipitation CO_2 is produced and the fluxes $F_{Ca^{2+}}$, $F_{CO_3^-}$ and $F_{CO_3^{2-}}$ are reversed. In the case (c) where the surface at $z = -\delta$ consists of calcite corresponding fluxes are also present. In all cases stoichiometry requires that each $CaCO_3$ released from the crystal lattice consumes one molecule of CO_2. Therefore in all three cases the condition holds

$$F_{Ca^{2+}} \cdot A = \int_{-\delta}^{\varepsilon} R_{CO_2}(z)\, dz \cdot A, \tag{3.1.1.6}$$

where A is the area of the calcite surface, and R_{CO_2} (z) is the conversion rate of CO_2 into H_2CO_3. Conversion of CO_2 occurs by the two simultaneous reactions (Kern 1960; Usdowski 1967).

$$H_2O + CO_2 \rightleftharpoons H_2CO_3 \tag{3.1.1.7}$$

$$CO_2 + OH^- \rightleftharpoons HCO_3^- . \tag{3.1.1.8}$$

The conversion rate is

$$R_{CO_2} = (k_1^+ + k_2^+[OH^-]) - (k_1^-[H^+] + k_2^-)[HCO_3^-] \tag{3.1.1.9}$$

where the concentrations denoted by [ ] are dependent on z, and the k are rate constants. The flux of Ca^{2+} from the calcite surface resulting from the heterogeneous surface reactions has been investigated by Plummer et al. (1978), who propose an empirical relation.

$$F_{Ca^{2+}} = k_1\,[H^+]_s + k_2[H_2CO_3^*]_s + K_3 - K_4 \cdot [Ca^{2+}]_s \cdot [HCO_3^-]_s, \qquad (3.1.1.10)$$

where $[\ ]_s$, denote activities of the corresponding species at the solid surface and k_1, k_2, k_3, are rate constants of the forward reactions, whereas k_4 denotes an effective rate constant resulting from all the elementary back reactions and therefore depends on the activity of $[H_2CO_3^*]_s = [CO_2] + [H_2CO_3]$. To calculate dissolution of precipitation rates the following mass transport equations have to be solved

$$-\frac{\partial[CO_2]}{\partial t} + D_M \frac{\partial^2[CO_2]}{\partial z^2}\; R_{CO}{}^2(z) \qquad (3.1.1.11)$$

$$-\frac{\partial[HCO_3^-]}{\partial t} + D_M \frac{\partial^2[HCO_3^-]}{\partial z^2} = -R_{CO_2}(z) \qquad (3.1.1.12)$$

$$-\frac{\partial[CO_3^{2-}]}{\partial t} + D_M \frac{\partial^2[CO_3^{2-}]}{\partial z^2} = 0; -\frac{\partial[Ca^{2+}]}{\partial t} + D_M \frac{\partial^2[Ca^{2+}]}{\partial z^2} = 0\,. \qquad (3.1.1.13)$$

The left-hand sides of these equations represent mass transport by diffusion and dispersion, expressed by the concentration gradients and the coefficient of molecular diffusion D_M, which together with the coefficient of hydromechanic dispersion D' forms the coefficient of hydrodynamic dispersion D ($D = D_M + D'$). The right hand sides represent chemical reaction rates in the solution. It is assumed that all reactions, except $H_2O + CO_2 \rightleftharpoons H_2CO_3$, are sufficiently fast, for the species as H^+, OH^-, HCO_3^- and CO_3^{2-} to be in equilibrium with each other.

A very important point is that the hydrodynamic conditions of the liquid surrounding the calcite determine the mass transport by changing the coefficient of hydrodynamic dispersion D (Bear 1972).

1. If the solution is stagnant or is flowing laminarly one has to use the coefficient of molecular diffusion D_M, which for all species involved is approximately of the same order of magnitude ($D_M \approx 10^{-5}$ cm²/s).
2. In the case of turbulent flow there is a region $0 \leq z \leq \varepsilon$, called the diffusion boundary layer. The thickness of ε can be determined from the theory of diffusional mass transfer in turbulent flow (Kay and Nedderman 1985) according to the expressions

$$\varepsilon = L_c \cdot a^{-1} \cdot Re^{-0.8} \cdot Sc^{-1/3}\,; Re = \frac{v \cdot L_c}{\nu}\,; Sc = \frac{\nu}{D_M}\,, \qquad (3.1.1.14)$$

where, depending on the geometry of the calcite sample $0.02 \leq a \leq 0.04$, L_c is a characteristic dimension (e.g. length of a flat plate, diameter of a tube). v is the velocity of the fluid in the bulk solution and ν is the kinematic viscosity. Within the diffusion boundary layer mass transport is effected by molecular diffusion. In the bulk of the fluid, however, mass transport is due to turbulent eddies which by their chaotic movement increase diffusion considerably, such that an effective coefficient of diffusion $D_{eff} = 10^3 \cdot D_M$ has to be used for D in Eqs. (3.1.1.11)–(3.1.1.13).

3. At extremely high Reynolds numbers Re, ε tends towards zero and in this ideal case one has to use $D = D_{eff}$ everywhere in the solution.

Therefore the case of laminar flow and the case of extreme turbulence are mathematically equivalent since they differ only by the magnitude in the constant D. Even then the solution of Eqs. (3.1.1.11)–(3.1.1.13) is complex and we will give only the results and their interpretation. The details of the model calculations are given by Buhmann and Dreybrodt (1985a,b) and by Dreybrodt (1988). The intermediate case (2) is discussed by Dreybrodt (1988).

3.1.1.3 Results of the Mathematical Model

Figure 3.1.1.2 shows dissolution rates calculated for case (a) (laminar flow or stagnant solution) as a function of the Ca^{2+} concentration in the bulk of the solution. The system is open to a CO_2-containing atmosphere of $p_{CO2} = 30$ Pa. The dissolution rates have been calculated for various thicknesses δ of the water layer, given in units of cm at the corresponding curves. Clearly there is a strong dependence of the dissolution rates on δ. Three regimes of dissolution mechanisms can be recognized from Fig. 3.1.1.2.

Regime 1 (control by CO_2-conversion):

For $\delta < 0.01$ cm the dissolution rates depend linearly on the thickness δ of the water layer. From Eq. (3.1.1.4) one sees that the dissolution rate $F_{Ca^{2+}}$ decreases with decreasing δ, because at sufficiently small δ the value of the integral, which gives the amount of CO_2 converted per second, becomes small and therefore controls the rate of $CaCO_3$ dissolution. At sufficiently small distance δ, no strong concentration

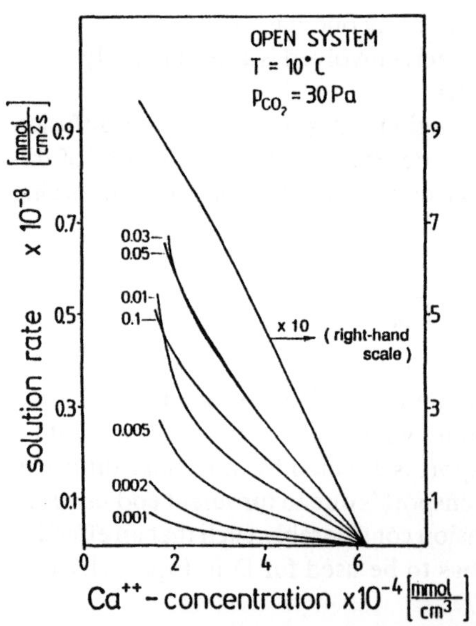

Fig. 3.1.1.2. Theoretical dissolution rates for laminar flow at various thicknesses δ of the water layer. Values of δ in cm are written to the curves. The *uppermost curve* is for fully turbulent flow and for $\delta \geq 1$ cm. (Buhmann and Dreybrodt 1985a). *Numbers on the right scale* refer to the uppermost curve, $p_{CO_2} = 30$ Pa, t = 10 °C

gradients can build up by diffusion. Therefore, the rate of CO_2 conversion is almost constant with respect to z, and from Eq. (3.1.1.4) the amount of CO_2 converted is proportional to δ and so is also $F_{Ca^{2+}}$, the dissolution rate.

Regime 2 (simultaneous control by diffusion and CO_2 conversion):

For 0.03 cm < δ < 0.05 cm the dissolution curves crowd into one common region. In this case a reaction zone close to the calcite surface builds up, with steep concentration gradients of CO_3^{2-} and H^+ ions. Most of the CO_2 conversion occurs in this region, the extension of which is given by

$$\lambda_1 = (D_M/k)^{1/2} \tag{3.1.1.15}$$

where k is the effective rate constant of CO_2 conversion. The length λ_1 represents the average distance a CO_3^{2-} ion released from the calcite surface can travel until it is converted to HCO_3^- by the instantaneous reaction $CO_3^{2-} + H^+ \rightleftharpoons HCO_3^-$. Since H^+ results from the conversion of $CO_2 \rightleftharpoons H^+ + HCO_3^-$, most of the CO_2 is affected. Therefore the thickness λ_1 determines dissolution rates and for not too large δ the reaction rates become independent of δ (Buhmann and Dreybrodt 1985a,b; Dreybrodt 1988).

Regime 3 (diffusion control):

The dissolution curve for δ = 0.1 cm drops below the curves of regime 2. In this case the slowest process is diffusion of Ca^{2+} ions into the bulk solution and it can be shown (Dreybrodt 1988) that the dissolution rates become inversely proportional to δ. Therefore in a large stagnant body of water, such as in a big subsurface cavern of karst areas, calcite dissolution rates become extremely low.

In all these three regimes the rate of the $CaCO_3$ release from the solid [Eq. (3.1.1.10)] is fast and therefore not rate-limiting. If one enhances diffusion by introducing turbulent motion and replaces D by $D_{eff} = 10^3 \cdot D_M$, then diffusion is no longer rate-limiting. If one also increases δ, (δ > 1 cm), the amount of CO_2 converted is sufficiently high not be rate-limiting. The dissolution rates are then determined entirely by the surface reaction rates given by Eq. (3.1.1.10). Due to the effective mass transport the activities at the surface are now equal to those in the bulk of the liquid. The uppermost curve in Fig. 3.1.1.2 shows the result. A tremendous increase in dissolution rates by about a factor of ten results, when turbulent flow occurs. This is of utmost geological importance. Summarizing the results shown in Fig. 3.1.1.2, one can approximate the dissolution rates in the region of Ca^{2+} concentrations above 20% of the equilibrium concentration ($[Ca^{2+}] \geq [Ca^{2+}]_{eq}$) by straight lines. Thus, a simple linear relation results which is given by the expression

$$F_{Ca^{2+}} = \alpha \cdot ([Ca^{2+}]_{eq} - [Ca^{2+}]) \tag{3.1.1.16}$$

and which holds also for precipitation if $[Ca^{2+}] > [Ca^{2+}]_{eq}$. This holds true as well for the closed system, in a temperature range from 1 to 40 °C and for p_{CO2} pressures from atmospheric p_{CO2} of 30 Pa up to pressures of 100 kPa CO_2. The numerical value of the kinetic constant α_k depends on the temperature, p_{CO2} on the type of flow, i.e. laminar or extremely turbulent ($\varepsilon = 0$), on the system conditions, i.e. open or closed system and on the thickness δ of the water layer covering the calcite surface (Buhmann and Dreybrodt 1985a,b).

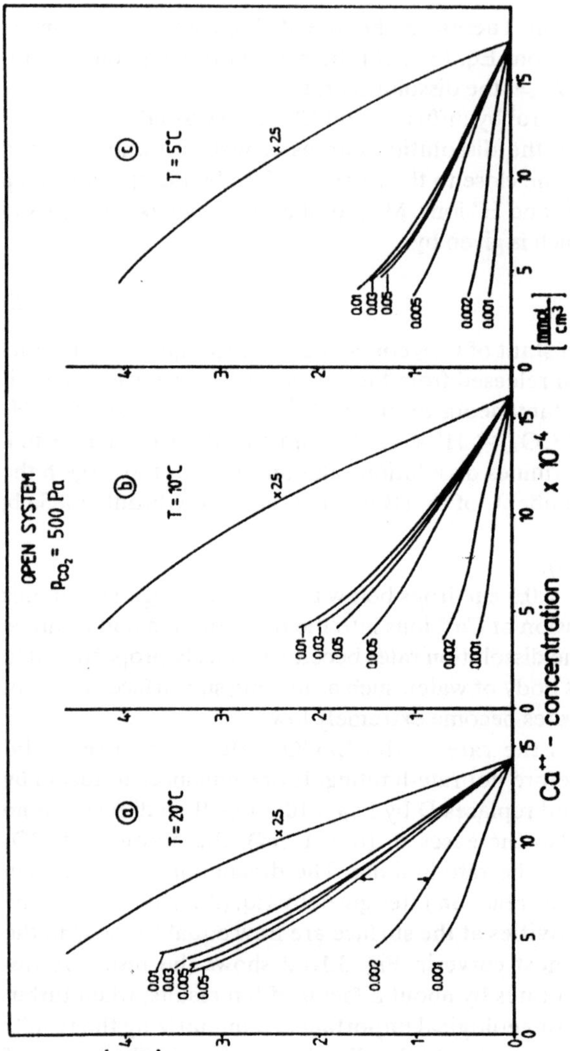

Fig. 3.1.1.3. Theoretical dissolution rates for laminar flow for different temperatures and film thicknesses. δ in cm is written on the curves. The *uppermost curves* are for fully turbulent flow. (Buhmann and Dreybrodt 1985a). p_{CO_2} = 500 Pa

Values of α_k are in the range of 10^{-5} to 10^{-4} cm/s. Tables of α_k and corresponding values of $[Ca^{2+}]_{eq}$ are given for the open and the closed system and for a variety of temperatures and CO_2 pressures by Buhmann and Dreybrodt (1985a,b). From these data dissolution and precipitation rates can be estimated for many geologically relevant situations. Figure 3.1.1.3 shows some results of dissolution rates for the open system at p_{CO_2} = 500 Pa and for different temperatures.

So far the influence of the diffusion boundary layer has been neglected. In this layer mass transport is controlled by molecular diffusion. Since the existence of this layer exerts a significant resistance to mass transport it can alter dissolution rates significantly. This is shown in Fig. 3.1.1.4, which gives dissolution rates for turbulent flow in the bulk of the liquid for a variety of thicknesses ε of the boundary layer. The thickness δ of the bulk of the solution is 1 cm in all cases. There is a decrease in dissolution rates with increasing ε and the curves crowd towards a limiting value $\varepsilon \approx 0.02$ cm, where the reaction is confined to the reaction zone of thickness $\lambda_1 \approx \varepsilon$. Further increase in ε exerts only minor influence. To measure dissolution rates as function of ε one uses the rotating disk technique. A calcite sample is mounted to the flat bottom of a body of cylindrical symmetry, so that it fits smoothly into this lower flat surface. The body is immerged into the solvent and rotated with a high angular speed. From the rotational speed the boundary layer thickness ε can be calculated (Pleskov and Filinovskii 1975). The dissolution rates predicted from the calculations are in agreement with experimental data (Herman 1982; Dreybrodt 1988).

Finally, it should be noted that the theoretical model is also applicable to dissolution of calcite in aquifers of porous media containing calcite grains (Baumann et al. 1985). In this case the amount of $CaCO_3$ dissolved per second in 1 cm^3 of the porous medium is given by equation

$$R_s = \frac{A}{V_p} \cdot \alpha_p \left([Ca^{2+}]_{eq} - [Ca^{2+}]\right), \tag{3.1.1.17}$$

where A is the surface area of the calcite grains and V_p is the pore volume of the medium. The quantity α_p is calculated from Eqs. (3.1.1.11) – (3.1.1.13) using boundary conditions for the closed system. The parameter δ + ε is replaced by the

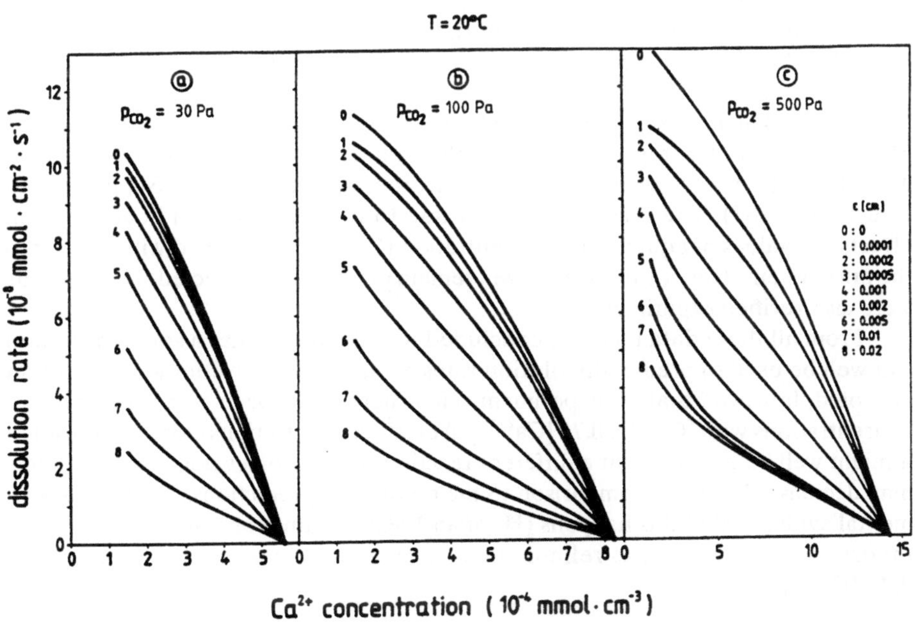

Fig. 3.1.1.4. Dissolution rates for turbulent flow in an open system for various p_{CO_2}. The *numbers on the curves* are related to the thicknesses ε of the boundary layer as denoted in panel **c**. Thickness of turbulent core δ = 1 cm. (Dreybrodt 1988). t = 20 °C

average grain size δ (cf. Fig. 3.1.1.1) and the coefficient of hydrodynamic dispersion has to consider hydrodynamic dispersion such that $D = \bar{v}\,\delta + D_M$, where $\bar{v}$ is the average flow velocity in the medium.

3.1.1.4 Experimental Verification of the Theory

To verify the theoretical predictions experimentally, the evolution of the Ca^{2+} concentration of a water layer in contact with a calcite surface were investigated under open or closed system conditions, respectively. For a water layer of thickness δ the time dependence of the average Ca^{2+} concentration is easily derived (Buhmann and Dreybrodt 1985a) according to equation

$$[Ca^{2+}]\,(t) = [Ca^{2+}]_{eq} \cdot [1 - \exp(-t/\tau)]\,;\, t = \delta/\alpha_k. \qquad (3.1.1.18)$$

From measurements, the time constant τ can be determined and compared to the theoretical value above.

The case of laminar flow is simulated by a stagnant water layer on top of the calcite surface. Turbulent flow is simulated by stirring water layers of thickness between 0.5 to 2 cm with a propeller (Buhmann and Dreybrodt 1985a,b; Buhmann 1984). Figure 3.1.1.5 shows results for the open system using various calcite samples ($p_{CO2} = 500$ Pa). The full lines are theoretically calculated values of τ. The results are in good agreement with the theory. Similar results were obtained for $p_{CO2} = 30$ Pa.

For turbulent layers also excellent agreement between experiment and theory is found (Buhmann and Dreybrodt 1985a). Experiments under closed system conditions also showed good agreement with the theory for both laminar and turbulent flow (Buhmann and Dreybrodt 1985b).

Since in karst systems most natural waters contain additional ions such as Na^+, Mg^{2+}, SO_4^{2-}, and Cl^-, one may look for their influence upon the dissolution rate. Therefore, the theory was extended by incorporating these ions (Buhmann and Dreybrodt 1987). The results show that due to ionic strength and common ion effects the values of equilibrium concentration $[Ca^{2+}]_{eq}$ are changed with respect to the pure system. The α_k values, however, remain practically unaffected. These results were also verified experimentally.

From all these data it can be concluded that the theory has a sound basis and can well be used to predict dissolution rates even for natural situations. It should be noted, however, that all experiments cited here, have been performed at Ca^{2+} concentrations with $[Ca^{2+}] < 0.7 \cdot [Ca^{2+}]_{eq}$. At Ca^{2+} concentrations close to saturation it might well be possible that a different reaction path becomes dominant and rates may decrease sharply in comparison to the theoretical predictions. There is experimental evidence that this happens (Herman 1982; Plummer and Wigley 1976). The theory can be regarded as reliable in all cases of undersaturated waters up to $0.7 \cdot [Ca^{2+}]_{eq}$.

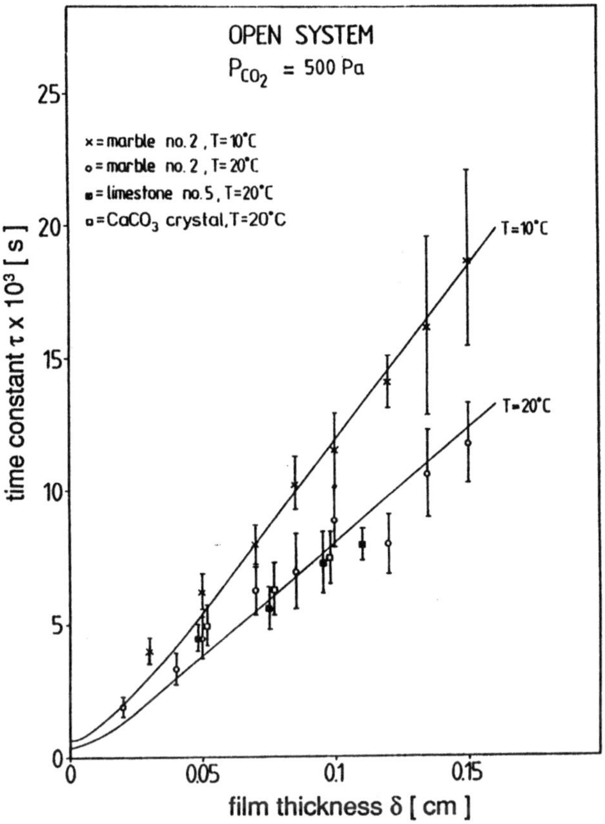

Fig. 3.1.1.5. Time constants τ of $CaCO_3$ dissolution measured for various film thickness at $p_{CO_2} = 500$ Pa (open system) for stagnant water layers. The *solid line* gives the results of the theoretical model. (Buhmann and Dreybrodt 1985a)

3.1.1.5 Dissolution and Precipitation Rates of Limestones in Natural Environments

Dissolution rates (denudation rates) by meteoric water on bare limestone in karst areas have been measured by a variety of methods. Jennings (1985) has summarized some of these data from the recent literature. Values range from 15–40 mm/1000 a.

Short-term (1 a) measurements of denudation on bare rocks are performed by use of the microerosionmeter (Trudgill et al. 1981). Reported values range from 5–30 mm/1000 a (Jennings 1985; Dreybrodt 1988). From Fig. 3.1.1.2 one obtains dissolution rates of 5×10^{-9} mmol/(cm^2•s) for laminarly flowing water sheets of a depth in the order of several tenths of a millimeter. Since rain water running down limestone surfaces remains far from equilibrium (Bauer 1964), the denudation rates were estimated from the dissolution rates in Fig. 3.1.1.2 at 50% saturation. From this a denudation rate of 50 mm/1000 a results. Assuming that the rainfall occurs only during 20 to 40% of the time, an annual retreat in bedrock of about 15 mm/1000 a is expected in accordance with observations. Note that a dissolution rate of 10^{-6} mmol/(cm^2•s) corresponds to a denudation rate of 8.5 mm/1000 a (Dreybrodt 1988).

The intensity of corrosion in cave rivers has been measured by a variety of methods. These are summarized by Dreybrodt (1988). At CO_2 concentrations of cave waters corresponding to p_{CO_2} values of about 100 Pa up to 400 Pa, typical bed rock retreat is in the range from 100 to 300 mm/1000 a, which is higher by a factor of 10 than denudation on bare rock. Average values are 150 mm/1000 a. In cave rivers flow is turbulent in any case. Therefore, Fig. 3.1.1.4 can be used to estimate rates of bedrock retreat. Assuming 50% saturation and fully turbulent flow one finds a value of 100 mm/1000 a. At boundary layer thicknesses of $\varepsilon \approx 0.02$ cm this decreases to 20 mm/1000 a.

The influence of hydrodynamic flow conditions on dissolution rates was recently demonstrated by Lauritzen (1986), who showed that in an isolated phreatic cave conduit dissolution rates are dependent on the discharge of the resurgence with maximal rates of approximately 100 mm/1000 a at high discharges. These examples show that observations in natural environments and model predictions are in good accordance.

Precipitation rates were measured in a calcite-precipitating stream (Usdowski et al. 1979) by immersion of a rectangular flat-shaped limestone sample from the increase in weight within a given immersion time. The Ca^{2+}, Mg^{2+}, Na^{+}, K^{+}, SO_4^{2-}, pH and alkalinity data of simultaneously collected water samples at the same locations were used to calculate calcite deposition rates by numerical solution of Eq. (3.1.1.11)–(3.1.1.13). The thickness ε was determined by immersion of gypsum plates of similar shape at the same locations. Since the dissolution of gypsum is controlled entirely by diffusion, ε could be calculated from the weight loss of these samples (Dreybrodt 1988) to range between 0.01 and 0.03 cm.

The newly developed theoretical model of calcite reaction kinetics defines the chemical, hydrodynamical and geometrical parameters determining dissolution and precipitation rates. Once these parameters have been measured for a specific natural situation, it is easy to estimate the resulting dissolution or precipitation rates. The results of the model are tabulated for a variety of situations (Dreybrodt 1988) and provide data to the geologist from which conclusions can be drawn for many natural systems.

3.1.2 Field Measurements and Laboratory Experiments of Calcite Dissolution Kinetics of Natural Porous Media

J. Baumann and H.D. Schulz

3.1.2.1 Background

Calcite dissolution processes were studied at two field sites with fluvioglacial deposits, one in the Segeberger Forst (Schleswig-Holstein) 40 km N of Hamburg, the other in Delhi, Ontario/Canada (Fig. 1.1, Table 1.1). At both sites the decalcification of the sediments could be measured as a function of calcite weathering. Laboratory experiments were carried out in order to measure calcite dissolution kinetics under boundary conditions close to those in nature. The experiments were made in systems open and closed to CO_2.

3.1.2.2 Field Investigations

Study Areas

The investigation site SEG1 (Jagen 223) is at the northwestern edge of the Segeberger Forst, a woody hilly area of 60 km^2 extension. The surface deposits are of fluvioglacial origin, consisting of medium to coarse-grained sands with varying amounts of gravel and stones. The unleached sediments contain up to 10% wt. carbonate material primarily from glacial and/or fluvioglacial reworked Cretaceous deposits. The average thickness of the unsaturated zone is about 7 to 10 m.

The study area in Canada is located approximately 7 km SE of Delhi (CDN) neat Trout Creek, a small stream within the Big Creek Drainage Basin. In the near surroundings of the site the topography is quite plain and the water table is about 12 m below the surface. The surface deposits consist of the Pleistocene Norfolk Sand, an eolian reworked clean medium-grained sand of lacustrine origin containing dominantly quartz and feldspar and up to 15% wt. calcite and 5% wt. dolomite (Reardon et al. 1979a,b). The thickness of this unit varies from some meters to tens of meters. The total thickness of Quarternary deposits at the study site is approximately 60 m. Underlying bedrocks are Devonian limestones (Barnett 1978).

At both sites many important hydrolygical and hydrogeological parameters are well known from earlier investigations (Albertsen 1977; Albertsen et al. 1980; Reardon et al. 1979a,b, 1980; Schulz and Albertsen 1981).

Sampling Methods and Analyses

At the study site Jagen 223 in northern Germany, two boreholes SEG1–1 and SEG 1–2 0.5 m apart from one another were hand-augered to depths of 12.3 m and 10.3 m in February 1983. In the deeper one a CO_2-measuring device was installed. This device consisted of a number of polyethylene tubes reaching to different depths. In the second hole a device for the in situ measurement of soil moisture was installed.

In March 1985 two boreholes CDN1 and CDN2 about 10 m apart from one another were augered to depths of 5.9 m and 6.1 m respectively at the study site Trout Creek. No field measurement devices were installed because a similar CO_2-measuring arrangement already existed nearby.

The CO_2 content of the ground air was measured with Dräger(R) CO_2-analyzing tubes. A glass syringe was used to extract 500 ml of ground air from the sampling device and then to squeeze it through the analyzing tube. The volumetric pore water contents were determined with a neutron moisture meter. Water and CO_2 content were measured monthly over a period of 1 year at the investigation site SEG1 (Jagen 223). At Trout Creek CO_2 analyses were performed only during March 1985.

Samples of sediment material were collected at depth intervals of 30 to 40 cm during drilling. In order to prevent considerable alteration of pore water chemistry due to temperature changes and extensive degassing, 400 to 500 g of the sediment material was centrifuged immediately after sample collection and a small part of the extracted pore water (5 ml) was used to determine pH and alkalinity.

The major anions (Cl^-, SO_4^{2-}, NO_3^-) were determined 1 day after sampling by IC. Acidified water samples were analyzed for the cations Ca^{2+}, and Mg^{2+} with AAS,

Na^+ and K^+ by ICP-AES. Determination of total organic and inorganic carbon was done by coulometric measurement using about 1 g of air-dried soil material.

Results

The data from the C_{inorg} analyses indicate a carbonate content for the unleached sediment material of about 20 to 25% wt. at Trout Creek and a maximum of 3% wt. at SEG1 (Jagen 223) (Figs. 3.1.2.1–3.1.2.2). In spite of the small distance (10 m) between the two boreholes at Trout Creek, carbonate minerals are leached to very different depths below surface: 1.1 m (not shown) and 4.3 m (Fig. 3.1.2.3), respectively. This may be due to different infiltration rates – borehole CDN2 is located within a small depression.

At the investigation site SEG1 (Jagen 223) the upper 4.3 m of the profile are carbonate-free, but in addition a second leached zone could be observed between 6 and 7.8 m below the surface and finally the deepest part of the profile is also decalcified (Fig. 3.1.2.1). A slightly higher organic content, silicified organic material and iron hydroxide grains indicate a paleosoil with its appropriate leaching zone in these parts of the profile.

All pertinent pore water analyses for site SEG1 (Jagen 223), which are relevant for the carbonate equilibria are given in Figs. 3.1.2.1 and 3.1.2.3. In spite of pore water extraction immediately after sampling, alteration of pH due to slightly CO_2 degassing could not be avoided (Pearson et al. 1978). Therefore, corrected pH values were calculated from HCO_3^- (alkalinity) and measured p_{CO_2} of the ground air (Figs. 3.1.2.2 and 3.1.2.4) using the equation:

$$pH = -\log K_{H_2CO_3^*} - \log K_{Water} - \log p_{CO_2} + \log[HCO_3^-], \qquad (3.1.2.1)$$

where $K_{H_2CO_3^*}$ and K_{Water} are equilibrium constants and brackets denote activities. With corrected pH values and pore water chemical data, saturation indices with respect to calcite and dolomite were calculated using a simple iterative computer programme similar to WATEQ (Truesdell and Jones 1974). The results of the computations are shown in Figs. 3.1.2.2 and 3.1.2.4.

For the study site CDN2 pore water 4.5 m below surface is slightly oversaturated for calcite (log SI 0.1–0.2; Fig. 3.1.2.4). In the borehole CDN1 (not shown) calcite saturation is attained at about 1.2 m depth.

The results from the Canadian study site show that calcite saturation equilibria and seepage rate determines the leaching rate of this mineral. Calculations taking into account a seepage rate of 5 m/a, a porosity of 0.38 and a pore moisture content of 9% vol. (Reardon et al. 1980) in combination with a saturation equilibrium of the pore water corresponding to a p_{CO_2} of 0.4% vol. (open system) result in a "flux of calcite" from the sediment of 11 g/($m^2 \cdot a$). This signifies that during the last 10 000 years at climatic conditions similar to today's approximately 4 m of sediment were decalcified. This value fits quite well with the observed depth of the calcite leaching zone.

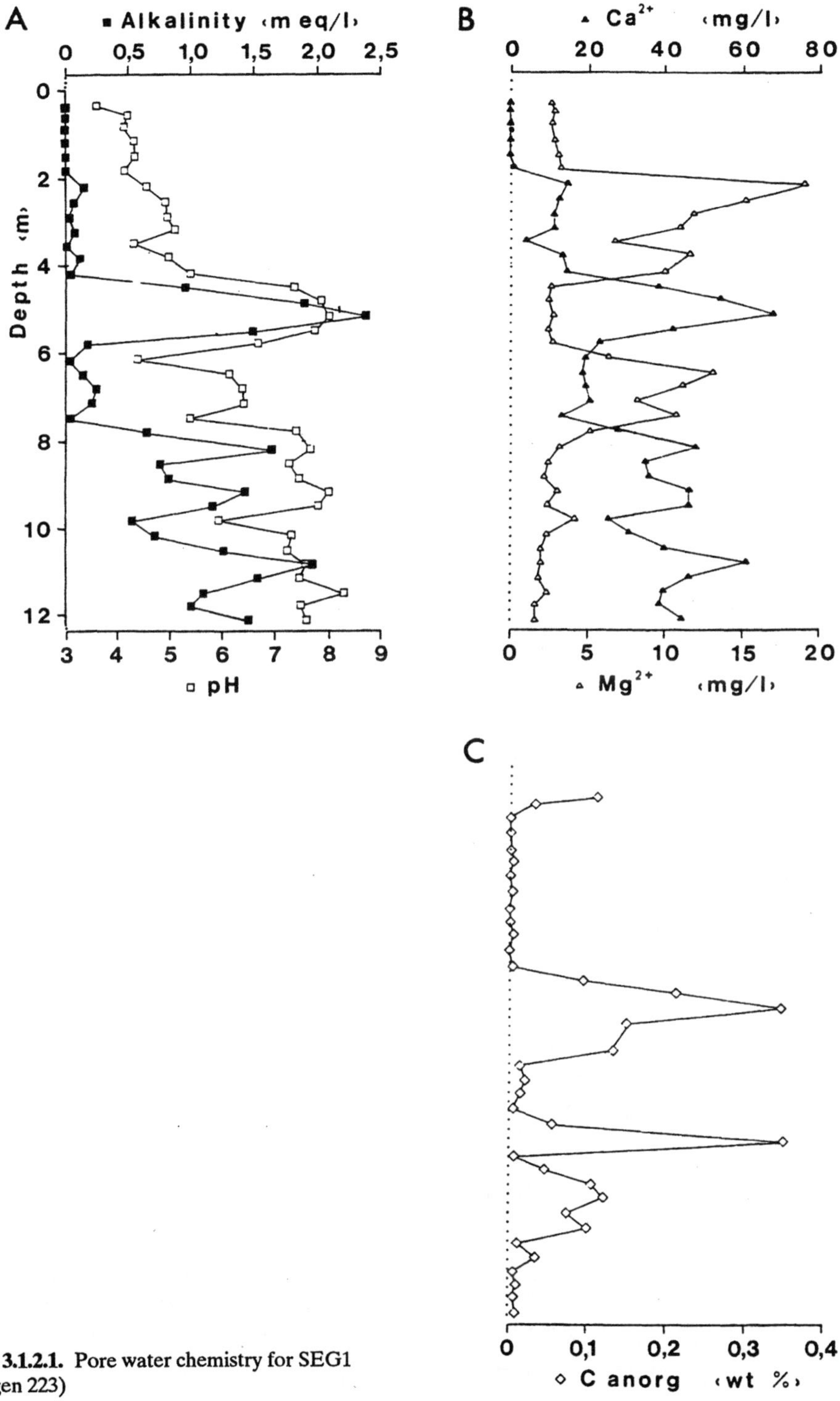

Fig. 3.1.2.1. Pore water chemistry for SEG1 (Jagen 223)

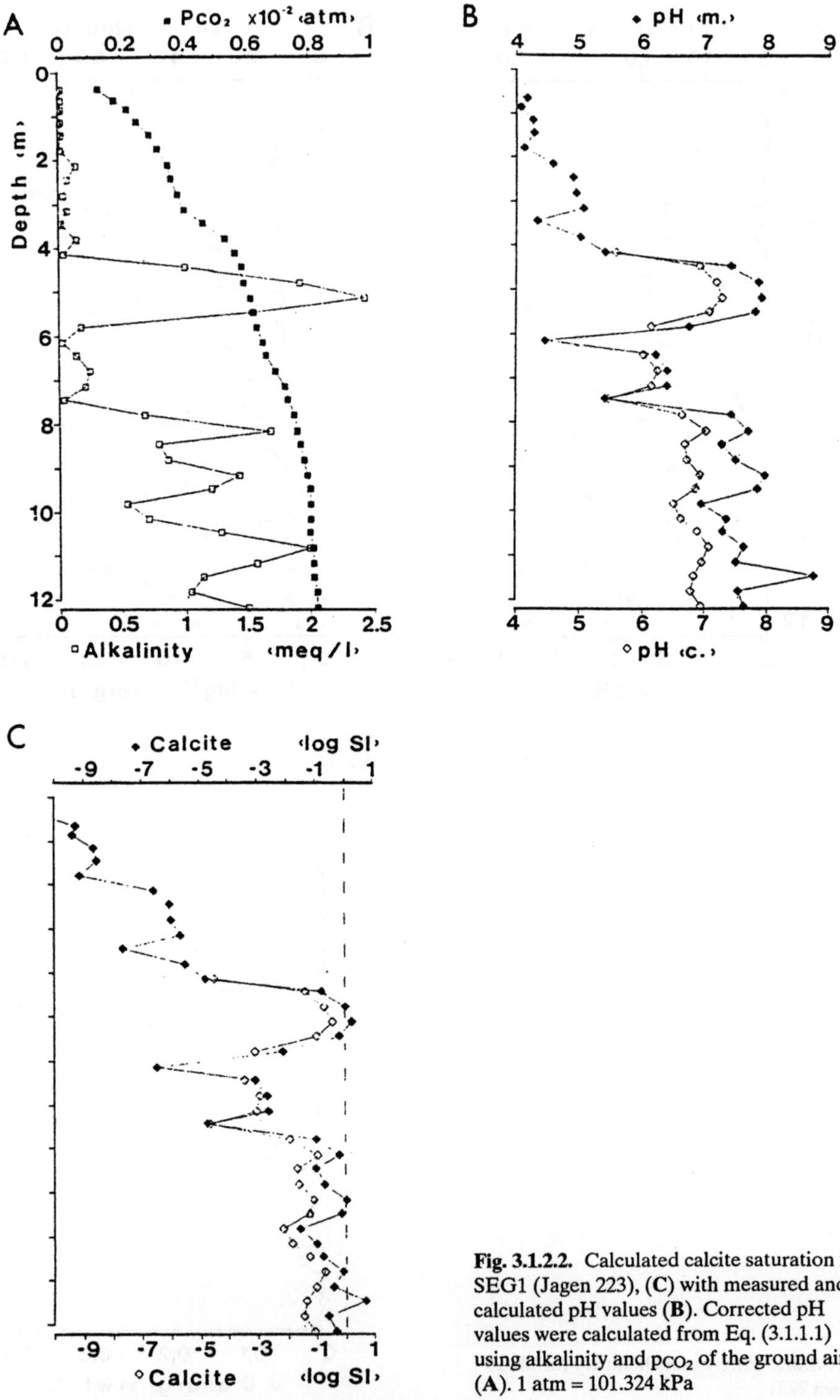

Fig. 3.1.2.2. Calculated calcite saturation for SEG1 (Jagen 223), (**C**) with measured and calculated pH values (**B**). Corrected pH values were calculated from Eq. (3.1.1.1) using alkalinity and p_{CO_2} of the ground air (**A**). 1 atm = 101.324 kPa

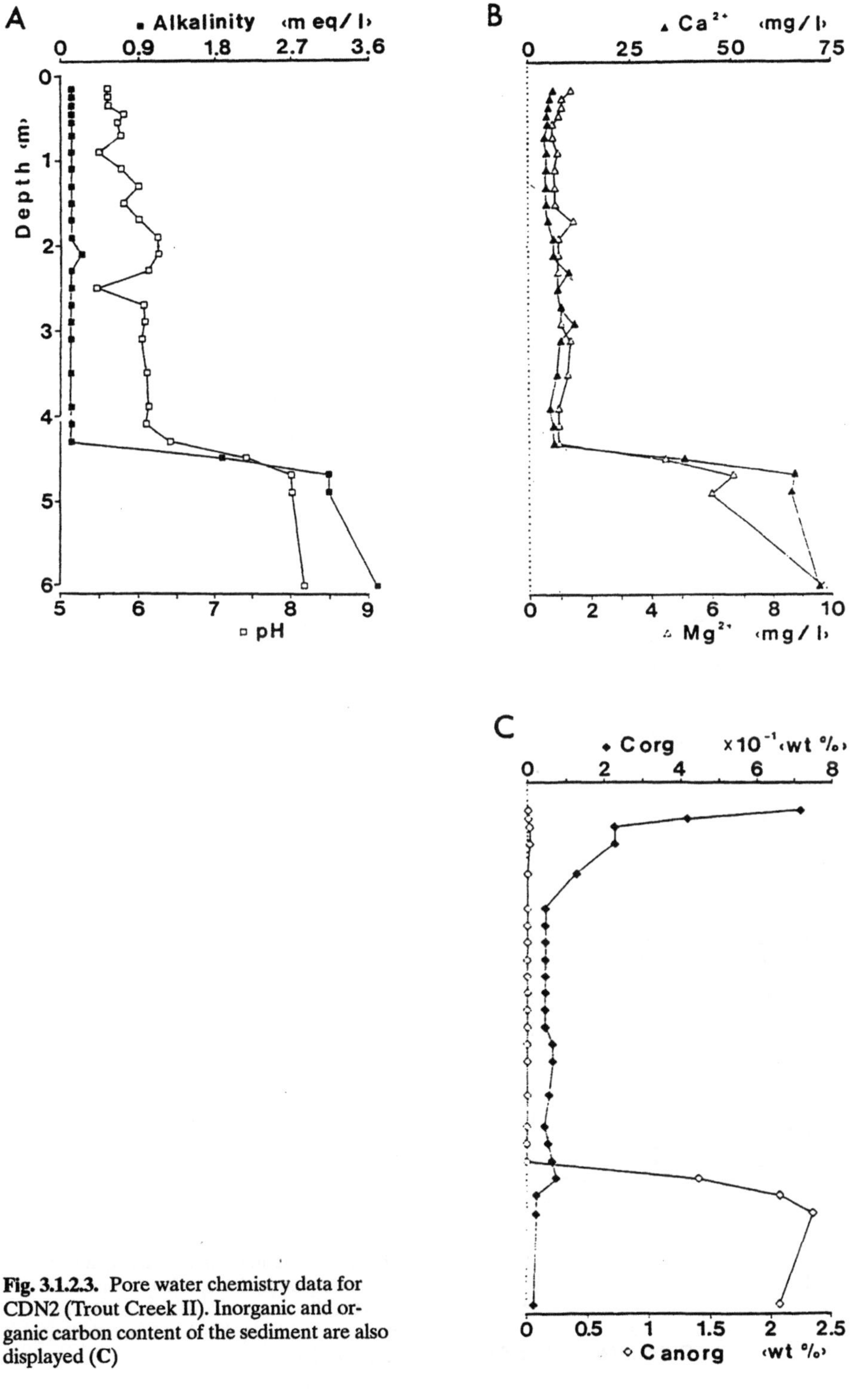

Fig. 3.1.2.3. Pore water chemistry data for CDN2 (Trout Creek II). Inorganic and organic carbon content of the sediment are also displayed (**C**)

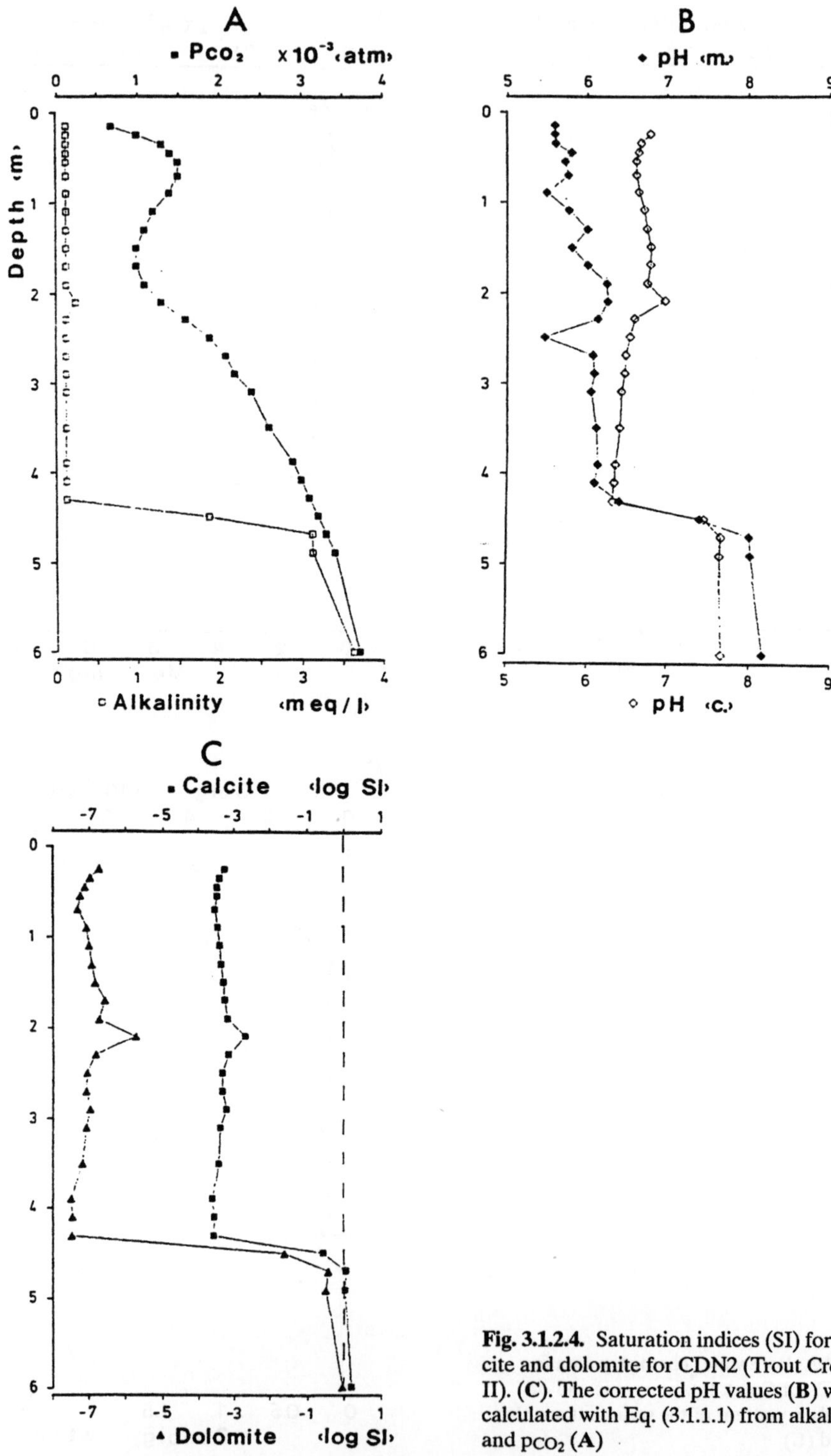

Fig. 3.1.2.4. Saturation indices (SI) for calcite and dolomite for CDN2 (Trout Creek II). (**C**). The corrected pH values (**B**) were calculated with Eq. (3.1.1.1) from alkalinity and p_{CO_2} (**A**)

3.1.2.3 Dissolution Experiments Under Open System Conditions

Experimental Design and Boundary Conditions

Different measurements of calcite dissolution kinetics under laboratory conditions have been published (Berner and Morse 1974; Plummer et al. 1978; Sjöberg 1978), but the boundary conditions for these experiments are usually not comparable to natural situations. Buhmann and Dreybrodt (1985a) have reported some laboratory experiments and a theory which is valid for the flow patterns in fractures. A modification of this theory is also applicable for hydrodynamic conditions in porous media (Baumann et al. 1985).

A schematic view of the experimental equipment which is similar to that used by Rauch and White (1977) is given in Fig. 3.1.2.5. A solution is circulated from a vessel through reaction tubes filled with calcite samples and back into the vessel, where it is re-equilibrated to a certain partial pressure of CO_2. For some experiments with very low flow velocities, a pre-concentrated solution is prepared in a second vessel using the same calcite material. Solution is then transferred to the gas equilibration vessel passing a 0.45 micron filter. In this case dissolution rates are determined only for certain presaturation conditions. All experiments have been performed at 25 ± 0.5 °C. Temperature and electrical conductivity were recorded as functions of time. The temperature-corrected conductivity in the system is directly correlated to the concentration of Ca^{2+}. From these data the dissolution rates were calculated as a function of the Ca^{2+} concentration of the solution.

All characteristic data of the experiments are given in Table 3.1.2.1. The p_{CO_2} in the gas equilibration vessel was maintained by bubbling a CO_2-N_2 mixture (presaturated with water vapour) at a constant rate of 2 l/min through the solution. From the vessel the solution was pumped with a peristaltic pump to the samples. Since degassing of the solution is possible, the gas mixture enters the samples with a somewhat lower p_{CO_2}. Once the Ca^{2+} concentration of the entering solution is in equilibrium with this actual pressure, no further increase of concentration in the system is possible. Therefore, the p_{CO_2} can be determined using the final saturation value of Ca^{2+} in the vessel.

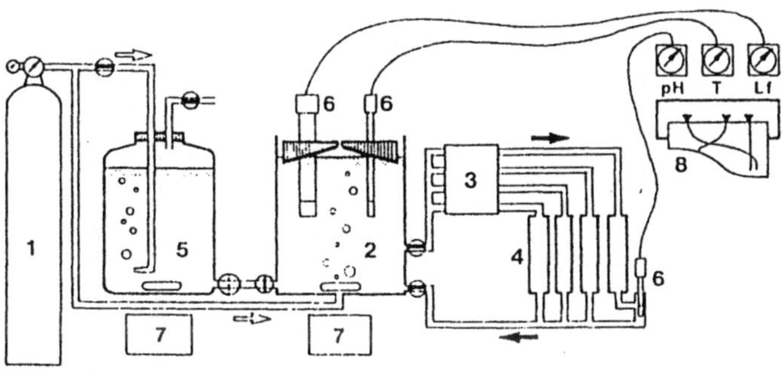

Fig. 3.1.2.5. Schematic view of the experimental setup to measure the calcite dissolution rates in porous media. *1* mixture of CO_2-N_2 gas; *2* gas re-equilibration vessel; *3* peristaltic pump; *4* tubes containing samples of calcite grains; *5* vessel to preconcentrate solution; *6* electrodes to measure pH, temperature and electrical conductivity; *7* magnetic stirrer, *8* recording equipment

Table 3.1.2.1. Boundary conditions of the laboratory experiments

Experiment	p_{CO_2} kPa	$p^s_{CO_2}$ kPa	V_{tot} cm^3	S cm^2	$\bar{v}$ cm/s	d cm	d_v cm	Carbonate material[a]	M_c g	[b]	L cm	Δ cm^2	*n*	n
0	97	75	130	70	1.15	0.045	0.016	Calcite II	1.2	(A)	4.0	0.196	0.40	1
1	97	75	130	70	1.15	0.045	0.016	Calcite II	1.2.	(A)	4.0	0.196	0.40	1
3	28	11	620	209	1.51	0.045	0.016	Calcite I	3.6	(A)	4.8	0.196	0.40	2
3a	28	11	620	209	0.031	0.036	0.013	Calcite I	2.86	(A)	4.2	0.196	0.40	2
4	9.5	5	640	139	1.51	0.045	0.016	Calcite I	2.4	(A)	4.0	0.196	0.40	2
5	0.9	0.6	500	263	0.47	0.036	0.013	Calcite I	3.6	(A)	1.2	0.636	0.40	3
7	0.9	0.8	500	14	0.006	0.028	0.01	Calcite I	0.15	(B)	0.15	0.636	0.40	1
8	0.9	0.8	600	62	0.004	0.036	0.02	Bryoz. Fx	0.27	(B)	0.2	0.636	0.45	3
9	0.9	0.8	600	62	0.005	0.036	0.013	Bryoz. Fx	0.27	(C)	5.5	0.636	0.40	3
10	0.9	0.8	600	45	0.006	0.039	0.02	Bryoz. SEG	0.3	(B)	0.15	0.636	0.45	3
11	0.9	0.8	600	45	0.006	0.039	0.014	Bryoz. SEG	0.3	(C)	5.5	0.636	0.40	3
12	0.9	0.8	600	45	0.002	0.039	0.014	Bryoz. SEG	0.3	(C)	5.5	0.636	0.40	3

[a] Fx = Faxe, SEG3 = Radesforder Berg.
[b] (A), (B), (C) cf. Fig. 3.2.1.6 mixture (C): ≈ 5% wt. $CaCO_3$.
p_{CO_2} = CO_2 partial pressure in the re-equilibration vessel.
$p^s_{CO_2}$ = CO_2 partial pressure at the inlet of the tubes.
V_{tot} = total volume of solution in the system.
S = total surface of the carbonate samples.
$\bar{v}$ = average flow velocity.
d = average particle diameter.
d_v = average diameter of pores.
M_c = total mass of carbonate material.
L = length of the sample.
Δ = area of tube cross-section.
n = porosity.
n = number of tubes.

Sample Preparation

In the experiments four different calcite materials were used (Table 3.1.2.1)

- well-crystallized, spary calcite (Calcite I); crushed material shows well-defined cleavage rhomboedrons;
- fine-grained, calcite (Calcite II); cleavage rhomboedrons are not as well defined as for "Calcite I";
- bryozoa from Cretaceous deposits near Faxe/Denmark (Bryoz.Fx); these deposits contain more than 90% bryozoa;
- Cretaceous bryozoa; separated from the fluvioglacial sands from the Segeberger Forst (Bryoz.SEG).

Non-biogenic calcites (Calcite I, Calcite II) were crushed. All materials were sized by wet sieving with deionized water with a pH between 8.5 to 9.0 after addition of NH_4OH. Calcite I and Calcite II were additionally washed with dilute HCl (0.01 M) for about 10 s, rinsed with deionized water, and finally with acetone. Subsequently all sample materials were dried at 100 °C.

Specific surfaces of the different sample materials and their porosites were estimated by a geometrical approach taking into account, e.g. packing and shape of grains (Baumann 1986). Geometrical parameters of all experiments are listed in Table 3.1.2.1.

Surfaces of all sample materials were checked by scanning electron microscopy before and after the experiments using a representative number of grains.

Results

Measured dissolution rates as function of the initial Ca^{2+} concentration are displayed in Figs. 3.1.2.7–3.1.2.8. Additionally, calculated rates from a kinetic theory (Baumann et al. 1985) are shown in Figs. 3.1.2.7 and 3.1.2.8 for the non-biogenic calcite. The calculated rates are in good agreement with the observed ones at high partial pressures of CO_2 and high flow rates (Exp. 0 and 1, Fig. 3.1.2.7). With decreasing p_{CO_2} and/or flow velocities, increasingly higher deviations between calculated and measured rates occur (Fig. 3.1.2.8).

The dissolution experiments with biogenic calcites were performed at very low flow rates and partial pressures. A flow velocity from 0.006 to 0.002 cm/s is common within the saturated zone and a p_{CO_2} of 800 Pa is also a typical value for ground air. For example Fig. 3.1.2.9 displays respective dissolution rates of the bryozoa material from Faxe/Denmark (Bryoz.FX.). The measured rates of experiment 8 are about 1.4 times higher than those of experiment 9. This difference results from the special arrangement of the calcite sample in a mixture with inert material (Fig. 3.1.2.6). This leads to an increasing effective length of diffusion from the reactive surface into pore space. This effect was also predicted from the theory (Baumann et al. 1985).

The saturation length (x_s), which is a very useful parameter to describe the combined mechanism of dissolution kinetics at the calcite surface and transport by diffusion and dispersion was calculated for all experiments from the equation:

$$x_s = \frac{-2 \cdot D}{v \cdot [1 - \sqrt{(1 + 4 \cdot \alpha_k S/V_p \cdot D/v^2)}]} \qquad (3.1.2.2)$$

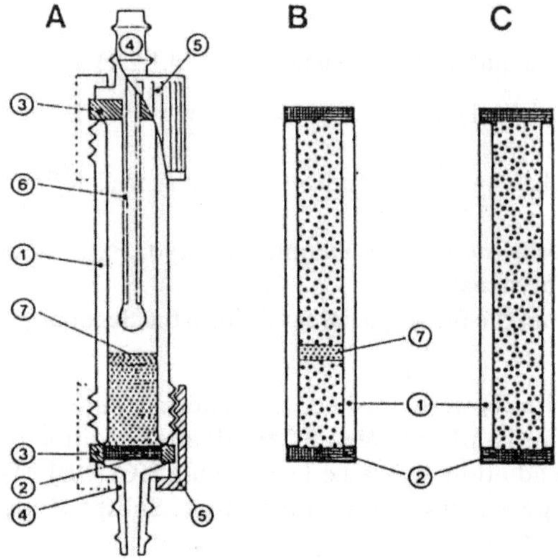

Fig. 3.1.2.6. Scheme displaying different arrangements of the calcite material (**A** layer; **B** thin layer within inert material; **C** mixture with inert material); **1** glass tube (I.D. = 9 mm, length 55 mm); **2** = filter plate (glass, pore size 0.05–0.1 mm); **3** packing (silicon); **4** tube connection (polyethylene); **5** stopper of the tube; **7** carbonate grain

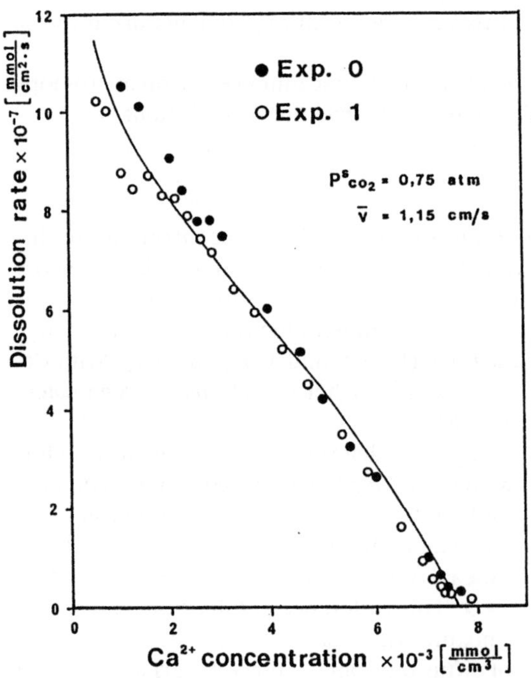

Fig. 3.1.2.7. Measured dissolution rates of porous calcite samples (*Exp. 0* and *1*). The *solid line* represents rates calculated by using a kinetic model. (Baumann et al. 1985)

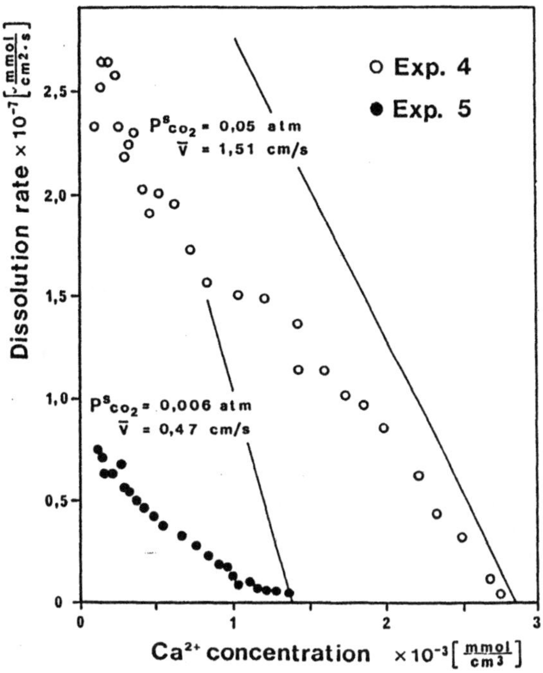

Fig. 3.1.2.8. Dissolution rates of calcite from *Exp. 4* and *5*. The *solid line* represents rates calculated by using a kinetic model. (Baumann et al. 1985)

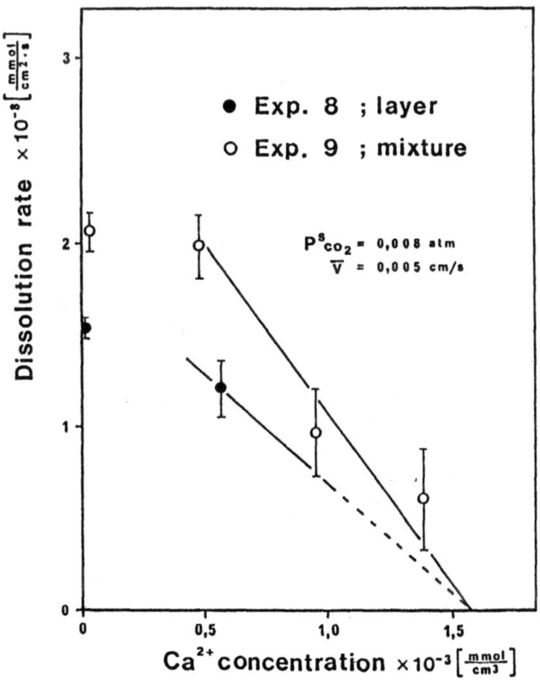

Fig. 3.1.2.9. Dissolution rates of the Faxe-Bryozoa (*Exp. 8* and *9*). The observed differences result from the arrangement of the material: once as a thin layer and second as a homogeneous mixture with inert grains

given by Palciauskas and Domenico (1976). This quantity describes the length of a flow path which is long enough to lead to a saturation of 63%. Parameters used in this formula are explained together with the results of the calculations in Table 3.1.2.2. The results clearly show that for experiments with low flow rates (Exps. 7–12 and 3a) the saturation length (x_s) is smaller than the total length of the sample (L) and in the experiments 7, 8 and 10 only on the order of the grain size (d). This implies that the solution leaving the sample should be approximately saturated with respect to calcite. In these cases the flux of dissolved calcite out of the sample is only determined by the respective saturation equilibrium and the flow rate.

SEM pictures taken from the spary calcite material (Calcite I) and from bryozoa (Bryoz.SEG) after dissolution clearly demonstrate that the entire surface was active during the dissolution processes. Nevertheless, the structured surfaces indicate regional differences of dissolution intensity.

Table 3.1.2.2. Measured and calculated parameters of the dissolution experiments

Experiment	S/V_p cm^{-1}	$\bar{v}$ cm/s	d 10^{-2} cm	D cm^2/s	$\bar{\delta}$ 10^{-2} cm	$[Ca^{2+}]_{eq}$ $\mu mol/cm^3$	L cm	α_k 10^{-4} cm/s	x_s cm
0	223	1.15	4.5	0.052	0.8	7.6	4.0	1.95	26.4
1	223	1.15	4.5	0.052	0.8	7.5	4.0	1.90	27.1
3	278	1.51	4.5	0.068	0.8	3.8	4.8	1.0	54.3
3a	317	0.031	3.6	0.001	0.7	3.8	4.2	0.8	1.2
4	222	1.51	4.5	0.068	0.8	2.8	4.0	1.6	42.5
5	287	0.47	3.6	0.017	0.7	1.3	1.2	3.1	5.3
7	370	0.006	2.8	0.00017	0.5	1.5	0.15	3.6	0.07
8	360	0.004	3.6	0.00014	1.0	1.5	0.02	3.9	0.05
9	15	0.005	3.6	0.00018	1.3	1.5	5.5	4.0	0.8
10	350	0.006	3.9	0.00023	1.0	1.5	0.15	4.0	0.07
11	11	0.006	3.9	0.00023	1.4	1.5	5.5	4.1	1.3
12	11	0.002	3.9	0.00008	1.4	1.5	5.5	3.6	0.5

S/V_p = quotient of total reactive surface of grains and total volume of pores ($V_p = L \cdot F \cdot n \cdot N$; Table 3.2.1.1).
$\bar{v}$ = average velocity of the solution through the sample.
d = average diameter of grains.
D = effective diffusion constant, corresponding to the longitudinal dispersion coefficient ($D = \bar{v} \cdot d$; Bear 1972).
δ = distance for the diffusion and/or dispersion transport from the calcite surface into the pore solution (experiments 0–8 and 10: $\delta = w/2$ for 100% calcite; experiments 9, 11 and 12: $\delta = w$ for a mixture of 5% wt. calcite with inert material).
$[Ca^{2+}]_{eq}$ = saturation concentration, calculated from the thermodynamic equilibrium equations (open system; $p_{CO_2} = p^s_{CO_2}$)
L = length of the sample.
α_k = kinetic constant; calculated for the experiments 0.1 and 5 from $\bar{\delta}$, $\bar{D}$ and $p^s_{CO_2}$ (after Buhmann and Dreybrodt 1984) and interpolated for the others (after Buhmann 1984).
x_s = saturation length.

3.1.2.4 Laboratory Measurement of the Saturation Length

Theory and Experimental Design

The theory of Buhmann and Dreybrodt (1985a,b) shows clearly that the saturation length is a very useful parameter to describe the kinetics of calcite dissolution with respect to all factors governing the entire process.

Thus, for a system closed with respect to CO_2, a relatively simple experimental setup is used to determine the saturation length in laboratory experiments. This parameter represents the length of a flowpath which is necessary to reach a certain degree of saturation with respect to calcite according to a specific flow velocity. In the experimental setup, which is shown in Fig. 3.1.2.10, the process of calcite dissolution can be simulated in the laboratory under conditions which compare well with those in a number of natural aquifers. The distribution of concentrations within the column containing the calcite-bearing material can be measured by taking small samples of the pore water at different positions of the column. The dissolution rate at a certain initial concentration can be calculated from the concentration at saturation and from a constant (k) which combines physical and chemical influences governing the process of calcite dissolution in natural material.

After a certain flow time, a steady state of calcium concentrations in the column is reached. In this situation the calcium concentrations in solution are only governed by two different velocities. The first one is the flow velocity of the water within the column – which is well known – and the second is given by all factors due to the calcite dissolution kinetics. From the distribution of calcium concentra-

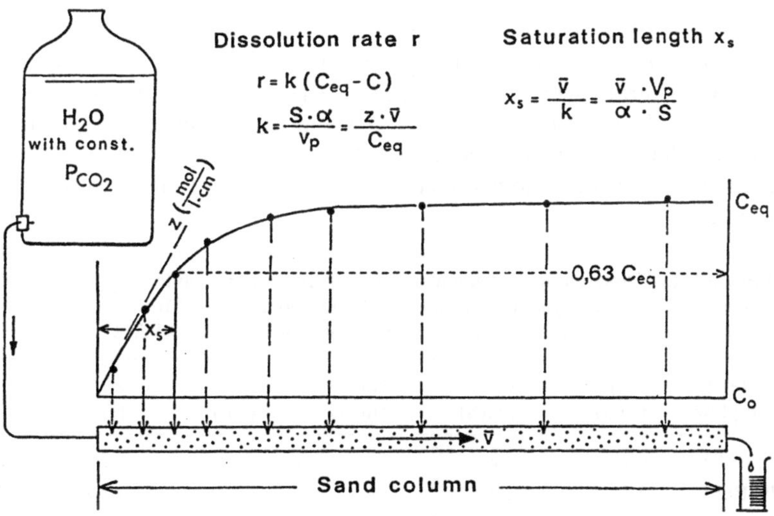

Fig. 3.1.2.10. Experimental setup for the measurement of calcite dissolution kinetics in a system closed with respect to CO_2. r dissolution rate [mol/(l·s)]; k constant factor of rate equation (l/s); ceq equilibrium concentration (mol/l); c concentration for which r is calculated (mol/l); S reacting surface area (cm^2); α slope of the specific rate function (cm/s); V_P pore volume (cm^3); z initial slope of the rate function [mol(1 cm)]; $\bar{v}$ average flow velocity (cm/s); x_s saturation length (cm)

tions within the column it is easily possible to calculate the characteristics of calcite dissolution kinetics using the equations given in Fig. 3.1.2.10.

The main advantage of this method is that for a calcite-bearing material parameters can be easily determined which are necessary to use the calcite dissolution kinetics in computer models with respect to flow velocities which are in the same range as those for ground water. The main disadvantage of this method is that most of the basic physical processes of calcite dissolution kinetics are hidden within the sum parameter *k*. For the present study experiments were carried out with natural materials and with synthetic mixtures of inert quartz grains and calcite and/or dolomite crystals. With respect to the synthetic materials with known reacting surfaces, it is possible to describe even the processes at the grain surface using the equations given in Fig. 3.2.1.10.

The input water for most experiments was a deionized water with a partial pressure of CO_2 adjusted to values close to those of natural waters. The p_{CO_2} within the storage vessel at the input side of the experiments was controlled by measuring the electric conductivity and/or the pH value. If the p_{CO_2} became too low, fresh CO_2 gas was bubbled through the water until the desired value was reached again. In some experiments input water with a bakcground of sodium sulfate in a concentration comparable to that of natural waters was used, in other experiments possible dissolution inhibitors were added.

Boundary Conditions and Results

Experiments with natural material will easily lead to the rates of calcite and dolomite dissolution, but will not give a better understanding of the dissolution processes as obtained in experiments with synthetic mixtures of inert quartz grains and carbonate crystals.

Table 3.1.2.3 lists as an example the boundary conditions and the results from some of the experiments carried out with a mixture of quartz grains and calcite crystals of grain sizes of about 1 mm. The reacting surfaces were geometrically estimated. The partial pressure of CO_2 for most experiments ranged from 1 to 6 kPa. These values are well comparable to those in ground water or only slightly higher than these. Also the flow velocities were lying in a range comparable to those of ground water.

Table 3.1.2.3. Dissolution experiments in one of the columns under closed system conditions with respect to CO_2 (Reacting surface of calcite 2340 cm^2; pore volume 510 cm^3; grain size 1–2 mm; column length 95 cm; total column volume 1500 cm^3; results shown in Fig. 3.1.2.11)

Date of sampling	p_{CO_2} (kPa)	$\bar{v}$ (cm/s)	c_{eq} (mg/l)	x_s (cm)	*k* (l/s)	α_k (cm/s)
11.09.85	6.18	2.2×10^{-3}	83	10	2.2×10^{-4}	4.8×10^{-5}
02.10.85	5.88	1.7×10^{-3}	78	7.5	2.3×10^{-4}	5.0×10^{-5}
07.10.85	5.57	4.0×10^{-3}	75	12.5	3.2×10^{-4}	7.0×10^{-5}
29.11.85	3.55	3.8×10^{-3}	47.5	14.5	2.7×10^{-4}	5.9×10^{-5}
12.12.85*	4.66	6.1×10^{-3}	62	18	3.4×10^{-4}	7.4×10^{-5}
11.02.86	6.38	3.1×10^{-3}	86	9.7	3.2×10^{-4}	7.0×10^{-5}

The results of the experiments can be compared with the calculated values by comparing the α_k-values of Tables 3.1.2.2 and 3.1.2.3. In both cases the α_k-values differ by not more than a factor of 2. If the different experimental setups and flow velocities of both experiments are taken into account, as well as the fact that in both cases the reacting surfaces were only geometrically estimated, this result can be taken as surprisingly good and as a proof for the significance of the values.

The measured distribution of calcium concentrations within the column of the experiments marked in Table 3.1.2.3 with an asterisk (Fig. 3.1.2.11) show quite clearly an exponential curve with concentrations close to saturation at the end of the column (the first 4.5 cm of the column are filled with an inert filter material). The saturation length can easily be calculated from this curve at the flow distance where 63% of the equilibrium concentration is reached.

Figure 3.1.2.12 shows an experiment with a column filled with a mixture of quartz grains and dolomite crystals. The reacting surface in this column was slightly larger than in the column of Fig. 3.1.2.11, and the flow velocity was slightly lower. The results show quite clearly in all dolomite experiments that the dissolution kinetics of dolomite is only slightly slower than that of calcite.

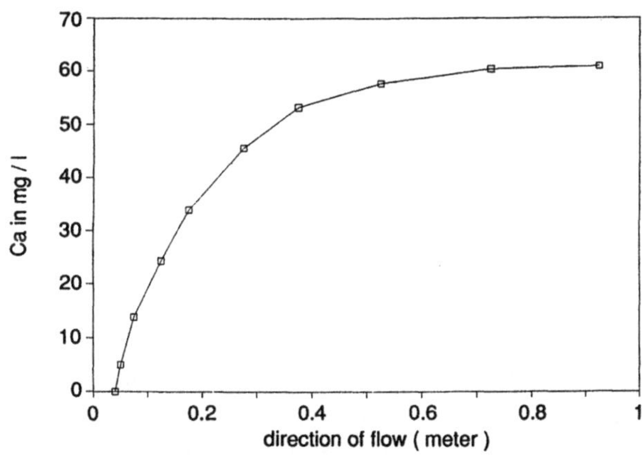

Fig. 3.1.2.11. Dissolution experiment in a column with a mixture of inert quartz grains and pure calcite crystals. This experiment is marked in Table 3.1.2.3 with an asterisk

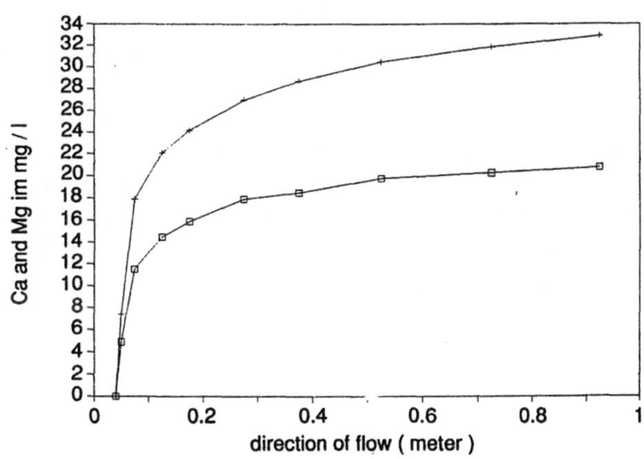

Fig. 3.1.2.12. Dissolution experiment in a column filled with a mixture of inert quartz grains and dolomite crystals

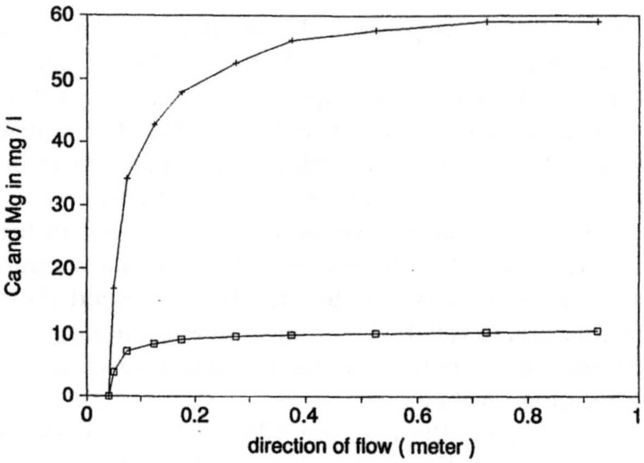

Fig. 3.1.2.13. Dissolution experiment in a column filled with a mixture of inert quartz grains with calcite and dolomite crystals

The fact that dolomite dissolution is an incongruent dissolution cannot be seen from these experiments. The calcium/magnesium ratio of the solution is nearly the same as that in the dolomite. Possible differences are beyond the accuracy of these measurements. That means that for most computer modelling of reactions in ground water the dolomite dissolution can be taken as congruent.

For the experiment in Fig. 3.1.2.13, a column with a mixture of inert quartz, calcite and dolomite grains was used. The distributions of calcium and magnesium concentrations in the column show that saturation for both minerals is attained. The calcium/magnesium ratio is similar to those of most natural ground water, where obviously the concentrations are governed by similar processes.

Influence of Dissolution Inhibitors

In some experiments input water with an addition of possible dissolution inhibitors [sulfate (50 mg/l), fluoride, copper, phosphate (2 mg/l)] was used. In the experiments with sulfate, fluoride and copper, no significant influence either on the equilibrium-concentration or on the dissolution kinetics could be observed. Only the experiments with phosphate as an inhibitor showed a significant influence on the dissolution process. The results of such an experiment are shown in Figure 3.1.2.14. From the saturation curve it becomes quite clear that initially the dissolution of calcite occurs normally until a calcium concentration of about 6 mg/l is attained where saturation for a second mineral (hydroxylapatite) is reached. This secondary phase now controls the concentration of calcium in solution.

Obviously, the inhibiting effect of phosphate reflects nothing else than a system controlled by two solid phases. It depends upon the phosphate concentration of the solution which of the solid phases controls the calcium concentration. It is presumed that this process is typical for most effects described under the name dissolution inhibition.

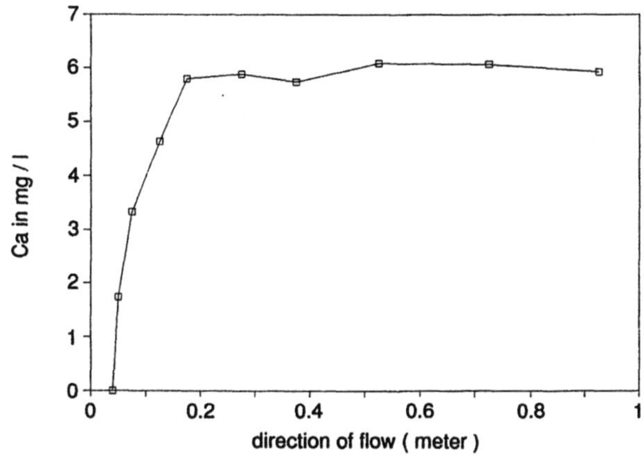

Fig. 3.1.2.14. Dissolution experiment in a column filled with a mixture of inert quartz grains and calcite crystals. The input solution had a concentration of 2 mg/l phosphate

3.2 Field Studies on Subsurface Water of Selected Sites

B. Merkel and J. Grossmann

3.2.1 Pore Water Sampling in Carbonate Terrains

3.2.1.1 Basic Annotations

Monitoring natural compounds and pollutants in pore and seepage water has become more common in the last decade. The increased activity is due to the knowledge that ground water quality has its origin in physico-chemical processes within the unsaturated zone. Pore water can be sampled by zero tension (lysimeters), by tension water samplers (suction cups), or by extraction of soil samples (centrifuge). Zero tension systems will sample water only if a gravitational pore water movement occurs. Thus, no water can be sampled in periods with a soil tension less than about 70 kPa. Furthermore, the natural soil profile and the exchange between the pore water and the ground air are disturbed by the installation. Thus, tension water samplers have been used more frequently. These systems use a partial vacuum to extract pore water into a porous cup which is placed into the soil. Thus, sampling can be done even if there is no gravitational water movement. However, pore water cannot be sampled if the capillary pressure in the soil is less than about 20 kPa. In this case pore water can be obtained for chemical analysis only by extraction of soil or rock samples.

Samples from lysimeters and suction cups extracted from identical soils under identical conditions differ significantly with respect to K^+, Mg^{2+}, Ca^{2+}, HCO_3^-, SO_4^{2-}, NO_3^-, pH, electric conductivity and DOC. Samples from zero suction samplers are characterized by higher electric conductivity and lower pH values due to an increased partial pressure of CO_2 in ground air in the lysimeter (Merkel 1983). The elevated CO_2 partial pressures partly originate from clipping gas exchange at the lysimeter bottom. Similar results have been reported by Haines et al. (1982).

3.2.1.2 Advantages and Disadvantages of Tension Sampling

Two important advantages of tension samplers are easy installation and sampling even during dry periods. However, there are some problems and disadvantages as well. Most of them are linked to the use within the carbonate-water-CO_2 system, which is sensitive to the CO_2 partial pressure.

Suction sampling may be preferred out of the following reasons:

- easy to install and easy to remove,
- nearly no interference with the geological environment
- useful for investigating spatial variability of soil water chemistry, and
- small sample volume means small interference with the chemical system.

On the other hand, potential disadvantages of tension sampling may be:

- sorption and ion exchange effects with the sampler material,
- loss of volatile compounds,
- bypassing of seepage water,
- increasing of seepage rate due to applied suction,
- sampling only possible with vacuum applied,
- depending on soil tension pores of different sizes are sampled,
- small sample volume (requires special equipment for chemical analysis).

Sorption and ion exchange as well as degassing effects during sampling will be discussed in detail below. The use of lysimeters in soil systems with macropores yields bypass effects. Such effects were found in a Tritium tracer experiment (Eichinger et al. 1984) and have also been reported by Barbee and Brown (1986). Water chemistry of samples drained from macropores may differ from those extracted from micropores due to kinetically controlled and surface reactions. These effects are related to the dynamics of the unsaturated flow and the diffusive exchange between macro- and micropores. Thus, sampling of pore water is greatly affected by an uncertain spatial variability of pore size distribution.

The hydraulic contact between soil pores and sampler is of great importance and is one of the main problems, in particular with the study of sandy gravel and/or in dry soils. The contact between sampling device and soil can be improved by adding sieved silt from the soil around the suction device. Some researchers have used powdered silicate to maintain a good contact. Talsma et al. (1979) pointed out that the applied differential pressure has only little influence on the flow rate towards the cup and the accumulation of water in the sampler.

Suction cups may be useful to investigate the spatial variability of soil water chemistry, which can differ significantly within some decimeters. On the other hand, a relatively large number of cups has to be used to obtain representative data. The reduced (under suction) pressure influences the potential field around the suction cup and thus the size of the drained soil volume as well as the chemical composition of the sample (Grossmann et al. 1987).

3.2.1.3 Pore Water Sampling with Suction Cups

Different systems of tension sampling have been designed and are displayed in Fig. 3.2.1.1 (Suarez 1986; Grossmann et al. 1987).

Pore Size of Suction Cup

If the pores of the suction cup are too large, the sampler will suck air instead of water. On the other hand, if the pore size is too small, macromolecules and microorganisms will not enter the sampler. The relation between suction head and pore size is given by Eq. (3.2.1.1):

$$\rho_c = -\frac{\sigma_w \cdot \cos \alpha_m}{\bar{d}_v \cdot g \cdot \rho_w} \cdot 10^3 \qquad (3.2.1.1)$$

where ρ_c = capillar potential (cm), σ_w = surface tension of water (N m^{-1}), α_m = moistening angle (0...180°), $\bar{d}_v$ = average pore diameter (mm), g = gravitational acceleration (m s^{-2}), ρ_w = water density (g cm^{-3}).

Using, for example, a pore size of 3 µm instead of 0.5 µm will lead to significantly higher concentrations of humic acids in the sample and simultaneously to increased heavy metal contents due to complexing. Thus, using suction cups with different pore sizes will result in data that are not comparable. For optimal conditions the pore size of tension samplers should have a uniform pore size distribution.

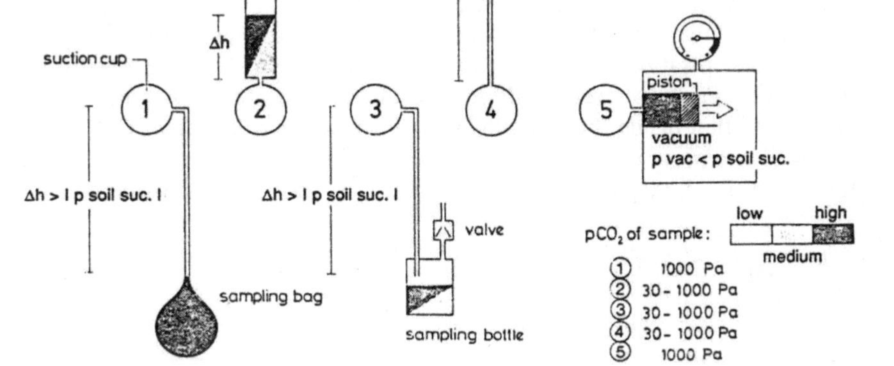

Fig. 3.2.1.1. Schematic representation of different systems of tension lysimeters. *1* Water column: water bag; 2 air space: shaft; 3 water column: bottle; 4 air space: bottle or air space container; 5 syringe/piston: syringe

In any case, membrane filtration with 0.45 or 0.2 μm has to be done to obtain comparable data, if trace metals are analyzed (Hem 1985).

Degassing Effects

The partial vacuum in the sample system degasses the CO_2 of the pore water and thus changes the chemical composition of the pore solutions via the various carbonate equilibria. The natural soil system is a more or less open system in many cases, while the suction cup is a closed system with respect to CO_2. The sampling procedure can be broken down into two parts. Part 1 is the transfer of pore water through the porous cup wall, which acts as an interface between the open and the closed system. Part 2 deals with the problem of storage conditions in the sampling device.

Transfer of pore water into the sampling systems requires a reduced pressure. This leads to a degassing of the water sample and to a new equilibrium pressure. Degassing, however, is limited by the kinetics of the CO_2 diffusion. The fundamental chemical reactions, which are implied, are:

$$CO_{2(gas)} \rightleftharpoons CO_{2(aq)} \tag{3.2.1.2}$$

$$HCO_3^- \rightleftharpoons CO_{2(aq)} + OH^- \tag{3.2.1.3}$$

$$Ca^{2+} + CO_3^{2-} \rightleftharpoons CaCO_{3(s)} \tag{3.2.1.4}$$

Reaction (3.2.1.2) has a half-life $T_{1/2}$ of 0.03 s, while reaction (3.2.1.3) has a half-life $T_{1/2}$ of 28 min (Gmelin 1973). Plummer et al. (1978) determined the forward and backward rate constants of calcite dissolution with respect to a stirred bulk liquid phase and single crystals. Kunz and Stumm (1984) found that nucleation of calcite will occur if water is supersaturated and the saturation index exceeds a certain level. The critical saturation index depends on the surface, catalytic effects and/or inhibitors such as phosphate (Reddy 1977).

Reactions (3.2.1.2) and (3.2.1.3) will establish a p_{CO_2} pressure in the sampling system which is lower than the partial pressure of CO_2 in the soil. In consequence, the $CO_{2(aq)}$ concentration in the water of the sampling system will be lower than in the pore water. Furthermore, the difference of the CO_2 partial pressure between ground air and that inside the sampling system leads to a diffusion of CO_2 through the cup wall, which is controlled by Fick's law. The transport of carbon species through the cup wall may be described by the convection-diffusion equation

$$\frac{\delta c_C}{\delta t} = D_M \cdot \underbrace{\frac{\delta^2 c_{CO_2}}{\delta x^2}}_{\text{diffusion}} + \underbrace{v \cdot \frac{\delta c_C}{\delta x}}_{\text{convection}} \tag{3.2.1.5}$$

where c_C = concentration of all carbon species (mol/cm^3); c_{CO_2} = concentration of $CO_{2(gas)}$ (mol/cm^3); D_M = coefficient of CO_2 diffusivity in water (cm^2/s); t = time (s); x = distance (cm); v = flow velocity (cm/s).

The pressure within the lysimeter will increase with the amount and flow of pore water into the suction cup. Sampling systems using a piston device or a bag strap will affect the sample less than sampling systems with a partial vacuum. For example, using sampling system type 4 (Fig. 3.2.1.1) and a sampling depth of 2 m, the total pressure in the lysimeter has to be reduced to about 70 kPa. Assuming a CO_2 partial pressure of 30 Pa in the atmosphere, an initial CO_2 partial pressue of 21 Pa in the lysimeter has to be seen in relation to the CO_2 partial pressure above the pore water (1 kPa). Thus, pore water, which is in equilibrium with a CO_2 partial pressue of 1 kPa will be equilibrated after being moved into the lysimeter with a partial pressure of 21 Pa. This leads to a significant degassing of the water and changes in the distribution of carbon species. The calculated calcite saturation index of the pore water with respect to the new partial pressure in this example was SI = 1.51. Thus, calcite precipitation may occur immediately (Kunz and Stumm 1984). To avoid those effects, the gas volume in the sampler should be as small as possible or the air space has to be filled before with ground air. However, this is quite difficult to achieve. From the above it follows further that pore water sampling from carbonate terrains should be done continuously, if possible, because degassing reaches a maximum if an empty cup starts to collect pore water.

Degassing Experiments

Carbonate ground water was equilibrated with respect to atmospheric CO_2 partial pressure. The calculated saturation index SI was about 8 at the beginning of the experiment. Tests were carried out applying different partial vacua and volumes to eight uniform samples of ground water for 14 days in a closed system (glass bottles). Some of the bottles were stirred three times a day. In most samples no precipitation of calcite occurred. Only in the stirred bottles with a total pressure of 20 kPa a significant decrease of calcium and hydrogen carbonate concentrations was detected (samples no. 8 and 9, Table 3.2.1.1). These tests showed that precipitation of calcite in a sample bottle does not occur if specific boundary conditions are given. No precipitation occurs even in stirred sampling bottles at 12 °C and a

Table 3.2.1.1. Results of a degassing experiment (duration 14 days, sample volume 50 ml, 12 °C); sample no. 1 is untreated, samples no. 6 to 9 have been stirred three times a day

Sample no.	1	2	3	4	5	6	7	8	9
p (kPa)	100	80	50	20	20	80	50	20	20
Volume (1)	0	2	2	2	8	2	2	2	8
Stirred	No	No	No	No	No	Yes	Yes	Yes	Yes
Measured									
SI	0.90	0.95	1.11	1.36	1.36	0.95	1.11	–	–
Ca^{2+} (meq/l)	1.96	1.96	1.96	1.96	1.95	1.97	1.98	1.25	0.75
HCO_3^- (meq/l)	2.59	2.57	2.59	2.52	2.57	2.59	2.61	2.39	1.88
Calculated									
Ca^{2+} (meq/l)	1.28	1.22	1.12	1.02	0.88	1.22	1.12	1.02	0.88
HCO_3^- (meq/l)	1.26	1.20	1.10	0.98	0.84	1.20	1.10	0.98	0.84

saturation index at about 1.11. The equilibrium concentrations of Ca^{2+} and HCO_3^- (Table 3.2.1.1) were calculated from the CO_2 partial pressure, taking into account the amount of CO_2 degassed into the sampling system.

In the course of the degassing experiment, pH values were measured as well as calculated from the CO_2 partial pressure, using a numerical model (Parkhurst et al. 1980; Grossmann 1988). The recorded pH values varied between 7.9 and 8.57, while calculated values were in the range of 8.9 to 9.4. The pH determination in seepage water is a well-known and serious problem with respect to the interpretation of the pH values of pore water samples, and thus the calculation of the saturation index. Both laboratory and field experiments indicate that the pH is subject to a significant alteration during the sampling procedure. An additional cause for pH variations may occur with microbiological activities in the sampling system.

Figure 3.2.1.2 displays measured Ca^{2+} – and HCO_3^- concentrations versus suction head in the soil of a laboratory lysimeter. There is no significant difference between the samples withdrawn at different pressures.

Sorption and Cation Exchange of Solutes by Suction Cups

Ceramics and sintered Al_2O_3 have been widely used for the construction of tension lysimeters, and recently sintered nylon, teflon, polyethylene and cellulose fibres have been tested for their suitability. In this context, it has been well established that leaching as well as sorption can significantly affect trace metal and phosphate concentrations (Grossmann et al. 1987).

New Al_2O_3 cups yield several milligrams per litre of Ca^{2+}, Mg^{2+}, Na^+ and HCO_3^- if pore water with pH values of about 7.5 is passed through them. It was found that after flushing the Al_2O_3 cups with HCl, deionized water and pore water, a solution passing through the lysimeter stayed unchanged with respect to both major ions and heavy metals.

In general, the materials of suction cups have to be hydrophile to maintain moisture and water flow through the lysimeter. In consequence, suction cups have a remarkable cation exchange capacity, which may cause significant differences in composition between the solutions collected in the lysimeter and the soil solution, especially with regard to trace metals. Soil and cup materials have a low to moderate

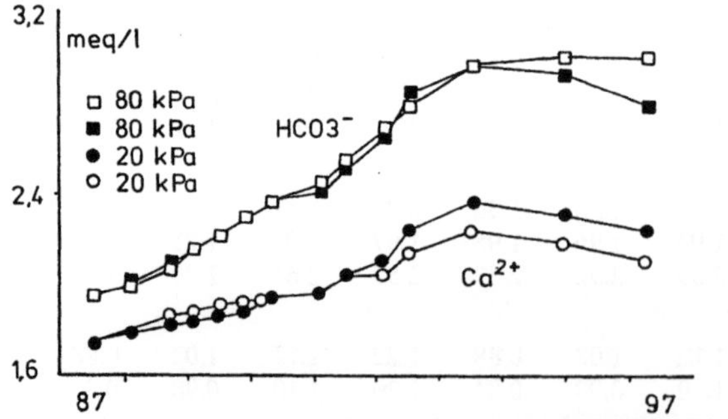

Fig. 3.2.1.2. Ca^{2+} and HCO_3^- concentrations of seepage water samples versus pressure head withdrawn by suction cups from a laboratory lysimeter filled with carbonate gravel. Vacuum applied: 20 and 80 kPa and air space: 1 and 8 l

preference for main cations, whereas heavy metals especially at trace levels are adsorbed selectively (Grossmann et al. 1988). Sorption is mainly controlled by the following factors:

- cation exchange capacity of the cup material;
- pH value of the pore solution,
- content of dissolved organic carbon,
- sampling rate.

Many attempts have been made to solve the problem of cation exchange capacity of suction cup material in using plastic material (Grossmann et al. 1987). Since hydrophile plastics have an exchange capacity as well, a final solution of the problem is not available. Figure 3.2.1.3 displays copper concentrations versus flow through lysimeters from laboratory experiments with Al_2O_3 and nylon cups. The copper concentration of the test solution was altered stepwise between 5 and 25 µg/l. Figure 3.2.1.3 indicates that nylon cups did not change the copper concentration while Al_2O_3 cups caused significant changes with regard to the pH value of the solution. Some alteration occurred even for nylon cups after adding 4 mg/l of HA. Figure 3.2.1.4 shows the copper concentrations while using both Al_2O_3 and nylon cups in a field experiment (Grossmann 1988). Although the real concentrations in the pore solution are not known, Fig. 3.2.1.4 implies that the nylon cups give more realistic results because the Al_2O_3 shows in most of the cases smaller peaks and/or measurable time lags.

At high pH values, the effect of cation exchange with respect to trace metals is stronger than at low pH (Fig. 3.2.1.3). Since in most carbonate pore waters the pH varies between 7 and 9, the sorption of trace metals by the suction lysimeter is a more serious problem than in acid soils.

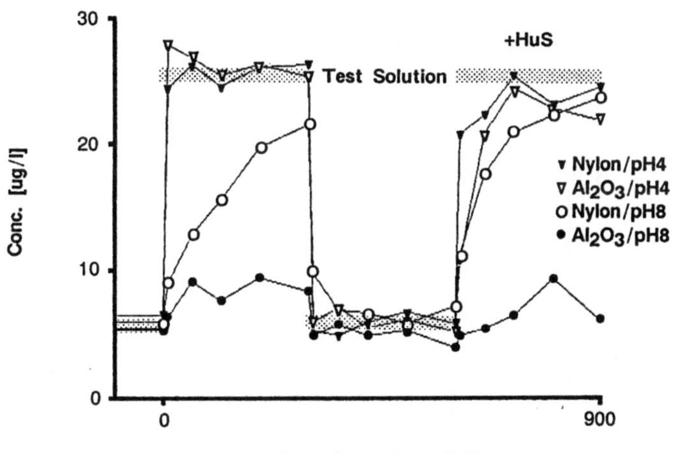

Fig. 3.2.1.3. Flush test with Al_2O_3 and nylon cups and a copper solution. The copper concentration was varied according to a step function with time. The *four lines* represent the copper content of the solution versus cumulative flow after passing Al_2O_3 or nylon cups (test solution at pH 4 and 8, with and without 4 mg/l HA)

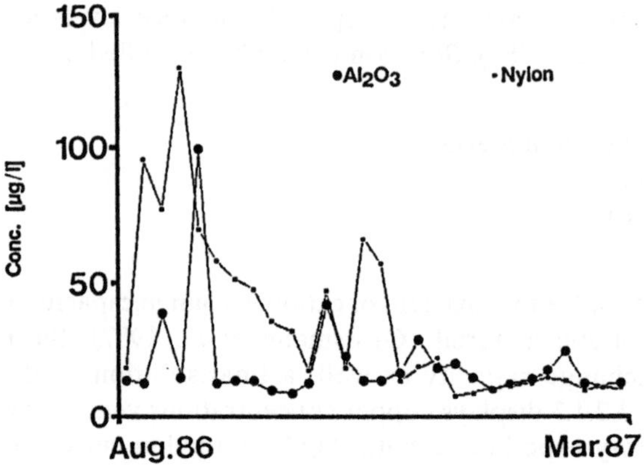

Fig. 3.2.1.4. Copper content of pore water versus time sampled with Al_2O_3 and nylon cups during a field experiment in carbonate gravel. (Grossmann 1988). Sampling depth: 2 m, pH: 8, DOC: 0.5–3 mg/l, flow rate: 0.1 to 1 ml/h

The capacity of humic acids to bind metal ions reduces cation exchange reactions at the lysimeter wall (e.g. Fe). On the other hand, humic compounds are adsorbed by the cup material, especially by ceramics, and increase the exchange capacity of the cup (e.g. Co, Ni, Pb).

Figure 3.2.1.5 shows that high flow velocities through the Al_2O_3 cup prevent the adjustment of the exchange equilibrium and thus reduce sorption. A step function for the input concentration from 2 to 25 µg/l and two flow rates (3 and 30 ml/h) were used. At a flow rate of 30 ml/h the solution in the cup was exchanged every 6 min, and every hour at a flow rate of 3 ml/h.

Figure 3.2.1.6 displays Zn concentrations and sample volume versus sample number. Again, nylon and Al_2O_3 cups were used in this field experiment (Grossmann 1988). Large differences between samples from nylon and Al_2O_3 lysimeters occurred and were significantly correlated with small sample volumes. This provides evidence for the effect of flow velocity or time in order to achieve cation exchange equilibria. Minor sorption at high flow rates was reported by Hansen and Harris (1975) as well.

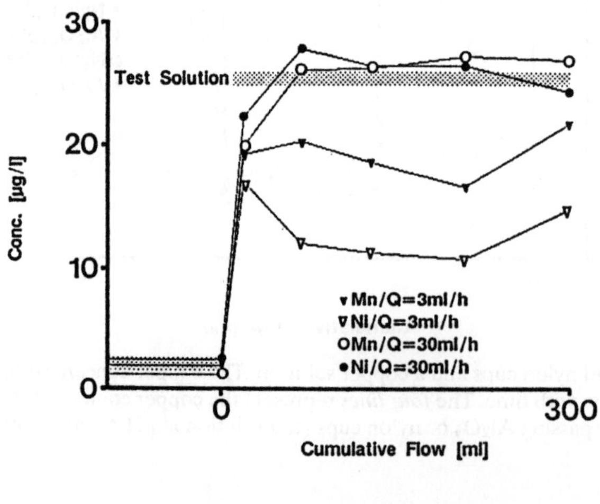

Fig. 3.2.1.5. Flush test with Al_2O_3 and nylon cups. The concentration of the test solution was changed according to a step function of time from 2 to 25 µg/l Mn and Ni. Flow rate: 3 and 30 ml/h

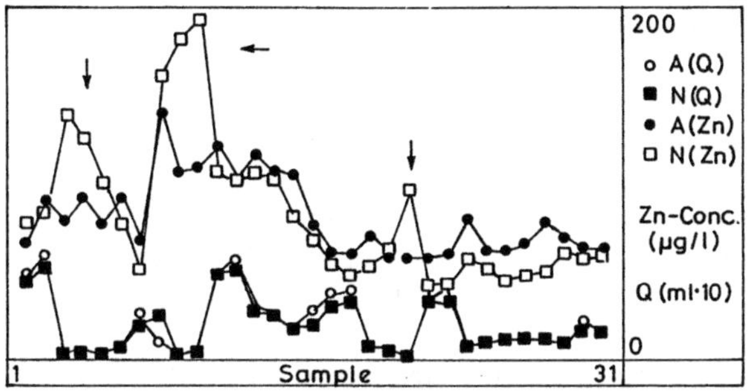

Fig. 3.2.1.6. Zn concentration and sample volume versus sampling number, measured during a field experiment (Grossmann 1988). Both nylon (*N*) and Al_2O_3 (*A*) cups were used at identical points. The time intervals of the sampling was about 14 days

Table 3.2.1.2 shows the behaviour of ten trace metals with respect to sorption by suction cups (Grossmann 1988). Because of the complexity of the situation, Table 3.2.1.2 can provide only a general overview. In cases of any doubt, laboratory tests should be carried out as close as possible to natural conditions. The results are, strictly speaking, only representative for the conditions under which they were obtained.

Pre-Treatment of New Suction Cups

New suction cups contain contaminants which may leach into the percolating solution. Thus, a cleaning and conditioning procedure is necessary before their installation and use (Grossmann et al. 1987). Ceramic cups should be flushed by 100 ml 1.0 M HCl and 0.5 l deionized water at a low flow rate of approximately 50 ml/h. An additional rinsing with 100 ml 1.1 M NaOH will remove organic pollutants. For plastic cups 20% of these quantities are sufficient.

To minimize sorption effects it is recommended to flush the cups with a solution similar to the soil solution. A pre-treatment with respect to the measurement of trace metals (concentrations < 100 µg/l) requires at least 5 l for ceramic cups (10 l for copper, iron or lead). For plastic cups 20% of these quantities are sufficient.

Table 3.2.1.2. Sorption of trace metals by ceramic and nylon suction cups in carbonate terrains (Grossmann 1988). Evaluation of several laboratory and field experiments (0 negligible sorption, + measurable sorption, ++ strong sorption; blank cells = information missing)

Cup material	Al	Be	Cd	Co	Cu	Fe	Mn	Ni	Pb	Zn
Al_2O_3	++	++	++	0	++	++	0	0	++	++
Ceramic	++		++		++				++	++
Nylon	+	+	0	0	+	++	0	0	++	0

3.2.1.4 Conclusions

Suction cups are simple devices to sample water from the unsaturated zone. The installation is easy and the impact on the investigated soil is moderate. Some problems of this method are clogging of cups, filter effects and impact on seepage water flow. In carbonate terrains with neutral to slightly alkaline pH values the sorption of solutants, especially trace elements, may be a serious problem. The use of cup materials with low cation exchange capacity partly solves this problem.

Users of suction cups should focus attention on the following points:

- The pore diameter of the cup should correspond to the standards of solid/liquid separation (Skougstad and Scarboro 1968). Otherwise the samples have to be filtered.
- New suction cups have to be cleaned with acid and deionized water. The pretreatment includes also a conditioning procedure with the natural water to be sampled.
- To achieve equilibrium between cup wall and soil an equilibration period of several weeks after the installation is recommended. Before valid data can be obtained at least 1 l of pore water should have rinsed through each cup (Grossmann 1988).
- To minimize exchange effects for the investigation of trace metals, e.g. Al, Be, Cd, Fe, Pb, and Zn, plastic cups should be used. Appropriate tests of the suitability of a cup should be carried out with conditioned cups.
- The pressure differential between lysimeter and soil should be as small as possible, while still permitting the collection of soil water.
- A piston device or a water bag in combination with a hanging water column will give the best results due to a minimum of air space in the sampling system.
- Sampling should be continuous if the pressure difference is obtained by a partial vacuum.
- pH measurements in soil water samples from carbonate terrains are affected by degassing, which causes changes in the carbonate equilibria. Humic acids and microbiological activities may influence pH measures as well. Thus measured pH values should not be used for chemical equilibria calculations. It is more reasonable to measure the CO_2 partial pressure of the ground air directly for geochemical modelling.

3.2.2 Variation of Inorganic Carbon in the Unsaturated Zone of a Carbonate Gravel System

L. Eichinger and B. Merkel

3.2.2.1 Background

In the unsaturated zone biological, chemical and physical processes control the formation and the decay of dissolved inorganic carbon (DIC), the dissolution and precipitation of carbonate minerals and the degassing of CO_2. These processes determine also the carbon isotope compositions of the DIC, the ground air CO_2 and the aqueous carbonate species. Hence, in determining ^{14}C ages of ground waters, one must consider the knowledge of the mass transfer and carbon isotope evolution in the unsaturated zone.

Detailed field investigations extending over a period of several years were carried out in the unsaturated and saturated zones of a carbonate gravel system (Munich Gravel Plain). These studies included hydrodynamic, hydrochemical and isotopical measurements of ground air, seepage and ground water. These investigations were aimed at defining the origin of the dissolved carbon in the ground water.

3.2.2.2 Areas of Investigation

The investigations were performed in the Quaternary carbonate gravel system of the Munich Gravel Plain. The gravel layer has a thickness from 10 to 60 m, while the unsaturated zone has a thickness from 10 to 50 m. The ground water depth varies between 1 and 10 m at maximum. The average ground water flow velocity is assumed to be in the range from 10 to 30 m/d. The Pleistocene carbonate gravels are covered with 30 to 50 cm of loamy soil. Ninety percent of the gravel consists of calcite and dolomite, the remainder being sandstones and crystalline material with interstitial lenses of sand and cemented gravel. Two different field sites were chosen in this area, where ground air could be sampled from different depths in order to compare the effect of different vegetation on the ^{14}C content.

Forest area: this area is covered with mixed forest. The average water content of the unsaturated zone was varied from 0.04 to 0.06 (volume ratio of water/soil), and the average seepage velocity, derived from ^{3}H measurements, was about 2 m/a (Eichinger et al. 1984). The thickness of the unsaturated zone in this study area is between 20 and 24 m and rises to about 50 m towards the recharge area. The saturated zone is only 1 to 3 m thick. The bottom of the aquifer consists of Tertiary silt layers with a very low permeability. The mean residence time of the ground water was calculated from a hydrological balance to be about 6 to 10 years. Ground air samples were taken from five depths between 3 and 20 m (Fig. 3.2.2.1).

Grass-covered area: this site is covered with grass and the unsaturated zone has a thickness of 10 to 13 m. The average water content was between 0.05 to 0.08

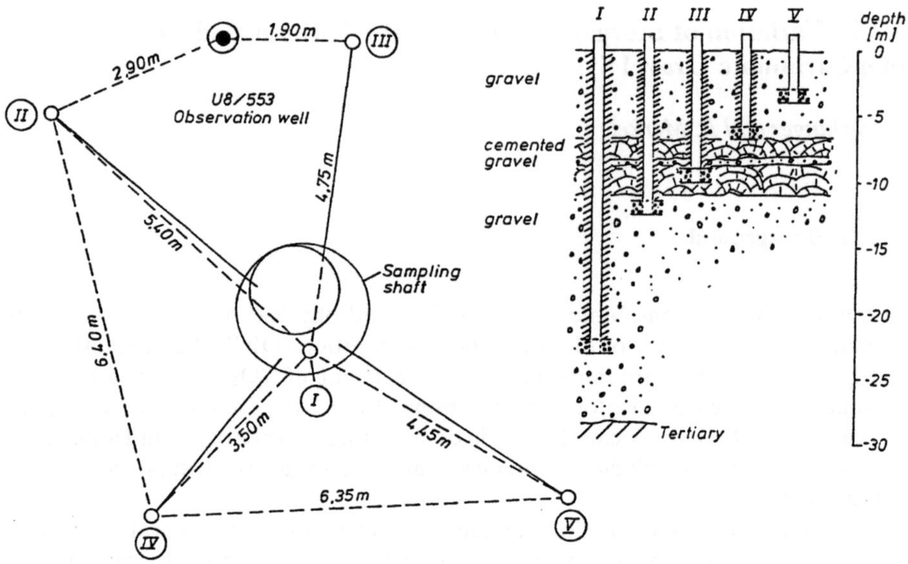

Fig. 3.2.2.1. Schematic representation of the ground air probe at the investigated forest site (MUC 4)

(volume ratio of water/soil). The average seepage velocity as derived above was about 2 m/a (Eichinger et al. 1984). An accessible research shaft (Fig. 3.2.2.2) was used to sample ground air and seepage water at depths from 0.2 to 9 m (Merkel et al. 1982). A computer-based data acquisition system was used to record continuously the CO_2 concentration of ground air, capillary pressure heads, ground temperatures and the electric conductivity of seepage water. Pore water was sampled with suction cups at different depths and analyzed weekly for major components and monthly for isotope contents (^{13}C, ^{18}O).

3.2.2.3 Results

Ground Air, Seepage and Ground Water

Measurements of CO_2 contents are used to estimate the amount of CO_2 production. The CO_2 partial pressure varied between 0.05 and 4 kPa in the ground air (Table 3.2.2.1). Figure 3.2.2.3 shows the variation of Ca^{2+}, Mg^{2+}, HCO_3^- and $\delta^{13}C$ (‰, PDB), while Fig. 3.2.2.4 displays the variations of the major ions with time at 6 m depth.

Measurements of CH_4 concentration show both production and decomposition of methane in the unsaturated zone. Figure 3.2.2.5 displays the CH_4 content of ground air versus depth. Methane, which diffuses into the top soil layer from the atmosphere (average atmospheric content: 2 vpm), is decomposed in the top soil almost totally. Curve b in Fig. 3.2.2.5 provides evidence for the assumption that at depths below 1 m a moderate CH_4 production takes place, although the entire unsaturated zone is supplied with oxygen. This means that methane productions take place in micropores due to the lack of oxygen supply. Figure 3.2.2.6 shows average

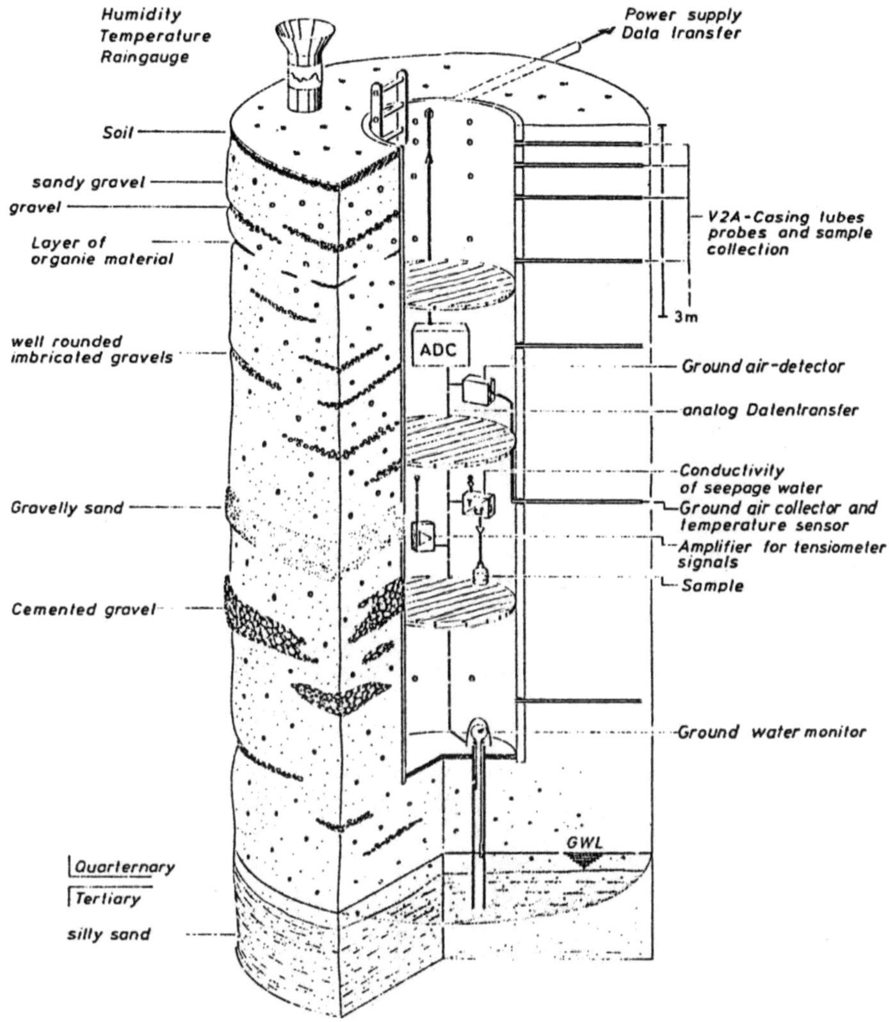

Fig. 3.2.2.2. Accessible research shaft at the grass-covered site area Munich-Großhadern (MUC 3)

concentrations of DOC, nitrate, ammonium and nitrate versus depth. Rain water infiltrates into the soil with an average nitrate concentration of 12 mg/l. This value is computed by accounting for evaporation. The nitrate decreases to values of 2 mg/l at 0.2 m depth, which seems to be due to microbially mediated reactions. At greater depths, however, organically bound nitrogen is converted to nitrate. This assertion is based on the inverse correlation of the DOC and nitrate concentration depth profiles (Fig. 3.2.2.6). The maximum nitrite concentrations occur in the zone of methane production and can be interpreted in the proposed manner.

Measurements of the ^{14}C content of the ground air carbon dioxide are shown in Fig. 3.2.2.7a. The ^{14}C values decrease from 121% modern to 106% modern at a depth

Table 3.2.2.1. Values for CO_2, Ca^{2+}, Mg^{2+} and HCO_3^- in different depths of the unsaturated zone (minimum, maximum and arithmetic mean)

Depths (m)	CO_2 kPa n=20 000	Ca^{2+} mg/l n=180	Mg^{2+} mg/l n=180	HCO_3^- mg/l n=180
0.2	0.053	38.5	4.5	120
	3.71	100.0	15.4	362
	0.58	68.5	9.1	250
0.5	0.10	40.0	4.5	133
	4.02	107.1	22.4	369
	0.95	68.4	11.3	260
1.0	0.14	35.8	7.9	127
	3.74	112.0	21.3	385
	1.05	60.3	11.7	325
2.0	0.20	36.4	8.5	104
	2.00	120.0	25.5	362
	1.01	61.0	14.3	242
3.5	0.20	37.3	5.0	129
	1.72	152.0	29.2	370
	0.92	61.8	17.5	259
6.0	0.21	40.0	9.4	187
	1.73	160.7	25.4	386
	0.75	59.5	17.9	264
9.0	0.22	35.7	10.2	140
	0.97	79.0	25.0	327
	0.40	51.4	17.4	227

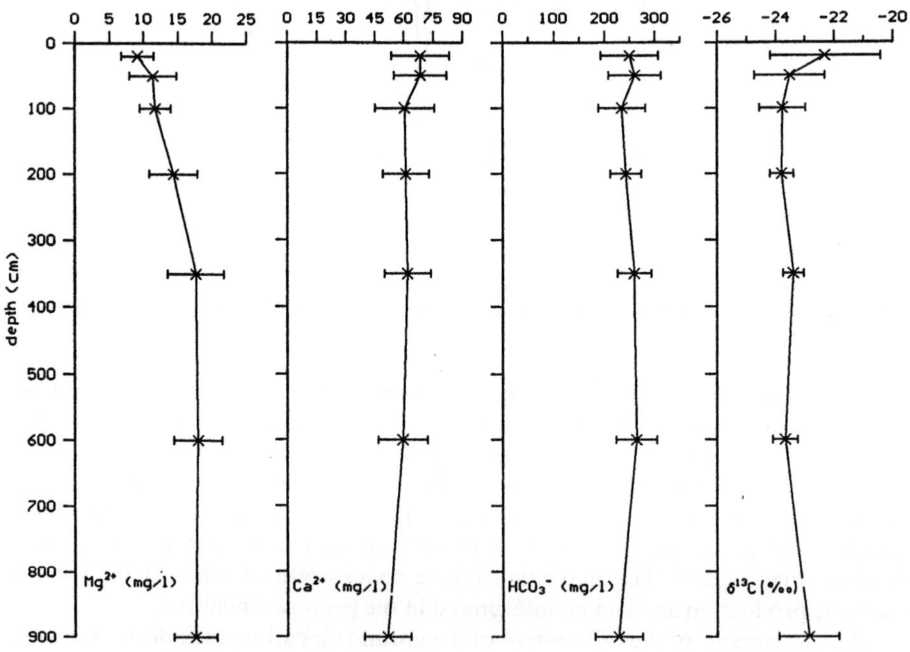

Fig. 3.2.2.3. Minimum, maximum, and average values of Ca^{2+}, Mg^{2+}, HCO_3^- and $\delta^{13}C$ versus depth

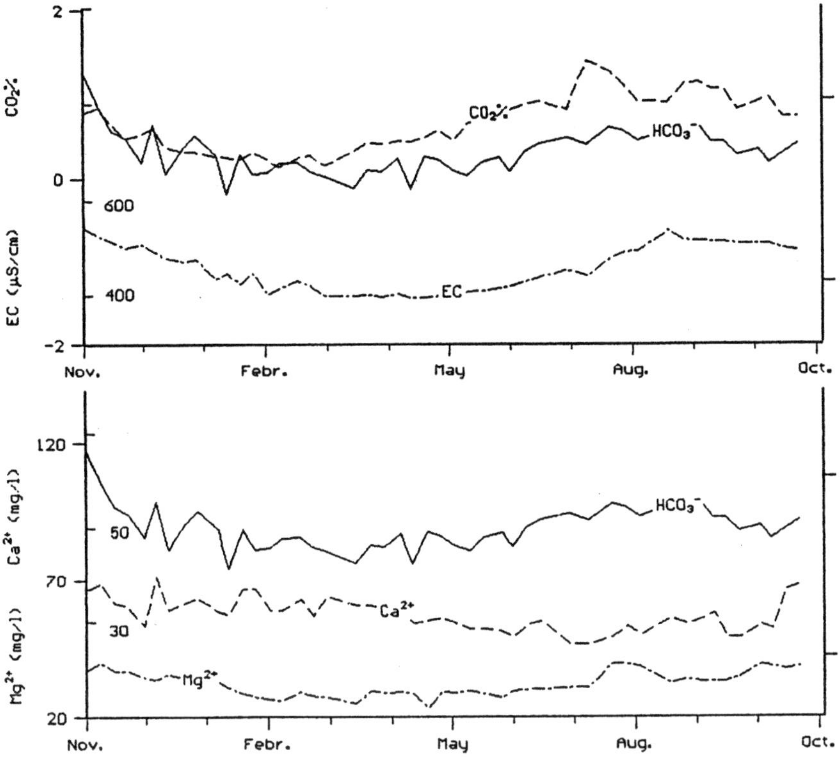

Fig. 3.2.2.4. Variation of CO_2, Ca^{2+}, Mg^{2+} and HCO_3^- contents at a depth of 6 m (grass-covered area; MUC 3)

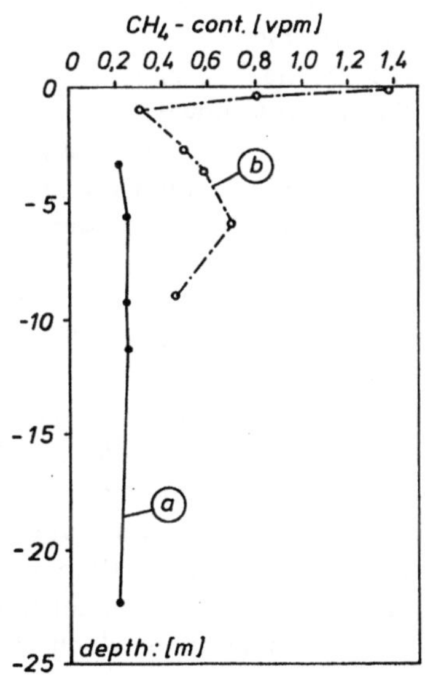

Fig. 3.2.2.5. Average CH_4 content in ground air versus depth. **a** Forest site area (MUC 4); **b** grass-covered site area (MUC 3)

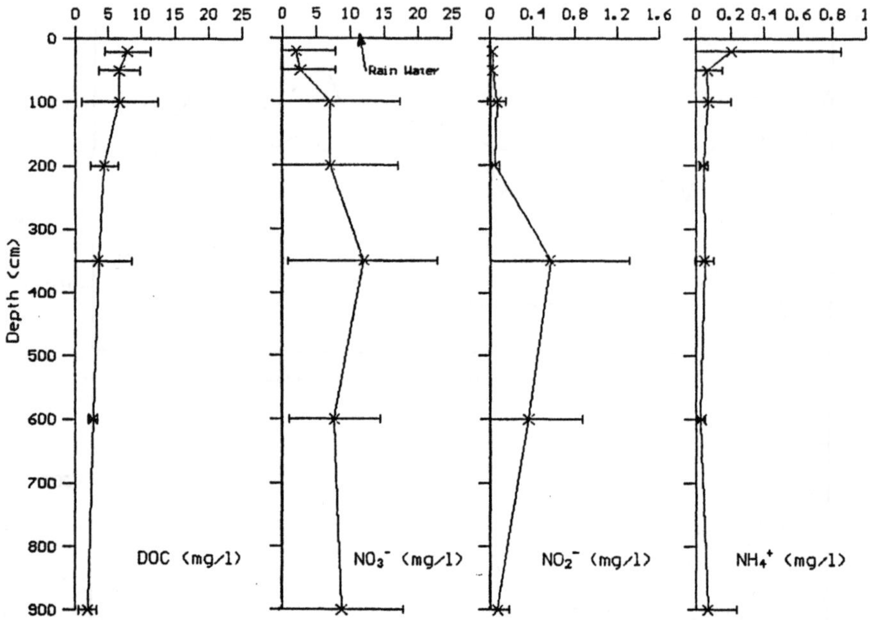

Fig. 3.2.2.6. DOC, nitrate, nitrite and ammonium concentration (mg/l) versus depth. Averages from a time period of 3 years

of 6 m. At a depth of 23 m 90% modern were found with no differences between the measurements in 1979/80 and those from 1983 to 1985. The difference of 15% modern at a depth of 6 m is related to the decreasing ^{14}C content of the atmospheric CO_2 (Dörr and Münnich 1986).

The ^{14}C content of the ground water at the investigation sites was 80 to 85% modern (Fig. 3.2.2.7a). This value does not fit to the values of the unsaturated zone, because the ^{14}C contents of the soil CO_2 were higher than those of the ground water. The 5 to 10% lower ^{14}C activity in the ground water, however can be explained by the increasing thickness of the unsaturated zone (up to 50 m) in the recharge area, where the ^{14}C contents of ground water samples are about 70% modern (Eichinger and Merkel 1982). This supports the assumption that the ground water is a mixture derived from different recharge areas. At the grass covered area the situation is quite similar. The ^{14}C content of the ground water is lower than the ^{14}C contents of the sampled ground air CO_2 from the unsaturated zone (Fig. 3.2.2.7b). This can be interpreted as before.

The $\delta^{13}C$ contents of ground air CO_2 of the two sites show significant differences (Fig. 3.2.2.8): $\delta^{13}C$ = –21.3‰ for the forest area and –23.4‰ for the grass covered area (averages over time and depth). All $\delta^{13}C$ values refer to the PDB standard. At the grass covered area the $\delta^{13}C$ profile of the ground air CO_2 shows maxima at depths of 0.2 m and 9 m (Fig. 3.2.2.8b). At the depth of 0.2 m the $\delta^{13}C$ values are probably influenced by diffusion from atmospheric CO_2 ($\delta^{13}C \approx -8$‰). Pore water samples which were taken between the surface and a depth of 9 m

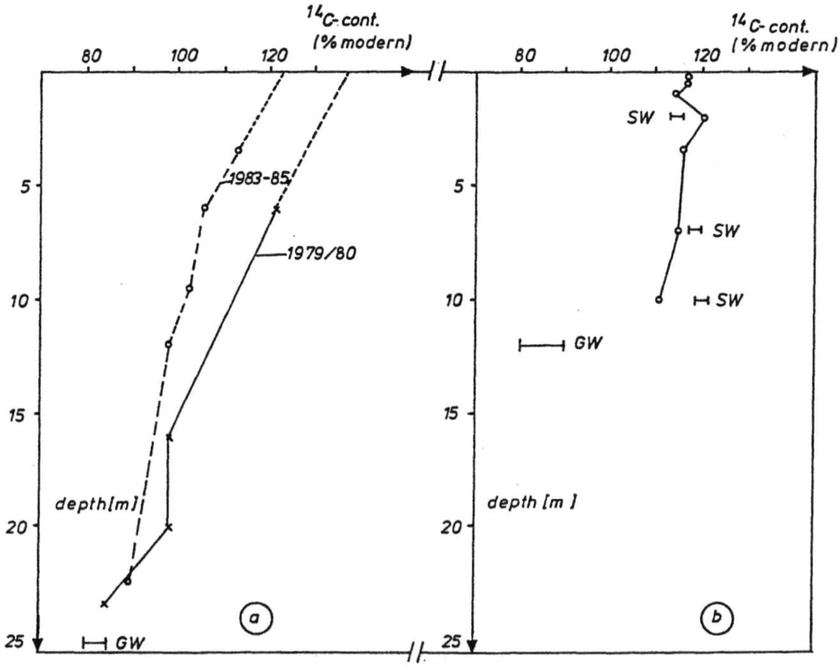

Fig. 3.2.2.7. Average ^{14}C content of ground air CO_2, seepage water (*SW*) and ground water (*GW*). **a** Study area forest (MUC 4), investigation in 1979/80 and 1983–1985, **b** Study area grass (MUC 3), investigations in 1983–1985

showed mean $\delta^{13}C$ values of –15.0 and –13.3‰ for DIC (Fig. 3.2.2.8b). The highest $\delta^{13}C$ values in DIC were measured at a depth of 9 m.

Ground water from different sites within the Munich Gravel Plain had ^{14}C contents of DIC between 70 and 122% modern. Mean residence times (MRT) between 1 and 15 years were estimated from measured ^{3}H contents. Because of the comparable hydrochemical composition of the ground water and similar $\delta^{13}C$ values of the DIC (–14.5 to –12.2‰, PDB) a similar initial ^{14}C content should be expected for all ground water samples. Since the ^{14}C content of atmospheric CO_2 has been higher in the past, it is to be considered that with increasing mean residence time of the ground water (Fig. 3.2.2.9) due to slower seepage or greater thickness of the unsaturated zone, the initial ^{14}C of the DIC has been higher. Initial values from 50 to 100% of the atmospheric ^{14}C content can be deduced comparing the atmospheric CO_2 and ^{14}C content of the DIC. However, this is valid only if there are no differences in the hydrochemical composition and the $\delta^{13}C$ of the DIC, and if an initial content calculated from hydrochemical models does not match with experimental results.

From measurements of the ^{85}Kr content in ground air and assuming equilibrium between ground air and pore water, an average seepage velocity of 2 to 3 m/a (Eichinger et al. 1984) was calculated for the forest site area. Using the results of the ^{14}C measurements in ground air CO_2 and the computed mean residence time of

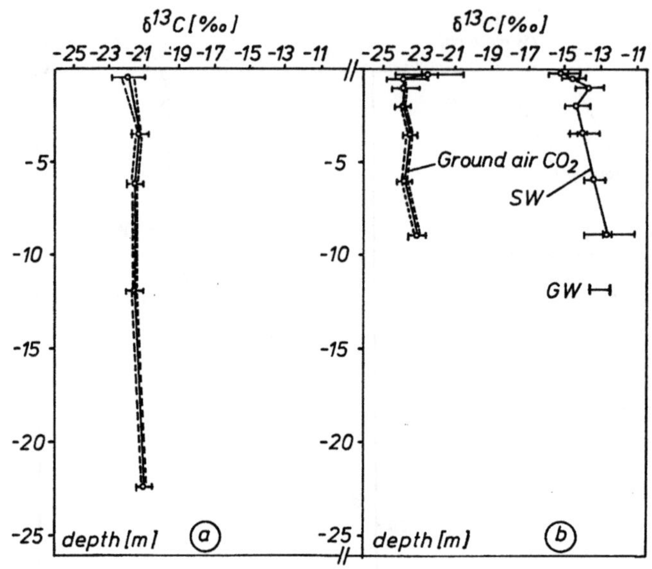

Fig. 3.2.2.8. Average $\delta^{13}C$ content (PDB) of ground air CO_2, seepage water (*SW*) and ground water (*GW*). $\delta^{13}C$-CO_2 values are given with standard deviation of the single measurement and mean value (*dashed lines*). **a** Study area forest (MUC 4), **b** Study area grass-covered site (MUC 3)

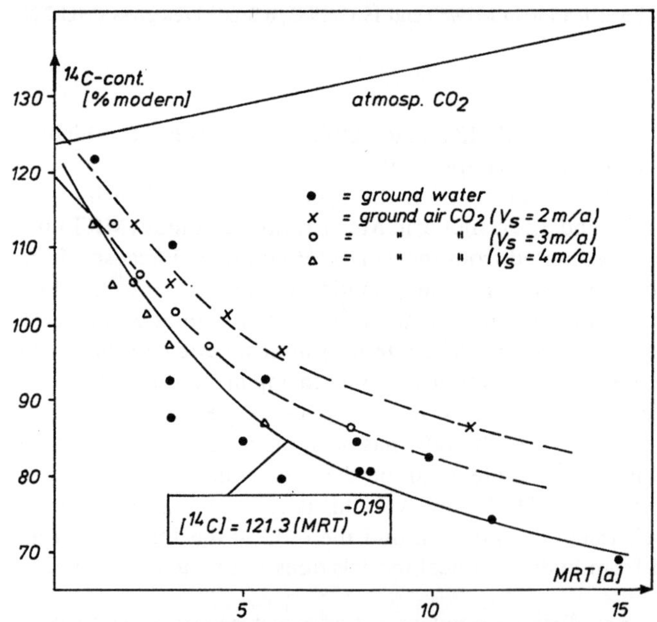

Fig. 3.2.2.9. ^{14}C content of ground water and of ground air CO_2 as function of mean residence time (MRT) compared to the ^{14}C content of atmospheric CO_2. The ground air samples are from different depths of the unsaturated zone (forest site, MUC 4). The MRT of the CO_2 is calculated for different seepage velocities

pore water, which was deduced from the ^{3}H and ^{85}Kr measurements, one obtains a relation between seepage velocity and ^{14}C content of ground air CO_2 (Fig. 3.2.2.9). A good correlation is obtained, if a seepage velocity of 4 m/a is assumed. Thus, the initial ^{14}C content of ground water samples strongly depends on the mean residence time in the unsaturated zone.

Carbon Isotope Exchange

Considerable variations of CO_2 partial pressure were observed down to a depth of 23 m. Thus, the seepage water even in greater depths proves to be both over-saturated and undersaturated with respect to calcite and dolomite. The ^{14}C content of ground air CO_2 is assumed to be in isotopic equilibrium with the DIC of seepage water if an open system is assumed. Thus, the significant ^{14}C decrease of ground air CO_2 with depth provides evidence for dissolution and precipitation of carbonate minerals in the unsaturated zone of the Munich Gravel Plain.

Due to the CO_2 maximum at 0.5 to 1 m depth and decreasing CO_2 concentrations with depth precipitation of carbonate minerals is predicted by thermodynamical modelling (Hoppe et al. 1987). Since precipitated carbonate minerals may be detected from C and O isotope contents, carbonate coatings from the surface of rock material were dissolved as well as the carbonate cement (Nagelfluh), using chemical and mechanical methods. The measurements of the $\delta^{13}C$ and the $\delta^{18}O$ values may be interpreted as an effect of mixing (Fig. 3.2.2.10). One end point

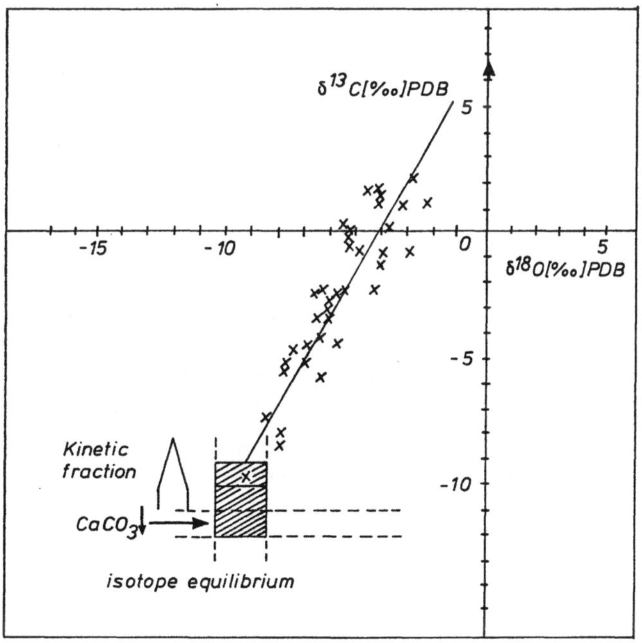

Fig. 3.2.2.10. $\delta^{13}C$ and $\delta^{18}O$ of carbonate minerals. The *hatched area* marks the $\delta^{13}C$ and $\delta^{18}O$ values of carbonates precipitating in isotopic equilibrium with seepage water under kinetic fractionation. $CaCO_3\downarrow$ = precipitated carbonate

represents the primary component close to $\delta^{13}C$ values of +1.0‰ and $\delta^{18}O$ values of –4.0‰. The other end point represents the precipitated calcite. The $\delta^{13}C$ and $\delta^{18}O$ values of this calcite may be deduced from the composition of the seepage water which are close to a $\delta^{13}C$ value of –10‰ and a $\delta^{18}O$ value of –10‰.

Samples of cemented gravels (Nagelfluh) from boreholes in the forest site area were prepared in order to determine their ^{14}C content using a special preparation line. Weak acid was added at a rate of 10 ml per day to the rock material in a closed system. The reaction vessel was filled with 10 l of distilled water and with about 10 kg of the sampled material. The produced CO_2 was flushed with N_2 carrier gas. In three out of six samples ^{14}C was detectable.

The amount of carbonate minerals precipitated in the forest site at a depth from 6 to 22 m was estimated from the average ^{14}C content of ground air CO_2, the mean DIC content in the seepage water (5.5 mmol/l), the mean ground air CO_2 concentration (0.5 kPa), a water content of about 0.05 (volume ratio water/soil) and a porosity of 0.3 to be 30 g/(m^2/a) $CaCO_3$, assuming a seepage velocity of 2 m/a, and to be 15 g/(m^2/a) $CaCO_3$ with a seepage velocity of 4 m/a. Assuming today's climatic conditions for the past 10 000 years, about 0.5 to 2% (of mass) of the total amount of carbonate has been precipitated from seepage water since the deposit of the Quaternary gravel.

3.2.2.4 Conclusion

In carbonate terrains the initial ^{14}C content of ground water is mostly a function of the mean residence time of seepage water in the unsaturated zone, which itself is a function of the thickness of the unsaturated zone and the average seepage velocity. Thus, the initial ^{14}C content is affected by the higher ^{14}C values of the past, and the ^{14}C content of the ground water DIC increases with increasing mean residence time. The decrease of ^{14}C values with depth in the unsaturated zone is due to carbon isotope exchange between ground air CO_2, DIC, and rock carbonate, and mainly to carbonate dissolution and precipitation reactions. Precipitation and dissolution are controlled by changes of CO_2 partial pressure, which were monitored even at greater depths.

The ^{13}C values of ground air CO_2, seepage water and ground water show that at all depths of the unsaturated zone the system is more or less open with respect to CO_2 (Eichinger 1987; Hoppe et al. 1987). ^{14}C content measurements at the forest area show that only minor CO_2 exchanges take place beween the soil zone and deeper layers of the unsaturated zone, which may be due to a cemented carbonate gravel which was encountered at a depth of about 5 to 7 m.

3.2.3 Isotope Geochemistry of the Subsurface Carbonate System in Sandhausen and Bocholt

H. Dörr, W. Leuchs, P. Obermann, W. Regenberg and C. Sonntag

3.2.3.1 Background

The evolution of the carbonate system in the unsaturated and saturated zone is controlled by the carbon flux from the ground air reservoir into pore solution and ground water. The carbon transfer (^{12}C, ^{13}C and ^{14}C) from soil CO_2 to the pore water depends on the partial pressure of CO_2 in the ground air and on the carbonate dissolution conditions (open or closed system).

At a fixed partial pressure of soil CO_2 carbonate dissolution under open system conditions causes a maximal carbon mass- and ^{14}C flux to the ground water, because aqueous CO_2 consumed by dissolution of ^{14}C-free solid carbonate is continuously replenished by modern soil CO_2, and carbon exchange between the total dissolved inorganic carbon species DIC (= $CO_{2,aq}$ + HCO_3^- + CO_3^{2-}) raises the $^{14}C/^{12}C$ isotope ratio of DIC. Isotope fractionation between gaseous CO_2 and DIC is considered by a ^{13}C correction of the $^{14}C/^{12}C$ ratios (Broecker and Olson 1961) to the present ^{14}C level in soil CO_2. The ^{13}C content of the dissolved carbonate species is also controlled by that of soil CO_2. Equilibrium isotope fractionation between gaseous CO_2 and dissolved carbon species causes DIC to be enriched in ^{13}C up to 10‰, depending on the mole fractions of HCO_3^- and $CO_{2,aq}$. The temperature dependence of the isotope fractionation is of minor influence.

Carbonate dissolution under closed system conditions leads to much lower concentrations than in case of the open system. Missing carbon exchange with the dominant soil CO_2 reservoir implies a decrease of the $^{14}C/^{12}C$ and an increase of the $^{13}C/^{12}C$ isotope ratio, due to stoichiometrical mixing of carbon derived from modern, isotopically light soil CO_2, and dead, isotopically heavy carbonates (Pearson and Hanshaw 1970; Mook 1976; Wigley et al. 1978; Fontes and Garnier 1979).

The identification of carbonate dissolution conditions from the isotopical composition of DIC in ground water could, however, be misleading if further carbonate dissolution and precipitation processes take place along the ground water flowpath. Additional dissolution may be caused by biogenic CO_2 produced by microbial oxidation of organic carbon in the aquifer (Sect. 5).

The present field study on the geochemical evolution of dissolved inorganic carbon and its isotopic composition has been carried out in a beech/pine forest at Sandhausen south of Heidelberg (SDH) (Fig. 1.1, Table 1.1, Sect. 4.2.3, this Vol.) and at the shallow Bocholt ground water system (Sect. 4.1.1. and 6.2.2, this Vol.). This study aimed to reconstruct the DIC content, and the $^{14}C/^{12}C$ isotope ratios of ground water from field measurements of the temporal and spatial variations of soil CO_2, and of the spatial distribution of the carbonate dissolution conditions in the recharge area of the investigated ground water system

3.2.3.2 Depth Distribution and Temporal Variation of Soil CO_2

Soil CO_2 is produced by microbial decomposition of soil organic matter and by root respiration. Figure 3.2.3.1 shows four typical CO_2 concentration profiles in the unsaturated zone at Sandhausen (for details of CO_2 production measurements see Dörr and Münnich 1987, 1990). The CO_2 concentration in ground air increases with depth down to the ground water table, which is at a depth of about 4 m below ground. The plotted curves are calculated from the theoretical steady state concentration profile

$$C(z) = c_o\,(1 - e^{-z/\bar{z}}) + c_{atm}, \qquad (3.2.3.1)$$

which is obtained from the diffusion equation under the assumption of a homogeneous soil and an exponentially decreasing CO_2 production rate $Q = Q_o\,e^{-z/\bar{z}}$. c_{atm} is the atmospheric CO_2 concentration, $c_o + c_{atm}$ is the concentration in depth $z >> \bar{z}$. The relaxation depth $\bar{z}$ is the depth at which the CO_2 concentration reaches $(1 - e^{-1})$ of its value at great depth. All curves in Fig. 3.2.3.1 are calculated with a relaxation depth $\bar{z}$ = 10 cm, which is in good agreement with the relaxation depth of the soil organic matter profile (Dörr and Münnich 1986), indicating a major contribution of microbial decomposition to the total soil CO_2 production. CO_2 respiration from living plant roots, which may reach a depth of 1 to 3 m, causes a slight increase of the CO_2 concentration between 50 and 350 cm in late spring and early summer. Calculation of the source strength from CO_2 concentration profiles (Dörr and Münnich 1990) give about 1% contribution from a depth below 50 cm to the total CO_2 production. A soil CO_2 concentration decrease due to carbonate dissolution is practically not observable for two reasons: (a) the annual CO_2 consumption by carbonate dissolution is only about 1% of the annual soil CO_2 production (Sect. 4.2.3, this Vol.) and

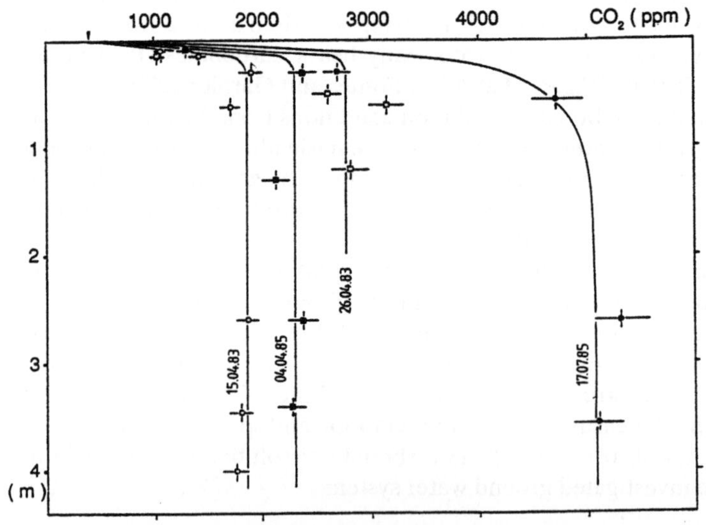

Fig. 3.2.3.1. CO_2 concentration profiles in ground air at Sandhausen/Heidelberg calculated from the theoretical steady-state profile. *Sampling dates* are given in the figure

(b) gaseous diffusion in ground air is fast enough to replenish consumed CO_2 almost instantaneously.

The latter is evident from a comparison of the time constants for carbonate dissolution and for replenishing CO_2 by molecular diffusion in ground air. The diffusion relaxation time $\tau_{diff} = L^2/(6D)$ (Carslaw and Jäger 1959), i.e. the time needed to restore steady state conditions over a length L = 1 m (maximum distance over which carbonate dissolution takes place) is τ_{diff} = 7 h (D = 0.1 cm^2/s). The time constant of carbonate dissolution $\tau_{diss} = I/R_c$, where I is the DIC inventory in the soil column, and R_c is the carbonate dissolution rate, is τ_{diss} 28 days ($I = c_{eq} \cdot L \cdot W = 3 \times 10^2$ mmol/cm^2, c_{eq} = 3 mmol/l at p_{CO_2} = 2500 ppm (vol.), 1 = 1 m, soil moisture w = 10% vol. and $R_c \approx 10^{-8}$ mmol/($cm^2 \cdot$ s) at a $CaCO_3$ content in the soil of about 10% (Plummer et al. 1979; Sect. 3.1.1, this Vol.). This rough estimate indicates that gaseous diffusion of CO_2 in ground air is not the limiting factor in carbonate dissolution processes.

Figure 3.2.3.2 shows the variation in soil temperature, soil CO_2 production and soil CO_2 concentration between 1983 and 1988. Soil CO_2 production is strongly correlated with soil temperature. The effect of soil moisture on CO_2 production is dominant only during extremely dry and wet periods. The annual cycle of the soil CO_2 concentration is much less pronounced than the annual cycle of soil CO_2 production. This is due to the fact that the CO_2 concentration depends not only on the CO_2 production rate, but also on the soil permeability, and thus on the soil moisture content. The effect of soil moisture on CO_2 production and soil permeability is discussed in detail by Dörr and Münnich (1987, 1990).

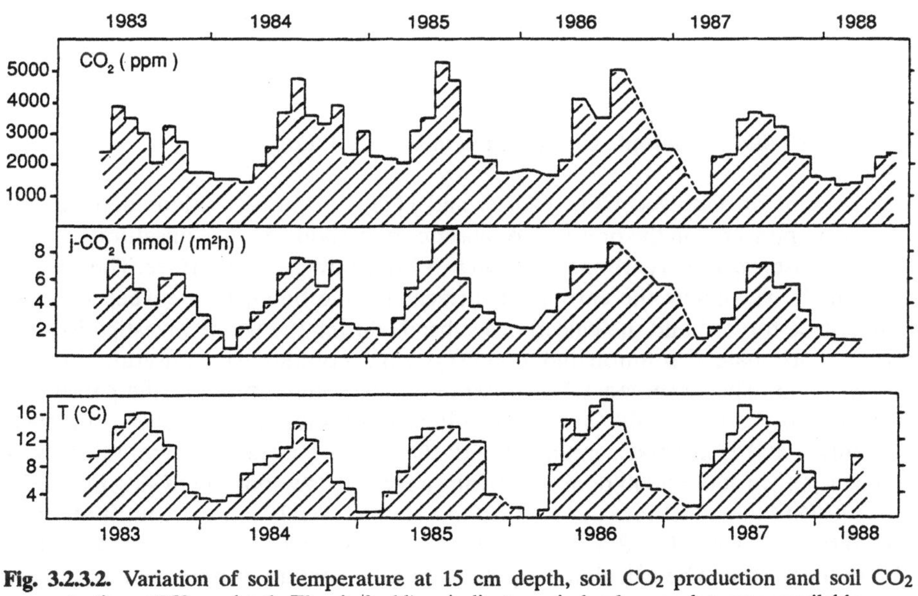

Fig. 3.2.3.2. Variation of soil temperature at 15 cm depth, soil CO_2 production and soil CO_2 concentration at 350 cm depth. The *dashed lines* indicate periods when no data were available

3.2.3.3 Isotopic Composition of Ground Air CO_2

The $^{13}C/^{12}C$ isotope ratio of ground air CO_2 is constant with depth and shows no significant temporal variations (Dörr and Münnich 1980). The average $^{13}C/^{12}C$ isotope ratio of ground air CO_2, obtained from seven depth profiles taken at Sandhausen between April 1985 and May 1986, is $\delta^{13}C_g = -20.2 \pm 1.1\%$ (PDB). The average composition of soil respiration CO_2 at this site, i.e. the CO_2 that diffuses out of the soil into the atmosphere, is $\delta^{13}C_{resp} = -24.3 \pm 0.9‰$ and agrees with the $\delta^{13}C$ of the decomposed organic material from which it is derived ($\delta^{13}C$ of the 1983/84 litter fall was $\delta^{13}C_{org} = -25.0 \pm 0.1‰$). Ground air CO_2 is richer in $\delta^{13}C$ than CO_2 diffusing out of the soil by about 4‰, due to a 4‰ difference in the molecular diffusion coefficients of $^{13}CO_2$ and $^{12}CO_2$. This isotope fractionation between ground air CO_2 and soil respiration CO_2, which has been already observed at other locations, is explained in more detail by Dörr and Münnich (1980).

The ^{14}C content of soil respiration CO_2 depends on the contribution of root respiration and microbial decomposition of soil organic matter to the total soil CO_2 production. While root respiration CO_2 can be assumed to be nearly identical with the atmospheric ^{14}C level, the ^{14}C content of decompositional CO_2 may, depending on the carbon residence time in the soil, be depleted by radioactive decay or be even enriched by the contribution of "bomb ^{14}C". The measured ^{14}C content of soil respiration CO_2 shows a seasonal cycle with a near atmospheric ^{14}C level in summer and a 10% lower ^{14}C content in winter (isotope fractionation effects are excluded by the normalization of measured $^{14}C/^{12}C$ isotope ratios to $\delta^{13}C = -25\%$ (Broecker and Olson 1961)). The observed annual variation of the ^{14}C content of soil respiration CO_2 is explained by the seasonal varying contribution of a fast (turnover time $\tau_t = 1$ a) and a slowly ($\tau_t = 100$ a) decomposing soil organic carbon reservoir (Dörr and Münnich 1986).

3.2.3.4 DIC and Its Isotopic Composition in Pore Solution

The amount of dissolved CO_2 (CO_{2aq}) and HCO_3^- in pore solutions depends on ground air p_{CO_2} and on the presence or absence of carbonates. The unsaturated zone at Sandhausen is free of carbonates down to 3.2 m below ground surface. The ground water table was below the carbonate surface (i.e. the boundary between the carbonate-free top soil and the calcareous subsoil) during the whole investigation, indicating carbonate dissolution under open system conditions in the unsaturated zone below 3.2 m. The observed DIC content in pore solutions above the carbonate surface of 0.4 ± 0.2 mmol/l (average of nine samples taken between April, 1984, and July, 1984) is in good agreement with the calculated 0.2–0.3 mmol/l dissolved CO_2 at a p_{CO_2} of 4000–5000 ppm (vol.) using a Henry's law constant of $-\log K_H = 1.27$ at 10 °C (Stumm and Morgan 1981). In this simple calculation the contribution of HCO_3^- to DIC of less than 8% at a measured pH between 4.2 and 5.2 (Sect. 4.2.3, this Vol.) is neglected. The calculated equilibrium DIC content in pore solution below the carbonate surface (assuming open system carbonate dissolution) during the winter period, when ground water recharge mainly occurs, is 3.0 mmol/l [average winter $p_{CO_2} = 2500$ ppm (vol.)].

The measured average ^{13}C content of the dissolved inorganic carbon species of $\delta^{13}C_{DIC} = -23.8 \pm 2.5‰$ is comparable with the expected value of −21.4‰ if an isotope fractionation of −1.1‰ between gaseous and dissolved CO_2 (Vogel et al. 1970) is considered. To calculate the $\delta^{13}C$ of DIC in pore solution below the carbonate surface (open system carbonate dissolution), the mole fractions of HCO_3^- and CO_{2aq} and a further isotope fractionation between dissolved CO_2 and HCO_3^- of 9.6‰ (Mook et al. 1974) have to be considered.

3.2.3.4 DIC, $\delta^{13}C$ and ^{14}C Depth Distribution in Ground Water

The evolution of the carbonate system and its isotopic composition in ground water has been studied in more detail in the shallow Bocholt aquifer. The geohydraulic and hydrochemical conditions at this site are described in Sections 4.1.1, 6.2.2, and 6.2.3 (this Vol.). The shallow ground water can be considered to be modern ground water in the sense of ^{14}C dating, i.e. having experienced only negligible ^{14}C loss by radioactive decay. Thus, ^{14}C can be used as an additional tracer in studying the evolution of the carbonate system in ground water.

Figure 3.2.3.3 shows vertical distributions of ^{14}C and $\delta^{13}C$ across the ground water system at the location of the multi-level wells BOC1 and BOC4 (Fig. 3.2.3.4). The ^{13}C and decreasing ^{14}C with depth in the upper part of the profile is caused by carbonate dissolution under closed system conditions, as can be seen from the increase of DIC (Fig. 3.2.3.5). ^{14}C contents of approximately 40 pmc only at 4 m below ground surface at BOC1 and at 10 m at BOC4 are extraordinarily low, since bomb tritium in this layer suggests the presence of bomb-^{14}C, equivalent to a soil CO_2 ^{14}C level of about 120 pmc. These observed ^{14}C contents are thus much lower than expected for carbonate dissolution under closed system conditions, which should yield a ^{14}C content of about 60 pmc, if the ^{14}C of ground air CO_2 is assumed to be 120 pmc. The $\delta^{13}C$ data ranging between −12‰ and −15‰ suggest additional carbonate dissolution caused by dead CO_2 from microbial oxidation of fossil organic carbon of the ground water.

At depths below 15 m the ^{13}C profiles practically coincide, and the ^{14}C profiles are close to approximately 65 pmc down to about 36 m, where a clay layer defines the base of the aquifer at BOC1. The BOC4 profiles extend further down to 49 m without reaching the aquifer base. Along this lower section of BOC4, ^{14}C decreases from 65 pmc to 46 pmc, whereas $\delta^{13}C$ increases from −15‰ to −13‰. This decrease of the $^{14}C/^{12}C$ ratio may partly be due to radioactive decay, since the estimated ground water ages are in the order of magnitude of some 10^2 to 10^3 years (Sect. 6.2.4, this Vol.).

Between a depth of 7 m and 15 m below ground surface ^{13}C is more depleted and the ^{14}C content is higher at BOC1 than at BOC4. This indicates a larger contribution of carbon species dissolved under closed system conditions in the recharge area of BOC4, while in the recharge area of BOC1 open system conditions seem to prevail. This is also indicated by a higher DIC content in the considered ground water layer at BOC1 than at BOC4 (Fig. 3.2.3.5).

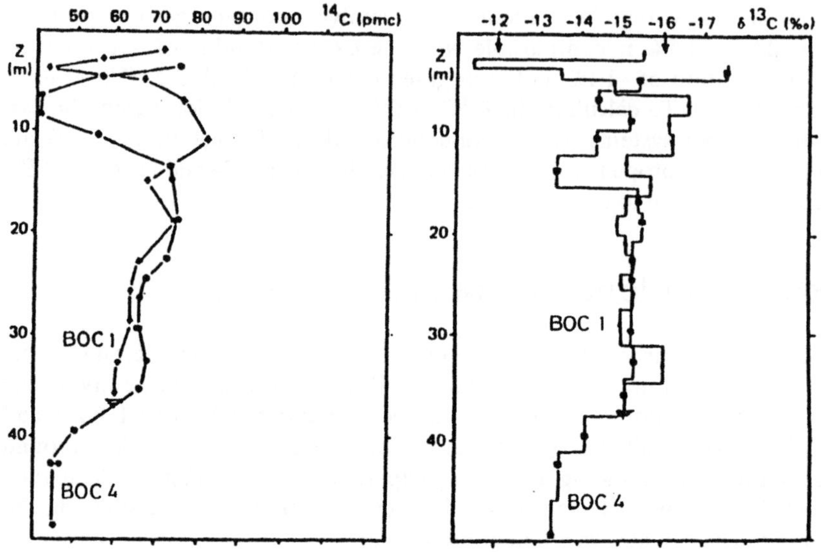

Fig. 3.2.3.3. Depth distribution of ^{14}C and ^{13}C at the wells BOC1 and BOC4. Vertical variations in the ^{13}C profiles are related to a spatial variability of carbonate dissolution conditions in the recharge area. The *arrows* indicate the expected limits of ^{13}C for carbonate dissolution under open ($\delta^{13}C$ = −16‰) and under closed ($\delta^{13}C$ = −12‰) system conditions

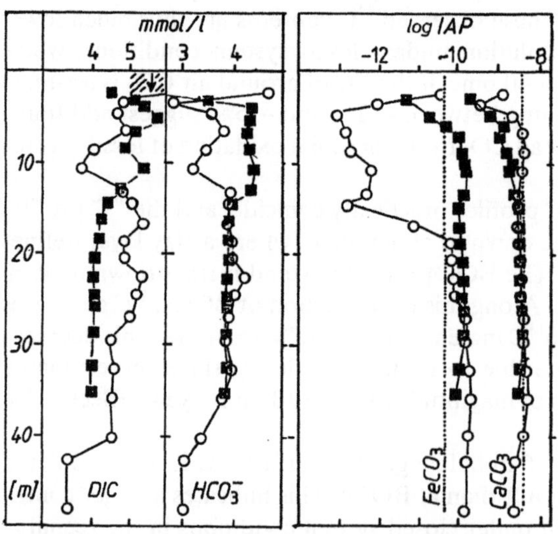

Fig. 3.2.3.4. Depth distribution of total dissolved organic carbon (DIC) and HCO_3^- and the ion activity products for $CaCO_3$ and $FeCO_3$. BOC1: *full squares*; BOC4: *open circles*

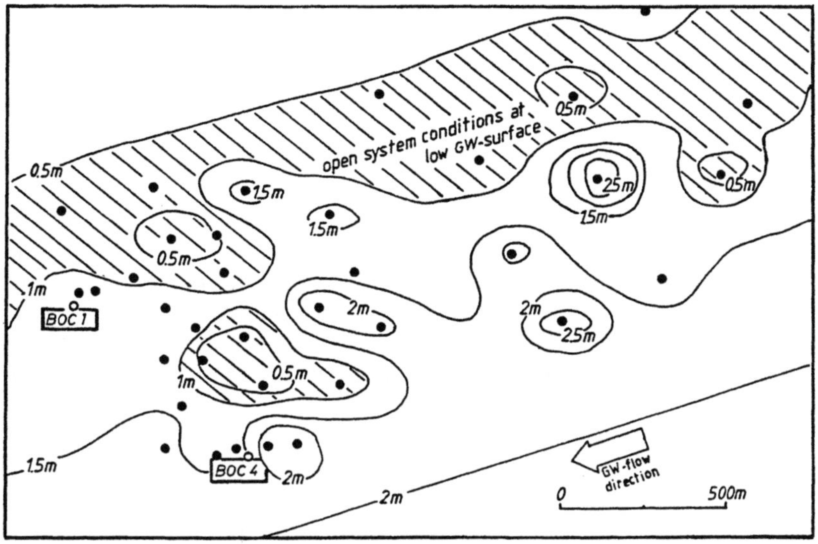

Fig. 3.2.3.5. Difference between the depth of the solid carbonate surface and of the ground water surface. *Positive numbers* indicate regions with closed system carbonate dissolution

3.2.3.6 Spatial Variability of Carbonate Dissolution Conditions

For reasonable model simulations of the carbon isotope profiles of BOC1 and BOC4 information on the geochemical conditions for carbonate dissolution in the recharge area upgradient from these wells is needed. This information was obtained from measurements of the soil CO_2 content and of the position of the carbonate- and ground water surface. Carbonate dissolution under open system conditions is expected to occur at those locations where the carbonate surface is situated above the ground water table, so that solid carbonate is exposed to gaseous soil CO_2. In the opposite case, when the carbonate surface lies well below the ground water surface, dissolution may occur under closed system conditions. Thus, the position of the carbonate surface in comparison to the ground water table is a proper parameter to identify locations, where open or closed system conditions may prevail.

Figure 3.2.3.4 shows a map which has been obtained by analysis of 35 spot measurements in the infiltration area. The positive numbers indicate closed system condition, i.e. the carbonate surface is found at greater depth than the ground water surface, whereas negative numbers denote open system conditions. This pattern is, of course, subjected to seasonal and climatic variations of the ground water surface and thus highly time dependent. Figure 3.2.3.4 refers to October 1985, when the ground water level was unusually high.

Nevertheless, Fig. 3.2.3.4 is very instructive. It shows clearly a change from small differences between carbonate and ground water surface upgrade of BOC1 to larger differences upgrade of BOC4. Due to temporal variations of the ground water table by about + 1 m, carbonate dissolution will occur more often under open system conditions (shaded areas in Fig. 3.2.3.4) in the recharge area of BOC1 than in the recharge area of BOC4.

This is confirmed by isotope measurements of four ground water samples taken close to the ground water surface in the recharge area in December 1984. The ^{14}C and ^{13}C content in ground water from regions where open system conditions prevail is 117 pmc and $\delta^{13}C = -19.2‰$, respectively. In regions where closed system carbonate dissolution occurs the ^{14}C and ^{13}C content in ground water is 57 pmc and $\delta^{13}C = -12.6‰$, respectively, as expected from carbon isotope models (Fontes and Garnier 1979).

The variation of ^{13}C and ^{14}C in the ground water between 7 m and 15 m below ground surface of the wells BOC1 and BOC4 can thus be related to spatial variations of the carbonate dissolution conditions in the recharge area and to time variations of the water table position. This is consistent with the DIC content (Fig. 3.2.3.5) at both wells. Soil CO_2 concentrations in the recharge area range from 0.9 to 2.7% vol. with a mean value of 1.5% vol. The samples were taken in October and December, and, by comparison with the annual variation of soil p_{CO_2} at Sandhausen, are supposed to represent the CO_2 concentration and isotopic composition of soil CO_2 during the whole period of ground water recharge. The DIC content of ground water, adjusted to the mean CO_2 content in ground air, is calculated to be 5.7 mmol/l assuming open system conditions and 1.5 mmol/l in case of closed system carbonate dissolution. The DIC contents at BOC1 and BOC4 range between these limiting values (Fig. 3.2.3.5), indicating a predominance of carbon species derived under open system conditions between 7 m and 15 m at BOC1, while at BOC4 closed system conditions should be more frequent.

$\delta^{13}C$ of soil CO_2, measured at seven sites in the infiltration area, varies between –19‰ and –29‰, showing a mean value of $\delta^{13}C_g = -23.5‰$, $\delta^{13}C$ of solid carbonates sampled close to the ground water surface lies between + 0.2‰ and 1.0‰ with an average $\delta^{13}C_c = 0.6‰$. The ^{14}C content of soil CO_2 is 117 pmc.

$\delta^{13}C$ of DIC expected from the isotopic composition of soil CO_2 and solid carbonate in the recharge area are $\delta^{13}C_{open} = -16‰$ and $\delta^{13}C_{closed} = -12‰$, respectively. These values are indicated by arrows in Fig. 3.2.3.3. $\delta^{13}C$ in ground water is closer to –16‰ than to –12‰, which is due to the predominance of carbon species dissolved under open system conditions. The contribution of carbon species dissolved under open system conditions is estimated from a mass- and ^{13}C balance to be 65% (BOC4) and 95% (BOC1) above 15 m and about 88% (BOC1 and BOC4) below 15 m.

The ^{14}C depth distribution at BOC1 and BOC4 cannot be explained by the spatial variability of the carbonate dissolution conditions alone. Commonly used model predictions for the initial ^{14}C content differ considerably from the measured ^{14}C profiles. The observed difference, however, could be related to an additional carbonate dissolution by oxidized fossil organic carbon. It is supposed, that this microbial oxidation of organic carbon is stimulated by an anthropogenic input of microbially decomposable chemical constituents like NO_3^- and SO_4^{2-} to the ground water.

But even in deeper and thus "unpolluted" ground water the ^{14}C content is influenced by oxidized dead organic carbon. Stoichiometric microbial oxidation of organic carbon associated with nitrate and sulfate reduction is estimated to produce a total of about 1.8 mmol/l dissolved carbon species in the upper part of the profile and about 1 mmol/l in depth below 10 m. The ^{14}C profile at BOC4 can be explained by a mixing of carbon species derived from carbonate dissolution under open and under closed system conditions, and of oxidized dead organic carbon (Dörr et al. 1987).

3.2.3.7 Conclusions

Measurements of the CO_2 concentration and its isotopic composition in ground air are used to determine the evolution of dissolved inorganic carbon species in pore solution and ground water. The vertical variation of the ^{14}C and ^{13}C content in the Bocholt ground water system is partly related to spatial variations of the hydrochemical conditions for the evolution of ground water carbonate equilibrium. The occurrence of open and closed system conditions for carbonate dissolution in the recharge area is identified by mapping the position of the carbonate surface in relation to the ground water table.

3.2.4 Application of Stable Carbon and Sulfur Isotope Models to the Development of Ground Water in a Limestone-Dolomite-Anhydrite-Gypsum Area

K.W. Schaefer and E. Usdowski

3.2.4.1 Background

The most abundant carbonate rocks, limestones and dolomites, are an important reservoir in the geochemical cycle of carbon. This cycle is strongly controlled by the mechanisms of surface and subsurface rock dissolution. In principle, rain water takes up biogenic CO_2 in the soil zone and reacts with the underlying carbonate rock. Taking calcite as an example the following overall reactions occur

$$CO_2 + H_2O \rightleftharpoons H_2CO_3 \tag{3.2.4.1}$$

$$CaCO_3 + H_2CO_3 \rightleftharpoons Ca^{2} + 2HCO_3^- . \tag{3.2.4.2}$$

Recently, sulfuric acid from atmospheric pollution is enhancing the dissolution according to the reaction

$$2CaCO_3 + H_2SO_4 \rightleftharpoons 2Ca^{2+} + 2HCO_3^- + SO_4^{2-} . \tag{3.2.4.3}$$

Garrels and Christ (1965) have pointed out that the reaction of carbonates with dissolved CO_2 is bracketed by two cases: in the open system the dissolution occurs at a fixed partial pressure of CO_2, whereas in the closed system water is saturated with CO_2 at a given partial pressure and reacts with the solid after its separation from the gas phase. In order to evaluate a hydrogeochemical system it is necessary to establish its position with respect to the limiting cases. Deines et al. (1974) have shown that the stable isotopes of carbon have to be considered for this purpose. These models have been refined by Kirchhoff (1985), and Schaefer and Usdowski (1987), who accounted for the influence of reaction (3.2.4.3) on the major element and carbon isotope composition of a carbonate ground water. The stable isotopes of sulfur are used to distinguish between sulfate derived from atmospheric pollution or the dissolution of anhydrite and/or gympsum.

3.2.4.2 Theory

With respect to the open system, the isotopic composition of the dissolved carbonate, $\delta^{13}C_T$, is controlled at any time during the dissolution of the solid by the isotopic composition of the gas phase, $\delta^{13}C_g$. This is expressed by the equation

$$\delta^{13}C_T = \alpha_T (\delta^{13}C_g + 10^3) - 10^3. \tag{3.2.4.4}$$

The overall partition coefficient, α_T, is defined by the equation

$$\alpha_T = \alpha_f \cdot x_f + \alpha_b \cdot x_b + \alpha_c \cdot x_c, \tag{3.2.4.5}$$

where α's and x's refer to the individual partition coefficients and mole fractions respectively, and the subscripts indicate chemical species, i.e. $f = H_2CO_3 + CO_{2(aq)}$, $b = HCO_3^-$, and $c = CO_3^{2-}$. As the mole fractions of the carbonate species depend upon the pH value of the solution, $\delta^{13}C_T$ can be calculated from measured pH and $\delta^{13}C_g$.

With respect to the closed system, the isotopic composition of the dissolved carbonate is a mixture of the isotopic compositions of the initial solution, $\delta^{13}C_o$, and of the dissolved solid carbonates, $\delta^{13}C_s$, in the proportion of the two carbon sources. This is expressed by

$$\delta^{13}C_T = \delta^{13}C_o \frac{\Sigma C_o}{\Sigma C} + \delta^{13}C_s \left(1 - \frac{\Sigma C_o}{\Sigma C}\right), \tag{3.2.4.6}$$

where $\Sigma C = (CO_2)_{aq} + (H_2CO_3) + (HCO_3^-) + (CO_3^{2-})$ is the dissolved carbonate at any time of the dissolution, and the subscript zero refers to initial conditions. The quantity ΣC is measured, and the quantity ΣC_o is calculated with the equation

$$2\Sigma C_o = \Sigma (2x_f + x_b) + (H^+) - (OH^-) + 2(SO_4^{2-})_a, \tag{3.2.4.7}$$

where (H^+) and (OH^-) are derived from pH, and $(SO_4^{2-})_a$ is the sulfate derived from acid rain (Schaefer and Usdowski 1987). This quantity is obtained with the equation

$$\delta^{34}S_T = \delta^{34}S_a \frac{(SO_4^{2-})_a}{(SO_4^{2-})_T} + \delta^{34}S_s \left(1 - \frac{(SO_4^{2-})_a}{(SO_4^{2-})_T}\right), \tag{3.2.4.8}$$

were $\delta^{13}S_T$, $\delta^{34}S_a$, and $\delta^{34}S_s$ are the measured sulfur isotope compositions of the ground water, the acid rain, the gypsum and/or anhydrite, and $(SO_4^{2-})_T$ is the sulfate concentration of the ground water.

3.2.4.3 Geohydrology

The study area is the Zechstein plateau between Osterode and Förste (FOR), located at the west border of the Harz mountains (Table 1.1, Figs. 1.1 and 3.2.4.1). It is bounded in the west and the east by deep valleys which meet each other in the north of the area. The plateau consists mainly of dolomitic limestones of the Zechstein-2 formation and is capped partially with loess. The fractured surface layers of the carbonate rocks are underlain by Zechstein-2 anhydrite which is about 500 m thick and which has been partially transformed into gypsum. Shallow ground water emerges at springs or in wells at fractures of the Zechstein-2 dolomitic limestone layers near the top or at the flanks of the plateau. The area is drained by the river Söse. Close to Förste several other springs emerge at the intersection of two tectonic lines of the rheno-hercynic fault system, running SSW-NNE and ESE-NWE, respectively. In this area water from the plateau mixes with brines which are thought to migrate from the region of Echte, located about 10 km northwest of Förste, according to unpublished reports.

3.2.4.4 Geochemistry

Major Components

Table 3.2.4.1 shows minimum and maximum values measured during a period of 1 year in 1984/1985. The springs Forellenteich (FOR1), Grafenquelle (FOR2), and Heilquelle (FOR3) were sampled at monthly intervals. Measurements at Feldbrunnen (FOR4) were made at intervals of 2 months, beginning with July 1984.

The FOR4 water is generated in the dolomitic limestone and emerges near the top of the Zechstein plateau. It is undersaturated to saturated with respect to calcite, undersaturated with respect to dolomite, and extremely undersaturated with respect

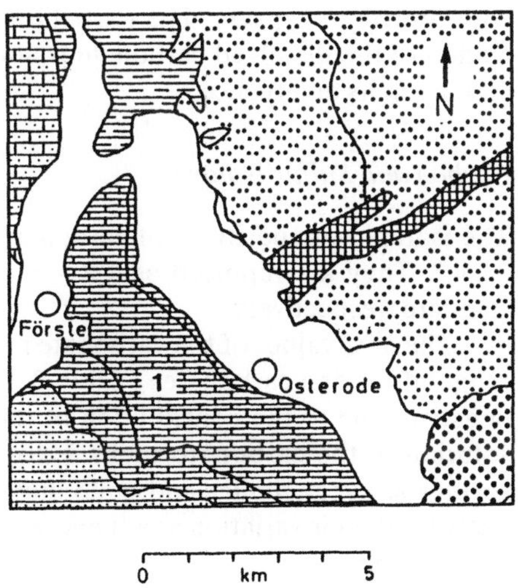

Fig. 3.2.4.1. The Zechstein plateau between Osterode and Förste (*1*). *Stippled areas* various silicate rocks of the Harz mountains. *Brick pattern* carbonate rocks (Jordan 1976). Location of springs: FOR1 is about 2 km south of Förste, and FOR2 and FOR3 are about 1 km north of Förste, FOR4 is in the area marked *1*

Table 3.2.4.1. Variations of major components in 1984/1985

mg/l	FOR1	FOR2	FOR3	FOR4
Ca^{2+}	394.8 - 413.8	260.5 - 272.5	482.0 - 595.2	92.7 - 96.2
Mg^{2+}	40.1 - 52.9	30.4 - 37.1	74.8 - 113.1	39.3 - 41.3
Na^+	5.6 - 9.1	140.8 - 174.6	1465.0 -2429.9	6.9 - 7.8
K^+	0.9 - 1.8	1.9 - 4.2	24.5 - 38.0	3.3 - 3.5
Cl^-	26.1 - 30.8	227.8 - 303.9	2224.7 -3897.6	23.8 - 27.7
SO_4^{2-}	843.0 - 866.1	511.2 - 545.8	1300.9 -1522.4	110.2 - 122.8
HCO_3^-	270.5 - 275.3	237.6 - 240.6	252.8 - 258.1	276.4 - 280.3
NO_3^-	10.3 - 37.0	11.3 - 37.0	<2.6	38.6 - 59.8
t(°C)	8.8 - 11.3	8.4 - 10.1	11.4 - 13.0	8.8 - 8.9
pH	7.29 - 7.45	7.19 - 7.35	7.19 - 7.29	7.23 - 7.37
SI_c	0.38 - 0.52	0.12 - 0.26	0.17 - 0.36	(-0.12)- 0.004
SI_d	(-0.02) - 0.09	(-0.28)- (-0.14)	(-0.04)- 0.06	(-0.28)- (-0.14)
SI_g	(-0.28) - (-0.26)	(-0.62)- (-0.60)	(-0.29)- (-0.24)	-1.52
-log p_{CO_2}	2.15 - 1.94	2.09 - 1.94	2.06 - 1.90	2.03 - 1.89

(SI_c, SI_d, and SI_g: saturation indices with respect to calcite dolomite, and gypsum).

to gypsum. The relatively high concentration of nitrate is due to the use of fertilizers in the catchment. The FOR1 water emerges at the eastern margin of Förste. Its chemical composition is similar to that of FOR4. However, the concentrations of calcium and sulfate are much higher, demonstrating the effect of the dissolution of the $CaSO_4$ minerals as the water migrates downward from the top of the plateau. The FOR3 water is a brine emerging from a borehole of a depth of about 25 m. It may be concluded that it is generated from the subsurface dissolution of Zechstein evaporites containing halite, sylvite, magnesium sulfate and chloride, and calcium sulfate. The FOR2 water appears to be a mixture of the brine with ground water from the plateau.

The FOR1, FOR2 and FOR3 waters are moderately supersaturated with respect to calcite, undersaturated to saturated with respect to dolomite, and undersaturated with respect to gypsum. The partial pressure of CO_2 corresponding to the concentration of the dissolved carbonate is about 1 kPa, showing that any water degasses CO_2 upon its emergence.

Carbon and Sulfur Isotopes

The isotopic compositions of carbon and sulfur are compiled in Tables 3.2.4.2 and 3.2.4.3. They are reported as $\delta^{13}C$- and $\delta^{34}S$ values relative to the PDB- and CD standard, respectively.

The $\delta^{13}C$ values of the carbonate rocks (limestone and dolomite) are within the range of marine carbonates (Table 3.2.4.2). Those of the dissolved carbonate are negative, due to the contribution of the isotopically light soil CO_2 during the dissolution of the carbonate rocks. Except for FOR3, the $\delta^{13}C$ values of the other springs are very similar and quite uniform. The distinctly more positive values for FOR3 and their variation as a function of time are reflecting the complicated history of this brine.

Table 3.2.4.2. $\delta^{13}C$ values (‰, PDB) of Zechstein-2 rocks, ground water and soil CO_2

	Average	Max	Min	No. of samples
Dolomite	5.9	6.4	5.3	8
Calcite	4.8	5.3	4.7	5
FOR1	−11.4	−10.8	−12.3	12
FOR2	−10.5	−10.0	−11.1	12
FOR3	− 7.7	− 4.4	− 8.9	11
FOR4	−11.9	−11.8	−12.0	3
Soil CO_2	−25.9	−24.0	−26.6	5

Table 3.2.4.3 $\delta^{34}S$ values (‰, CD) of Zechstein 1–4 rocks, ground water, rain and snow

	Average	Max	Min	No. of samples
Zechstein				
1, gypsum, anhydrite	11.2	11.4	9.9	7
2, gypsum	10.7	11.4	9.9	4
3, gypsum	11.9	12.4	11.5	6
4, gypsum	11.5	–	–	1
Water				
FOR1	11.0	11.4	10.6	12
FOR2	11.0	11.3	10.8	12
FOR3	11.2	11.6	10.9	12
FOR4	8.3	8.5	8.2	3
Rain, snow	4.5	6.0	3.0	6

The $\delta^{34}S$ values of the gypsum and anhydrite rocks vary little (Table 3.2.4.3). Those of the FOR1, FOR2 and FOR3 ground waters are very similar to the values of the host rocks. The $\delta^{34}S$ values of the more shallow FOR4 ground water reflect an isotopic mixture between the atmospheric sulfate and the dissolved solids. From the values of Table 3.2.4.3 and an average sulfate concentration of 115.8 mg/l, it can be calculated that about 50 mg/l or 44% of the sulfate is derived from the atmosphere. Similar results have been obtained by Kirchhoff (1985). These results are comparable to those from chemical balance computations using rain, pore and ground water compositions which yield about 66 mg/l atmospheric sulfate calculated by Kirchhoff (1985), from the data of Schulz (1970), Beese et al. (1975), and Kühn and Weller (1977).

3.2.4.5 Model Considerations

The results of model calculations are shown in Figs. 3.2.4.2 and 3.2.4.3. It can be seen that in none of the cases the dissolution of the carbonate rocks occurs under open system conditions. With respect to FOR1 and FOR2 the closed system is most nearly attained. The more shallow FOR4 shows intermediate conditions, as expected, with an estimated contribution of the closed system of about 70%. This demonstrates that the conditions of carbonate dissolution are changing as water

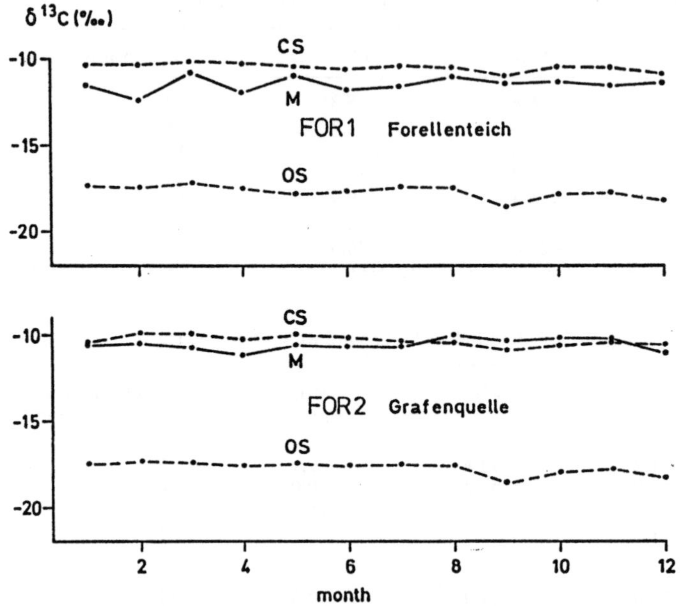

Fig. 3.2.4.2. Annual variations of the δ^{13}C-values of the dissolved carbonate in 1984. *CS* and *OS* closed system and open system model calculations, respectively. *M* measured values

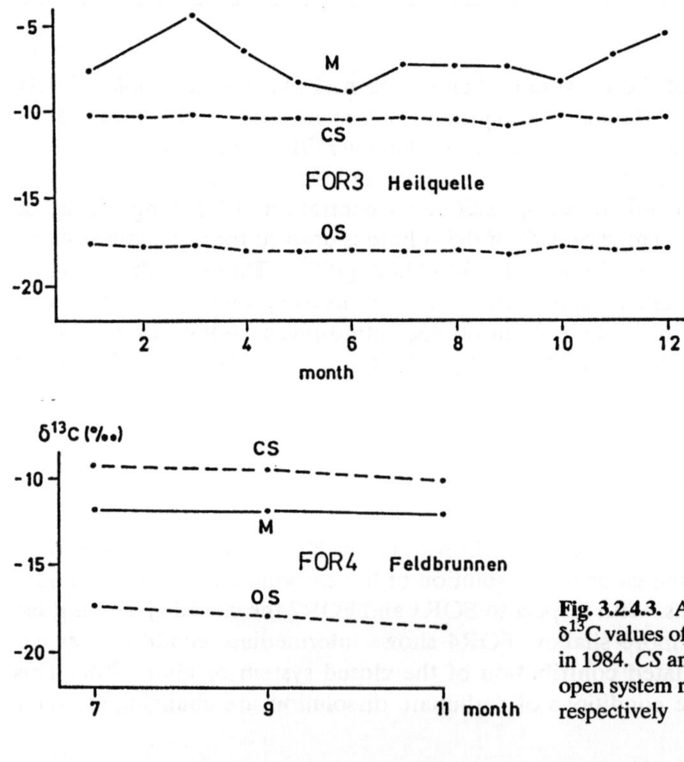

Fig. 3.2.4.3. Annual variations of the δ^{13}C values of the dissolved carbonate in 1984. *CS* and *OS* closed system and open system model calculations, respectively

migrates downward in an aquifer. The FOR3 brine gives anomalous results. The measured $\delta^{13}C$ values are beyond those of the limiting cases. It should be pointed out that in the calculations with respect to this water the $\delta^{13}C$ values of the above mentioned Zechstein-2 carbonates have been used. As a matter of fact the recharge area of the brine, and consequently the isotopic composition of the carbonate rock in question, are not known. There is, however, some evidence that isotopically heavy carbonates are dissolving in the subsurface of the recharge area, as shown for example by the calcite-depositing spring of Westerhof near Echte, where the $\delta^{13}C$-values of the ground water are very similar to those of the brine (Usdowski et al. 1979).

In conclusion: the refined models of carbonate dissolution may be applied to hydrogeochemical systems if all the required data are available.

3.2.5 A Dissolution Front at the Contact of Sands to Marly Limestone Aquifers

H.R. Langguth and R. Schulz

3.2.5.1 Hydrogeology of the Seffent Springs Catchment

The permanently flowing Seffent springs (from the Latin: septem fontes, seven sources), are formed by two groups of sources (AAC6 and AAC7). Their catchment area, covering 8 km^2, is situated at the western margin of Aachen (Fig. 3.2.5.1), the springs being located in its northern part. The southern part of the catchment is covered by the Aachen forest, the northern part by the campus of the Technical University Aachen (RWTH Aachen). The substratum of the catchment area is mainly formed by gently dipping Upper Cretaceous sands and marly limestones (Fig. 3.2.5.1). The Seffent and the Senserbach faults are the major tectonic elements in the catchment, which limit the Klinikum/Königshügel graben in the north and the south. At the Seffent fault, which runs exactly between the two source groups of the Seffent springs (AAC6 and AAC7 on Fig. 3.2.5.1), the younger Cretaceous limestones are in contact with the 8-m-uplifted Cretaceous sands of the northern block. At the Senserbach fault, 2 km to the south, the uplifted Upper Cretaceous sands are in direct hydraulic contact with younger Cretaceous limestones (Breddin et al. 1960; Albers 1976; Knapp 1978; Kalterherberg 1982).

The ground water contour lines in February 1984 (Fig. 3.2.5.1) indicate a ground water flow generally from south to north, with a small northwest turn of direction in the neighbourhood of the Klinikum towards the Seffent springs, which are naturally the only permanent ground water outflow of this drainage basin. From the sandy aquifer in the south a steady ground water flow enters with an average of 32 l/s at the Senserbach fault into the marly limestone aquifer in the north (unpublished data of H. Breddin and H.R. Langguth 1972). Here, the ground water table drops about 20 m over a horizontal distance of about 500 m and continues in northwestern direction beginning at the 182 m contour line with the normal hydraulic gradient known from other regions in the chalks and limestones (Fig. 3.2.5.1).

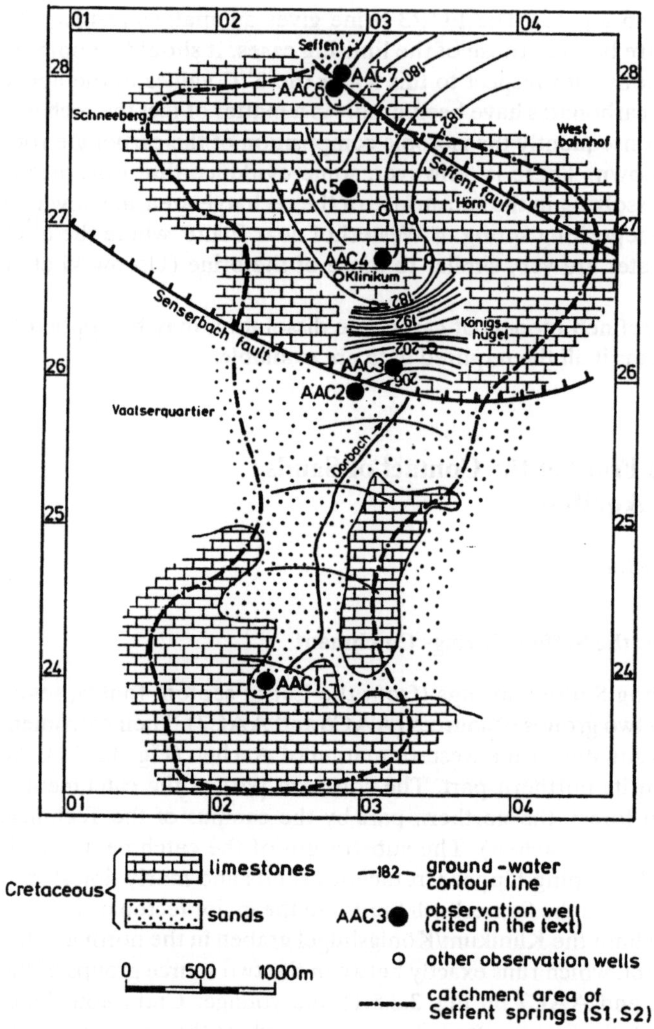

Fig. 3.2.5.1. Location and geology of the Seffent springs catchment close to Aachen

A small naturally ephemeral brook, the Dorbach, originates in a small source group in the southern part (AAC1 in Fig. 3.2.5.1). This brook is influent south of the Senserbach fault, but effluent north of it due to the water table drop mentioned above. Here, the Dorbach infiltrated completely in former times over a distance of about 250 m. Nowadays the water passes this area in an artificial canal.

The fluctuation of the ground water levels at AAC3 and AAC5 as well as the discharge of the source group AAC6 situated in the fault zone and the northern marly limestone aquifer domain are highly correlated to the precipitation.

3.2.5.2 The Hydrogeochemistry

The hydrogeochemical data of the ground water differ in the sandy aquifer and in the marly limestone aquifer (Table 3.2.5.1). The development and typical chemical changes of silicium, SO_4^{2-}, HCO_3^- and aggressive CO_2 along the ground water flow path can be shown by the geochemical cross-section between the Dorbach spring (AAC1) and the Seffent springs (AAC6), which practically follows a ground water flow line (Fig. 3.2.5.2).

The observation points AAC1 and AAC2 are located in the sandy aquifer. The concentration of the major components in AAC1 is typical for the shallow ground water of the sandy aquifer, being largely influenced by local infiltration. The spring water is undersaturated with respect to $CaCO_3$ indicated by an average content of 14.5 mg/l aggressive CO_2, and characterized by a silicium content fluctuating around 2.2 mg/l, and by moderate concentrations of SO_4^{2-} (42 mg/l) and HCO_3^- (62 mg/l). After a subsurface flow length of 2000 m in the sandy aquifer, the well water AAC2 shows considerably higher concentrations: Si 4.9 mg/l, SO_4^{2-} 168 mg/l, and HCO_3^- 224 mg/l. The HCO_3^- concentration is close to that in the limestone aquifer with some remaining aggressive CO_2. Obviously, the water quality of well AAC2, which is representative for the deeper sandy aquifer and the transition sandy aquifer/marly limestone aquifer, is influenced by some limestone locally overlying the sandy aquifer (Fig. 3.2.5.1).

The hydrochemical dates of the well AAC3 in the marly limestone aquifer about 230 m north of the Senserbach fault are characterized by generally lower SO_4^{2-} (51 mg/l), but higher HCO_3^- contents (280 mg/l; Table 3.2.5.1). The higher HCO_3^- contents coincide with a diminution of aggressive CO_2 (average value at well AAC3: 3.6 mg/l). The concentration of silicium in AAC3 (8.7 mg/l) is almost twice that of AAC2 (4.9 mg/l). The concentrations in the well AAC5, about 1200 m north of AAC3, are even higher, mostly due to HCO_3^- (AAC5: 349 mg/l, AAC3: 279 mg/l). Practically the same concentrations as in AAC3 are observed in the water of group AAC6 of the Seffent springs, about 625 m further to the north. The Si, SO_4^{2-} and HCO_3^- concentrations are only slightly increased here, and the water of AAC6 is saturated with respect to $CaCO_3$.

The hydrochemical composition north and south of the Senserbach fault is characterized by relatively higher SO_4^{2-} concentration in the south coinciding with a

Table 3.2.5.1. Characteristic chemical parameters of selected observation wells. Average values from March 1983/March 1984 (in mg/l)

Observation well	AAC1	AAC2	AAC3	AAC5	AAC6
Aquifer type	Sand		Marls and limestones		
Si	2.2	4.9	8.7	10.2	12.2
SO_4^{2-}	42	168	51	68	69
HCO_3^-	62	224	279	349	322
$CO_{2\ aggr.}$	14.5	15.1	3.6		0.0

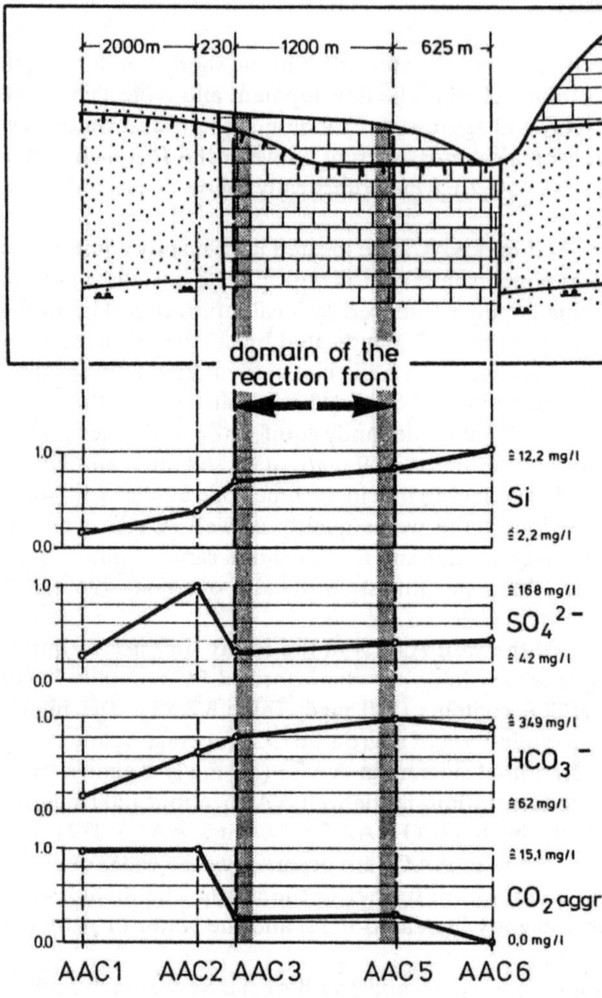

Fig. 3.2.5.2. Schematic N-S hydrogeological cross-section (AAC1-AAC6). The chemical data are average values for March 1983 to March 1984

lack of $CaCO_3$ in the sandy aquifer and consequently lower HCO_3^- concentrations, whereas relatively higher HCO_3^- concentrations in the north are connected with lower SO_4^{2-} contents. Relatively high permanent contents of aggressive CO_2 are found in the south, whereas they decrease or even disappear episodically in AAC5 and permanently in AAC6 in the north.

The chemical data are extensively evaluated in Section 6.3.

3.2.5.3 Formation of a Reaction Front

The water of the sandy aquifer is rich in CO_2 and does not attain CO_2-$CaCO_3$ equilibrium due to the lack of $CaCO_3$. North of the Senserbach fault the incoming aggressive ground water from the south attacks the argillaceous (marly) limestones on the fissure planes. The remaining clay material coats the fissures and suppresses further attack by aggressive water. The fissure diameters are successively diminished by clogging, and consequently the hydraulic permeability decreases. In response to this, the hydraulic gradient rises abruptly, forming the subterranean "cascade" of the ground water table mentioned above.

This dissolution or reaction front moves slowly northwards since the last glaciation. Its actual position is about 500 to 600 m north of the Senserbach fault. In the periods of high ground water levels the ground water preferentially flows in less altered and hence higher permeable layers, thus carrying aggressive water faster to the north and creating a transition zone, in which aggressive CO_2 can occasionally be observed, e.g. at AAC5 1200 m north of the Senserbach fault. Beyond the reaction zone the ground water has a rather stable chemical composition.

3.2.5.4 Conclusions

The horizontal flow of CO_2-aggressive ground water passing from a sandy aquifer through a fault into a marly limestone aquifer causes corrosion phenomena, indicated by changes in the hydrochemical composition and by the formation of a dissolution front, in which $CaCO_3$ is dissolved mainly in fissure planes. The argillaceous constituents of the limestone remain in place, thus coating the fissure planes, finally clogging the fissures, diminishing the permeability and thus prohibiting further $CaCO_3$ dissolution. This reaction front moves slowly into the limestone aquifer, and is visible on a ground water contour map by its condensed contour lines, indicating high hydraulic gradients.

3.2.6 Carbonate Rock Dissolution Under Intermediate System Conditions

J. Michaelis

3.2.6.1 Background

Under humid climate conditions carbonate rocks are dissolved essentially by biogenic CO_2 from the soil zone. Garrels and Christ (1965) have shown that this process is bracketed by so-called open and closed system conditions, and Deines et al. (1974) have demonstrated that the stable isotopes of carbon may be used in order to distinguish between these limiting cases. Modifications and extended versions have been presented by other authors (e.g. Wigley 1975; Reardon and Fritz 1978; Wigley et al. 1978; Michaelis et al. 1984). Other models take into account the additional reaction of carbonate rocks with sulfuric acid from atmospheric pollution, using the stable isotopes of sulfur in order to make the necessary distinction between anthropogenic sulfate and sulfate which originates from the dissolution of calcium sulfate minerals (Kirchoff and Usdowski 1985; Schaefer and Usdowski 1987). However, virtually none of these models or field studies has been concerned with the evolution of carbonate ground water, which develops from strong undersaturation to saturation with respect to calcite under conditions between the limiting cases. The present study shows that chemical and isotopic analyses of soil water as a function of depth may be used in order to deduce the boundary conditions of carbonate rock dissolution in such intermediate systems.

3.2.6.2 Study Area

The study area Mackenröder Spitze/Göttinger Wald (MSP) is located about 7 km east of Göttingen on top of a Muschelkalk plateau (Table 1.1, Fig. 1.1). The area has been closed to the public for several years, and has been reserved for research projects. It is completely covered with dense forest, mostly beech about 100 years old. The soil is mull rendzina and terra fusca rendzina. It is underlain by limestone of the Muschelkalk formation. The annual rainfall in the area is about 700 mm (Andres 1984). The water emerges at springs at various sources on the western flank of the plateau.

Over a 2 years' period the composition of the pore water at depths of 5, 30 and 100 cm, as well as that of the so-called stemflow (i.e. water trickling down from trees) has been analyzed at monthly intervals for major components and the stable isotopes of carbon and sulfur. The water of a spring, the Seckborn (MSP1), has been measured at various occasions and the mineralogy of the soil and the limestone has been studied. Sampling procedures and analytical methods have been described in detail by Michaelis (1987).

3.2.6.3 Results

Soil and Rock Material

To a depth of about 20 cm the soil consists essentially of quartz, feldspar (mainly K-feldspar), illite and chlorite. The grain size fraction < 2 mm contains 1–5% calcite in contrast to the fraction > 2 cm, containing 95% calcite. Below that depth the amount of quartz decreases to about 15% while that of calcite increases to about 75%. At about 90 cm depth the limestone is encountered. This rock is about 95% calcite with 5% non-carbonate material (quartz and clay minerals of the above type). Neither dolomite nor $CaSO_4$ minerals can be detected by optical and X-ray methods. According to Füchtbauer (1950), the limestone has CaO/MgO ratios of about 100:2 (CaO = 50–53%, MgO = 0.9–1.1%, CO_2 = 40–43%).

The isotopic composition of the limestone lies between $\delta^{13}C = 2.01$ and 2.75‰, PDB (average: 2.45‰). Nielsen and Rambow (1969) report $\delta^{34}S = 20.0 \pm 1$‰ CD. Several measurements of the soil CO_2 during the sampling period gave an average $\delta^{13}C = -23.0$‰, PDB.

Input Solutions (Stemflow)

The characteristic parameters which are necessary for the following mass-balance calculations are summarized in Table 3.2.6.1. For complete analytical data see Michaelis (1987). The composition of the stemflow varies widely as a function of time, as generally observed. Figure 3.2.6.1 shows the variation of sulfate as an example. During two long-lasting smog periods in January and February 1985, respective maximum values of 210.7 and 160.8 mg/l were measured. The corresponding pH values were 2.47 and 2.75. These were the lowest values observed over the whole period of 2 years. It should be pointed out that the variation of the $\delta^{34}S$ values of the stemflow is much less than that of the other parameters.

Table 3.2.6.1. Mean values and maximum variations at Mackenröder Spitze/Göttinger Wald in 1984–1986

	pH	Ca^{2+} (mg/l)
1	3.52 (2.74–4.27)	11.2 (2.4–44.9)
2	6.21 (5.51–6.82)	26.2 (11.6–49.7)
3	7.42 (6.71–7.95)	48.4 (39.7–58.5)
4	4.35 (7.08–8.62)	100.9 (66.5–129.1)
5	7.57 (7.30–7.67)	112.6 (108.6–117.0)
	SO_4^{2-} (mg/l)	$\delta^{34}S$ (‰ CD)
1	56.7 (19.1–210.7)	1.42 (–0.02–2.59)
2	34.5 (16.7–62.3)	2.12 (0.46–3.50)
3	32.7 (24.0–44.5)	2.80 (1.91–3.56)
4	42.0 (31.2–66.7)	3.64 (3.10–4.44)
5	35.8 (33.3–37.5)	3.05

1 = Stemflow.
2 to 4 = pore solutions at 5, 30, and 100 cm depth, respectively.
5 = MSP1 water.

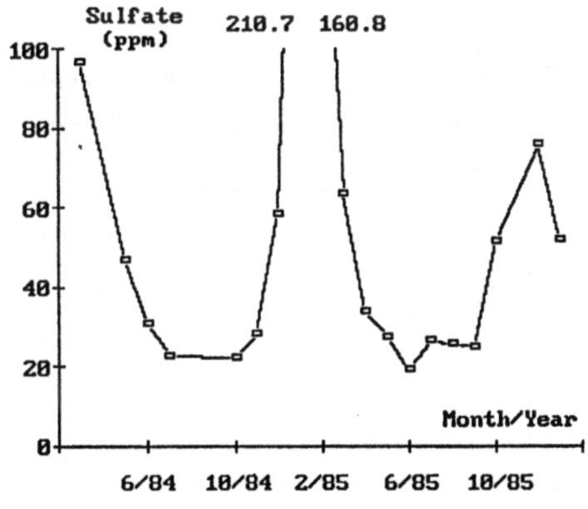

Fig. 3.2.6.1. Sulfate concentrations of stemflow Mackenröder Spitze, March 1984-February 1986

Pore Solutions

The concentrations of Ca^{2+}, and HCO_3^- as well as pH increase as a function of depth, and the usual statistical scattering of data which occurs as a function of time levels out with increasing depth (Table 3.2.6.1, Fig. 3.2.6.2). At a depth of 100 cm the solutions are approximately saturated with respect to calcite. The concentrations of Mg^{2+} = 0.5–3.9 mg/l, Na^+ = 0.9–8.8 mg/l, and Cl^- = 0.3–16.3 mg/l are approximately the same as that of the stemflow. However, K^+ which is removed in the biological cycle decreases to 0.3 mg/l. It should be noted that the concentration of sulfate decreases to a depth of 30 cm, but increases below. A very similar pattern has been observed by Hauhs and Dise (1989). This behaviour is attributed to the absorption and desorption and/or the formation and decomposition of aluminium-sulfate complexes at various depths (Van Breemen 1973; Ulrich 1984).

Output Solution

The output solution (MSPI water) is a typical ground water for limestone areas. This water is approximately saturated with respect to calcite. Concentrations of components not listed in Table 3.2.6.1 are Mg^{2+} = 5.8 mg/l, Na^+ = 2.8 mg/l, K^+ = 0.8 mg/l, HCO_3^- = 315.2 mg/l, and Cl^- = 6.7 mg/l. Comparison with the pore water at 100 cm depth shows that the pore water migrating into the saturated zone emerges virtually unchanged at the spring.

3.2.6.4 Evolution of the Ground Water

Chemical Reactions

The chemical development of the pore solutions shows that the dissolution of calcite proceeds as a function of depth until at 100 cm approximate saturation with respect to that mineral is attained. In the hydrogeochemical system under considera-

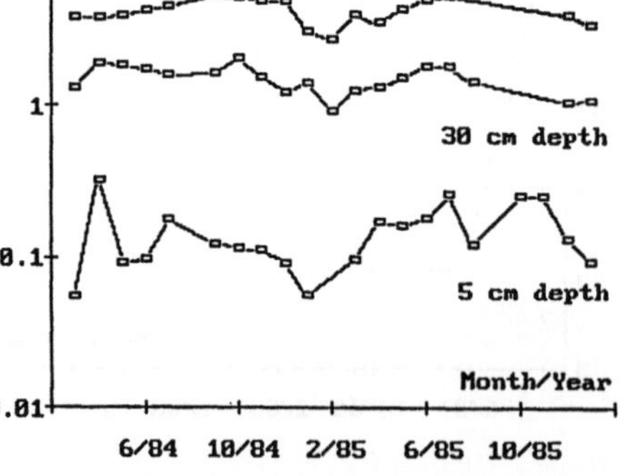

Fig. 3.2.6.2. Hydrogen carbonate concentrations of pore solutions Mackenröder Spitze, March 1984-February 1986

tion two parallel reactions are controlling this process: the reaction of calcite with dissolved biogenic CO_2 of the soil zone

$$CaCO_3 + H_2CO_3 \rightleftharpoons Ca^{2+} + 2HCO_3^- \quad (3.2.6.1)$$

and the reaction of the solid with anthropogenic sulfuric acid from air pollution

$$2CaCO_3 + H_2SO_4 \rightleftharpoons 2Ca^{2+} + 2HCO_3^- + SO_4^{2-}. \quad (3.2.6.2)$$

The concentration of the acid is indicated by the sulfate content of the stemflow (Table 3.2.6.1). In order to quantify the proportion of the two reactions, mass balances for sulfate, calcium and carbonate have to be calculated. Further, the concentration of sulfate which is derived from the dissolution of the finely dispersed $CaSO_4$ minerals according to the reaction

$$CaSO_4 \rightleftharpoons Ca^{2+} + SO_4^{2-} \quad (3.2.6.3)$$

has to be considered.

Sulfate Balance

In the study area the sulfate balance at any depth is given by the expression

$$T = A + G + D, \quad (3.2.6.4)$$

where T is the measured total molar sulfate concentration. The quantity A is the concentration of infiltrated rain water sulfate which is obtained dividing the sulfate concentration of the stemflow by the so-called deposition factor Q, which has a value of 1.35 for the study area. The quantity G is the concentration of sulfate according to reaction (3.2.6.3), and D is the change of concentration due to the combined effects of adsorption and desorption of sulfate by clay minerals and/or the formation and decomposition of aluminium-sulfate complexes (Michaelis 1987). Using the isotope balance

$$T\,\delta^{34}S_T = A\,\delta^{34}S_A + G\,\delta^{34}S_G + D\,\delta^{34}S_D \quad (3.2.6.5)$$

and the pertinent data the quantities A, D, and G can be calculated for any depth (Michaelis 1987). The result shows that the concentration of sulfate produced according to reaction (3.2.6.3) increases from zero at 5 cm, where the soil is free of Ca-sulfate, to 4 mg/l (0.04 mmol/l) at a depth of 100 cm.

Calcium Balance

Calcium is produced by reactions (3.2.6.1) to (3.2.6.3) and has attained a total concentration $Ca_{(T)} = 2.52$ mmol/l at a depth of 100 cm (Table 3.2.6.1). All these reactions occur in the subsurface. However, reaction (3.2.6.2) takes place at still another site of the system: the surface of the vegetation covering the study area. In dry periods limestone dust is blown off from parts of the Muschelkalk plateau and settles on the leaves and stems of the trees. On rainfall, these particles dissolve according to Eq. (3.2.6.2), causing a calcium concentration $Ca_{(2a)} = 0.21$ mmol/l in the stemflow. From the stoichiometry of the reaction it follows that this concentration is equivalent to a sulfate concentration of 0.5 $Ca_{(2a)}$. The total input of the anthropogenic sulfate is A = 0.44 mmol/l (stemflow/Q, Table 3.2.6.1). Thus, the input

Table 3.2.6.2. Ca^{2+} produced by the various chemical reactions as indicated below and in the text

	mmol/l	%
$Ca_{(1)}$	1.598	63.5
$Ca_{(2a)}$	0.207	8.2
$Ca_{(2b)}$	0.667	26.5
$Ca_{(3)}$	0.045	1.8
$Ca_{(T)}$	2.517	100.0

$Ca_{(1)}$ = Eq. (3.2.6.1). ($Ca_{(2a)}$ = Eq. (3.2.6.2) surface.
$Ca_{(2b)}$ = Eq. (3.2.6.2) subsurface.$Ca_{(3)}$ - Eq. (3.2.6.3).

of sulfuric acid is $R = A - 0.5\ Ca_{(2a)} = 0.34$ mmol/l. This yields a calcium concentration $Ca_{(2b)} = 2\ R = 0.67$ mmol/l) from Eq. (3.2.6.2) in the subsurface. According to Eq. (3.2.6.3), the calcium from the dissolution of $CaSO_4$ minerals is equivalent to $G = 0.04$ mmol/l (obtained from the sulfate balance) so that $Ca_{(3)} = G$. The calcium due to reaction (3.2.6.1) is $Ca_{(1)}$. This quantity is calculated from the mass balance

$$Ca_{(T)} = Ca_{(1)} + Ca_{(2a)} + Ca_{(2b)} + Ca_{(3)} = Ca_{(1)} + 2A + G. \qquad (3.2.6.6)$$

The calcium concentrations of the various reactions and their percentages are summarized in Table 3.2.6.2. At a depth of 100 cm, 64% of the calcium is derived from the reaction of $CaCO_3$ with dissolved biogenic CO_2. The reaction of $CaCO_3$ with sulfuric acid on the surface of the hydrogeochemical system produces 8%. In the subsurface the same reaction yields 27%, and 2% of the calcium is due to the dissolution of $CaSO_4$ minerals.

Carbonate Balance

In analogy to calcium, the mass balance for hydrogen carbonate is given by Eq. (3.2.6.7):

$$HC_{(T)} = HC_{(1)} + HC_{(2a)} + HC_{(2b)}\ , \qquad (3.2.6.7)$$

where $HC_{(T)}$ is the total HCO_3^--concentration, $HC_{(1)}$ is that produced according to Eq. (3.2.6.1), and $HC_{(2a)}$ and $HC_{(2b)}$ are concentrations due to Eq. (3.2.6.2) on the surface and in the subsurface of the system, respectively. From the stoichiometry of the reaction follows that $2\ HC_{(1)} = Ca_{(1)}$, and $HC_{(2a)} + HC_{(2b)} = Ca_{(2a)} + Ca_{(2b)} = 2A$. Combining these expressions with Eqs. (3.2.6.6) and (3.2.6.7) leads to Eq. (3.2.6.8):

$$Ca_{(T)} = 0.5\ HC_{(T)} + A + G, \qquad (3.2.6.8)$$

where A and G have the same meaning as before. Figure 3.2.6.3 shows that the measured calcium hydrogen carbonate relation is virtually identical with Eq. (3.2.6.8), obtained from the mass balances. The observed slope is 0.49, and the intercept is 0.56 versus A + G = 0.48. The small difference may be attributed to calcium in rain water.

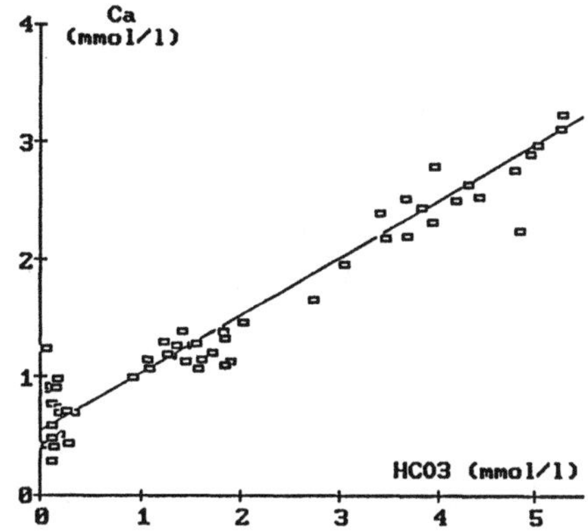

Fig. 3.2.6.3. Ca^{2+}/HCO_3^- ratios of pore solutions Mackenröder Spitze, March 1984-February 1986. The linear regression yields: $[Ca^{2+}] = 0.49 \cdot [HCO_3^-] + 0.56$ (mmol/l); $r^2 = 0.94$

3.2.6.5 Boundary Conditions for Limestone Dissolution

The conditions of limestone dissolution are limited by two cases. In the open system the dissolution occurs at a constant partial pressure of CO_2, established by the continuous production of biogenic CO_2 in the soil zone. Under closed system conditions, water saturated with CO_2 in the soil is separated from the gas phase and reacts in the deeper subsurface. Both cases may be expressed in terms of the ratio of the gas volume (V_g) to the volume of the solution (V_l). With respect to the open system this ratio is $V_g/V_l = \infty$, whereas it is $V_g/V_l = 0$ under closed system conditions. In general, the conditions of the $CaCO_3$ dissolution change from open to closed as the water migrates from the soil into the underlying rock, filling the pores more and more along its path. This intermediate case in which the ratio of the volume of gas to the volume of solution changes from infinity to zero is expressed by $0 < V_g/V_l < \infty$. Model considerations for intermediate systems have been presented by Schaefer and Usdowski (1987), and Michaelis (1987). The basic equations for intermediate conditions are

$$\delta^{13}C_T \Sigma C = \delta^{13}C_s[Ca^{2+}] + \delta^{13}C_o \Sigma C_o + \delta^{13}C_{g,o} \cdot (p_{CO_2,o} V_g/V_l \cdot R \cdot T) - \delta^{13}C_g (\Sigma C x_f \cdot V_g/K_{CO_2} \cdot R \cdot T \cdot V_l) \quad (3.2.6.9)$$

and

$$\Sigma C = \{2 \Sigma C_o + 2p_{CO_2,o} V_g/R \cdot T \cdot V) - H^+ + K_W/[H^+] + 2[SO_4^{2-}]\}/ /2x_f (1 + V_g/K_{CO_2} \cdot R \cdot T \cdot V_l + x_b), \quad (3.2.6.10)$$

where $\delta^{13}C_T$ = total dissolved carbonate, $\delta^{13}C_s$ = solid carbonate, $\delta^{13}C_o$ = initial solution carbonate, $\delta^{13}C_{g,o}$ = initial gas carbon, $\delta^{13}C_g$ = remaining gas carbon, C, C_o = total dissolved carbonate at any time, initially, x_f, x_b = mole fractions of $H_2CO_3^*$ (= CO_{2aq} + H_2CO_3), HCO_3^-, $[SO_4^{2-}]$ = anthropogenic sulfate, R = gas constant, T = absolute temperature, K_{CO_2} = solubility constant for CO_2, K_W = ion product of water, $p_{CO_2,o}$ = initial partial pressure of CO_2.

These equations may be used to calculate $\delta^{13}C_T$ at a given V_g/V_l – ratio as pH increases upon the dissolution of calcite. Figure 3.2.6.4 shows some examples. On the other hand, the V_g/V_l-ratio can be deduced from the chemical and isotopic data of the system. From the present data it can be calculated that on average, the evolution of the water occurs at a V_g/V_l ratio of 13 (Curve B, Fig. 3.2.6.5), and that 90% of all solutions are bracketed by curves A and C (Michaelis 1987). This compares well to the measured $\delta^{13}C$- and pH values in the water of Seckborn spring, which drains part of the study area, and to other springs (marked Grone and Weende in Fig. 3.2.6.5) of the Muschelkalk formation near Göttingen.

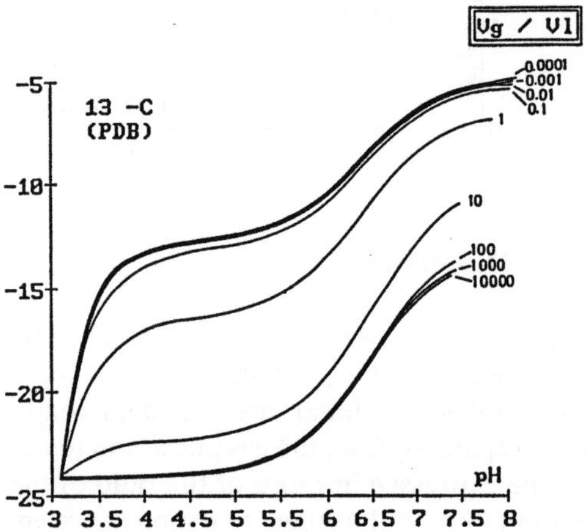

Fig. 3.2.6.4. Theoretical $\delta^{13}C$-pH relations during the evolution of carbonate ground waters at various V_g/V_l ratios (0.0001 to 10 000)

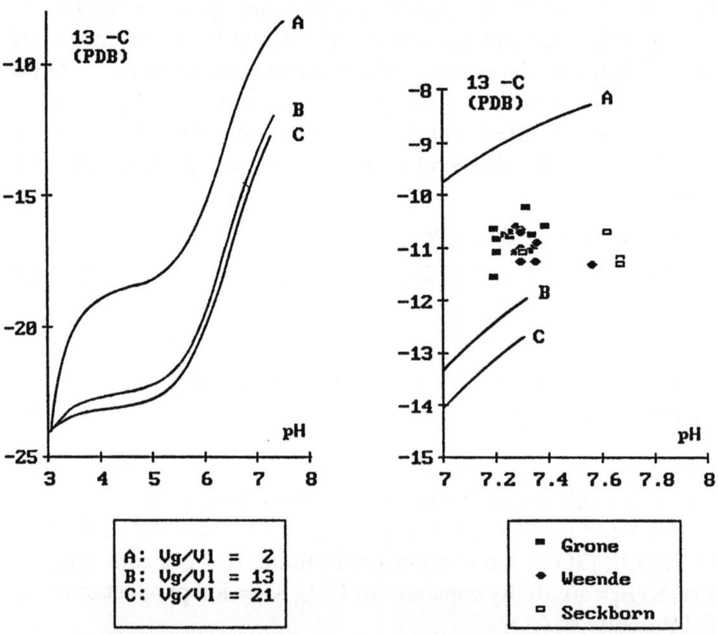

Fig. 3.2.6.5. Evolution of carbonate ground water at Mackenröder Spitze (curves *A, B, C*). *B* Average V_g/V_l ratio for all analyses. *A* and *C* Boundaries for 90% of the cases. The *symbols* represent measured $\delta^{13}C$- and pH values of calcite-saturated springs in the area

3.3 Alteration in Karst Systems

3.3.1 Mineralogy and Hydrogeochemistry of the Gympsum Karst of Foum Tatahouine, South Tunisia

W. Smykatz-Kloss, H. Hötzl and H. Kössl

3.3.1.1 Background

The purpose of the present paper is the study of mineralogical and hydrogeochemical changes during karstification of sulfate-rich rocks in semi-arid regions. From preliminary investigations on semi-arid gypsum karst environments in Libya, it emerged that intensive processes of dissolution and precipitation lead to remarkable changes in the mineralogical composition of the karst and foreland rocks, in the geomorphological appearance of the region, and to an enrichment of evaporitic minerals in the foreland of the Mesozoic escarpment of the Libyan Jabal Nefusa (Kosa and Smykatz-Kloss 1978; Kosa 1980; Smykatz-Kloss 1980; Goebelbecker 1985).

The area of Foum Tatahouine has been chosen for a detailed study of karstification processes because of the evident and close connection between the karst area and the playa where the dissolved material is redeposited by evaporation (Figs. 3.3.1.1 and 3.3.1.2). It was not possible to quantify the hydrogeochemistry of the area because of lack of rainfall during the campaigns. Therefore, this paper stresses the observed mineralogical changes along the path of transportation: the comparison of the karst and playa mineralogy allows the reconstruction of dissolution and precipitation processes on the base of physico-chemical principles.

3.3.1.2 Geology and Observed Karst Phenomena

The geology of the region is relatively well known (Domergue et al. 1952; Busson 1967; Perthuisot 1976). An Upper Triassic to Middle Jurassic sequence of calcium sulfates and intercalated dolostones of at least 500 m total thickness is exposed E of the Jabal Dahar. It covers an area of some hundred km^2. Dissolution and erosion have resulted in smooth and rounded hills and deep erosion channels (Fig. 3.3.1.3). Many karst features can be observed, although they are much less clearly visible than in the comparable gypsum karst area of Bir-Al-Ghanem at the foothills of the Jaba Nefusa in Libya.

Precipitation occurs only during a few weeks in the winter. The annual rainfall is in the range between 200 and 280 mm and shows that the region is semi-arid.

The length of the few caves occurring in the karst area (A, Fig. 3.3.1.2) is not more than 300 m. Swallets and resurgences are numerous. The main water flow during the short rain period occurs on the surface (in wadis). But there is subsurface water flow, as can be seen from holes filled with brines in the centre of the playa. In

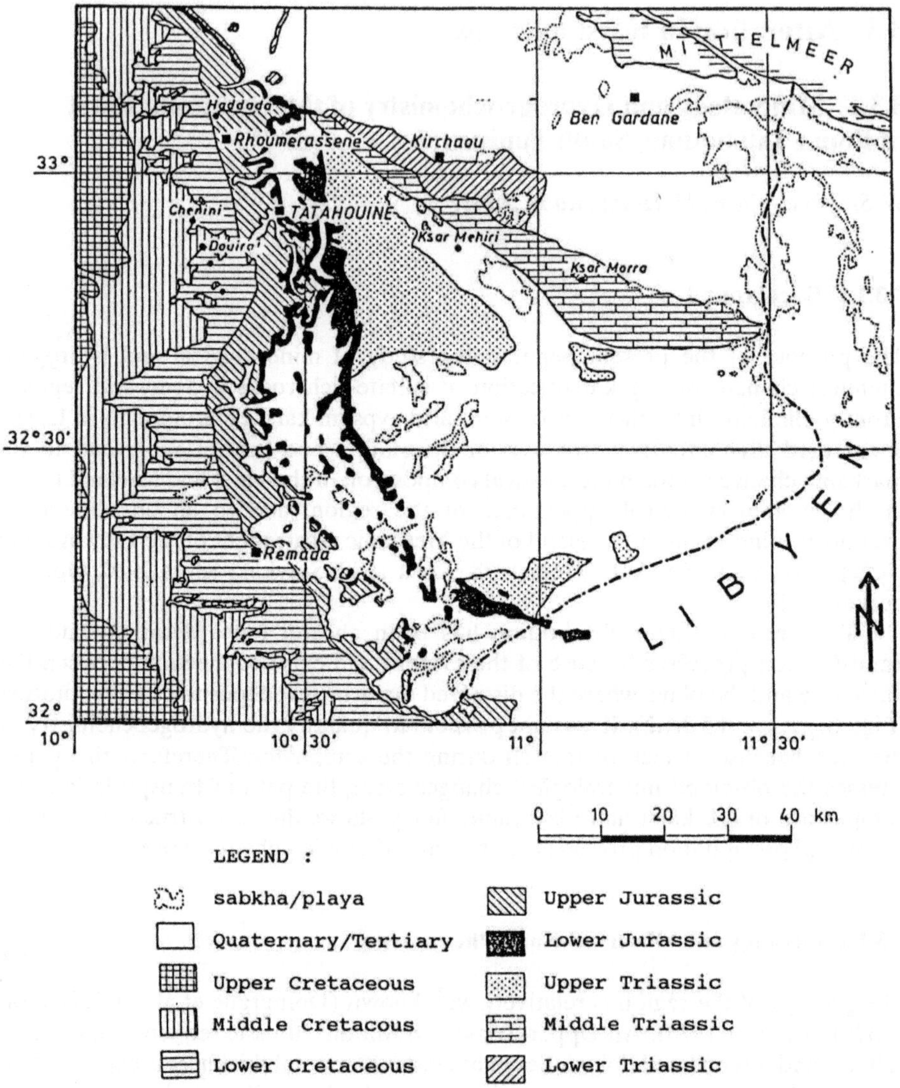

Fig. 3.3.1.1. Geology of the area of Foum Tatahouine, South Tunisia

fact, there are two levels of subsurface water flow which are related to marly-clayish layers (Hötzl et al. 1984; Kössl 1988).

All the surface and subsurface waters of the studied area east of Foum Tatahouine flow into the playa Oum el Krialate (TN1; Fig. 3.3.1.3), where extensive evaporation takes place. Thus, the immense richness in minerals found in this playa is mainly due to evaporation in a nearly closed basin, which is bordered by a barrier of rocks in the east (Ouedi el Assel, Fig. 3.3.1.3).

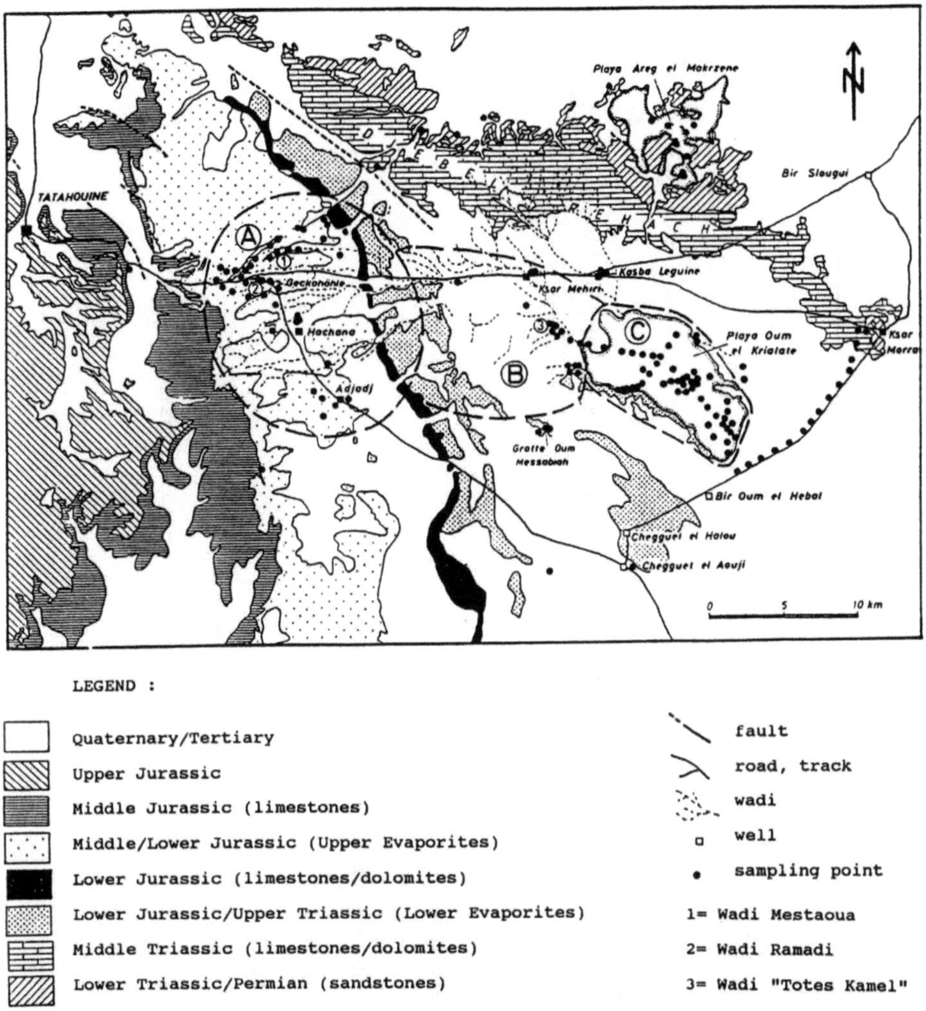

Fig. 3.3.1.2. The study area: *A* gypsum karst; *B* transport region; *C* playa Oum el Krialate (TN1)

3.3.1.3 Hydrogeochemistry

The karst waters contain between 1700 and 2500 mg/l TDS. The cation and anion proportions (Table 3.3.1.1, Fig. 3.3.1.4) differ in dependence on the composition of the rocks in the recharge area. The proportions of the ions change with increasing distance from the recharge area (A). The playa waters show TDS between 3000 and 6000 mmol/l. Especially the amounts of dissolved Na^+, SO_4^{2-} and Cl^- have increased in the playa waters. The direction of the water flow is W → E (SE) (Fig. 3.3.1.5). The southern part of the playa shows a Cl^- concentration of 1800 mmol/l, which is more than twice the concentration of SO_4^{2-} (800 mmol/l).

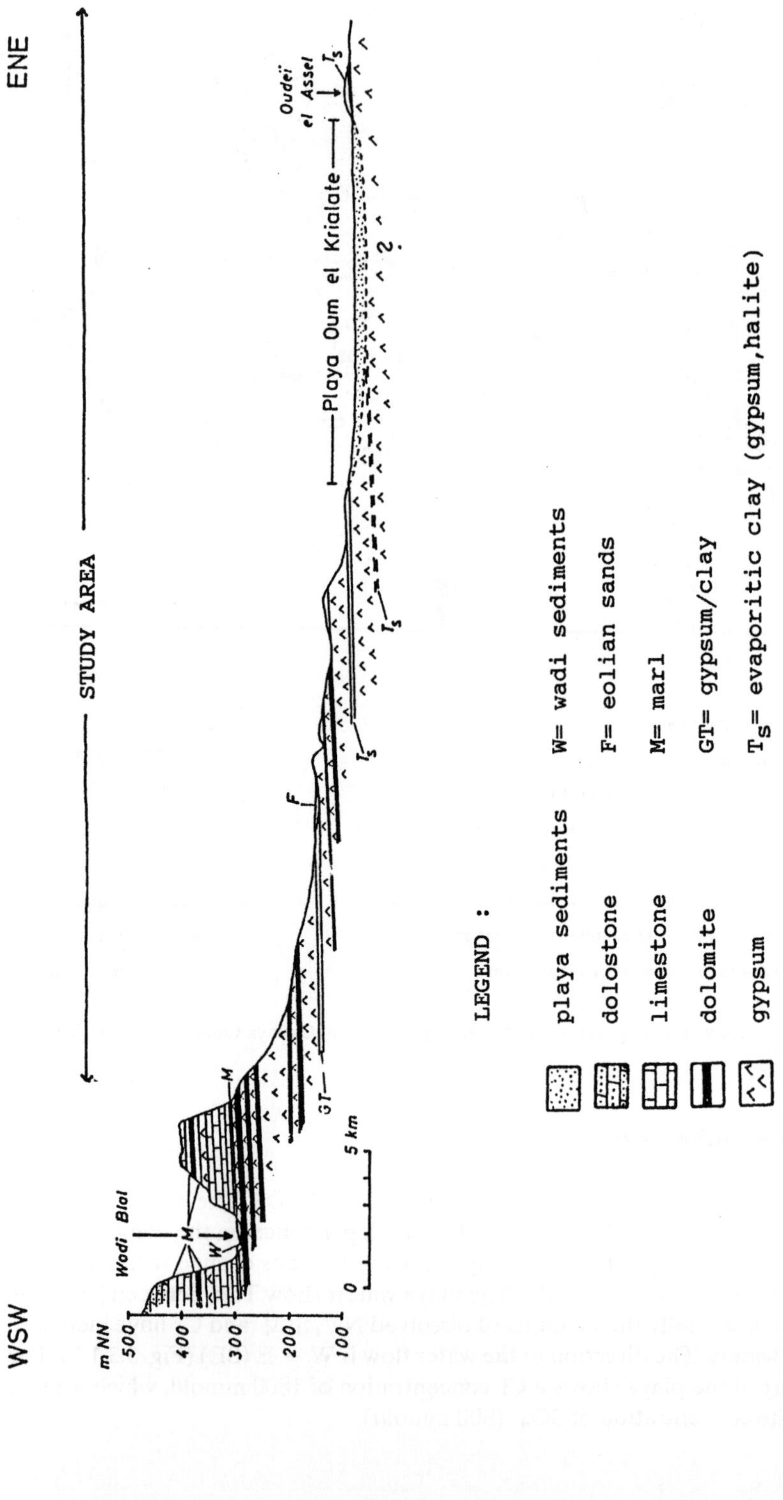

Fig. 3.3.1.3. Geological profile of the study area

Table 3.3.1.1. Cation and anion proportions of the karst waters, Wadi Ramadi; TDS: 100 mmol/l

Cations	Ca^{2+}	Mg^{2+}	Na^{+}
($\Sigma = 100$)	95	4	1
Anions	SO_4^{2-}	HCO_3^-	Cl^-
($\Sigma = 100$)	78	18	4

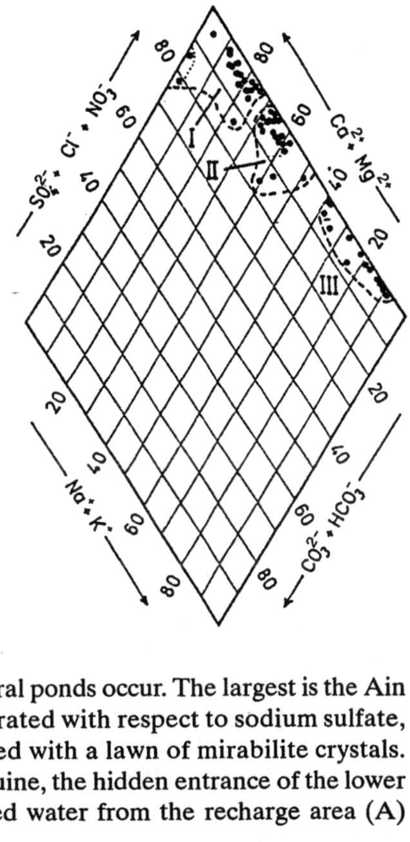

Fig. 3.3.1.4. Chemical composition of the water samples (PIPER-diagram). *I* waters from the karst area (A); *II* waters from the area north of the karst area; *III* playa waters

In the central part of the playa some natural ponds occur. The largest is the Ain el Leguine. The waters of the ponds are saturated with respect to sodium sulfate, and in fact the bottom of the ponds is covered with a lawn of mirabilite crystals. After pumping the water from the Ain el Leguine, the hidden entrance of the lower water flow level could be seen which supplied water from the recharge area (A) during the field work (3 weeks).

3.3.1.4 Mineralogy

The mineralogy of the karst area is very simple. Gypsum and anhydrite occur, the latter only in deeper parts of the rocks. Subordinate celestite ($SrSO_4$) forms clear blue crystals, some cm in size. The main mineral gypsum shows a large variety of types including white, clear, pink and grey, black (coloured by organic matter) and green crystals (due to inclusions of Mg-rich smectite). Thick layers of gypsum and anhydrite alternate with thin layers of dolostones. The dolomite of the carbonaceous layers shows the same weathering phenomenon which has been observed in the

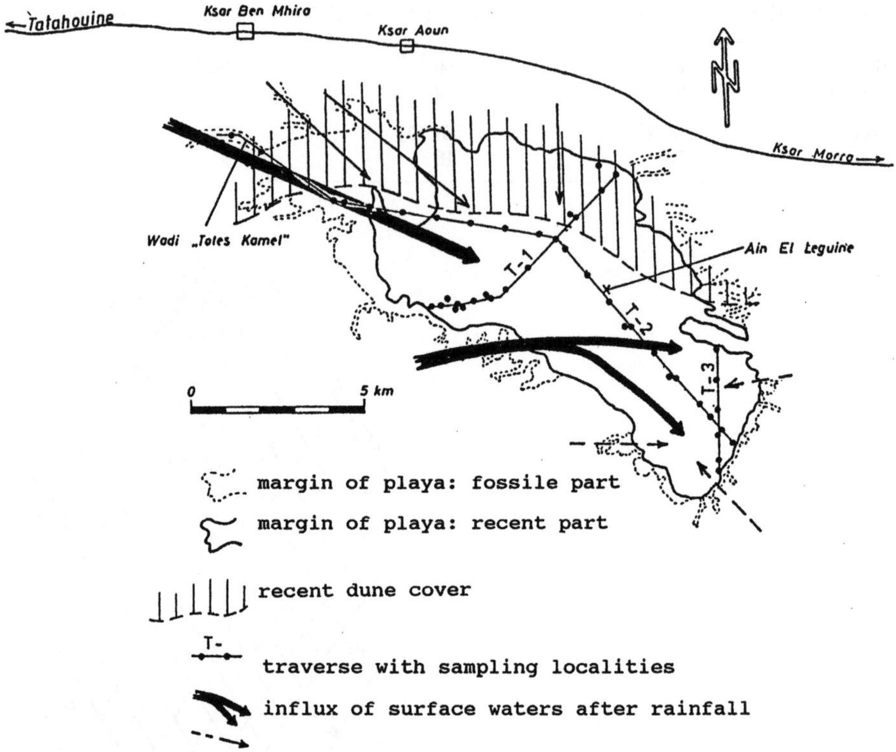

Fig. 3.3.1.5. Water flow into the playa Oum el Krialate

Jabal Nefusa and the Jabal Dahar (Sect. 3.3.2, this Vol.; Kössl 1988), e.g. a transformation of well-ordered, stoichiometric dolomite into a disordered, non-stoichiometric "proto-dolomite". Small amounts of calcite, quartz, halite, illite, albite and pyrite are present in the area (A) as well. At some spots protected from wind and rain, efflorescences of halite, hexahydrite, bloedite and bassanite occur (Smykatz-Kloss et al. 1985). Occasionally, newly formed thenardite, glauberite and jarosite (near the primary pyrites) can be observed as well on the rock surfaces of the region (A).

Approaching the playa, e.g. entering the area (C) (Fig. 3.3.1.2), striking changes in mineralogical composition occur. Using the WATEQ-2 programme to calculate the saturation indices (SI) from the karst and playa waters for the main sulfates and carbonates, it emerges that these minerals are stable for all of the three regions: the karst area (A), the transportation area (B) and the playa (C, Fig. 3.3.1.2). The saturation indices of some other evaporite minerals also increase from the recharge area (A) to the playa (C) (Table 3.3.1.2).

For many of the evaporite minerals which occur in the playa Oum el Krialate (Table 3.3.1.3) SI cannot be computed because thermodynamic data are not available. From the concentrated brines of the playa more than 30 sulfates, chlorides, carbonates and even silicates have been formed. Corresponding to the differences in

Table 3.3.1.2. Saturation indices (SI), calculated using the WATEQ-2 programme

Mineral	SI for areas A + B (= karst + transport, see Fig. 3.3.1.2)	SI for area C (= playa)
Dolomite	+ 0.741	+ 1.900
Calcite	+ 0.551	+ 0.499
Aragonite	+ 0.431	+ 0.327
Gypsum	+ 0.303	+ 0.654
Anhydrite	+ 0.042	+ 0.394
Epsomite	- 2.74	- 1.02
Nahcolite	- 4.13	- 2.19
Mirabilite	- 4.74	- 0.13
Halite	- 5.41	- 1.86
Thenardite	- 5.92	- 1.30

Table 3.3.1.3. Sulfates, carbonates and chlorides occurring in the Playa Oum el Krialate

Gypsum	$CaSO_4 \cdot 2\,H_2O$
Bassanite	$CaSO_4 \cdot 1/2\,H_2O$
Anhydrite	$CaSO_4$
Glauberite	$Na_2Ca(SO_4)_2$
Polyhalite	$K_2MgCa_2(SO_4)_4 \cdot 4\,H_2O$
Syngenite	$K_2Ca(SO_4)_2 \cdot H_2O$
Mirabilite	$Na_2SO_4 \cdot 10\,H_2O$
Thenardite	Na_2SO_4
Hexahydrite	$MgSO_4 \cdot 6\,H_2O$
Kieserite	$MgSO_4 \cdot H_2O$
Bloedite	$Na_2Mg(SO_4)_2 \cdot 4\,H_2O$
Loeweite	$Na_{12}Mg_7(SO_4)_{13} \cdot 15\,H_2O$
Picromerite	$K_2Mg(SO_4)_2 \cdot 6\,H_2O$
Langbeinite	$K_2Mg_2(SO_4)_3$
Kainite	$K_4Mg_4Cl_4(SO_4)_4 \cdot 11\,H_2O$
Rozenite	$FeSO_4 \cdot 4\,H_2O$
Szomolnokite	$FeSO_4 \cdot H_2O$
Calcite	$CaCO_3$
Aragonite	$CaCO_3$
Dolomite	$CaMg(CO_3)_2$
Magnesite	$MgCO_3$
Nahcolite	$NaHCO_3$
Vaterite	$CaCO_3$
Halite	$NaCl$
Sylvite	KCl
Carnallite	$KMgCl_3 \cdot 6\,H_2O$

cation and anion proportions, the evaporites occur in several zones (Figs. 3.3.1.6 and 3.3.1.7). The central, eastern and southern parts of the playa are covered by an evaporitic crust which increases in thickness from the western margin (2–3 mm) to the centre of the playa (20–30 mm). Very common and expected evaporites are found such as gypsum, halite, hexahydrite, thenardite, bloedite, calcite, aragonite and (proto-)dolomite. Other but rarely observed minerals are bassanite, K-feldspar,

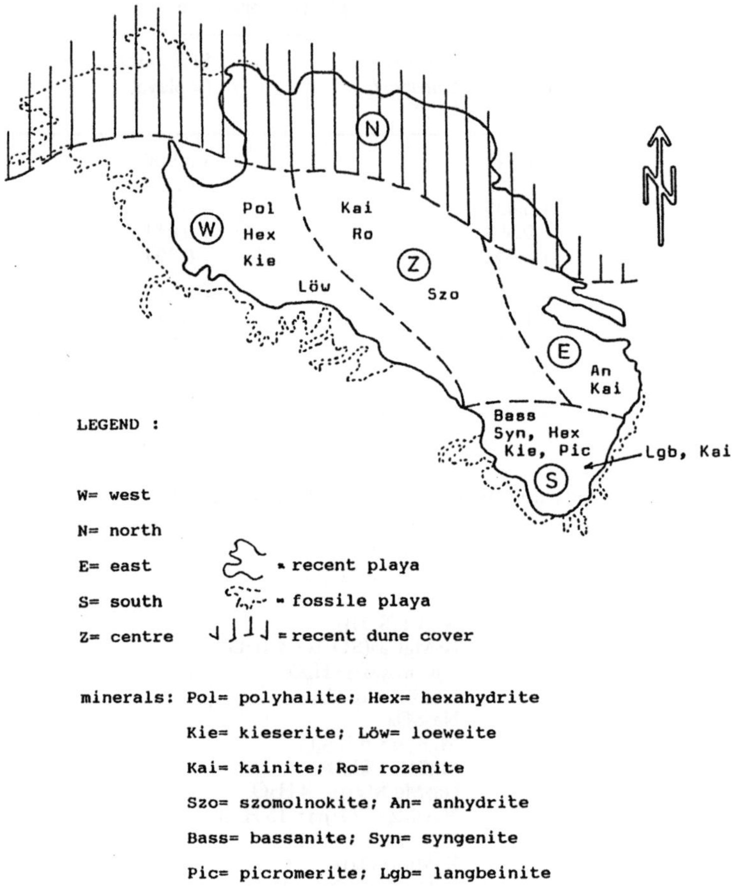

Fig. 3.3.1.6. Distribution of the sulfates in the playa

Mg-smectite, polyhalite and nahcolite. Additional minerals which occur in traces in the playa Oum el Krialate and which – to the knowledge of the authors – have never been described before from playa crusts are langbeinite, loeweite, vaterite, rozenite and szomolnokite.

All these minerals are very fine-grained, but sometimes idiomorphic, except for mirabilite, which forms very large crystals, but only if in contact with the brines of the ponds. Thus, on the bottom of the Ain el Leguine a lawn of mirabilite crystals up to 50 cm in length occurs. Removed from the brine, the mirabilite crystals are transformed within minutes into a fine-grained white powder of water-free sodium-sulfate thenardite.

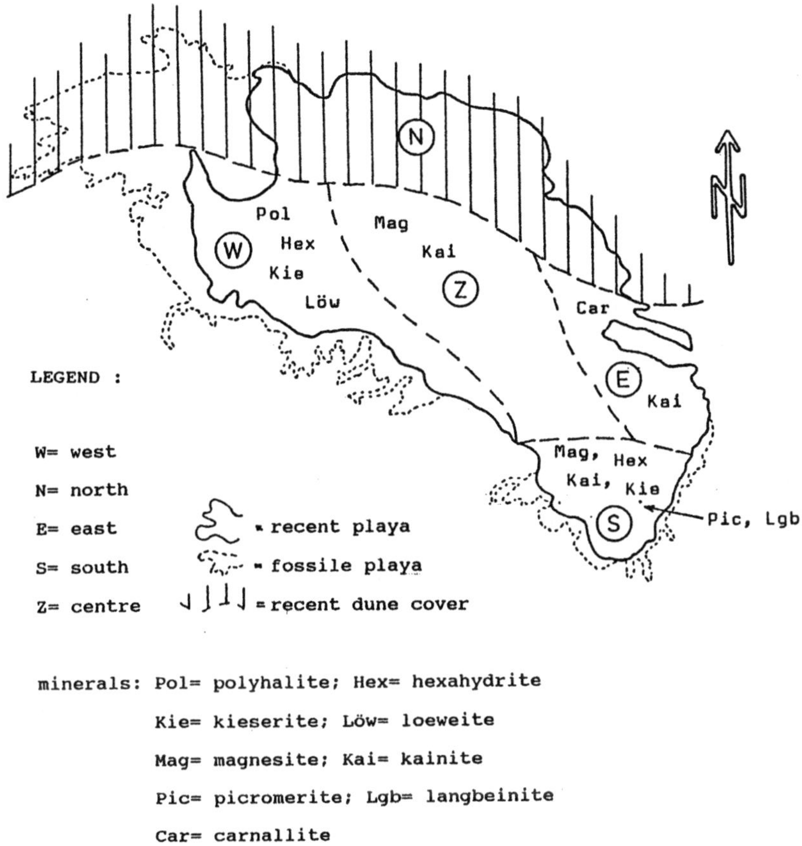

Fig. 3.3.1.7. Distribution of the magnesium minerals in the Playa Oum el Krialate: notations as in Fig. 3.3.1.6

3.3.1.5 Discussion and Conclusions

The hydrogeologically closed system of the Foum Tatahouine karst, including the Playa Oum el Krialate, makes it possible to follow the transformation of the primary sulfates, carbonates and silicates into the evaporites of the playa. The composition of the karst waters does not change remarkably from karst to playa. The TDS are relatively constant between 120 and 160 mmol/l. The karst waters are supersaturated with respect to the main sulfate and carbonate minerals.

Evaporation in the playa causes an increase of TDS in the playa waters. The reason for the low Ca^{2+} content of these waters and for the enrichment of Na^{+} and Cl^{-} is the continous precipitation of gypsum. Consequently, with increasing evaporation a sequence of minerals is observed in the order gypsum and other Ca-bearing sulfates, Mg- and Na-Mg sulfates (hexahydrite, bloedite, loeweite) and finally Na sulfates (mirabilite, thenardite). These sulfates of the western playa zone are accompanied by dolomite, calcite and aragonite and by halite, which is the only chloride mineral.

The waters flow from W to E and SE. This means that the western playa zone displays the first (= oldest) evaporites. When the waters continue flowing to the SE, e.g. after the formation of the Ca evaporites, they are more concentrated in Na^+, K^+ and Cl^-. The eastern and especially the southern part of the playa are the areas where the alkali sulfates and chlorides are formed (kainite, syngenite, picromerite, langbeinite; sylvite, carnallite). They are accompanied by the carbonates magnesite, nahcolite and – very rarely – vaterite. The occurrence of some of these minerals is surprising because they are thermodynamically metastable. The reason for the occurrence of so many minerals and for the unusual and metastable parageneses is the hydrologically closed system: Any solution from the karst area reaches the playa. According to Braitsch (1971), it is possible to follow the "crystallization paths" of evaporating solutions by means of Jaenecke diagrams (e.g. K_2-Mg-SO_4; Goebelbecker 1985; Sect. 3.3.2, this Vol.). From these diagrams it emerges that the only thermodynamically stable sulfates (besides gypsum) should be thenardite (mirabilite) or bloedite. The fact that both minerals occur together and that they are associated with numerous other minerals (Table 3.3.1.3) points to a disequilibrium situation. This means the evaporation must have occurred rather fast.

The very unusual occurrence of the Fe^{2+} sulfates rozenite and szomolnokite can be explained by the primary association of pyrite and organic matter in some marly layers of the karst area. The presence of organic matter during weathering and transport keeps the iron in the di-valent stage so that it may be transported into the playa. There, the formation of $FeSO_4 \cdot 4\,H_2O$ (e.g. rozenite) and of $FeSO_4 \cdot H_2O$ (szomolnokite) takes place when the playa waters evaporate. The semi-arid environment favours the formation of Fe-, Mg- and Na sulfates poor in water: rozenite and szomolnokite instead of melanterite ($FeSO_4 \cdot 7\,H_2O$), hexahydrite ($MgSO_4 \cdot 6\,H_2O$) and kieserite ($MgSO_4 \cdot H_2O$) instead of epsomite ($MgSO_4 \cdot 7\,H_2O$), and thenardite (Na_2SO_4) instead of mirabilite ($Na_2SO_4 \cdot 10\,H_2O$).

3.3.2 Dedolomitization and Salt Formation in a Semi-Arid Environment

W. Smykatz-Kloss and J. Goebelbecker

3.3.2.1 Background

The climate of the Jabal Dahar area (South Tunisia) is semi-arid, and rainfall is limited to a few weeks during winter. The walls of the escarpments of the Jabal Dahar, rising to several 100 m above the Jafara plain, are the places where the rain falls essentially, while the regions south and west of them are more arid. The rocks of the Jabal Dahar are Upper Jurassic to Lower Cretaceous dolomitic limestones (Sect. 3.3.1, this Vol.). Evidently the small annual precipitation between 220 and 280 mm causes intensive weathering: after the rainy period numerous secondary products are formed on the walls of the escarpment, on its foothills and in the foreland. This study aimed to evaluate rock weathering and the formation of new minerals in the area of the Jabal Dahar (TN2) as a case study for hydrogeochemical processes and mineral transformations in dolomitic rocks under semi-arid conditions.

3.3.2.2 Field Studies

Rock, soil and water samples (from wells and one cistern) were collected during three periods in winter 1982/83 and 1983/84 and in autumn 1983. Furthermore, a few surface water samples from wadis could be taken.

The five rock profiles investigated consist of porous Jurassic (Ksar Hadada, Ghoumrassen) and Cretaceous (Chenini) limestones with varying amounts of dolomite. A compact and nearly dense chert layer occurs on the top of the limestones as a protective cover against weathering. But this cover exhibits numerous fissures and clefts which permit the descent of rain waters (Goebelbecker 1985).

3.3.2.3 Dedolomitization

The rocks are limestones, which have been partially dolomitized during the late diagenetic stage. This is deduced from the incomplete dolomitization, the partly idiomorphic development of the dolomite crystals (Fig. 3.3.2.1), the occurrence of

Fig. 3.3.2.1. Idiomorphic protodolomite + calcite ("rice grains"); SEM photo (1 *bar* = 100 μm). Sample R1-ad

dolomite edges around primary calcite, the occurrence of zoned dolomite crystals (Fig. 3.3.2.2), the iron content (up to 2.2 cation-% in the dolomite structure), and the paragenesis with K-feldspar, microcrystalline quartz (chert) and illite (Swett 1968).

The predominant process of the (sub-) recent weathering is the dedolomitization. The composition and structural properties of the mineral dolomite change from the upper part of the profiles downwards. The dolomite is stoichiometric [$Ca_{1.0}$(Mg, Fe)$_{1.0}$ $(CO_3)_2$] in the upper nine-tenths of the profiles. while it is non-stoichiometric [$Ca_{1.08}$(Mg, Fe)$_{0.92}$$(CO_3)_2$] within 5–20 m from the bottom. Figure 3.3.2.3 shows the chemical composition of well-ordered standard dolomites and protodolomites from the Jabal Dahar. The non-stoichiometric dolomite exhibits different degrees of structural disorder which can be seen in X-ray diagrams and from differential thermal analyses, so that it must be classified as a protodolomite (Fig. 3.3.2.4). From the remarkable decrease of the decomposition temperatures (Goebelbecker 1985), it must be concluded that the Jabal Dahar protodolomites are disordered essentially with respect to the cation layers (Lippmann 1973).

The formation of protodolomite is the first step of dedolomitization of well-ordered dolomite. Optical and electron optical investigations show that the dolomite is partly dissolved, preserving the primary forms (rhombohedra, Fig. 3.3.2.1) and is stepwise replaced by freshly precipitated protodolomite, which has a smaller Mg^{2+} and Fe^{2+} content.

The dedolomitization is enhanced by zoned dolomite crystals with an iron-rich core. SEM and microprobe analyses of nearly 100 zoned, iron-bearing dolomites show in about 50 cases that the dissolution starts at zones rich in iron, e.g. in the centre of the rhombohedral planes or at the interface between an inner (iron-rich) and outer (iron-poor) zone (Fig. 3.3.2.2). The dedolomitization is supported by high Ca^{2+}/Mg^{2+} ratios (Munn and Jackson 1980) or by high $CaSO_4$ and NaCl contents (Braitsch 1971) of the weathering solutions due to presence of the pertinent minerals.

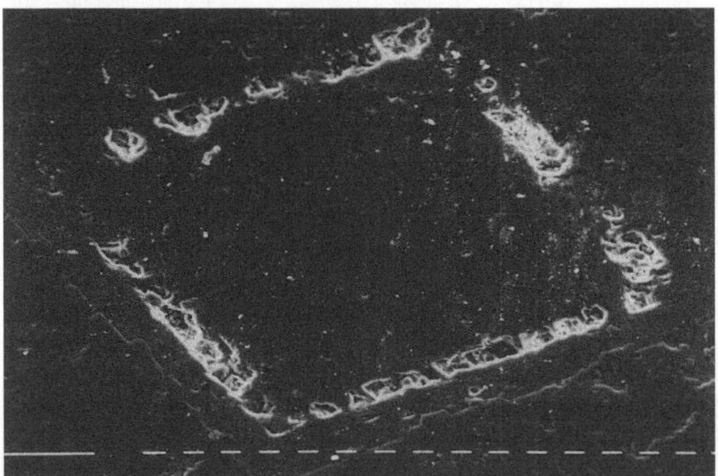

Fig. 3.3.2.2. Zoned protodolomite rhombohedron with holes at the "interface" rim/core dolomite. SEM photo (*1 bar* = 10 μm). Sample R1aa chert matrix

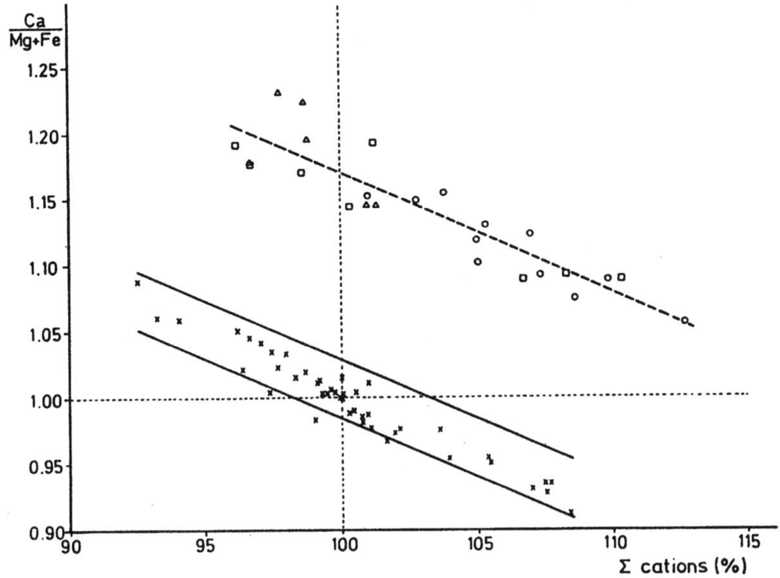

Fig. 3.3.2.3. Ca/(Mg + Fe) ratio versus the Σ cations of well-ordered standard dolomites (x) and disordered protodolomites from the Jabal Dahar (Δ, □, ○). Microprobe measurements

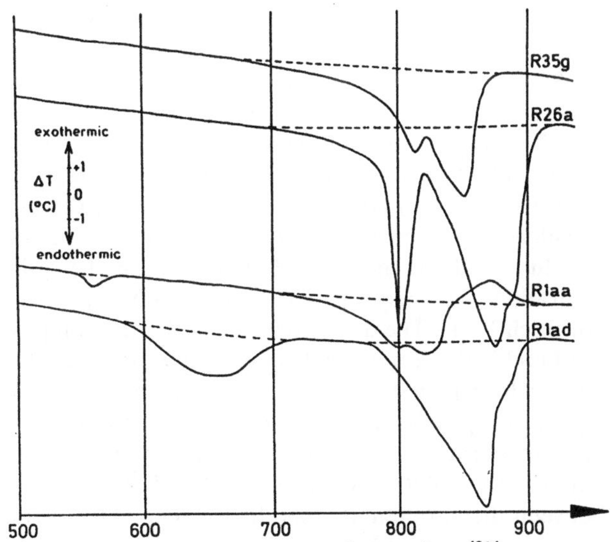

Fig. 3.3.2.4. Different thermal analyses of protodolomites. Sample R 26 a: nearly well ordered. R1aa and R35-g: disordered mainly with respect to the Ca layer. R1ad: extremely disordered with respect to the Mg layer

Dedolomitization on or near rock surfaces has been described from different localities (Al-Hashimi and Hemingway 1973; Frank 1981).

Protodolomite is one product of dedolomitization. Associated minerals are calcite, gypsum, hexahydrite, goethite, bloedite and halite, which form partially efflorescences (hexahydrite, bloedite), partially "rice grains" (Fig. 3.3.2.1), or flowstones (calcite), or encrustations (goethite) at the lower part of the profiles.

From a mineralogical point of view, the formation of protodolomite by weathering is of especial interest. Generally, protodolomite is found as a precursor of well-ordered dolomite in recent sediments (Usdowski 1967; Lippmann 1973; Magaritz et al. 1980). The present study shows (to the knowledge of the authors for the first time) that the mineral protodolomite is the successor of well-ordered dolomite, forming as a metastable product of dedolomitization. As a consequence, the reported occurrence of non-stoichiometric, Ca-rich dolomite in old sedimentary rocks (Füchtbauer 1962; Füchtbauer and Goldschmidt 1965; Reeder 1981) may be re-interpreted as formed by (sub-) recent weathering processes. From the thermodynamic point of view (Goldsmith and Heard 1961; Carpenter 1980), the conservation of "primary" Ca-rich dolomites in Mesozoic or Paleozoic rocks seems implausible.

3.3.2.4 The Formation of Evaporite Minerals in the Foreland Soils

Newly formed minerals are observed in and on the rocks of the escarpments and in the soils of the foreland. In addition to the aforementioned minerals (hexahydrite, gypsum, halite and bloedite), some dehydration products of H_2O-rich sulfates appear in the soils: thenardite ($NaSO_4$), leonhardtite ($MgSO_4 \cdot 4H_2O$), bassanite ($CaSO_4 \cdot 1/2H_2O$) and – rarely – anhydrite (with respect to bassanite from arid soils of other regions see Akopodje 1985, and Smykatz-Kloss et al. 1985).

The weathering solutions which reach the foreland are saturated with respect to $NaCl$, $CaSO_4$, $MgSO_4$ and Na_2SO_4. Using the data of D'Ans (1933) and the methods of Braitsch (1971) and Drouby et al. (1976), the evaporation sequence may be calculated from solubilities and compositions of the weathering solutions. The stability fields of the Na-K-Mg sulfates in the quinary system of sea water at saturation with respect to $CaCO_3$, $CaSO_4$, and $NaCl$ are shown in Fig. 3.3.2.5 (Braitsch 1971). The diagram includes also the composition of 30 foreland waters of the Jabal Dahar (open circles) and a hygroscopic salt sample which is a mixture of hexahydrite ($MgSO_4 \cdot 6H_2O$) and bischofite ($MgCl_2 \cdot 6H_2O$). Thermodynamic equilibrium provided thenardite or bloedite should be precipitated from the solution upon evaporation in addition to gypsum and halite. Precipitation of thenardite enriches the solution with respect to Mg^{2+}, so that it should pass the boundary between thenardite and bloedite. Consequently, thenardite should be re-dissolved and bloedite should be precipitated. But this will only happen in the case of ideal thermodynamic equilibrium and if the precipitate stays in contact with the solution. Normally, both conditions will not be fulfilled in the case of capillary transport of the solutions in playas or sabhkas. The parageneses of evaporite minerals in the foreland soils represent disequilibrium conditions.

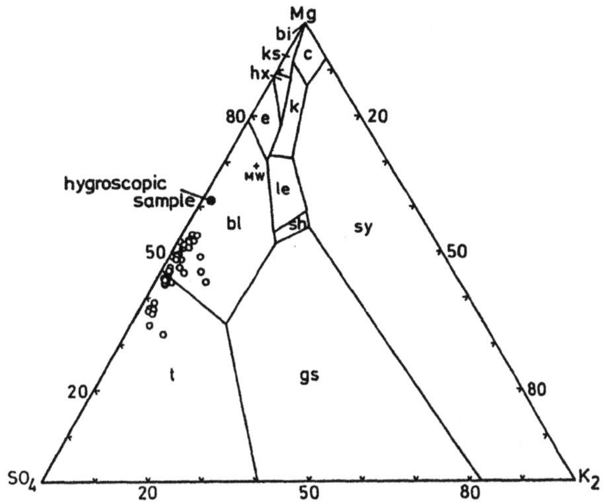

Fig. 3.3.2.5. Plots of water analysis in the Jaenecke diagram (quinary seawater system, 25 °C). Abbreviations: *bi* bischofite; *bl* bloedite; *c* kainite; *ks* kieserite; *le* leonite; *sh* schoenite; *sy* sylvite; *t* thenardite; SMOW standard mean ocean water

The temperature of the solutions was between 25 and 30 °C. This means that the stable Na sulfate is thenardite (not mirabilite), and the expected Mg sulfates are the less hydrated minerals kieserite ($MgSO_4 \cdot H_2O$) and leonhardtite ($MgSO_4 \cdot 4H_2O$). The main minerals in the foreland soils are thenardite and hexahydrite ($MgSO_4 \cdot 6H_2O$). Leonhardtite occurs in traces and kieserite not at all. The precipitation of hexahydrite requires a temperature of 30–40 °C (Braitsch 1971).

3.3.2.5 Conclusions

The observed weathering phenomena lead to the following conclusions.

- Under semi-arid conditions the main weathering process in the dolomitic limestones of the Jabal Dahar is dedolomitization.
- The first step of dedolomitization is the formation of a non-stoichiometric, Ca-rich dolomite (= protodolomite).
- The presence of $CaSO_4$ and NaCl supports the process of dedolomitization.
- The dissolution of (Fe-bearing) dolomite and of protodolomite leads to the formation of calcite (rice grains, flowstones), goethite (encrustations) and hexahydrite (efflorescences) and to a solution rich in sodium, magnesium and sulfate.
- The evaporation of this solution in the foreland occurs under disequilibrium conditions and leads to the formation of meta-stable evaporite minerals (thernadite, bloedite, hexahydrite, leonhardtite, bischofite, gypsum, halite).

3.3.3 Transformation Processes in Paleokarst Sediments and Chemistry of Modern Waters in the Aladag Region, Turkey

M. Cevrim and W. Echle

3.3.3.1 Background

The formation of vertical extended karst systems and the diverse fabrics of their various fillings are results of varying hydrodynamical and physico-chemical conditions in the hydrographical karst zones. Paleokarst systems are particularly suitable for study and interpretation of mineralogical reactions. Such investigations have been carried out in a small (12×4 km^2) area of the Aladag Mountains (TR) at 1600–3000 m altitude, in the Southern part of the Zamanti Region, Middle Taurus (Fig. 3.3.3.1). In this area the geology is characterized by alpine nappe tectogenesis (Tekeli 1980). The main tectonic units consist of sedimentary rocks, in particular limestones of Devonian up to Cretaceous age. They are often bordered by small ophiolithic zones. Intensive karstification since the Triassic and later, facilitated by local fracture zones (Wippern 1964; Imreh 1965; Cevrim 1984), took place in Permian and Jurassic pure limestones. The genesis of several Zn-Pb mineralizations related to an Eocene erosion surface is probably also connected with this paleogeographical evolution (Cevrim 1984; Cevrim et al. 1986).

The actual hydrological conditions in this area are independent of the inactive paleokarst system. The geochemical behaviour of spring and surface waters has been examined in relationship to the lithology and weathering processes. Along overthrust boundaries of nappes, perennially active springs exist. Due to prevalent

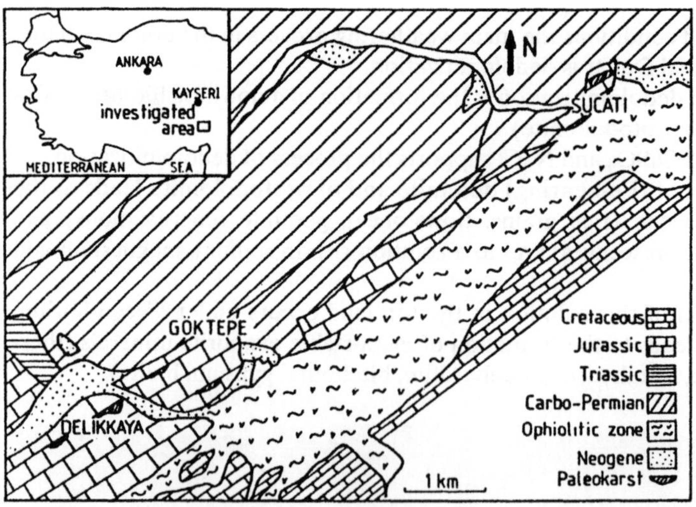

Fig. 3.3.3.1. Geological map of the investigated area in the Aladag Mountains

limestone terrain and intensive karstification, many springs and surface waters in the highland cease to flow in the late summer.

3.3.3.2 The Paleokarst Fillings

Composition and Origin of the Karst Sediments

The sedimentary, mostly allochthonous filling of the karst cavities exhibits the results of diverse weathering processes, which began in the various source rocks, continued during surface transport and in the karst environment. The grain size of the mostly detrital cavity fillings ranges from clay up to pebble. The coarsest fraction consists of friction detritus. The grain sizes of the clastic constituents diminish continuously from top to bottom of the karst filling. Due to intensive weathering, the original mineralogical composition can seldom be recognized. In the fillings, detrital and authigenic minerals can be distinguished.

The detrital cavity filling consists of breccias and splinters of limestone, fragments of quarzite, chert, radiolarite, and bauxite. Besides grains of quartz, amphiboles, pyroxenes, feldspars, micas, and other minerals altered to serpentine, minerals like septochlorite, smectite, kaolinite, mixed layers and illite have been observed. Chromite, magnetite, hematite and rutile are present in minor amounts.

Authigenic minerals are pyrite, marcasite, sphalerite, galena, smithsonite, hemimorphite, cerussite, limonite and gels of Fe-Al hydrous oxides. Besides precipitated calcite and aragonite, opal and chalcedony have been found. Zinc-alumosilicates and a Zn-Al-OH-CO_3 phase are other remarkable authigenic minerals (Cevrim and Echle 1988).

The detrital silicates and heavy minerals are essentially derived from the ophiolithic rocks. Only a small portion is from the residue of the karstified or eroded limestones (Cevrim 1984). Rendzina-like soils covering Malmian and Permian limestones are characterized by the same set of detrital minerals which can be found in the paleokarst sediments (Cevrim 1984). The karst filling took place in the post-Cretaceous, in particular during Eocene time (Cevrim 1984).

Depositional Accumulation and Weathering Processes in the Karst Environment

The investigation of different types of karst sediments was facilitated through numerous fresh exposures in the paleokarst. In the open-pit Zn-Pb mine of Delikkaya for example the fossil karst is exposed over 150 m in vertical extent (2850–2700 m a.s.l.) and shows a relatively complete hydrographic zoning according to Civijic (1918), Bernard (1973), Bögli (1978) and others (Fig. 3.3.3.2).

A typical feature of the cavity fillings is alternating bedding of detrital silicatic and chemically precipitated calcareous material. A periodic filling progress (Cevrim 1984) is indicated by changes in grain-size distribution from clayey to sandy material, geopetal fabric, interbedding, and also by changes of mineralogical and chemical compositions. All this may be observed over small distances from millimetres to decimetres. Table 3.3.3.1 summarizes the mineral assemblages at different depths of the paleokarst exposures at Delikkaya. These observations are of

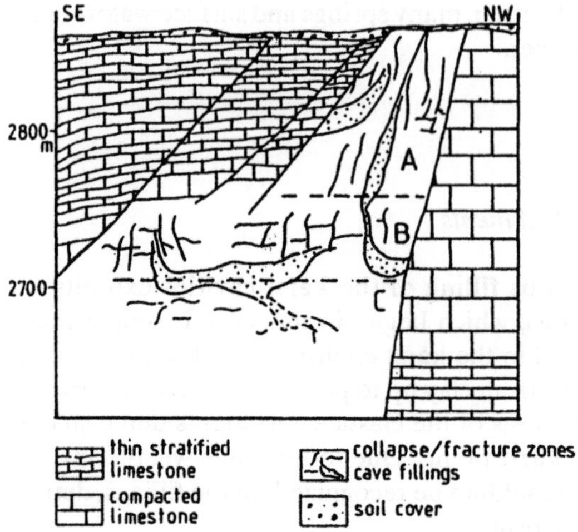

Fig. 3.3.3.2. Hydrographical zones of the paleokarst at Delikkaya (schematically). *A* percolation zone; *B* circulation zone; *C* stagnation zone

little help in order to establish relationships between mineral assemblages and particular hydrographic zones in the karst. Grain size distribution and roundness types are also of little help in this regard. The reason is that allochthonous and heterogenous materials of different origin and weathering state are transported and redeposited several times in the karst cavities. In a single thin section poorly weathered grains of olivine occur close to strongly altered olivines and amphiboles, and relatively fresh orthoclases and pyroxenes occur close to extremely decomposed plagioclases and other detrital silicates.

The sediments in karst cavities reflect their dependence on periodically changing hydrodynamic and physico-chemical conditions and on the karst topography. This indicated by strong variations with respect to the nature of the fillings in adjacent karst tunnels, joints and caves (Fig. 3.3.3.3). At the same level, mineral

Table 3.3.3.1. Semi-quantitative mineralogical composition (XRD) of selected karst sediments from different depths of the paleokarst at Delikkaya

Sample	Q	C	F	M	K	Sc	S	Ch	Sm	Gt	OM	Altitude (m)
DK 11	++	+++		+	++	(+)		(+)		++		2865
DK 13	+	+++		+	+					+++	+	2845
DK 23	(+)				(+)	+++	(+)	(+)		++	++	2825
AT 3	(+)		(+)	(+)	++	++	(+)	(+)	+	+	(+)	2805
DK 35	++	++		(+)	++			+		++	(+)	2755
DK 38	+	+			(+)	++	(+)			+	+++	2735
AT 1	+			++	++		(+)			++	+	2715
AT 6	(+)	+		+	++	(+)	(+)	+	+	+	(+)	2715
AT 8	++		(+)	+	(+)	+	(+)		+	+	+	2715

Q = quartz; C = calcite; F = feldspar; M = mica/illite; K = kaolinite; Sc = "septochlorite"; S = serpentine; Ch = chlorite; Sm = smectite; Gt = goethite; OM = Zn-Pb ore minerals (mainly smithsonite, hemimorphite, cerussite).
+++ Major components, ++ major-minor components, + minor components, (+) trace.

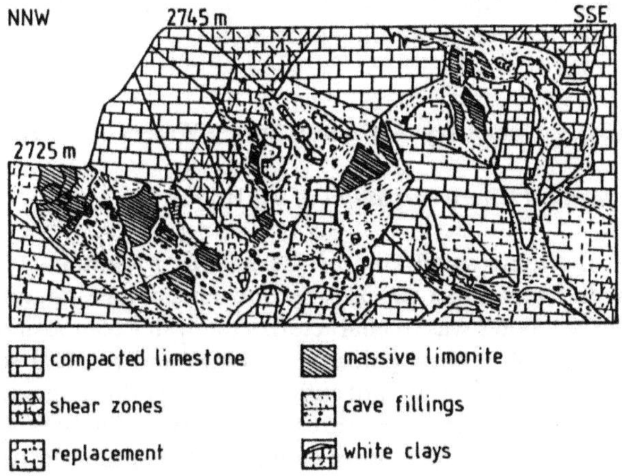

Fig. 3.3.3.3. Cross-section of a paleokarst exposure in the Zn-Pb open pit at Delikkaya

assemblages range from clayey green fillings with septochlorite and authigenic sulfides to cave clays with accretions of limonite, kaolinite-halloysite, Zn-carbonate, and even to kaolinite-smectite. These observations indicate an overlap of reducing phreatic and oxidizing vadose conditions. In some places the lateral variation of the karst fillings is furthermore due to an abrupt change of the limestone facies. At Delikkaya, stagnation zones with reducing conditions were created in low depth due to damming up by tectonically tilted, platy and clayey-bituminous limestones. The pure massive limestones in larger horizontal distances contain coarser-sized loamy fillings and exhibit the conditions of the percolation zone.

The open-space and cave fillings in the upper paleokarst levels with better percolation frequently have very heterogenous grain-size distributions and mineralogical compositions. Limonite, kaolinite and illite predominate and demonstrate a high weathering intensity.

The sediments of the deep and shallow ground water zones of the exposed paleokarst show better sorting, higher amounts of clay and are characterized by formation of speleothems. Serpentine and chlorite, which contribute to the typical green colour of these sediments, must be considered to be weathered phases of detrital ophiolite components. Reducing Fe- and Mg-rich slightly alkaline solutions generated by weathering processes, lead to the formation of smectite, berthierine and sulfides. In mineralized zones, a high zinc supply has caused the formation of Zn-berthierine and Zn-smectite (Cevrim and Echle 1988).

Further characteristic weathering products can be observed in the karstified limestones, where solutions reacted with limestone at cave rims and in fractured zones. White clays are formed by the reaction of Si- and Al-bearing waters (involved in weathering of detrital fillings) with decomposing limestone. Their mineral assemblage consists of amorphous phases, allophane, para-alumohydrocalcite, halloysite and meta-halloysite. During progressive alteration they were transformed into kaolinite and gibbsite. The high zinc content of solutions caused by weathering of the karst-hosted ores leads also to the formation of a Zn-Al-OH-CO_3 phase, a new mineral of the sjögrenite-pyroaurite group (Cevrim and Echle 1988) and Zn-Al-silicates (fraipontite, Zn-berthierine). The karst fillings exhibit numerous overlaps of the various weathering processes.

3.3.3.3 Chemistry of Modern Spring Waters of the Aladag Region

Chemical Characterization of the Waters

In the same small area of the Aladag Mountains (Fig. 3.3.3.1) studied for karst phenomena (Sect. 3.3.3.2), hydrochemical investigations were carried out. The aim of the studies was to characterize the water compositions and to show influences of changing lithology, paleokarst fillings and residual soil on the ion input into the waters. In this area, not influenced by anthropogenic contamination, most of the springs arise along overthrust boundaries of nappes and fractured tectonical structures. Water samples were collected at 21 springs in June 1983. At ten of the springs a re-sampling took place in October 1983. Simultaneously, 11, and later 1 surface water samples were collected. In October 1983, nearly all other springs, as well as the surface waters, had ceased to flow. Immediately after sampling total hardness, alkalinity, acidity and the amount of chloride were determined by titrating with Visocolor Testbesteck (Machery-Nagel). A portion of each water sample was preserved with concentrated HNO_3 and subsequently analyzed by atomic absorption spectroscopy (AAS Perkin Elmer 5000) for major cations (Ca^{2+}, Mg^{2+}, Na^+, K^+) and for selected base metals (Zn, Fe, Pb). Tables 3.3.3.2 and 3.3.3.3 show analytical spring water data for the major components.

The temperatures of the spring waters ranged from 6 to 12 °C and those of the surface waters from 8 to 15 °C. At the same locations water temperatures did not change from June to October. All spring waters are weakly alkaline, with pH values ranging from 7.2 to 8.2 according to their content of earth alkalines and hydrogen carbonate. The alkalinity values ranged from 1.7 to 5.3 meq/l, those of the acidity between 0.2 and 0.8 meq/l. Regarding the ionic balance, the sulfate content (not measured) should be very low. The total hardness varied between 2.2 and 5.4 meq/l the Ca^{2+} contents between 20 and 85 mg/l, and the Mg^{2+} contents between 1.2 and 28.0 mg/l. High concentrations represent waters from areas with basic magmatites. Low concentrations are related to dilution in the fissured rocks during recharge periods. The slightly to distinctly higher total cation content of spring waters in October compared with the samples collected in June (Fig. 3.3.3.4) and the increased Mg/Ca ratios at the locations TR19 and TR21 can be explained by shallow ground water exchange. The small change in June and October at location TR1 indicates a deeper ground water system.

The temperatures of the surface water varied between 2 and 15 °C, the pH between 7.75 and 8.30, the total hardness (Ca^{2+} + Mg^{2+}) between 2.4 and 3.8 meq/l, the alkalinity between 2 and 3 meq/l, and the acidity between 0.2 and 0.6 meq/l. The respective concentrations ranged between 10 and 68 mg/l Ca^{2+}, between 1.9 and 20.5 mg/l Mg^{2+}, between 0.08 and 2.0 mg/l Na^+ and between 0.10 and 0.88 mg/l K^+.

Table 3.3.3.2. Properties of spring water collected in June (J) and October (O) 1983

					meq/l			mg/l			
No.	Location	Date	T	pH	TH	Alk	Ac	Ca^{2+}	Mg^{2+}	Na^{+}	K^{+}
TR1	Gökoluk	J	8.0	7.90	3.0	4.6	0.4	30	2.8	0.92	0.30
TR1	Gökoluk	O	7.5					41	2.8	0.90	0.33
TR2	Sarioluk	J	9.0	8.10	3.2	3.2	0.4	43	11.0	1.70	0.13
TR2	Sarioluk	O						47	28.0	2.12	0.54
TR3	Sarioluk	J	8.0	7.50	3.6	3.6	0.6	63	6.0	1.00	0.64
TR3	Sarioluk	O	8.0	7.60		3.9	0.5	67	5.2	1.10	0.50
TR4	Böcülü	J	12.0	8.15	2.8	2.4	0.6	37	8.0	1.34	0.10
TR4	Böcülü	O						46	11.9	2.60	0.36
TR5	Stimaoluk	J	5.0	7.60	2.4	2.1	0.6	40	3.7	1.30	0.54
TR6	Aygirpinar	J	8.0	7.80	3.0	2.6	0.8	35	11.1	0.56	0.23
TR7	Karatas	J	8.0	8.10	3.0			39	7.4	0.72	0.28
TR7	Karatas	O	8.5	7.90				37	13.0	1.00	0.39
TR8	Tasoluk	J	9.0	7.85	3.4	3.1	0.6	56	3.5	1.24	0.41
TR9	Kösk	J	9.0	8.00	2.6	2.2	0.4	38	4.1	0.62	0.35
TR10	Disdöken	J	3.0	8.00	2.2	1.7	0.4	20	8.9	0.36	0.24
TR11	Fadimekadin	J	7.0	7.90	2.6	2.2	0.4	43	2.6	0.72	0.47
TR11	Fadimekadin	O	7.5	7.60		2.4	0.4	45	2.8	0.96	0.44
TR12	Tuzlupinar	J	12.5	7.95	5.4	5.3	0.8	68	15.1	3.26	0.09
TR13	Tuzlupinar	J	8.0	8.00	3.8	3.7	0.6	50	11.7	1.86	0.36
TR14	Tuzlupinar	J	7.0	8.20	3.8			46	7.7	2.24	0.64
TR15	Tuzlupinar	J	7.0	8.10	3.8	3.2	0.4	51	7.1	1.46	0.55
TR16	Kösoglu	J	8.0	8.00	4.4	3.8	0.6	60	8.6	1.18	0.16
TR16	Kösoglu	O	8.0	7.60				62	9.6	1.80	0.27
TR17	Köskyayla	J	6.0	7.20	3.0	2.0	0.2	31	6.3	1.24	0.67
TR18	Köskyayla	J	10.0	7.60	5.0	4.0	0.7	54	24.5	2.40	0.66
TR19	Köskyayla	J			5.0	4.4		65	11.5	2.00	0.12
TR19	Köskyayla	O	10.5	7.35		4.2		50	21.9	2.00	0.37
TR20	Sarioluk	J			2.8	2.2	0.4	46	1.2	0.52	0.16
TR20	Sarioluk	O						50	1.4	0.74	0.14
TR21	Sucati	J	11.0	7.80				55	1.7	0.86	0.37
TR21	Sucati	O	12.0	7.90		3.8	0.6	85	9.4	2.76	0.22

T = Temperature (°C), TH = total hardness, Alk = alkalinity, Ac = acidity

Table 3.3.3.3. Averaged properties of spring waters related to geological units of investigated area

	mg/l				meq/l				
	Ca^{2+}	Mg2+	Na^{+}	K^{+}	TH	Alk	Tl[a]	pH	
TR8, TR9, TR20	47	2.9	0.8	0.3	2.8–3.6	2.5	2.6	7.9	Carbo-Permian (limestone, shale, quarzite)
TR10	20	8.9	0.4	0.2	2.2	1.6	1.8	8.0	Triassic (shale)
TR21	55	1.7	0.9	0.4	3.0	n.d.	2.9	7.9	Jurassic (limestone)
TR1, TR3, TR14, TR15	48	5.9	1.4	0.5	3.0–4.0	3.0	2.9	8.1	Jurassic/Cretaceous
TR12, TR13, TR18, TR19	59	15.7	2.4	0.3	3.8–5.4	4.5	4.4	7.9	(limestone/ophiolite)
TR5, TR6, TR7	37	6.8	0.8	0.3	2.4–3.0	2.2	2.4	7.8	Cretaceous
TR11, TR16, TR17	44	5.4	0.9	0.4	2.4–4.4	2.5	2.2	8.0	(ophiolite/limestone)

[a]Tl = total cation load.
Other abbreviations see Table 3.3.3.2.

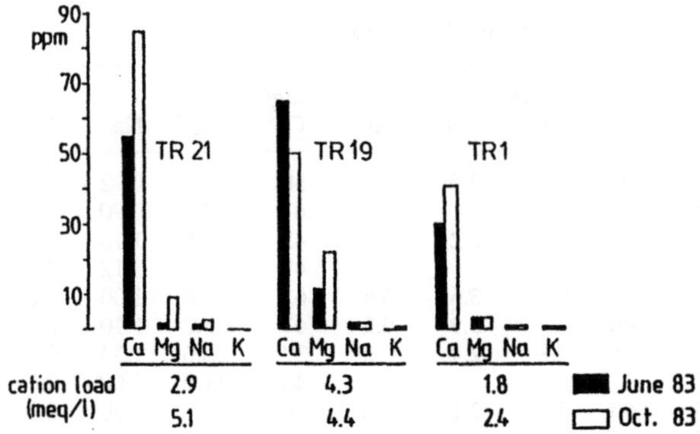

Fig. 3.3.3.4. Major cation concentrations of water sampled in June and October, 1983, from three springs in the Aladag Mountains

Lithological Control of the Water Chemistry

The absolute concentrations reported in Table 3.3.3.2 and summarized in Table 3.3.3.3 indicate that the chemical composition of the spring waters varies only little. However, the relative values of major cations show more pronounced differences and mainly reflect the influence of lithological factors within the catchment areas. Figure 3.3.3.5 shows meq-% values for Ca^{2+} and Mg^{2+}. Spring waters arising from limestones as well as comparable surface waters contain > 80 meq-% Ca^{2+}. Triassic shales and basic magmatites (ophiolite zone) contribute to the high relative amount of Mg^{2+} (> 20 meq-%) in the respective waters. Ca^{2+} released by lateritic weathering of ophiolithic rocks is responsible for the relatively high Ca^{2+} content of the spring waters. The relatively high amount of Mg^{2+} in some waters draining dolomite-free karstified limestones is due to the weathering of the Mg-bearing karst fillings and erosion soils. Both are in part results of the weathering of ophiolithic nappes (Sect. 3.3.3.2).

The concentrations of Na^+ and K^+ in the spring waters and in comparable surface waters indicate also a lithological control (Fig. 3.3.3.6). Waters flowing through basic igneous rocks are characterized by high relative amounts of Na^+. Waters from clayey sediments of the Upper Devonian, Carbo-Permian and Triassic contain more K^+. High average values of K^+ and increased K^+/Na^+ ratios of the spring and surface waters in Jurassic limestones are (similar to Mg^{2+}) the result of weathering and eluation of paleokarst fillings.

Zn, Pb, and Fe in the waters were measured in order to show the influence of the paleokarst-related Zn-Pb mineralization on the local ground water composition. Water collected in June had more than 20 µg/l Zn and 100 µg/l Fe in the immediate proximity of the ore deposits. In October the values were distinctly higher (20–50 µg/l Zn; up to 900 µg/l Fe), especially in low discharge springs and in surface waters draining the karstified Jurassic limestones. Pb was not detected by flame AAS. These low metal contents are thought to be a result of ion exchange processes between water and clay minerals and the incorporation of metal cations in silicates, carbonates and oxides formed in the karst environment (Sect. 3.3.3.2).

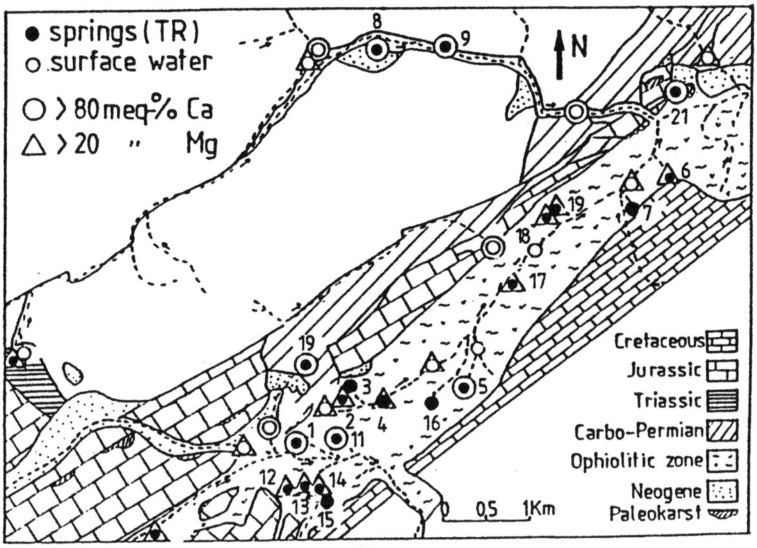

Fig. 3.3.3.5. Ca^{2+} and Mg^{2+} content of spring and surface waters in lithologically different drainage basins

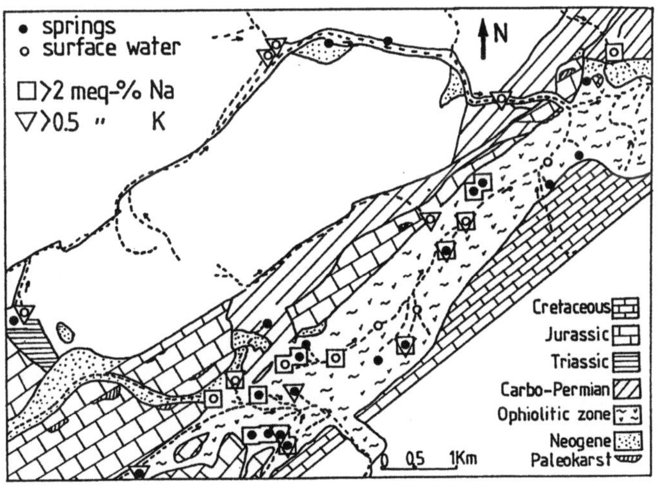

Fig. 3.3.3.6. K^+ and Na^+ content of spring and surface waters in lithologically different drainage basins

3.3.3.4 Conclusions

The petrological and mineralogical studies of post-Cretaceous paleokarst sediments in Malmian limestones reveal that the predominant detrital silicates and heavy minerals are essentially derived from ophiolithic rocks. Weathering processes resulted in the formation of authigenic sulfides, clay minerals, zinc-alumosilicates and Zn-takovite in the hydrographic karst zones.

The modern spring and surface waters in this area are hydrogen carbonate waters with low total concentrations. Nevertheless, a dependence of the water chemistry on the lithology and mineralogy of the drainage area does exist. Waters flowing through limestones predominantly contain Ca^{2+}. Mg^{2+} and Na^{+} are relatively enriched in waters from ophiolites and Mg^{2+} and K^{+} dominate in waters flowing through shales. The low concentrations of Zn, Pb and Fe in waters from the immediate proximity of Zn-Pb ore deposits are the result of effective "filtering" by ion exchange and metal fixation in silicates, carbonates and oxides formed in the karst environment.

4 Silicate Systems

4.0 Introduction

G. Matthess

Silicates are major components of the Earth's crust both in magmatic and metamorphic rocks and in sedimentary rocks. Igneous and metamorphic rocks are at or near the Earth's surface in about 20% of the land surface. They are covered in the remaining parts of the present continental surface by 5×10^8 km^3 of sedimentary rock (in geosynclinal areas 75% vol. with an average thickness of 10 km; in platform areas 25% vol. with an average thickness of 1.8 km), within which silicate sediments are predominant (shales, clays < 77%, sandstones, graywackes, arkoses > 15%; Wedepohl 1969). For the interaction between water and rocks of the same mineralogical composition, the same principles hold independently of their nature as crystalline rocks, consolidated or unconsolidated sedimentary rocks. Silicates and especially the feldspars make up nearly 80% and 57.9% of the Earth's crust, respectively (Correns 1940). Wedepohl (1969) reports 41% plagioclase, 21% K-feldspar, 6% amphibole, 4% biotite, 4% pyroxene, 0.06% olivine and 21% quartz for the intrusive rocks of the upper crust. The importance of silicate weathering for substance fluxes is indicated by the total annual chemical load delivered from the continents to the oceans of 39.3×10^{14} g (29.9 g/(m^2.a). With respect to components of special interest here, the dissolved masses annually delivered by rivers to the ocean are quoted as follows: Al 0.003×10^{14} g, Na^+ 2.07×10^{14} g and SiO_2 4.26×10^{14} g (Garrels and Mackenzie 1971), which correspond to fluxes of Al 2.3×10^{-3} g/(m^2.a) = 0.09 mmol/(m^2.a), Na^+ 1.57 g/(m^2.a) = 52.4 mmol/(m^2.a) and SiO_2 3.24 g/(m^2.a) = 53.9 mmol/(m^2.a). Investigations in the research programme were focussed on the unsaturated and saturated zones in sand and gravel aquifers, where the hydraulic conditions were relatively easy to establish.

Oxidation and reduction are geochemically important processes, which together with the hydrogen ion activity (pH) determine the solubility of many substances. The presence of oxidizing or reducing agents controls these processes going on in the weathering crust of primary rocks and in the porous systems of unconsolidated sediments. Major oxidizing components in subsurface systems are atmospheric oxygen, and the oxidized states of dissolved ions or solids like Fe^{3+}, Mn^{3+}, SO_4^{2-}, NO_3^-, H^+ and $Fe(OH)_3$. Reducing conditions are caused by oxygen-consuming substances, particularly by organic matter, and in the presence of Fe^{2+}, S^{2+}, NO_2^-, NH_4^+, H_2 and $Fe(OH)_2$. Utilization of oxidized species in aquifers in the presence of reduced dissolved organic matter apparently follows a sequence in the order of their decreasing negative values of free enthalpy change at pH 7. The oxidized species are reduced in the following order: dissolved oxygen, nitrate, manganese (IV) oxide, ferric hydroxide, sulfate, dissolved CO_2 (i.e. HCO_3^-), and dissolved nitrogen (Stumm and Morgan 1970; Champ et al. 1979). Occurrence of such a sequence has been observed in ground water flow systems by Jackson and Inch (1980), Edmunds (1977), Edmunds et al. (1984), Golwer et al. (1970, 1976) and Obermann (1982).

The redox reactions in sandy and gravelly aquifers were studied at Bocholt (Leuchs and Obermann, Sect. 4.1.1, this Vol.), Fuhrberger Feld near Hanover (Böttcher et al., Sect. 4.1.3), in a shallow ground water system near Bensheim/Hessen (Ebhardt and Fritsch, Sect. 4.1.4), in a lake shore zone in Berlin (Brühl et al., Sect. 4.1.5), in a leakage system in the syncline of Dorsten (Northrhine-Westphalia) (Hoffmann et al., Sect. 4.1.6) and in a deep aquifer system at Hamburg (Löhnert et al., Sect. 4.1.7). A major result is the proof of the possible use of iron sulfides as reducing agents for nitrate, first detected at Fuhrberger Feld (Sect. 4.1.3) and confirmed at Bocholt (Sect. 4.1.1). Details of the redox reactions were studied using isotopes. Among these the investigations using nitrogen and oxygen isotopes as indicators of nitrification and denitrification processes are to be mentioned (Schmidt et al., Sect. 4.1.2).

Field studies in silicate/water systems showed that the weathering behaviour of feldspars and quartz could be described in some cases by determining the saturation indices of the pore solution in contact with these minerals. This was obvious when corrosion or precipitation effects could be observed by SEM inspections of the mineral grains, which coincided with negative or positive saturation indices, indicating undersaturation or oversaturation of the respective solutions (Ohse et al. 1984). However, discrepancies between theoretical predictions and the real findings occur. The reason for this may be incomplete or poor chemical data, inadequate water samples, inadequate thermodynamic input constants for the dissolved and the solid species, the presence of organic or inorganic coatings which reduce the contact surfaces between mineral grains and water, the neglect of organic complex-forming species in the geochemical models used, and finally kinetic effects, which in the given hydrologic system produce non-equilibrium conditions. Laboratory experiments on the dissolution behaviour of feldspar show a characteristic parabolic curve of concentration of dissolved components vs. time (Helgeson et al. 1984). This finding is usually explained by diffusion through a coating totally covering the mineral grain as the rate-limiting step. The coating is assumed to be caused by precipitates (precipitate-protection layer theory; Wollast 1967; Helgeson 1971) or by leaching of Na, K, Ca or Si from the alumosilicate structure (residual layer theory; Correns and Engelhardt 1938; Luce et al. 1972; Paces 1973; Busenberg and Clemency 1976; Chou and Wollast 1984; Goossens et al. 1989). The surface reaction hypothesis (Berner 1981; Dibble and Tiller 1981; Aagaard and Helgeson 1982) considers the fast dissolution of small crystallites from the crushed minerals as cause for the initially higher reaction rates. Furthermore, the precipitation of secondary phases covering the grain surfaces may explain the parabolic curve (Holdren and Adams 1982). However, the same macroscopic effect may be produced by the blocking of the reactive surfaces by mineralorganic complexes (chemisorption) (surface coordination model; Furrer and Stumm 1986).

Consequently, special field and laboratory studies dealt with some aspects of these open questions.

Mobility of aluminium species play an important role in weathering processes. The vertical distribution of aluminium species in soil profiles at forest sites in the Solling, Wingst, Hils hills, and in Fuhrberger Feld were studied. It could be shown that organic complexes play an important role in the transport behaviour of aluminium (Dietze and Ulrich, Sect. 4.2.1). Seasonal and vertical variations of Na^+, Ca^{2+}, SO_4^{2-}, Al and pH in the unsaturated and saturated zone of a sandy aquifer at a forest ecosystem near Heidelberg could be related to ion exchange reactions. The local silicate weathering rate was derived from mass balances (Jacob et al., Sect. 4.2.3). Interaction between pore solutions and clay minerals were studied at two soils profiles on Saalian tills and fluvioglacial sand in Hamburg. The changes observed mainly deal with exchange of Ca^{2+} and K^+ and with the formation of coatings and cements by amorphous Fe-, and Mn precipates (Sztuka and Valeton, Sect. 4.2.2).

The interrelation between the hydraulic flow system and the rates of chemical reactions determines whether equilibrium between water and rock may exist. Considering the overall knowledge on silicate weathering rates, it is to be expected that equilibrium will often not be attained in near-surface water bodies. However, for a better understanding of the processes involved and of their kinetics, laboratory investigations were performed using feldspars from three pegmatites and glaciofluvial sand from the Segeberger Forst south of Kiel (Schenk et al., Sect. 4.3.1). These results were used together with data from Petersen (Sect. 2.2.5) in the evaluation of field data from three sites in the Segeberger Forst (Matthess et al., Sect. 4.3.2), and indicate an important role of organic substances in the weathering processes.

4.1 Redox Reactions in the Subsurface

4.1.1 Anoxic Reaction Zones in an Aquifer Influenced by Increasing Nitrate and Sulfate Contents

W. Leuchs and P. Obermann

4.1.1.1 Degradation of Organic Matter in the Environment

Under the conditions of the Earth's surface organic matter is decomposed by microorganisms into stable solids, solutes or gases dependent on the geochemical environment. This is connected, as mentioned before, with the sequential reduction of oxygen, nitrate, manganese oxide, iron hydroxide and sulfate, which is characteristic for the hydrochemical evolution of many ground waters. The different inorganic electron acceptors are often used in a distinct succession along the ground water flow path and the ground water depth respectively according to decreasing standard free reaction enthalpies (Table 4.1.1.1).

However, the sequence of reactions listed in Table 4.1.1.1 can be more or less complete depending on the concentrations of reactive inorganic and organic electron acceptors, depth of aquifer and ground water flow velocity. Ground water systems, in general, exhibit low contents of organic substances which are relatively resistant to microbial attack (e.g. Westrich and Berner 1984). Thus, the reaction sequence is mostly incomplete and is extended over the entire depth of the aquifer. In contrast, ocean and lake sediments rich in reactive organic substances show a complete degradation sequence within the upper decimetres of the sediment profile (e.g. Claypool and Kaplan 1974).

The present study on the vertical zoning of such a microbially mediated degradation sequence was carried out in a shallow unconfined aquifer system in the lower Rhine valley near Bocholt (BOC), Northrhine-Westfalia (Fig. 4.1.1.1). The hydrostratigraphy at the site is 30 to 50 m of sand and gravel with corresponding hydraulic conductivities between 8×10^{-5} m/s and 1×10^{-3} m/s; today the ground water flows with a mean apparent flow velocity of about 25 m/a in a western direction.

4.1.1.2 Sampling and Analytical Procedure

For sampling purposes the boreholes BOC1 and BOC4 to BOC7 were dry-drilled down to the aquiclude, a clay of Miocene age with a hydraulic conductivity $< 10^{-7}$ m/s (for a three-dimensional plot of the hydrogeological conditions see Fig. 6.2.4.2, this Vol.). Sediment samples were collected in plastic bags and immediately flooded with nitrogen gas before sealing. The samples were freeze-dried and stored at –30 °C prior to chemical and mineralogical analysis to avoid oxygenation.

In the sediments, organic carbon as a measure of the organic substances was determined by coulometry as the difference between total carbon (O_2-combustion at 1200 °C) and inorganic carbon content (acid treatment).

Pyrite sulfur was determined by coulometric titration after oxidation to sulfur dioxide at 950 °C (S_{950}). Other co-oxidized sulfur compounds, e.g. iron monosulfides and organic sulfur, occur only in negligible amounts, their contents being

Table 4.1.1.1. Standard reaction enthalpies (ΔG_r^o) for bacterial redox reactions. Data taken from Berner (1980) and Thauer et al. (1977)

	Reaction	ΔG_r^o [kJ/mol]
O_2 Reduction	(1) $CH_2O + O_2 \rightarrow CO_2 + H_2O$	- 475
NO_3^- Reduction	(2) $CH_2O + \frac{4}{5}NO_3^- \rightarrow \frac{2}{5}N_2 + \frac{4}{5}HCO_3^- + \frac{1}{5}CO_2 + \frac{3}{5}H_2O$	- 448
Mn(IV) Reduction	(3) $CH_2O + 3CO_2 + H_2O + 2MnO_2 \rightarrow 2Mn^{2-} + 4HCO_3^-$	- 349
Fe(III) Reduction	(4) $CH_2O + 7CO_2 + 4FE(OH)_3 \rightarrow 4Fe^{2-} + 8HCO_3^- + 3H_2O$	- 114
SO_4^{2-} Reduction	(5) $CH_2O + \frac{1}{2}SO_4^{2-} \rightarrow \frac{1}{2}H_2S + HCO_3^-$	- 77
CO_2 Reduction	(6) $2H_2 + \frac{1}{2}CO_2 \rightarrow \frac{1}{2}CH_4 + H_2O$	- 66
CH_4 Fermentation	(7) $CH_2O \rightarrow \frac{1}{2}CO_2 + \frac{1}{2}CH_4$	- 58

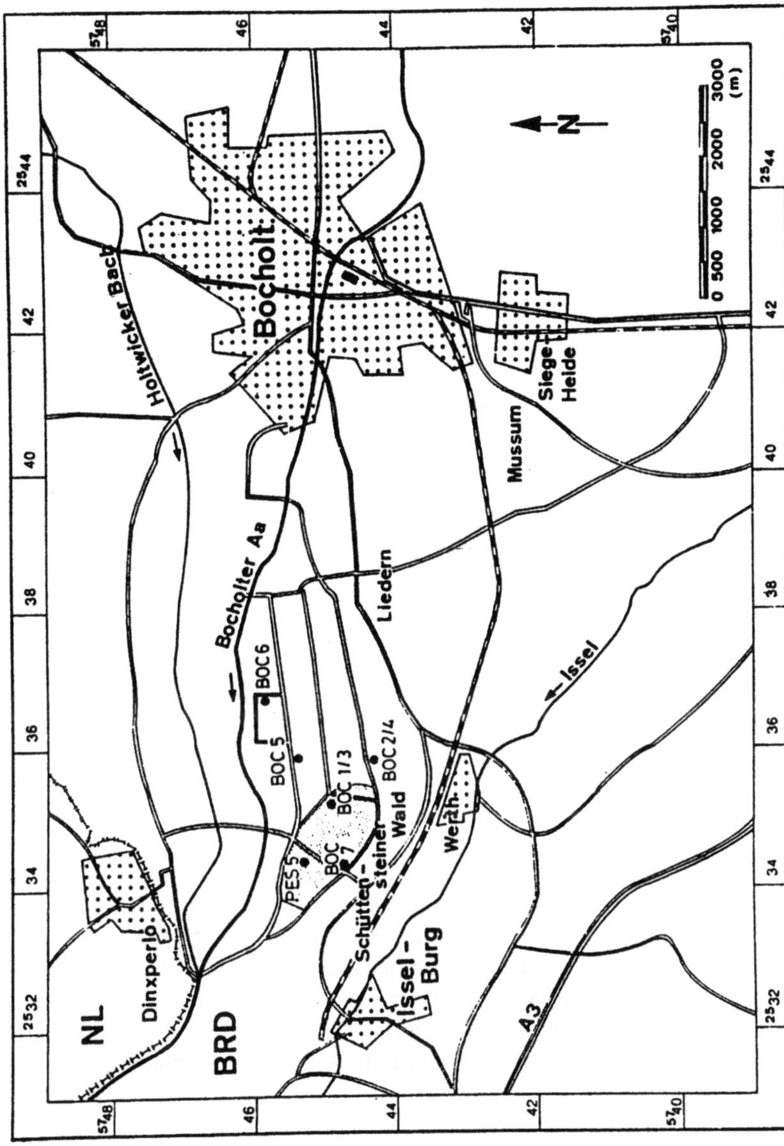

Fig. 4.1.1.1. Study area Bocholt aquifer system

one to two orders of magnitude lower than those of pyrite sulfur. Sulfur bound in iron monosulfides was analyzed separately as hydrogen sulfide after distillation with dilute hydrochloric acid (Berner 1964).

For sampling of ground water from different depths at a site where new multi-levels wells were being constructed using an approach similar to Atakan et al. (1974b). The wells consist of up to 20 single sampling points attached to a central PVC pipe at different depths, each connected to the ground level by PVC tubes

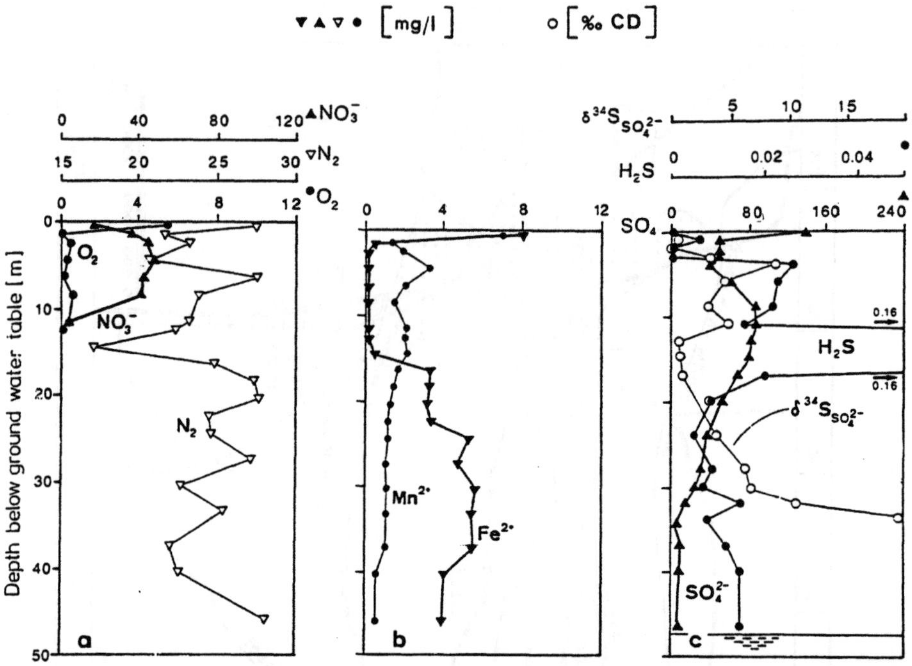

Fig. 4.1.1.2. Depth distribution of concentrations. **a** Oxygen, nitrate, and nitrogen (well BOC4). **b** *Manganese and ferrous iron* (well BOC4). **c** Sulfate and dihydrogen sulfide, and $\delta^{34}S$ of sulfate (versus CD troilite) at well BOC7 (measurements, Prof. Dr. Puchelt/Karlsruhe)

(Leuchs 1988). To avoid vertical mixing of ground water during the sampling procedure all sampling points were pumped simultaneously with the same pump while ground water from only one point was passed through the sample flask. This handling guaranties a high depth resolution (Fig. 4.1.1.2). The chemical ground water composition was determined with standard methods: pH and E_H by electrometry, O_2 by electrometry or GC, N_2 by GC, CO_{2total}, HCO_3^-, and SO_4^{2-} by titrimetry, NO_3^-, H_2S_{total} and Fe^{2+} by photometry, Ca^{2+} and Mn_{total} by AAS (Leuchs 1988).

4.1.1.3 Reaction Zones

All reactions or reaction products mentioned in Table 4.1.1.1 could be identified in the aquifer systems. The redox reactions or groups of them occur in distinct vertical zones or small and isolated areas of the aquifer, which is evident from the sedimentological and hydrochemical data given below. The respective microbial populations are described in Section 5.3.2 (this Vol.).

Consumption Zone of Oxygen, Nitrate and Manganese Oxide

Water in the uppermost portion of the saturated zone has low dissolved oxygen concentrations due to oxygen consumption in the unsaturated zone. The highest content of 5.5 mg/l is found at the ground water table (Fig. 4.1.1.2a). Further depletion within 5 m below the ground water table corresponds to nitrification, oxygen reduction or iron hydroxide precipitation.

Nitrate reduction starts simultaneously at the water table but continues into greater depths than oxygen consumption (Fig. 4.1.1.2a). The distribution of nitrate is very similar in all wells (e.g. Fig. 4.1.1.2a). The concentrations are highest (50 mg/l) within the first 5 m below the water table and decrease with depth. Thus, nitrate cannot be detected at 12 m below the water table. The depth distribution of nitrate is influenced by three processes:

- nitrification near the ground water table,
- variations in nitrate input concentrations,
- denitrification by bacterial nitrate reduction to nitrogen gas.

The influence of denitrification is poorly reflected by the depth distribution of dissolved nitrogen (Fig. 4.1.1.2a). Presumably the dissolution of soil gaseous nitro-

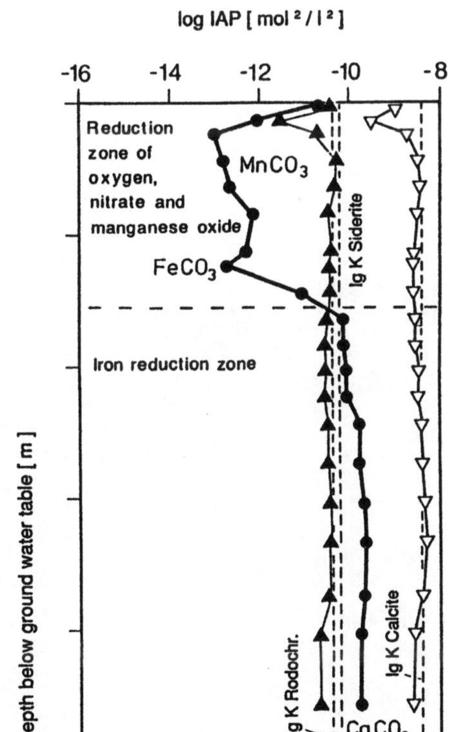

Fig. 4.1.1.3. Depth distribution of SI of various carbonates (*dashed lines*) at well BOC4, calculated with WATEQF. (Plummer et al. 1976)

gen of the ground air at the water table, which is in equilibrium with the different partial pressures over the recharge area, dominates over the nitrogen originating from nitrate reduction along the ground water flow path. However, evidence for denitrification is given by increase of the ratio N_2/Ar with depth and a positive correlation between $\delta^{15}N$ of nitrate and nitrogen gas (Sect. 4.1.2).

Elevated concentrations of manganese ions (about 4 mg/l) are found within the nitrate reduction zone, indicating two competing reactions at the same depth (Fig. 4.1.1.2b). The competition results from various types of bacteria present.

Zone of Iron Hydroxide Reduction

The reduction zone of nitrate and manganese oxide is characterized by very low concentrations of ferrous iron (< 0.2 mg/l). After complete removal of nitrate,

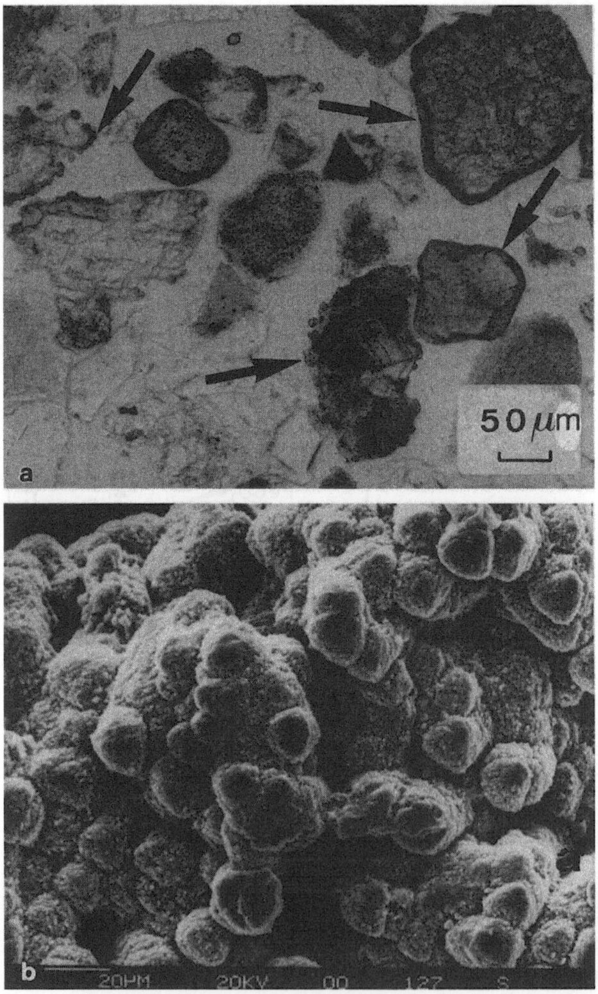

Fig. 4.1.1.4. Photomicrographs of a thin section (**a**) and SEM samples (**b-d**). **a** Overgrowths of authigenic siderite on sedimentary calcite grains. **b** Framboidal siderite with the composition $Fe_{0.78}\ Ca_{0.13}\ Mn_{0.09}\ CO_3$ (EDAX)

ferrous iron increases to values of about 5 mg/l below 15 m depth (Fig. 4.1.1.2b), indicating increased activity of iron-reducing bacteria. It is likely that in the presence of nitrate, iron-reducing bacteria cannot compete successfully with nitrate (or manganese)-reducing bacteria because nitrate reduction yields lower ΔG_r values (about –450 kJ/mol) than iron reduction (ΔG_r about –120 kJ/mol).

The elevated concentration of ferrous iron in the uppermost sampling point (oxygen and nitrate reduction zone) is due to the dissolution of iron hydroxides in acidic ground water percolating through carbonate-free sediments and subsequent reduction of ferric to ferrous iron.

Whereas in the nitrate/manganese reduction zone the formation of rhodochrosite may be expected from thermodynamic calculations (Fig. 4.1.1.3), siderite was identified in the iron reduction zone controlling ferrous iron concentration in the aqueous phase.

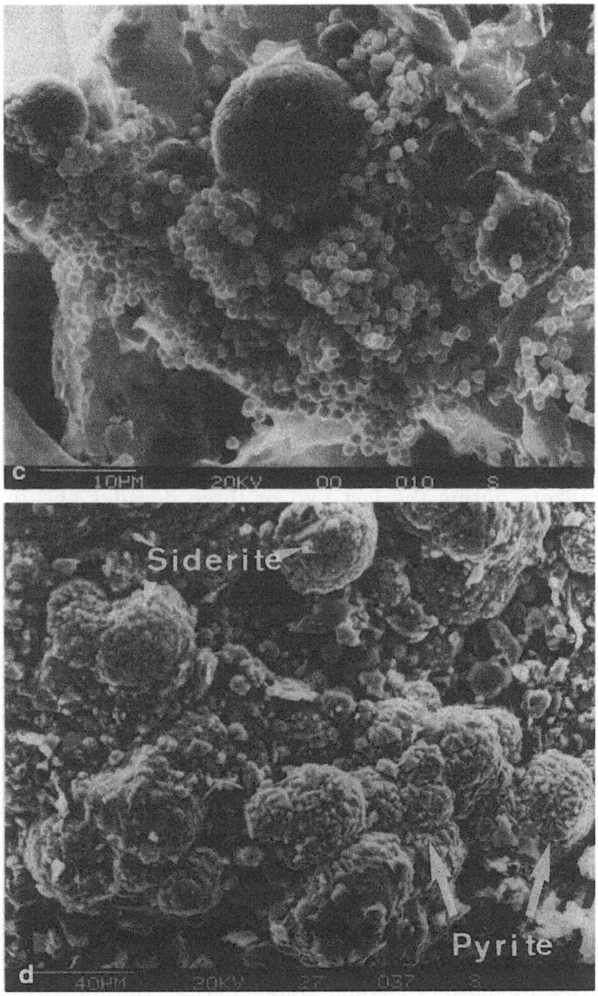

Fig. 4.1.1.4. **c** Framboidal and euhedral pyrite from a wood surface. **d** Framboidal and euhedral pyrite replacing corroded formerly formed siderite

Table 4.1.1.2. Composition of authigenic siderites suggested by energy dispersive X-ray analyses

Hole	Depth below ground water table	Mean composition	Number of measures	Kind of sample	Reaction zone
BOC1	8	$Fe_{0.74}\,Ca_{0.15}\,Mn_{0.11}\,CO_3$	3	Fine sand	Iron
BOC1	12	$Fe_{0.80}\,Ca_{0.16}\,Mn_{0.04}\,CO_3$	2	Fine sand	hydroxide
BOC1	25	$Fe_{0.60}\,Ca_{0.30}\,Mn_{0.10}\,CO_3$	1	Fine sand	reduction
BOC7	33	$Fe_{0.82}\,Ca_{0.15}\,Mn_{0.03}\,CO_3$	8	Wood debris	Relic of
BOC7	44	$Fe_{0.85}\,Ca_{0.10}\,Mn_{0.05}\,CO_3$	4	Wood debris	methanogenesis

Framboidal siderite forms coatings preferably upon sedimentary calcite grains (Fig. 4.1.1.4a,b). This indicates that siderite is authigenic. Energy dispersive x-ray analysis (EDAX) showed that this authigenic siderite contains calcium and manganese carbonate impurities of up to 20 mol% (Table 4.1.1.2).

The solubility control by the observed mixed carbonate phase is confirmed by the SI depth distribution, the ion activity products (IAP) of siderite, rhodochrosite and calcite compared to the solubility constants ($\log K_{siderite} = -10.34$, $\log K_{rhodochrosite} = -10.45$, $\log K_{calcite} = -8.41$ at ground water temperature of 10 °C). Below the nitrate reduction zone the ground water remains slightly supersaturated (SI > 0) with respect to siderite, saturated (SI = 0) with respect to calcite and slightly undersaturated (SI < 0) with respect to rhodochrosite (Fig. 4.1.1.3).

Occurrence of Sulfate Reduction Environments

The ground water sulfate concentration decreases continuously from about 150 mg/l at the water table to values below 10 mg/l between 35 and 50 m depth (Fig. 4.1.1.2c). This distribution is due to increased usage of nitrate and sulfate containing fertilizers during the last two or three decades and to increased oxidation of iron monosulfides (amorphous FeS, mackinawite) by nitrate in the upper parts of the aquifer. Penetration of young sulfate from fertilizers into the deep and old ground water, which has a MRT of about 300 years at the base of the wells (Sect. 6.2.6), results from vertical transversal dispersion, a process which has not yet smoothed (balanced) the concentration gradient.

Additionally, bacterial sulfate reduction takes place along the sulfate depth profile, as is suggested by elevated dissolved sulfide concentrations and increasing $\delta^{34}S$ values of sulfate with depth (Fig. 4.1.1.2c). Dissolved sulfide is found not only at greater depth but also at lower ground water levels containing oxygen or nitrate. Obviously, sulfate reduction occurs within micro-environments in small lenses or layers of wood and peat embedded in the aquifer, which offer higher levels of degradable organic matter than the surrounding sediment. Iron sulfides should precipitate in the presence of iron within these micro-environments as a consequence of sulfate reduction, similar to the precipitation of a metastable iron monosulfide (e.g. mackinawite) in anoxic marine sediments, which then recrystallizes to thermodynamic stable pyrite (Berner 1970). This assumption could be verified by sulfide determinations and scanning electron microscope investigations (SEM) from isolated wood samples. Concentrations of pyrite sulfur (S_{950}, Fig. 4.1.1.5a) and iron monosulfide sulfur (S_{FeS}, Fig. 4.1.1.5b) are depicted as a function of the content of organic substances in Fig. 4.1.1.5. Compared to the organic-poor sandy sediments (< 2% wt. organic carbon), the wood material (about 50% wt. organic carbon) is characterized by a maximum concentration of S_{950} (about 4% wt.) and S_{FeS} (about 0.1% wt.).

Replicates of the same wood sample have different sulfide contents. The wood surface is probably coated by isolated patches of iron sulfide. SEM observations confirm that framboidal FeS or FeS_2 and euhedral FeS_2 (Fig. 4.1.1.4c) cover a minor part of the wood surface. The sulfide overgrowths probably reflect zones of different resistance against microbial degradation in wood debris.

Methanogenesis

Pyrite of wood samples from deeper aquifer parts sometimes replaces previously formed siderite (Fig. 4.1.1.4d and Table 4.1.1.2). Presumably this siderite was precipitated under methanogenic conditions, when the deep old ground water was sulfate-free. Methanogenesis (CO_2 reduction or acetate fermentation) is inhibited in aerobic ground water environments. However, methanogenic bacteria (thermophilic) and dissolved methane were detected in deep ground waters (Beeman and Suflita 1987; Ghiorse andWilson 1988), even as deep as 1200 to 1800 m (Olson et al. 1981).

Under the influence of infiltration and rapid downward movement of anthropogenic sulfate the older methanogenic environment has continuously been replaced by sulfidic conditions followed by the transformation of siderite into pyrite.

Dissolved methane was measured in well BOC7. Its concentration varies between 0.015 and 2.3 mg/l. The maximum concentrations occur in the intermediate section of the ground water body. It is assumed that this methane is formed recently after the complete removal of sulfate in few organic micro-environments.

4.1.1.4 Development in Time

The Bocholt aquifer system is characterized by the reduction of oxygen, nitrate and manganese in the upper part (maximum depth today 15 m below the ground water table) and by reduction of iron hydroxide at greater depth. Sulfate reduction occurs in wood and peat micro-environments over the entire depth of the aquifer. Presumably methanogenesis is restricted to a few micro-environments where sulfate could be completely degraded. The levels and depth intervals of the different redox reactions depend on the kind and the amount of infiltrating inorganic electron acceptors.

About 300 years ago, when the oldest ground water present today was recharged, only low amounts of nitrate and sulfate infiltrated. Consequently, both nitrate and sulfate were totally removed within the upper portions of the aquifer, allowing for methanogenesis and siderite formation in deeper parts. Today a thick zone of iron reduction and siderite precipitation occurs almost over the entire aquifer depth in the organic-poor sandy environment (< 2% wt. organic C).

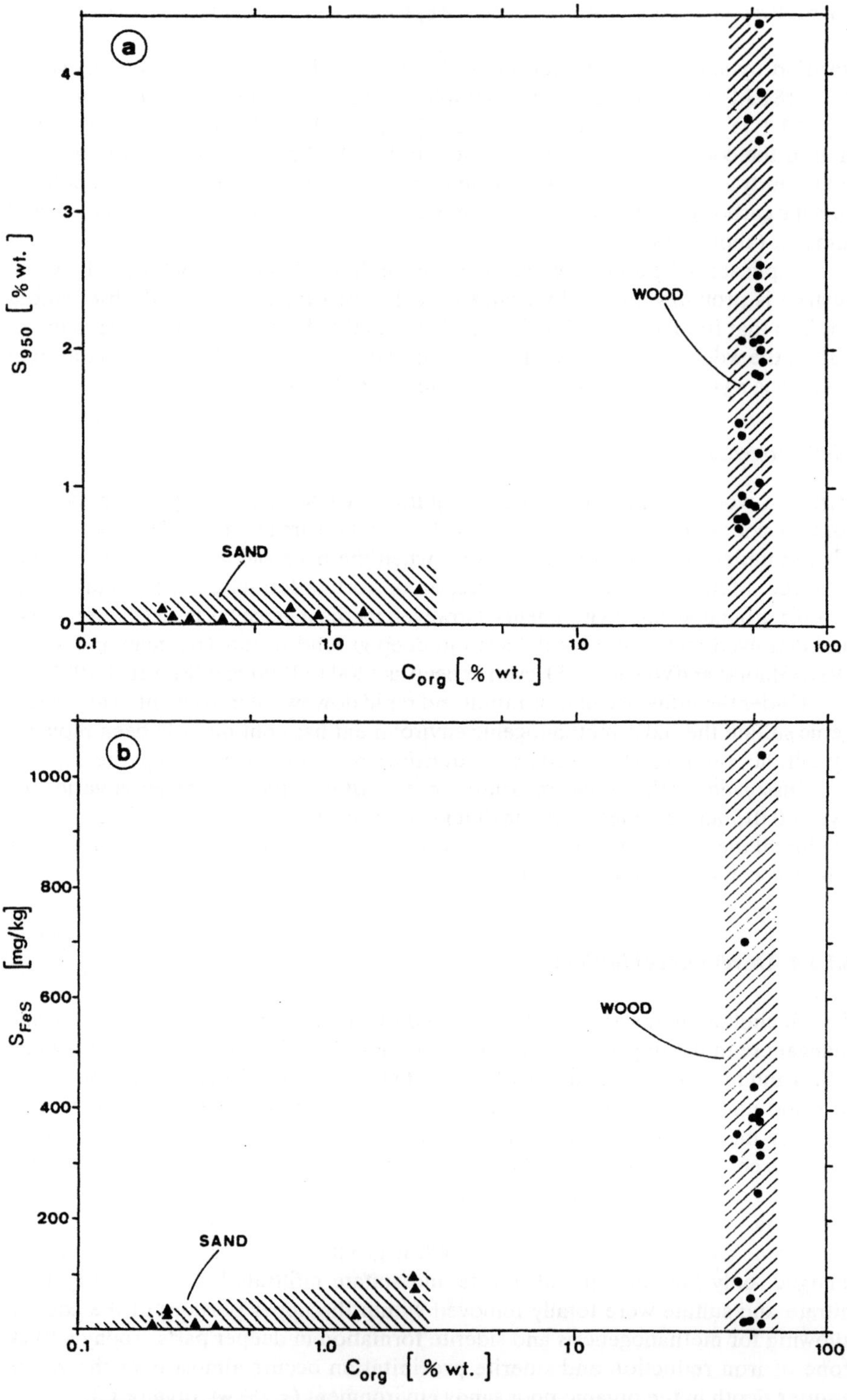

Fig. 4.1.1.5. Relation between concentrations of solid sulfide (**a** pyrite-sulfur S_{950}; **b** iron monosulfide-sulfur S_{FeS}) and organic carbon content

The increased application of fertilizers in the last two or three decades has resulted in a nitrate and sulfate contamination of the shallow ground water. Thus, surplus nitrate and sulfate penetrates into formerly nitrate- and sulfate-free deep ground water. As a consequence, the iron reduction zone continuously moves downwards, because the microbial community prefers nitrate as electron acceptors for thermodynamic reasons. Under the changed geochemical conditions authigenic siderite becomes unstable and is most likely transformed into thermodynamically stable rhodochrosite.

The replacement of siderite by pyrite takes place in organic micro-environments because at depth sulfate-reducing bacteria successfully compete with methane-forming bacteria. At the base of the aquifer, methanogenic siderite relics are still preserved, indicating the very recent downward shift of the sulfate-reducing environment.

Further reduction of the different subsurface electron acceptors will cause a systematic and irreversible consumption of the degradable organic material in the sediments. It will also affect sorbed organic matter that has been accumulated by the entrainment of soluble soil humic substances over thousands of years preferably in the shallower aquifer sediments. When the available solid organic matter will be depleted, the capacity of the system for any further degradation will only depend on the amount of coinfiltrated dissolved organic substances. However, this carbon source will only allow for the reduction of much lower concentrations of electron acceptors than are present in the aquifer today.

4.1.1.5 Effects on pH, E_H and Dissolved Inorganic Carbon

The microbial reaction sequence is accompanied by increasing pH-values (max. pH = 7.7) and decreasing E_H values (min. E_H = +20 mV) with depth (Stumm and Baccini 1978; Froelich et al. 1979). Two groups of ground water samples can be distinguished in a pH-E_H diagram, depending on their nitrate content (Fig. 4.1.1.6).

The E_H of nitrate-containing samples does not show any statistically significant pH dependence in the pH range 6.5 to 7.5 (E_H about +200 mV). In contrast, the redox potential of nitrate-free ground water decreases from +200 mV at a pH value of 6.0 to +20 mV at a pH of 7.7. This correlation may partly be buffered by the redox couple $Fe(OH)_3/Fe^{2+}$ having a steeper slope than the measured pairs of values. Furthermore, the redox couple $S(+0)/H_2S$, which, according to Berner (1963), characterizes sulfidic environments could also contribute to the measured composite redox potential of nitrate-free ground water. The E_H of this redox couple has nearly the same pH dependence as the measured values but is in the negative E_H range between pH 6 and 8 (see also Böttcher and Strebel 1985c).

With regard to carbonate equilibria (i.e. precipitation and dissolution of a solid phase), the theoretical shift in pH, E_H and total CO_2 (for depth distribution see Leuchs 1988) can be deduced from the charge and electron balance of each reaction listed in Table 4.1.1.1 (Froehlich et al. 1979). However, carbonate dissolution under varying open and closed conditions in the recharge area (Sect. 3.2.3) superimposes the effect of the microbial reactions in the upper ground water body.

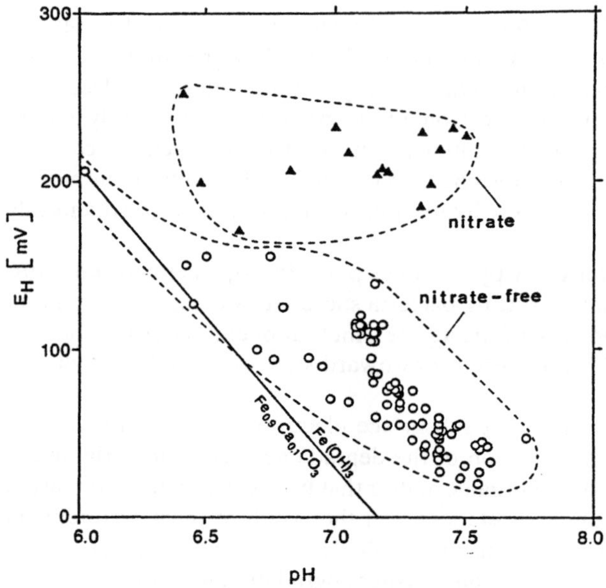

Fig. 4.1.1.6. pH-E_H diagram of all ground water samples

On the other hand, the gradual decrease in CO_2 and HCO_3^- with depth below 20 m can be explained by increasing oxidation of organic carbon by ferric iron and precipitation of Ca-(Mn-) siderite (Leuchs 1988).

4.1.2 Nitrogen and Oxygen Isotopes as Indicators for Nitrification and Denitrification

H.-L. Schmidt, S. Voerkelius and A. Amberger

4.1.2.1 Background

Nitrogen-containing compounds of different oxidation states occur in most soil layers. Nitrogen is applied to the soil as fertilizers in form of urea, NH_4^+ or NO_3^-, and from here it enters the ground water as NO_3^- or the atmosphere as NH_3, N_2 or N_2O. All these substances are correlated among each other within the natural nitrogen cycle, in which, especially in the soil, oxidation and reduction processes connect the different nitrogen pools through different pathways (Fig. 4.1.2.1). Therefore the identification of the origin of a given substance and the determination of nitrogen fluxes and balances become extremely difficult.

For corresponding investigations no suitable radiotracers exist, and the use of stable isotope tracers is largely limited by the interference of their natural occurrence. However, any transfer between the different nitrogen pools implies isotope effects: in general, nitrogen compounds containing heavy isotopes of nitrogen or

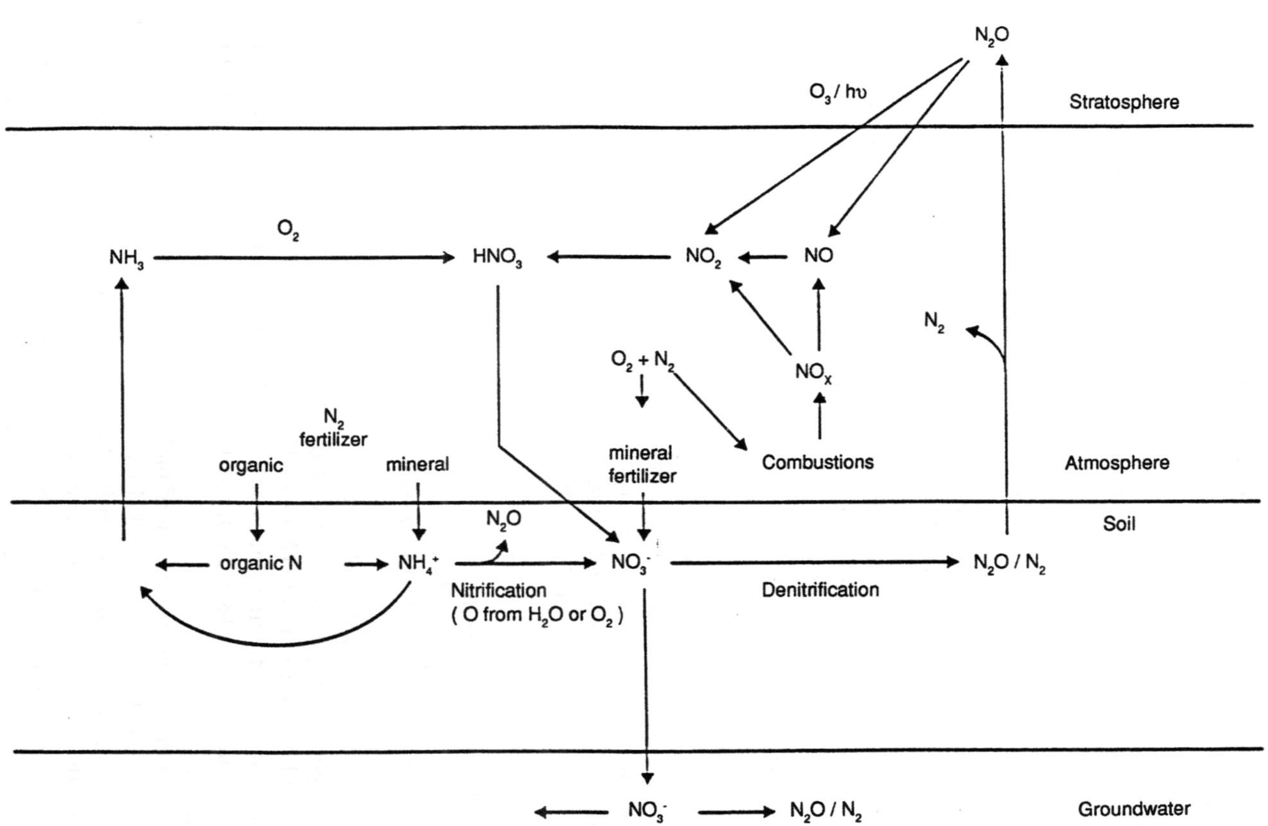

Fig. 4.1.2.1. Biological nitrogen cycle indicating the origin of N-compounds and their transfer between atmosphere, soil and ground water. Main processes with isotope discrimination: NH_3 evaporation, denitrification, nitrification, combustion and probably ozone cycle

oxygen are subject to a slower turnover than their light isotopomers. As a consequence of these isotope effects, variations of the natural isotope abundances between substrate and product pools are observed, the proper extent of which depends on the reaction type and the actual turnover. The knowledge of these intercorrelations in principle permits assignment of nitrogen compounds to their origin and turnover on the basis of their isotope abundances (expressed as δ-values). In the case of nitrogen, this fact has several times been used for the assignment of NO_3^- using $\delta^{15}N$ values (Mariotti and Létolle 1977; Karamanos and Rennie 1980; Flipse et al. 1984). However, because of the very complex interdependencies between the various nitrogen pools already mentioned, the interpretation of such results implies large uncertainties. Yet the evidence and information of isotope ratio determinations can be considerably enhanced if the $\delta^{15}N$ values are supplemented by the $\delta^{18}O$ values (Amberger and Schmidt 1987). As the oxygen atoms in the substances can only originate from H_2O or O_2, representing two pools with largely different $\delta^{18}O$ values, an assignment to these pools and to the synthetic pathway can easily be performed.

This has been carried out so far for NO_3^-, but can also be applied to other nitrogen compounds. On the basis of the mechanism of their formation and turnover, and taking into account the isotope effects accompanying the corresponding reactions, general principles for the assignment of the origin of these compounds can be established. In the present study, this will be shown for NO_3^-, N_2O and N_2 as the main products of nitrification and denitrification in soil. Experimental details are given elsewhere (Amberger and Schmidt 1987; Voerkelius 1989).

4.1.2.2 Results and Discussion

Factors Determining the Isotope Composition of NO_3^- and N_2O

N_2 from air ($\delta^{15}N = 0‰$) is the sole nitrogen source in the industrial production of NO_3^-; the oxygen atoms predominantly originate from air-O_2 ($\delta^{18}O = +23.5‰$), and only to a small extent from H_2O ($\delta^{18}O = -10.5‰$). As a matter of fact, the isotope composition of synthetic nitrate, as calculated on the basis of this prediction ($\delta^{15}N = 0 \pm 2‰$, $\delta^{18}O = +18 \pm 2‰$), has been found in corresponding samples (Table 4.1.2.1).

Investigations on the mechanism of biological nitrification (Aleem et al. 1965; Hollocher et al. 1981; Andersson and Hooper 1983; Hollocher 1984; DiSpirito and Hooper 1986) suggested that NO_3^- arising from this process contains one O-atom from O_2 and the other two O-atoms from water. By oxygen exchange between the intermediates and water even the O-atom from O_2 may get lost. On this basis $\delta^{18}O$ values in NO_3^- of $< +0.8‰$ $[(1 \times 23.5 + 2 \times -10.5)/3]$ are expected, while the $\delta^{15}N$ values of NO_3^- may vary within large limits, as the starting material originates from different N sources. This expectation could be confirmed in a fundamental investigation (Amberger and Schmidt 1987). Nitrification NO_3^- had a $\delta^{18}O$ value of $-1.5‰$, while its $\delta^{15}N$ value was determined by that of the N source and by the isotope effect of the nitrification process. Possible corresponding oxygen isotope effects, which would lead to ^{18}O-depleted NO_3^-, will probably not become evident due to oxygen exchange reactions of intermediates with water.

Table 4.1.2.1. Isotope abundances of nitrogen compounds depending on their origin. $\delta^{15}N$-values as AIR standard, $\delta^{18}O$-values as SMOW standard. Sources: O_2: $\delta^{18}O = +23.5$‰, H_2O: $\delta^{18}O = -10.5$‰, N_2: $\delta^{15}N = 0$‰

Compound	Origin	$\delta^{15}N$[‰]		$\delta^{18}O$[‰]	
		Expected	Measured	Expected	Measured
NO_3^-	Techn. synth. 0		−2...+ 2	+18 ± 2[a]	+17...+23
	Nitrification	<−10...+10	− 30...+10	+ 1[b]	−1.5
	Rain water	0	− 6...+ 1	>+23.5	+50...+60
N_2O	Denitrification	>0	−	>>0	+36...+50[d]
	Nitrification	>0	−	0...+2	+22[d]
N_2	Air	0	0	−	
	Denitrification	− 3...+15[c]	− 5...+ 2	−	

[a]Calculated on the base of: $N_2 + 2.5O_2 + H_2O \rightarrow 2\ HNO_3^-$.
[b]Basing on H_2O as the main oxygen source.
[c]Assuming a shift of 18‰ rel. to the original NO_3^-.
[d]After Yoshinari & Wahlen (1985).

Isotope effects must also be the reason for a secondary enrichment of the heavy nitrogen- and oxygen isotopes in NO_3^- during denitrification. In our laboratory the in vitro kinetic isotope effects on the nitrate reductase reaction found were $k_{14}/k_{15} = 1.03$ and $k_{16}/k_{18} = 1.015$ (Olleros 1983). These values were the basis for the "denitrification slope" in a $\delta^{15}N/\delta^{18}O$-diagram permitting the assignment of NO_3^- to origin and pretreatment (Fig. 4.1.2.2). The application of this diagram to samples of known origin confirmed its suitability, with the exception of NO_3^- from water of forest soils. However, as will be explained in the following, this nitrate is probably formed from nitrogen oxides in the atmosphere.

Corresponding reflections can be made concerning the isotope composition of N_2O. Nitrous oxide can be formed by nitrification and by denitrification (Fig. 4.1.2.1). Its main production in soils has been observed under conditions on the border between aerobic and anaerobic environments. According to our present knowledge, in the course of denitrification O-atoms from NO_3^- are not only removed by reduction, but may also partially be removed by exchange with water. Hence even the remaining O atom in N_2O, one of six from NO_3^-, may not necessarily originate from this starting material. However, if no exchange occurs, this oxygen atom should be enriched in ^{18}O due to isotope effects of the denitrification. During the first step of the nitrification one O-atom from O_2 is introduced into NH_3. If this is not lost during the subsequent reactions, $\delta^{18}O$ values close to +23.5‰ are to be expected.

Wahlen and Yoshinari (1985) and Yoshinari and Wahlen (1985) found very high mean ^{18}O enrichments in N_2O from the atmosphere ($\delta^{18}O$ = +45‰). On the other hand, N_2O formed by nitrification attained only a $\delta^{18}O$ value of +22‰, which is in line with the origin of the O atom from O_2, and excludes oxygen exchange with water during formation, therefore other sources of highly ^{18}O-enriched N_2O must exist in the atmosphere. In fact, N_2O formed under conditions of denitrification showed $\delta^{18}O$ values near +36‰, and N_2O produced by combustion processes is reported to be enriched up to 58‰ (Wahlen and Yoshinari 1985).

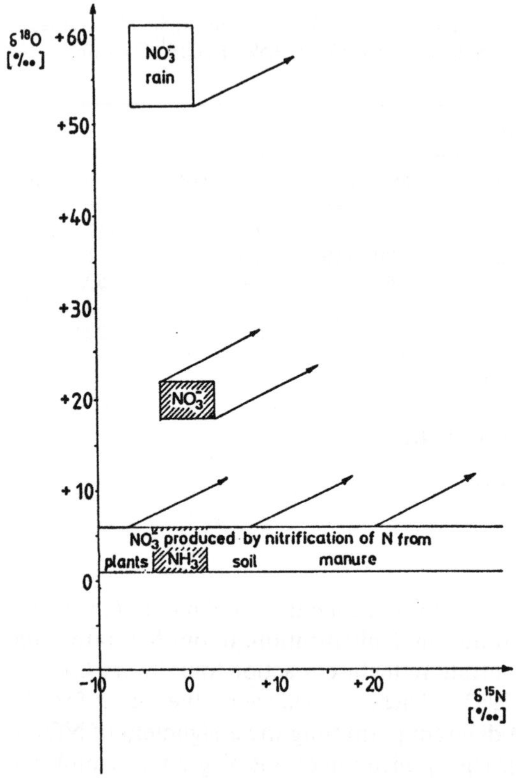

Fig. 4.1.2.2. $\delta^{15}N$- and $\delta^{18}O$ values of NO_3^- as a function of origin consumption. Slope of shift from ratio of isotope effect for nitrogen and oxygen during denitrification. (After Olleros 1983 and Amberger and Schmidt 1987). □ Mineral fertilizers

In order to obtain more details about denitrification, the isotope effects of the in vitro reduction of NO_2^- to N_2O were investigated. With the fungus *Fusarium solani* a fractionation factor for nitrogen 1.018 was found, i.e. a depletion of ^{15}N in the product N_2O relative to the starting material NO_2^-. On the other hand, the heavy oxygen isotope was enriched in N_2O with regard to NO_2^-, indicating a large intramolecular oxygen isotope effect of the process. The ^{18}O enrichment found in the product N_2O simultaneously indicates that no significant exchange of oxygen atoms takes place between the intermediates and water during the reaction.

In an additional experiment, N_2O formed in a soil under conditions of denitrification (flooding with water after addition of glucose) was collected. The gas showed $\delta^{18}O$ values up to +50‰, and it was largely enriched with ^{18}O as compared to the NO_3^- from which it had been formed. These preliminary results confirm the assumption that N_2O formed by denitrification is more enriched with heavy oxygen than that originating from nitrification. This must be due to isotope effects of the NO_3^- reduction. In addition the contribution of an isotope effect during the reduction of N_2O to N_2 could be shown. This effect is, depending on the concentration of N_2O, between 1.005 and 1.027 (Yoshida 1984; Voerkelius 1989).

Application of the Results for Origin Assignments of Nitrogen Compounds in Pore- and Ground Water

The $\delta^{15}N$ values of NO_3^- from pore water of farming areas showed, as expected, correlations to the fertilizers applied: NO_3^- in extracts from soil collected at various depths under a farming area which, over many years, had been fertilized exclusively with mineral fertilizers ($\delta^{15}N$ = −2 to +2‰) showed only minor ^{15}N enrichments. In extracts from a corresponding soil which had additionally been fertilized with manure, relatively high NO_3^- concentrations and ^{15}N enrichments up to +13 δ units were measured.

Drinking water of the city of Freising, originating from areas with mixed agricultural use, had NO_3^- concentrations near 60 mg/l with $\delta^{15}N$ values of +12 ± 2‰. This is compatible with its formation by nitrification of organic fertilizers. On the other hand, in the drinking water from Augsburg, which is collected under forest and agricultural areas with preferably mineral fertilization, lower NO_3^- concentrations (25–40 mg/l) and $\delta^{15}N$ values between +3 and +10‰ were measured. This and the $\delta^{18}O$ values of this NO_3^- indicated its origin from nitrification of mineral fertilizers and by mineralization.

The isotope analyses of NO_3^- from ground water samples of a drilling hole near Bocholt (Sect. 4.1.1, this Vol.), for which E_H values (below 250 mV) and other data suggested reducing conditions indicative of denitrification showed – contrary to the expectation – decreasing $\delta^{15}N$ values with decreasing NO_3^- concentration (Fig. 4.1.2.3).

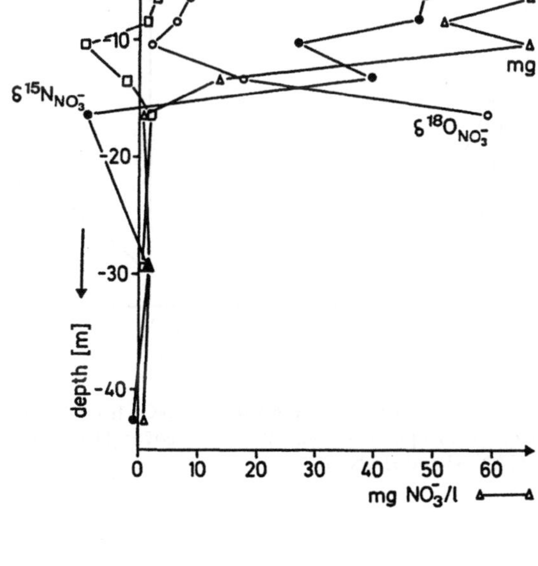

Fig. 4.1.2.3. Concentrations and δ-values of NO_3^- from water of multi-level well BOC4 and $\delta^{15}N$ values of dissolved N_2 depending on depth

So far this result cannot be explained, unless different NO_3^- sources are supposed. On the other hand, the $\delta^{18}O$ behaved precisely as expected from the isotope effect for the denitrification: while the NO_3^- concentration decreased, it largely increased.

In this context, the origin of N_2 dissolved in the ground water could be, at least in part, assigned to the denitrification process. The depth profile of the $\delta^{15}N$ values of the dissolved N_2 followed with a shift of +18‰, that of the NO_3^- in the same sample (Fig. 4.1.2.3). This proves its formation from nitrate, e.g. by denitrification. Correspondingly, in the dissolved gas the ratio of the concentrations of dissolved N_2 to Ar was higher than that which is found in water in equilibrium with air. This is further proof for the origin of the dissolved N_2 from a source different from air.

Other water samples, provided by O. Strebel, had been collected from different depths of the saturated zone in the Fuhrberger Feld (Sect. 4.1.3, this Vol.). This area is well characterized concerning the profiles of the ground water flux, the redox potentials, the NO_3^- concentrations and other data (Böttcher et al. 1985). The strong decrease of the NO_3^- concentration with depth (Fig. 4.1.2.4) coincides with other data indicative of a denitrification process. With decreasing NO_3^- concentration, the $\delta^{15}N$- and $\delta^{18}O$ values of the remaining anion increased, and the ratio of the enrichments of the two isotopes agreed well with the slope calculated from the isotope effects measured in the vitro experiments (Olleros 1983).

As before, the water samples from forest soils did not fit the diagram conceived for nitrate originating from mineral fertilizer and organic nitrogen either in the NO_3^- concentration nor the δ-values.

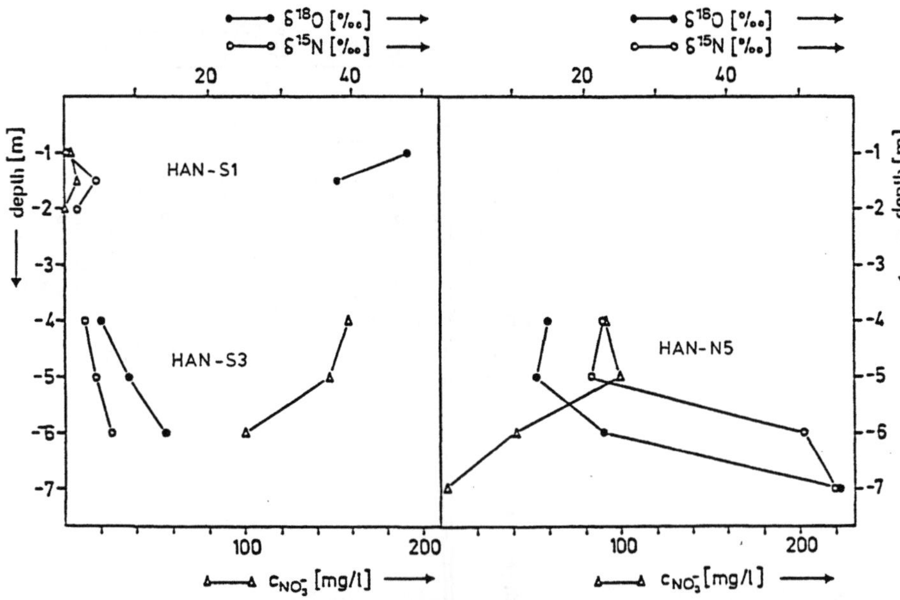

Fig. 4.1.2.4. Concentrations and δ-values of NO_3^- from ground water from different depths of the Fuhrberger Feld aquifer system: under forest cover (HAN-S1), and from predominantly agricultural areas (HAN-S3 and HAN-N5)

These samples are characterized by a very low NO_3^- concentration, rather low $\delta^{15}N$ values, but extremely high $\delta^{18}O$ values (Fig. 4.1.2.4). Similar isotope contents have only been found for N_2O. This gas is one of the most long-lived atmospheric emissions. In part it is converted to HNO_3 in the ozone cycle of the stratosphere. Yet the contribution of N_2O to NO_3^- in rain is not very important (Crutzen 1983). The bulk of the atmospheric HNO_3 originates from NO_x of combustions, which is oxidized, at least in part, by tropospheric ozone. For stratospheric ozone a heavy enrichment of ^{18}O has been found, but not for tropospheric ozone (Mauersberger 1981). However, as a matter of fact, NO_3^- from several samples of rain water showed $\delta^{18}O$ values between +52 and +61‰. Therefore it must be assumed that so far unknown processes are the reason for the ^{18}O enrichment in this rain HNO_3^-. The origin of the nitrate in water from forest soils must be due to the acid rain, and the diagram on the correlation between δ-values and origin of NO_3^- (Fig. 4.1.2.2) can be completed by an area for a third important source, NO_3^- from the atmosphere.

Thus it is now possible to assign several nitrogen compounds to their origin and pretreatment on the basis of their isotope abundances. In addition, the means of natural isotope abundance determination can be better used as a tool for the elucidation of bacterial and inorganic processes in the nitrogen cycle, especially for the measurement of nitrogen turnover in the saturated and unsaturated zone.

4.1.3 Redox Conditions and Microbial Sulfur Reactions in the Fuhrberger Feld Sandy Aquifer

J. Böttcher, O. Strebel and W. Kölle

4.1.3.1 Study Area and Methods of Investigation

The ground water recharge area of the Fuhrberger Feld waterworks in the Wietze lowlands northeast of Hanover (HAN; Fig. 1.1, Table 1.1) consists of an unconfined aquifer of about 20–30 m thick unconsolidated sands and gravelly sands. These Quaternary sediments contain in uneven distribution small lignitic pebbles with a predominant diameter of less than 5 mm, yielding an average of about 0.1–1.0% wt. organic carbon. Larger lignitic wood residues are rare. Furthermore, reduced sulfur compounds are present in the sediment material in different amounts. The whole aquifer is virtually completely oxygen-free, traces of oxygen (≈ 0.1 mg O_2/l) can only be measured near the ground water surface. The aquifer is underlain by clays and clayey marls of Cretaceous age (Lillich et al. 1973). In this aquifer substantial microbial reactions take place which strongly influence the quality of the extracted well water. In this section the respective reaction mechanisms and their environmental prerequisites in the aquifer are identified. The quantitative aspects of the microbial solute transformations (input-output balance of solutes, modelling of transformations) are dealt with in Section 6.2.5.

In the recharge area of one extraction well, defined from piezometer level contour maps and covering about 1000 ha (Fig. 6.2.5.1 in Sect. 6.2.5), 13 multi-level samplers with 15 to 30 m depth were installed approximately following a flow line. The use of the miniscreen method (Pickens et al. 1977a,b; Snelting 1979) allowed the measurement of the depth function of major ground water solutes with a resolution in depth of mostly 1 m (Böttcher et al. 1985). The samples were collected four times in 1983 and 1984 through filter tubes (12 mm diameter, tubing 5 mm diameter) to the soil surface by pumping with a simple vacuum hand pump.

The measured depth functions with respect to time and space were analyzed on the basis of the vertical-plane two-dimensional ground water flownet (Fig. 4.1.3.1).

4.1.3.2 Depth Functions of Major Solutes in Ground Water, Controlling Reactions

The depth function of nitrate, sulfate, hydrogen carbonate, pH and redox potential of multi-level well HAN-N10 is a typical example (Fig. 4.1.3.2). This well is located in an extensive area of arable land. However, the ground water between 6 and 10 m in depth was additionally affected by the input of untreated sewage in its recharge area (prior to the installation of a sewage treatment plant in 1972 for the village of Fuhrberg). Figure 4.1.3.2 reveals the presence of two ground water portions with different chemical properties. In the upper portion down to about 7 m

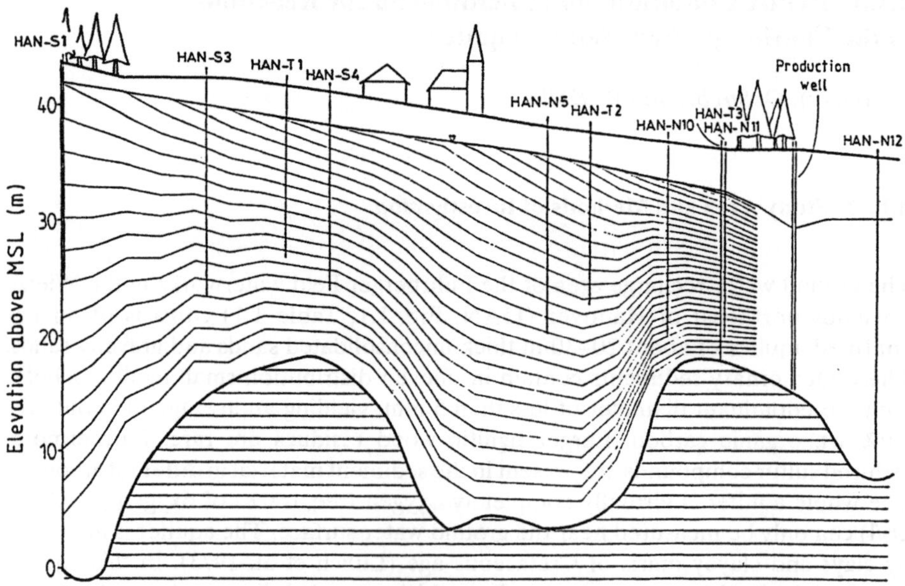

Fig. 4.1.3.1. Vertical-plane two-dimensional ground water flownet, and position of multi-level wells and of extraction well 1 in the investigated area. (After Böttcher et al. 1985)

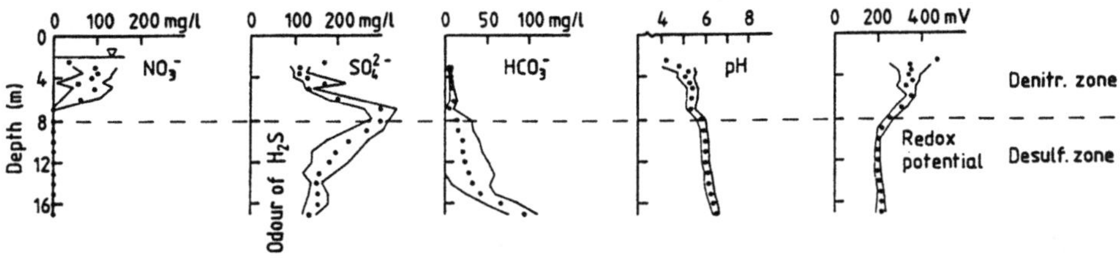

Fig. 4.1.3.2. Depth function of major solute concentrations, of pH and E_H in the ground water (mean values ± standard deviation) of the multi-level well HAN-N10 in an extensive area of arable land

depth, nitrate concentration and E_H decrease (from > 100 to close to 0 mg NO_3^-/l and from > 400 to about 300 mV, respectively), whereas sulfate concentrations increase significantly. In the lower portion, below about 7 m depth, sulfate concentration and E_H decrease (from about 300 to < 200 mV), but the hydrogen carbonate concentration increases significantly and H_2S can be smelled. However, this solute depth function does not hold for the whole investigated area. It depends mainly on the specific solute input with the ground water recharge (Sect. 6.2.5) due to different land use, e.g. under extensive areas of forest another solute depth function can be observed (Böttcher et al. 1985).

The upper ground water portion with its decrease in nitrate and simultaneous increase in sulfate concentrations represents a zone of intensive bacterial denitrification, where particles of reduced sulfur compounds are used as electron donors (Kölle 1982). The denitrification follows the reaction equation

$$5\,FeS_2 + 14\,NO_3^- + 4\,H^+ \rightarrow 7\,N_2 + 10\,SO_4^{2-} + 5\,Fe^{2+} + 2\,H_2O \qquad (4.1.3.1)$$

In this equation, ferrous disulfide is used as a possible reduced sulfur compound (Sect. 4.1.3.3). Reaction products are gaseous nitrogen, sulfate and ferrous iron ions. Furthermore, ferrous iron released by this reaction, may be involved in a second type of denitrification reaction (Kölle et al. 1985):

$$5\,Fe^{2+} + NO_3^- + 7\,H_2O \rightarrow 5\,FeOOH + 0.5\,N_2 + 9\,H^+ \qquad (4.1.3.2)$$

The lower ground water portion, which is practically free of nitrate, represents a zone of bacterial sulfate reduction using solid particles of organic carbon as electron donors (Kölle 1982):

$$SO_4^{2-} + 2\,C + 2\,H_2O \rightarrow 2\,HCO_3^- + H_2S \qquad (4.1.3.3)$$

The released H_2S then reacts with ferrous iron and some trace metals, and the resulting sulfides are precipitated due to their very low solubility products.

4.1.3.3 Bacterial Denitrification in the Upper Ground Water Zone

The bacterial denitrification according to Eq. (4.1.3.1) in the upper ground water zone is proven by several independent approaches (isotope fractionation of nitrate-nitrogen and nitrate-oxygen in the residual nitrate, see Sect. 4.1.2 and Böttcher et al. 1990a; reaction kinetics and quantification, see Sect. 6.2.5). It is a first-order reaction (Böttcher et al. 1989) conditioned on certain prerequisites, which are the presence of nitrate-reducing bacteria, a sufficient stock of microbially oxidizable reduced sulfur compounds, and the presence of certain redox/pH conditions (Sect. 4.1.3.5). The limited reserves of reactive reduced sulfur compounds of the upper ground water zone are gradually depleted by this denitrification reaction.

Tentative microbiological investigations of sediment and ground water samples verified the presence of the autotrophic bacterium *Thiobacillus denitrificans* (P. Werner 1985, unpubl. report), which may use nitrate in the absence of molecular oxygen taking reduced sulfur compounds as electron donors. Typically 10^5 organisms/g sediment and $> 10^4$ organisms/ml ground water were found.

Reduced sulfur compounds can be found in the aquifer in the inorganic material as well as in the lignitic pebbles, solid sulfate is not present in the sediment. The maximum "sulfate formation capacity" (SFC) of the sediment was chemically determined as total sulfur content. The lignitic pebbles were quantitatively separated from the sediment to determine the "lignite-specific" maximum SFC (Kölle et al. 1985, 1987, 1990; Strebel et al. 1985). However, only a portion of the maximum SFC of the sediment is microbially oxidizable. This portion was characterized by a standardized laboratory test. It is defined by the formation of sulfate-S in about 8 months, in a laboratory batch experiment exposing fresh sediment material to a nitrate solution, after the sulfate formation has reached very small rates (Kölle et al. 1985).

Figure 4.1.3.3 shows the measured depth function of the respective SFC of the sediment samples, their "maximum SFC" and their maximum "lignite-specific" SFC from several drillholes in the study area. Under arable land, significantly lower SFC values (average about 0.1 kg sulfate-S/m^3 sediment) are found in the upper portion of the profile down to 6–8 m depth, representing about the denitrification zone (Fig. 4.1.3.2), than below this zone (average about 0.4 kg sulfate-S/m^3 sediment). This may be attributed to a previous consumption of the SFC in this denitrification zone. Furthermore, Fig. 4.1.3.3 indicates that the SFC is controlled mainly by the "non-lignite-specific" sulfur compounds, since apparently no relation exists between the measured SFC and the "lignite-specific SFC". However, the lignitic pebbles play an important role in the lowering of the redox potential. Finally, Fig. 4.1.3.3 reveals some appreciable differences between measured and "maximum SFC". These differences may be attributed to less oxidizable and rather stable reduced sulfur compounds. The incomplete and differing transformation of the total stock of reduced sulfur compounds (= "maximum SFC") is caused by several factors: chemical and mineralogical composition, degree of crystallization, form and particle size of the respective compounds (specific surface) (Kölle et al. 1987, 1990). A survey found mineral grains of pyrite, marcasite and framboidal melnikovite, furthermore sulfidic materials were observed as poorly crystallized or X-ray amorphous crusts on or as cement between mineral grains.

The bacterial denitrification according to Eq. (4.1.3.2) is still debated in microbiological literature; however, there is considerable evidence of it in the investigated aquifer. However, it was not possible to relate the reaction with certain bacterial species. When simultaneous denitrification according to Eqs. (4.1.3.1) and (4.1.3.2) occurs, the respective bacteria compete for the nitrate. Furthermore, reaction (4.1.3.1) consumes H^+ ions, whereas reaction (4.1.3.2) produces H^+ ions.

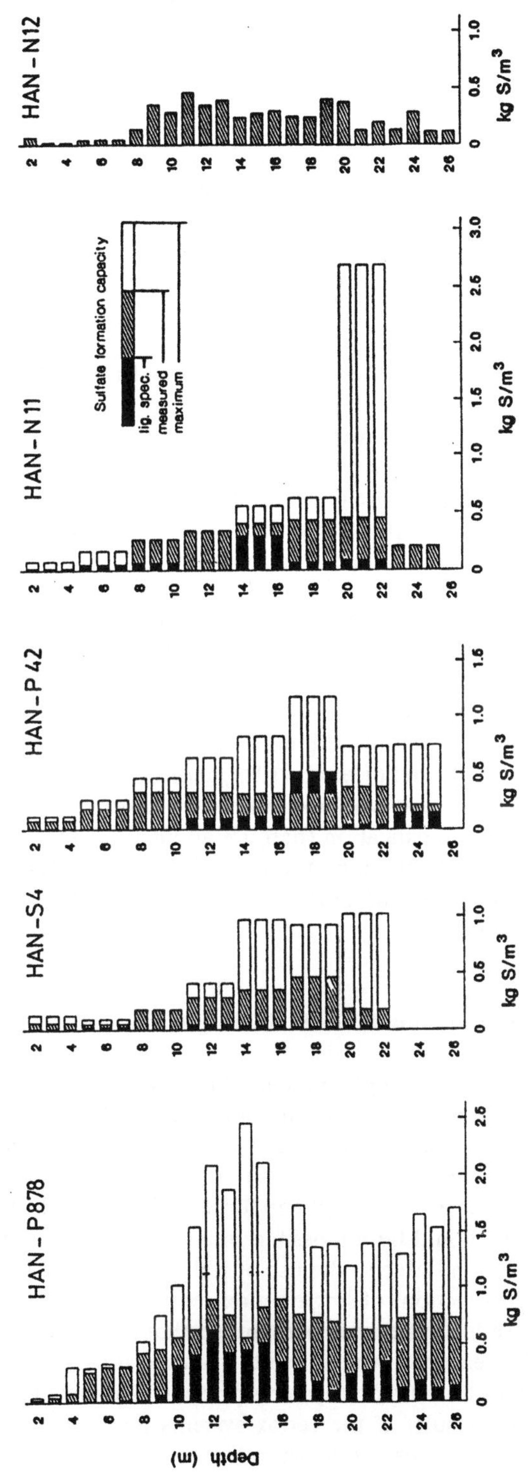

Fig. 4.1.3.3. Depth functions of the sulfate formation capacity (SFC) in kg sulfate-S/m^3 sediment at several drill holes near Fuhrberg: "maximum SFC", "lignite-specific" SFC and measured SFC (laboratory experiment)

In the poorly buffered upper ground water zone, a certain feedback of both reactions is to be expected, which results in an overall H^+ balance (Kölle et al. 1985, 1987, 1990), so that no specific pH change is connected with the denitrification (Fig. 4.1.3.2, pH in 4–7 m depth).

Regarding the depth function of solute concentrations at the sampling well HAN-N10 (Fig. 4.1.3.2), an additional aspect must be included. The ground water in the depth from 6 to 10 m originates from a recharge area which until 1972 had nitrate concentrations significantly above the level of arable land conditions (Sect. 4.1.3.2). Therefore, after denitrification, the measured resulting sulfate concentration at 6 to 10 m depth is accordingly high.

4.1.3.4 Bacterial Sulfate Reduction in the Lower Ground Water Zone

The microbial sulfate reduction according to Eq. (4.1.3.3) in the lower portion of the aquifer is proven by several independent approaches: by a noticeable odour of H_2S, the decrease of sulfate concentration and the increase of hydrogen carbonate concentration with depth and with ground water age, respectively (Fig. 4.1.3.2), and by the presence of isotope fractionation for sulfate-sulfur and sulfate-oxygen, i.e. increase of the heavier ^{34}S and ^{18}O isotopes in the residual sulfate with decreasing sulfate concentration (Strebel et al. 1990).

The respective heterotrophic bacteria were not identified, but important prerequisites for reaction according to Eq. (4.1.3.3) are fulfilled, such as the presence of microbially oxidizable solid organic carbon (lignitic pebbles), virtually nitrate-free ground water and the necessary E_H/pH conditions (Sect. 4.1.3.5). The results described in Böttcher et al. (1989) indicate that the sulfate reduction follows first-order kinetics, but with a significantly lower rate (half-life about 76–100 a) than that of the denitrification (half-life about 1–2.3 a) (Sect. 6.2.5.3). In any case, in the lower ground water zone the limited reserves of reactive solid organic carbon are decreased by the sulfate reduction, and the reserves of reduced sulfur compounds are increased.

4.1.3.5 Redox Conditions in the Denitrification and Desulfurication Zones

In Fig. 4.1.3.4 all E_H/pH data measured (Böttcher and Strebel 1988d) in the investigated area are attached to the nitrate and sulfate reduction zones. The data of the nitrate reduction are, with only a few exceptions, below the stability line of NO_3^-/NO_2^-; however, all those measured for the sulfate reduction are well above the stability lines for SO_4^{2-}/H_2S and for S_0/H_2S_{aq}, respectively. Thus, the observed sulfate reduction should be impossible in a thermodynamic point of view. This apparent contradiction is due to the use of stability lines in Fig. 4.1.3.4, which are only valid for redox reactions of single redox couples. However, in ground water a mixture of several redox systems is present, so that a mixed potential (Stumm and Morgan 1981) is measured, which results from the superposition of the individual polarization curves for all redox systems present. The mixed potential can be approximated with good results, if the redox systems involved are known and their individual polarization curves are measured.

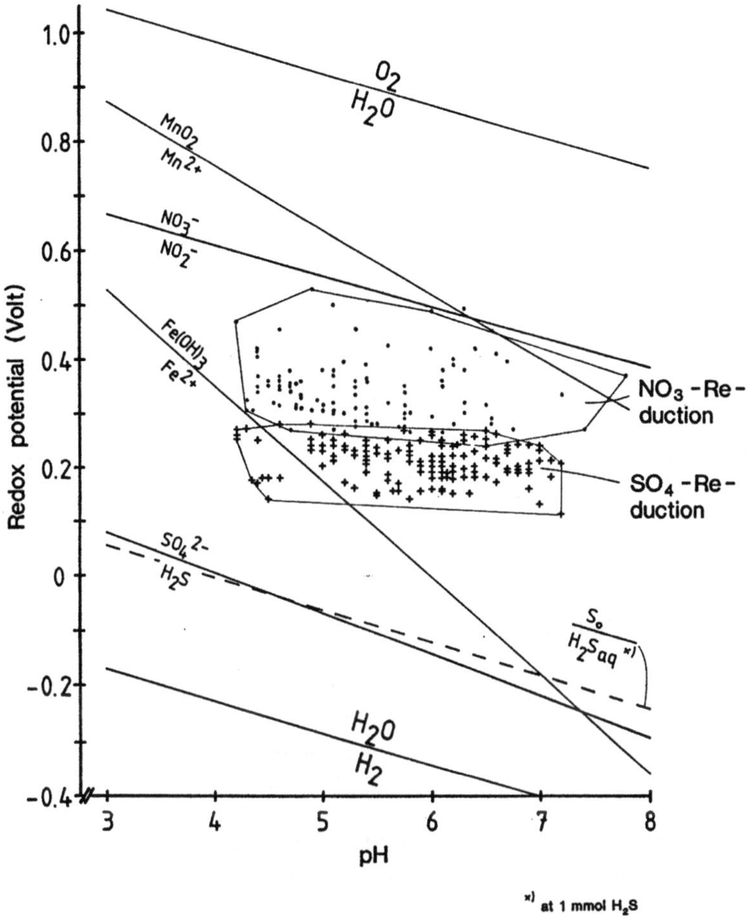

Fig. 4.1.3.4. EH/pH diagram of the redox characteristics of denitrification and sulfate reduction in the ground water at the investigated area and stability lines of single redox couples. (Böttcher and Strebel 1985c)

The results of respective measurements and calculations for the Fuhrberger ground water are represented in Fig. 4.1.3.5 (details in Böttcher and Strebel 1985c). The calculated mixed potential in this figure is referred to the presence of the following redox systems: in the denitrification zone the redox couples MnO_2/Mn^{2+}, $Fe(OH)_3/Fe^{2+}$, NO_3^-/NO_2^- and organic compounds (≈14 mg DOC/l), in the sulfate reduction zone MnO_2/Mn^{2+}, $Fe(OH)_3/Fe^{2+}$, S_o/H_2S_{aq} and organic compounds (≈10 mg DOC/l). The mixed potentials calculated in this way fit well with the measured E_H/pH stability field in the case of denitrification, but less in the case of the sulfate reduction. This may be due to complexation reactions between organic ligands with Fe- and Mn-ions, which could not be considered in the calculations. Nevertheless, the apparent contradiction between measured data and theoretical prediction (Fig. 4.1.3.4) is practically eliminated.

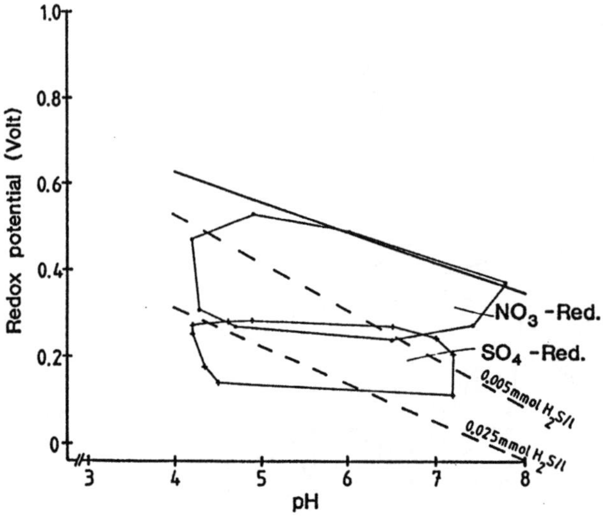

Fig. 4.1.3.5. pH dependency of the calculated mixed potentials in the denitrification zone (*solid line*) and in the sulfate reduction zone (*dashed line*) in the ground water of the investigated area and measured EH/pH ranges for denitrification and sulfate reduction (Fig. 4.1.3.4). (Böttcher and Strebel 1985c)

4.1.3.6 Concluding Remarks

In major parts of the investigated ground water recharge area of an extraction well an upper denitrification zone and a lower sulfate reduction zone can be observed in the aquifer. In these two zones microbially mediated reactions occur, controlling important ground water quality parameters such as the concentrations of NO_3^-, SO_4^{2-} and Fe^{2+}. However, the quality of the raw water extracted by the well is controlled by the solute input with ground water recharge and the transformation of solutes in the aquifer in the whole catchment area of this production well (Sect. 6.2.5).

4.1.4 Influence of Fine-Grained Cover Beds on the Chemistry of Shallow Ground Water

G. Ebhardt and P. Fritsch

4.1.4.1 Topic and Site of Investigation

The influence of the unsaturated zone on the chemistry of the uppermost ground water is significant, especially concerning the redox state and the carbonate system. Both are of considerable importance for the use of ground water, because they control the content of many aqueous components including nitrate, alkalinity, calcium, magnesium, iron and others. Within the area of the ancient Neckar river along the Odenwald, a sharp difference has been found between the oxidized, oxygen- and nitrate-bearing ground water underlying eolian sands and the reduced, divalent iron- and manganese-containing ground water underlying clayey deposits (Al-Azawi and Ebhardt 1983). This redox change is coupled with an increase in carbon dioxide and hydrogen carbonate. The investigation aimed at more precise information on this relationship by hydrochemical and sedimentological investigation of the overlying strata, with special attention to loamy sediments.

The investigated area at Hähnlein near Bensheim (BEN; Fig. 1.1, Table 1.1) is essentially an east-west strip of 3 km length across the flood bed of the ancient Neckar river (Fig. 4.1.4.1). There are only small differences in altitude between flat dunes in the east and the flood plains with shallow former river beds. Another slightly higher platform with flood deposits follows to the west. The whole area is underlain by permeable sand and gravel with intercalations of silty layers, which cannot be traced over a long distance. The base of this uppermost aquifer is made up of the so-called Upper Clay in about 40 m depth. The ground water flows generally from east to west towards a large well gallery. The depth of ground water table below surface ranges between 3 and 4 m in the dune area, 0–1 m in the ancient river beds and 1.5–4 m in the flood plain. At high ground water levels a local drainage system is active, during the summer ground water is withdrawn by irrigation wells.

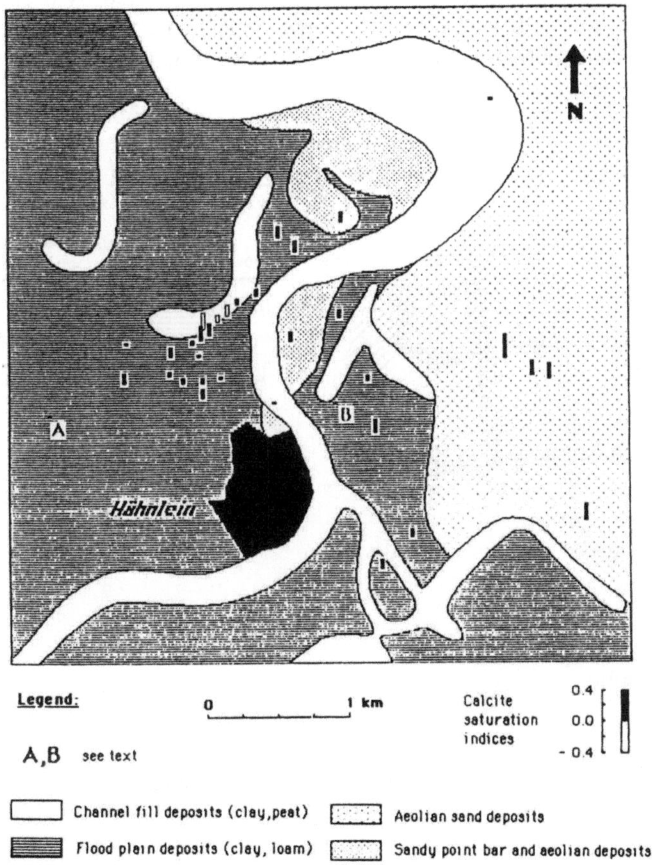

Fig. 4.1.4.1. Geological map of the investigated area with $SI_{carbonate}$ of shallow ground water

Both facts are not relevant to the overall water balance. The western part of the area and parts of the dunes are covered with forest, the rest is used by agriculture, in the deepest parts as meadows. The climate is rather warm and dry (about 700 mm annual precipitation, with a mean annual temperature of 9 °C).

The dunes of the eastern part consist of typical eolian fine- to medium-grained sands with narrow grain size distributions (Kupfahl et al. 1972). There is a rather poor soil development towards brown soils (cambisols in the FAO system of Soil Taxonomy), carbonate is removed to several m depth, but normally not down to the ground water table.

Within the alluvial flat the aquifer is overlain by one to several m of fine-grained deposits (Fig. 4.1.4.2) containing 30 to 50% clay and 5 to 20% sand. The clay mineral composition is varying, mostly dominated by kaolinite and illite, but also containing smectites and chlorite. Generally the cover sediments and their sandy substratum are rich in carbonate nearly up to the Earth's surface. Close to their base carbonate is enriched up to 50%. Correspondingly high pH values are found there.

Due to their young age the soils are poorly developed, pelosols prevail with more or less gley or pseudogley features (i.e. vertisol-like regosols and gleysols). The ground water influence has been diminished by drainage and in the west by ground water withdrawal. In some areas the uppermost clayey layer is primarily free of carbonáte, resulting in low pH values (Fig. 4.1.4.3).

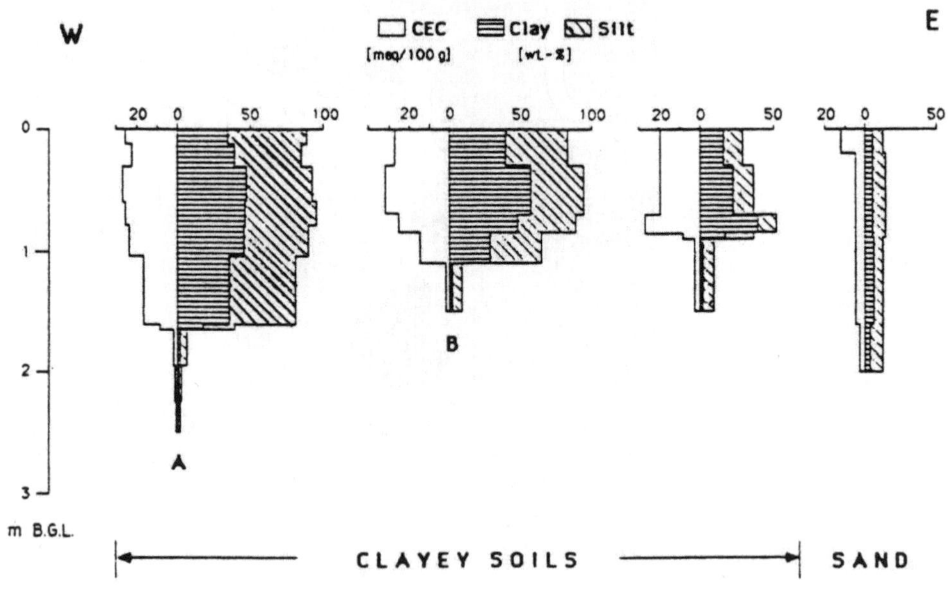

Fig. 4.1.4.2. CEC, soil horizons and grain size in a W-E sequence

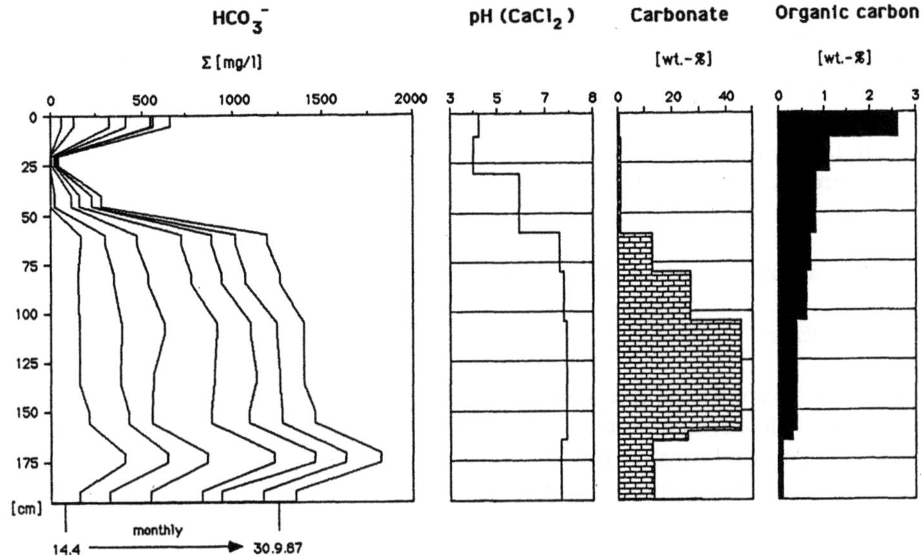

Fig. 4.1.4.3. Cumulative hydrogen carbonate concentration and soil properties in the clay profile of site BEN-A (forest)

4.1.4.2 Methods

The ground water analyzed was taken from 15- to 20-m-deep irrigation wells and from some shallow hand-made wells. O_2, pH and conductivity were measured in the field in a flow cell. When these values remained constant, samples were taken which were titrated with 0.1 N HCl and 0.1 N NaOH for HCO_3^- and CO_2 respectively, Fe, Mn, Na, K, Ca, Mg were measured in the laboratory by AAS, NO_3^- and NH_4^+ by photometry, SO_4^{2-} gravimetrically and Cl^- by titration.

Ground air was sampled from boreholes which were closed from the atmosphere, by suction with a hand pump into a glass tube and fused at both ends. After cracking the capillary end in the laboratory, the gas was immediately sucked into a gas-tight syringe and injected into a GC for analysis of O_2 and CO_2.

Tritium and ^{14}C values of ground water, $\delta^{13}C$ values of ground water, ground air and soil carbonates were measured by Dr. H. Dörr at the Heidelberg Institut für Umweltphysik.

The extraction of pore water from the clay soils was carried out by the immiscible liquid displacement method (Kinniburgh and Miles 1983) using trifluoroethane as displacent in a refrigerated high-speed centrifuge. For this purpose 30-mm diameter soil cores were taken on 2 × 2 m plots down to a depth of 2 m (soil A) and 1.5 m (soil B). After immediate freezing the core tubes in liquid nitrogen, the samples were stored in a freezer until laboratory treatment. Further preparation was carried out in a glove box with nitrogen atmosphere. Five-centimetre segments of the cores were taken in distances of 15 or 20 cm, distributed among 3–4 50-ml centrifuge tubes and subsequently centrifuged at 15 000 rpm for 2 h. The procedure yielded 1–5 ml pore solution.

After pH determination with a micro glass electrode, HCO_3^- was measured immediately by backtitration (0.01 N NaOH) of the sample previously acidified with 0.001 N HCl. NH_4^+ was subsequently analyzed with an ion-sensitive electrode. After membrane filtration (0.45 μm) all anions were determined with an ion chromatograph. Divalent iron was measured photometrically using bipyridine, all other cations were analyzed with AAS.

4.1.4.3 Carbonate-Carbon Dioxide System

Ground Water

Figure 4.1.4.4 shows the frequency distributions of the concentrations of CO_2 and HCO_3^- in shallow ground water under sandy and clayey strata and within a transition zone characterized by a narrow change in ground water chemistry. Ground water under loamy strata is characterized by higher concentrations of these components as well as by corresponding hardness and pH values (Al-Azawi and Ebhardt 1983). An evaluation of the chemical data with the thermodynamic program WATSPEC (Wigley 1977) found that most of the waters are supersaturated to a small extent in relation to calcite (Fig. 4.1.4.1). This is consistent with the ubiquitous presence of carbonates. Slightly undersaturated waters have been found in very shallow boreholes near old river beds. Saturation indices are about zero with respect to dolomite, a significant supersaturation with respect to siderite could not be explained.

Ground Air

In 1984/85 carbon dioxide (Fig. 4.1.4.5) and oxygen contents in the ground air and soil moisture in several profiles were determined monthly. In sandy soils the CO_2 content varied between 0.2 and 2%. In the alluvial clays, where sometimes the gas

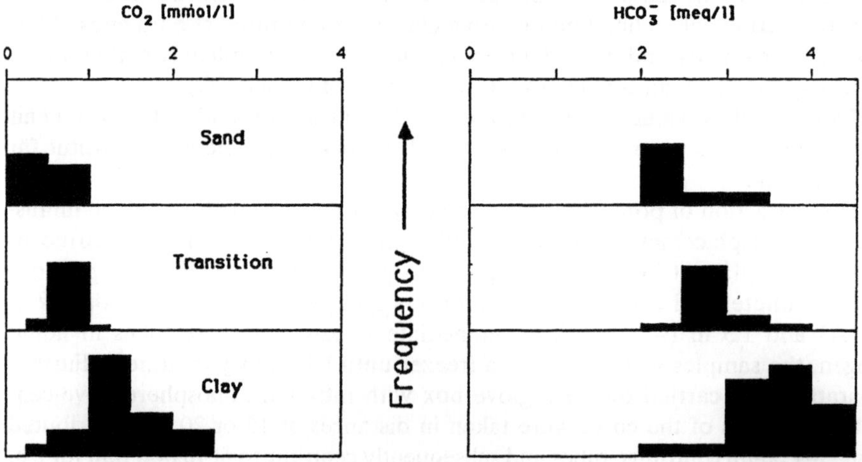

Fig. 4.1.4.4. Frequency distribution of carbon dioxide and hydrogen carbonate in shallow ground water

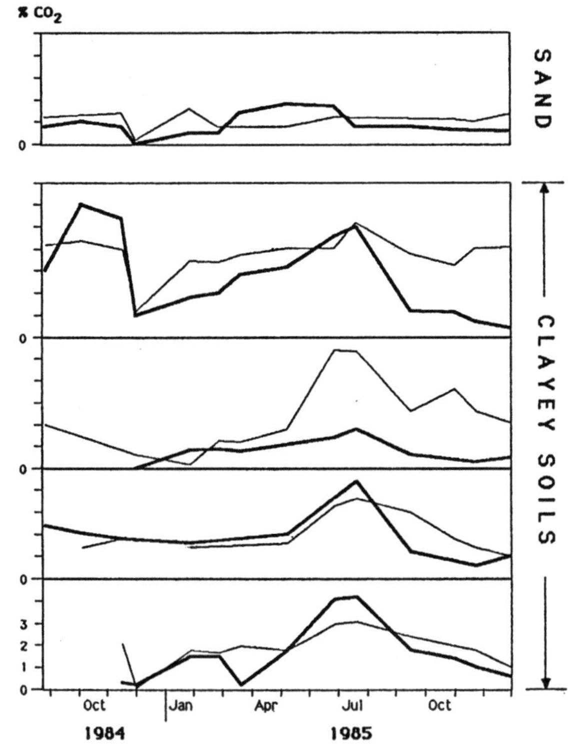

Fig. 4.1.4.5. CO_2 content in the ground air; *thick lines* near surface; *thin lines* close to water table

sampling was impossible because of soil water saturation, values between 2 and 12% were found. Besides the annual cycle, the rather irregular variations were influenced by the soil moisture, i.e. by precipitation. The oxygen deficiency, compared with the atmospheric oxygen content, lay in the same magnitude as the carbon dioxide and showed about the same variation.

Isotopes

Isotope determinations allow the following conclusions, with respect to the place of carbonate solution and the genesis of the carbonate concretions mentioned above.

- The $\delta^{13}C$ values of the ground air are as usual, reaching about –25‰ PDB.
- The $\delta^{13}C$ values in the ground water are –13.8 to –15.5‰ PDB, indicating that carbonate is dissolved within an open system. The difference with respect to ground air is essentially given by the partition factor $CO_{2(gas)}/HCO_3^-$. This is consistent with the observation of carbonates above the ground water table, and indicates that carbonate dissolution takes place in the unsaturated zone.
- The ^{14}C ages are recent, the tritium contents (30 to 50 TU) indicate a normal subhorizontal ground water flow with no ascending flow of old water from deeper strata.

- The results of ^{13}C determinations of the solid carbonates of two alluvial profiles (BEN-A and BEN-B; Fig. 4.1.4.6) make it possible to quantify the relation of primary (sedimentary) and secondary (precipitated from ground water) carbonates by a simple mixing calculation on the base of the ground water ^{13}C-value. The concretions are concentrated in the horizons primarily rich in carbonate and are located within an ancient zone of ground water fluctuation.

Model Considerations

At present only simple estimates of the concentrations and input of the inorganic carbon components in the seepage water are available. In Fig. 4.1.4.7 the CO_2 content of ground water – calculated from HCO_3^-, pH and temperature – is compared with that in water in equilibrium with measured ground air. Especially under sandy cover ground water contains more CO_2 than expected from ground air concentration. This can be explained by a generation of carbon dioxide within the ground water. This assumption is supported by an equivalent oxygen deficit; however, it

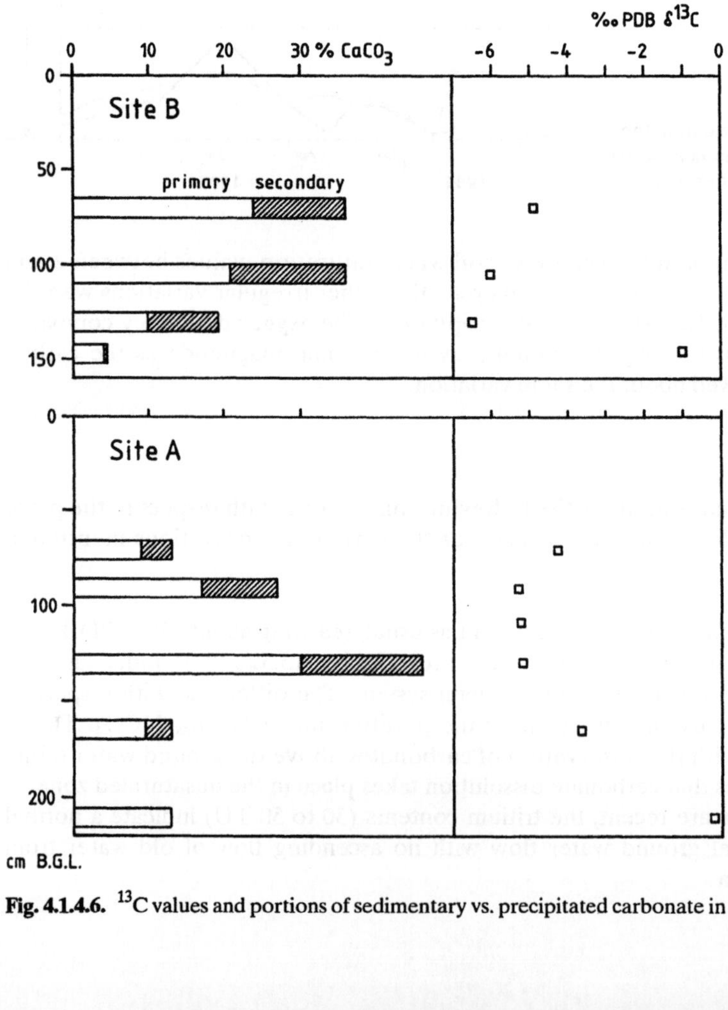

Fig. 4.1.4.6. ^{13}C values and portions of sedimentary vs. precipitated carbonate in two alluvial profiles

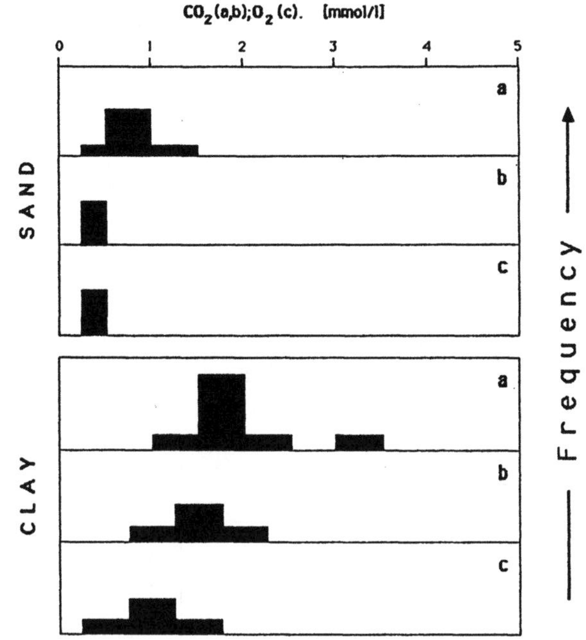

Fig. 4.1.4.7. Calculated concentration of CO_2. **a** In ground water; **b** in equilibrium with ground air; **c** ground water oxygen deficit in relation to ground air

is in contradiction to the open system concept indicated by the carbon isotopes. An alternative explanation might be the existence of local horizons with restricted gas communication, which, however, were not detected in the investigated profiles. Furthermore, the input and mean concentrations of CO_2 and HCO_3^- were estimated from calculated concentrations in seepage water and ground water recharge assuming equilibrium between seepage water, ground air and solid calcite (Fig. 4.1.4.8). Ground water recharge has been estimated both from observed runoff of the nearby lysimeter Alsbach and calculated from a climatic water balance with the method of Renger et al. (1974). Whereas the former shows runoff over the first half of the summer 1985, the calculated recharge ended abruptly with the end of March, regardless of different field capacities used.

Keeping the strong seasonal variation of CO_2 in the ground air in mind, it is not surprising that the calculations based on the lysimeter runoff yielded markedly higher inorganic carbon input than that based on the climatic water balance (Table 4.1.4.1). A further source of error is to be expected in the possibility of still higher CO_2 contents in such phases of ground water recharge in which gas sampling was partially impossible. This might be responsible for the fact that all calculated concentrations are lower than those measured in the field, but these higher values are not unrealistically far from those.

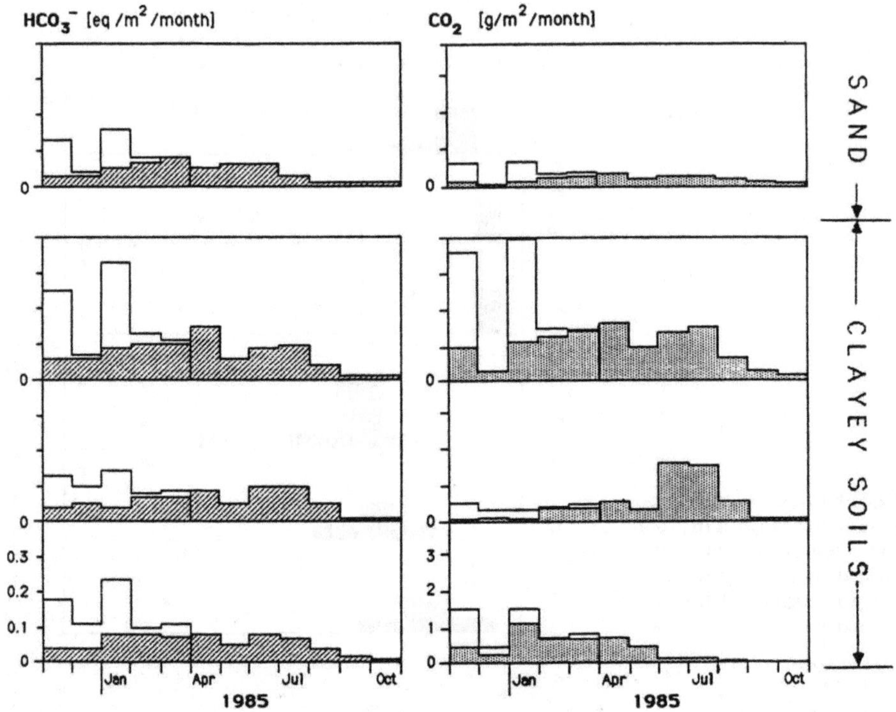

Fig. 4.1.4.8. Seasonal variation of CO_2^- and HCO_3^- input calculated from ground water recharge and ground air composition. *Thick lines* from lysimeter runoff; *hatched* from water balance

Table 4.1.4.1. CO_2^- and HCO_3^- ground water input and mean concentration of inorganic carbon species calculated from recharge and ground air composition. Recharge taken from lysimeter runoff or calculated from climatic water balance

		Input [mmol/(m^2.a)]		Concentration (mmol/l)	
		Lysimeter recharge	Calculated recharge	Lysimeter recharge	Calculated recharge
CO_2	Sand	43	54	0.4	0.3
	Clay	190	140	1.4	1.0
HCO_3^-	Sand	520	480	4.0	3.4
	Clay	750	690	5.9	4.9

4.1.4.4 Solute Concentration and Element Balances

The investigations focussed on the heavy clay soils in the research area, mainly influencing element transformation and solute transport to ground water, and on the relationships between lithologic parameters and the adsorbed and dissolved cation concentrations. Unlike the relatively well-known physical properties of heavy clay soils such as swelling or shrinking, pedal structures and rapid water movement along macropore systems, little information is available on clay interstitial water chemistry. The reason is the difficulty in extracting sufficient water to carry out full chemical analyses. However, the immiscible liquid displacement method provided good results (Sect. 4.1.4.2).

The sampling sites were chosen in the central floodland area (Fig. 4.1.4.1), soil BEN-A in forest land (*Quercus robur*, *Carpinus betulus*), soil BEN-B in grassland, both sites undercovered by sand, which was encountered in the deepest part of the boreholes.

At site BEN-A the ground water table has been lowered by ground water extraction to a depth of 4 m below the surface, whereas at site BEN-B the ground water table recently rose up to 1.1 m below the surface because of a relatively wet spring in 1987. The samples were taken every 2 weeks or monthly throughout the vegetation period 1987.

Figures 4.1.4.3, 4.1.4.9–4.1.4.11 show examples of cumulative solute concentration profiles as a function of depth, with the outer boundary line featuring mean distributions. For comparison with these data, some soil parameters and the portion of cation types on total exchange capacity are plotted. The different solute distribution in both sites are mainly due to different land use. Although unfertilized plots were chosen, the former arable land site BEN-B still contains considerable amounts of anthropogenic carbonate in the naturally carbonate-free upper horizons down to 50 cm depth (Fig. 4.1.4.9). Therefore the exchange capacity near the surface is almost saturated with calcium, while in the deeper parts of the profile the magnesium increases relatively. Similarly, the curve of HCO_3^- is relatively uniform, without seasonal variations, but slightly decreasing with depth. The highest concentrations coincide with the zone of maximum CO_2 production near the Earth's surface. The

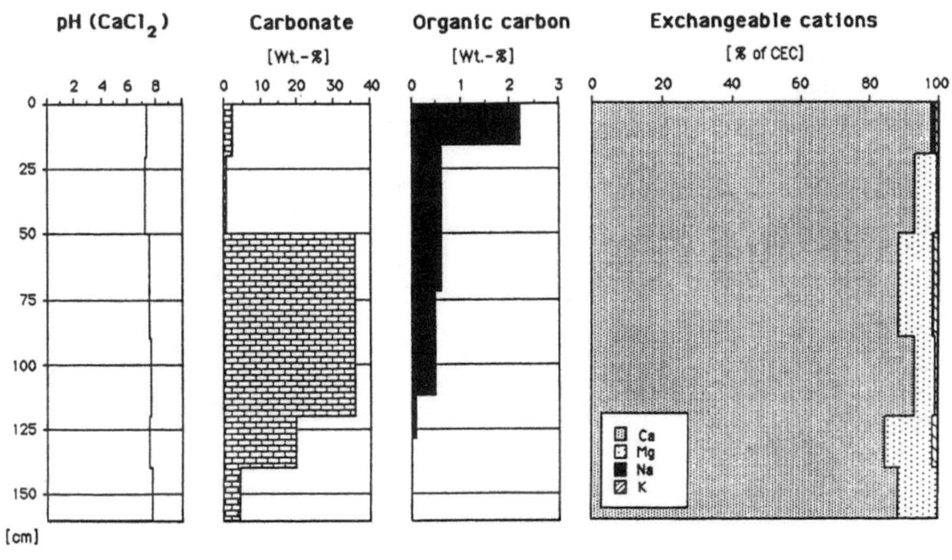

Fig. 4.1.4.9. Soil properties in the clay profile at site BEN-B (grassland)

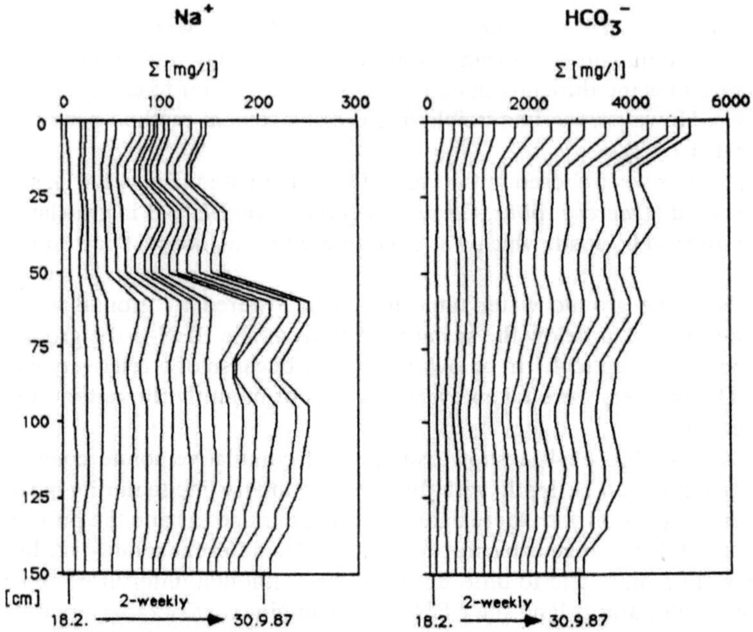

Fig. 4.1.4.10. Cumulative solute concentration profiles of site BEN-B

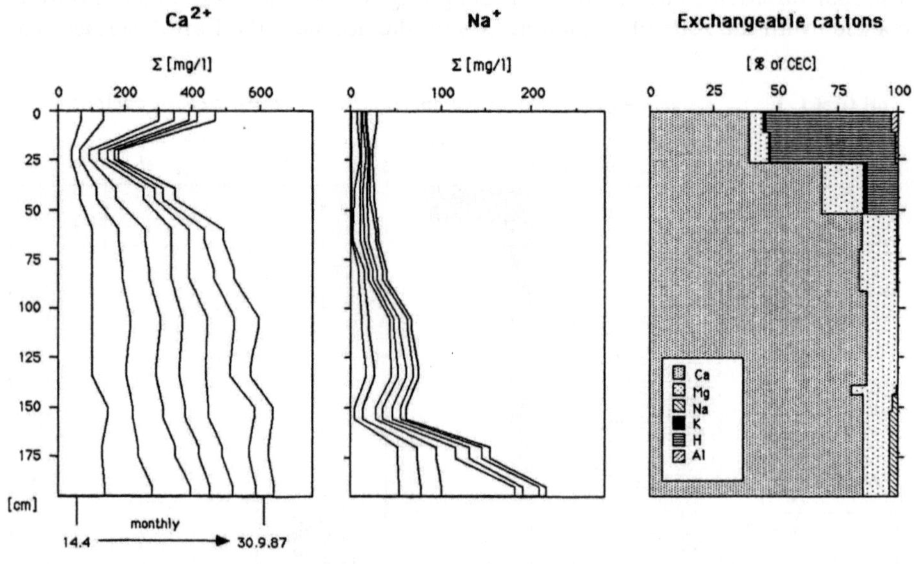

Fig. 4.1.4.11. Cumulative cation concentration and CEC at site BEN-A

mean HCO_3^- content in the ground water-saturated sand of the deepest sample segment reached only 3.1 meq/l compared with about 6 meq/l in the shallow ground water wells.

Considering the poorly adsorbable sodium as a tracer, the depth-concentration profiles show maximum mean concentrations at 60 cm below surface, which is similar for nitrate and sulfate, which are not plotted here. This fact seems to be connected with a main root zone in this horizon above the subjacent stratum, which is cemented by precipitated carbonates. The solute concentrations in April were rapidly lowered in May by leaching, and in the case of nitrate by plant uptake and denitrification. The wet spring and summer with small evaporation presumably prevented the soil solution to be concentrated, whereas Cl^- reached high contents at the end of July, presumably caused by K-fertilizer drifted from a neighbouring corn field.

Forest soil BEN-A (Figs. 4.1.4.3, 4.1.4.11) shows completely different cation exchange capacity patterns. Acid rain largely effects the cation exchange proportions. Hydrogen ions dominate in the sorption sites of the upper horizons and amount still about 12% in 25–50 cm depth. The occurrence of aluminium in these horizons is caused by the low pH ($CaCl_2$) range of 4.7 (Ah) to pH 4.5 in the subsequent horizon, which is only partially similar to the pore solution measurements, even taking the methodical difference $pH_{(CaCl_2)}$ to $pH_{(H_2O)}$ of +0.5 pH into account.

The HCO_3^- distribution shows the effect of buffering processes after acid precipitation (Fig. 4.1.4.3). Deposited protons will first of all be buffered by dissolving carbonate dust, which causes in addition to carbon dioxide production the HCO_3^--input. However, cation exchange forms the major soil buffer system in the clay soils investigated. Furthermore, the rapid decrease of HCO_3^- concentration at 20–25 cm depth indicates, that the input rate of H^+ exceeds the rate of the exchange mechanism. This is most likely a problem of accessible surfaces in these structured soils, which allow bypassing of water through larger pores. The increasing HCO_3^- content especially between 50–60 cm depth can be explained by dissolution of $CaCO_3$ in the subsoil. The Ca^{2+} concentration profiles confirm these assumptions, except that the concentration peak of HCO_3^- in 1.75 m depth is not reflected by an equivalent peak in Ca^{2+} concentration.

The distributions of chloride, nitrate and sulfate show similar tendencies to the sodium curve plotted here. The higher transpiration in the forest area should cause an increasing solution concentration down the soil profile. However, this cannot explain the very sharp increase at 150 cm depth. Compared to the investigations of Jacob (1987) at the Sandhausen site (Sect. 4.2.3) with sandy soils, but similar precipitation conditions, considerably higher mean solute concentrations could be observed for Na^+ (10 times) and Cl^- (4–5 times) in the deeper part of the profile.

Nitrate, which did not exceed 1 mg/l in Heidelberg soil solution, reached concentrations in the clay soils up to about 350 mg/l at a depth of 2 m, presumably because of extensive N mineralization due to clearing of trees. Considering the reduced ground water conditions, surprisingly no reduction of the oxidized components of the seepage water took place during its passage through the clay soil as already indicated by the high nitrate content of the pore solutions. Only in the first 5 cm were Fe^{2+} (0.4–5.5 mg/l) and NH_4^+ (1.2–4.4 mg/l) found in significant concentrations.

The results of the pore water analysis were used to calibrate a two-phase layer model predicting element distribution and output to ground water, considering rapid water movement down macropores (Fritsch 1990).

4.1.4.5 Conclusions

The investigations on the relation between sedimentology and pore water and gas content of fine-grained cover sediments and the chemical content of the subjacent ground water gave consistent results concerning the CO_2 system. The inorganic carbon species in the ground water can be referred – with a certain deficit – to carbon dioxide content of the ground air. The isotopic data indicate that carbonate dissolution occurs predominantly in the open system of the unsaturated zone.

The geologically indicated assumption of an influence of heavy clay soils on the reduced ground water could not be verified. With the exception of the uppermost horizon, only oxidized species were found in the pore water within the unsaturated zone. The element distribution in the pore solution is influenced by atmospheric acidification and could be referred to soil properties, mainly exchange capacity and carbonate content. A layer model for solute movement in structured soils, including the effect of an heterogeneous pore system, probably may help to explain the more concentrated pore solution in the deeper part of this soil.

4.1.5 Hydrogeochemical Processes During the Passage of Surface Water and Ground Water Through Genetically Different Organic Sediments

H. Brühl, A. Moschick and H. Verleger

4.1.5.1 Targets

Ground water which passes aquifers with intercalations of organic rich mud and peat, as well as bank filtrates which flow through organic sediments will undergo characteristic chemical changes. This water usually is reduced, low in sulfate and less mineralized than the original water; however, it contains dissolved organic substances that may cause problems for the treatment of drinking water. The respective microbial metabolic processes and the water-rock interactions were investigated with five column experiments using two genetically different organic sediments and two natural waters of different composition as well as deionized water simulating the passage through organic sediment layers abundant in the Pleistocene aquifers of northern Germany. These investigations focussed on the behaviour of the major components and the reactions of humic and fulvic acids during the passage through the organic sediments (Moschik 1990).

Table 4.1.5.1. Main components of the pore solution of the sediments and of the percolated water of the column experiments

meq/l	Ca^{2+}	Mg^{2+}	Na^{+}	K^{+}	HCO_3^-	Cl^-	NO_3^-	SO_4^{2-}
B-PEAT1	5.70	0.86	1.20	0.25	6.17	1.72	0.01	0.73
B-MUD2	5.24	1.04	2.41	0.30	6.84	2.11	0.08	0.39
B-SW	9.85	1.21	4.35	0.12	7.20	2.59	0.03	5.56
B-GW	2.74	0.63	0.31	0.04	3.22	0.42	0.04	0.29

4.1.5.2 Performance of Column Experiments

For experiments the following sediments were used.

- A medium decomposed phragmites peat from the Riemeisterfenn/Berlin (B-PEAT), from a depth of about 1 m.
- A recent silty fine sand mud from the lake Südparkteich/Berlin (B-MUD) taken from 60 cm below the bottom.

The sediments were transported and processed in a nitrogen atmosphere immediately after sampling to preserve the natural chemical equilibrium. Subsamples were instantaneously packed into acrylic glass columns for the experiments B-PEAT 1.1 and B-MUD 2.1 and B-MUD 2.3; the rest of the samples were deep frozen and stored for the experiments B-PEAT 1.2 and B-MUD 2.2. The experiments B-PEAT 1.1 and B-MUD 2.1 were performed using a comparatively strongly mineralized oxygen-containing surface water B-SW (Table 4.1.5.1), the experiment B-PEAT 1.2 and B-MUD 2.2 involved a weakly mineralized reduced ground water B-GW, both from Berlin.

The water was percolated under steady-state conditions with an apparent velocity of 3 to 6 cm/d until the ion concentrations of the percolate remained nearly constant. Thus, the solution of the specified yield of pore space (here called pore volume) had been exchanged five to six times. During a following period of stagnation of about 30 days, equilibrium between sediment and pore solution was allowed to evolve. In experiment B-MUD 2.3 the flow was continued with oxygen-saturated distilled water until some of the ions could no longer be detected in the exchanged water. The exchanged water leaving the columns was collected and sampled in fractions of 100 to 200 ml for chemical analyses.

4.1.5.3 Investigations on Components and Chemical Properties of the Sediments

The mineral matter is composed of quartz and feldspar. Moreover, pyrite and small quantities of calcite were detected in the peat, the mud contained calcite, some pyrite, vivianite and other unidentified iron compounds.

The cation exchange capacity (CEC) of the sediments depends mainly on the amount, the age and the origin of the organic matter (Fig. 4.1.5.1, CEC/Org-ratio). In relation to the CEC, the amount of the exchangeable bound alkaline-ions found

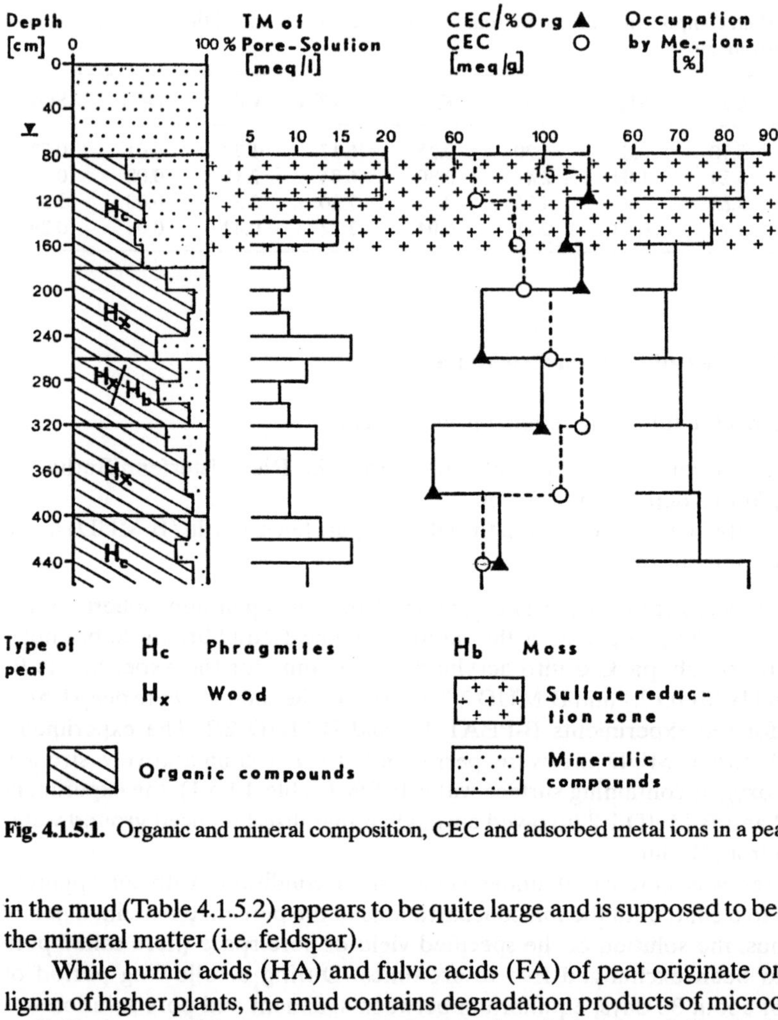

Fig. 4.1.5.1. Organic and mineral composition, CEC and adsorbed metal ions in a peat profile

in the mud (Table 4.1.5.2) appears to be quite large and is supposed to be bound onto the mineral matter (i.e. feldspar).

While humic acids (HA) and fulvic acids (FA) of peat originate only from the lignin of higher plants, the mud contains degradation products of microorganisms as well (Jackson 1974), consequently the humic substances of the mud are characterized by higher nitrogen and hydrogen contents and lower carbon contents (Table 4.1.5.3). The total acidity of the peat depends mainly on carboxyl groups, but in the mud phenolic OH groups contribute 10 to 50% of the total acidity.

Table 4.1.5.2. Composition, cation exchange capacity (CEC) and adsorbed metal ions in the sediment samples

Sample	H_2O % wt.	TOC % d.wt.	CEC	Ca^{2+}	Mg^{2+}	Na^+	K^+	Fe^x	Al^{3+}+H^+
			meq/100 g						
B-PEAT1	60	51	87.4	61.5	3.86	1.14	0.68	0.47	19.75
B-MUD2	47	10	28.4	12.7	0.72	0.67	0.88	0.63	12.80

Table 4.1.5.3. Elemental composition, total acidity (TA) and some functional groups of HA and FA

	C	H	N	O	TA	COOH	phenol OH
	%				meq/g		
Soil humic acids (SHA)							
B-PEAT	50.6	4.6	3.1	41.7	4.53	3.66	0.87
B-MUD	48.5	4.9	3.8	42.8	7.14	3.27	3.87
Soil fulvic acid (SFA)							
B-PEAT	47.8	4.2	1.8	46.2	5.19	5.01	0.18
B-MUD	45.9	4.4	2.1	47.6	4.74	4.05	0.69
Aquatic fulvic acid (AFA)							
B-PEAT	48.4	4.0	1.3	46.3	6.54	6.50	0.04
B-MUD	49.3	4.5	1.2	44.9	4.89	4.39	0.50

Although HA and FA of the mud differ in elemental composition and many properties from those of the peat, their structures seem to be largely similar, as far as it can be inferred from IR spectra (Fig. 4.1.5.2).

All preparations show the distinct absorption bands at 3400 cm^{-1} (caused by H-bonded OH) and 1630 cm^{-1} (aromatic C=C). Differences of the single categories of substances are indicated by decreasing intensities of the absorption bands at 2920 cm^{-1} (aliphatic C-H) and 1050 cm^{-1} (polysaccharide like substances) in the order SHA-SFA-AFA. In the same order, the intensities of the bands at 1720 cm^{-1} (COO-ion) and 1380 cm^{-1} (aliphatic C-C and phenlic C-O) increase. The SHA preparations show absorption bands that are not observed in the SFA- and AFA preparations. Following Stevenson (1982) the bands at 3050 cm^{-1} are considered as aromatic hydrocarbons, those between 1540 and 1515 cm^{-1} as nitrogen-containing compounds and those between 1150 and 1050 cm^{-1} as polysaccharide-like substances.

4.1.5.4 Results of the Column Experiments

The experiments aimed to analyze the process of equilibrating and the dynamic equilibrium that evolves, when organic sediments are passed by surface and ground water which differs from the original pore solution in mineralization and composition. On the basis of the development of ion concentrations during the passage and of the ion balance, the reactions involved in the process of equilibrating could be identified and the sorption characteristics of organic sediments could be inferred. Furthermore, the influence of the mineralization of the pore water on the quantity and the quality of the dissolved organic compounds was evaluated.

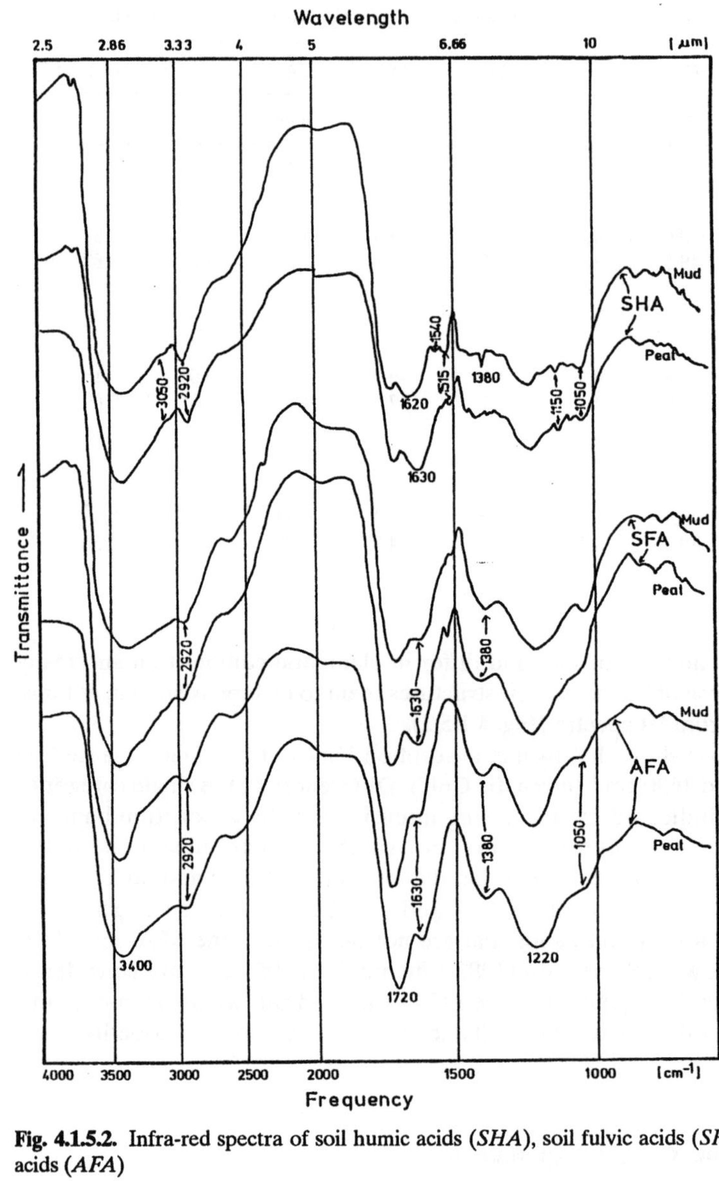

Fig. 4.1.5.2. Infra-red spectra of soil humic acids (*SHA*), soil fulvic acids (*SFA*) and aquatic fulvic acids (*AFA*)

The Process of Equilibrating

The process of equilibrating can be divided into three phases (Fig. 4.1.5.3):

During the first phase, constant ion concentrations marked the displacement of the pore volume of original pore solution. The second phase was characterized by fast linear changes of the ion concentrations with increasing contact time. Ion adsorption (IA) in the experiment with the strongly mineralized water and ion

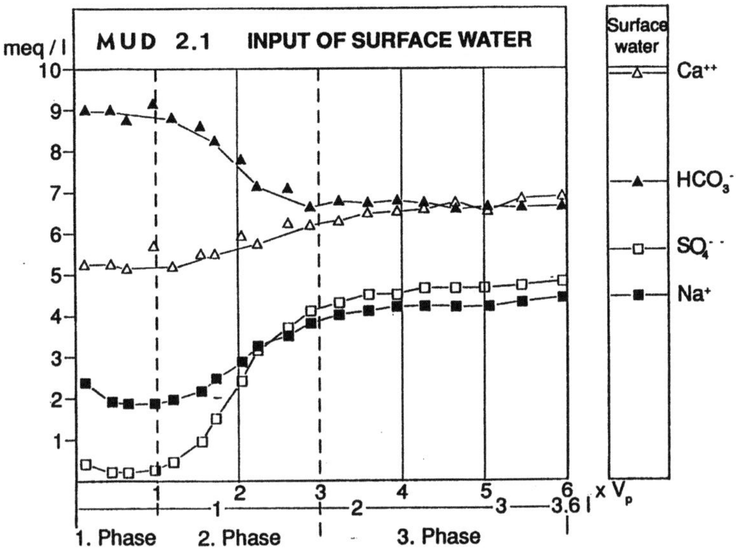

Fig. 4.1.5.3. Concentration development of selected ions in the pore solution of the mud percolated by a surface water with high TDS. The unit amount of exchanged water is equal to one exchange of the solution in the pore volume V_P

desorption (ID) in the experiment with the weakly mineralized water took place to adapt the sediment to the changed ionic strength of the new pore solution. Ion adsorption and desorption were almost completed after two additional exchanges of the solution in the pore volume, since at the end of the second phase more than 75% of the solutes of the percolated water could pass the sediment column.

The fast adaptation to the ionic strength occurred by preferential adsorption or desorption of alkaline ions, as was evident in the development of the alkaline earth/alkaline ion equivalent ratios (EA) in Fig. 4.1.5.4. In the experiment B-PEAT 1.1, the EA of the pore solution, having a higher value than the percolated water, remained high at first. In the experiment B-PEAT 1.2, the EA ratio even decreased due to the preferential desorption of alkaline ions until it returned to the higher value of the percolated water in the third phase.

During the third phase the ion concentrations approached in an exponential curve the mineralization and composition of the percolated water. As an example, the Na^+ concentration curve is given (Fig. 4.1.5.3). By ion exchange reactions the sediments adapted to the altered ion equivalent ratios of the pore solution. The duration of this phase depends on the differences between the percolated and the original solution. After five exchanges of pore volume of solution, almost a dynamic equilibrium in the experiments B-PEAT 1.1 and B-MUD 2.1 was attained. The equivalent ratios in the experiment B-PEAT 1.2 differed too much to be balanced within the duration of the experiment (Figs. 4.1.5.3 and 4.1.5.4).

The dynamic equilibrium which was attained after the fifth exchange of the solution in the pore volume of the experiments B-PEAT 1.1 and B-MUD 2.1 deviated from the ideal case without any interaction between percolated water and

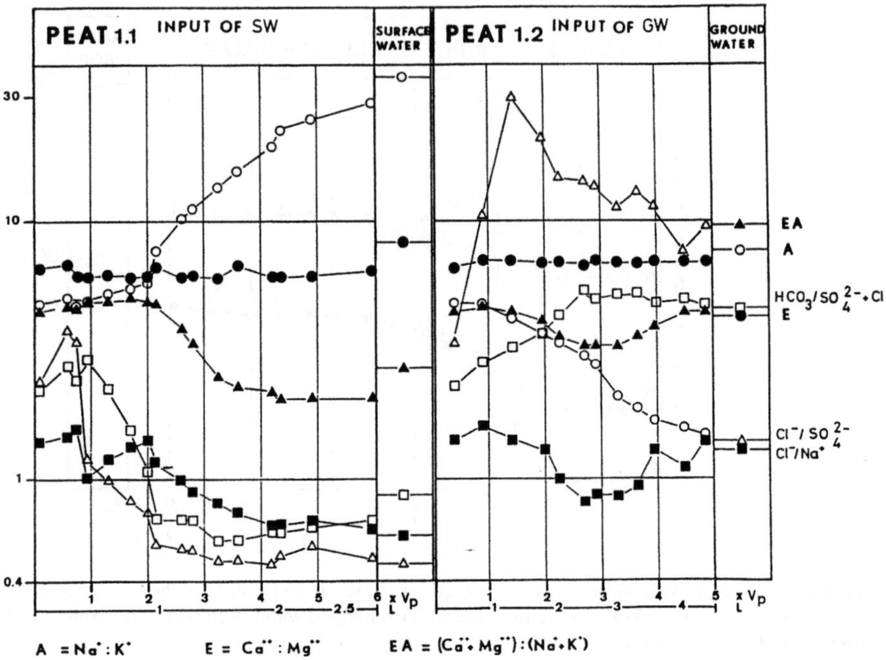

Fig. 4.1.5.4. Development of the ion-equivalent ratios in the exchanged water of the peat during the flow of the ground water with high TDS

the sediments. Particularly the passage and adaptation of Ca^{2+}, SO_4^{2-} and HCO_3^- is hindered, obviously by microbial metabolism.

The oxygen content of the water used for the experiments B-PEAT 1.1 and B-MUD 2.1 was exhausted by the oxidation of reduced species within a few centimetres' pathway in the column, which led to a redox potential (E_H) of –200 mV to –250 mV halfway down the column. Under these conditions sulfate was reduced. Reduction processes consume protons (Benade 1954) so that the redox environment within the sulfate reduction zone is characterized by low E_H- and high pH values. Under these conditions carbonate precipitates. These processes became evident in the mud after the first exchange of the pore volume of solution (flat Ca^{2+} concentration curve in Fig. 4.1.5.3). In the peat the precipitation of calcium carbonate is indicated by increasing amounts of retained Ca^{2+} after the third pore volume of solution had passed through the column (Table 4.1.5.4). During the percolation of the reduced ground water in the experiment B-PEAT 1.2, HCO_3^- was produced in an equivalent amount to the reduced SO_4^{2-}, without a precipitation of calcium carbonate, whereas in the experiment B-MUD 2.2 calcium carbonate was precipitated. The peat could obviously release more protons than were needed for the reduction of the small amounts of SO_4^{2-}, so that HCO_3^- remained dissolved.

Table 4.1.5.4. Balance of composition after three changes of the solutions in the pore volume and at the end of the flow. Negative values refer to removed components due to adsorption, precipitation or degradation, positive values to added components due to dissolution, desorption or mineral formation. The transformations are standardized to meq/kg dry sediment

Sample	Ca^{2+}	Mg^{2+}	Na^{+}	K^{+}	Cl^{-}	HCO_3^{-}	SO_4^{2-}
B-PEAT 1.1							
Exch. 2–3x	–3.30	+0.28	–3.58	+0.13	–0.15	–3.25	–3.02
Exch. 3–6x	–6.16	+0.12	–1.78	+0.35	+0.15	–6.86	–0.96
B-MUD							
Exch. 2–3x	–6.77	–0.40	–1.89	+0.38	–0.96	–3.56	–4.10
Exch. 3–6x	–7.33	+0.33	–0.22	+0.12	–0.39	–4.45	–2.22
B-PEAT 1.2							
Exch. 2–3x	+5.46	–0.27	–2.48	+0.53	+2.19	+5.68	–0.50
Exch. 3–5x	+0.47	–0.61	+0.52	+0.42	+0.37	+0.74	–0.40
B-MUD 2.2							
Exch. 2–3x	–0.16	–0.32	+1.06	+0.21	+1.60	–0.31	–0.43

The Sorption Characteristics of the Sediments

Organic sediments are characterized by high cation exchange capacities. The chemosorptive bonds with metal ions and especially with protons are very stable under natural conditions. The amount of ions retained or released from the sediment after the third exchange of the pore volume of solution (after the adaption of the sediment to the new pore solution to a large extent and before the precipitation of carbonates) attained not more than 1 to 1.7% of the CEC, although the exchange sites had been adapted to solutions with the double or half concentration. Obviously the ionic strength of natural fresh waters is not capable of altering substantially the ratio of protons and metal ions on the exchange sites.

Vertical profiles from the peat bog with influent conditions, however, show that the exchange sites are almost completely occupied by metal ions within the sulfate reduction zone (Fig. 4.1.5.1). This means that the changes of the ion distribution on the exchange sites are primarily based on interactions of microbially induced reduction processes with the resulting alterations of the electrochemical environment.

Adsorption of anions was proven at high pH values. It occurred to compensate the charge in the pore solution when carbonates precipitated. Desorption of anions in those experiments with the weakly mineralized water showed a decrease in sorption energy in the order of PO_4^{3-} >> CO_3^{2-} > SO_4^{2-} > Cl^-. The release of PO_4^{3-} and Cl^- was associated with an increase of the DOC content and/or a quality change of the dissolved organic substances; a sorptive bond to specific organic compounds can be presumed.

The Behaviour of the Dissolved Organic Matter

The percolates leaving the columns always exhibit light yellow to brown colours due to dissolved organic substances measured as dissolved organic carbon (DOC). These are composed of 95–98% FA, 1–3% HA and 1–2% other organic substances.

The dissolved organic substances can be characterized by the ratio of UV light absorption at 254 nm and the DOC content (A254/DOC). The ratio increased in presence of the more complexly structured organic acids. High values of A254/DOC are usually associated with a high buffer capacity of the solutions. Due to the absorption of visible light by HA, the A436/A254 ratio rises with increasing contents of HA in the solution.

The solubility of the organic substances is mainly influenced by the earth alkaline/alkaline ratio (EA) of the pore solution; its total mineralization, however, is less important. Therefore the DOC content of the weakly mineralized percolates was similar at the beginning and at the end of the experiments (Fig. 4.1.5.6). Increasing DOC content was always connected with a decreasing EA of the pore solution. These changes occurred during the percolation of a weakly mineralized water with preferential desorption of alkaline ions in the second phase of equilibration, and when the pore solutions were depleted of earth alkaline ions due to precipitation of carbonates.

Quantity and composition of the dissolved organic substances are equally influenced by the EA as well. The release of more complexly structured FA (rising A254/DOC-ratio in Fig. 4.1.5.5) and more HA (rising A254/A436-ratio in Fig. 4.1.5.6) occurred on the above-mentioned occasions with decreasing EA of the pore solution.

The interpretation of the behaviour of organic substances is complicated by their slow dissolution kinetics. All experiments showed after the fifth exchange of

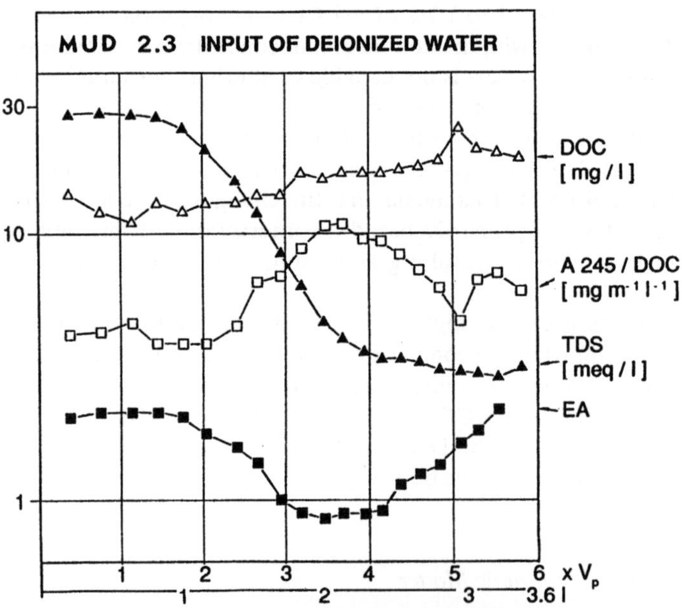

Fig. 4.1.5.5. Developments of TDS- and DOC contents in the exchanged water of the mud during the flow of deionized water. The earth-alkaline/alkaline ratio (*EA*) depends on variable release of earth-alkaline- and alkaline ions, the A254/DOC ratio is a quality feature for the DOC

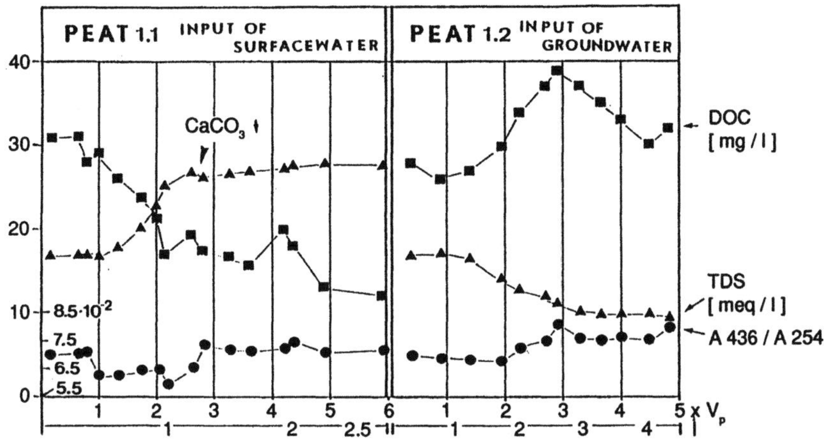

Fig. 4.1.5.6. Developments of the TDS- and DOC contents in the exchanged water of the peat during the flow of the surface water with high TDS and the ground water with low TDS. The A436/A254 ratio is an indicator for the presence of dissolved HS

the pore solution a substantial release pulse of organic substances that differed qualitatively from the general trend (Fig. 4.1.5.5). This release pulse of less complexly structured organic substances was presumably produced by the change of the pore solution.

4.1.5.5 Conclusions

Hydrochemical interactions of the major inorganic constituents of infiltrating surface and percolating ground water occur in different organic sediments. In column experiments, the adjustment of an equilibrium between solid (peat, mud) and liquid phase after replacing the original pore solution by a highly mineralized surface water (B-SW), a low mineralized ground water (B-GW) and distilled water was observed. Both sediments displayed similar hydrochemical properties although their humic and fulvic acids showed distinctive, genetically caused differences in elemental composition and in their functional groups. The difference in hydrochemical behaviour is primarily based on different quantities of the available organic material and the mineral composition of the sediments.

The sorptive bonds of metal ions and protons with the organic material are stable enough that no more than 1 to 2% of the existing exchange sites react to doubling or halving the ion concentration of the pore solution by ion adsorption or desorption.

The TDS composition of the pore solution is more effectively changed by microbially mediated processes. This happens directly by the reduction of the oxygen-containing dissolved compounds, indirectly by the change of chemical equilibria due to the reducing and proton-lacking environment. This causes mainly the precipitation of carbonates, the increase of sorption gains to the organic material, and adsorption of anions.

The percolates dissolved predominantly fulvic acids from the sediments (> 95% of the DOC). Quantity and quality of the dissolved organic substances are mainly influenced by the earth alkaline/alkaline ratio of the pore solution, less by their ionic strength. Quality and quantity changes of the dissolved organic matter caused by changes of the inorganic water composition are retarded.

4.1.6 Hydrochemical Phenomena in the Dorsten Leakage System

M. Hoffmann, H.R. Langguth and J. Larue

4.1.6.1 Target and Site of Investigation

From a review of the published material (Langguth and Voigt 1980), it is evident that the leakage effect has been demonstrated only in an indirect way by means of pumping tests in the aquifer itself or by the determination of the hydraulic parameters of the semi-permeable bed (aquitard). Little is known about the hydrogeochemistry during continuous leakage. This was investigated in the hydrogeological entity of the Upper Cretaceous Haltern sands near Dorsten (DOR), Northern Ruhr district (Figs. 1.1 and 4.1.6.1, Table 1.1) (Meiners 1985; Hoffmann 1986; Hoffmann et al. 1987). It is formed by unconfined and confined aquifer units (multi-layer system characterized by leakage phenomena).

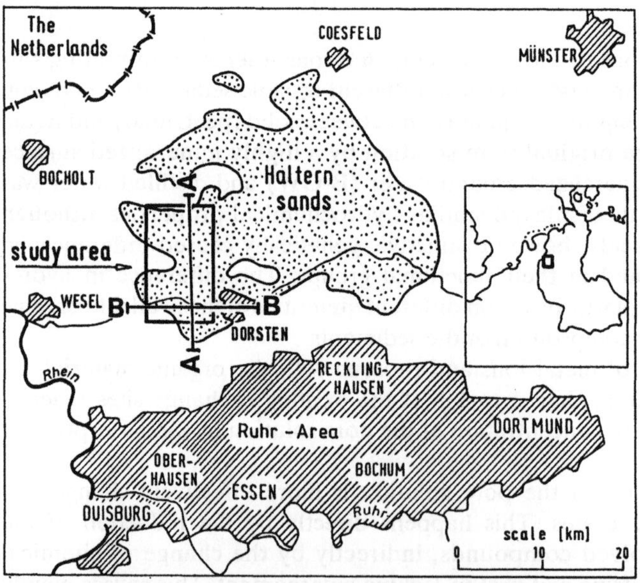

Fig. 4.1.6.1. Geographical location of the study area (*A-A, B-B* cross-sections of Fig. 4.1.6.2)

At the suggestion of the authors, the local ground water supply agency (Rhenish-Westphalian Water Supply Society Mühlheim/Ruhr) agreed to drill observation wells screened in the aquitard. Thus, it was possible to observe directly the hydraulic behaviour in the confining bed while pumping in the main aquifer.

Additionally, the hydrochemical characteristics of the water in the confining bed could be described and correlated with the chemistry in the main aquifer.

4.1.6.2 Geology and Hydrogeology

The geological structure and the location of the study area are characterized as follows (Fig. 4.1.6.2; Hilden and Suchan 1974; Völtz 1979):

- The base of the hydrogeological system is formed by impervious layers of Emscher marls (Coniacium to Santonium), which, overlain by the sands of Haltern and Recklinghausen (Upper Santonium), form the main aquifer (40 m < thickness < 120 m). Their hydraulic conductivity ranges from 10^{-4} to 10^{-3} m/s.
- In slightly undulating synclines a semi-permeable layer (aquitard) is observed with a hydraulic conductivity of maximum 10^{-6} m/s (Bottrop marls; Santonium/ Campanium).
- The aquitard is overlain in the domains of the alluvial valley of the Lippe by a Quaternary sandy aquifer of up to 12 m thickness.

Due to the general structure of the main aquifer in the form of flat synclines and anticlines, the natural ground water recharge essentially takes place in the central part (Marl anticline). There the main aquifer is exposed at the surface. The unconfined ground water flows from here generally southward in the direction of the Dorsten syncline (semi-confined conditions).

4.1.6.3 Multi-Level Wells and Pumping Tests

In the Dorsten syncline a well field with 48 wells at present pumps 20 million m^3/a. At the suggestion of the authors, the management of the ground water plant agreed to drill and screen two multi-lvel wells DOR1 and DOR2 in the aquitard of the Bottrop marls. They are close to the production wells 10 and 31. Their design is given in Fig. 4.1.6.3. The depth of the screen below surface (z) is added in metres (e.g. observation well DOR2, depth 22 m = DOR2.22). The screens are installed in sequences which have higher contents of fine sand, recognized by cuttings and logging, thus monitoring zones with an assumed higher permeability. Only the screen DOR2.22 has been intentionally placed in a dense marly sequence.

The way the pressure heads in the aquitard react to variations in the pressure head in the underlying main aquifer was observed by means of pumping tests

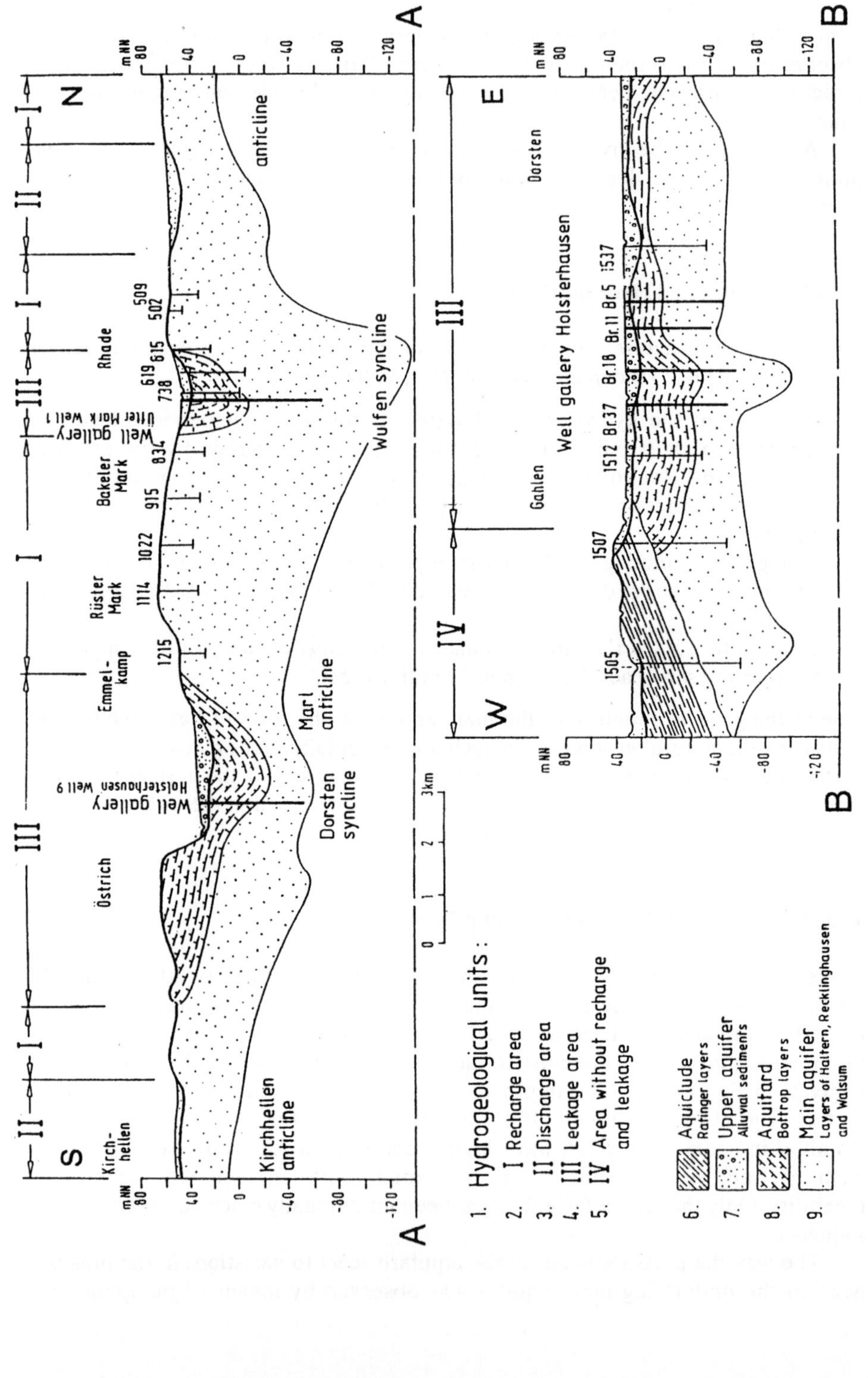

Fig. 4.1.6.2. Geological S-N and W-E sections in the study area

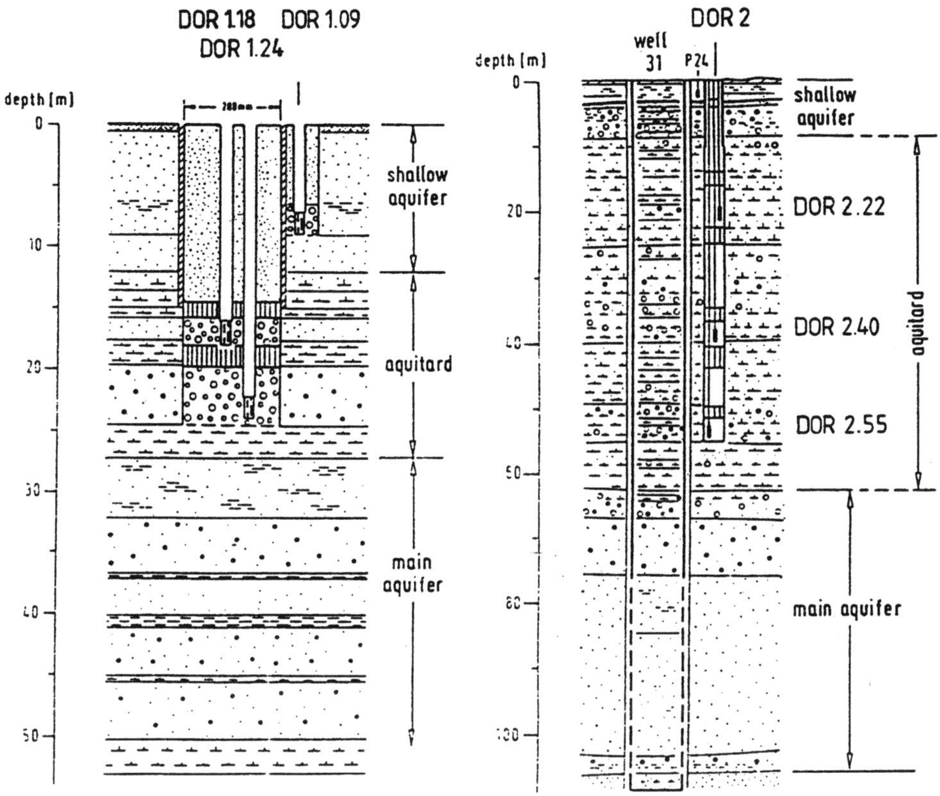

Fig. 4.1.6.3. Cross-section of the multi-level wells DOR1 and DOR2

(Fig. 4.1.6.4), whose details are not discussed here. They proved the occurrence of both Noordbergum (Neuman and Witherspoon 1969; Verruijt 1969) and Rhade (Langguth and Treskatis 1987) effects. Thus, the unsteady-sate flow conditions typical for leakage aquifer systems have been demonstrated by direct observations.

4.1.6.4 The Mineralization of Water Along the Flow Path

The chemical composition of the water in the aquifer system was determined from 1983 to 1985. In 1984 samples were collected from 94 observation wells by complete chemical analyses. These data allow the description of time variations and the geochemical processes along the flow paths of ground water.

The areal distribution of the calcite saturation index (SI_c), calculated with the geochemical programme WATEQF (Plummer et al. 1976), shows in a synoptical way the interdependance of the ground water geochemistry of each hydrogeological unit of the aquifer system (Fig. 4.1.6.5). In the recharge area, the Marl anticline area in the north, ground water is undersaturated (aggressive) with

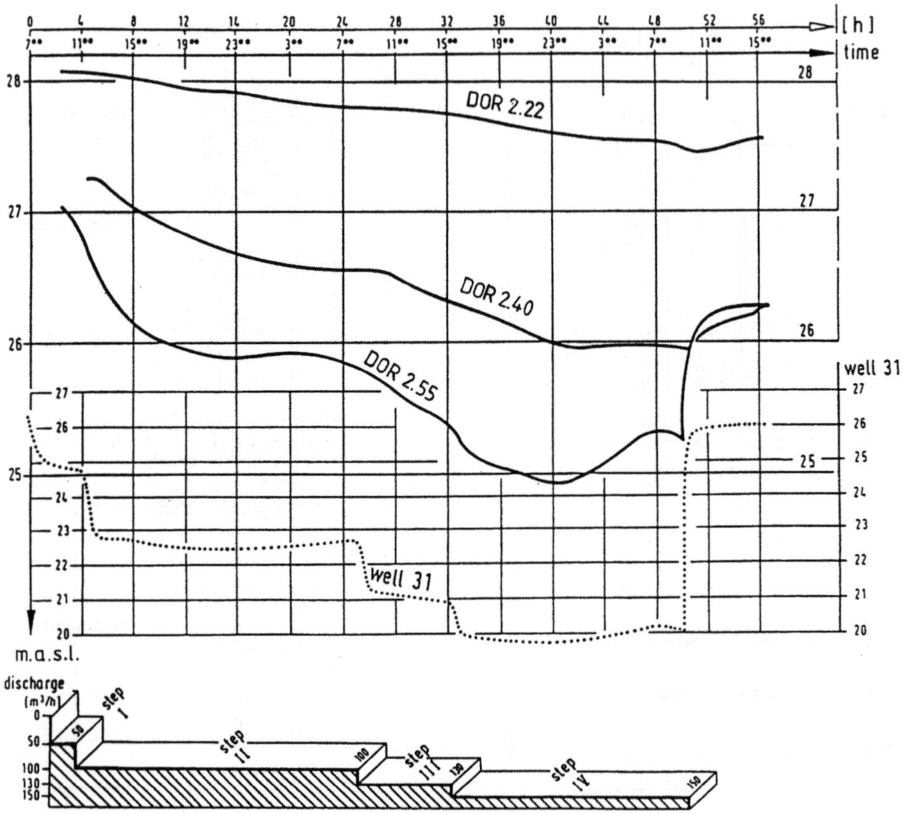

Fig. 4.1.6.4. Hydraulic heads of well 31 and of the observation wells DOR2.22, DOR2.40 and DOR2.55 in the aquitard during the pumping test at certain discharges

respect to calcite ($SI_C = -1.0$). On its flow path to the south, ground water approaches equilibrium with calcite (SI_C about -0.25). This is particularly valid for the transition zone towards the Dorsten syncline. Finally, in the surrounding area of the Holsterhausen pumping gallery, all the water shows SI_C near zero, thus indicating a $CaCO_3$-CO_2 equilibrium. Consequently, the dominating process in the flow direction of ground water from the recharge area in the Marl anticline to the southern discharge area in the Dorsten syncline is the dissolution of calcite.

Eight other parameters have been treated in the same way: TDS, pH, O_2, Na^+, Cl^-, Ca^{2+}, HCO_3^- and SO_4^{2-}. They permit analogous interpretations. The results are summarized in Table 4.1.6.1.

In some cases, the coincidence of geological boundaries with correspondant variations in geochemistry is not as clear as in the case of SI_C. But this is understandable considering the nature of these dynamic processes, which evolve somewhat independently of the hydrostratigraphy.

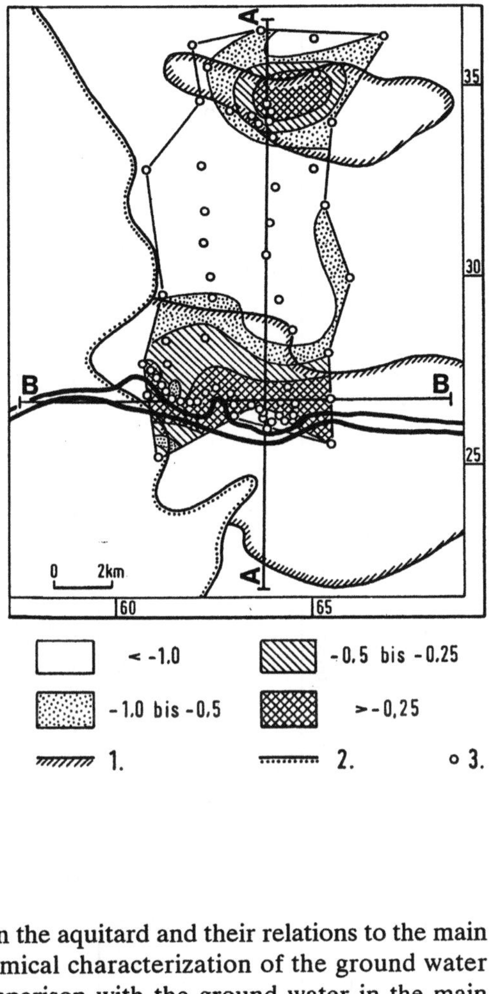

Fig. 4.1.6.5. A real distribution of saturation indices for calcite (SI_c) in the main aquifer. *1* Boundaries of confining beds; *2* Boundary of Tertiary cover; *3* observation well

4.1.6.5 Hydrogeochemical Results

Besides the hydrodynamic processes in the aquitard and their relations to the main aquifer, the data allowed the geochemical characterization of the ground water circulating in the aquitard and a comparison with the ground water in the main aquifer. There are significant differences concerning the concentrations and the chemical type:

- The waters in the main aquifer have very low to intermediate TDS (80 to 500 mg/l). They belong to the Ca-HCO_3 type.
- The waters in the aquitard have considerably higher TDS (800 to 1500 mg/l). They belong to the Ca-SO_4-HCO_3^- and the Ca-SO_4-type.

Mineralogical studies by Prof. Echle and Dr. Ing. Gussone of the Institute for Mineralogy and Petrology, RWTH Aachen, and thermodynamic calculations using the model WATEQF (Plummer et al. 1976; Pagenkopf 1978) show that this is due to the dissolution of calcite in the main aquifer and calcite and gypsum in the aquitard. So regional discrepancies in the above-mentioned hydrogeochemical types, namely Ca-CO_3 and Ca-HCO_3-SO_4, in the ground water of the main aquifer indicate the attribution of leakage water from the aquitard, in general, simultaneously accom-

Table 4.1.6.1. Typical concentration intervals, evolution of chemical composition in ground water flow direction, origin of main components, dominant geochemical processes and other parameters of the main aquifer

Hydrogeological domain	TDS epm	pH	O_2 ppm	Na^+ epm	Cl^- epm	Ca^{2+} epm	HCO_3^- epm	SI_C	SO_4^{2-} epm
	Typical concentration intervals								
Syncline of Wulfen	3 to 9	7 to 7.5	3 to 9	<0.5	>1.0	2 to 4	1 to 3	–0.5 to ± 0	0.5 to 1.5
Anticline of Marl	<3	<6.5	6 to > 9	<0.5	<0.5	<2	<1	<–1.0	<0.5
Syncline of Dorsten	6 to >9	About 7.5	<3 to 6	<0.5	0.5 to 1.0	2 to 4	1 to 3	–0.5 to ± 0	<0.5 to 1.5
	Geochemical behaviour in direction of ground water flow								
Syncline of Wulfen	Constant	Locally >7.5	Not uniform	Constantly low	Stable	Stable	In the centre slightly raised	Aggressive up to saturated	Not uniform
Anticline of Marl	Locally slightly raised	Increasing southerly	Constantly high	C o n s t a n t l y low	low	low	Constantly low	Constantly aggressive	Constantly low
Syncline of Dorsten	Increasing southerly	Mostly <7.5	Decreasing towards SW	Stable increasing locally	Not uniform, in places raised	Increasing towards W and SE	Increasing towards SE	Transit N to S from less aggressive to saturated	Clearly raised in places

General direction of ground water flow ↓

Table 4.1.6.1. (*Continued*)

Hydrogeological domain	TDS epm	pH	O_2 ppm	Na^+ epm	Cl^- epm	Ca^{2+} epm	HCO_3^- epm	SIc	SO_4^{2-} epm
	Origin and dominant geochemical process								
Syncline of Wulfen	Dissolution of calcite	Dissolution of calcite	Lowered by oxidation	Infiltration	Infiltration/man-made	Dissolution of calcite	Saturation of calcite continuously attained	Undersaturation in respect to calcite	Infiltration
Anticline of Marl	Infiltration	Dissociation of CO_2 from air in the soil	Air in the soil	Infiltration	Infiltration	Infiltration/dissolution of calcite	Infiltration/dissolution of calcite	Undersaturation in respect to calcite	Infiltration
Syncline of Dorsten	Dissolution of calcite leakage locally	Dissolution of calcite	Lowered by oxidation	Infiltration, locally ion exchange	Infiltration/man-made	Dissolution of calcite	Dissolution of calcite	Towards S saturation of calcite	Infiltration/leakage

General direction of ground water flow →

panied by increasing TDS. Hence the leakage can also be proved by chemical means.

The following three geochemical processes can be discriminated on a Ca-SO_4 plot (Fig. 4.1.6.6):

1. dissolution of calcite,
2. dissolution of calcite and gypsum,
3. addition of aquitard water to main aquifer water by leakage.

Dissolution of Calcite in the Main Aquifer

This is reflected by the fact that the calcium concentration increases without any significant variation in the sulfate content. The calcite equilibrium is attained when the following values of Ca^{2+} and pH are reached: 3 to 3.5 meq/l Ca^{2+} (60 to 70 mg/l, pH 7.5).

Dissolution of Gypsum in the Aquitard

Due to its higher solubility gypsum is still dissolved, after calcite equilibrium has been reached. The dissolution of gypsum increases the calcium content in the water, exceeding the solubility product of calcite. Consequently calcite can precipitate as a secondary product. This well-known fact is indicated in Fig. (4.1.6.6) by the slightly hyperbolic curves of the sulfate and calcium content, which slowly approaches a straight line of gypsum dissolution.

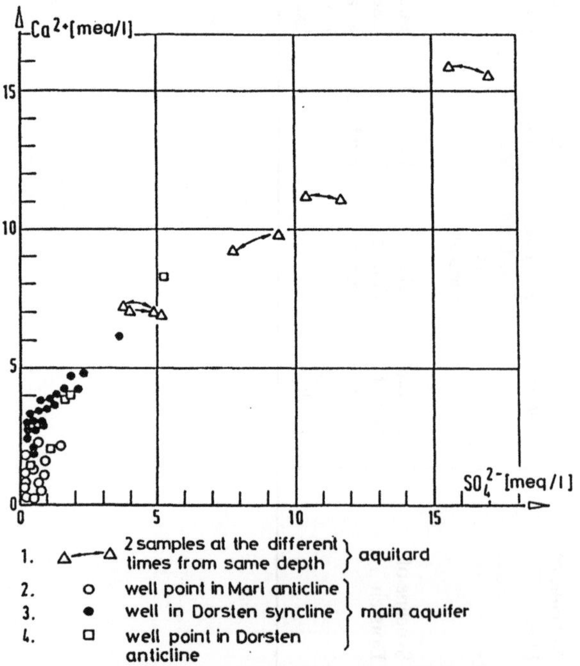

Fig. 4.1.6.6. Calcium-sulfate ratio of ground water in the aquitard and the main aquifer. *1* Chemical analysis at different times from the same depth; *2* samples from unconfined areas; *3* samples from production wells under confined conditions; *4* samples from confined areas

Addition of Aquitard Water to Main Aquifer Water by Leakage

An addition of leakage water to the main aquifer forms a straight line of mixing of the original ground water of $Ca\text{-}HCO_3$ type in the main aquifer and the higher mineralized $Ca\text{-}SO_4$ water in the aquitard as distal members.

In the case of aquitard water it is not established whether the most concentrated member has been found by the given arrangement of the screened sections of the multi-level wells.

The available data at the given sampling depths indicate that the rate of leakage water to the main aquifer differs locally. Using the contrasting values of the distal members (Fig. 4.1.6.6), it is only possibly to be sure of a leakage input when its percentage exceeds 5% of the horizontal ground water flow.

The leakage effect could also be proved by isotopic data. Tritium, deuterium and 18-oxygen were measured by Dr. L. Eichinger in the water of the multi-level wells and the corresponding production wells. The samples were taken directly at the well head producing and using a small submersible pump at the multi-level wells. In the latter case sampling took place after the stabilization of the specific electric conductivity of the water.

The δ^2H- and $\delta^{18}O$ data of the aquitard and the aquifer show the normal environmental values of the study region. Thus, they do not contribute to arguments for the direction of leakage flow.

The tritium values range from 6.4 to 45.2 TU (Table 4.1.6.2), which shows that all samples contain some water portion which has been recharged to the aquifer by precipitation falling after 1953.

The sample from well 31 gives the lowest value of 6.4 TU. Considering the hydrogeological situation at this site this can be explained by the mixing of tritium-free ground water from the main aquifer with tritium-containing leakage water. It is difficult to prove the real reference depth of the sample, for well 31 is screened from 75 to 105 m below surface. In Fig. 4.1.6.7 a reference depth of 95 m b.s. has been assumed.

An almost linear decrease of tritium with depth is observed if all samples from well 31 and the associated multi-level wells in the aquitard are plotted together (Fig. 4.1.6.7). This indicates that the water in the aquitard becomes increasingly young in an upwards direction. Therefore, relatively young ground water (residence

Table 4.1.6.2. Contents of stable isotopes (δ^2H and $\delta^{18}O$) and tritium in wells and multi-level wells. (Holsterhausen Water Supply Westphalia/Germany)

Sampling well	Depth (m)	δ^2H ‰	$\delta^{18}O$ ‰	Tritium (TU)
Well 10	55	–53.6	–7.89	35.8
DOR1.18	18	–52.7	–7.62	45.2
DOR1.24	24	–54.5	–7.81	43.1
Well 31	95	–52.9	–7.56	6.4
DOR2.22	22	–52.0	–7.48	44.0
DOR2.40	40	–52.5	–7.68	26.6
DOR2.55	55	–52.4	–7.50	31.4

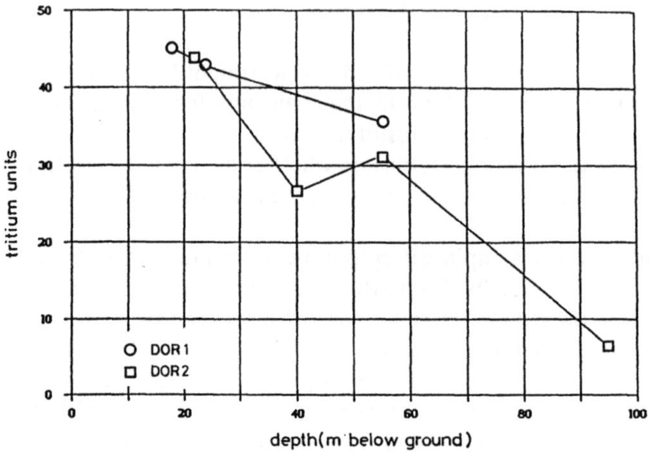

Fig. 4.1.6.7. Tritium contents versus depth in multi-level wells DOR1 and DOR2

time < 7 a) percolates as leakage water through the aquitard from the upper aquifer and mixes with older ground water which originates in the recharge area of the main aquifer (residence time > 30 a).

Due to the irregular pumping regime of the well gallery, mixed water could possibly penetrate into the basal parts of the aquitard. This secondary mixing would explain the deviation of tritium values from the straight lines in Fig. 4.1.6.7. Otherwise, the straight line shows that the geological matrix of the aquitard reacts like a quasi-homogeneous stratum.

In well 10 (35.8 TU) and the neighbouring multi-level wells DOR1.18 and DOR1.24, relatively high tritium contents (about 45 TU) are observed. The assumed reference depth here is 55 m (position of screen between 33 and 77 m below surface). The high tritium contents can be explained by the smaller thickness of the aquitard and its more sandy composition.

In summary, both the nature of the aquitard water and the tritium contents identify leakage through the aquitard (Fig. 4.1.6.3a,b) and suggest differences in the leakage rate.

4.1.7 Hydrogeochemical Processes in the Hamburg Deep Aquifer System

E.P. Loehnert, W. Bauhus and C. Sonntag

4.1.7.1 Study Subject: the Aquifer System

Geologically the Hamburg area (Fig. 4.1.7.1), which covers approximately 1000 km^2, is part of the North German plainlands and is underlain by an unconsolidated, multi-

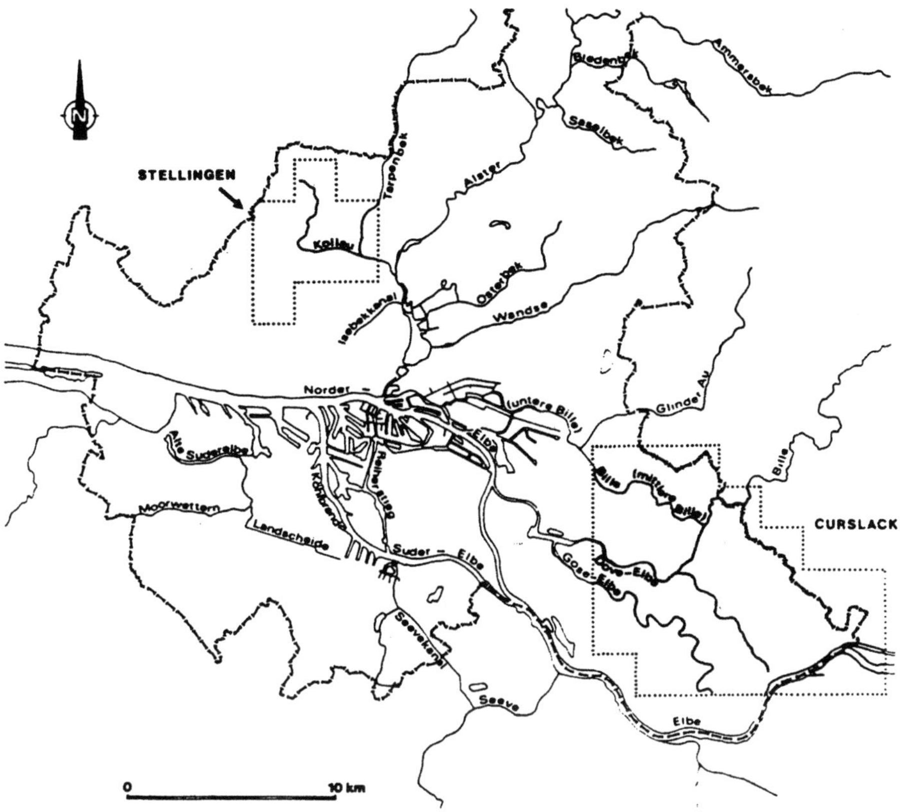

Fig. 4.1.7.1. Location map. Case-study areas Stellingen and Curslack are marked

layered aquifer system of Tertiary (Miocene, Pliocene) and Wuaternary (Pleistocene) age (Fig. 4.1.7.2). Deep aquifers (down to a maximum depth of 500 m below surface) include (Loehnert 1985, 1987):

1. Miocene Lower and Upper Lignite Sands, separated by the so-called Hamburg Clay and underlain by marine clays/silts of the Vierlande stage (Lower Miocene)
2. Pleistocene coarse-grained glacial-fluviatile deposits in a channel system cut into the Tertiary.

The thickness of the single aquifer units varies in the range of decameters, while channel deposits, consisting of sand and gravel, reach a few hundred metres. Hydraulic conditions of the deep aquifer system are sketched in Loehnert (1987); aquifers are mainly confined, and the Elbe valley serves as a major discharge area.

The proper investigation of the mineralogical composition of the aquifers and the structural position of the system has been hampered because only drill cuttings from boreholes and wells are available. Nevertheless, a stratigraphic correlation using geophysical borehole logging has been possible.

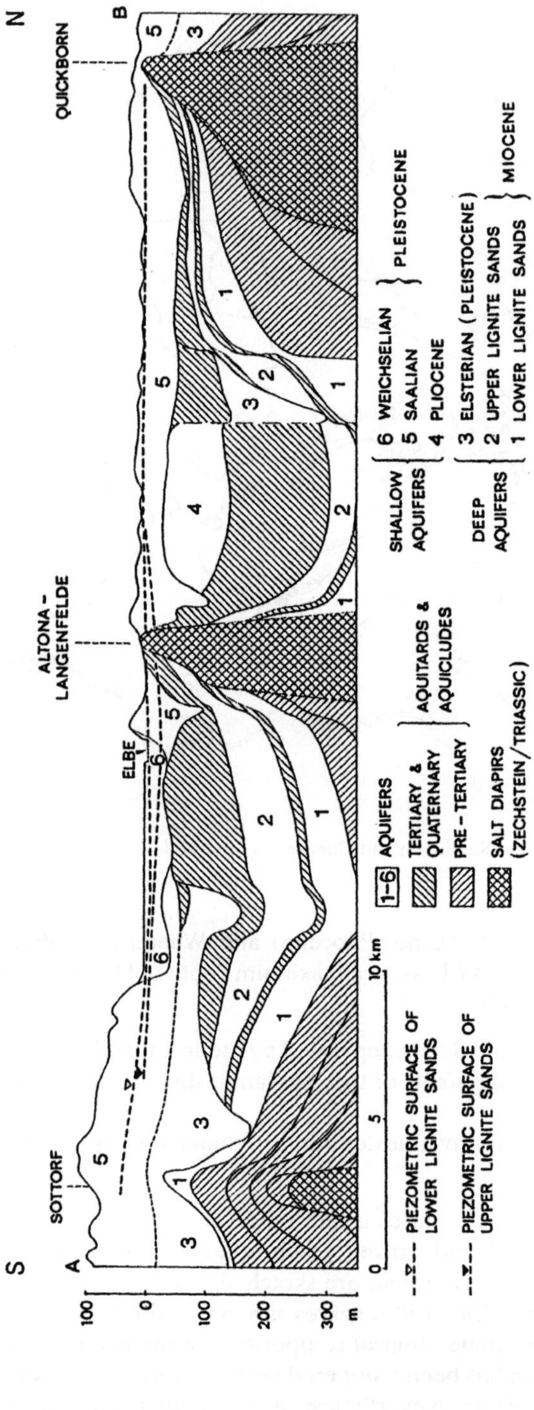

Fig. 4.1.7.2. Hydrogeologic cross-section of the Hamburg aquifer system. (Loehnert 1985)

Organic matter, in the form of lignite seams, humic silts and coatings on quartz grains in brownish-coloured sands, occurs frequently in Miocene aquifers. Pleistocene aquifers, on the other hand, contain regularly calcium carbonate unless they are made up of locally transported Tertiary material without components of nordic (granitic) origin, i.e. transported by glaciers. Thus, the mineralogic aquifer composition reflects differences in the depositional environment and geological age (climate). These differences are expected to result in differing liquid/solid phase interactions in the system.

The two case study areas (HAM, Figs. 1.1 and 4.1.7.1, Table 1.1) differ with respect to the structural position of deep aquifers (Loehnert et al. 1986; Bauhus 1988):

In the Curslack area the system is flat-lying and shallow, max. 150–200 m below ground. The salt diapir of Reitbrook with its top at 800 m depth hardly affects the horizontal bedding of aquifer layers. Upper Lignite Sands and younger confining beds were eroded in the Elbe valley. Water balance calculations confirm that ground water recharge takes place mainly to the north, outside the valley, which itself provides a discharge region. Upwardly discharging ground water is partially saline, especially in the Pleistocene channel aquifers.

The Stellingen area is controlled by some N-S-aligned salt domes, one of them outcropping at the surface. Aquifer occurrence, thickness and structure is determined by syn- and postgenetic halokinetic movement and subrosion of evaporites which lie partially in contact with deep aquifers. As a result, ground water salinization occurs in a much more complicated manner as in Curslack since former (fossil) and on-going salt solution processes have to be taken into account.

4.1.7.2 Metods of Investigation

Ground water head measurements were performed while taking water samples. These were taken by pumping from piezometers and from production wells. The network of sampling points is fairly dense and well distributed over the above-mentioned case-study areas. Approximately 100 complete hydrochemical analyses were executed, most of them by Bauhus (1988). Besides major cations and anions, some 30 determinations of trace elements (Li, Sr, Ba, Br and I) were done. After balancing equivalents of cations vs. anions, results were plotted automatically in Piper-Schoeller diagrams using programmes by M. Hannemann et al. (pers. commun.).

During sampling, physico-chemical parameters such as temperature, specific electrical conductivity, pH, E_H, dissolved oxygen, H_2S and CO_2 were measured on the spot. Some of the readings were used to calculate thermodynamic equilibria, among others saturation indices to assess water/rock reactions (Plummer et al. 1976; Wigley 1977). Environmental isotopes were applied to provide a time scale for hydrochemical processes and to support interpretation based on hydraulic and chemical data. About 140 measurements of tritium, deuterium and oxygen-18 from deep ground waters are available in addition to 50 of carbon-13 and carbon-14. Some 20 sulfur-34 data, mainly of dissolved sulfate, were also obtained. The latter are to be compared with values of shallow, unconfined aquifers which were obtained recently (Bauhus and Loehnert 1987; Hund 1988).

In 1986/87, towards the end of this project, a limited number of samples were analysed for their bacterial content (Nehrkorn et al. pers. commun.). The proof of the presence of sulfate-reducing microorganisms was particularly valuable in connection with sulfur-34 findings. However, no clear tendency towards reduction was established due to the incalculability of pumping effects.

4.1.7.3 Results

Hydrochemistry

General

The system is under reducing conditions: dissolved oxygen is mainly zero (in all cases < 1 mg/l), and E_H varies between –200 and +370 mV, with values above +185 mV restricted to the Curslack area. In particular the Stellingen deep ground waters yielded hydrogen sulfide gas, and iron sulfide incrustations of the aquifer matrix are also common here. The pH ranges between 6.9 and 8.9, the latter values being associated with ground waters affected by cation exchange.

The research focussed on hydrochemical processes along the fresh/saline water interface. Consequently, these two extreme kinds of ground water are to be considered first.

Fresh ground waters with TDS of 0.3 g/l occur down to a maximum depth of 400 m in the Hamburg region. They are predominantly of the Ca-HCO_3 type. Cation exchange results in Na-(Cl-) HCO_3 types. Subsurface exchange conditions in the North German plainland were summarized by Loehnert (1970). Distinctly sodium-dominated fresh waters are found close to the salt water contact. The cation exchange effect may be assessed by applying the Schoeller's base exchange index I_{BA} (Schoeller 1962):

$$I_{BA} = \frac{(Cl^- - (Na^+ + K^+))\ meq/l}{(SO_4^{2-} + HCO_3^-)\ meq/l} \qquad (4.1.7.1)$$

Calculated values for Stellingen waters are plotted against total concentration of ions in Fig. 4.1.7.3. Negative indices are limited to max. 240 meq/l and with above 25 meq/l restricted to Tertiary aquifers. Na^+ is largely associated either with HCO_3^- or with SO_4^{2-}.

Maximum salinity measured was about 100 g/l, i.e. almost threefold the concentration of ocean water. Whereas saline waters of Curslack are NaCl-dominated, those of Stellingen commonly contain considerable amounts of $CaSO_4$.

Dissolution and Origin of Salts

In the Stellingen case-study area both rock salt and gypsum were/are available for dissolution by ground water. Highest salinity was encountered in sandy-silty perme-

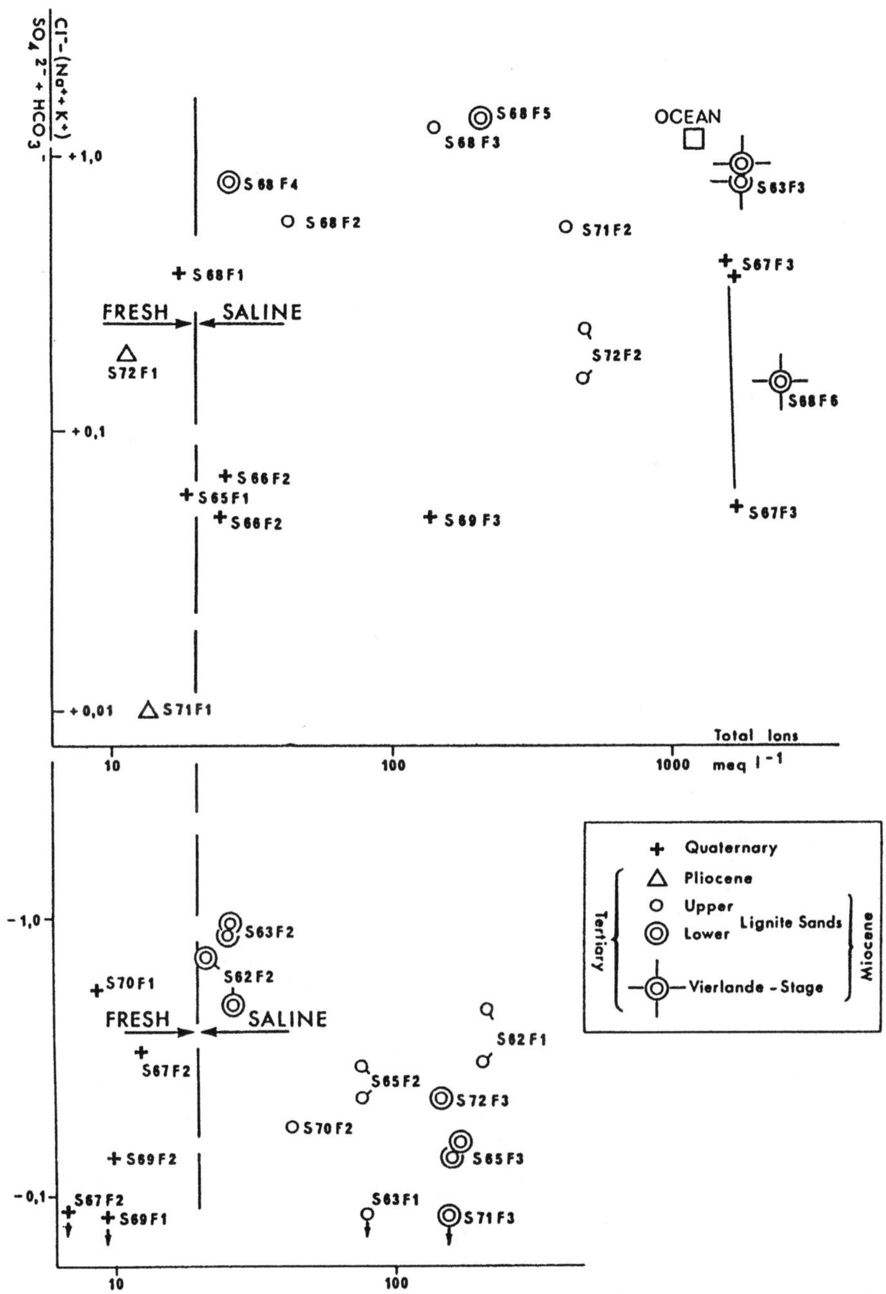

Fig. 4.1.7.3. Schoeller Base Exchange Index vs. total ions for Stellingen ground waters

able layers of the Vierlande stage at 592–615 m b.g.l. and in Pleistocene channel deposits at 392–396 m b.g.l., in all cases with positive I_{BA} (Fig. 4.1.7.3). These waters differ with respect to the Ca + Mg content and Ca:Mg ratios, the meq values of the latter are sometimes <1, thus approaching ocean water (Ca:Mg = 0.19) in marine Lower Miocene layers. Sulfate appears reduced with respect to ocean water (meq $Cl:SO_4$ max. four times that of ocean). Hence if connate water were assumed to occur, its primary character was completely altered among others by sulfate reduction and cation exchange (total amount of Ca + Mg diminished but Ca:Mg ratio increased). According to Bauhus (1988), saline ground waters in the Stellingen area tend to be almost saturated in gypsum but undersaturated in halite. The tendency to reach equilibria differs obviously between Tertiary and Quaternary aquifers. This fact is even more obvious in Curslack saline waters which show extreme undersaturation in gypsum in Pleistocene channel aquifers. Salinity amounts to 30 g/l TDS, though evaporites are not present for direct dissolution. It has long been recognized that older Tertiary and Cretaceous rocks underlying the aquifer system contain brines of Na-Ca-Cl or Na-Cl types (Loehnert and Paluska 1977). The results of Bauhus (1988) strongly suggested hydraulic interconnection by presenting straight mixing lines between an "oil field brine" (TDS 175 g/l) and saline ground waters of the aquifer system under consideration; chloride and trace elements such as Li^+ and Sr^{2+} are evidently only subjected to dilution in ground waters ascending in the Elbe valley (Fig. 4.1.7.4).

Carbonate Hydrogeochemistry

Cation exchange and sulfate reduction observed in the Hamburg deep aquifer system are closely interrelated with CO_2 production and carbonate dissolution. This will be shown later in the context of carbon-13 variation.

Tertiary aquifers, in particular, regularly contain organic matter, from which carbon dioxide is produced through oxidation. At the same time equilibrium CO_2 is

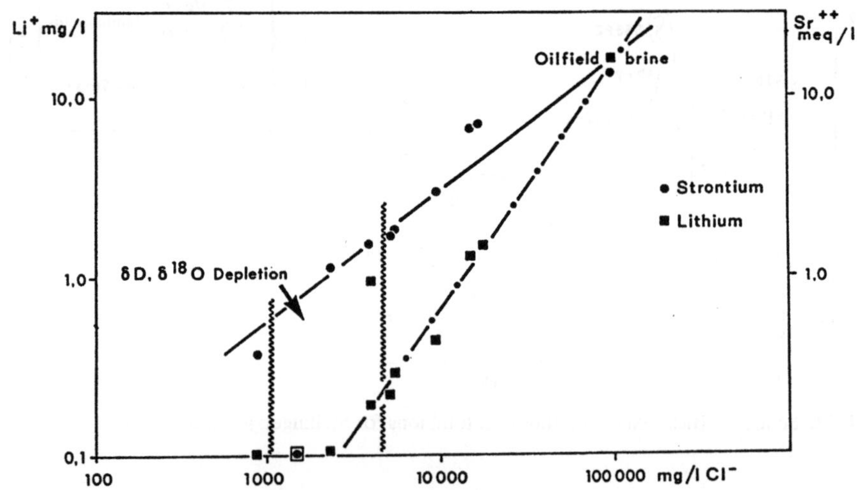

Fig. 4.1.7.4. Chloride vs. lithium and strontium in Curslack ground waters

released while Na^+ is brought into solution instead of Ca^{2+} by cation exchange. Eventually, sulfate reduction processes result in an increase in HCO_3^- particularly under closed system conditions at deeper levels in an anoxic environment. Highest hydrogen carbonate amounts (max. 14 meq/l) are encountered in the marine silts of the Vierlande stage while the normal range is from 3–5 meq/l (with pH 7 up to 8). These waters are slightly over-saturated with respect to calcite. Tendency to undersaturation with respect to calcite and other minerals is developed in Quaternary as compared to Tertiary aquifers (Loehnert et al. 1986; Fig. 7). A relatively short residence time seems to be responsible for this tendency, which is supported by the occurrence of tritium.

Saturation equilibria differ completely in the Curslack area where waters show clear tendencies of calcite and dolomite supersaturation in Quaternary channel deposits whereas Tertiary aquifer waters display equilibria or slight undersaturation of these minerals. Ascending ground waters obviously play a decisive role in controlling carbonate hydrogeochemistry here. If they are supposed to mix with waters of different carbonate composition, extreme supersaturation is being produced (Wigley and Plummer 1976). This effect was simulated by Bauhus (1988) applying the equilibrium model PHREEQE (Parkhurst et al. 1980). The calculated results are in good agreement with actual readings.

Environmental Isotopes

Age Structure

Carbon-14 dating was attempted with little success for reasons given in Subsection 4.1.7.4. A number of carbon-14 age correction methods (Ingerson and Pearson 1964; Tamers 1967; Tamers and Scharpenseel 1970; Fontes and Garnier 1979) were employed. The assumption of an initial C-14 activity of 85% is evidently too high. Deep-seated saline ground waters in either case-study areas show frequently C-14 activities <10% and sometimes close to zero. Thus, most of them are of Pleistocene age.

A grouping is much evident for stable isotopes D and O-18 in deep ground waters which follow mainly the meteoric water line (Loehnert and Sonntag 1981) except the "oil field brine". Frequency distribution of oxygen-18 combined with tritium is shown in Fig. 4.1.7.5. δO-18 varies between –7.2 and –9.6‰ and elevated tritium values (max. 45 TU) are restricted to waters relatively enriched in O-18. In agreement with authors in other European investigation areas, group I with depleted O-18 and D is regarded as recharged under cold climatic conditions, possible in a Weichselian interstadial time period. Ground waters of this group differ as to their depth and chemistry but they are exclusively encountered in Miocene aquifers. This is in agreement with the occurrence of confining clay layers which prevent current recharge to these aquifers.

While in Curslack group I waters are saline, in Stellingen both fresh and saline waters belong to this isotopically depleted group. "Fresh" means, however, also distinct cation exchange character (Na-Cl-HCO_3 type), with lowest I_{BA} between –0.5 and –0.9 (Fig. 4.1.7.3). Waters of group I interfinger laterally with stronger saline types, all of them with increased sulfate, where sometimes meq/l $SO_4^{2-} > Ca^{2+} + Mg^{2-}$. Most of these waters belong to group II according to their O-18 and D content.

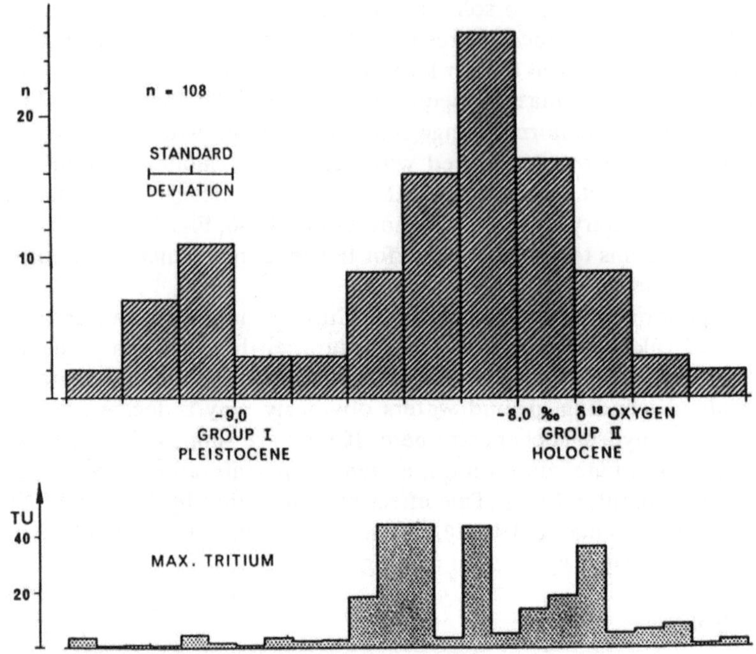

Fig. 4.1.7.5. Frequency distribution of δoxygen-18 and tritium (maximum TU values per 0.1‰ O-18 steps) in Hamburg deep ground waters

For Curslack NaCl waters of group I have lower TDS than group II. Waters of group II are SO_4^{2-} free and contain elevated concentrations of Mg^{2+}. It is the water of group II which constitutes the link to "oil field brines" (Fig. 4.1.7.4).

Stable Isotopic Composition of Dissolved Matter

The variation of carbon-13 is linearly related to HCO_3^- in both areas (Fig. 4.1.7.6). This relation is less distinct in Curslack where fresh waters of a supposed starting composition appear to begin with a lower HCO_3^- content. In both areas highest δC-13 is approx. –3‰, decreasing within the common mixing range of soil CO_2 and rock carbonate. Increasing hydrogen carbonate along with C-13 depletion as shown in Fig. 4.1.7.6 is explained by secondary CO_2 generation within the aquifer system, namely by dissolution of calcite and dolomite, cation exchange, oxidation of organic material and microbial sulfate reduction. These processes overlap each other in a complicated manner. Carbonate dissolution can be excluded even in fossiliferous marine Miocene deposits (S 68 VI in Fig. 4.1.7.6a). C-13 depletion occurs, on the other hand, in Curslack well waters which contain tritium. This is an exception, however, and deep ground waters usually contain dissolved CO_2 originating from the oxidation of organic matter by sulfate reduction.

Microbially catalyzed sulfate reduction was proved by sulfur-34 measurements (Fig. 4.1.7.7). Readings often fall in the range of Zechstein evaporites (Nielsen and Ricke 1964) although Triassic salts may also be subject to dissolution in the study

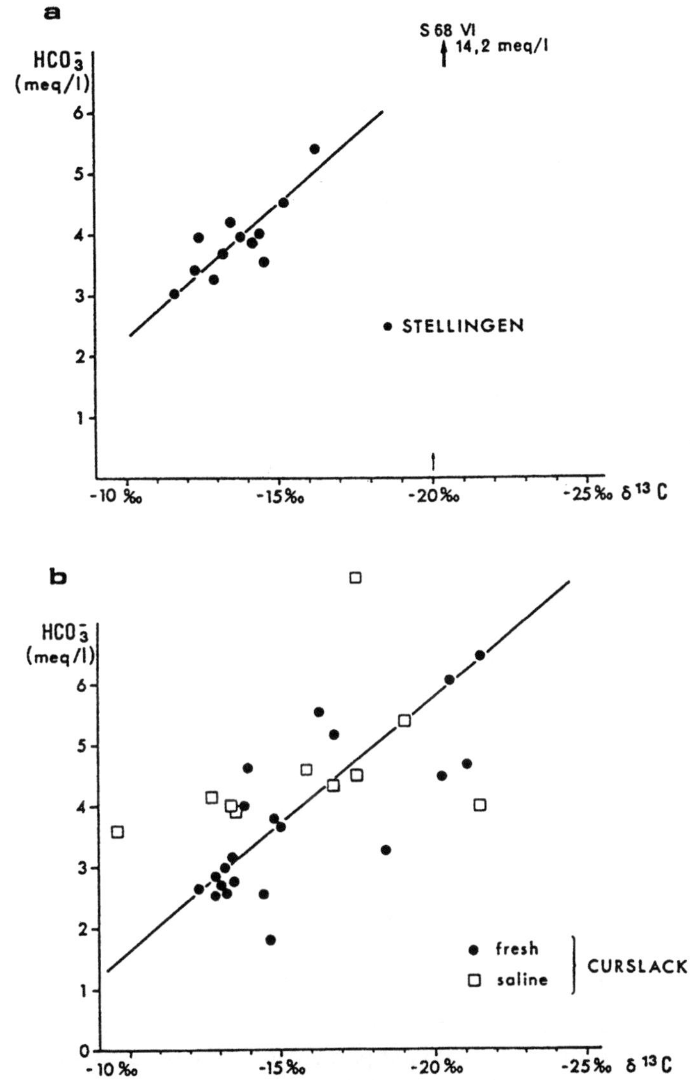

Fig. 4.1.7.6. Relationship of hydrogen carbonate/δcarbon-13 in Stellingen (**a**) and Curslack (**b**) ground waters

area. All deep ground waters have positive S-34 values and some of them are more positive relative to evaporites, independent of the sulfate amount in solution. The relative enrichment of δS-34 is due to isotope fractionation resulting from microbiologic sulfate reduction (Puchelt 1967; Puchelt and Nielsen 1967). These processes appear to take place preferentially within the fresh/saline water transition zone of both case-study areas.

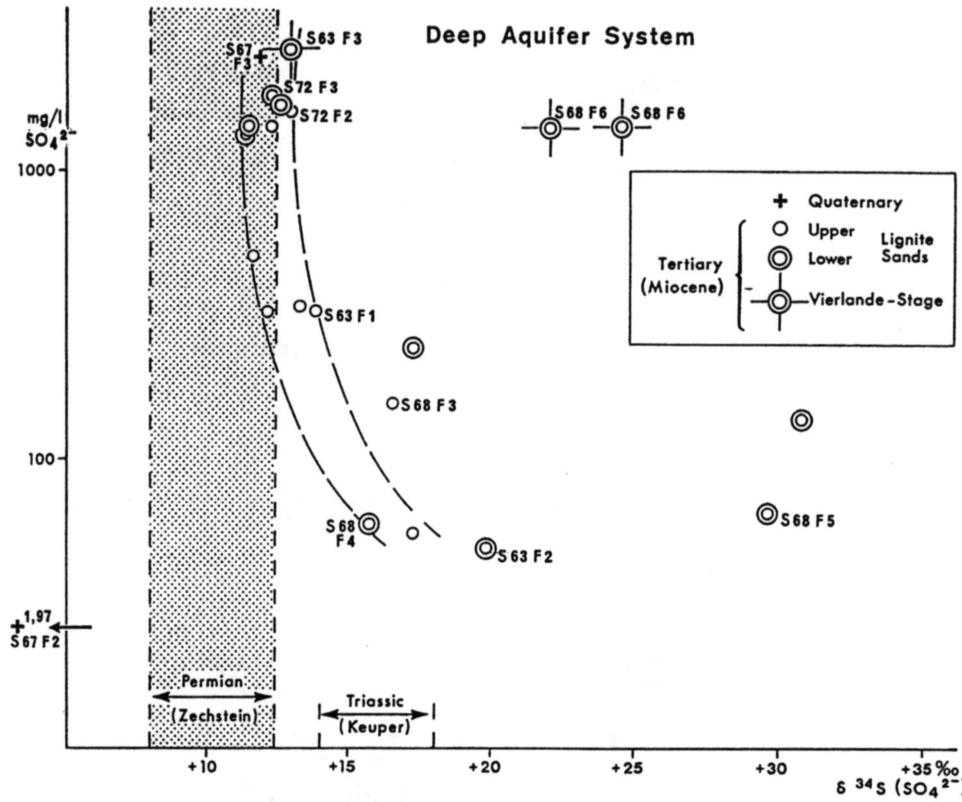

Fig. 4.1.7.7. Plot of sulfate (log scale) vs. δ sulfur-34 (of dissolved sulfate) mainly in Stellingen ground waters

4.1.7.4 Discussion and Conclusion

Environmental isotopes yielded three generations of deep ground water

1. "Palaeowaters" of Pleistocene age, depleted in oxygen-18 and deuterium, free of tritium, which do not participate in the recent hydrologic cycle
2. Deep ground water with the same stable isotope composition as the shallow aquifer system, i.e. recharged under climatic conditions similar to Recent, no tritium
3. Deep ground water with O-18 and D contents similar to those in (2) but with tritium.

1. "Paleowaters" are almost exclusively found in Tertiary aquifer basins originating from halokinetic subsidence. These ground waters were recharged after withdrawal of the inland ice sheet when the rising sea water level caused a water table recovery as well. This origin includes, in Tertiary aquifers of the Stellingen area, a fossil fresh/salt water contact, with fresh waters of Na-Cl-HCO_3 base exchange type. In the Curslack area, type (1) waters are represented by Na-Cl saline waters which are poor in Sr and Li (Fig. 4.1.7.4).

2. Waters without tritium occur in both areas with higher salinity interfingering with paleowaters (1). Sulfate-rich types are supposed to have been mobilized by increased gypsum dissolution resulting in a 200-m-deep subrosion depression on the top of the salt barrier. In Curslack saline waters of this generation have elevated Mg-(Ca-) concentrations either through mixing as mentioned earlier or due to cation exchange (action of NaCl waters upon calcareous rocks). These waters constitute the link with deep-seated brines.

3. Waters with tritium are superimposed on type (2). Given favourable hydrogeologic conditions, these waters might penetrate to approx. 150 m depth (Bauhus and Loehnert 1987). In other cases tritium contents are induced artificially, either by local overexploitation or by artificial recharge using surface waters. In any case, tritium is the ideal tracer indicating that some of the ground water was recharged in the past 30 years. With this time scale in mind, the following conclusions about hydrogeochemical processes may be made.

In the Curslack area ground water velocities were calculated based on carbon-14 contents, the result being about 0.2 m/a at a deeper level (100–150 m) and 1.4 m/a at a shallower level (50–100 m) of the aquifer system. Hydrochemical changes along flowpaths crossing the salt/fresh water boundary could be assessed, and cation exchange and sulfate reduction processes resulting in carbon dioxide generation (increase in p_{CO_2}) are evident (Bauhus 1988: Figs. 64–66). Saturation indices of calcite and dolomite are, however, not affected by these processes.

The hydrochemical evolution of the Stellingen area is similar. However, the structural position of the aquifer system is far more complicated there, due to active salt diapir movements and evaporite dissolution at different time periods (post-Middle Miocene to Holocene) in addition to hydraulic influences of deeply cut Pleistocene channels. Consequently, there is a strong variation in aquifer depth, temperature gradients and depth to the saline/fresh water boundary. The latter occurs twice in some sections but with lateral transitions into hydrochemically and isotopically differing ground waters. Flow velocities are definitely lower, therefore it was termed a "stagnant system" unlike the "dynamic system" at Curslack (Loehnert et al. 1986).

4.2 Rock-Water Interaction

4.2.1 Aluminium Speciation in Acid Soil Water and Ground Water

G. Dietze and B. Ulrich

4.2.1.1 Aluminium in Subsurface Aquatic Systems

One consequence of changes in soil chemistry due to acid deposition is the mobilization and transport of aluminium from the edaphic to the aquatic environment (Cronan and Schofield 1979; Nilsson and Bergkvist 1983; Ulrich and Matzner

1983; David and Driscoll 1984; Seip et al. 1984). The continuing soil acidification is characterized by a decrease in base saturation and a large amount of exchangeable Al^{3+} and hydroxy-Al ions which are at present the predominant forms of acidic cations in forest soils (Ulrich 1989). The pH values of soil solutions range between 3.0 and 4.4. The temporary storage of aluminium by precipitation of $AlOHSO_4$ presents a further source of available aluminium for the soil water system (Matzner 1988). The increased solubility of aluminium in the soil can be affected directly by decreasing pH and also by an altering humification behaviour of soil organic matter. The latter means an elevated level of dissolved organic compounds which form strong complexes with aluminium moving downward into mineral horizons (podsolization processes; Stobbe and Wright 1959; Dawson et al. 1978; Buurman and van Reeuwijk 1984). In a comparative analysis of seepage waters and ground water samples from different forest soil profiles the distribution pattern of inorganic and organically bound aluminium species (labile and non-labile forms) at varying depths has been examined. Many concepts and analytical procedures for the determination of chemical species of aluminium already exist (Campbell et al. 1983; Driscoll 1984; La Zerte 1984; Ares 1986; Backes and Tipping 1987; Hodges 1987; Shotyk and Sposito 1988). The most widely used method for the fractionation between the inorganic and organic forms of aluminium is the cation exchange column method developed by Driscoll (1984). This method combined with other procedures is also applied in this work. The single inorganic monomeric aluminium species such as Al^{3+} and $AlOH^{2+}$ are calculated by a mathematical model for coupled chemical equilibria with established equilibrium constants (data not presented) (Prenzel 1986).

Sampling sites were selected to represent different parent materials and vegetation, so the different sorption characteristics of mineral soils, the different origins of water- soluble soil organic matter, and the distribution profile of sample pH with increasing depth could be taken into account.

4.2.1.2 Materials and Methods

Pore water samples were collected from tension lysimetes (Meiwes et al. 1984) which are located in several forested sites in W. Germany, by handauger-slitprobe technique and a multi-level well (Strebel and Böttcher 1985). A total of six soil profiles were sampled (Table 4.2.1.1). The samples were filtered through 0.45 μm membrane filters (Sartorius) and stored in a refrigerator at 4 °C.

4.2.1.3 Analytical Procedures

Aluminium fractionation was accomplished using a semi-automated ion exchange column method earlier reported by Driscoll (1984) which was combined with dialysis procedures and cation exchange experiments in the batch mode.

Table 4.2.1.1. Site description, time of sampling and soil characteristics

Solling spruce (plot SOL1 F_1, Mayer 1981)					
Time of sampling	August until November 1985				
Depth of sampling/cm	0	10	20	40	80
Horizon	Oh/Aeh	Bv1	Bv1	SBv2	IIBv3
Soil type	Podsolized brown earth				
Parent rock	Lower Triassic sandstone with Loess cover				
Solling beech (plot SOL2 B_1, Mayer 1981)					
Time of sampling	August until November 1985				
Depth of sampling/cm	0	10	20	40	80
Horizon	Oh/Aeh	Bv	Bv	Bv	IIBv
Soil type	Podsolized brown earth				
Parent rock	Lower Triassic sandstone with Loess cover				
Wingst spruce (WIN distr. 28, Büttner and Lamersdorf 1985)					
Time of sampling	December 1985				
Depth of sampling/cm	35	100			
Horizon	Bhs	Bv			
Soil type	Podsol				
Parent rock	Quicksand over fluvioglacial sand				
Hils beech (HIL1 distr. 39, Raben 1985)					
Time of sampling	August 1985				
Depth of sampling/cm	20	40	80		
Horizon	Ahe/Bsh	Bv	(Cv)/Bv		
Soil type	Podsolized brown earth				
Parent rock	Mud of Hilssandstein with Loess cover				
Hils spruce (HIL2 distr. 109, Raben 1985)					
Time of sampling	August 1985				
Depth of sampling/cm	20	40	80		
Horizon	Ae,	Ae	Bsh		
Soil type	Podsol				
Parent rock	Hilssandstein				
Fuhrberger Feld (HAN) (Strebel and Böttcher 1985)					
Zone of sampling	A1-A6,B6	E2,E4	F3	H2,H5-H7	
Depth of sampling/cm	100	170	250	100	
Horizon	Gr	Gr	Gr	Gr	
Time of sampling	February 1986				
Multi-level well	HAN-S1	HAN-S3			
Depth of sampling/cm	200	500			
Horizon	Ground water				
Time of sampling	February 1985				
Soil type	Podsol to gley-podsol				
Parent rock	Quaternary cover sand				
Vegetation	A-F, H6 pine H2 birch (mixed woodland) H5 black alder H7 spruce (forest plantation area)				

Column and Batch Ion Exchange Experiments

The ion exchange experiments were performed with Amberlite IR-120 (analytical grade, 20–50 mesh, Serva) in its protonated form. A 6-ml bed of strongly acidic cation exchange resin was used in a 0.7-cm diameter column fitted to a peristaltic pump (Gilson Minipuls 8). Before packing the column, the resin was slurried in distilled water and washed intensively. The minimum flow rate was 4 ml/min per ml of exchanger bed volume. The resin was replaced after about ten samples have been cation-desalted. The eluate was analyzed for aluminium concentration by flame AAS (acetylene-N_2O flame). If aluminium concentrations were below 1 mg/l, the graphite furnace procedure was applied (Heinrichs et al. 1985). Further details of the method have been described by Driscoll (1984) and Dietze (1985).

Dialysis Procedure

Dialysis tubings of Spectrapor, 3500 molecular weight cutoff (3500 MW) were employed. The tubings were initially cut to a length such that they could be filled with 5 ml sample volume. Then they were carefully soaked and rinsed several times for 2 h in double distilled water. After being filled with the sample and sealed, the bags were immersed in plastic beakers containing an aqueous slurry of cation exchange resin (Na-form, 0.1 g/ml sample volume). The ratio of the solution volumes outside to inside the bag was 10 to 1. For homogenizing, the beakers were shaken. After each hour the outside solution was removed, discarded, and replaced with the same amount of double distilled water, while the resin was kept inside. Repeating this procedure five times, the tubes were unclipped and the contents were analyzed for aluminium and dissolved organic carbon (DOC). DOC contents were measured by a total carbon analyzer (Beckman).

4.2.1.4 Theory and Calculations

The aluminium fractions which are operationally defined on filtered samples (Fig. 4.2.1.1) include: total dissolved aluminium (Al_t) determined by atomic absorption, labile (mainly inorganic) aluminium (Al_{ex}) that can be exchanged to a cation exchange resin, and moderate-labile (organically bound) aluminium (Al_{nonex}) that remains in the effluent. Labile aluminium is obtained by subtraction, Al_t-Al_{nonex}. Furthermore, by dialysis procedure inorganic and organic forms of aluminium (Al_D) which pass through the 3500-MW membrane is separated from aluminium (Al_R) that is bound to organic solutes which have a higher molecular size and therefore cannot penetrate a 3500-MW membrane. The amount of aluminium (Al_B) which is not replaceable by the cation of cation exchanger in the batch mode is classified as the non-labile-complexed fraction. According to previous experiments (Dietze 1985), it was assumed that Al_B belongs to the higher molecular weight fraction Al_R. A fractionation scheme is given in Fig. 4.2.1.1.

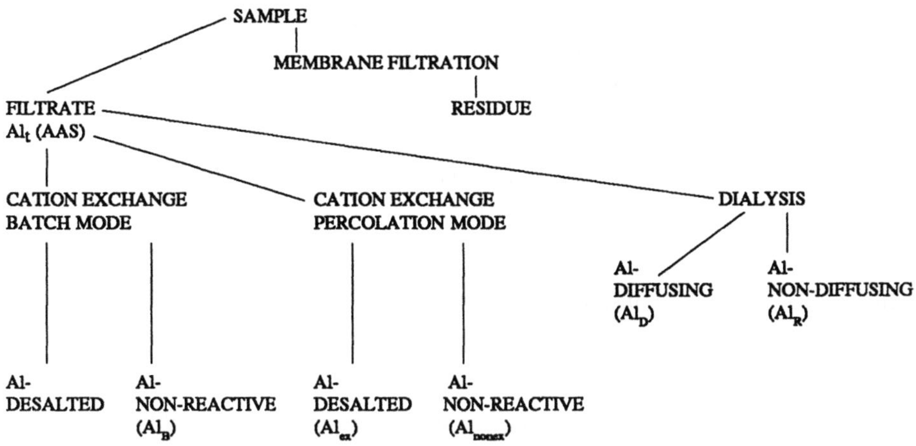

Fig. 4.2.1.1. Operational definitions used in the cation exchange and dialysis procedures for aluminium specification. (Dietze 1986)

The following cases are distinguished:

1. The calculated sum of desalted and non-diffusing aluminium exceeds total aluminium concentration in sample solution:

$$Al_{ex} + Al_R > Al_t. \qquad (4.2.1.1)$$

Dialysis gives values for organically bound aluminium similar to those from the sulfonate ion exchange column method. But the results by dialysis are generally higher. Cation exchange in the percolation mode occurs in proportion to the sum of the free metal concentration and the metal held in very labile complexes. In direct contact to the resin the high negatively charged field of the exchange resin can polarize the metal complexes, thus they dissociate and aluminium enters the resin phase while the organic ligands pass through. This reaction path is prevented by separating the resin phase and the sample by a membrane. According to this fact, the difference between the amounts dialyzed and exchanged aluminium can be attributed to the fraction of very labile complexes that cannot penetrate a 3500-MW membrane (Al_x).

A systematic source of error can be an overestimation of inorganic aluminium, since diffusing aluminium can also be partially bound to low molecular weight organic compounds passing the membrane.

The criterion for separation is simply that the various solution components diffuse through the membrane at different rates. Independent diffusion of the anions and cations does not occur. This observation is attributed to the fact that the electroneutrality condition in the inside and outside solution for all components must be fulfilled (Donnan 1911; Bull 1951). Its initial concentration level and the nature of anions present in the solution influence the diffusion of aluminium. Thus, the retention of aluminium was found to increase with increasing sulfate concentration in solution. This finding is in agreement with that reported by La Zerte (1985). Dialysis procedure against a slurry of cation exchange resin should improve the diffusion behaviour of inorganic aluminium and insure its complete separation after 5 h.

The advantage of this approach is that weak complexes can be determined and perturbations of chemical equilibria are avoided. Only a small amount of protons actually enters the inside solution, the pH of the samples changes by about a half unit during the dialysis.

The kinetically moderate labile complexed quantity can be taken as the difference between the non-reactive portions in the column and batch ion exchange methods:

$$Al_{nonex} - Al_B = Al_{y>3500} \tag{4.2.1.2}$$

In this case the fraction $Al_{y>3500}$ belongs to the higher molecular weight species. The sum made from all four fractions is equal to the initial aluminium concentration in sample solution.

2. The calculated sum of desalted and non-diffusing amounts of aluminium is below the total aluminium concentration of sample solution:

$$Al_{ex} + Al_R < Al_t \text{ and} \tag{4.2.1.3}$$

$$Al_R > Al_B \tag{4.2.1.4}$$

In this rare case one obtains kinetically moderate labile complexed metal species which belong to the low molecular weight species:

$$Al_t - (Al_{ex} + Al_R) = Al_{y<3500} \tag{4.2.1.5}$$

Inorganic aluminium corresponds then to Al_{ex}. The very labile aluminium fraction Al_{ex} is set inevitably to zero (possibly a systematic analytical error). The higher molecular weight fraction $Al_{y>3500}$ can be calculated from

$$Al_R - Al_B = Al_{y>3500} \tag{4.2.1.6}$$

4.2.1.5 Results and Discussion

The speciation has been calculated as described above. All data can be understood from Tables 4.2.1.2 to 4.2.1.4. The distribution of Al species in the pore solutions in percent can be seen from Table 4.2.1.5. In all cases the extreme and moderate labile forms of organic complexes predominate. As expected, the different kinds of soil materials exhibit a different feature in aluminium transport.

For the sandy soils (Hils/Wingst) that are characterized by a high water permeability and a low sorption capacity, dissolved aluminium is largely associated with organic compounds throughout the whole profile depth. Transformation by precipitation or dissolution processes and redistribution of Al species may occur simultaneously as overlapping steps (Buurman and van Reeuwijk 1984). The samples taken from the catchment Fuhrberger Feld (HAN), a sandy shallow aquifer of about 20 m depth, are characterized by various sampling zones (A to H). The ground water level of these sites declines distinctly with decreasing distance to the

Table 4.2.1.2. Analysis of soil solutions for pH, aluminium and dissolved organic carbon

Point of sampling	Depth (cm)	pH	Al_t (mg/l)	DOC_t (mg/l)
Solling Spruce (SOL1)				
SOL1/4a	0	3.03	1.60	36.0
SOL1/4b	0	3.05	2.30	29.0
SOL1/1	10	3.26	6.20	19.5
SOL1/2	10	3.60	4.80	14.0
SOL1/4	20	3.75	10.40	11.0
SOL1/5	20	4.01	8.60	8.6
SOL1/7	40	4.21	10.00	4.5
SOL1/9	40	4.20	10.90	4.5
SOL1/10	80	4.10	9.20	4.5
SOL1/11	80	4.13	10.20	5.0
SOL1/1	10	3.45	9.10	15.0
SOL1/4	20	3.89	11.00	6.0
SOL1/7	40	4.12	9.00	4.5
SOL1/10	80	4.14	10.30	4.5
Solling beech (SOL 2)				
SOL2/3a	0	3.50	1.90	135.0
SOL2/3b	0	3.44	1.05	57.5
SOL2/1	10	3.73	3.10	13.0
SOL2/7	40	4.33	4.70	5.0
SOL2/9	40	4.23	6.10	4.0
SOL2/1	10	3.41	5.10	18.0
SOL2/4	20	4.00	4.10	9.0
SOL2/7	40	4.21	4.50	4.0
SOL2/10	80	4.18	5.10	4.0
Wingst spruce (WIN)				
WIN/28	35	3.99	18.24	26.0
WIN/28	100	4.15	21.11	11.5
Hils beech (HIL1)				
	20	3.86	1.20	40.5
	40	4.40	0.61	19.5
	80	4.28	5.83	9.0
Hils spruce (HIL2)				
	20	3.49	1.56	60.0
	40	3.46	1.77	50.5
	80	3.48	1.67	65.5

Table 4.2.1.2. (*Continued*)

Point of sampling	Depth (cm)	pH	Al_t (mg/l)	DOC_t (mg/l)
Fuhrberger Feld (HAN)				
A1	100	4.0	22.73	38.5
A2	100	4.0	13.40	37.5
A3	100	4.0	14.17	49.5
A4	100	4.2	13.22	6.5
A5	100	4.2	16.97	6.0
A6	100	4.4	4.15	19.0
B6	100	3.9	19.71	29.5
E2	170	5.4	2.92	33.0
E4	170	4.3	10.66	24.5
F3	250	4.2	33.54	16.5
H2	100	4.2	7.21	22.0
H5	100	4.8	0.32	8.0
H6	100	4.6	0.80	17.0
H7	100	4.0	25.02	12.0
HAN-S1	200	4.17	6.60	23.0
HAN-S3	500	4.23	4.10	18.0

Table 4.2.1.3. Column and batch ion exchange experiments, analytical measurements

Point of sampling	Depth (cm)	Al_{nonex} (mg/l)	Al_B (mg/l)
Solling spruce (SOL1)			
SOL1/4a	0	0.395	0.023
SOL1/4b	0	0.360	0.049
SOL1/1	10	0.294	0.022
SOL1/2	10	0.354	0.021
SOL1/4	20	0.228	0.019
SOL1/5	20	0.225	0.025
SOL1/7	40	0.140	0.010
SOL1/9	40	0.069	0.016
SOL1/10	80	0.016	0.006
SOL1/11	80	0.051[a]	0.021
SOL1/1	10	0.336	0.123
SOL1/4	20	0.231	0.023
SOL1/7	40	0.050	0.022
SOL1/10	80	0.189	0.038
Solling beech (SOL 2)			
SOL2/3a	0	1.635	0.258
SOL2/3b	0	0.616	0.050
SOL2/1	10	0.318	0.025

Table 4.2.1.3. (*Continued*)

Point of sampling	Depth (cm)	Al_{nonex} (mg/l)	Al_B (mg/l)
SOL2/7	40	0.040	0.010
SOL2/9	40	0.016	0.010
SOL2/1	10	0.498	0.023
SOL2/4	20	0.252	0.010
SOL2/7	40	0.019	0.011
SOL2/10	80	0.022	0.017
Wingst spruce (WIN)			
WIN/28	35	1.750	0.204
WIN/28	100	1.710	0.021
Hils beech (HIL1)			
	20	0.116	0.035
	40	0.348	0.023
	80	0.144	0.020
Hils spruce (HIL2)			
	20	0.164	0.051
	40	0.935	0.029
	80	0.172	0.077
Fuhrberger Feld (HAN)			
A1	100	1.785[b]	0.847
A2	100	1.281[b]	0.781
A3	100	1.869[b]	0.968
A4	100	0.306[b]	0.120
A5	100	0.396[b]	0.061
A6	100	0.777[b]	non-tested
B6	100	1.023[b]	non-tested
E2	170	0.376[b]	non-tested
E4	170	0.528[b]	non-tested
F3	250	0.825[b]	non-tested
H2	100	0.474[b]	non-tested
H5	100	0.138[b]	non-tested
H6	100	0.285[b]	non-tested
H7	100	0.715[b]	non-tested
HAN-S1	200	0.403	0.099
HAN-S3	500	0.299	0.201

[a]Ion exchange resin in its saturated sodium form.
[b]For altering acidity of the resin H^+/Na^+-ratio was 1.

Table 4.2.1.4. Dialysis procedures, analytical measurements

Point of sampling	Depth (cm)	Al_R (mg/l)	DOC_R (mg/l)
Solling spruce (SOL1)			
SOL1/4a	0	1.045	29.0
SOL1/4b	0	0.660	20.0
SOL1/1	10	0.057	7.0
SOL1/2	10	0.016	9.2
SOL1/4	20	0.022	4.8
SOL1/5	20	0.378	1.5
SOL1/1	10	1.491	5.0
Solling beech (SOL 2)			
SOL2/3a	0	1.596	129.0
SOL2/3b	0	0.748	23.0
SOL2/1	10	0.010	10.5
SOL2/1	10	0.036	9.2
SOL2/4	20	0.564	1.9
Wingst spruce (WIN)			
WIN/28	35	4.020	13.5
WIN/28	100	3.520	6.0
Hils beech (HIL1)			
	20	0.727	33.0
	40	0.153	16.5
	80	2.320	4.0
Hils spruce (HIL2)			
	20	1.050	52.5
	40	1.302	41.0
	80	1.281	57.0
Fuhrberger Feld (HAN)			
A1	100	1.659	11.5
A2	100	1.260	11.0
A3	100	1.806	15.5
A4	100	0.006	1.5
A5	100	0.026	3.0
A6	100	0.693	11.5
B6	100	1.218	10.5
E2	170	0.360	15.5
E4	170	0.177	5.0
F3	250	1.176	3.5
H2	100	0.066	4.5
H5	100	0.092	5.0
H6	100	0.130	11.0
H7	100	0.162	3.0
HAN-S1	200	2.60	6.0
HAN-S3	500	1.80	6.5

Table 4.2.1.5. Distribution patterns of inorganic and organically bound aluminium species at varying depths

Point of sampling	Depth (cm)	Al_{inorg} / Al_t0.01	Al_x / Al_t0.01	$Al_{y<3500}$ / Al_t0.01	$Al_{y>3500}$ / Al_t0.01	Al_B / Al_t0.01
Solling spruce (SOL1)						
SOL1/4a	0	34.4	40.6	0.0	23.3	1.4
SOL1/4b	0	71.3	13.0	0.0	13.5	2.1
SOL1/1	10	95.3	0.0	3.8	0.5	0.4
SOL1/2	10	92.6	0.0	7.0	0.4	0.4
SOL1/4	20	97.8	0.0	2.0	0.0	0.2
SOL1/5	20	95.6	1.8	0.0	2.3	0.3
SOL1/7[a]	40	98.6	–	–	–	0.1
SOL1/9[a]	40	99.4	–	–	–	0.2
SOL1/10[a]	80	99.8	–	–	–	0.2
SOL1/11[a]	80	99.5	–	–	–	0.1
SOL1/1	10	83.6	12.7	0.0	2.3	1.4
SOL1/4[a]	20	97.9	–	–	–	0.6
SOL1/7[a]	40	99.5	–	–	–	0.2
SOL1/10[a]	80	98.2	–	–	–	0.1
Solling beech (SOL2)						
SOL2/3a	0	16.0	0.0	0.0	72.5	13.6
SOL2/3b	0	28.8	12.6	0.0	53.9	4.8
SOL2/1	10	89.7	0.0	9.5	0.0	0.8
SOL2/7[a]	40	99.2	–	–	–	0.4
SOL2/9[a]	40	99.7	–	–	–	0.3
SOL2/1	10	90.2	0.3	9.1	0.0	0.5
SOL2/4	20	86.2	7.6	0.0	5.9	0.2
SOL2/7[a]	40	99.6	–	–	–	0.7
SOL2/10[a]	80	99.6	–	–	–	0.3
Wingst spruce (WIN)						
WIN28	35	78.0	12.5	0.0	8.5	1.1
WIN28	100	83.3	8.6	0.0	8.0	0.1
Hils beech (HIL1)						
	20	39.3	51.1	0.0	6.8	2.9
	40	42.5	21.5	32.2	0.0	3.8
	80	60.2	37.3	0.0	2.1	0.3
Hils spruce (HIL2)						
	20	32.7	56.8	0.0	7.2	3.3
	40	26.4	20.7	0.0	51.2	1.6
	80	23.3	66.4	0.0	5.7	4.6
Fuhrberger Feld (HAN)						
A1	100	92.1	0.0	0.6	3.6	3.7
A2	100	91.0	0.0	0.2	3.6	5.8
A3	100	86.8	0.0	0.4	5.9	6.8

Table 4.2.1.5. (*Continued*)

Point of sampling	Depth (cm)	Al_{inorg}	Al_x	$Al_{y<3500}$	$Al_{y>3500}$	Al_B
		$Al_t 0.01$	$Al_t 0.01$	$Al_t 0.01$	$Al_t 0.01$	$Al_t 0.01$
A4	100	97.7	0.0	2.3	0.0	0.9
A5	100	97.7	0.0	2.2	0.0	0.4
A6	100	81.3	0.0	2.0	16.7	–
B6	100	93.8	1.0	0.0	5.2	–
E2	170	87.1	0.0	0.5	12.3	–
E4	170	95.0	0.0	3.3	1.7	–
F3	250	96.5	1.0	0.0	2.5	–
H2	100	93.4	0.0	5.7	0.9	–
H5	100	57.4	0.0	14.2	28.4	–
H6	100	64.5	0.0	19.3	16.2	–
H7	100	97.1	0.0	2.2	0.6	–
HAN-S1	200	60.6	33.3	0.0	4.6	1.5
HAN-S3	500	56.1	36.6	0.0	2.4	4.9

[a]Without dialysis procedure; $Al_{inorg} = Al_{ex}$.

multi-level well (HAN-S1) within the recharge area (Strebel and Böttcher 1985). The speciation results also give evidence of a downward transfer of aluminium complexes to B and lower horizons. However, dissolved aluminium still remains in solution in such a deep part of the profile. This means that inorganic as well as organic forms of aluminium pass the Bsh horizon without being involved in precipitation or sorption processes. The differing behaviour within the sampling zone H may be caused by the change of vegetation (see site description) or higher pH values of samples. Although the more likely reason was to be the smaller Al/DOC-ratio of these samples. The samples taken from the well (HAN S3) indicate that dissolved aluminium complexes, especially with small organic molecules, are transported within the recharge area down to 5 m depth.

The two podsolized brown earths in the Solling research area (SOL1 and SOL2) are especially of clayey textured parent materials which provide, in view of its excellent sorption characteristics, a strong interaction behaviour with the solution phase. In the deeper parts of Bv1 horizons (> 20 cm) Al is present predominantly in inorganic forms. Dissolved total organic carbon decreases steadily down from the Oh/Aeh horizon to the Bv1 horizon, while the total aluminium content increases sharply from the top of the Bv1 horizon. Thus two possible mechanisms can be suggested. On one hand, the preferential immobilization of organic compounds due to precipitation, sorption, polymerization processes at the top of the B horizon, and on the other hand the further release of aluminium to soil solution in considerable amounts, which means a pH control to aluminium solubility. At this moment there is no real proof if the soluble Al complexes dissociate or if aluminium is partially co-precipitated with organic matter. Considering earlier experiments, the latter is probable (König et al. 1986). Considerable amounts of dissolved aluminium were analyzed in seepage water samples of the O horizon before having contact with mineral soil (Dietze 1985). This fact indicates an Al accumulation mechanism in the organic layer.

The pH profiles do not show a significant effect on aluminium speciation. The influence of pH seems to be overlapped by sorption processes.

4.2.2 Mineral-Pore Water Interaction in Two Soil Types on Pleistocene Sediments at Hamburg

F. Sztuka and I. Valeton

4.2.2.1 Sites of Investigation

The development of soil profiles under similar environmental conditions – such as climate, morphology and vegetation – is generally controlled by the properties of the parent material (e.g. carbonate content, quantity and properties of clay minerals, content of organic carbon, porosity, sedimentary textures, pattern and intensity of drainage) as well as by the interaction of minerals and pore solutions. These interactions were studied in two soil profiles in the Hamburg area (Fig. 1.1, Table 1.1 – a luvisol on calcareous fine-grained and clayey till at Beckedorf (HAM1; Fig. 4.2.1.1) and a podsol on coarse-grained fluvioglacial sand at Volkspark/Hamburg (HAM2; Fig. 4.2.2.2), both soils developed on sediments of the Saalean age (Drenthe stage; Sztuka 1985). The evaluation of the chemical composition of the pore water and the mineralogical composition of the solids, especially the secondary minerals, was aimed at providing a better understanding of the interactions which produce different soils and their typical composition of secondary minerals.

4.2.2.2 Methods

Pore water samples were collected in the field and laboratory with lysimeters, porous suction cups, centrifuges and pressure filtration. Furthermore, pore solutions were produced by "equilibrating" 1 unit deionized water with 0.8 units dry soil within 24 h (Ulrich and Khanna 1971). The water samples were usually very small (15–35 ml). Therefore the analyses of dissolved Ca^{2+}, Mg^{2+}, Sr^{2+}, $Fe^{2+(3+)}$ and Mn^{2+} were performed with AAS (Perkin-Elmer 305 B), of Al, PO_4^{3+}, NH_4^+ and SiO_2 by photometric methods (Perkin-Elmer 550 SE UV/VIS), of F^-, NO_2^-, NO_3^-, Cl^- and SO_4^{2-} by IC methods (Dionex 2120 i) and Na^+ and K^+ by flame photometry. If 500–1000 ml samples were available, standardized analytical methods were used (DEV 1971–1988). The temperature and the oxygen content were measured with the combined O_2-temperature probe Oxi 01, Mod. WTW-Weilheim, pH with the temperature-corrected probe (Ingold) Mod. 521, WTW-Weilheim, E_H with a Pt-AgCl electrode (Ingold).

In the solid phase the exchangeable cations were determined by treatment with ammonium acetate at pH 7 and measurement of the exchanged cations in the solution. The CEC was determined by exchange of the fixed ammonia by KCl at pH 2.5 and distillation after Kjeldahl. Base saturation (BS) is calculated as percentage of the sum of exchanged cations (Na^+, K^+, Ca^{2+}, Mg^{2+}) at the CEC. Adsorbed H^+ and Al^{3+} were measured by extraction with KCl and titration of one aliquot with NaOH to pH 8.2. One aliquot of this extract was titrated with NaF to determine the exchangeable Al as $[AlF_6]^{3-}$ (Schlichting and Blume 1966). The H_2O-rich X-ray amorphous Fe- and Mn oxides (Fe_o, Mn_o) were extracted with oxalic acid ammonium oxalate solution (Schwertmann 1964), the "free" or alkaline-soluble Si and Al by successive treatment with NaOH and HCl (Hashimoto and Jackson 1960), and the pedogenic iron- (Fe_d) and manganese oxides (Mn_d) by dithionite-citrate treatment (Mehra and Jackson 1960). Total chemical composition was determined by X-ray fluorescence analysis.

The primary (rock-forming) minerals and the secondary minerals, especially the clay minerals, were identified using optic and X-ray methods. The clays were semi-quantitatively characterized by X-ray diffractrometry (Vortisch 1982), but due to its low clay content, the clay in profile HAM2 had to be enriched before measurement.

Saturation indices (SI) of relevant mineral phases were calculated from the pore water analyses using the programme WATEQF (Plummer et al. 1976). These values were compared with the mineralogical results.

4.2.2.3 The Properties of the Two Soils

Due to the different sedimentological properties of the parent material, the soil-forming processes, especially the interactions between minerals and pore solutions during pedogenesis, yielded partly similar, but characteristic features in the two soil profiles. Grain size distributions, contents of organic carbon and $CaCO_3$, porosities, and the indicators for the mobility (Fe_o, Fe_d, molar ratio Si/Al and Mn_o) of the two profiles HAM1 and HAM2 are visualized in Figs. 4.2.2.1 and 4.2.2.2, the distribution of the exchangeable cations, the CEC, the *BS* and the clay mineral composition in Figs. 4.2.2.3 and 4.2.2.4. The chemical data of the pore waters are listed in Tables 4.2.2.1–4.2.2.2.

4.2.2.4 Interactions Between Minerals and Pore Solutions

Typical precipitates found in the pores of comparable soil horizons are indicative for similar geochemical environments of these soil layers. The precipitates are either X-ray amorphous oxidic, hydroxidic or carbonate coatings, or newly formed minerals or their diagenetic transition states. The secondary minerals often fill the pore space as cement. The specific geochemical interactions between the pore solutions and the solid phases can often be described with the help of the saturation index (SI). These interactions are described for iron and manganese precipitates, secondary clay minerals and carbonates.

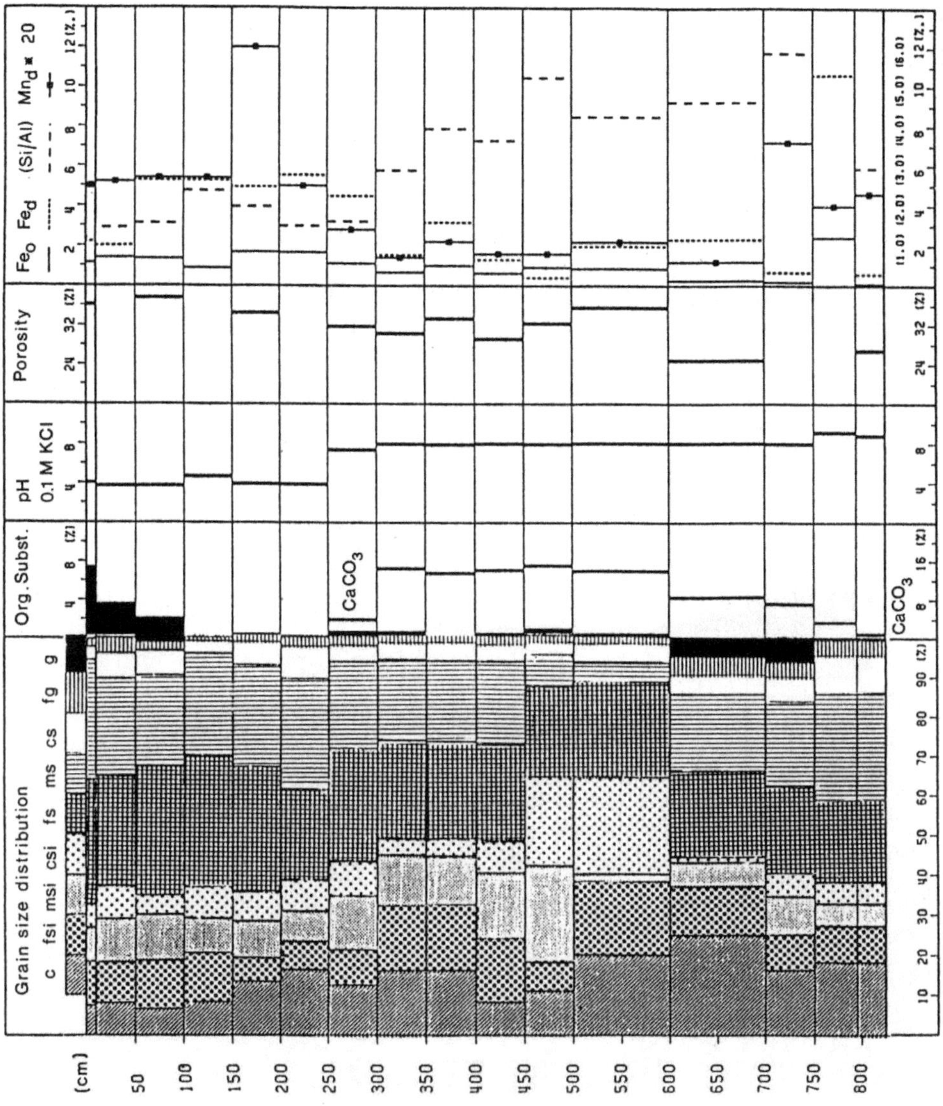

Fig. 4.2.2.1. Relation between depth of soil and grain size distribution (c - clay, si - silt, s - sand, g - gravel; prefix means f = fine, m = middle, c = coarse), carbonate, organic carbon, pH, porosity, q/f'sp-ratio, main clay mineral, solubility and mobility of Fe_d, Si/Al, and Mn_d at site HAM1. (Sztuka 1985)

Iron- and Manganese-Containing Precipitates

Parent minerals for iron in the given sediments are the iron-containing silicates, like biotite, pyroxenes, amphiboles and olivine, and for manganese the manganese-containing minerals, like biotite and amphiboles.

Typical coatings of Fe precipitates are found on quartz grains, on which locally Fe aggregates have grown into the pore space. Such aggregates are typical for the

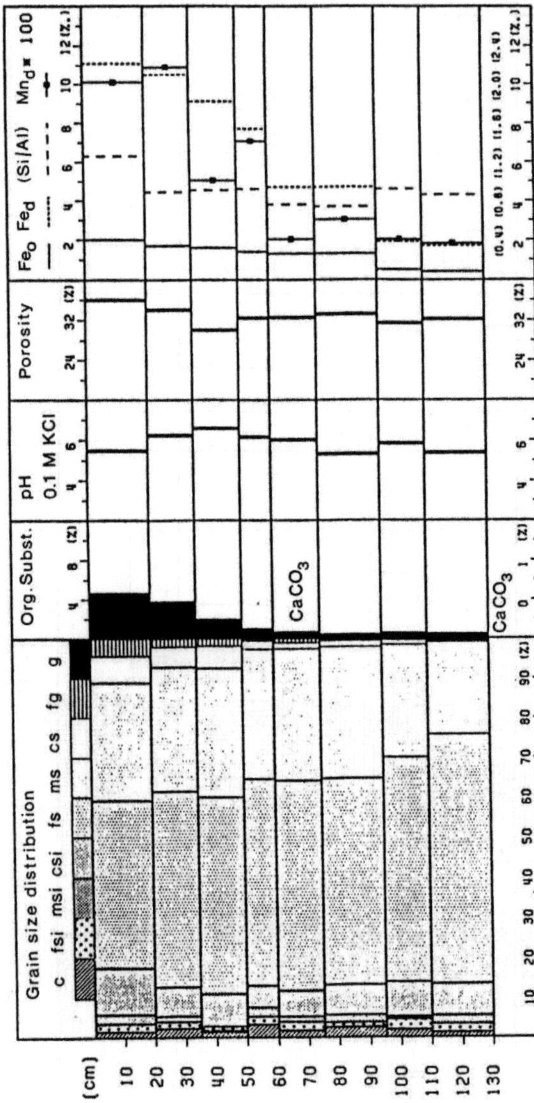

Fig. 4.2.2.2. Relation between depth of soil and grain size distribution (c - clay, si - silt, s - sand, g - gravel; prefix means f = fine, m = middle, c = coarse), carbonate, organic carbon, pH, porosity, q/f'sp-ratio, main clay minerals, solubility and mobility of Fe_d, Si/Al, and Mn_d at site HAM2. (Sztuka 1985)

rusty patches which can be observed macroscopically. These iron-enriched patches are often surrounded by iron-depleted fallow zones. Aggregates of Fe- and Mn oxides on surfaces of clay minerals are observed in the stagnation layer of HAM1. The precipitates range from H_2O-rich X-ray-amorphous phases to crystalline phases, which is indicated by their activity ratios Fe_o/Fe_d and Mn_o/Mn_d. This activity ratio, according to Scheffer and Schachtschabel (1982), is an indicator for the crystallinity of the precipitates. From the ratios in Tables 4.2.2.3 and 4.2.2.4, it can be deduced that poorly crystallized phases are predominant in the two profiles. The precipitates are characteristically distributed in the soil horizons. In the luvisol

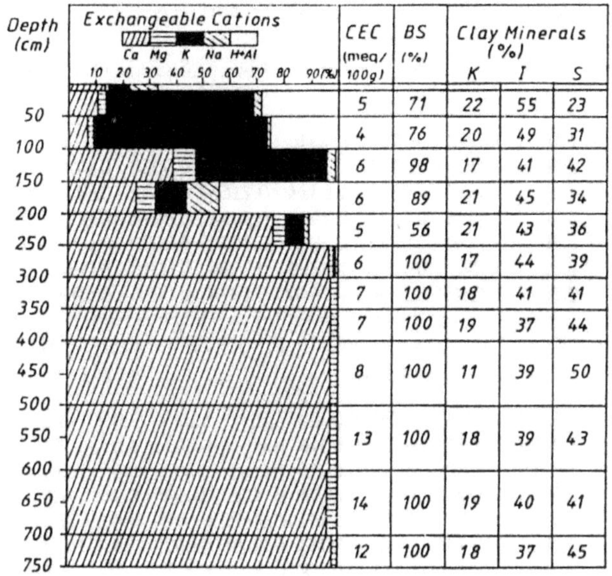

Fig. 4.2.2.3. Cation exchange capacity (*CEC*), base saturation (*BS*) and clay minerals (*K* kaolinite; *I* illite; *S* smectite) at site HAM1. (Sztuka 1985)

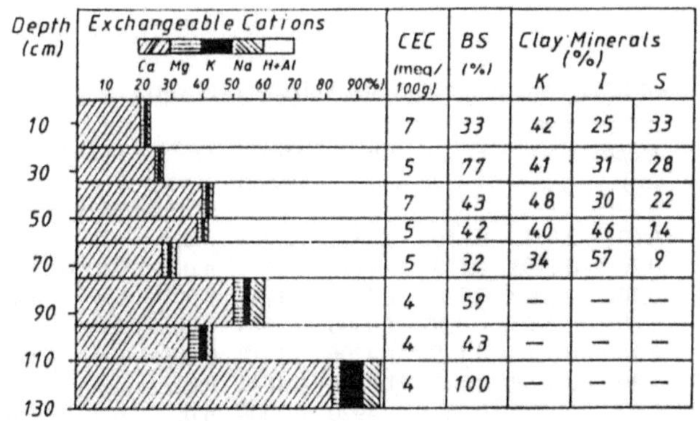

Fig. 4.2.2.4. Cation exchange capacity (*CEC*), base saturation (*BS*) and clay minerals (*K* kaolinite; *I* illite; *S* smectite) at site HAM2. (Sztuka 1985)

(HAM1), precipitates of Mn are found in the lower part of the profile ($BS_{fe,t}$), whereas those of Fe, Mn and Al are present mainly in the uppermost top soil (Ah-SE), due to retarded percolation (Fig. 4.2.2.1). They are explained by increasing E_H values in carbonate-free, well-aerated zones.

In the well-drained profile HAM2 no horizon-specific iron enrichments are found. Fe_o decreases continuously from 0.2‰ at the top to 0.09‰ at the base of the profile. The Mn precipitates indicate a higher mobility: a first enrichment with 0.21‰ Mn_o in the upper soil zone is followed by a depleted zone underlain by a second maximum with 0.01 and 0.12‰ Mn_o (Table 4.2.2.4). The solubility and mobility of iron and manganese, indicated by the Fe_o/Fe_d – and Mn_o/Mn_d ratios, are much higher in the podsol (HAM2) than in the luvisol (HAM1) (Tables 4.2.2.3 and 4.2.2.4).

The SI of the pore solutions indicate in the upper parts of profile HAM1 an undersaturation with respect to amorphous $Fe(OH)_3$, whereas goethite and hematite

Table 4.2.2.1. Pore waters of profile Beckedorf (HAM1)

No.	Depth (m)	Na^+	K^+	Ca^{2+}	Mg^{2+} (meq/l)	Sr^{2+}	$Fe^{2+(3+)}$	Mn^{2+}	Al	SiO_2	HCO_3^-
1	1.10	0.7177	0.1790	0.1040	0.2680	0.0040	0.0740	0.0120	1.0960	0.0910	0.0001
2	2.40	0.9526	0.1201	2.2008	1.1080	0.0052	0.4740	0.0698	0.8044	0.0879	0.0899
3	4.00	0.5655	0.0365	1.9424	0.2166	0.0022	0.0144	0.0006	0.0077	0.0935	0.3499
4	6.20	0.5088	0.0843	2.3984	0.3694	0.0018	0.0084	0.0011	0.0098	0.1028	1.2995
5	7.20	0.4480	0.0854	2.3802	0.3178	0.0022	0.0286	0.0045	0.0311	0.1298	1.9994
6	8.30	0.5567	0.1048	1.3394	0.2396	0.0018	0.0002	0.0002	0.0077	0.0661	1.3196
7	9.50	0.3742	0.1534	1.2640	0.1974	0.0016	0.0006	0.0002	0.0309	0.0324	1.3552

No.	Cl^-	NO_3^-	SO_4^{2-} (meq/l)	PO_4^{3-}	F^-	Zn^{2-}	O_2 (mg/l)	pH	E_H (mV)	Conduct. (μS/cm)	Temp. (°C)	Ionic strength
1	1.5290	0.2193	1.5580	0.0006	n.d.	n.d.	3.9	4.22	435	192	5.8	0.00843
2	1.3410	0.1161	3.6424	0.0046	n.d.	0.0038	4.3	4.45	430	416	6.1	0.01599
3	1.0718	0.0780	1.8238	0.0024	0.0016	0.0020	6.8	6.85	310	272	6.0	0.00806
4	0.5923	0.0293	2.0910	0.0074	0.0058	0.0028	8.2	7.66	225	255	5.8	0.00967
5	0.4580	0.0177	1.7160	0.0084	0.0044	0.0024	9.4	7.77	240	302	5.8	0.00924
6	0.2838	0.0189	0.9910	0.0068	0.0029	0.0673	7.9	7.42	315	232	5.8	0.00583
7	0.3102	0.0064	0.9372	0.0018	0.0018	n.d.	7.3	6.88	295	194	5.7	0.00549

Table 4.2.2.2. Pore waters of profile Volkspark (HAM2)

No.	Depth (m)	Na^+	K^+	Ca^{2+}	Mg^{2+} (meq/l)	Sr^{2+}	$Fe^{2+(3+)}$	Mn^{2+}	Al	SiO_2	HCO_3^-
1	0.00	0.0713	0.0506	0.0518	0.0232	n.d.	n.d.	0.0000	0.1432	0.0127	0.0499
2	0.20	0.1191	0.0130	0.4179	0.0574	n.d.	0.0107	0.0021	0.0911	0.0313	0.5249
3	0.70	0.1022	0.0204	2.4286	0.1068	0.0018	n.d.	0.0000	0.0270	0.0156	1.3997
4	1.00	0.9425	0.0987	1.0506	0.2648	0.0015	n.d.	0.0014	0.0170	0.0482	0.9248
5	2.20	0.6803	0.1152	1.4206	0.1912	0.0014	n.d.	0.0104	0.0172	0.0531	1.0747
6	2.30	0.5193	0.1419	1.6816	0.4842	0.0012	n.d.	0.0000	0.0107	0.0572	1.4746
7	2.40	0.5772	0.2511	1.4920	0.2608	n.d.	n.d.	0.0000	0.0121	0.0410	1.0210
8	4.00	0.2270	0.0690	0.9804	0.1874	n.d.	n.d.	0.0060	0.0062	0.0265	0.3749

No.	Cl^-	NO_3^-	SO_4^{2-}	PO_4^{3-} (meq/l)	F^-	Zn^{2-}	O_2 (mg/l)	pH	E_H (mV)	Conduct. (μS/cm)	Temp. (°C)	Ionic strength
1	0.4987	0.0016	0.0770	0.0009	n.d.	0.0120	7.28	4.87	385	25	10.2	0.00114
2	0.3592	0.0066	0.2280	0.0010	0.0021	0.0110	7.02	6.57	315	68	9.3	0.00193
3	0.1851	0.2661	1.7280	0.0017	0.0006	0.0089	6.40	7.02	298	206	8.9	0.00840
4	0.3102	0.0999	1.1996	0.0015	0.0011	0.0054	7.90	7.36	256	198	8.7	0.00576
5	0.3420	0.0254	1.3425	0.0009	0.0022	0.0000	6.80	7.44	266	n.m.	8.6	0.00643
6	0.4897	0.2499	1.4156	0.0012	n.d.	0.0021	8.11	7.54	276	223	8.2	0.00781
7	0.3525	0.0387	1.4996	0.0022	0.0018	n.d.	6.24	7.38	198	n.m.	8.2	0.00691
8	0.3889	0.1306	0.8580	0.0031	0.0009	n.d.	6.31	6.84	125	154	8.2	0.00434

Table 4.2.2.3. Pedogene oxides in profile Beckedorf (HAM1)

No.	Depth (m)	Mn_o (‰)	Mn_d (‰)	Fe_o (‰)	Fe_d (‰)	Fe_o/Fe_d	Si_{aq} (‰)	Al_{aq} (‰)	Si_{aq}/Al_{aq}
1	0.10	0.18	0.23	1.14	2.20	0.51	4.61	3.86	1.19
2	0.50	0.20	0.26	1.40	2.01	0.69	4.53	3.12	1.45
3	1.00	0.22	0.27	1.35	5.26	0.26	4.41	2.81	1.56
4	1.50	0.24	0.27	0.88	5.27	0.17	3.29	1.38	2.38
5	2.00	n.m.	0.60	1.69	4.94	0.34	3.89	1.96	1.98
6	2.50	0.21	0.25	1.66	5.53	0.30	4.53	3.03	1.49
7	3.00	0.08	0.14	1.11	4.49	0.25	2.37	1.48	1.60
8	3.50	0.03	0.07	0.66	1.54	0.43	1.73	0.60	2.88
9	4.00	0.04	0.11	1.00	3.16	0.32	2.09	0.53	3.94
10	5.00	0.06	0.08	0.62	1.31	0.47	1.71	0.47	3.64
11	6.00	0.06	0.08	0.92	0.40	0.14	2.25	0.43	5.23
12	7.00	0.09	0.11	0.87	1.99	0.44	2.25	0.53	4.25
13	7.50	0.03	0.06	0.26	2.32	0.11	1.34	0.29	4.62
14	7.60	0.24	0.36	1.32	22.20	0.06	2.23	0.38	5.87
15	7.75	0.01	0.02	0.20	0.69	0.29	0.54	0.04	13.50
16	7.95	0.08	0.15	2.43	10.60	0.22	0.74	0.23	3.22
17	8.05	2.85	3.60	17.00	47.51	0.36	2.41	0.13	18.50
18	8.25	0.21	0.23	0.11	0.58	0.19	0.56	0.19	2.94
Crust		1.27	1.45	0.20	1.28	0.16	0.46	0.02	23.00
Crust		0.028	0.07	0.11	0.61	0.18	0.74	0.05	15.00

Table 4.2.2.4. Pedogene oxides in profile Volkspark (HAM2)

No.	Depth (m)	Mn_o (‰)	Mn_d (‰)	Fe_o (‰)	Fe_d (‰)	Fe_o/Fe_d	Si_{aq} (‰)	Al_{aq} (‰)	Si_{aq}/Al_{aq}
1	0.20	0.021	0.101	0.20	1.10	0.18	3.82	3.04	1.26
2	0.35	0.009	0.109	0.17	1.05	0.60	2.63	2.94	0.89
3	0.50	n.d.	0.049	0.16	0.91	0.18	1.17	1.28	0.91
4	0.60	0.005	0.069	0.14	0.77	0.18	1.93	2.09	0.92
5	0.75	n.d.	0.022	0.13	0.47	0.28	2.30	3.01	0.76
6	0.95	0.010	0.027	0.13	0.47	0.28	2.17	2.93	0.74
7	1.10	0.012	0.021	0.05	0.19	0.26	2.86	3.11	0.92
8	1.30	0.007	0.018	0.04	0.17	0.25	1.97	2.27	0.86

are stable in the whole profile and siderite, hausmannite, $Mn(OH)_3$, $Mn(OH)_2$, $MnSO_4$, manganite, and rhodochrosite are unstable. This is consistent with the observed distribution of precipitates and depletion of the primary iron- and manganese-containing constituents due to weathering processes.

Clay Minerals

The pore water constituents controlling the formation of clay are aluminium and silica (Scheffer and Schachtschabel 1982), which stem mainly from alumosilicates, e.g. feldspars, biotite and primary clay minerals. The concentration of silica depends mainly on the presence of amorphous silica, whereas the quartz proves to be thermodynamically stable. In soils and sediments with periodic changes in the input and drainage intensity of the aquatic phase, periodical stagnation of water and E_H-pH variations, and seasonal variation in bioproduction, changing equilibrium conditions between the following phases are developed:

initial clay mineral ⇄ aquatic phase ⇄ secondary clay mineral.

Thus, secondary clay minerals occur as transformed and newly formed crystallites besides the primary clay minerals. However, secondary clay minerals could not be detected by selective decomposition (see also Iwanoff 1981).

In the luvisol (HAM1), appreciably higher portions of clay and silt fraction are present than in the podsol (HAM2) (Figs. 4.2.2.1 and 4.2.2.2). In the profile HAM1 illite (55–44%) dominates over about equal portions of smectite and kaolinite (17–22% each). The clay minerals in profile HAM2 are characterized by a dominance of kaolinite over illite in the A and B horizons, a decrease in smectite content with depth, and a lack of any clay minerals at 80 cm depth. Alteration of clay minerals in the different soil environments produced contrasting values of *BS* and cation composition in the interlayer position. The depth distribution of the clay minerals in the two profiles shows the expected positive correlation between CEC and the contents of illite and smectite. In both profiles the BS rises from low values in the top soil (71 or 33%, respectively) to 100% in the C-horizon.

The exchangeable cations in the unweathered part of HAM1 consist mainly of potassium, whereas below the BS_t horizon calcium dominates (> 90%), consistent with the high content of calcium carbonate in the rock (Fig. 4.2.2.1) and of calcium ions in the pore solutions (Table 4.2.2.1). Due to these facts, the base saturation is high, and calcium and potassium hold mainly in the upper soil the exchange positions, whereas sodium and aluminium are of minor importance.

In the uppermost 1.10 m of profile HAM2, the BS is low (32–59%) and the adsorbed H^+ and Al of the clay minerals are correspondingly high. In the decalcified weathering zone the main exchangeable cations are H^+ and Al^{3+}, whereas the concentrations of exchangeable calcium varies between 20 and 55%, and those of magnesium, potassium and sodium are generally low. Together with the higher mobility of iron (Table 4.2.2.4), this type of CEC and the availability of aluminium for interlayer positions proves the more active disaggregation of smectite in this profile.

The SI of the pore solutions indicate in the profile HAM1 an undersaturation with respect to gibbsite and amorphous $Al(OH)_3$, whereas those from the profile HAM2 are completely supersaturated with respect to $Al(OH)_3$. At profile HAM2 the water is saturated or supersaturated with respect to gibbsite over the whole profile, with the exception of the uppermost 20-cm layer.

In the upper zone of HAM1 the pore solutions are undersaturated with respect to albite and anorthite, and approach saturation in deeper parts of the profile. Illite and smectite are thermodynamically unstable in the upper parts of the profile, whereas kaolinite is stable in the whole profile. This is consistent with the observed distribution of precipitates and newly formed minerals, which is connected with a breakdown and partial disappearance of primary minerals (Ohse 1983; Ohse et al. 1984), and a secondary generation or mineral alteration due to weathering.

The solubility and mobility of aluminium is much higher in the podsol (Table 4.2.2.4) than in the luvisol (Table 4.2.2.3). Besides the well-known pH dependency of the aluminium and silicium content, the mobilization of aluminium and – to a lower extent of silicium – also depends on the complexation by humic and fulvic acids (Sects. 2.8 and 4.2.1, this Vol.).

Carbonates

Both profiles, despite different parent carbonate content (HAM1 15.8% carbonate, HAM2 0.04–0.21%), show a decalcification of the upper soil horizons, which is combined with lower pH values in the pore solutions than those in the parent sediment. This is consistent with the observed distribution of the SI of the pore solutions, which indicate in the upper parts of both profiles an undersaturation with respect to calcite. In the lower part of the carbonate-poor profile HAM2 saturation is approached, whereas in the profile HAM1 the pore solution is saturated in the carbonate rich till layer.

The carbonates in the profile HAM1 could be discriminated into primary detrital carbonate of Cretaceous and Paleozoic origin (calcite with traces of strontium), secondary micritic precipitates in the pore space and on the grain surfaces as early diagenetic cement, and secondary compact carbonate crusts of 15–45-cm thickness containing 93% sand, formed at the transition zone between the fine-grained to coarser sediments. Compact carbonate precipitates are also found as veins in faulted zones.

The observed presence of strontium in the precipitates and in the pore solutions agrees with results of Valeton and Khoo (1980). According to Cornelius (pers. commun. 1982), the ratio of Ca to Sr in carbonates may be locally used as an indicator of secondary precipitation by ground water.

4.2.2.5 Conclusions

From these evaluations it can be deduced that the measurement and evaluation of pore water composition offers a tool for the understanding of kinetics of weathering processes in different profile zones due to the interactions between the pore water and the degree of single mineral phases (Sect. 4.3.1, this Vol.). For example, the mobility of Fe- and Mn oxides, SiO_2 and $Al(OH)_3$ are indicative of intermediate (transient) forms of weathering and mineral formation.

4.2.3 Subsurface Hydrochemical Reactions in the Sandhausen Forest Ecosystem

H. Jacob, W. Regenberg and C. Sonntag

4.2.3.1 The Experimental Site

The natural and the anthropogenically induced element input from the unsaturated zone into the ground water influences the composition and quality of the ground water. Therefore, the spatial and temporal variations of the hydrochemical conditions in the unsaturated zone of a forested area were studied to determine element fluxes through the unsaturated zone into the ground water (Jacob 1987).

The experimental site is located in a forest (Schwetzinger Hardt near Sandhausen) in the Rhine valley about 7 km southwest of Heidelberg (SDH; Fig. 1.1, Table 1.1). The soil is developed on river sediments (mostly sands with a small fraction of gravel) which are rather homogeneous (Regenberg 1987). In depths greater than 3.5 m the content of carbonates ranges between 5 and 10% wt. The strata above this level are decalcified. The ground water table sank steadily from 3.5 m depth in 1983 to about 5 m in 1987 and rose again to 4.4 m in 1988.

The mineralogical composition of the sediments is dominated by quartz and quarzites ranging between 80 and 95% wt. The rest are feldspars with a maximum of 15% and mica up to 3%. Other minerals were found in trace quantities only. Due to the absence of clay minerals and the homogeneity of the soil, this site offers a rather simple system with respect to hydrological and hydrochemical studies.

Pines (*Pinus silvestris*, 120–150 a) and beeches (*Fagus silvatica*, 20–40 a) grow in the surroundings of the experimental site. The forest floor has practically no vegetation cover. The humus layer is between 10 and 20 cm thick. With respect to the geological and mineralogical situation this site is typical for two thirds of the whole forest (Regenberg 1987). The rest of the area is covered by sand dunes which often contain carbonates up to the ground surface.

4.3.2.2 Sampling and Analytical Methods

Samples of percolating water and precipitation samples (bulk deposition) were collected on a weekly basis from 1984 to 1988. The content of the major ions (Cl^-, NO_3^-, SO_4^{2-}, H^+, K^+, Ca^{2+}, Mg^{2+} and Al) was determined by IC, pH electrode, AAS and ICP-AES. The uncertainty of the data is about ± 15% for the H^+ concentration and between ±5% and ±10% for the other ions. The AAS- and ICP measurements were performed at the Institute Fresenius, Taunusstein (Sect. 2.3, this Vol.).

For the sampling of pore water from different layers, a planked pit (2.35 m long, 1.35 m wide and 3.3 m deep) was installed in spring 1983. The influence of the pit on the adjacent soil was minimized by a plastic sheet which separates the 5-cm-thick wooden planks and the soil. Measurements of the soil temperature and the CO_2 concentration in the ground air showed no significant influence of the pit at distances greater than 50 cm.

Tubes with integrated porous cups (type: Ceratzki 1971) were installed horizontally in the soil between 33 and 258 cm depth. The total length of such a suction facility was 65 cm. The porous cups (manufacturer: Haldenwanger, Berlin) are mainly composed of Al_2O_3 (99.5%). They are cylindrical with an outer diameter of 1.3 cm and a length of 10 cm. Their mean pore diameter of about 0.6 µm (Mayer 1971) causes a bubbling pressure potential of 4900 hPa. With a low pressure of 500–600 hPa the pore water was sucked through the porous cups into glass bottles.

The use of Al_2O_3 porous cups is connected with some problems (Hansen and Harris 1975; Grossmann et al. 1985). In samples from porous cups which had been in the soil longer than 2 years, Al concentrations up to 35 mg/l were found. The values of other ions also increased, and the pH was much lower than originally. As new porous cups in the same soil layers yielded samples with four to five times lower solute content, one must assume that the hydrochemical environment around the old cups was strongly disturbed (Jacob 1987). Therefore, one should be careful with the interpretation of the data when the Al concentration and the concentrations of other ions rises drastically and the pH value is much lower than it was originally.

4.3.2.3 Results

Percolating Water

Time Series and Depth Profiles of Major Ions in the Pore Water

Model calculations (Jacob 1987) predict a seasonal variation of the ion concentrations in the pore water which is due to variations of the moisture content and also to variations of the element input and the nutrient uptake by plants. The concentrations should be highest in autumn or winter and lowest during the infiltration period in spring. The amplitude of this variation should decrease with increasing depth in the soil. Figure 4.3.2.1 shows the time series of the pH and the Na^+-, Ca^{2+}- and SO_4^{2-} concentrations in the pore water at two different depths for the years 1985 to 1988. The extremely low canopy throughfall in 1985 (1984: 482 mm, 1985: 328 mm, 1986: 498 mm, 1987: 427 mm) caused a very low moisture content of the soil during some months of the year. Thus, it was not possible to extract pore water samples in this period. Also from September to December 1987 the soil was too dry to obtain water samples from the upper layers (e.g. at 45 cm depth). These data gaps may hide considerable seasonal variations of the solute concentrations.

The annual mean concentrations of Na^+, Ca^{2+}, SO_4^{2-} and the pH (Fig. 4.3.2.2) show an increase, while the Al content decreases with depth. As the Cl^- contents were influenced by contamination, sodium was taken as a reference element. This is justified by the almost equal values of the sodium input by the canopy throughfall and the downward flux of sodium at 258 cm depth. This means that sources and sinks of sodium in the soil between the surface and 258 cm depth equalize each other.

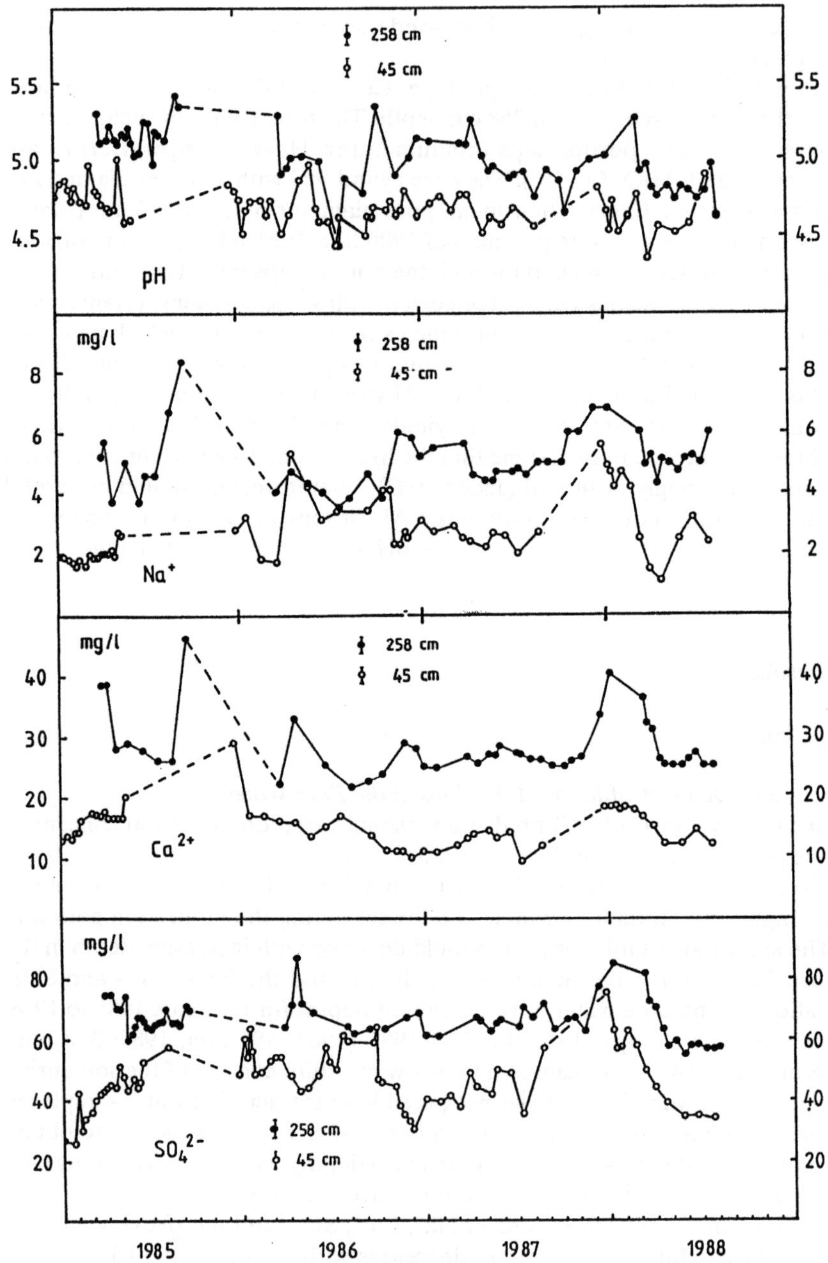

Fig. 4.2.3.1. Time series of the pH, and the contents of Na^{+}, Ca^{2+} and SO_4^{2-} in the SDH pore solutions at 45 and 258 cm depth for the years 1985 to 1988

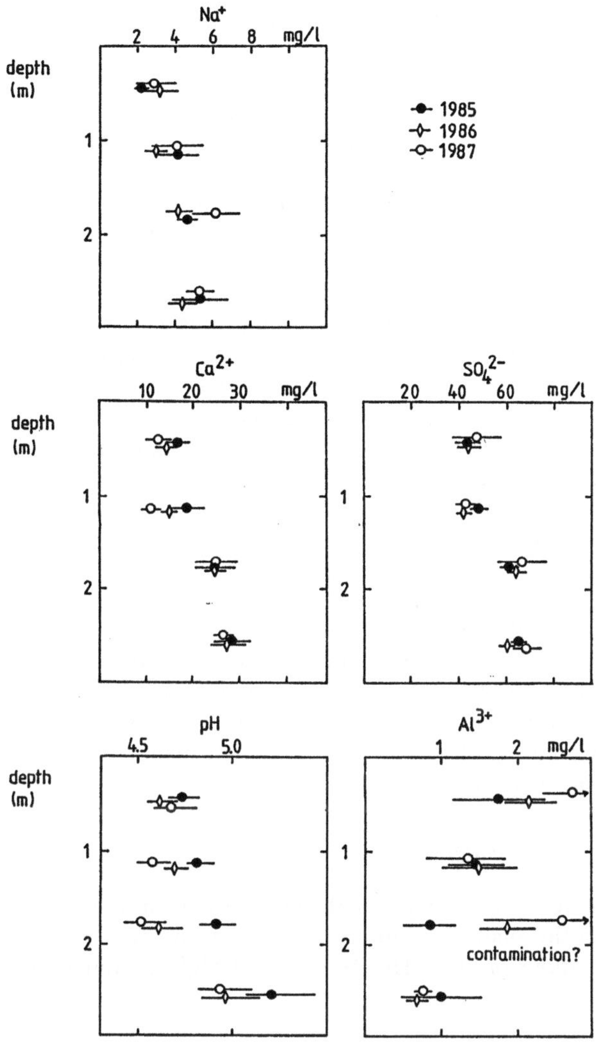

Fig. 4.2.3.2. Depth profiles of the Na^+, Ca^{2+}, SO_4^{2-} and Al^{3+}-concentrations and the pH in the SDH pore solutions. Mean values for 1985, 1986 and 1987 are given. Bars indicate the standard deviation

The SO_4/Na ratio and the Ca/Na ratio in the pore water do not change significantly between 45 cm and 258 cm depth. Thus, the turnover of these elements obviously occurs mainly in the upper 40 cm of the soil. Below this depth the element fluxes are nearly constant. The Al/Na ratio, however, sharply decreases with depth while the pH increases. This leads to the conclusion that Al ions which were released in the uppermost layers must have been partly immobilized in the deeper layers of the soil. (The apparent minimum in the pH data of 1986 and 1987 in 180 cm depth and the corresponding increase of the Al concentration is probably due to a contamination of the samples by old porous cups, as mentioned above. In 1988 new porous cups were installed at the same depth which yielded samples with a higher pH and Al concentrations ranging only between 0.4 and 1.0 mg/l).

Acidification of the Soil and Buffering Processes

The pH values of the pore water vary from 4.3 to 5.4. Values higher than 5.0 were found only at 258 cm depth. Values between 4.2 and 5.0 indicate that ion exchange is the main buffering process against the acidification of the pore water (Ulrich 1985). Since the clay content of the soil is negligible, the organic matter has to be considered as the predominant ion exchanger. Similar to the vertical distribution of the organic matter, the exchange capacity sharply decreases with depth. The ion exchange is practically restricted to the upper 50 cm of the soil. The *BS* on the exchanger drops from 27% in 0–5 cm depth to 22% in 5–10 cm and to 6% below 10 cm depth. Calcium forms about 80% of the stored nutrient elements.

The local contribution of silicate weathering to the buffering process is small. It is estimated to be lower than 17 meq $H^+/(m^2 \cdot a)$ (Jacob 1987). A buffering of the H^+-input (medium value: 53.2 meq/$(m^2 \cdot a)$) by ion exchange will lead to a progressive loss of nutrients (especially Ca^{2+}) which are washed out into the ground water. This was illustrated by an "event" in May and June 1984 when an unusually high H^+ input due to heavy acid rains caused a rise of the Ca^{2+} concentration in the pore water at 45 cm depth from less than 10 mg/l to about 80 mg/l. This Ca^{2+} pulse decreased again within 3 weeks (Jacob 1987).

The vertically integrated quantity of exchangeable Ca^{2+}-, Mg^{2+}-, and K^+ ions is about 3500 meq/m^2. If an annual H^+ input of 50 meq/m^2 is assumed it would take about 70 years to remove these components completely, when other buffering processes, the pH dependence of the actual exchange capacity or an additional input of nutrients by weathering or fertilization are not taken into account. A continuing high input of acids will result in detrimental losses of nutrients from the soil within the next decades.

Ground Water

A comparison between the ion concentrations in the pore water at 258 cm depth and in the shallow ground water (5–6 m depth, Table 4.2.3.1) shows that the concentrations of SO_4^{2-}, HCO_3^-, Ca^{2+} and Mg^{2+} are much higher in the local ground water

Table 4.2.3.1. Concentrations (in mg/l) of major ions in the SDH pore solution (at 2.6 m depth) and in the shallow SDH ground water (between 5 and 6 m)

	Pore solution	Ground water
Cl^-	12 – 19[a]	8 – 12
NO_3^-	< 1	< 1
SO_4^{2-}	60 – 80	90 – 140
HCO_3^-	< 10	210 – 280
Na^+	4 – 8	7 – 9
K^+	1 – 4	1 – 4
Ca^{2+}	25 – 40	100 – 130
Mg^{2+}	2.5 – 4	8 – 11

[a]Contamination cannot be excluded.

than in pore solution. The surplus of the last three ions is due to the dissolution of carbonates which takes place in the calcareous layers below 3.5 m at an annual rate of 290 ± 40 mmol/m^2. This estimate is based on a mean annual ground water recharge of 150 mm. The carbonate dissolution occurs mainly under open system conditions, i.e. in the unsaturated zone under the influence of soil CO_2. Only 1% of the annual CO_2 production in the soil is necessary for the dissolution of about 300 mmol/m^2 carbonates. At the experimental site the CO_2 content of ground air varies seasonally between 2000 and 5000 ppm (Sect. 3.2.3, this Vol.). Estimates of carbonate dissolution at fixed CO_2 pressure yield equilibrium-concentrations of HCO_3^- in soil solution between 165 and 215 mg/l. The HCO_3^- concentrations in ground water are slightly higher. This surplus is probably due to carbonate dissolution in the saturated zone, where free CO_2 from the decomposition of organic matter contributes to the dissolution.

It is more difficult, however, to define the origin of the surplus of about 100 mmol/(m^2 a) SO_4^{2-} which would be necessary to explain the ground water concentration, as no gypsum could be found in the profile. Possibly FeS, which was formed in formerly water-saturated layers under reducing conditions, is oxidized to sulfate in the now unsaturated zone after the water table had dropped.

4.3 Reaction Kinetics

4.3.1 Experimental Methods for Determining Dissolution Rates of Silicates – a Comparison

D. Schenk, G. Matthess, A. Dahmke and A. Petersen

4.3.1.1 Rate-Controlling Parameters

The thermodynamic description of chemical weathering processes of feldspars indicate very often non-equilibrium or only partial equilibrium conditions between pore solution and solid phases both in laboratory weathering experiments and in natural systems. This phenomenon is mainly due to the formation of secondary phases, especially of clay minerals, which impedes the establishment of a total equilibrium between feldspars and pore solutions at low temperatures. The rate of approaching the partial equilibria is controlled by such factors as chemical composition, crystallographic properties and specific reactive surfaces of the solid phases, temperature, pH and saturation state of the pore solution, presence of inhibitors in the pore solution, contact time between solid phases and pore solution and the ratio between volumes of solid phases and pore solution. The experimentally derived silicate dissolution rates vary over more than one order of magnitude, due to the large number of parameters and to the type of experimental setups used.

4.3.1.2 Methods

Rock Materials

The weathering experiments were performed with three feldspar rich rocks (Sect. 2.2.5, this Vol.). These materials were selected because they contain several feldspar components, which seemed advantageous for the application of the laboratory-derived reactions rates to natural systems, and – even more important – because these materials could be provided in sufficiently large amounts for the experiments (especially for the column experiments). The rock materials were characterized by thin section microscopy, X-ray diffractometric analyses, X-ray fluorescence analyses and microprobe analyses. More details of rock petrography, of the pretreatment steps, the determination of specific surfaces, the analytical procedures and the methods of evaluation of the rates are documented in Dahmke (1988).

Experimental Setups

Three different experimental setups were used for the laboratory measurement of dissolution kinetics of rocks, namely batch experiments, fluidized bed reactor experiments and column experiments (Schenk et al. 1987, 1989). All these experiments were performed at a continuously recorded room temperature of 23 ± 2 °C.

Batch Experiments
The experimental setups consisted of 0.5 l polyethylene bottles filled with suspensions of various amounts of ground rock in 0.5 l solvent (deionized water and 1.0, 0.1 and 0.01 mM solutions of CsCl). Depending on the grain size fraction used, the suspended quantities were: 3 g (fraction < 0.063 mm), 5 g (fraction 0.063–0.1 mm), 10 g (fraction: 0.2–0.4 mm) and 20 g (fraction: 0.8–2 mm). The experiments were performed without any pH adjustment and with a pH adjusted by buffering solutions or by CO_2 flushing. To achieve a continuous mixing of the suspensions as well as to minimize the abrasion of rock material during the shaking process, two different shaking methods depending on the grain sizes were used. The rock-water mixtures with the coarse fractions were rotated in an overhead shaker with a period of 30 turns per hour. The suspensions of the fine-grained fraction, being less sensitive to abrasion, were shaken on a horizontal shaker. At distinct time intervals, about 20-ml samples of the homogenized suspension were withdrawn, filtered and analyzed. In the coarse fractions the amount of water withdrawn was replaced by the respective amount of deionized water to maintain the rock/water ratio. The dilution caused by this procedure over the whole observation time (30 d) is calculated to be 4% at maximum.

Fluidized Bed Reactor Experiments
The experimental set consists basically of the inflow unit and the reaction unit (Fig. 4.3.1.1). In the inflow unit the deionized water is conditioned by a control unit to certain CO_2 partial pressures or certain pH values, respectively. The reaction unit is described in extenso by Chou and Wollast (1984). The amount of albite, anorthite and orthoclase rock materials used in the reaction unit were in all cases 5 g of the grain size fraction 0.063–0.1 mm. The inflow from the conditioning unit was started

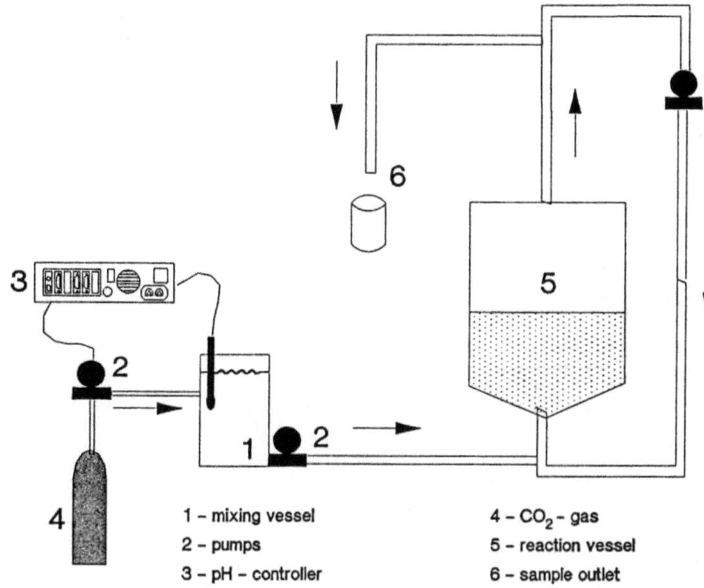

Fig. 4.3.1.1. Experimental setup of a fluidized bed reactor

3–5 days after the addition of the solid phase, thus excluding the effects of transient processes, such as exchange reactions (Busenberg and Clemency 1976) and the dissolution of adsorbed microcrystallites (Holdren and Berner 1979). For most of the experiments the inflow water was conditioned to a CO_2 partial pressure of $10^{-3,5}$ simulating atmospheric conditions, or to an CO_2 partial pressure of 10^{-2} simulating ground air conditions. Furthermore, several pH-buffered solutions (pH range of 4 and 6) were added to keep the pH constant during the experiment or to investigate the effects of an abrupt pH change in the aqueous solution. The samples were collected by a time-controlled fraction sampler (sampling time: 3600–10 000 s). The constancy of the inflow and outflow rates were controlled by weighing.

Column Experiments

The column experiments consist of an inflow unit (see previous section) and Plexiglas, columns (tubes) filled with rock materials of the grain size fraction 0.4–0.8 mm and a porosity of 0.4, through which water with a controlled CO_2 partial pressure was flowing continuously. Valves were installed between inflow and outflow in the Plexiglas columns to facilitate the collection of samples. At the 1-m columns, the first sampling valve was installed 3 cm from the inflow, the next three followed in equal distances of 5 cm, the remaining valves were arranged at 43 cm, 58 cm, 73 cm and 93 cm distance from the inflow. Samples were taken every 3 or 4 weeks. Between these sampling events, the volumes of the outflowing solutions were measured, from which the mean flow velocity $\bar{v}$ was determined. Furthermore, the outflowing solutions at the end of the column were analyzed in changing time intervals. The column samples were collected when flow velocity and concentration of the solution were constant. For sampling, the end of the column was closed and

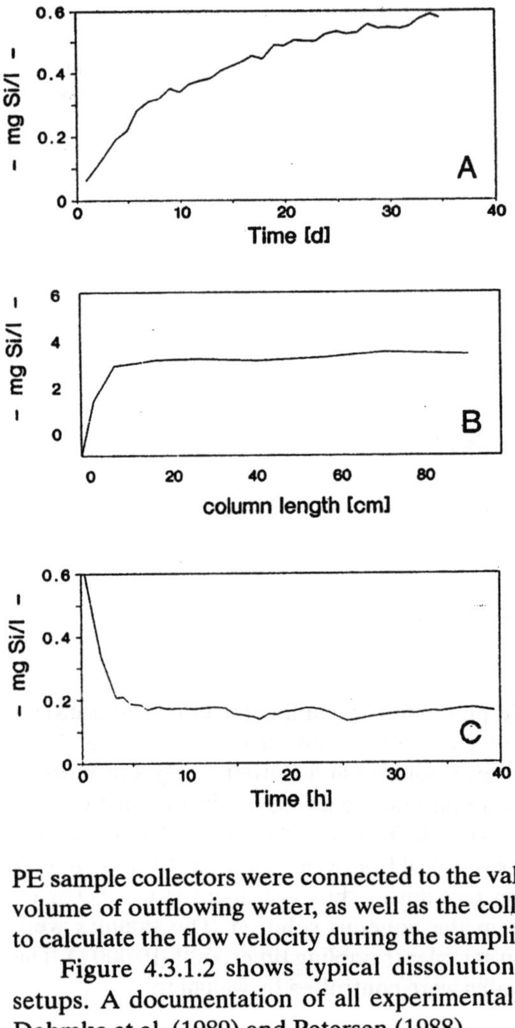

Fig. 4.3.1.2. Typical concentration vs. time curves for the three experimental setups. **A** Batch experiments. **B** Column experiments. **C** Fluidized bed reactor

PE sample collectors were connected to the valves, beginning with the last one. The volume of outflowing water, as well as the collecting time, were measured in order to calculate the flow velocity during the sampling procedure.

Figure 4.3.1.2 shows typical dissolution curves for the used experimental setups. A documentation of all experimental results is given by Dahmke (1988), Dahmke et al. (1989) and Petersen (1988).

4.3.2 Field Studies on the Kinetics of Silicate Minerals/Water Interaction

G. Matthess, A. Petersen, D. Schenk and A. Dahmke

4.3.2.1 Location

The evaluation of kinetic data of weathering processes in glacial outwash sediments is based on field studies, performed by the Institute of Geology and Paleontology,

Kiel University, in the Segeberger Forst (SEG), 40 km north of Hamburg (Fig. 1.1, Table 1.1) (Pekdeger 1977; Albertsen et al. 1980; Holthusen 1981; Ohse 1983; Ohse et al. 1984).

4.3.2.2 Analytical Method

The weathering processes and rates in the investigated area were studied in the profiles of three bore holes (SEG1–1, SEG1–2, SEG2; Fig. 4.3.2.1) by calculating specific fluxes of dissolved material through the boundaries between the horizons. This is obtained as a product of water fluxes determined with classical hydrological tools and the solute content of pore solutions from different depths. The rock samples were taken with an Eijkelkamp drilling rig. Pore solutions were obtained by centrifugating the core material in a cooled centrifuge at a temperature of 10 °C and stored after conservation with HCl or trichlorotrifluoroethane (TTE). The pore waters were analyzed with an ICP-AES (Perkin-Elmer, model ICP/6000) for cations and with HPLC (Dionex 2110i) for anions. The binding forms of the cations were determined with a sequential leaching procedure using solvents of increasing strength (Table 4.3.2.1). This method gives semi-quantitative information on the accessibility of substances within the solid phase for weathering processes which cannot be achieved by chemical bulk analyses only. Details of the drilling, sampling and preparation of the samples, and chemical data are given in Brend'amour (1986).

4.3.2.3 Geology

The Quaternary (Saalian) glacial outwash sediments of this forest area are primarily determined in their mineralogical composition by the crystalline glacial drift originating from Scandinavia. The glaciers also took up carbonate materials (bryozoa) of

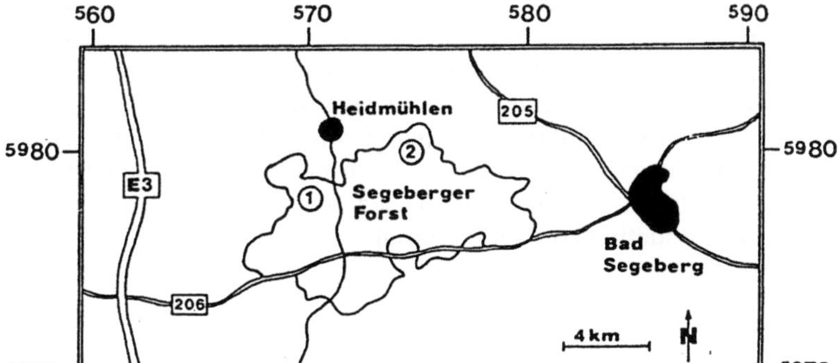

① Investigation area with boreholes SEG1 and SEG 3-11

② Borehole SEG 2

Fig. 4.3.2.1. Location of field site

Table 4.3.2.1. Methods for binding form analyses (sequential method)

Extraction step	Reagents
Water-soluble, exchangeable	In 1M ammonium acetate (pH 7)- $SrCl_2$ (pH 8.1)
Moderate acids, carbonatic	Monochlorine acetic acid (20%)
Reduction (1st step)	0.1 M Hydroxolamine-hydrochlorid
Reduction (2nd step)	Oxalic-acid/ammonium oxalate (pH 3.25)
Reduction (3rd step)	0.3 M Sodium citrate 1 M $NaHCO_3$, Sodium-dithionite
Strong acids	HCl (30%)

Cretaceous age, montmorillonite of Tertiary age (Eocene) and preSaalian Pleistocene deposits. Hence the nature of the deposits varies depending on the amount of admixed material of Cretaceous, Tertiary and Pleistocene origin and on the state of their degradation.

The Saalian sands form a flat hill rising 15–35 m above the surrounding sandy plain of Weichselian age. The water table is encountered between 7 and 10 m below ground level (SEG1–1 and SEG1–2), but the distance to the water table in some topographically elevated localities may reach about 30 m. For example, the borehole SEG3 at Radesforder Berg site penetrated 32 m of Saalian glacial outwash sands intercalated with three thin till layers at depths of 9–10 m, 24–25 m and 27–29 m below ground level. The water table was encountered at 31.2 m depth. The seepage velocities, calculated from the vertical distribution of tritium, Na^+ and Cl^- (Eichinger and Schulz 1984) range from 0.5 to 0.6 m/a.

4.3.2.4 Petrology

Major sand components are quartz (86% vol.), K-feldspar 4% vol.), plagioclase (3% vol.) and mica (2% vol.). The rest (5% vol.) consists of heavy minerals such as goethite, pyroxene, amphibole, cerussite, smithsonite, ilmenite, rutile and zircon, carbonate materials (bryozoa) and some undefined organic material and amorphous phases. A bulk X-ray analysis of the non-carbonate material indicates a predominance of sodium (Na_2O 0.59% wt.) over calcium (CaO 0.24% wt.). In the upper part of the profile (1–9 m) illite dominates among the clay minerals (Fig. 4.3.2.2). From 2 to 18 m depth montmorillonite increases, while illite and kaolinite decrease. Below 20 m montmorillonite decreases, while illite dominates again and kaolinite shows a small increase. Down to 15 m below ground level the sediments are decalcified. From 17 m to 31 m the sediments contain a few percent calcium carbonate and undefined organic material.

4.3.2.5 Geochemistry

The pore solutions close to ground level have total dissolved solid concentrations between 100 and 200 mg/l, whereas those from 20–30 m depth are characterized by distinctly higher solute contents (250–500 mg/l) due to increased Ca^{2+}, HCO_3^-, SiO_2, Cl^-, Na^+ and K^+ concentrations.

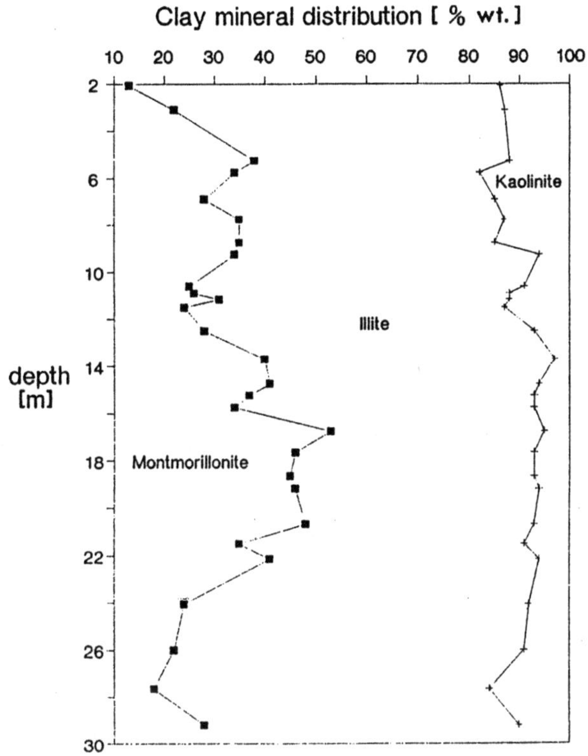

Fig. 4.3.2.2. Depth profile of clay mineral distribution

The silica concentrations of the pore water increase continuously from the Earth's surface to a depth of 3 m below surface. The silica concentration decreases slightly with depth either by precipitation or – under the influence of the ground water – by dilution. In correlation with the decalcification depth, the pH rises from 5 to 8. The ion exchange capacities range from 1.5 to 2.0 meq/100 g. The amount of sorbed calcium rises generally at the calcite dissolution depth beginning a few metres above this reference level (Fig. 4.3.2.3).

The evaluation of the results of the binding form analyses for relevant cations has to keep in mind that a small part of the primary minerals may also be dissolved within the different leaching steps. However, a general tendency of decreasing amounts of weakly bound cations versus depth can be detected (Fig. 4.3.2.4). These amounts of easily accessible components, e.g. aluminium, in the solid phase of a given volume element are always larger than the total amount of dissolved substance in the aqueous phase of the same volume element.

The CO_2 concentrations of ground air were measured at different times of the year using Draeger test tubes. Below 10 m depth the CO_2 concentration of about 1.1% is fairly constant throughout the year (Albertsen et al. 1980).

For calculations of the total amount of ions – dissolved and in solid form – the density, seepage velocity and soil moisture were taken from Albertsen et al. (1980), Ohse (1983) and Eichinger and Schulz (1984). These data, which are the base for further calculations, are given in Table 4.3.2.2.

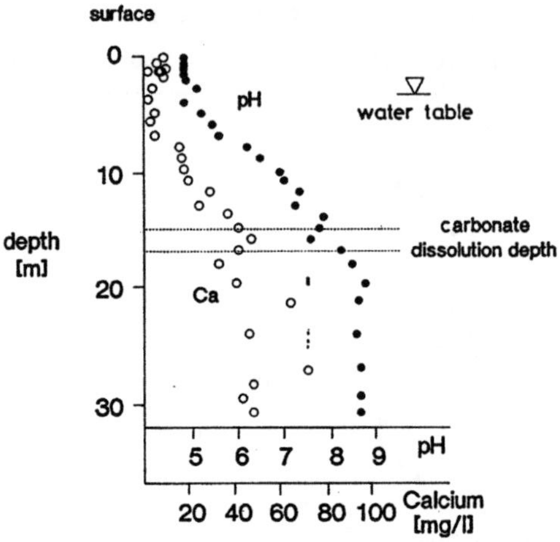

Fig. 4.3.2.3. pH and amount of exchangeable Ca^{2+} vs. depth. (Ohse 1983; Schulz and Baumann 1985)

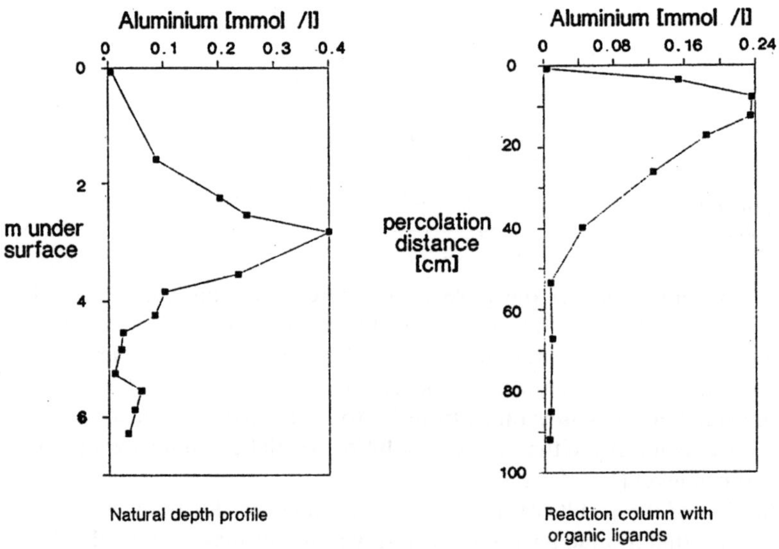

Fig. 4.3.2.4. Al concentration in reaction column compared with a natural weathering profile

Table 4.3.2.2. Sedimentological and hydrological parameters for mass balances

Density	1.7 g/cm^3
Porosity	0.37
Specific surface	400 cm^2/g
Feldspar content	5.3–10.4% vol.
Seepage velocity	200 mm/a
Precipitation	700–750 mm/a
Soil moisture	3–10% wt.

4.3.2.6 Thermodynamic Calculations

The thermodynamic speciation and saturation calculations with the computer programme WATEQF (Plummer et al. 1976) and PHREEQE (Parkhurst et al. 1980) indicate calcite undersaturation down to 17 m and quartz saturation from 3 m down to calcite dissolution depth. The pore solutions from deeper horizons are slightly oversaturated with respect to quartz, and generally undersaturated with respect to amorphous SiO_2 with decreasing intensity towards depth. The feldspars and some primary layered silicates are unstable under near-surface conditions, and dissolution and transformation to clay minerals and amorphous substances may take place. The pore solutions are undersaturated with respect to K-mica (muscovite) down to 6–7 m below ground level. Physical weathering of mica in connection with potassium release and inclusion of hydrated cations in peripheral parts of the sheets would favour the formation of hydromuscovite and illite. However, the calculations show, that illite is also unstable in the upper part of the profile. Kaolinite proves to be the thermodynamically most stable phase over the whole profile (Ohse et al. 1984).

Consistent with the thermodynamic predictions the SEM investigations show a general decreasing weathering alteration of primary mineral surfaces with increasing depth. However, in detail these model predictions do not cover all observed changes in the upper part of the water/rock system. The deviations from theory are caused by kinetic effects and by the formation of secondary minerals such as clay minerals and undefined amorphous phases, which are not adequately considered in the models used. Thermodynamic data, suitable for the prediction of dissolution and precipitation behaviour of secondary phases are rare and do not correspond to non-perfect, non-ideal natural minerals with undefined surface energies. In the upper part of the profile amorphous Si-Al phases and organic layers cover the mineral grains to some extent, thus diminishing direct contacts between pore solutions and silicate grains. The existence of organic and inorganic layers (coatings) on mineral grains observed by SEM investigations may lead to incorrect saturation indices of pore solutions and to reduced weathering rates. The thermodynamic models furthermore do not take into account the existence of organic compounds which may influence the equilibrium state (Sect. 2.2.5, this Vol.). The described corrosion and precipitation phenomena have developed in situ since the last glaciation (about 10 000 a). Formation of clay mineral zones could not be detected with certainty.

4.3.2.7 Weathering Rates Derived from Field Studies

The mass balances, calculated by Brend'amour (1986) from the transport rates of the main cations, generally show positive values indicating a permanent removal of dissolved rock components with the ground water outflow from the investigated area. The sodium content of the pore solution originates from the input by precipitation and from the release from feldspars and clay minerals in the sander. According to Schulz (1972), the ratio $[Na^+]/[Cl^-]$ in the local precipitation is nearly constant. Therefore it should be possible to determine the contribution of weathering of silicates to the total sodium content of the pore solution by subtracting the input of

Na^+ due to the atmospheric precipitation. Assuming no other sources for Cl^- (Schulz 1972) the sodium input can be calculated from the Cl^- concentration of the pore solution, when the average Na^+/Cl^- ratio in the rain water is known. Using this approach for an estimated Na/Cl ratio of 1 for pure NaCl or of 0.85 for sea water, negative differences of sodium concentrations have been calculated. It could not be clarified whether these negative values are results of an over-estimation of Na^+ input with rain or of any unknown sinks for sodium in the system, which would result in a smaller ratio. In comparison to the sodium input by precipitation the amount of sodium released from silicates is relatively small. No statistically significant separation between the rain water input of sodium and the enrichment due to weathering could be achieved for the profiles SEG1–2 and SEG2. Only profile SEG1–1 gives an overall sodium release rate in the order of 10^{-17} mmol/(g · s) as a mean value for the whole profile.

Albertsen et al. (1980) calculated sodium depletion rates for the whole Segeberger Forst with input/output comparisons to be in range of 54 000 kg/a Na^+. From the rock volume of 1.8×10^9 m^3 assuming a density of 1.7 t/m^3 a Na-dissolution rate of 2.5×10^{-15} mol/(g · s) can be calculated as a mean value for the whole area. This calculation does not differentiate between saturated and unsaturated zones and neglect the decreasing weathering intensity with depth.

As the dissolution rates in the laboratory experiments are calculated from Si concentrations, a direct comparison with Si dissolution rates from field studies is a major task. However, Si concentration in the pore water are controlled by the solubility of both amorphous SiO_2 and quartz. The thermodynamic calculations show that the pore solution 3 m depth below ground surface downwards is in equilibrium with both solid phases. It is therefore impossible to calculate rates from the variation of concentration in this part of the profile. Only above 3 m are the pore solutions undersaturated with respect to these species and the observed gradient enables a preliminary calculation of the silica release rates.

The Al concentration is also controlled by reversible dissolution-precipitation reactions with amorphous alumina or alumosilicates. Positive mass balances for the given volume elements can only be used for calculating dissolution rates when the Al concentration under the given pH value is far away from saturation or precipitation of Al is prevented by complexation with organic ligands. This apparently holds only for pore solutions close to the Earth's surface. According to thermodynamic calculations, 80–95% of the dissolved aluminium must be complexed by organic ligands. The relatively unstable polymeric organo-complexes are destroyed very fast if the supply of free organic ligands decreases with increasing distance from the soil in which the organic substances are mainly produced. This can be seen in the decrease of the aluminium concentration downwards from 3 m below surface (Fig. 4.3.2.4; Petersen 1988).

The dissolution rate calculated from the aluminium concentrations in the pore solutions down to 2.8 m below surface (SEG2) [1.3×10^{-15} mol/(s · cm^2)] differs only slightly from those derived from silica concentrations. This indicates a nearly congruent dissolution process which is consistent to laboratory results (Sect. 4.3.1).

Weathering rates based on Al concentrations can be calculated furthermore by the analyses of the binding forms. The input of Al into the system by rainfall is neglectable. Its outflow from the profile below the layers with notable C_{org} content cannot be larger than the average water flow rate times the maximum Al solubility under the given pH value. Much higher amounts of Al remain in the system as secondary Al phases, which can be detected by the binding form analyses and normalised to total amounts per volume element a rate of 4.9×10^{-16} mol/(g · s) Al is calculated assuming a weathering time of 10 000 years, which is realistic and consistent with local geological history. During this time the upper part of the profile was decalcified. This process was connected with a lowering of the pH values, and with an increase of weathering rates and of Al leaching rates due to increased Al solubilities.

4.3.2.8 Comparison Between Field and Laboratory Data

The comparison of experimental and field data is biassed by the assumption that the observed aluminium and silica release in the field is derived exclusively from the reactive surfaces of feldspars, whereas the calculations neglect the amorphous and microcrystalline silica phases of the outwash material and the presence of complex forming organic substances because their thermodynamical properties are not sufficiently known.

A better comparison of the field data with the laboratory rates takes into account that the outwash material consists only of 7% vol. feldspar. According to Dahmke (1988), the average specific surface of the outwash sand is about 400 cm^2/g. If a stoichiometric factor of 2.78 and 1, respectively, is used to convert Si and Al dissolution rates into feldspar dissolution rates, the calculated dissolution rates (Table 4.3.3.3) agree reasonably well with those derived from experiments considering the simplified assumptions in the evaluation and the uncertainties of the measurements (Dahmke 1988; Petersen 1988).

The results from column experiments with organic ligand solutions support the field observations that between pH 4 and 8 the dissolution behaviour and maximum solubility of aluminium in natural systems depends mainly on the physico-chemical properties and the transport behaviour of the organic ligands (Fig. 4.3.2.4).

Table 4.3.2.3. Weathering rates derived from field studies and laboratory experiments

Field data

Method	Element Release rate mol/(g·s)	Calculated feldspar dissolution rate mol/(g·s)
Na/Cl balance	1.0×10^{-17} Na	–
Total Na balance	2.5×10^{-15} Na	–
Si concentration profile (–1.5 m)	1.1×10^{-11} Si	1.4×10^{-13}
Al concentration profile (–1.5 m)	1.3×10^{-15} Al	4.6×10^{-17}
Al accumulation	4.9×10^{-16} Al	1.7×10^{-17}

Experimental data (Sects. 2.9 and 4.3.1, this Vol.)

Dissolution rates at different pH and different solvents corrected for feldspar dissolution rates		Dissolution rates for the experimental variations (calculated for Si-release, not for feldspar dissolution rate)				
	250 to 800 h lin. regression [mol/(s·cm^2)]	pH value	Grain size (mm)	Si-release rate [mol/(cm^2 s)]		
Albite		*Fixed pH value*				
Aqua dest.				Anorthite	Albite	Orthoclase
pH 4.5	$4.5 \times 10^{-16} \pm 4 \times 10^{-17}$					
pH 6.0	$1.5 \times 10^{-16} \pm 1 \times 10^{-17}$	4	0.063–0.1	4.0×10^{-15} $\pm 1.6 \times 10^{-16}$	3.0×10^{-15} $\pm 2.8 \times 10^{-16}$	2.3×10^{-15} $\pm 1.8 \times 10^{-16}$

Table 4.3.2.3. (*Continued*)

	250 to 800 h lin. regression [mol/(s·cm^2)]	pH value	Grain size (mm)	Si-release rate [mol/(cm^2 s)]		
				Anorthite	Albite	Orthoclase
Cit						
pH 4.5	$6.6 \times 10^{-16} \pm 5 \times 10^{-17}$	6	0.063–0.1	2.7×10^{-15} $\pm 3.3 \times 10^{-16}$	3.0×10^{-15} $\pm 2.8 \times 10^{-16}$	2.1×10^{-15} $\pm 3.1 \times 10^{-16}$
pH 6.0	$3.9 \times 10^{-16} \pm 1 \times 10^{-16}$					
VPy						
pH 4.5	$5.2 \times 10^{-16} \pm 2 \times 10^{-17}$	8	0.063–0.1	2.3×10^{-15} $\pm 2.2 \times 10^{-16}$	3.1×10^{-15} $\pm 4.3 \times 10^{-16}$	2.0×10^{-15} $\pm 4.2 \times 10^{-16}$
pH 6.0	$1.7 \times 10^{-16} \pm 3 \times 10^{-17}$					
		10	0.063–0.1	5.4×10^{-16} $\pm 1.2 \times 10^{-16}$	2.2×10^{-16} $\pm 2.4 \times 10^{-16}$	1.1×10^{-15} $\pm 3.4 \times 10^{-16}$
Orthoclase						
Aqua dest.		*Free pH value*				
pH 4.5	$2.7 \times 10^{-16} \pm 3 \times 10^{-17}$					
pH 6.0	$1.2 \times 10^{-16} \pm 3 \times 10^{-17}$			Anorthite	Albite	Orthoclase
Tart		7.5–8.5	<0.063	1.6×10^{-15} $\pm 1.4 \times 10^{-16}$	7.9×10^{-16} $\pm 7.1 \times 10^{-17}$	1.3×10^{-15} $\pm 7.5 \times 10^{-17}$
pH 4.5	$2.7 \times 10^{-16} \pm 3 \times 10^{-17}$					
pH 6.0	$2.7 \times 10^{-16} \pm 3 \times 10^{-17}$					
		7.5–8.5	0.063–0.1	2.2×10^{-15} $\pm 1.7 \times 10^{-16}$	1.2×10^{-15} $\pm 1.0 \times 10^{-16}$	1.5×10^{-15} $\pm 6.9 \times 10^{-17}$
Cit						
pH 4.5	$6.9 \times 10^{-16} \pm 3 \times 10^{-17}$					
pH 6.0	n.d.	7.5–8.5	0.2–0.4	9.3×10^{-16} $\pm 1.8 \times 10^{-16}$	4.2×10^{-16} $\pm 1.3 \times 10^{-16}$	1.2×10^{-15} $\pm 9.9 \times 10^{-17}$
VPy						
pH 4.5	$5.2 \times 10^{-16} \pm 1 \times 10^{-17}$	7.5–8.5	0.8–2.0	5.5×10^{-15} $\pm 6.1 \times 10^{-16}$	4.3×10^{-15} $\pm 6.1 \times 10^{-16}$	4.0×10^{-15} $\pm 5.3 \times 10^{-16}$
pH 6.0	$1.4 \times 10^{-16} \pm 3 \times 10^{-17}$					

5 Microbiology

5.0 Introduction

P. Hirsch

Microorganisms are present nearly everywhere, probably due to their small size as well as to their great metabolic diversity. Surprisingly, knowledge of the pristine ground water environment was almost completely lacking until the last decade. Vague concepts such as the microbial "self-purification" ability of ground water were not supported by detailed and systematic studies. Interest was mainly for practical purposes, such as in the study of the microflora of oil fields.

Wolters and Schwartz (1956) carried out the first systematic investigation of the microflora of uncontaminated samples of the saturated zone. They were followed by Kuznetsova (1961), Kramarenko (1962), Farkasdi et al. (1969) and Golwer et al. (1972). The early 1980's saw a rapid increase of interest in natural ground water organisms: Fuksa (1981), Marxsen (1981, 1982), Hoos and Schweisfurth (1982), Ladd et al. (1982), Ehrlich et al. (1983), Ventullo et al. (1983), Wallis and Ladd (1983), Wilson et al. (1983), Balkwill and Ghiorse (1985), and Selenka et al. (1986), to name only a few.

While some researchers were more interested in special microbial groups of deeper aquifers (Olson et al. 1981), others were fascinated by the microbial diversity in underground samples (Ghiorse and Balkwill 1982; Hirsch and Rades-Rohkohl 1983a,b, 1988). Sterile sampling methods had to be developed (Hirsch and Rades-Rohkohl 1983b). A recent review on the microbial ecology of the terrestrial subsurface deals with all such questions (Ghiorse and Wilson 1988).

When the research goals were formulated for this DFG priority programme, it was clear from the very beginning that microorganisms play an important role in many of the processes to be studied here. At this time information on numbers, types and distribution of aquifer microorganisms was almost lacking. This state changed rapidly due to a strong cooperation between hydrogeologists, geochemists and microbiologists, which is reflected in the data of this section as well as in other sections of this Volume.

The research projects described in Chapter 5 were partially focussed on the microbiology of specific sites, but partially also concerned some general problems of subsurface microbiology.

A literature review on morphological diversity in ground water samples had revealed 24 different genera (30 species names). They are documented combined with the results obtained in this priority programme by Hirsch et al. (Sect. 5.1.1). Indeed, the ground water environment contained a rich and very diverse, mostly oligotrophic microfloa: Gram-negative bacteria were mainly found in the free aquifer water, while the sediment contained more Gram-positive ones. Eukaryotes, (filamentous fungi, yeasts, protozoa) were exclusively found in the upper 10 m. Tremendous variations occurred due to local conditions.

In a critical evaluation of methods employed in ground water microbiology, data from four laboratories (three locations) were combined and recommendations formulated (Hirsch et al., Sect. 5.1.2). Most microbiological work should be initiated at the sample site to avoid changes of numbers and diversity of organisms in the samples. Thus, facilities to carry out these experiments should be available next to the boring sites.

The types and concentrations of organic nutrients in the aquifer environment were studied by Selenka and Hack (Sect. 5.1.3) using new sampling procedures to avoid contaminating the sample equipment. From bacterial counts and measurements of the average cell volumes it could be concluded that, in their case, only 0.1% of the total organic carbon (TOC) in ground water, and 0.01% of TOC in the sediment could have been due to microbial biomass. Humic acid fractions definitely must have contributed to the microorganisms' nutrition.

The review of physiological properties of ground water microorganisms by Hirsch (Sect. 5.2.1) and based on a previous description for the Bocholt and Segeberger Forst test sites (Kölbel-Boelke and Hirsch 1989) considers the distribution of "physiotypes" between the aquifer water and the particulate sediment. It indicates a physiologically diverse microflora of the saturated zone environment: among the ground water isolates there were always some strains that could utilize (in vitro) at least one of the many carbon sources tested. The surprising independence of ground water isolates from growth temperatures is explained by the constant ground water temperature regime.

Gottfreund and Schweisfurth† (Sect. 5.2.2) concerned themselves with the bacterial influence on manganese oxidation states, using an *Arthrobacter* sp. isolate. Organic substances or chelating agents decreased the rate of Mn (II) oxidation, while phosphate inhibited it completely without interfering with growth in the presence of a suitable C-source.

Tuschewitzki et al. (Sect. 5.2.3) studied microbial utilization of humic substances in laboratory batch cultures or in fixed-bed reactors with continuous addition of humic acids. Batch cultures did not allow growth of the test strains with humic acids, but the availability of phosphorus or nitrogen was reduced in their presence. Humic substances stimulated growth of test bacteria in fixed-bed reactors provided with glass beads to stimulate the quartz grains of the subsurface sediment.

The distribution and numbers of different physiological groups in the sediments of a profile in the Sandhausen forest ecosystem (Weyandt and Schweisfurth, Sect. 5.3.1) demonstrate the inhomogeneity of the subsurface aquifer system. Microhabitats totally different were assumed to be situated adjacent to each other. A sample consisting of many different microhabitats will thus only render average data. These authors also determined in vitro, "activities" of ground water microorganisms, employing fluoresceine-labelled substrate analogues.

The distribution of physiological types in the Bocholt aquifer system was studied by Kölbel-Boelke and Nehrkorn (Sect. 5.3.2) employing miniaturized test systems and statistical data analyses. A great diversity was found with profound differences between neighbouring wells as well as within one well and at different depths. Likewise, the determination of in vitro activities demonstrated this diversity.

The investigations of the Segeberger Forst aquifer system included initial studies on sampling methods and optimal sample treatment (Hirsch and Rades-Rohkohl 1988), on the microorganisms' interactions with surfaces and with each other (Hirsch and Rades-Rohkohl 1983b) and on the distribution of microorganisms between the free aquifer water and the solid sediment phase (Kölbel-Boelke and Hirsch 1989). Work on morphological and physiological properties demonstrated, just as in the Bocholt aquifer system study, the great diversity of microorganisms (Hirsch and Rades-Rohkohl, Sect. 5.3.3).

Finally, a very preliminary investigation of the unsaturated zone of a Quaternary gravel aquifer is presented by Alexander et al. (Sect. 5.3.4). Experiments in coarse limestone gravel were especially difficult to carry out.

All of these microbiological investigations have demonstrated a dire need for more information on microbial activity in the microenvironment. New experimental techniques will have to be designed to study the individual bacterium (or a colony) down within the aquifer/sediment environment to understand what the organism really does or can do in situ (Hirsch and Rades-Rohkohl 1990). An illustration (Fig. 5.0.1) depicts how individual microorganisms probably live within their microenvironment. It demonstrates the close proximity of aerobic and anaerobic locations, and the presence in free water as well as on mineral or humus surfaces.

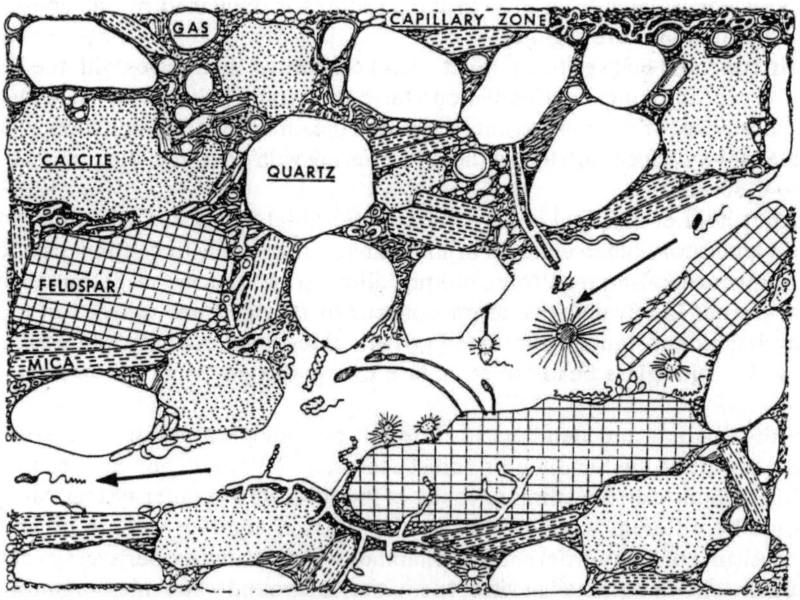

Fig. 5.0.1. Conceptual drawing of the microbial environment in a porous aquifer. All microorganisms shown have been found in water or sediment samples. The picture demonstrates the diversity of microhabitats, often with aerobic and anaerobic conditions right next to each other. For better recognition the microorganism sizes have been increased

Although the dimensions may not be correct, one could imagine, in this way, what a microbe's subsurface life might be like.

5.1 Characterization of the Natural Subsurface Environment

5.1.1 Morphological and Taxonomic Diversity of Ground Water Microorganisms

P. Hirsch, E. Rades-Rohkohl, J. Kölbel-Boelke and A. Nehrkorn

5.1.1.1 The Morphological Approach

There are several approaches to studying and describing morphological and species diversities of ground water microorganisms. One could concentrate on different habitats such as free ground water, the water present in wells, the interstitial water running off from sediment samples of the saturated zone, the sediment itself or the locations where organisms live attached to the well's walls. The water in the wells is often even covered by microbial neuston. Also, it can be expected that different depths in the aquifer might have different microbial assemblages (Kölbel-Boelke 1988). Obviously, the methods of study will reveal different microorganisms: one could employ direct microscopic techniques on the samples using light phase contrast microscopy or electron microscopy. A morphological study of pure culture isolates grown on different media will result in additional information. A more recent approach centres on direct chemical analysis of samples to find species- or group-specific indicator substances (White et al. 1983).

Another approach would be to look for ecologically relevant properties: attachment, motility, storage of reserve materials, pigmentation, spore formation, aggregation, formation of polymer, predatory activity, etc.

Finally, one could be interested in the taxonomical or phylogenetic position of ground water organisms and their presence in various subsurface habitats: where are viruses, bacteria, fungi, protozoa, etc? A complete description of the ground water microbiota should include all of these approaches. Unfortunately, such complete information may not be possible because of the system's complexity.

Wolters and Schwartz (1956) first devoted their interest to ground water samples from pristine environments. They cultivated microorganisms from well water, using especially developed (but still fairly rich) media. Bacteria were isolated, and attempts were made to determine their taxonomic position. Well water was also studied by Hässelbarth and Lüdemann (1967), Sládecková (1972) and Lüpkes (1974), who all employed microscopy of water samples only. Isolation of 500 pure cultures from continuously pumping deep wells was reported by Stetzenbach et al. (1986); these authors published a few genus names. Well water was also investigated by Cullimore and McCann (1977) who were mostly interested in iron- and manganese-oxidizing bacteria. Cultivation of most observed forms was reported.

Morphological diversity of ground water organisms was studied in greater detail by Hirsch and Rades-Rohkohl (1983a,b, 1988b), who observed more than 100 different "morphotypes" (MT) in their ground water and sediment samples. Descriptions of the fine structure of sediment organisms were published by Ghiorse and Balkwill (1983), and Balkwill and Ghiorse (1985), who also noted differences in colonial morphologies. Bone and Balkwill (1988) gave a more detailed account of the distribution of colony types in various depths of soils as well as of unsaturated or saturated subsurface sediments. They found that the total number of different colony types decreased with depth when medium PTYG or a 5% dilution of it were used. This medium contained, per liter, 5 g each of peptone and trypticase (BBL), 10 g each of yeast extract and glucose, as well as 0.6 g $MgSO_4 \cdot 7H_2O$ and 0.07 g $CaCl_2 \cdot 2H_2O$; agar-15 g. Samples collected in January 1985 contained, throughout the profile, 36 different colony types: at 0.05 m depth there were 22 types, at 0.2 and 1.3 m 10 types and at 7.6 m only 4. In the saturated sediment they did not find streptomycetes or eukaryotes, nor were there endospore-forming rods or non-branching filaments. These authors also applied direct microscopic techniques such as the method of Waid (1973). Single cocci and small straight rods were found in all sediment samples of the saturated zone. Coccoid rods were missing from the lowest zone (8.4 m), and large cocci were observed only in a sediment zone rich in gravel (7.6 m). Below the water table (2.8 m) they observed only three to five different MT per sample. Using soil extract agar (SSA) which lacked the addition of organic carbon other than was present in the 100 ml/l of soil extract, these authors observed a much larger colony diversity. This points to the fact that a large proportion of ground water microorganisms can be considered to be oligotrophic.

A thorough study of microbial diversity in deep aquifers (25 °C) was carried out by Balkwill (1989). A large number of samples collected from three sites down to 265 m depth were processed. Most samples contained on PTYG agar (or 1% PTYG agar) 11–40 different colony types. Diversity varied from one depth to another but did not decrease with depth. Of more than 3500 isolates a total of 1200 strains were characterized morphologically. Of these, 1.6% were fungi, 3.5% resembled streptomycetes, and the other 94.9% were mainly rods (81%) but also pleomorphic (9%), cocci or short rods (9%), or filaments (1%). Of these, 86% were Gram-negative. Sediments from permeable aquifer zones had more morphological diversity than sediments from zones high in clay. The presence of various microbial groups in deep aquifers has been registered by Sinclair and Ghiorse (1989).

General discussions of these matters can be found in Ghiorse and Wilson (1988) or Albrechtsen (1988). Table 5.1.1.1 summarizes observations on the reported presence of microorganism types or genera in the ground water environment.

The present paper describes observations made on general morphology and diversity of ground water and sediment microorganisms found at the Bocholt (BOC) and Segeberger Forst (SEG) aquifer systems. Parts of this work have been published previously (Hirsch and Rades-Rohkohl 1983a,b, 1988b; Kölbel-Boelke 1988; Kölbel-Boelke et al. 1988a).

Table 5.1.1.1. Microorganisms observed in ground water and sediment: data from literature

Prokaryotes	Reference[a]	Prokaryotes	Reference[a]
Achromobacter sp.	12,26	Manganese oxidizers	14
A. cycloclastes	26	Manganese reducers	14
A. eurydice	26	*Methanobacterium* sp.	2
A. iophagum	26	*M. thermoautotrophicum*	20
A. liquefaciens	26	Methanogenic bacteria	20
A. superficiale	26	*Mathanosarcina* sp.	2
Acinetobacter sp.	12,24	*Micrococcus* sp.	4,8,26
A. calcoaceticus	8,10	*M. conglomeratus*	26
A. lwoffii	17	*M. luteus*	8
Actinomyces sp.	8	*M. roseus*	8
Aerococcus sp.	8	*M. varians*	6
A. viridans	8	*Micromonospora* sp.	8
Aeromonas sp.	4,8,17	*Moraxella* sp.	8,10,12,24
Alcaligenes sp.	8,9	*Naumanniella* sp.	7
A. denitrificans	9	Nitrate-ammonifyers	14
Arthrobacter sp.	1,8	Nitrifiers	11
A. globiformis	8	*Nocardia* sp.	8,26
A. flavens	8	*Ochrobium* sp.	7
Bacillus sp.	4,6,12	Oligotrophs	11
B. cereus	17	*Pasteurella* sp.	10
B. licheniformis	17	*Planococcus citreus*	8
B. megaterium	17	Pseudomonas sp.	1,4,6,8,9,12,21
B. subtilis	17	*P. acidovorans*	4,8
Bacterium sp.	26	*P. alcaligenes*	4,8,24
B. fulvum	26	*P. aeruginosa*	9
Beggiatoa sp.	18	*P. cepacia*	4,8,10
B. alba	23	*P. diminuta*	4,8,10,17
B. minima	23	*P. fluorescens*	4,6,8,10,17,21
Branching filaments	1	*P. mallei*	8
Brevibacterium sp.	1	*P. maltophilia*	4
Caulobacter sp.	12	*P. mendocina*	6
Cellulomonas sp.	8	*P. pseudoalcaligenes*	4,10
Chromobacterium sp.	1,8	*P. pseudomallei*	4
C. lividum	6	*P. putida*	9,17
C. typhiflavum	10	*P. putrefaciens*	4,8
Clonothrix sp.	7	*P. stutzeri*	4,9
Cocci	5,11	*P. testosteroni*	4
Coliforms	4	*P. thomassii*	4
Coryneforms	4	*P. vesiculare*	4
Crenothrix sp.	7	*Rhodococcus rhodochrous*	15
Denitrifiers	14	*R. ruber*	15
Desulfotomaculum sp.	2	Rods	5,11
D. nigrificans	20	*Siderocapsa* sp.	7,13
Flavobacterium sp.	8,17,26	*Siderococcus* sp.	13
F. aquatile	8,26	*Sphaerotilus* sp.	7
F. breve	8	*S. natans*	23
F. devorans	8,26	*Staphylococcus* sp.	4,8
F. sevanense	26	*Streptomyces* sp.	8
Gallionella ferruginea	13,23	Sulfur oxidizers	14
G. minor	13	Sulfate reducers	14,20,25
Halobacteroides acetoethylicus	22	*Thiobac. denitrificans*	16
Hydrocarbon oxidizers	3	*T. ferrooxidans*	7,16
Hyphomicrobium sp.	8	*T. thiooxidans*	16

Table 5.1.1.1. (*Continued*)

Prokaryotes	Reference[a]	Prokaryotes	Reference[a]
Iron precipitating bacteria	14	*T. thioparus*	16
Iron reducers	14	*Thiothrix tenuis*	23
Leptothrix sp.	7,12,13	*Toxothrix trichogenes*	13,23
L. ochracea	13,23	*Xanthomonas* sp.	8
Eukaryotes	**Reference[a]**	**Eukaryotes**	**Reference[a]**
Fungi (mycelial)	14,23	*Epistylis subterranea*	19
Fungi (yeasts)	14,18	*Cothurnia ruthae*	19
Flagellates (colourless)	23	*Scyphidia husmannii*	19

[a]1 Balkwill and Ghiorse (1985);
2 Beeman and Suflita (1987);
3 Bogdanova (1965);
4 Bonde (1985);
5 Bone and Balkwill (1988);
6 Buchanan-Mappin et al. (1986);
7 Cullimore and McCann (1977);
8 Dott et al. (1982);
9 Ehrlich et al. (1983);
10 Gehlen et al. (1985);
11 Ghiorse and Balkwill (1983);
12 Gounot et al. (1987);
13 Hässelbarth and Lüdemann (1967);
14 Hoos and Schweisfurth (1982);
15 Ivshina et al. (1981);
16 Kramarenko (1962);
17 Ladd et al. (1982);
18 Leuchs (1988);
19 Lüpkes (1974);
20 Olson et al. (1981);
21 Parker and James (1985);
22 Rengpipat et al. (1988);
23 Sládecková (1972);
24 Stetzenbach et al. (1986);
25 van Beek and van der Kooij (1982);
26 Wolters and Schwartz (1956).

5.1.1.2 Materials and Methods

Staining reactions, microscopic preparations, and light phase contrast (LPM) or transmission electron microscopy (TEM) have all been described (Hirsch and Rades-Rohkohl 1983a,b, 1988b; Kölbel-Boelke et al. 1988b). For SEM thin slices of the inner surface of a PVC tube (sampling tube SEG8, Segeberger Forst; Fig. 5.1.1.1) were fixed immediately in 2% glutaraldehyde when the tube was pulled up after 2.5 years of exposure. Dehydration, critical point drying and gold-plating followed standard procedures. The preparations were viewed at 20 kV with a Cambridge SEM.

5.1.1.3 Results

Observations at the Bocholt Aquifer System

A total of 2700 isolates were obtained by statistical methods from 60 sediment and ground water samples (Kölbel-Boelke et al. 1988a). Of these nearly 72% were Gram-negative and the rest stained Gram-positive or -variable. Major representatives were Gram-negative, straight rods (52.2%), Gram-positive coryneform or

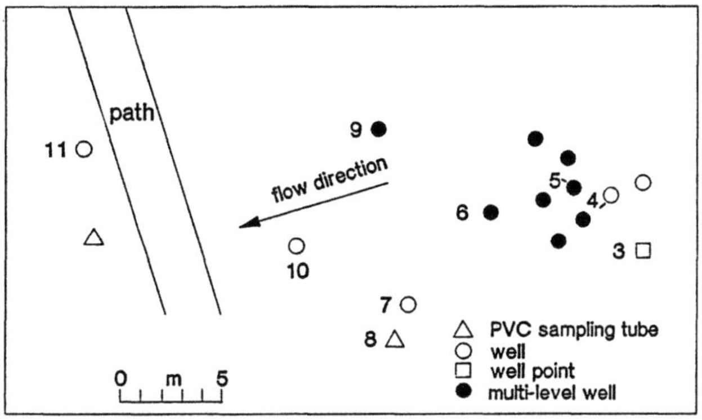

Fig. 5.1.1.1. Wells at experimental site Segeberger Forst

Arthrobacter-like bacteria (17.2%), Gram-negative organisms of variable forms (9.5%), Gram-positive coccoid rods or cocci (8.6%), and other bacteria (12.5%); gliding and filamentous forms: *Cytophaga* group, Gram-negative cocci, *Caulobacter* and *Bacillus* spp. and a few actinomycetes. More than half of the isolates were motile, mostly by flagella.

A comparison was made of the microorganisms from different bore holes and great differences in community structure were observed (Kölbel-Boelke 1988). *Caulobacter* spp. were isolated from six of the seven bore holes investigated, and gliding forms came only from 4 of these seven sites. Some of these differences might have been attributable to different constructions of the wells and filters. But a comparison of different depths from one bore hole revealed likewise striking differences in the community structure of all samples (Kölbel-Boelke 1988). When water communities were compared with those from sediment samples, differences were likewise highly significant. Most water communities had less than 11% Gram-positive organisms; their proportion in one well was as low as 5.8%. The water samples yielded more Gram-negative straight or curved rods, while sediment samples had more *Arthrobacter*-like bacteria, more *Cytophaga* types, and Gram-positive rods, usually spore formers. Motility was widely spread among water isolates (61.2%), while sediment organisms were less often motile (44.6%). Pigmentation was more common among sediment bacteria, and these showed generally more morphological variation than the water strains. It can be assumed that a higher percentage of attaching organisms would exhibit more cell polarity, induced by the presence of a surface.

Taxonomical identification on the genus level, basing exclusively on morphology, was only possible for a few isolates (*Caulobacter, Cytophaga/Flexibacter, Bacillus, Micrococcus, Staphylococcus, Streptomyces*). For the majority of isolates the additional determination of physiological characters was absolutely necessary for the classification into genera or species. Although many strains could be attached to certain genera or species, others remained unidentified (Sect. 5.3.2, this Vol.).

Observations at the Segeberger Forst Aquifer System

Phase contrast light microscopy was employed initially to study morphotype diversity in SEG samples; the methods used were described previously (Hirsch and Müller 1986). A total of 107 morphotypes could be recognized. Of these 80 were prokaryotes and 27 eukaryotes. The largest proportions of prokaryotes were cocci, rods, spore formers and caulobacters; among eukaryotes the flagellates were most frequent (Figs. 5.1.1.2 and 5.1.1.3). Several of these MT were so characteristic that they could be identified even to the genus level. Of course, confirmation of this must await further studies of physiological and other properties. Table 5.1.1.2 lists

Fig. 5.1.1.2. Ground water and sediment morphotypes shown as drawings: prokaryotes found in BOC or SEG wells

1 Cocci, < 0.5 µm;
2 cocci, > 0.5 µm, ± polymer;
3 diplococci, ± polymer;
4 cocci, tetrades w.polymer;
5 streptococci ± polymer;
6 cocci, clumping;
7 rods, short to coccoid;
8 rods, short to cubical;
9 rods, minute, Ø < 0.5 µm;
10 rods ± polymer;
11 rods w.dark granules;
12 rods w.external grains;
13 rods, light refractile;
14 rods, corn cob-shaped;
15 rods, filled w.crystals;
16 rods, capsule w.Fe-oxides;
17 rods, cells w.Fe-oxides;
18 rods, w.large irreg. capsule;
19 rods, pleomorphic;
20 rods, aggregating, w.polymer;
21 rods in chains (*Leptothrix*);
22 rods, chains in sheath;
23 filaments, long, flexible;
24 filaments, short, pointed;
25 rods, pointed, in chains;
26 semi circles (*Microcyclus*);
27 minute, bdellovibrious;
28 curved rods (*Vibrio*-like);
29 spirilla, short ± polymer;
30 spirilla, long;
31 rods, bent, bundles, polymer;
32 *Gallionella* spp.;
33 *Nevskia* spp.;
34 coryneforms ± polymer;
35 rods, tapering (rockets);
36 vibrios w.pointed ends;
37 vibrios, small, parallel;
38 *Leptospira* spp.;
39 spirochetes, large turns;
40 spirochetes, long;
41 rods, spores spherical;
42 rods, spores elongated;
43 rods, pointed, w.spores;
44 *Clostridium* spp.;
45 rods, wavey w.spores;
46 *Seliberia*, thin, long;
47 *Seliberia* w.stalks;
48 clubs, straight, attached;
49 *Planctomyces*, egg-shaped;
50 *Planctomyces limnophilus* ?;
51 *Planctomyces*, pear-shaped
52 stalked, slender, budding bacteria ("Henricia");
53 cocci, small, budding;
54 cocci > 5 µm, budding;
55 rods, short, budding;
56 *Blastobacter*, buds lateral;
57 *Blastobacter*, buds terminal;
58 rod chains, budding;
59 rods, rosettes, budding;
60 *Pirellula* spp.;
61 *Pirellula*, pointed;
62 *Prosthecobacter*, slender;
63 *Prosthecobacter*, stout;
64 *Prosthecobacter*, bacteroid;
65 *Caulobacter*, bacteroid;
66 *Caulobacter*, cells long;
67 *Caulobacter*, limonoid;
68 *Caulobacter*, bent rods;
69 *Caulobacter*, vibrioid;
70 *Caulobacter*, curved;
71 *Caulobacter*, stalks twisted;
72 *Thiodendron* spp.;
73 *Stella* spp.;
74 *Prosthecomicrobium* spp.;
75 *Hyphomicrobium* spp.;
76 *Hyphomonas* spp.;
77 *Mycobacterium*-like rods;
78 *Actinomyces*, short branches;
79 *Nocardia* spp.;
80 actinomycete, highly branched

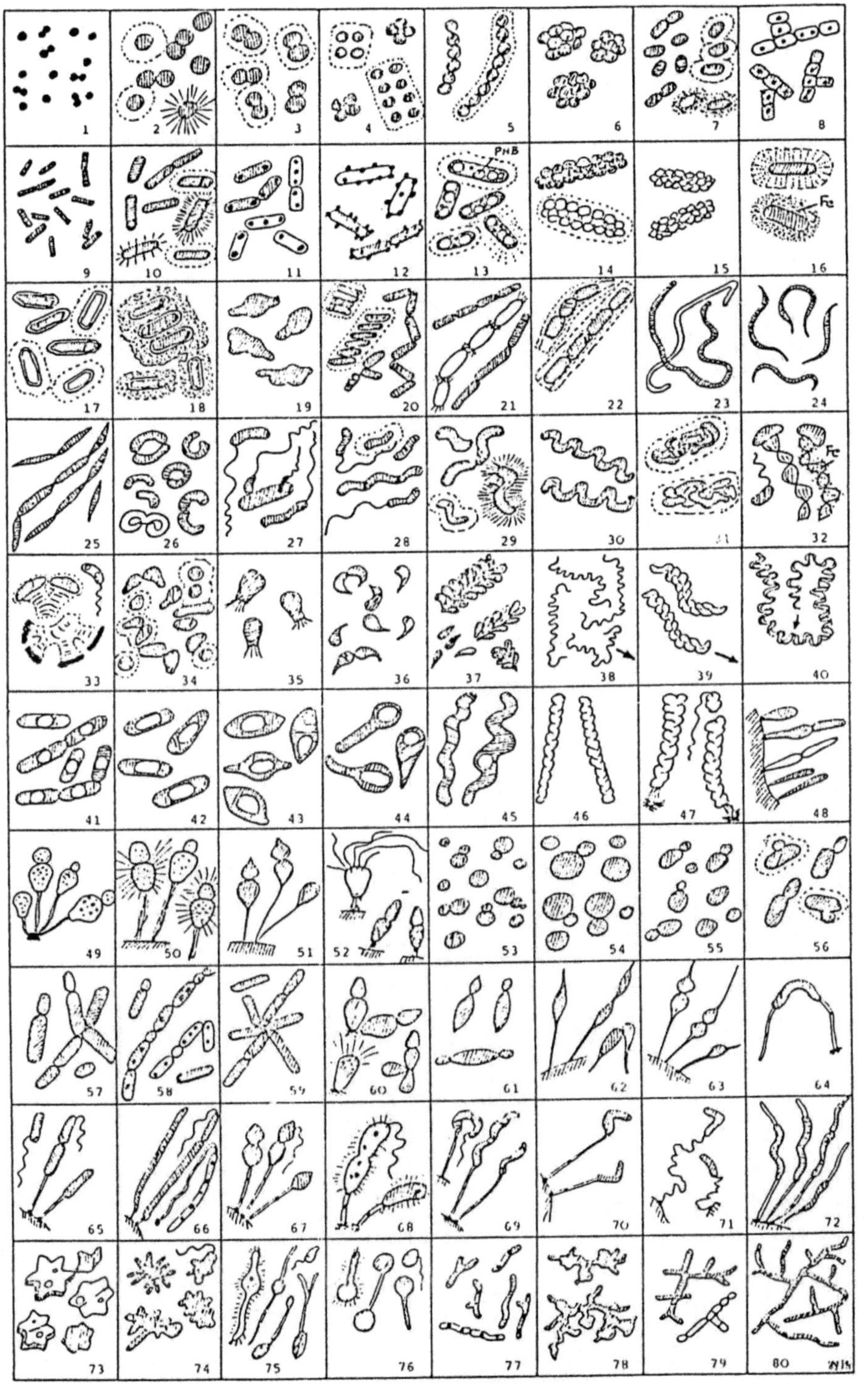

these genera and some of their distinctive properties. Also indicated here is their origin, i.e. if they were observed in GW, in interstitial water samples or in the sediment samples (SS). As can be seen, 20 of the morphotypes occurred in all three sample types or 17 at least in both, the ground water and sediment. Only three MT would be specific for sediment, and eight for ground water. It should be mentioned

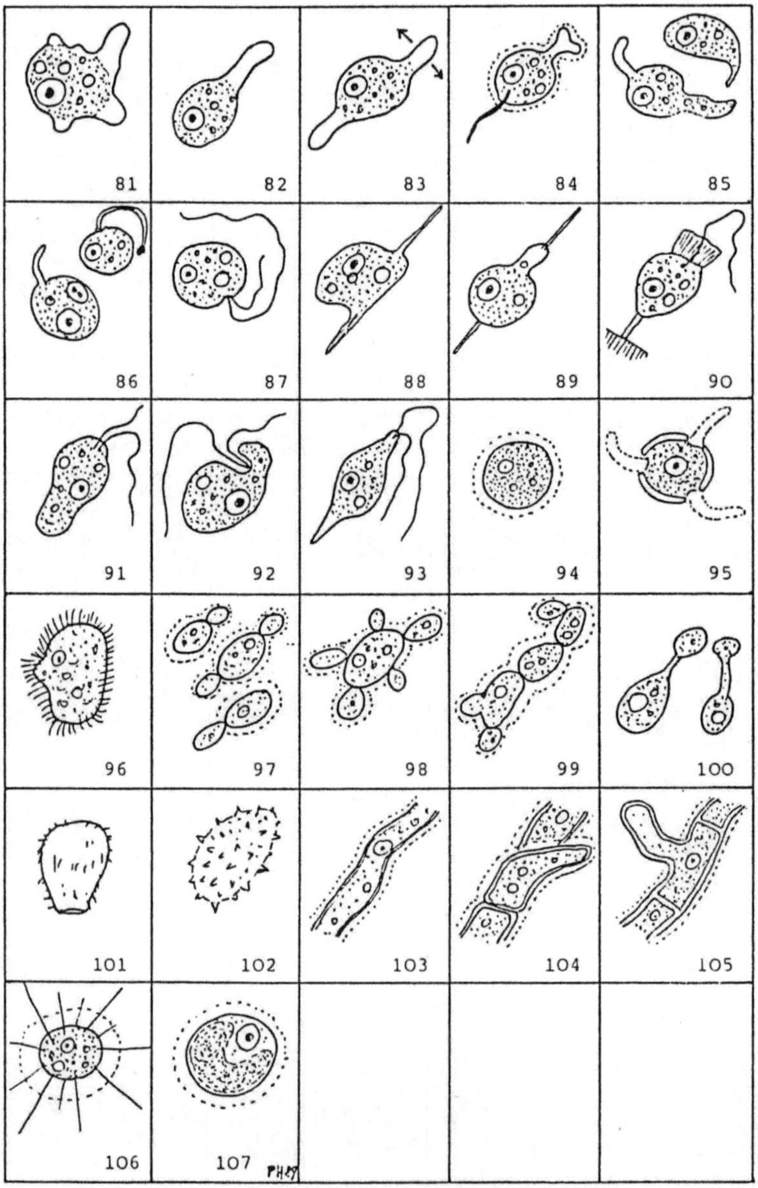

that many of the more unusual organisms were only observed after much prolonged storage of water samples or enrichments.

The high numbers of MT were unexpected. Several rather unusual forms were found. A rod-shaped bacterium had extracellular granules (MT 12, Fig. 5.1.1.2), another one was packed with light refractile storage granules which resembled poly-β-hydroxybutyric acid. Deposition of iron oxides was observed frequently (MT 16, 17, 22, 32). Also quite common were large polymer capsules (MT 2, 3, 4, 5, 7, 10, 13, 14, 16, 17, 18, 20, 29, 31, 34, 56). The unusually high proportion of attaching bacteria can be explained by the normal surface growth of these organisms within the sediment (MT 32, 37, 47, 48, 49–52, 57, 59, 60, 62–72). Also high was the number of budding forms (MT 49–58, 60, 61, 73–76). Finally the frequent observation of aggregates (of members of the same species) should be mentioned (MT 4, 5, 6, 8, 18, 20, 21, 31, 33, 37, 57–60, 72).

The use of the TEM brought further information that allowed the observation of two additional MT: *Agrobacterium*, like *A. polysphaeroideum* with rod-shaped cells that resemble an ear of corn (MT 14) and rod-shaped cells that were packed with small crystals (*Achromatium?*). Also revealed were the structures of many of the capsula polymers. These were seen best on phosphotungstic acid (negatively) stained preparations. There were furs of long, flexible "fimbriae", or of short to very long and rigid spines, amorphic, finely granular layers, or iron oxide-encrusted thick granular structures. The TEM also resolved the structure of "stalks" seen so often with these bacteria. *Gallionella* (MT 32) stalks were twisted bands which consisted of a large number of flagella-like filaments. *Nevskia* (MT 33) stalks were laterally excreted polymer. Stalks of *Seliberia* sp. (MT 47) and those of *Planctomyces* (MT 50) and "Henricia" (MT 52) were composed of twisted fibers that had been excreted from one cell pole. "Stalks" of *Prosthecobacter*, *Caulobacter* and *Thiodendron* were true prosthecae (Staley 1968), i.e. they were composed of cell wall, membrane and cytoplasm.

Some observations were made on the distribution of MT. The diversity found in SEG samples resembled that of other test fields, although those were less intensively studied. Most MT found in SEG samples occurred also in Haltern (Westfalia)

Fig. 5.1.1.3. Groundwater and sediment morphotypes shown as drawings: eukaryotes from BOC or SEG wells

81 Amoeba, w. pseudopodia;
82 amoeba w. 1 lobopodium;
83 amoeba w. 2 lobopodia;
84 mastigamoeba, spherical;
85 mastigamoeba, pointed;
86 flagellate, 1 flagellum;
87 flagellate, 2 flagella;
88 flagellate, irregular;
89 flagellate, bottle-shaped;
90 choanoflagellate;
91 flagellate (*Bodo* sp.?);
92 flagellate, anisokont;
93 flagellate, euglenoid;
94 protozoan cyst;
95 cyst with three pores;
96 ciliate;
97 yeasts w.polar buds;
98 yeasts, multiply budding;
99 yeasts, mycelial;
100 yeasts, buds w.sterigma;
101 fungal spore, drop-shaped;
102 fungal spore, oval;
103 fungus, hyphae w/o septa;
104 fungus, septate, false branches;
105 fungus, septate, true branches;
106 chrysophyte (?);
107 chlorophycean alga

Table 5.1.1.2. Bacterial genera identified in Segeberger Forst samples and some of their morphological properties

Bacteria	Grain stain	Staked	Prosthecate	Hyphae produced	Budding	Glycocalix	Sheath present	Aggregating	Chains formed	Storage granules	Fe-precipitation	Swarmers	Attaching	Motile	Habitat in which organism was found
Actinomycete I (with short branches)	+									+					W,P,S[a]
Actinomycete II (many long branches)	+									+					W, S
Agrobacterium polysphaeroideum (?)						+				+			+		W
Arthrobacter sp.	+					±									W,S
Bacillus sp. I (spores spherical, median)	+									+					W,P,S
Bacillus sp. II (spores ellipsoidal)	+														W,P,S
Bacillus sp. III (cells pointed on one pole; spores round)	+														S
Bactoderma sp. (probably a myxobacterium)	–							+							W, S
Bdellovibrio sp.	–													+	W,P,S
Blastobacter sp. I (rods with polar buds)	–				+			±		+		+	+	+	W,P,S
Blastobacter sp. II (short rods with lateral buds)	–				+							+	+	+	W,P,S
Blastobacter sp. III (rods with suppolar buds)	–				+							+	+	+	W
Caulobacter sp. I (short rods)	–		+					±				+	+	+	W,P,S
Caulobacter sp. II (long rods, often short prosthecae)	–		+					±				+	+	+	W,P,S
Caulobacter sp. III (cells small, limonoid)	–		+					±				+	+	+	W,P,S
Caulobacter sp. IV (cells stout, slightly bent)	–		+					±				+	+	+	W,P,S
Caulobacter sp. V (cells thin, vibrioid)	–		+					±				+	+	+	W
Caulobacter sp. VI (cells curved)	–		+					±				+	+	+	W, S
Caulobacter sp. VII (cells bent, small, prosthecae twisted)	–		+					±				+	+	+	S
Clostridium sp. (cells terminally swollen)	+														W, S
Cytophaga sp. (cells flexible, small, pointed)	–														W, S
Gallionella sp. (cells bean-shaped; stalks broad, ± Fe)	–	+									+	±	+	æ	W
Hyphomicrobium sp. (hyphae furry with short pili)	–		+	+	+	+		±		+	±	+	±	+	W,P,S
Hyphomonas sp. (cells spherical, hyphae partly furry)	–		+	+	+							+		+	W
Leptospira sp. (cells long, thin, ends with hooks)	–													+	W,P,S
Leptothrix sp. (cells short to long rods, sheath present)	–						+		+		+		±	±	W,P,S
Microcyclus sp. (cells forming semicircles or pretzels)	–					±									W,P,S
Mycobacterium sp. (cells small, often branched, granules)	+									+					W, S
Nevskia sp. (cells thin, bent, with lateral polymer)	–	+				±		±				±		±	W, S
Pirellula sp. I (cells pointed on attached pole)	•[b]				+			±				+	+	+	W, S
Pirellula sp. II (cells pointed bipolarly)	•				+			±				+	+	+	W,P,S
Planctomyces sp. I (cells egg-shaped, without bristles)	•	+			+							+	+	+	W,P,S
Planctomyces sp. II (cells egg-shaped, many bristles)	•	+			+	+						+	+	+	W,P,S
Planctomyces sp. III (cells piriform, pointed on prox. pole)	•	+			+			±				+	+	+	W, S
Prosthecobacter sp. I (cells slender, prosthecae long)	–		+			±							+		W,P,S
Prosthecobacter sp. II (cells short, pear-shaped)	–		+			±							+		S
Prosthecomicrobium sp. (prosthecae rather short)	–		+	±	+				±			+		+	W, S
Seliberia sp. I (cells without stalks, long, thin, twisted)	–												+		W, S

Table 5.1.1.2. (*Continued*)

Bacteria	Grain stain	Staked	Prosthecate	Hyphae produced	Budding	Glycocalix	Sheath present	Aggregating	Chains formed	Storage granules	Fe-precipitation	Swarmers	Attaching	Motile	Habitat in which organism was found
Seliberia sp. II (cells with stalks, thicker, twisted	–	+			+							+	+	+	W, S
Spirillum sp. I (cells vibrioid, short)	–					+			±	+				+	W, S
Spirillum sp. II (cells long, more than 2 turns)	–									+				+	W,,P,,S
Spirochete I (cells short, large turns, narrow)	–													+	W
Spirochete II (cells very long, large loose turns)	–													+	W
Stella sp. (cells rather small, star-shaped)	–		+		+	±				+					W
Thiodendron sp. (cells vibrioid, long prosthecae)	–		+	+		±		+					+		W,P,S
Vibrio sp. (cells short)	–								±					+	W,P,S

[a]W Free aquifer water, P – pore water (Interstitial water), S – sediment.
[b]These organisms do not react normally with Gram stain, probably because they lack a peptidoglycan layer.

or Dornach (Munich), albeit in different proportions. Only in a Dornach sample was a colourless eukaryote which resembled a chrysophycean alga.

There were also no significant differences in the MT diversity of individual wells of the SEG testfield when well water was pumped off prior to sampling. There were differences, however, when the microflora of water from stainless steel wells was compared with that of multi-level wells with plastic tubing. The well water of PVC tubing wells had a significantly higher proportion of pigmented bacteria and attaching forms (data not shown).

Within a given well, there were variations with depth, which resulted from inhomogeneities of the sediment profile. Samples from clay horizons had other and fewer MT than those from gravelly or sandy layers. Generally, eukaryotes were mainly present in samples from the upper 10 m of the water-saturated subsurface profile, while prokaryotes were found in all depths of the profile.

A detailed comparison between organisms of the free ground water, of the interstitial water, and of the sediment was made in case of the Bocholt and the Segeberger Forst aquifer systems (Kölbel-Boelke and Hirsch 1989). Although these comparisons were made mainly on the basis of physiological properties of pure culture isolates, some general differences were observed among them. Sediment (biofilm-) isolates were more often Gram-positive, contained more pigmented bacteria as well as those which were stimulated by HA preparations, and were less often motile. Additional information on the presence of eukaryotes in these three habitats is shown in Table 5.1.1.3. All eukaryotes were found in aerobic ground water, about one-half of them was also present in sediment samples. Most common were amoebae and flagellates.

Isolation of pure cultures was attempted for as many of the morphotypes as possible. However, the growth and enrichment conditions supported only some of them, mostly rods and cocci. In some cases preliminary identification steps led to

Table 5.1.1.3. Eukaryotic morphotypes observed in Segeberger Forst samples and their presence in various subsurface habitats

Morphotype	MT no.[a]	Aquifer water		Pore water[b]		Sediment	
		$+O_2$	$-O_2$	$+O_2$	$-O_2$	$+O_2$	$-O_2$
Amoeba, several short lobopodia	81	+	–	+	–	+	–
Amoeba, one lobopodium	82	+	–	–	–	–	–
Amoeba, two thin lobopodia	83	+	–	–	–	–	–
Mastigamoeba, spherical	84	+	–	–	–	–	–
Mastigamoeba, pointed, euglenoid	85	+	–	–	–	+	–
Flagellate, spherical, two flagella	87	+	–	–	–	–	–
Choanoflagellate, single, stalked	90	+	–	–	–	+	–
Flagellate, two flagella (*Bodo* sp.)	91	+	–	+	–	+	+
Flagellate, anisokont	92	+	–	–	–	–	–
Flagellate, euglenoid	93	+	–	–	–	–	–
Protozoan cyst, spherical	94	+	–	–	–	+	–
Cyst with three pores	95	+	–	–	–	–	–
Ciliate	96	+	–	–	–	+	–
Yeast, polar (bipolar) buds	97	+	–	–	–	+	–
Yeast, multiply budding	98	+	–	–	–	–	–
Yeast, mycelial	99	+	–	–	–	–	–
Yeast, buds with sterigmata	100	+	–	–	–	+	–
Fungal spores	101/102	+	–	–	–	+	–
Fungus, hyphae without septa	103	+	–	–	–	–	–
Fungus, septate, false branches	104	+	–	–	–	–	–
Fungus, septate, true branches	105	+	–	–	–	+	–
Alga (chlorophycean)	107	+	–	+	–	–	–

[a]See Fig. 5.1.1.2.
[b]Interstial water of sediment samples.

genus names listed in Table 5.1.1.4. Further characterization and identification procedures are in progress.

5.1.1.4 Discussion

The present data were obtained largely by directly observing "morphotypes" in samples, in enrichments or in pure cultures. The MT concept and its applicability have been discussed before (Hirsch and Rades-Rohkohl 1988b). The most important requirement for such a study is to optimally prepare the subsamples and to use LPM with the highest possible resolution. The agar slide technique of Pfennig and Wagener (1986) allowed us to prepare optimal slides for microscopy.

Despite all the limitations, working with MT will provide some preliminary information that helps to understand the diversity of a given habitat. In the case of stored ground water samples, it is quite likely that most, if not all MT present could be detected during repeated observations. This certainly was not the case with sediment. It is quite difficult to predetermine the proportion of microorganisms dislodged from the particle surfaces; a reasonable estimate is that only about one

Table 5.1.1.4. Pure cultures of aerobic bacteria obtained from Segeberger Forst ground water and sediment samples and some preliminary identifications

Type/Genus	No. of strains	Type/Genus	No. of strains
Cocci, various sizes	11	Vibrios, small forms	2
Short rods, various sizes	32	*Spirillum* spp.	1
diplo forms	1	*Blastobacter* (budding rods)	3
thick, in chains	1	*Prosthecomicrobium* spp.	2
encapsulated	2	*Caulobacter* spp.	20
pointed ends	6	*Thiodendron* sp.	1
pleomorphic	2	*Planctomyces* sp.	1
cells bent	2	*Hyphomicrobium* sp.	8
		Arthrobacter sp.	3
Rods, medium sizes	88	Coryneform bacteria	2
minute to small forms	13	*Bacillus* sp., terminal spores	3
very thin	19	central spores	2
long, thin filaments	2	*Mycobacterium* sp.	1
small, bent forms	5	*Nocardia* sp.	2
branched, thin rods	2	Yeasts, various forms	46
thick, large forms	33	Fungi, filamentous, ± septa	32
thick bent rods	25	Algae, unicellular, green	1
large, pointed rods	37		
long or very large rods	7		
parallel rods	1		
rosette forming rods	4		
encapsulated rods	6		
pleomorphic rods	4		
Total: 433 strains			

third of all forms present can be obtained in suspension by the techniques known. The question thus needs to be asked whether any MT perhaps never detach and thus are overlooked. The same may hold true for those organisms which are present only in very low numbers and which therefore usually do not become frequent enough to be seen in a subsample. Storing the samples for several months and repeatedly investigating these may overcome such problems, at least partially.

Identification of MT to the genus was, of course, very preliminary. "Correctness" of such identifications often rests only on the fact that so far there have not been any studies on the organisms' physiology, chemotaxonomy and molecular properties, so only one species of the genus is known and this one has been described on the basis of morphology. *Gallionella, Nevskia, Thiodendron* etc. are such cases.

The microscopic study revealed a high proportion of attaching forms. This reflects the observation that most subsurface organisms live on sediment surfaces where microbial numbers are found to be one to two orders of magnitude higher than in free water (Kölbel-Boelke et al. 1988a). There may be higher substrate concentrations on particle surfaces with attached macromolecules such as capsular polymers, flagella, fimbria, sacculi or just HS. In such a generally oligotrophic environment, aggregation is probably also advantageous to younger cells. The frequent occurrence of budding forms may be a direct consequence of the attached life, which leads to cell polarity and to the necessity of producing mobile offspring that may get away from the immediate neighbourhood of the parent cell.

A comparison of the data from the Bocholt aquifer system with those from the Segeberger Forst aquifer system showed general agreement with respect to morphotype diversity, although studies at the Bocholt site concentrated mainly on physiological diversity. The more frequent presence of Gram-positive bacteria in sediments and the predominance of Gram-negative, motile rods in the ground water samples seem to be generally true for these and many other ground water locations. Different wells harboured different communities, which was well documented in the case of the Bocholt aquifer system (Kölbel-Boelke 1988; Kölbel-Boelke et al. 1988a); to a certain degree this also held true for the SEG wells. Differences between samples from different wells of one site are partially explained by the types of well-constructing material employed. But this may also be due to differences in sediment texture and composition. The latter certainly also explains the different communities observed in various depths of one well. There may be clay lenses, or the sediment texture and particle sizes or organic substrate composition may vary.

In conclusion, the present studies on two aquifer systems have shown that the subsurface environment – when unpolluted – harbours microbial communities of great morphological diversity. Physiological diversity may be even greater, as has been shown by Kölbel-Boelke et al. (1988a,b). This indicates the possiblity of multiple interactions and interdependencies among the components of these subsurface communities, with special abilities to utilize (degrade) more complex substrates, as long as these occur in low concentrations.

5.1.1.5 Conclusions

A literature search for ground water microorganisms revealed 24 genera or 30 named species. Most of these identifications rested on morphological observations.

Microscopic morphotype analyses were performed on samples from two unpolluted North German test fields: the Bocholt and the Segeberger Forst aquifer systems. In both cases free aquifer water contained mainly Gram-negative, motile rods, while Gram-positive pigmented bacteria were more prevalent in sediment samples. There was much variation in the microflora of different bore holes (wells) of one site; part of this variation may have been due to different construction materials, other parts were probably due to differences in underground texture and nutrient present in situ.

A detailed morphotype analysis was made in Segeberger Forst subsurface samples. Of 107 recognized morphotypes, 80 were prokaryotes and 27 eukaryotes. Most of the prokaryotes were cocci, rods, spore formers and caulobacters (prosthecate and attached bacteria). Among the eukaryotes the flagellates were most frequent. A high proportion of subsurface bacteria was covered with thick furs of flexible or rigid fimbriae (spines). Also common were other types of polymer capsules. An unusually high proportion of subsurface bacteria were attaching or aggregating forms, probably a consequence of nutrient deficiency in many of these samples. The high frequency of budding forms may be a consequence of the predominantly attached life which induces cell polarity.

Distributional studies showed that 20 of 48 bacterial morphotypes were present in the free ground water, in interstitial water as well as in sediment; an additional 17 showed the same distribution but could not be found in interstitial water. Only eight morphotypes were found just in ground water, and three were restricted to sediment samples. The high morphological diversity in these subsurface communities is paralleled by at least an equal physiological diversity. This is taken as an indication for a high potential of the subsurface microbiota to utilize (degrade) a wide variety of even large molecular substrates, provided these are offered at low concentrations.

5.1.2 Methods of Studying Ground Water Microbiology: Critical Evaluation and Method Suggestions

P. Hirsch, E. Rades-Rohkohl, J. Kölbel-Boelke, A. Nehrkorn, R. Schweisfurth[†], *F. Selenka and A. Hack*

5.1.2.1 Aim

To collect ground water samples from a well or a spring, or sediment samples by drilling a hole into the ground water zone is fairly easy; however, it proves to be a crucial task to obtain meaningful information on the natural ground water microflora. One may carry out microbial cell counts and all sorts of determinations, but all external contaminations during the sampling processes must be excluded. Furthermore, the microbiological methods should be suitable to detect all or at least the majority of the organisms present, and to determine not only their potential properties but also the in situ activities. This section discusses the experience gathered in the course of studying undisturbed microbial ground water communities at three aquifer systems in Germany. Although the four research groups in bochum, Bremen, Homburg and Kiel used somewhat different approaches, the methods employed and the general experiences gained were very similar. Therefore, suggestions are made for appropriate procedures in the investigation of the ground water microflora.

5.1.2.2 Critical Description of Methods

Selection of Sampling Sites

Investigations on ground water microorganisms are favoured by the observation that seasonal differences have not been observed in the deeper ground water zones. This may be due to the usually slow changes of chemical water composition and the exponential damping of the temperature fluctuations with growing distance to the Earth's surface (Matthess 1990).

Any study of the ground water microbiota in the saturated zone should include both the sediment and the water, as substantial differences have been found between the microflora of these two habitats (Kölbel-Boelke and Hirsch 1989). However, both cannot be sampled at the same time. The drilling and construction procedures of a well usually result in severe disturbances of the well's environment, which are indicated by unnaturally high cell counts in the first water samples. Total and viable cell counts will decrease until after approx. 1 year an equilibrium with the surrounding aquifer is re-established: this is the moment when microbiological analysis can begin. It can be checked by collecting well water samples as described below and by periodical viable cell count (colony forming units, CFU) determinations after the well construction.

The microflora differs even in adjacent wells, sometimes with respect to cell numbers (Kölbel-Boelke et al. 1988a), sometimes with respect to morphological and physiological diversity (Hirsch and Rades-Rohkohl 1988a; Kölbel-Boelke et al. 1988a). The reasons for that may be differences in the ground water flow (rate, direction), or different aquifer materials and textures (sand, gravel, clay lenses, peat layers, wood pieces etc.). Consequently, more than one well should be studied, and the hydrogeological regime including the water quality should be known, i.e. pH, redox potential, degree of aerobiosis, the concentration of organic carbon as well as that of the major cations and anions.

Establishment of Wells; Collection of Sediment Samples

Most hand augers will not reach the ground water zone, thus larger drilling devices will have to be used. Lubricating liquids containing organic carbon such as methylcellulose ("tylose") should not be applied, as these will contaminate the sediment samples and the subsurface milieu, as indicated by increased cell numbers and changes in the microbial diversity. Augers often heat up because of friction. Optimally, after precoring with an auger, a new auger with a smaller diameter should be employed which contains sterile plastic tubing. The plastic tube containing the core can be taken with only little disturbance of the layering. After opening the tube with a sterile scalpel, the core can be sampled with sterile spoons after removal of the outer core layers (Hoos and Schweisfurth 1982; Gottfreund et al. 1984; Hack et al. 1984). Care should be exercised not to mix the sometimes very thin layers, which can vary greatly in microbial composition. The plastic tube employed should be free of plasticizers which could be utilized as substrates or are possible inhibitors of microorganisms. In most cases sediment samples of the saturated zone will release some water by gravity. This interstitial water should be tested immediately for microbial diversity and activity, which often resemble that of the sediment. Attached microorganisms in the sediment most likely produce motile (or at least mobile) offspring which can be found in the interstitial water.

As the interstitial water runs off, the emptied pores within the core will fill with oxygenated air. Thus, the sediment samples, previously microaerophilic or even anaerobic, will become increasingly aerobic. Therefore, sampling for microaerophiles or anaerobes must be done rapidly, as well as measurements of pH, redox potential, temperature, sediment gases etc. Therefore it is not advisable to transport the core to the laboratory before microbiological sampling. All sampling and sample processing should be done immediately after the core has been retrieved. A search for strictly anaerobic forms should begin by immediate inoculation of anaerobic media in flasks under a constant stream of O_2-free nitrogen and possibly after addition of a reducing compound and a redox indicator. This way true anaerobiosis can be established and monitored.

Transportation to the laboratory, even over ice, has been found to reduce microbial diversity within a few hours, although total and/or viable cell counts could increase (Hirsch and Rades-Rohkohl 1988a). A total alteration of the relative composition of the samples' microflora and also the cell numbers can be foreseen in case of prolonged storage.

Microbial cell counts are usually related to the sediment dry weight. Therefore, the water content of the sediment at the time of sampling has to be measured. Subsamples for this determination should be taken immediately after the retrieval of sediment cores, but that is hampered by the uncontrolled loss of interstitial water. This can be overcome by measuring the water content of a saturated parallel core.

Processing of Sediment Samples

For the investigation of aerobic microorganisms the sediment samples can be transferred into sterile, wide-mouthed flasks. Subsamples can then be inoculated into preweighed culturing vessels. The vessels are weighed again, and sterile medium is added. In order to study morphological or physiological diversity (or to obtain isolates) enrichments should be made with media that do not contain much more organic carbon than is present in the sediment. In most unpolluted sites the low C-content calls for oligotrophic media such as PYGV (Staley 1968), dilutions of it (PYGV/100, or for medium P (Wolters and Schwartz 2956). Other authors prefer soil extract media (Beloin et al. 1988). Incubation should be carried out as closely as possible to the ambient ground water temperature and in the dark, as many oligotrophic bacteria can be inhibited by light. Enrichment cultures should be scanned for morphotypes (and subcultured) after 1 and 4 weeks as well as after 6 and 12 months; many slow-growing microorganisms (especially protozoa) will not immediately develop to population sizes that can be observed.

For total and viable cell counts in the sediment the samples can be treated as suggested by Ghiorse and Balkwill (1982). The procedure consists of suspending a known amount of sediment in sterile ground water (taken from the same depth as the sample), or in sterile 0.1 N Na-pyrophosphate solution, or in sterile 0.85% NaCl. This suspension is then treated with an electric mixer and allowed to settle for about 2 min. Then subsamples are taken and diluted with sterile ground water for viable cell counts. The dilutions are immediately spread onto the corresponding agar plates and these are transported to the laboratory in plastic bags over ice. Incubation was described above. Evaluation of the viable counts should occur after 1, 2, and 6 weeks: slow-growing Gram-positives will often develop late. Other subsamples are filtered for total acridine orange direct cell counts (AODC) according to Ghiorse and Balkwill (1983, 1985).

Another technique, suggested by A. Hack and successfully employed by Kölbel-Boelke et al. (1988b), consists of releasing the cells from the sediment by suspending 25 g (wet wt.) in 100 ml sterile tap water and of adding 20 g sterile glass beads (4 mm Ø). The suspension is then shaken for 30 min at 160 agitations per minute. The resulting suspensions may be diluted before filtration through membrane filters. Staining is accomplished with amido-black 10B, the microorganisms are then destained with acetic acid and counted.

Viable cell counts were determined by Kölbel-Boelke et al. (1988b) by filtering the suspensions through sterile 0.2-μm cellulose nitrate membranes and placing these on a modified P-medium. In all such experiments variability of cell numbers within one sample will be observed: it is therefore mandatory to inoculate enough parallel plates (up to five) for sufficient statistics.

Most of these methods for sediment analysis assume that the treatment releases the majority of microorganisms that live attached to sediment particles. That this is not the case, at least for unsaturated zone samples, was shown by Balkwill et al. (1977): at the most about 30% of the organisms could be detached from the particle surfaces. Up to now the remaining attached microflora has not been studied satisfactorily. An analysis of the morphotype diversity without enrichment and/or isolation can be made by employing the method of Waid (1973). A small amount of sediment (pea-sized) is placed in a watch glass and water or polyvinyl pyrrolidone (Bone and Balkwill 1986) is allowed to flow over it, beginning from the bottom. The resulting thin polymer pellicle which now floats on the surface is then transferred to a glass slide or electron microscope grid and can be examined for morphotypes. Sediment samples can also be studied with the SEM using standard procedures of fixation, dehydration, critical point drying and electron microscopy. In another technique, suspensions are prepared from samples after a standing time of some months; these can be viewed either directly in the light- or electron microscope, or inoculated into sterile media for eventual investigation. For light microscopy the preparations are made on agar-coated glass slides according to Pfennig and Wagener (1986).

Collection and Processing of Water Samples from Wells

It is necessary to decide on the topic of the study, the "real" aquifer microflora, the microorganisms living free in the well water, or those living attached to the well wall (or plastic tubing's wall). Even the surface of the well water is often colonized by neuston microorganisms: a thin pellicle develops which can be removed by touching the surface off with a disk covered with plastic household foil.

The actual sampling can be done in various ways, depending on the kind of well. Stainless steel wells of sufficient inner width can be sampled with an immersible pump; pushing up the well water is often preferred to sucking it up, as the underpressure can remove dissolved gases. The pump should at least be disinfected by immerging it (and all the necessary tubing) in 70% ethanol and washing off the ethanol with sterile ground water. Multi-level wells with plastic tubing are more difficult to sample. Here the tube mouths should be disinfected, preferentially after installation of a stainless steel tap with stopcock. Sampling of such wells was described by Hirsch and Rades-Rohkohl (1983a,b; Sect. 5.1.1). Sampling for analysis of anaerobes requires a constant flow of anaerobic gas at the mouth of the tubing to avoid oxygenation of the sample (Sect. 5.3.3).

Investigation of the free ground water requires sufficient prepumping and discarding of the prepumped water to avoid collecting the microflora of the well water, which is practically removed after pumping off about 10 well water volumes. This can be monitoried by determining viable cell counts. The Bocholt group has prepumped >500 l per well. Too vigorous prepumping may dislodge the attached organisms from the well wall, and will draw water from above and below the sampled aquifer horizon. The latter drawback can be overcome by simultaneous sampling of several horizons with the appropriate pressure, e.g. by using a multi-channel valve or a vacuum box (Atakan et al. 1974b; Leuchs 1988).

The microflora adhering to inner well surfaces can be studied in various ways. Glass slides with or without agar layers containing nutrients can be hung into the well water and retrieved after removal of the neuston layer. Exposure times vary according to the temperature, but 1–2 weeks are sufficient. The slides are dried rapidly and viewed under the LPM. For best optical conditions the slides are remoistened with one small drop of filtered well water on each end and cover slips are placed on them. As long as the drying was effective there is no need for special fixation procedures. For electron microscopy one can expose 300-mesh nickel or stainless steel grids which are fastened to glass slides with a 0.25–0.35% formvar film (Hirsch and Pankratz 1970). To avoid the floating off of the film and grids, the former is secured to the slide with autoclaveable electric tape. After harvesting (as described above), the grids are rapidly dried, shadowed at low angle or stained negatively with phosphotungstate. Electron microscopy should be done at 60 kV, taking care that the electron beam does not tear the film through local overheating. Exposure of plastic household foil will also result in attachment of well organisms. After drying, these foil pieces can be dehydrated, gold-coated and viewed under the SEM.

Organisms adhering to the walls of plastic tubes (as in the multi-level wells) can be studied after the tubing has been withdrawn. The inner surface is carefully cut off with sterile scalpels and the sections are treated as in standard procedures for SEM. This kind of preparation will provide an excellent view of the distribution pattern of microorganisms, revealing – in most cases – a patchy distribution. All of these methods that involve attachment of microorganisms must be interpreted with great caution since the type of surface and its structure (charge, nutrient content, preexposure etc.) determine the types and numbers of microorganisms that will attach (Hirsch 1972).

It was stressed already above that inoculations for the determination of viable cell numbers and for morphotype studies or for isolation attempts should be ideally performed on site, immediately after sample retrieval. Samples that will be used for total cell counting after staining with AODC will have to be fixed immediately, e.g. with formaldehyde (2% v/v, final concentration). In waters rich in Fe^{2+} an addition of 0.01 M EDTA will avoid the precipitation of ferric hydroxide which could otherwise clogg membrane filter pores (Kölbel-Boelke 1987).

Transportation and Further Treatment of Water Samples

For transportation, aerobic samples should be kept in glass flasks with screw caps, and the flasks should be only half-filled. Anaerobic samples should be filled completely, leaving an approx. pea-sized gas bubble of N_2 or argon over the liquid as a pressure buffer. Transportation to the laboratory should be immediate, preferentially over ice packs. The use of dry ice is not recommended, as it has been found that the developing CO_2 will alter the diversity of the microbial community (Galikowski and Hirsch 1989).

Water samples intended for cell counting should not be stored, as has already been pointed out for the sediment samples. Within a few hours changes occur in the species composition: some organisms will disappear, others will multiply, sometimes to very high numbers. The result may be an increase in total and viable cell counts, but a decreased diversity.

An excellent way of studying diversity of ground water morphotypes is to simply store the samples in the dark at the ambient ground water temperature. After several weeks or months, subsamples are taken from the thin film of bottom- or wall growth. These are studied by light- and electron microscopy, e.g. after negative staining of the preparations (Hirsch and Rades-Rohkohl 1983a,b, 1988b). This technique does not allow the study of all possible morphotypes because of the changes occurring during storage, but the majority of attaching and oligotrophic microorganisms can be seen in this way.

Enrichments and Isolations

For the investigation of many different physiological types enrichments will have to be prepared. This can be done by inoculating various media (liquid or solid) with different pH, nutrient concentrations, types of C- or N sources, varying amounts of vitamins etc. Incubation can be varied with respect to temperature, especially low temperatures. Incubation time should be prolonged in any case since many organisms from ground water require long lag periods before they develop to numbers large enough to be seen. Gram-positive bacteria often grow on agar plates only after 4–6 weeks. Eukaryotic microorganisms will usually multiply only after an appropriate population of food organisms has developed. In liquid cultures this may take several months.

For isolations, a method was successfully used which employs sterile tooth picks. Colonies on agar plates are selected randomly or accordingly to pigmentation, size, surface structure etc. Each colony is touched lightly with a sterile tooth pick and the material is transferred to a distinct and numbered square on a "master agar plate". The rest of the material on the tooth pick is then spread on a numbered part of a glass slide and dried and scanned for interesting organisms by placing a small droplet of water on the colony smear and by using phase light microscopy (1000 x). If an interesting morphotype is found, the corresponding square on the masterplate can be purified further. The advantage of this method is that a large number of colonies can be scanned in a comparatively short time.

For statistical purposes, all colonies on an agar plate may be picked or a portion of them at random without selecting for special colony types (Kölbel-Boelke et al. 1988b). Further purification of the isolates is carried out on an appropriate agar until only one colony type remains per isolate. Restreaking on another agar medium will provide an additional assurance of purity, as does microscopic examination.

Pure cultures from ground water can be kept at 4–9 °C agar slants, as lyophils (better), or suspended in plastic caps in medium with the addition of glycerol and shock-frozen at –80 °C.

Liquid enrichment cultures can also serve to quantify cell numbers of distinct physiotypes. Thus, dilution series of the liquid samples are prepared using selective media and the most probable number procedure (MPN) is applied. Although this method is not too accurate, it could render data on the frequency of sulfate- or nitrate reducers, sulfur oxidizers, nitrifyers etc.

Characterization of Pure Cultures

Tests for the characterization of pure cultures can be done according to the types of organisms found and by employing standard methods (Skerman 1967; Gerhardt et al. 1981). However, when the standardized tests are performed at distinctly higher temperatures than those occurring in the natural habitat investigated, and when psychrophiles or psychrotrophic bacteria are involved, the tests will have to be carried out at ambient ground water temperatures (Hirsch and Rades-Rohkohl 1988a). In the absence of distinctly psychrophilic bacteria the identification tests could be carried out at 20 °C with the advantage of shorter incubation or reaction times (Kölbel-Boelke et al. 1988b).

Many of the ground water organisms are adapted to low nutrient conditions; they are oligocarbophilic or oligonitrophilic. This should be kept in mind when tests are made for the utilization of C- or N sources. Parallel controls without the corresponding C- or N source should test the possibility that the organisms acquire the missing nutrients from the air (gas phase). Ground water isolates should only be characterized by meaningful tests, thus excluding draught resistance or UV resistance, etc. Very valuable information includes the ability to break down humic substances, to carry plasmids, or to live facultatively anaerobic by using NO_3^- as an electron acceptor. Generally, it has been found that the scope of physiological capabilities among ground water isolates is nearly as wide as it is in surface microorganisms (Hoos and Schweisfurth 1982; Kölbel-Boelke et al. 1988a,b; Sect. 5.3.3).

Where large numbers of isolates have to be characterized, the use of microtitre plate techniques has been successful (Dott and Thofen 1980; Moaledj 1984; Kölbel-Boelke et al. 1988b; Hirsch et al. 1989; Sect. 5.3.3). Sterile conditions are mandatory, as is the incubation in a moist atmosphere such as in a plastic box. One problem exists with the microtitre plate technique: if it is to be time-saving, all 96 wells of a plate have to be inoculated and all strains to be tested have to be in an active state of growth. Since the rates of growth are quite different, the test strains have to be prepared according to their standard rates of growth. Evaluation of microtitre plate data can occur by various methods such as using colour reactions, fluorescence of 4-methylumbelliferon-coupled substrate analogues, or simply by detecting the amount of resulting growth (turbidity), possibly with a mirror or a special type of photometer.

The gathering of large data bases will often require computer use and evaluation with special programmes. Comparison of the microbiota of several wells was made by Kölbel-Boelke et al. (1988a) as well as by Hirsch et al. (Sect. 5.3.3).

Determination of in Situ Activities

All information on physiological capabilities gathered in the laboratory from pure cultures reflects only a potential; it does not mean that these reactions in fact occur in situ (Kölbel-Boelke et al. 1988a). Relatively few experiments have been made so far to describe or measure the actual activity within the ground water environment. As activity has been defined in several ways, different possibilities exist to determine it in situ. The presence of dividing cells in the sediments and attached on the walls of plastic well tubing has been interpreted as activity by Ghiorse and Balkwill (1983), Balkwill and Ghiorse (1985) and Hirsch et al. (1989).

Another indication of in situ activity is the observation of a complete sediment colonization comparable to that of natural samples at 12 weeks' exposure of sterilized sediment samples in a well (Hirsch and Rades-Rohkohl 1990; Sect. 5.3.3).

The proportions of respiring cells have been determined by Beloin et al. (1988) and Marxsen (1988), who tested for reduction of the artificial electron acceptor, 2-(p-iodophenyl)-3-(p-nitrophenyl)-5-phenyltetrazolium chloride (INT). This procedure was carried out on acridine orange-stained cells which have been used to determine total cell counts (Beloin et al. 1988), but the specificity of this procedure still needs to be tested. Microbial growth rates in sediment samples taken out of the profile were determined by Thorn and Ventullo (1988) by testing for (^{3}H)-thymidine incorporation. The presence and activity of exoenzymes in ground water or sediment samples can be determined with the use of fluorescent substrate analogues (Weyandt and Schweisfurth 1989). Here and in all other cases the activity was determined in samples that had been removed from the ground water environment.

5.1.2.3 Conclusions

Painstaking considerations are necessary for the selection of the sampling site, the sampling and processing methods, and the water and sediment samples themselves:

- Suitable are "aged" water wells with a well-known geological and screening profile, that have been constructed in a natural ground water system without lubricating drilling fluids and organic compounds.
- Sediment samples should be collected by coring performed with sterile plastic tubing.
- When ground water samples are taken from distinct depths, e.g. by multi-level wells, approx. 10 well volumes of the water present in the well have to be pumped off first, to remove those organisms that live within the well water or attached to the well walls. To avoid mixing of water from different horizons, the water should be extracted simultaneously from all screens.
- Transportation of water or sediment samples to the laboratory results in changes of microbial numbers and diversity. Therefore, processing of the samples, i.e. measurements and subculturing, should be done immediately on the site. Here, a facility to carry out microbiological work under sterile conditions should be established. This means the necessity of access for electricity, bench space and protection from air movement etc.

- Enrichments and isolations should be made with oligotrophic media, as most natural ground water environments are low in organic carbon. Enrichments should be cultivated at ground water ambient temperatures. Evaluations for morphotype diversity or for subculturing should be made several times for up to a period of 1 year. Culture collections can be kept on slants, as lyophils, or as frozen cultures in glycerol containing media in small caps.
- Characterization of pure cultures can be made by employing well-known standard procedures, but incubation shoud always be at ambient ground water temperature. Special attention is directed towards using microtitre plate techniques which allow the handling of large numbers of strains. In this case it is important to start with all strains on a plate in a state of simultaneous active growth. The evaluation of research data obtained from studying subsurface sediment samples should be related to sediment dry weight, but quantitative data of microbial numbers in sediment samples can not be compared directly to numbers found in water samples.
- Methods which allow the determination of in situ activities are still needed. Colonization experiments can be performed with sterilized sediment confined in membrane filter chambers and placed in well water, but these only reflect activities of the microflora within the well.

5.1.3 Organic Substances in Ground Water and Sediments and Their Relationship to Microorganisms in a Sandy Aquifer

F. Selenka and A. Hack

5.1.3.1 Background

The bacterial cell counts in the saturated zone vary in order of magnitude from 10^4 to 10^7/g d.wt. in the sediment, and in the corresponding ground water between 10^3 and 10^7/ml. Their numbers vary significantly with depth. This behaviour is due to the physico-chemical state of the sediments and the corresponding water, the structure of sediments and nutritional conditions. The organisms in question cover their demand for energy and carbon in two fundamentally different ways.

- In the case of lithoautotrophic microorganisms, reduced inorganic substrates such as H_2S, H_2, CO, Fe^{2+}, NH_3, S^0, Se, and Mn^{2+} are used, whereas the carbon, necessary for the formation of new organic substances, is fixed by assimilation of CO_2 (Ehrlich 1978; Schlegel 1985).
- Mixotrophic and chemoorganotrophic microorganisms use the deposited organic material in sediments and the dissolved organic materials in ground water as energy and carbon sources.

The organic compounds are present as fossil materials (plant residues, wood, peat, lignite and humic substances), recently deposited material from the top soils (humic substances, carbohydrates, organic nitrogen compounds, lipids and others) and newly formed metabolites and decay products of the entire microflora. Depending on the agricultural utilization of the covering soils, the distance to the soil surface, the age of sediment layers and ground water, the velocity of ground water flow, the type and quantity of inorganic substances in ground water (NO_3^-, SO_4^{2-}, S^0 etc.) and the redox conditions, there is a marked variability in composition and amount of organic substances in sediments. The organic matrix, i.e. the total organic material, is bound to the sediments by adsorption or deposited as particles. Fossil plant residues and humic substances are dominant, followed by extracellular polymeric substances (EPS) and carbohydrates or organic compounds with high nitrogen content.

In the Bocholt aquifer system (Fig. 1.1, Table 1.1, Sect. 4.1.1) organic substances in ground water and corresponding sediments were estimated as total organic carbon (TOC), total organic nitrogen (TON), extractable humic substances (HS), and carbohydrates associated with humic substances (CH_{HS}). These data were compared to bacterial counts and in a few cases to bacterial activities, thus giving information about distribution and turnover conditions of organic material in sediments and ground water in the depth range of 1 to 50 m.

5.1.3.2 Materials and Methods

Sediment and water samples were collected at seven Bocholt sampling sites named BOC1 to BOC7 and treated as described in Section 4.1.1.

Sampling of Sediment Material

Among different procedures for sediment sampling, drilling with coring devices gave the best results. In this case the natural stratification of the sediment was maintained and contamination of the sample by ground water in the drilling tube could be avoided to some extent. In addition, the drilling equipment had to be free from oil, fat and adhering soil particles from other sediment layers, especially those from the surface layers. Therefore, the equipment was thoroughly cleaned at the beginning and, if necessary, repeatedly during the drilling process.

Additional problems arose when sampling under anaerobic conditions was required. The sediment material was transferred to glass tubes which were immediately filled with inert gases such as nitrogen or argon.

Cryothermic drilling procedures and sample conservation methods also seem to be suitable, but are problematic when microbiological investigations have to be performed.

Sampling of Ground Water

Ground water samples can be acquired by means of sucking pumps or submerged forcing pumps. The sucking method renders minimum contamination of the sample by the pump and by the tubing materials, provided the pump is connected downstream of the sampling equipment. Use of submerged pumps, especially mobile ones, brings about the risk of contamination by pump, tubing material and cross-contamination while transferring the equipment from one well to another.

To avoid inadequate results, sampling has to be done under stady-state conditions. This means constancy of concentrations of the substances of interest. By changing the pumping rate to higher yields during sampling, withdrawal of water from additional compartments of the sediments in the neighbourhood of the filter elements will take place. Because concentrations of organic compounds are not necessarily equally distributed in the interstitial water, especially in the vicinity of a well, significant variations of the concentration of organic substances may occur in the sample. Moreover, additional mobilization of particles may cause varying TOC concentrations of the samples.

The construction material of wells (synthetics or stainless steel) has no recognizable effect on the main parameters i.e. TOC, HS, carbohydrate concentration, nitrogen content, provided the well has been cleaned thoroughly and the sampling rate is held constant and the sampling quantity is optimized. In the case of specific organic substances, especially non-polar compounds, a sorptive effect of the well construction material cannot be excluded.

5.1.3.3 Results

Improvement of Sampling Techniques for Ground Water

The total organic carbon concentration in water samples was used to estimate the quality of ground water sampling because this is influenced clearly by differences in construction materials of the wells and by variations in pumping rates. When ground water is pumped simultaneously from all filter elements of a multi-level well, the probability of hydraulic short cuts is low (Leuchs 1988). A well made of polyvinylchloride (PVC), polyethylene (PE) and polypropylene (PP) materials contained about 0.2 mg/l more TOC in ground water from the well vicinity than in water from more distant compartments. This could be seen from the decrease of TOC concentration with longer pumping time. This was most likely due to a release of organic substances from the plastic material even a year after the well construction.

Half a year later the TOC concentrations in ground water from the well's vicinity were lower (Fig. 5.1.3.1a). There was no longer any difference in the TOC concentration between the plastic and the steel well constructions (Fig. 5.1.3.1a,b). When 300 l ground water were pumped from a filter element, the TOC content in the water sample was constant in both cases. At the end of the first sampling cycle the pumping rate was increased to 7.5 l/min. This also resulted in temporarily lower TOC concentrations at the beginning, probably caused by the same activation process of TOC-depleted compartments adjacent to the well as mentioned above (Figs. 5.1.3.1a,b). The lower TOC concentrations were probably caused by intensified microbial decomposition of organic substances with oxygen penetrating into the aquifer by diffusion via sampling tubes and filter elements. These observations demonstrated once more that organic material is decomposed in subsoils if suitable oxidants are present.

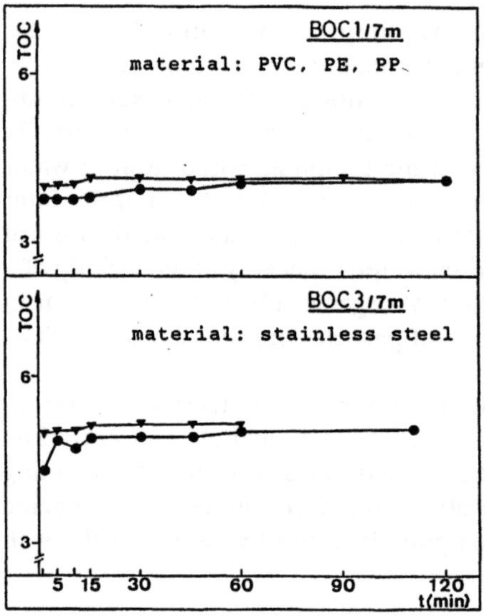

Fig. 5.1.3.1. Comparison of TOC contents (mg/l) in ground water from wells BOC1 (synthetics) and BOC3 (stainless steel) in 7 m depth, and dependence of TOC concentrations on pumping (l/min) and time (min). Sampling cycle 1 ●: production 5.0 l/min; sampling cycle 2 ▼: production 7.5 l/min, immediately after cycle 1; *PVC* polyvinylchloride; *PE* polyethylene; *PP* polypropylene. Sampling date: 24.07.1985; well BOC1 constructed 19 months ago; well BOC3 constructed 9 months ago

Organically Bound Carbon and Nitrogen in Sediment Samples and in Extracts of Humic Substances from Sediments

The concentration of org.C in sediment ranged from 3600 mg/kg near the surface of the aquifer examined to 160 mg/kg in greater depth. Particles of fossil wood or lignite caused TOC concentrations in the range of 650 to 1000 mg/kg d.wt. (Fig. 5.1.3.2).

A percentage of 45 to 65% of the TOC in sediments from 1–2 m below the surface could be extracted with 0.2 N NaOH. In lower depth this portion decreased below 5% of the total organic carbon in the sediment (Fig. 5.1.3.3). The same was true for strata near the ground water surface with high contents of nitrate in ground water.

The distribution of org.N in sediment, in the total extractable fraction and in the FA fraction was in accordance with the distribution of the TOC in these samples.

The molar C/N ratio in sediment layers without fossil plant material was about 16 in untreated sediment samples (Fig. 5.1.3.3) and in extracts from sediments. In extracts from fossil plant material the C/N ratios were about 22. However in sediment layers with ground water of high NO_3^- concentration the C/N ratios dropped to about 10 and in sediment extracts even to about 2 to 5.

→

Fig. 5.1.3.2. TOC in sediment of bore-holes BOC3, BOC4, BOC5 and BOC7. Dependence of TOC (mg/kg d.wt.) on depth. ● fossil organic material; ▼ ground water table

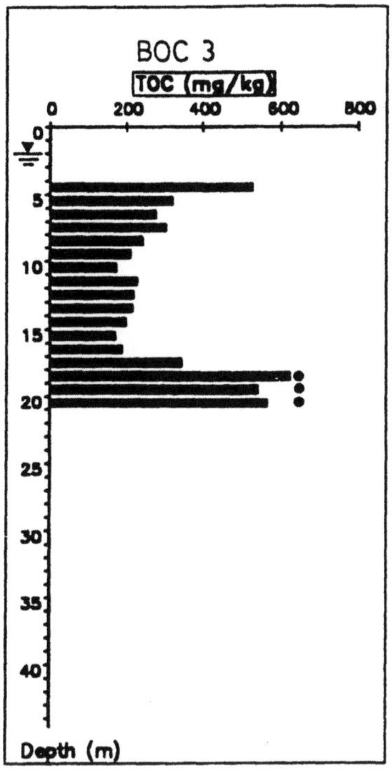
BOC 3
TOC (mg/kg)
0 200 400 600 800
0 5 10 15 20 25 30 35 40
Depth (m)

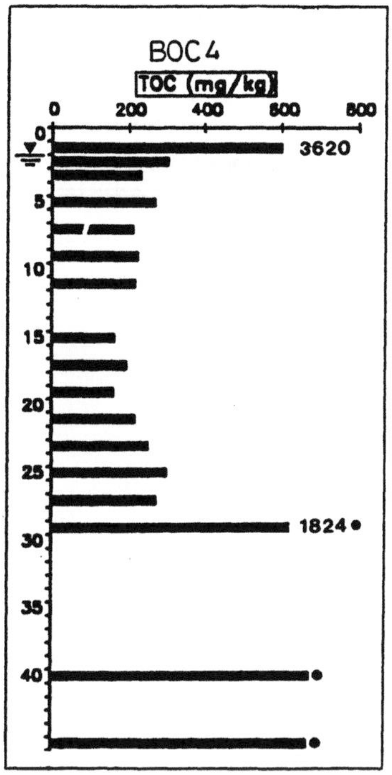
BOC 4
TOC (mg/kg)
0 200 400 600 800
3620
1824
0 5 10 15 20 25 30 35 40

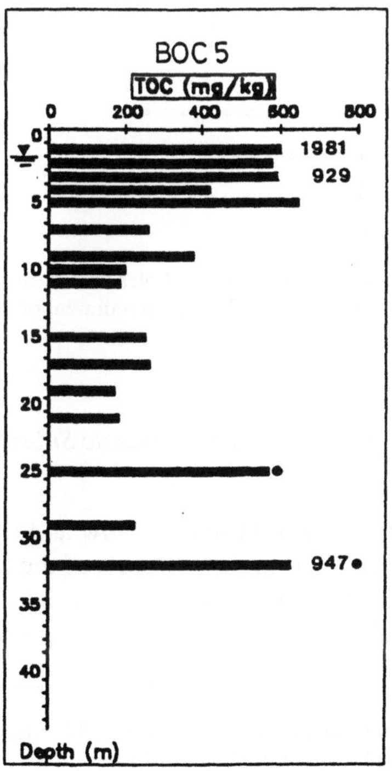
BOC 5
TOC (mg/kg)
0 200 400 600 800
1981
929
947
0 5 10 15 20 25 30 35 40
Depth (m)

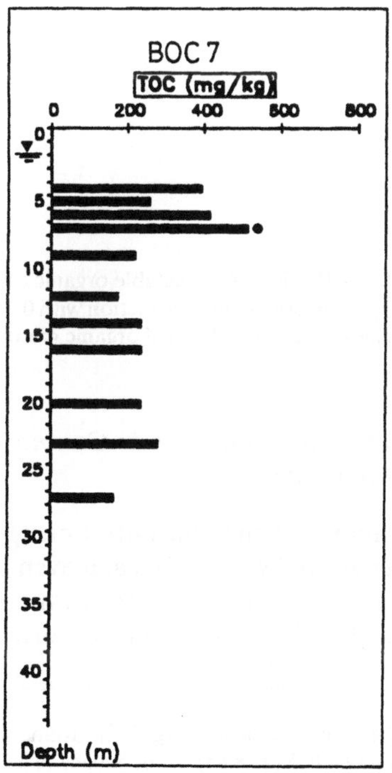
BOC 7
TOC (mg/kg)
0 200 400 600 800
0 5 10 15 20 25 30 35 40
Depth (m)

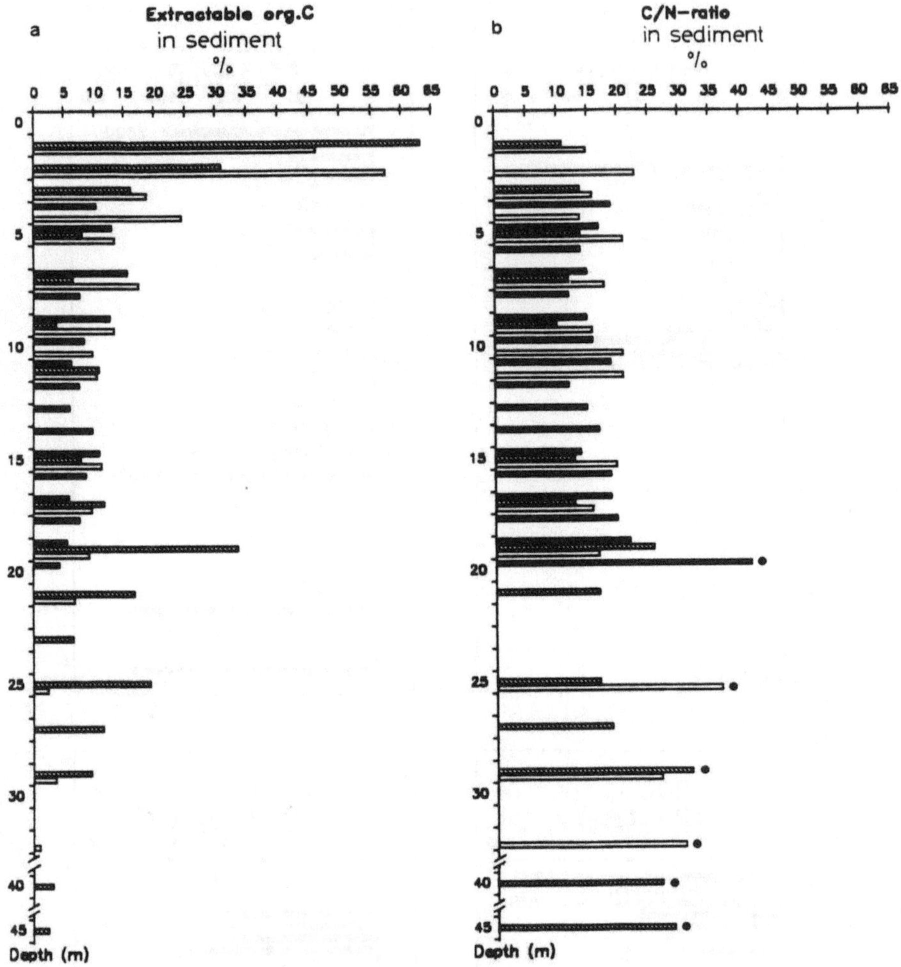

Fig. 5.1.3.3. **a** Portion of extractable organic carbon in sediment from bore-holes BOC3, BOC4, and BOC5 depending on depth. Extraction with 0.2 N NaOH. **b** Molar C/N ratio in sediment of the three wells. Signatures as in a. ● Fossil organic material

Organically Bound Carbon in Ground Water and in Extracts of Humic Substances from Ground Water

In ground water the concentration of org.C was found to be as low as 17.6 mg/l near the ground water table and even lower (2 mg/l) in deeper parts of the aquifer (Fig. 5.1.3.4). Sediment zones with many inclusions of organic material did not cause higher TOC values in the corresponding water. This behaviour is different

Fig. 5.1.3.4. Organic carbon (org. C) in ground water from wells BOC1 (7.8.85), BOC3 (7.8.85), BOC4 (5.8.85) and BOC5 (27.8.85). Dependence of TOC (mg/l) on depth. ▼ Ground water table

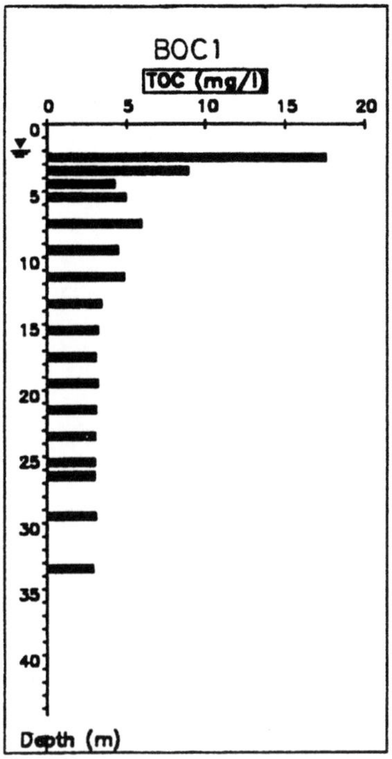
BOC1
TOC (mg/l)
0 5 10 15 20
0 5 10 15 20 25 30 35 40
Depth (m)

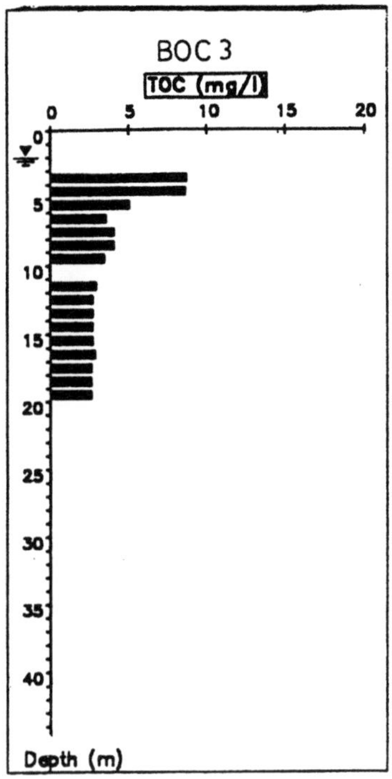
BOC 3
TOC (mg/l)
0 5 10 15 20
0 5 10 15 20 25 30 35 40
Depth (m)

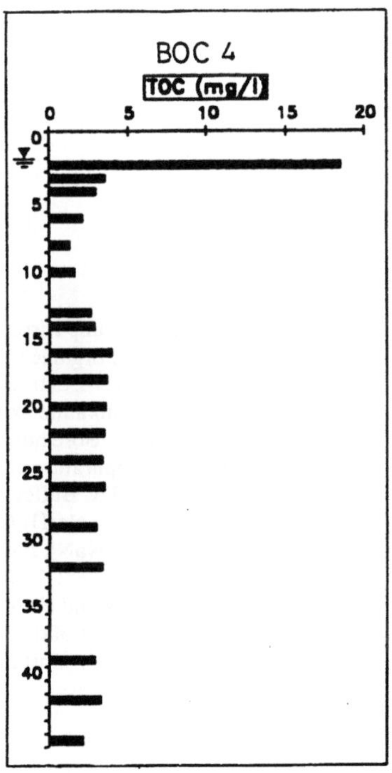
BOC 4
TOC (mg/l)
0 5 10 15 20
0 5 10 15 20 25 30 35 40

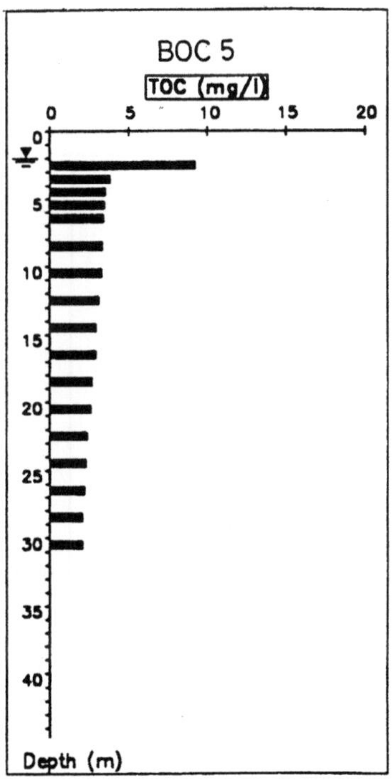
BOC 5
TOC (mg/l)
0 5 10 15 20
0 5 10 15 20 25 30 35 40
Depth (m)

from water seeping into the ground as cited in the literature. The organic substances were enriched in vadose water that passed horizons rich in organic materials (Pekdeger 1977). In the Bocholt field 60–70% of the TOC in ground water could be extracted with XAD-2. There was no dependence on depth. C/N ratios in water and in humic substance extracts from water reached 25–30 near the ground water table and about 18 below that. This was in agreement with the values found in sediment and in sediment extracts (Fig. 5.1.3.5).

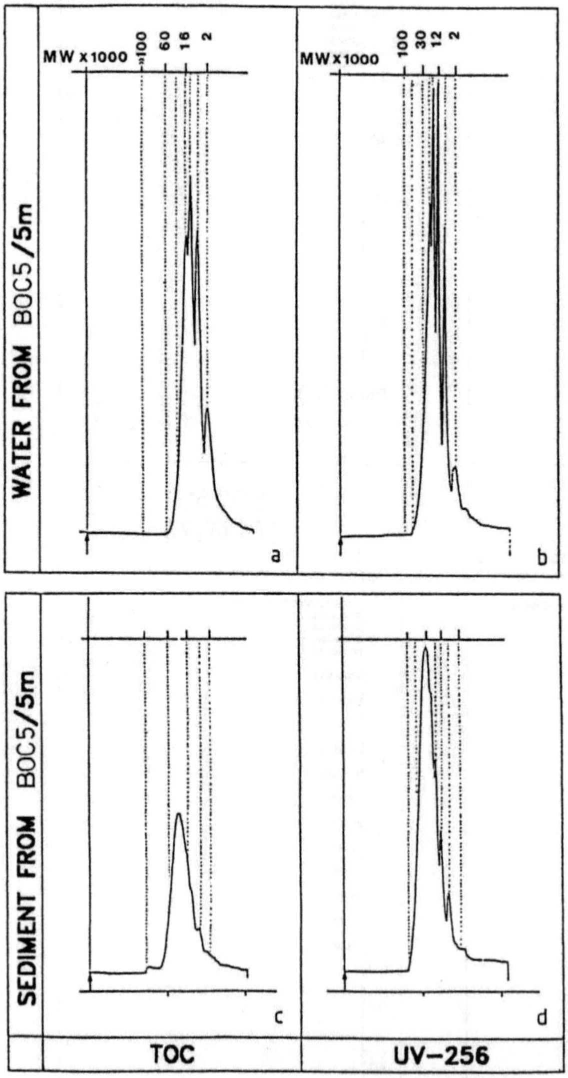

Fig. 5.1.3.5. Comparison of the distribution of organic substances in XAD-2 extracts from water and an alkaline extract from sediment. Bore hole BOC5, 5 m depth. Separation column: TSK G 3000 SW. Buffer: 0.7 g KH_2PO_4 + 1.1 g $Na_2HPO_4 \cdot 2 H_2O$ + 0.002 g NaN_3; Flow: 1 ml/min; detection: UV_{256} (**b** and **d**) and TOC (**a** and **c**); *MW* molecular weight of dextrans used as reference substances

Distribution of Organic Substances Between the Solid and Aqueous Phase

Only a small proportion of the organic substances in the aquifer was dissolved in the aqueous phase; on an average 500 to 800 times more org.C was bound to the sediment. In layers with fossil plant material we found up to 5000 times higher concentrations of org.C in sediment than in the corresponding water.

Carbohydrates Associated with Humic Substances

In sediment and ground water the concentrations of carbohydrates that were associated with humic substances (CH_{HS}) decreased with depth, absolutely as well as with regard to HS concentrations.

The presence of fossil plant residues caused an increase in the carbohydrate contents in the HS extracts from sediments, but this was limited to the strata involved. The distribution of carbohydrate carbon in HS extracts from sediment resembled that of extracts from water. The total "CH_{HS} carbon" near the ground water table was determined to be 10%, while at depth below 5 m it was at most 2% and in some cases only 0.2%. Acidic hydrolysis of HS samples from the Bocholt aquifer and from Geiseke (valley of the river Ruhr near Dortmund) gave rise to the same monomeric carbohydrates: the bulk was represented by arabinose, fucose, galactose, glucose, mannose, rhamnose and xylose. This was in accordance with findings of other authors and from other localities (De Haan and De Boer 1978) and it is also the case for the total carbohydrate fraction from river water (Ittekkot et al. 1982).

After hydrolysis, glucose, galactose and rhamnose were the most abundant monomeric carbohydrates in HUS extractions in our examinations.

The composition of the monomeric carbohydrate pool in sediment and water after hydrolysis was relatively uniform in all strata and at all examined sampling sites (Fig. 5.1.3.6). This implies that the carbohydrates observed were integrated into humic material by similar processes and most likely originated from similar sources.

Bacterial Counts and Biomass in Ground Water and Sediment

Microscopic total bacterial counts in the Bocholt aquifer system ranged from 10^6/ml near the ground water table (about 2 m) to 10^4/ml in depths below 25 m (Fig. 5.1.3.7).

Viable bacterial counts determined on P-Agar (Wolters and Schwartz 1956) by cultivation for 14 days at 14 °C were lower than microscopic counts by a factor of 10^2 to 10^3. In sediments bacterial counts were made after removal of the organisms from the sediment by shaking for 30 min with glass beads (size: Ø 4 mm), the most efficient method when compared with ultrasonic or ultraturrax techniques.

Between 10^4 and 1.2×10^6 bacteria/g d.wt. of sediment were observed microscopically. Extreme variations existed between the total counts of sampling sites within the top 20 m of the aquifer. The differences were sometimes noted in layers no more than 1 or 2 m apart from each other. It could be shown that most of the microbial population thus cultivated consisted of facultative anaerobes. Colony

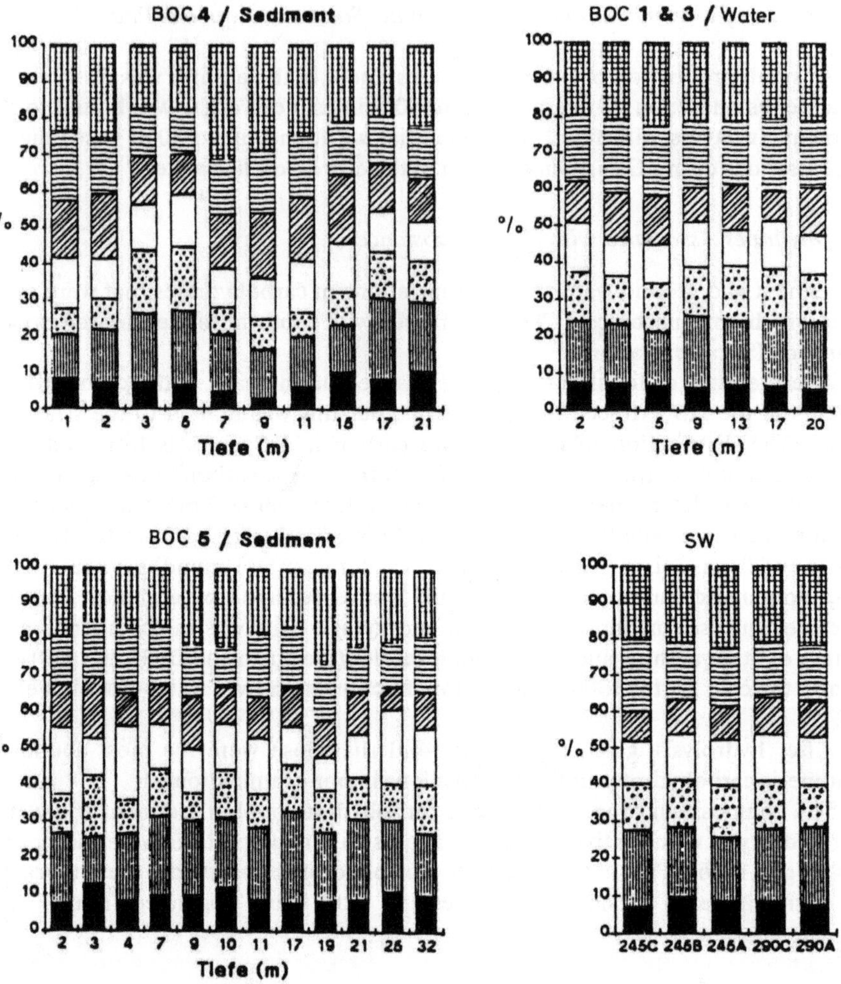

Fig. 5.1.3.6. Percent distribution of monosaccharides in hydrolyzed FA extracts of sediment and water from the Bocholt aquifer system (*BOC*) compared to extracts from bank filtrate of the Hengsensee (*SW* river Ruhr near Dortmund). ⊞ Glucose; ⊟ galactose; ▨ mannose; □ xylose; ▩ fucose; ◫ rhamnose; ■ arabinose

counts were generally lower by the factors of 10 to 10^4. There were no significant differences in colony counts when cultivated under aerobic or anaerobic conditions.

The carbon content of microorganisms is relation to their volume has been calculated by Watson et al. (1977) as 1.21×10^{-13}g C/μm^3 and by Ochiai and Nakajima (1985) as 7×10^{-14}g C/μm^3 of bacterial volume. The average volume of bacterial cells in the present investigation was 0.25 μm^3. Using these data it is found that only 0.1% of the TOC in water and 0.01% of the TOC in sediments in these studies could have been due to microbial biomass.

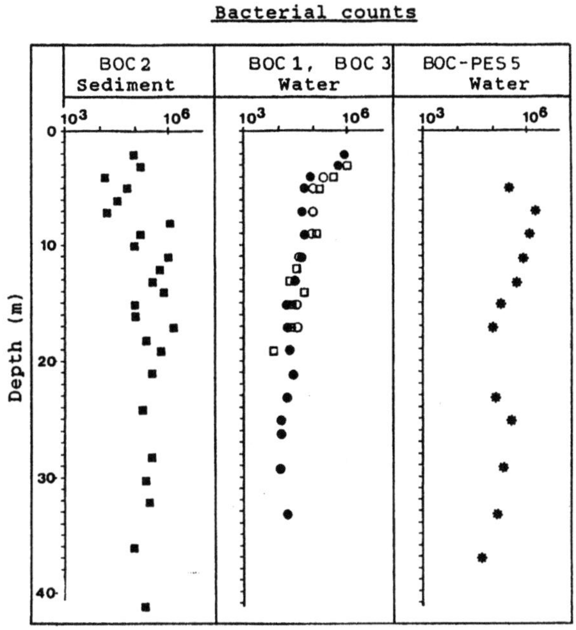

Fig. 5.1.3.7. Total bacterial numbers/ml water in the all filtered well BOC-PES 5 (11.7.83 *), in the multi-level wells BOC1 (synthetics; 29.8.84 ● and 31.1.85 ○), and BOC3 (stainless steel; 31.1.85 □), compared to the total bacterial number/g d.wt. in sediment from borehole BOC2 (Dec. 1983 ■)
The well BOC-PES 5 is totally filtered and is situated about 1000 m northwest of sampling sites BOC1 and BOC3 (See Sect. 4.1.1). Water samples were taken from BOC-PES 5 by packer technique (Obermann 1976) from BOC1 and BOC3 by simultaneous pumping of all filter elements

5.1.3.4 Humic Substances and Fossil Plant Material as a Source of Nutrition for Microorganisms

The amount of extractable organic substances in sediment and water was outstandingly small in strata with high concentrations of nitrate. However, because the concentration of TOC in sediment decreased far less; it seems that it was the extractable fraction of organic substances that enhanced microbial activity, whereas particular or clay-bound organic material was less reactive.

Oxygen and nitrate in ground water are consumed rather quickly, and in depths below 4 m bacterial utilization of sulfate occurs. Leuchs (1985) demonstrated the occurrence of microbial metabolic processes using sulfate by considering isotope fractionation of sulfur (Sect. 4.1.1). The elimination of sulfate, however, was slower than the elimination of oxygen and nitrate (Leuchs 1985). Compared with HS, the reduction of CH_{HS} in the aquifer is generally favoured, indicating an overall microbial activity.

The total number of microorganisms in ground water was highest in strata near the ground water table, namely 10^6/ml, and it decreased slowly with depth to reach 10^4/ml. This correlates with the decrease of TOC in ground water from 17.6 mg/l to 2 mg/l. This means that DOC originating from the top soils was a main source of nutrition for microorganisms in the upper parts of the aquifer. High bacterial counts and elimination of organic substances of higher molecular weight near the ground water surface is in accordance with the observation of Meyer et al. (1977), who found that in black water systems microbial growth was better supported by organic substances with molecular weight $\geq 10^4$ than by those with molecular weight between 10^3 and 10^4.

In sediment the correlation between org.C content and bacterial number was less clear than in water. In contrast to the findings in ground water, bacterial counts were observed to range from 10^4 to > 10^6/g d.wt. of sediment. This may be due to microhabitats of the microbial population on the surface of organic particles, as pointed out by Jörgensen (1977), Kölbel-Boelke (1987), Leuchs (1985, 1988). These microhabitats may also be responsible for the presence of some strict anaerobes even in oxygenated areas.

5.1.3.5 Conclusions

Humic substances bound to mineral surfaces take part in sorption and binding of a wide spectrum of organic and inorganic compounds. At the same time, as shown above, they can be sources of energy for the autochthonous and allochthonous microorganisms in aquifers. Humic substances can be slowly decomposed by microorganisms as long as electron acceptors are present in sufficient quantities. For example, O_2, NO_3^- or SO_4^{2-} may serve as electron acceptors. It is therefore, reasonable to assume that loss of humic substances will impair the aquifers' ability in binding and degrading pollutants to a marked degree.

5.2 Microbial Activities

5.2.1 Observations on the Physiology of Microorganisms from Pristine Ground Water Environments

P. Hirsch

5.2.1.1 Background

Information on the physiology of ground water microorganisms is still rare. The few publications concerned with this matter describe properties of cultures that came originally from the aquifer environment; but physiological tests were normally made with pure cultures in the laboratory. Typical examples may even consist of tests for the utilization of various substrates offered at unnaturally high concentrations. Other information comes from viable cell-counting experiments performed with specific media such as the number of sulfate reducers which was obtained from growth on anaerobic lactate medium. Table 5.2.1.1 lists such general properties and the corresponding references.

Recently, Ghiorse and Wilson (1988) discussed the mineralization of a variety of organic compounds in laboratory experiments or in microcosms. Ground water- or sediment microorganisms were capable of utilizing glucose, arginine, glutamate, benzoate, some volatile fatty acids, glycolate, methanol or tertiary butyl alcohol. Mineralization of pesticides or xenobiotics such as phenol, toluene, chlorobenzene,

Table 5.2.1.1. Some selected general properties of ground water microorganisms described in the literature

Property	Reference
Oligotrophy, nitrification	Ghiorse and Balkwill (1983)
Denitrification	Kramarenko (1962), Hoos and Schweisfurth (1982), Ehrlich et al. (1983)
Nitrate-ammonification	Hoos and Schweisfurth (1982)
Hydrocarbon oxidation	Bogdanova (1965)
Methanogenesis	Olson et al. (1981), Beeman and Suflita (1987) Ghiorse and Wilson (1988)
Iron precipitation	Hässelbarth and Lüdemann (1967), Kramarenko (1962), Sládecková (1972), Cullimore and McCann (1977), Hoos and Schweisfurth (1982), Gounot et al. (1987)
Iron reduction, manganese oxidation or reduction	Hoos and Schweisfurth (1982), Ghiorse and Wilson (1988)
Sulfide oxidation	Kramarenko (1962), Sládecková (1972), Cullimore and McCann (1977), Leuchs (1988)
Sulfur oxidation	Sládecková (1972), Hoos and Schweisfurth (1982)
Sulfate reduction	Olson et al. (1981), Hoos and Schweisfurth (1982) van Beek and van der Kooij (1982), Beeman and Suflita (1987), Ghiorse and Wilson (1988)

2-chlorophenol, p-chlorophenol, 2-4-dichlorophenol, 2,4,6-trichlorophenol, nitrilo-acetic acid or m-cresol apparently does occur at various rates. Other observations describing formation of extracellular microbial polymers and storage compounds such as polyalcanoic acids need to be mentioned also.

5.2.1.2 Physiological Data from the Sandhausen Site

Weyandt and Schweisfurth (Sect. 5.3.1) investigated the microbiology of a subsurface profile located in the Schwetzinger Hardt near Sandhausen (SDH). Physiological groups of microorganisms occurring below the 5.3 m deep water table were identified with the plate-counting technique and employing specific media. They located (and counted) aerobic and anaerobic protein-ammonifiers, nitrate-reducers, thiosulfate- and sulfur-oxidizers and sulfate reducers. These authors also noted phosphatase activity in various depths of their water-saturated sediment.

5.2.1.3 Microbial Physiology of Bocholt Isolates

A thorough study of potential physiological properties of isolates from the Bocholt aquifer system (BOC) was carried out by Kölbel-Boelke and Nehrkorn (Sect. 5.3.2, this Vol.). These authors obtained 2700 bacterial strains and tested them for their ability to utilize (degrade) 58 substrates of many different chemical classes. Each of these 58 substrates was degraded by at least one member of each depth or well community. The most frequent physiological properties of all isolates were the excretion of phosphatase, the degradation of Tween 40, Tween 60, DNA, Tween 20, lecithin, galactose, mannose, glucose, gelatin, arabinose and xylose; 45–64% of all test strains had such capabilities. Also commonly utilized were: fructose, maltose, saccharose, melizitose, raffinose, acetate, citrate, formate, lactate, malonate, tartrate, oxalate, propionate, succinate, glutamate, arginine, lysine, ornithine, phenylalanine, inulin or xylan. The polymers, starch, chitin or glycogen, were more frequently degraded than xylan or cellulose. All in all, this demonstrates the enormous variability of physiological properties among the Bocholt ground water or sediment microorganisms. There were great differences also between the communities found in different but adjacent wells or depths.

Selenka and Hack (Sect. 5.1.3) investigated the same Bocholt sites and remarked that most of their isolates were facultatively anaerobic. These authors also observed microbial sulfate reduction in depths below 4 m, and they concluded that the presence of specific microhabitats for strict anaerobes would be responsible for the activity of sulfate reducers in oxygenated areas.

5.2.1.4 Observations on the Physiology of Isolate from the Segeberger Forst Test Area (SEG)

Employing various different enrichment, isolation and observation techniques, Hirsch et al. (Sect. 5.3.3) established a morphotype table and identified at least nine genera where common knowledge permits some guesses concerning physiological properties of these organisms (Table 5.2.1.2). All of these were oligotrophic, the majority probably oligocarbophilic. In some cases a chemolithotrophic life with CO-

Table 5.2.1.2. Possible physiological properties of selected bacterial genera found in Segeberger Forst ground water or sediment and deduced from general knowledge of the corresponding genera (Table 5.1.1.2)

Genus	Oligo-trophy	CO-oxida-tion	H2-oxida-tion	Faculta-tively auto-trophic	Utiliz. of CH_3OH	Utiliz. of amino-acids	Utiliz. of cellu-lose aerobic	Utiliz. of C-1 compds.	Fe^{2+} oxidation Fe^{3+} precipitat.
Bactoderma	+	+	•	+	•	•	•	•	•
Blastobacter	+	•	•	+	+	•	•	•	•
Caulobacter	+	•	•	•	•	+	•	•	•
Cytophaga	+	•	•	•	•	•	+	•	•
Gallionella	+	•	•	+?	•	•	•	•	+
Hyphomicrobium	+	•	•	•	+	(+)	•	+	±
Leptothrix	+	•	•	•	•	•	•	•	+
Microcyclus	+	•	+	+	+	•	•	•	•
Seliberia	+	+	+	+	•	•	•	•	•

or H_2 oxidation appears possible. Utilization of C-1 compounds and/or oxidation of ferrous iron are further properties predicted for such genera.

Additional statements can be made. Most strains (= 85%) grew well between a pH of 5–8; most isolates (67%) were eurythermal, i.e. they grew between +4 °C and 30–43 °C. Only 15 pure cultures could be considered to be psychrophilic. Organic acids were well utilized; the growth with sugars or sugar alcohols was concentration-dependent. Several strains were capable of cleaving the aromatic ring of protocatechuic acid. There was manganese oxidation as well as Mn-reduction, iron oxide precipitation and nitrate reduction.

5.2.1.5 Discussion and Conclusions

The data and observations presented here show the general diversity of "physiotypes" in the ground water environment. Apart from xenobiotics, it appears that nearly every compound tested was utilized by at least some ground water- or sediment isolates. On the other hand, all investigators stress the differences between communities of even adjacent wells or of different depths in one particular well. Obviously, the rather stable but oligotrophic conditions in the subsurface environments have selected for highly structured and interdepending microbial communities. Physiological differences between the sediment biofilm microflora and the freely floating organisms have been described by Kölbel-Boelke and Hirsch (1989).

An unexpected result was the temperature independency of most isolates. As the ground water temperature is rather constant, a high degree of adaptation to the local temperature would have been logical.

Finally, it should be stressed again that all data obtained from laboratory experiments with pure cultures will only describe potential capabilities, but they will not necessarily reflect the in situ reactivity of a given organism. The microenvironment determines the type and degree of physiological activity, of course, only within the realm of the organism's genetic potential.

5.1.2 Formation and Transformation of Manganese Oxidation States by Bacteria

J. Gottfreund and R. Schweisfurth†

5.2.2.1 Background

Since the first description of bacterial involvement in manganese oxidation (Beijerinck 1913), many experiments have been made to understand the ecological meaning of manganese oxidation and to describe the biological background of its occurrence (Bromfield 1956; Jung and Schweisfurth 1979). Recently, a dependency of manganese oxidation on CO_2 fixation has been found (Kepkay and Nealson

1987). The following mechanisms of microbiological manganese oxidation have been considered:

1. Complex bound manganese (Mn) was oxidized faster than Mn(II)-ions if the chelating agent of the Mn(II)-complex is decomposable by microorganisms and can be substituted by OH^--ions:

$$\text{Mn(II)-complex} + H_2O \xrightarrow{\text{bacteria}} Mn(OH)_2 + HCO_3$$

This reaction includes:

- the dissociation constant of Mn(II)-complexes
  $\text{Mn(II)-complex} \longrightarrow \text{Mn(II)} + \text{chelating agent } (K_1)$
- the pK_s (pK_B) value of the chelating agent

$$\text{chelating agent}^{x-} + y\ H_2O \longrightarrow \text{H-chelating agent} + z\ H_2O\ (K_2).$$

This means that the manganese oxidation would proceed more slowly with a stronger Mn(II)-complex. If the chelating agent is the base of a weak acid, a strong manganese oxidation is to be expected as can be seen from the equilibrium constant K_2.

2. Mn(II) serves the destruction of H_2O_2 during the bacterial metabolism (Dubinina 1978)

$$H_2O_2 + \text{Mn(II)} \xrightarrow{\text{catalase}} MnO_2 + 2\ H^+$$

3. By the oxidation of Mn(II) the bacteria form ATP (Ehrlich 1976)

$$\text{Mn(II)} \longrightarrow \text{Mn(III) (?)} \longrightarrow \text{Mn(IV)}.$$

$$\text{ADP} \curvearrowright \text{ATP}$$

During the reduction and mobilization of Mn(IV) the microorganisms are very important. Through pH/E_H change they can create an environment where Mn(IV) is reduced (Hem 1963). Other bacteria can be cultivated which reduce Mn(IV) under aerobic and anaerobic conditions. According to the literature, two mechanisms of microbiological Mn(IV) reductions have to be differentiated:

- Mn(IV) serves the elimination of H_2O_2 which has been formed during bacterial metabolism. This reaction can take place spontaneously or catalytically by means of enzymes (peroxidase)
- Mn(IV) serves as a sink for unneeded $NADH_2$ if there is an excess of decomposable organic carbon.

Further details of these investigations were published in Gottfreund and Schweisfurth (1983), Gottfreund et al. (1983, 1985a,b,c), Schweisfurth and Schauermann (1984).

5.2.2.2 Material and Methods

A pure culture of *Arthrobacter* sp. was isolated from 64 m depth of an unconsolidated sediment layer of a reduced aquifer (Hoos and Schweisfurth 1982). The outcome of biochemical and morphological tests indicated a relationship with *Arthrobacter* spp. (Davies and Newton 1969). All experiments with this *Arthrobacter* sp. were carried out on a shaker at 27 °C in a 1000-ml Erlenmeyer flask containing 200 ml of nutrient solution. The determination methods for Mn(II), Mn(III), and MN(IV) have been described previously (Gottfreund and Schweisfurth 1983).

5.2.2.3 Results

Manganese Oxidation with a Pure Culture of an Arthrobacter sp.

During all tests with *Arthrobacter* sp., the following values were determined for pH and E_H (Fig. 5.2.2.1) and for Mn(II), Mn(III) and Mn(IV) (Fig. 5.2.2.2). The microbial manganese oxidation began at the end of the growth phase. A further experiment consisted of adding KCN to the nutrient solution; it demonstrated that the oxidation of Mn(III) to Mn(IV) could take place without bacterial activity (Fig. 5.2.2.3). In further tests the microbiological oxidation rate of Mn(II) has been determined by the dependency on its binding state. The following results were obtained: Mn(II)-lactate > Mn(II)-acetate > Mn(II)-Cl_2/$MnSO_4$ > Mn(II)-citrate > Mn(II)-oxalate.

The dependency of manganese oxidation on its binding state, as well as the type of the chelating agent, could definitely be shown for this *Arthrobacter* sp. During the experiments a Mn(II)-oxalate complex was used where oxalate was not decomposable. A general dependency of manganese oxidation on decomposable chelating agents could not be found.

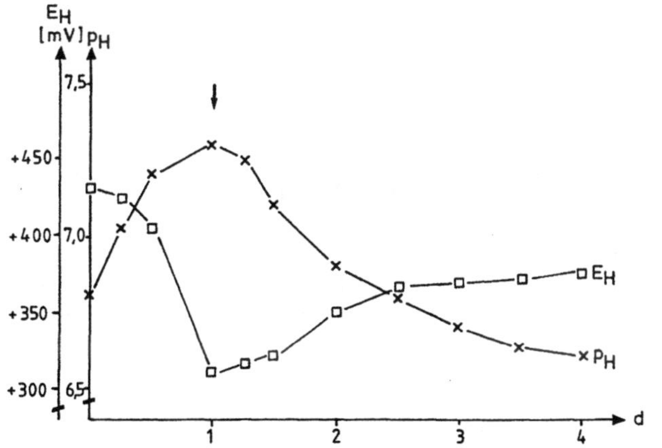

Fig. 5.2.2.1. Changes of E_H, and pH values in the culture during growth of *Arthrobacter* sp. and manganese oxidation

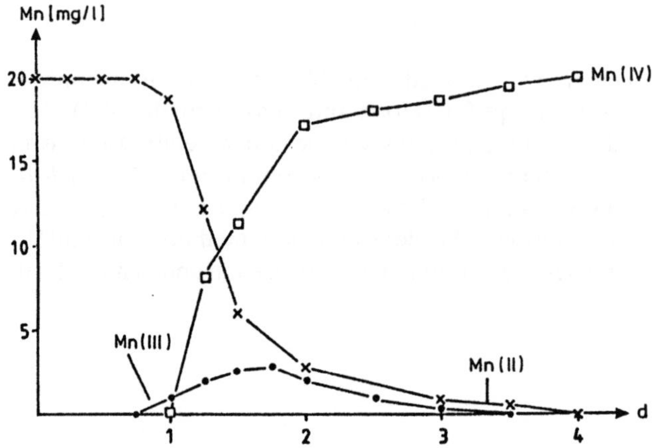

Fig. 5.2.2.2. Changes of the concentrations of Mn(II), Mn(III) and Mn(IV) in the medium of *Arthrobacter* sp.

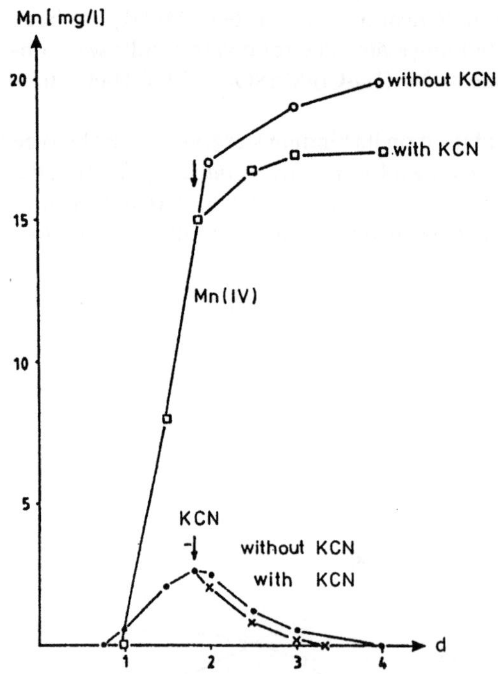

Fig. 5.2.2.3. Changes of the concentrations of Mn(III) with and without KCN (↓ = addition of KCN)

Very stable manganese complexes are formed with ethylendiamintetraacetate (EDTA) and nitrilotriacetic acid (NTA). Both chemicals did not serve as a carbon source for this *Arthrobacter* sp. The manganese oxidation depended on the concentration of both chelating agents (Fig. 5.2.2.4).

During the drilling of an 18-m-deep borehole near Hanover, loose sediment samples were collected and were used for the extraction of FA and HA. In the presence of both acids, the microbial manganese oxidation was dependent on the concentration of FA and HA (Fig. 5.2.2.5). The manganese oxidation was significantly inhibited when both compounds had been sterilized at 121 °C. Therefore, in later experiments the acids were filtered through a membrane filter.

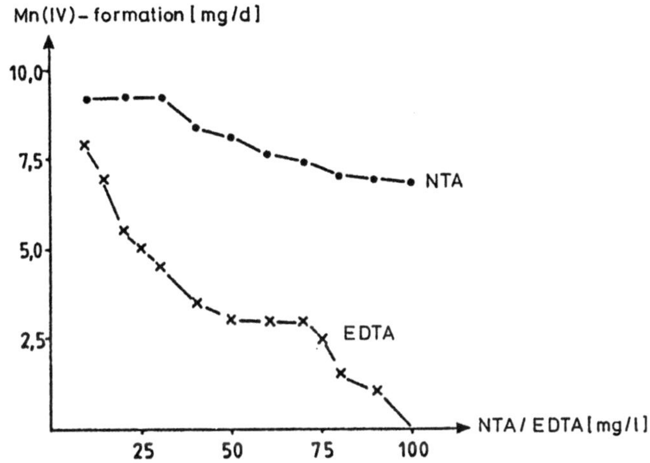

Fig. 5.2.2.4. Dependence of the manganese oxidation rate on the concentration of EDTA and NTA in the presence of 20 mg/l Mn(II)

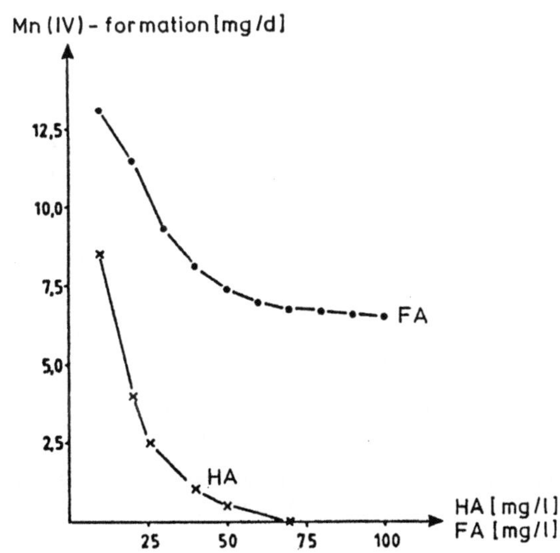

Fig. 5.2.2.5. Dependence of the manganese oxidation rate on the concentration of FA and HA by *Arthrobacter* sp.

As an inorganic chelating agent of Mn(II), phosphate was examined because of its influence on the microbiological manganese oxidation. In the presence of 10 mg/l P, orthophosphate inhibited manganese oxidation; diphosphate exerted this effect already at a concentration of 8.4 mg/l P. There was no proof of manganese oxidation in the presence of hexametaphosphate, trimetaphosphate and ultraphosphate. The growth of *Arthrobacter* sp. was not influenced by these conditions.

Manganese Reduction with a Pure Culture of a Pseudomonas sp.

The *Pseudomonas* culture originated from a sediment sample obtained from a depth of 16 m. The tests for biochemical classification revealed that this pure culture was a strain of *Pseudomonas*. Initially experiments to demonstrate manganese reduction were made under aerobic conditions with 1000-ml Erlenmeyer flasks which were shaken at 27 °C.

During manganese reduction initially Mn(III) was produced (Fig. 5.2.2.6). This Mn(III) was then reduced to Mn(II) by this *Pseudomonas* sp. After the addition of KCN to the nutrient solution during manganese reduction it could be seen that the major part of arising Mn(III) was reoxidized to Mn(IV) (Fig. 5.2.2.7). In a further experiment it was shown that Mn(III) could be reduced to Mn(II) by this *Pseudomonas* sp.

In the following experiment a condition was chosen where the bacteial activity was physically separated from the location containing solid manganese oxide. This was achieved by using membrane filters (pore size 0.05 μm). Under these conditions Mn(IV) was also reduced (Fig. 5.2.2.8). In contrast to the manganese oxidation, manganese reduction took place concomitantly with growth (Fig. 5.2.2.9). During these experiments the concentration of oxygen fell at times to zero, which was possibly caused by the strong metabolic activity of the bacteria. This gave the impression that manganese reduction would be an anaerobic process. Therefore, it was also tested for manganese reduction under strictly anaerobic conditions.

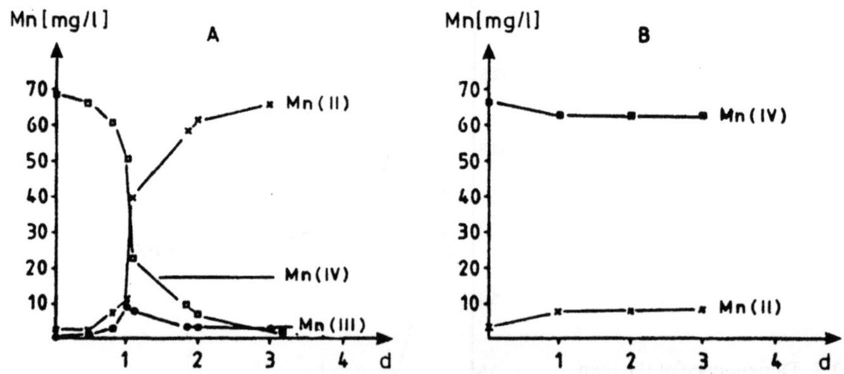

Fig. 5.2.2.6. Change of concentrations of Mn(II), Mn(III) and Mn(IV) in the nutrient solution during Mn(IV) reduction by *Pseudomonas* sp.

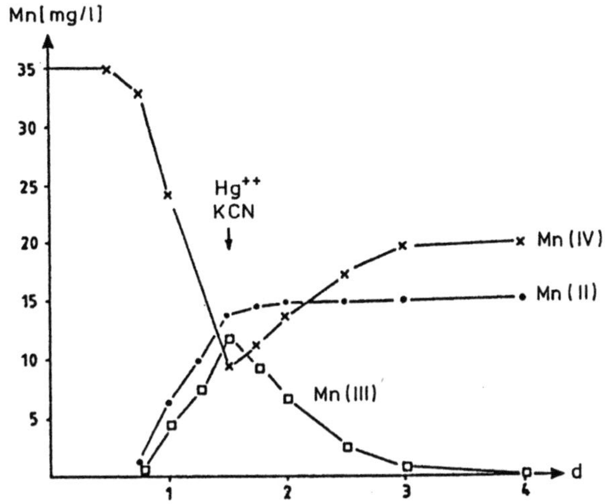

Fig. 5.2.2.7. Change of the concentrations of Mn(II), Mn(III) and Mn(IV) after addition of KCN or Hg^{2+} to the nutrient solution in the presence of Mn(IV) during manganese reduction by *Pseudomonas* sp.

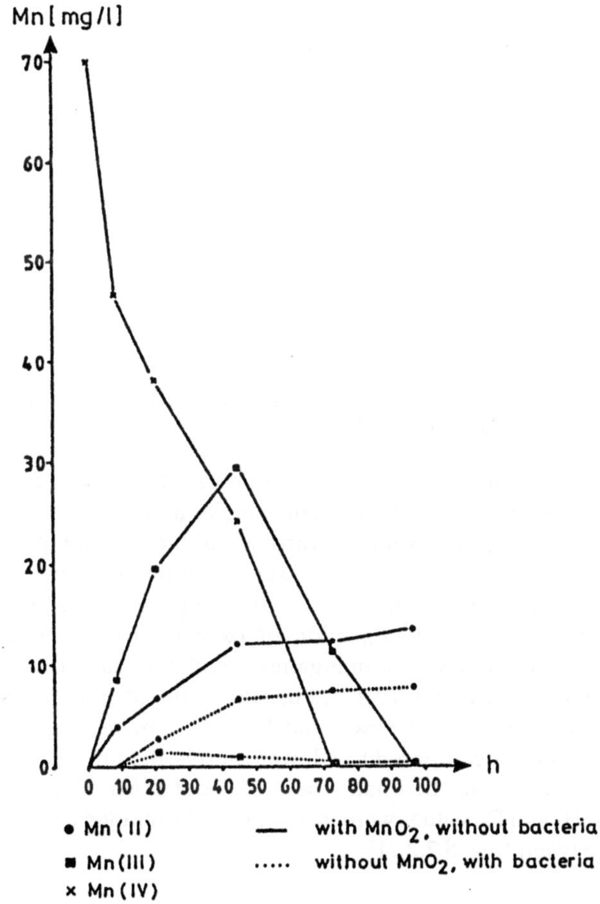

Fig. 5.2.2.8. Changes of concentrations of Mn(II), Mn(III) and Mn(IV) during manganese reduction

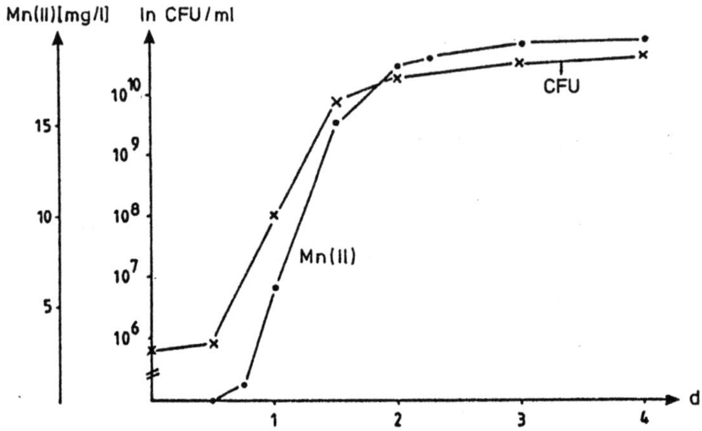

Fig. 5.2.2.9. Changes of concentrations of Mn(II) and the CFU of *Pseudomonas* sp. during manganese reduction

Under anaerobic conditions, but in the presence of Mn(IV), the growth of this *Pseudomonas* sp. was not determined. Under these conditions growth and Mn(IV) reduction took place only if nitrate, nitrite or $Fe(CN)_6^{3+}$ were added to the nutrient solution.

Under aerobic conditions (on a shaker) the manganese reduction was examined to study its dependency on carbon sources (glucose, fructose, glycerin, sodium lactate) (Table 5.2.2.1). More Mn(IV) was reduced to Mn(II) in the presence of high concentrations of organic carbon. But doubling the quantity of the carbon source did not lead to a comparable reduction rate: more Mn(IV) was reduced to Mn(II) per mol organic carbon in the presence of lower concentrations of organic carbon.

The Manganese Cycle in an Unsaturated Soil Horizon and in a Saturated Sediment Layer

Twelve different media were used to cultivate manganese-oxidizing bacteria. Altogether 225 soil and sediment samples and 30 water samples were microbiologically and chemically examined throughout the research project.

In a naturally reduced sandy aquifer system near Hanover, 10^2 to 10^3 manganese-oxidizing and 10^2 to 10^5 manganese-reducing bacteria per 10 g were observed. After the injection of oxygen-enriched water, the CFU increased to a maximum of 8×10^5 manganese-oxidizing bacteria and 1×10^6 manganese-reducing bacteria per 10 g. Thus, the ratio of manganese-oxidizing to manganese-reducing bacteria was about 1 in the samples originating from the oxygen-treated aquifer as compared to a lower ratio in the uninfluenced samples (Fig. 5.2.2.10). In those samples that had a ratio of about 1, Mn(III) and Mn(IV) could be found; the pH- and E_H values were in a range that excluded a chemical oxidation of Mn(II) with oxygen (Fig. 5.2.2.11).

Table 5.2.2.1. Highest amounts of reduced Mn(II) from Mn(IV) in dependence of the concentration of organic carbon and of the organic carbon source

	(mg/l)					
C-source	0.2	0.5	1.0	2.0	3.0	4.0
Glucose	9.5	13.0	15.0	18.0	19.0	20.0
Fructose	0.0	0.5	5.3	13.2	19.0	21.0
Glycerin	4.5	6.5	10.3	12.7	15.0	16.1
Na-lactate	0.0	0.5	8.9	11.2	13.4	14.7

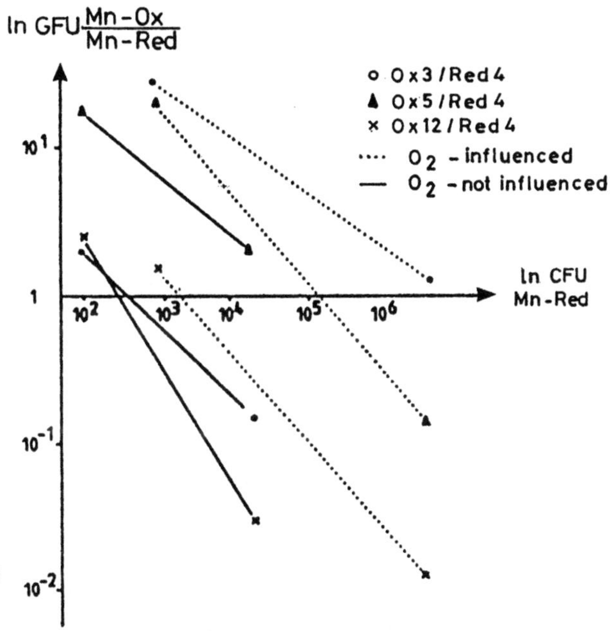

Fig. 5.2.2.10. Changes of the ratio of manganese oxidizing to reducing bacteria after injection of oxygen-enriched water into the aquifer. Designation of culture media: Ox 3/ Red 4; Ox 12/ Red 4. Ox media for cultivation of manganese-oxidizing bacteria; Red media for cultivation of manganese-reducing bacteria

5.2.2.4 Discussion

By the physiological tests, Mn(III) was shown to be an intermediate product of the manganese oxidation and reduction. Without microbiological presence and activity, the finding of Mn(III) in unconsolidated sediments cannot be explained.

The experiments with *Arthrobacter* sp. demonstrated that the utilization of an organic chelating agent is not a condition for the oxidation of Mn(II). Furthermore, there are no relationships between the chemical stability of the Mn complex and the microbial oxidation rate. The general statement that complex-bound manganese would be easier to oxidize by microorganisms, could not be verified (Möse and Brantner 1966). The first step of the microbial manganese oxidation is the attachment of the manganese complex or ions to the cell surfaces. With the addition of strongly chelating agents such as Cu^{2+} or Hg^{2+} or K^{+}, this attachment does not occur

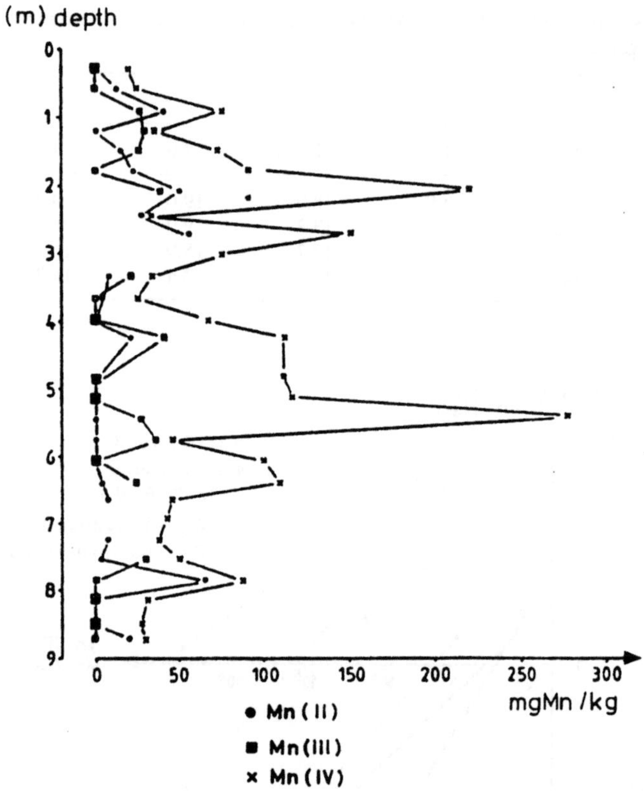

Fig. 5.2.2.11. Concentrations of Mn(II), Mn(III) and MN(IV) in unconsolidated sediment samples from oxygen-influenced area. ● Mn(II); ■ Mn(III); x Mn(IV)

and, therefore, manganese oxidation is not initiated. For analytical and methodical reasons, it could not be proven whether the attached Mn(II) was oxidized together with the ligand or only as an ion; further, it could not be observed where the manganese oxidation took place: e.g. within the cells or outside of them.

The mobilization and reduction of Mn(IV) to Mn(II) is a reaction which takes place outside the cells; Mn(III) must be chelated in a complex which was produced by the *Pseudomonas* sp. For this strain, manganese reduction already started under aerobic conditions and ceased under strictly anaerobic conditions. But in the presence of nitrate, Mn(IV) reduction started again under anaerobic conditions; Mn(IV) and nitrate were probably reduced at the same time.

The experiments with the pure culture demonstrated that manganese can be reduced microbiologically and that it can be oxidized under the same E_H and pH conditions. A regulation of both reactions could take place with the concentration of utilizable organic carbon. According to these experiments, manganese reduction was possible in the presence of oxygen and of small quantities of organic carbon; Mn(IV)-reduction was promoted in the presence of high concentrations of organic carbon; under these conditions the manganese oxidation was decreased.

The examination of aquifer samples has shown that manganese-reducing and -oxidizing bacteria could be cultivated under reduced and oxidized conditions; neither can be used alone to determine the redox conditions of a soil. On the other hand, if the numbers of both types were known, one could determine whether manganese oxidation or reduction takes place in situ.

The finding of Mn(II) in unconsolidated sediments has a particular meaning. Further tests have shown that Mn(III) can be determined in manganese deposits of wells. A catalytic activity of Mn(III) during the oxidation of Mn(II) to Mn(IV) has to be accepted.

Bacteria already produce Mn(III) under E_H/pH conditions, where a chemical production is very unlikely.

5.2.3.5 Conclusions

- A strain of *Arthrobacter* sp. started the oxidation of Mn(II) to Mn(III) at the end of the growth phase. In the presence of Mn(II) and oxygen, Mn(III) was converted to Mn(IV).
- The presence of organic substrates decreased the rate of Mn oxidation: Mn lactate > Mn acetate > Mn citrate. Mn lactate and acetate were easier to oxidize than $MnCl_2$ or $MnSO_4$. Mn(II) oxalate was very hard to oxidize, and oxalate could not be used as a carbon source by the *Arthrobacter* sp.
  The bacterial manganese oxidation decreased in the presence of high concentrations of chelating agents (EDTA, NTA, HA, FA). Phosphates inhibited the manganese oxidation but not the growth of the strain.
- During the manganese reduction, by the strain of *Pseudomonas* sp., Mn(III) was produced as an intermediate, metastable product. Mn(III) was reduced to Mn(II). For the reduction of Mn(IV) it was not necessary that the bacteria were in contact with MnO_2. Aerobic strains as *Pseudomonas* sp. reduced Mn(IV) under anaerobic conditions only in the presence of nitrate or nitrite.
- When oxygen was artificially added to a reduced Mn(II) containing aquifer, the CFU of manganese oxidizing and reducing bacteria in unconsolidated sediments increased by 10^2 to 10^3. The ratio of the CFU of manganese-oxidizing over those of reducing bacteria was determined for the reduced and the artificially oxidized aquifer samples. The data show that the reduced area had a ratio of under 1, and in the oxidized area the ratio was most of the time over 1.

5.2.3 Interactions Between Humic Acids and Microorganisms

G.-J. Tuschewitzki, B. Langer and H. Otremba

5.2.3.1 Background

In comparison with surface water, oligotrophic ground water with low levels of organic substrates is characterized by low counts of microorganisms. Growth and reproduction are often essentially limited by the deficiency of utilizeable carbon sources. As humic substances (HS) are quantitatively the largest carbon reservoir of many ground waters, it was questioned whether microorganisms use HS for growth and reproduction in ground water as they do in surface water (Tranvik and Höfle 1987).

Furthermore it is not known in which way the characteristics of humic acids (HA) influence growth of microorganisms in ground water.

The properties of humic substances, to complex organic and inorganic compounds, may reduce the low nutrient content of ground water to a much lower extent. In this way the microbial growth will be restricted. On the other hand, substrates near the surface may be complexed by HA, transported into deeper areas where, by releasing these substances, microbial growth will be enhanced (Hessen 1985; Shimp and Pfänder 1985).

These possibilities of interactions between microorganisms and HS were examined with two sets of experiments: first by batch cultures with selected bacterial strains and media of varying composition and second by experiments with fixed-bed reactors (*FBR*) which were continuously supplemented with ground water and HA. The latter set of experiments showed conditions more corresponding to those in nature.

5.2.3.2 Materials and Methods

The influence of humic acids on reproduction of microorganisms was examined by growth experiments in batch cultures with *Pseudomonas fluorescens* DSM 84 and other species also occurring in the ground water: *Pseudomonas testosteroni* DSM 38, *Pseudomonas aeruginosa* DSM 1117, *Pseudomonas palleronii* DSM 63, *Pseudomonas stutzeri* DSM 50227 and *Zoogloea ramigera* DSM 287.

The medium (100 ml in 259 ml bottles) used for all experiments contained the following ingredients per liter ddW (Drews 1976, modified): TRIS/HCl (pH 7.0), 0.05 mol; $(NH_4)_2SO_4$, 1000 mg; $CaCl_2$, 50 mg; $MgSO_4 \times 7H_2O$, 580 mg; $FeSO_4 \times 7H_2O$, 2 mg; KH_2PO_4, 6.8 mg; $Na_2HPO_4 \times 12H_2O$, 17.9 mg; trace elements solution 1.0 ml (Drews 1976).

Depending on the test conditions, organic compounds like sodium acetate, glucose, 4-hydroxybenzoic acid, vanillic acid or humic acids were added to the medium in various amounts. Two types of HA preparations were used: commercial humic acids from FLUKA (HA-Fl) and humic acids (HA-HAN) which were eluted from the ion exchangers used for preparing drinking water in the Fuhrberger Feld (Sect. 4.1.3).

The concentrated solutions were membrane filtered (0.45 μm pore size). Furthermore, concentrations of the other compounds of the medium such as N-, P- or mineral substances were reduced in order to examine the effects of HA under growth limiting conditions.

Batch cultures were inoculated with cells from precultures (final conc. 1%, OD_{405nm} 0.3–0.6) so that at the beginning each culture contained about 10^5–10^6 CFU/ml. Cultures were incubated in a rotary shaker water bath at 30 °C for up to 400 h. Growth of batch cultures was observed by following the optical density (OD) at 405 nm. In some experiments CFU numbers were determined by plating portions of 0.1 ml on "standard-plate count"-agar (tryptone 4 g, yeast extract 2.5 g, glucose 1 g, agar 15 g, distilled water ad 1000 ml; incubation 3 days at 30 °C). To facilitate comparison of different experiments, the OD of the first measurement was subtracted from the following measurements of that experiment.

A model system imitating conditions of the aquifer was used for further experiments. Five FBR (Fig. 5.2.3.1) of glass (volume 4.5 l) contained glass beads (4 mm diameter) and were fed with 5–10 l/h ground water, which came directly from the aquifer (water table at about 10 m depth). The ground water was free of organic pollutants and contained about 0.4 mg dissolved organic C/l. The surface area of the glass beads was 1.6×10^3 cm^2 and the reactors were charged with HA (FLUKA, 0.01% in 0.1 N NaOH, final concentration in the percolate 0.0001–0.001%). By plating on agar the HA concentrate appeared to be sterile. The pH of the ground water was 6.7–6.9 and that of the percolate 6.7–7.1. The temperature was between 10.5 and 11.5 °C. The *FBR* were observed for 6 months. The microbial flora of the inflowing ground water and of the percolates was determined qualitatively and quantitatively. Thirty to 50 isolates were tested of the ground water (inflow) and of each reactor.

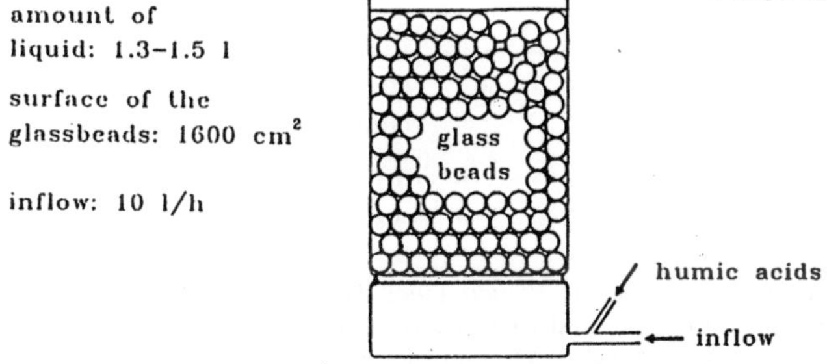

Fig. 5.2.3.1. Fixed-bed reactor (*FBR*) as used in the experiments (inflow = ground water)

These bacterial strains were classified in six groups by morphological and biochemical criteria: (1) pseudomonads and similar Gram-negative rods, (2) Gram-negative pigmented rods of the *Flavobacterium* group and similar ones, (3) Gram-negative facultatively anaerobic rods, (4) Gram-positive non-sporulating pleomorphic aerobic rods with coryneform morphology, (5) Gram-positive aerobic organisms with endospores and (6) other bacteria.

5.2.3.3 Results

Batch Experiments

Figure 5.2.3.2 shows six growth diagrams of *P. fluorescens* DSM 84. The mineral medium (MM) without any carbon source (negative control) showed only a very slight increase (up to 0.05 in OD at the beginning of the experiment. The addition of 0.1% acetate to the mineral medium (MM+AC, positive control) led to a steep increase of OD (0.6 after 15 h). The addition of 0.01% HA (MM+0, 01% HA, FLUKA) showed nearly the same result as the negative control.

Curves 4, 5 and 6 demonstrated the ability of *P. fluorescens* DSM 84 to convert aromatic compounds. Both the mineral medium with and without acetate was supplemented either with 4-hydroxybenzoic acid (Hy, 0.01 and 0.02%) or with 4-hydroxybenzoic acid and humic acids. Cultures with 4-hydroxybenzoic acid showed in comparison with the negative control dose-dependent higher OD. The combined addition of 4-hydroxybenzoic acid and humic acids, as well as acetate with 4-hydroxybenzoic acid and humic acids, did not show further increase in OD

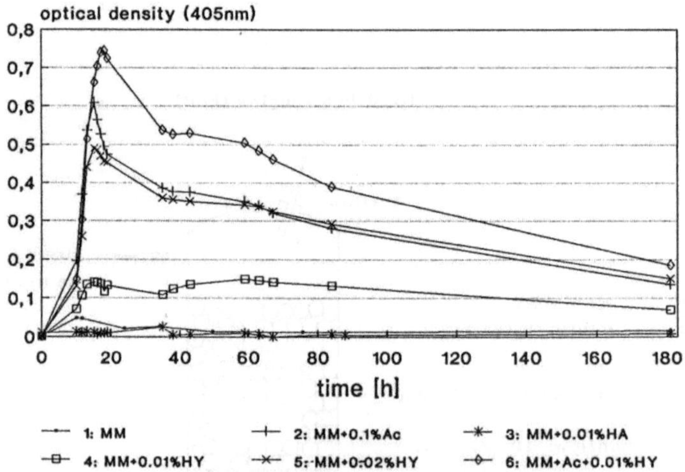

Fig. 5.2.3.2. Growth of *Pseudomonas fluorescens* DMS 84 in mineral medium (1:MM), in mineral medium supplemented with acetate (2:MM+0.1%AC), in mineral medium supplemented with 0.01% HA (3:MM+0.01%HA-Fl), in mineral medium supplemented with 4-hydroxybenzoic acid (4:MM+0.01%Hy, 5:MM+0.02%Hy, 6:MM+0.01%Hy)

than the same batches without humic acids (the results are therefore not shown in Fig. 5.2.3.2). Vanillic acid was not utilized under the given test conditions.

All of the other bacteria reacted in the same manner as *P. fluorescens* DSM 24 (strains tested: *P. testosteroni* DSM 38, *P. aeruginosa* DSM 1117, *P. palleroni* DSM 63, *P. stutzeri* DSM 50227 and *Zoogloea ramigera* DSM 287).

In a second set of batch experiments HA with acetate (0.1%) was tested with respect to the content of growth limiting nutrients such as N compounds and phosphate or other minerals salts. All previous experiments contained 0.1 mM PO_4^{3-} in the medium. In the following experiment (Fig. 5.2.3.3) in tests no. 1 and 2 the PO_4^{3-} concentration was decreased to a growth-limiting concentration of 0.01 mM. The P-limited growth attained an OD maximum of 0.15, however, under addition of HA (HA-Fl, 0.01%) 0.25. Both curves with and without humic acids (nos. 3 and 4) with 0.1 mM PO_4^{3-} reached an OD maximum of 0.62 and 0.67, respectively, depending on the acetate concentration.

The effect of HA on growth of microorganisms under N-limitation is shown in Fig. 5.2.3.4. The basic medium was used without $(NH_4)_2SO_4$ but supplemented with acetate (0.1%) and HA (0.01% HA-Fl or 1.0% HA-HAN). For a period of 90 h only a slight increase in OD (max. 0.03 after 90 h) could be observed without both N and humic acids. In comparison, cultures with HA showed an increase in OD: up to 0.07 units with HA-Fl (0.01%) and 0.36 units after 90 h with HA-HAN (1.0%).

Fixed-Bed Reactors

The changes of CFU numbers in both, inflow and outflow of the *FBR* charged with HA supplemented ground water are shown in Fig. 5.2.3.5. The ground water (inflow) used for feeding the reactors had 0–14 CFU/ml over a period of 200 d.

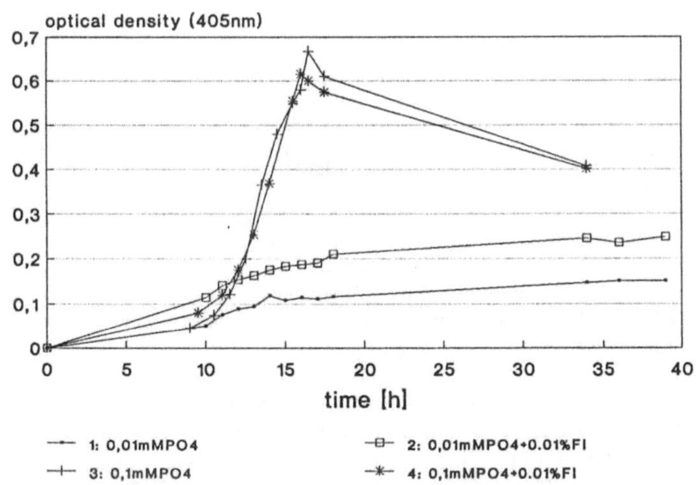

Fig. 5.2.3.3. Growth of *Pseudomonas fluorescens* DSM 84 in mineral medium with 0.1 (no. 3, no.4) and 0.01 (no. 1, no. 2) mM PO_4^{3-}, supplemented with HA (4: 0.1 mM PO_4^{3-} +0.01% HA-Fl; 2: 0.01 mM PO_4^{3-} +0.01% HA-Fl) and acetate (1%; no. 1,2,3,4)

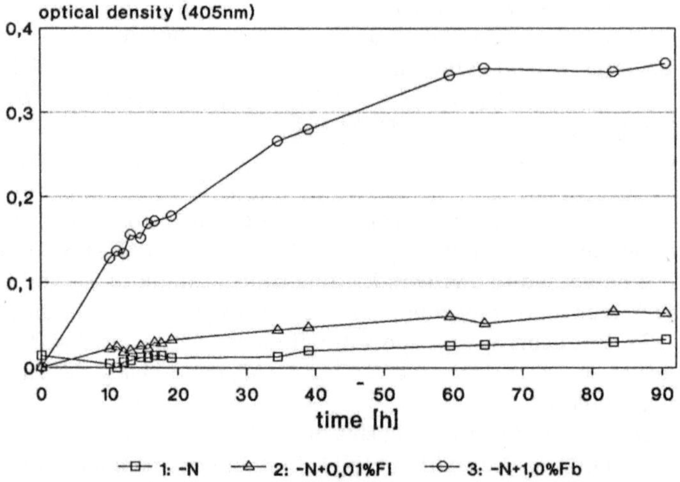

Fig. 5.2.3.4. Growth of *Pseudomonas fluorescens* DSM 84 in mineral medium without added ammonium (**1**), or supplemented with HA (**2** -N+0.01% HA-Fl; or **3** -N+1.0% HA-HAN)

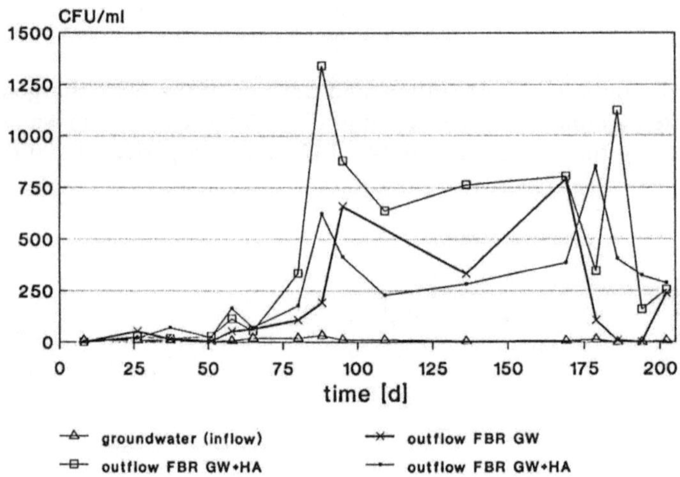

Fig. 5.2.3.5. Development of colony forming units in ground water (inflow) and percolates of *FBR* [one *FBR* with GW (outflow *FBR* GW), the other ones with ground water and HA (*FBR* GW+HA)] over a period of 200 d

Distinctly more microorganisms were present in all reactor percolates. After 50 to 80 days with CFU/ml ranging from 50 to 300, the CFU numbers increased to between 250 and 1300. The percolate of the reactor without HA (outflow *FBR* GW) was sampled after 90, 130 and 160 d; now the CFU were as high as those of the reactors with HA (outflow *FBR* GW+HA). Thus, the results of colony counts in the drains varied widely.

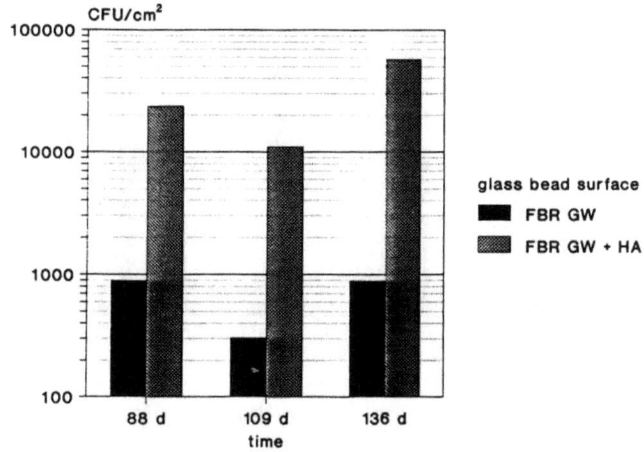

Fig. 5.2.3.6. Colony-forming units (CFU) on glass bead surfaces of the fixed-bed reactors fed with ground water (*FBR* GW) or fed with ground water supplemented with HA (*FBR GW*+HA) at three sampling dates

Completely different colony counts were observed on the surface of the glass beads. Figure 5.2.3.6 shows the comparison of glass beads from reactors with and without HA on three sampling dates (88, 109 and 136 d). Colony counts from glass beads of reactors fed with HA (*FBR* GW+HA; 7.8×10^3–7.8×10^4 CFU/cm2) were 1–2 log numbers higher than those from reactors without HA (FBR GW; $3 - 9 \times 10^2$ CFU/cm^2).

Comparing the distribution of bacterial counts in whole *FBR* it could be shown that 90–99% of all microorganisms were located on the glass bead surfaces and only 1–10% in the water.

About 200 bacterial strains were isolated from ground water (inflow) and from the percolates of the *FBR*; they were grouped according to morphological and biochemical characteristics as mentioned above (Fig. 5.2.3.7). The largest part (approximately 40%) represented pseudomonads and similar rods (group 1) found in the outflow of 3 *FBR*. In the ground water inflow these comprised only 15%. A large part of Gram-positive non-sporulating strains (35–45%; group 4) was observed in both, the reactors fed with HA and from the inflowing ground water but not from the reactor solely fed with ground water (without HA). Organisms of group 2 (*Flavobacterium* and similar ones) were found with less than 10% in all samples, bacteria with endospores could not be isolated. However, Enterobacteriaceae and related organisms were found in all samples. The majority of group 1 strains belonged to *Pseudomonas fluorescens* and those of group 4 to actinomycetes.

5.2.3.4 Discussion

Humic substances have been identified as reservoirs for secondary production in lakes (Hessen 1985; Tranvik and Höfle 1987). These conclusions were based on

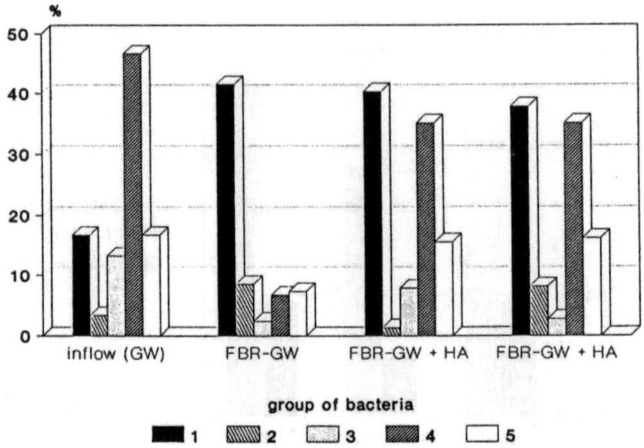

Fig. 5.2.3.7. Distribution of bacterial strains of fixed-bed reactors within different morphologic and taxonomic groups. Group *1* pseudomonads and similar Gram-negative strains; *2* Gram-negative pigmented rods like *Flavobacterium* and similar ones; *3* Gram-negative facultatively anaerobic rods; *4* Gram-positive non-sporulating rods with pleomorphy or coryneform morphology; *5* others

estimations of biomass, which were calculated by microscopic determinations of cell volumes and multiplication with a conversion factor (average mg C/μm^3 of cell volume).

The significance of humic substances as substrates in the seepage and ground water was still largely unknown. The results of these growth experiments (batch cultures with different media) and of the *FBR* fed continuously with ground water and HA gave strong evidence for a growth stimulation by HA in the ground water.

Multiplication of microorganisms in the batch experiments in the presence of all necessary minerals and HA could not be found as regards the OD measurements (not even with hydroxybenzoate as a primer). But the experiments with incomplete media showed on the other hand that both phosphorous and nitrogen, which are complexed within HA, will exert a strong impulse on growth, if these elements are growth-limiting and other growth substrates are present. Substances complexed by HA such as phosphate and nitrogenous compounds allow for microbial growth and reproduction in an environment which is naturally devoid of these growth limiting elements.

The experiments where *FBR* were fed with ground water simulated the conditions in an aquifer. The highly diverging colony counts in the outflow of *FBR* did not allow a clear decision whether there was a difference between reactors with or without HA. But the glass bead surface colonization in the *FBR* with HA was clearly greater than without. These results show the importance of surfaces as controlling factors in the colonization of low nutrient biotopes. The increase in CFU in all *FBR* (with or without HA) may be a result of the surface adsorption of substrates (ZoBell 1943).

The colonization of the surfaces in the *FBR* with HA indicates that microorganisms within a biofilm could be more versatile than free-floating microorganisms with respect to using recalcitrant HA (Atkinson and Fowler 1974; Molin and Nilsson 1985). The higher counts of pseudomonads and actinomycetes in the *FBR* outflow with HA indicate that there possibly may be two groups of microorganisms involved in the use of HA in ground water. In other biotopes two such different groups have been recognized (De Haan 1974; Gonzales et al. 1986; Martin et al. 1976).

5.3 Microbiology of Selected Location

5.3.1 Subsurface Microbial Activities in the Sandhausen Forest Ecosystem

R. Weyandt and R. Schweisfurth†

5.3.1.1 Microorganisms as a Final Step in the Degradation of Organic Matter

The translocation and transformation of biomass can be considered as hydrochemical processes in the natural water cycle of terrestrial systems. They are controlled by biotic and abiotic factors. The present investigations dealt with the biogeocoenosis of a forest ecosystem near Sandhausen/Heidelberg (SDH; Fig. 1.1, Table 1.1; Sect. 4.2.3) as an adaption to the local abiotic conditions. Generally, it is understood that the complexity of the processes occurring, the interrelations, and interdependencies that can be observed increase with the size of the investigated compartments. In the ecosystem under consideration here, the biotic components include the following groups of organisms: human beings, plants (trees and herbal bushes), birds and small mammals, invertebrates in soil (nemathelmintes, annelida, arthropoda), protozoa, algae, fungi and bacteria. The contributions of these organisms differ thoroughly in quality and quantity depending on the properties of the respective habitats. However, all of these groups contribute to the translocations and transformations of the living substance. The present analysis of mainly microbiological processes does not include most of the specific activities of the macroorganisms. Therefore, it is not a total overview but a contribution to the understanding of some specific phenomena.

5.3.1.2 Methods

The present report, which is a part of a study of distribution and activities of microorganisms in a subsurface profile (Weyandt 1990), discusses only the results gained from core samples collected down to a depth of 10 m (July 1987) and down to a depth of 1 m (December 1987), respectively. There is still no detailed information available about correspondence or variation of different sites; however, more recent investigations of the same area confirm the results.

Table 5.3.1.1. List of the culture media used for investigating bacterial colony-forming units (CFU)

Culture medium	Abbreviation	Remarks and references
P-agar	P	Copiotrophic (Kanazawa and Filip 1986) to oligotrophic agar (Wolters and Schwartz 1956)
Litter-extract agar	EXT	Organic substance used which was typical for this site (Müller 1965)
Tryptose-glucose-meat extract agar	TGE	Copiotrophic agar medium (Baumgarten and Leretzow 1969)

The investigations focussed on examinations of the soil and sediment samples, whereas the corresponding water samples were not available. Besides an investigation of different culture methods for physiological groups and colony-forming units of bacteria (Table 5.3.1.1) and fungi, the biochemical oxygen consumption (*BOC*), different enzymatic activities in soil, and the abiotic parameters were determined: pH value, organic carbon content and moisture content. A description of the methods employed and parameters is found in Weyandt and Schweisfurth (1989, Sect. 5.1.2; for the CO_2 data in the soil see Sects. 4.2.3 and 5.2.3).

5.3.1.3 Description of the Site

The investigated forest ecosystem, located in the Schwetzinger Hardt near Sandhausen (Sect. 4.2.3), consisted of pines (*Pinus silvestris*), beech (*Fagus silvatica*) and a few scattered oak trees (*Ouercus robur*). Moss, herbage and bush layers were almost absent. The forest soil was covered with litter. The subsoils are sandy and gravelly fluvial sediments (Jacob 1987). At the time of sampling, the water table was approximately at 5.30 m depth. During the last years, the ground water table had dropped significantly due to the extraction of ground water by local water-works. The shallow ground water flows relatively slowly (40 m/a). The litter as well as the following lower layers were strongly acid (Fig. 5.3.1.1). The roots developing between the litter and the upper layer rich in humus could be explained by better nutritional conditions at the soil surface; here the oxygen supply and release of CO_2 were better. Also the release of metal ions (especially aluminium) due to the strong acidification could force the plants to root in horizons rich in organic substances. Furthermore, because nitrification in acid soils is almost suppressed (Bewley and Stotzky 1983), plants must take up especially ammonia to supply their nitrogen demand. This process, too, enhances the acidification of the plant root habitats by the release of protons (Neitzke 1984). These specific conditions of the site are detrimental for the growth of pines and beeches, which are otherwise known as acid-tolerant (Lindquist 1932).

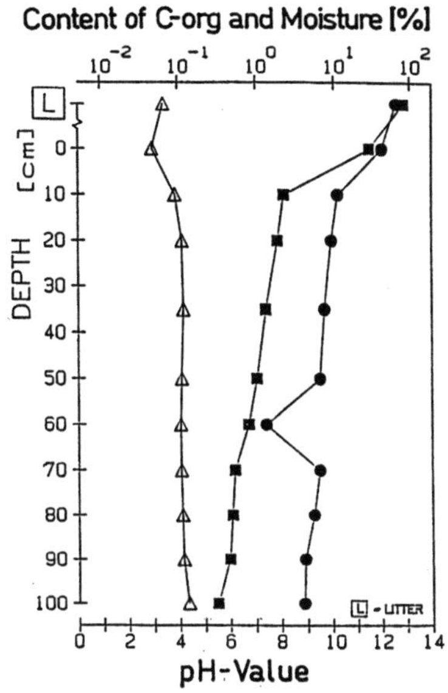

Fig. 5.3.1.1. Depth distribution of pH value (Δ), organic carbon (■) and moisture content (●) in the upper 1 m of the profile

5.3.1.4 Results

The significance of the upper soil layers, including the litter, as contact and border between the terrestrial system and the atmosphere and the particularities resulting from it justify a separate study of the upper 1-m zone (the soil sensu strictu), as compared to the lower parts of the profile down to a depth of 10 m.

Soil Microbiology

Distribution of Bacteria and Fungi
Despite a low pH value, the highest bacterial counts and those of CFU of fungi were found in the litter (Fig. 5.3.1.2), the bacterial counts being higher than those of the fungi. In the following 20–35 cm, the counts of bacteria and fungi decreased by more than two orders of magnitude and remained nearly constant between 60 and 100 cm depth, although there were slight variations, depending on the intensity of rooting at the respective site. In 1 m depth the CFU of fungi were below 1000/g dry soil, whereas bacterial counts ranged between 10^4 and 10^5/g, indicating a greater diminution of fungal CFU with depth.

Distribution of Different Physiological Groups in the Soil
Figures 5.3.1.3 and 5.3.1.4 show the distribution of various physiological groups. The numbers of organisms active in the nitrogen cycle (microorganisms with protein-ammonification potential $\pm O_2$, nitrate reducers or denitrifiers) or those contributing to the sulfur cycle (SO_4^{2-} and SO_3^{2-} reducers, S^o and $S_2O_3^{2-}$ oxidizers) corresponded to a large extent to the distribution profiles described above. No distinction was made between bacteria and fungi since liquid media were used to cultivate various physiological groups.

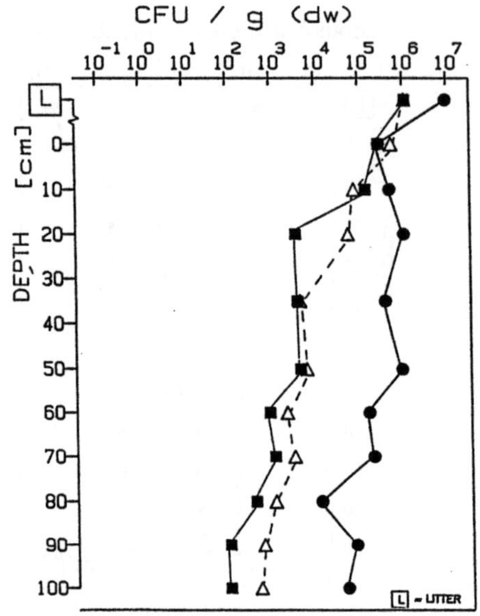

Fig. 5.3.1.2. Depth distribution of colony-forming units (CFU) of fungi (Δ) as well as of aerobic (●) and anaerobic (■) heterotrophic bacteria in the upper 1 m

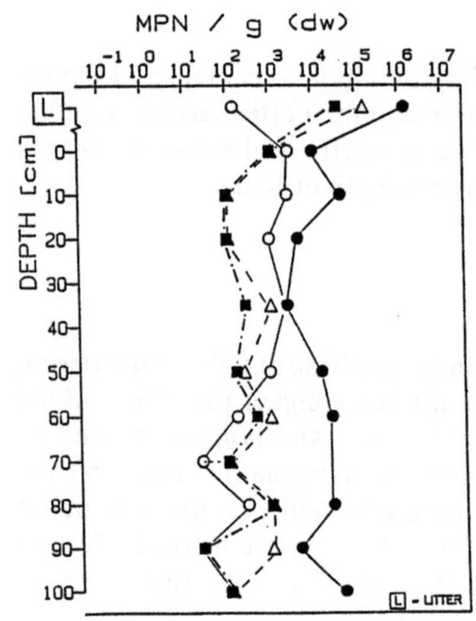

Fig. 5.3.1.3. Depth distribution of selected physiological groups of microorganisms in the upper 1 m: aerobic protein-ammonifiers (●); anaerobic protein-ammonifiers (○); denitrifiers (■); and nitrate reducers (Δ). *MPN* most probable number

Distribution of Enzyme Activities in the Soil Profile and Their Correlation with Bacterial and Fungal Numbers

The quantitative depth distribution of different microorganisms does not allow conclusions to be drawn on their momentary physiological activity, but it allows an estimation of their general participation in substrate transformation. Furthermore, soil enzymatic activities give some indications about more or less special and actual substrate turnover processes.

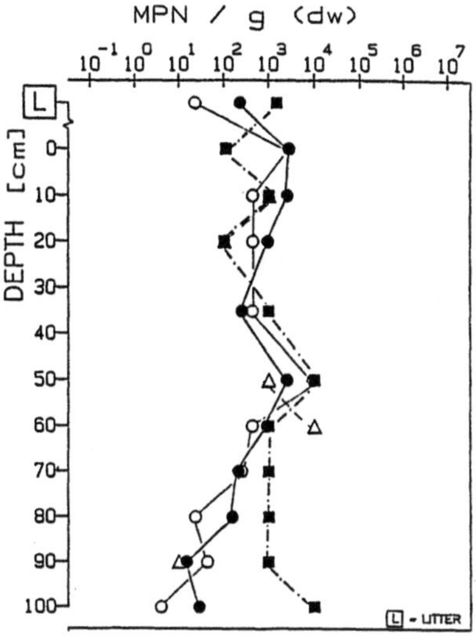

Fig. 5.3.1.4. Depth distribution of selected physiological groups of microorganisms representing the sulfur cycle: $S_2O_3^{2-}$ oxidizers (■); S^0 oxidizers (Δ); SO_4^{2-} reducers (○); SO_3^{2-} reducers (●); *MPN* most probable number

The term "enzymatic activity" in soil includes deliberately a possible participation by plant or animal enzymes. Therefore, the activities cannot clearly be related to specific bacteria, fungi or other microbial groups. Nevertheless, as a first step the activity of all organisms found in a given sample was determined. This approach excludes misinterpretations by adaptations of the organisms or by induction of enzymes as they result from biochemical reactions in vitro.

Figure 5.3.1.5 shows the profiles of pectinase-, cellulase-, and xylanase activity within the soil. The activities of catalase, saccharase and acid phosphatase (not represented) corresponded to this distribution pattern. These data revealed that the litter rich in humic substances was the layer with the highest detectable activity. The

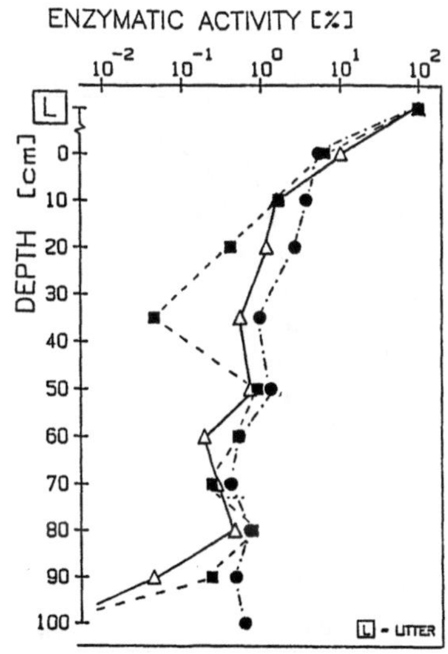

Fig. 5.3.1.5. Relative activities with depth of pectinases (●), cellulases (Δ) and xylanases (■) related to that of the litter (set to 100%)

activities of esterases (Schnürer and Rosswall 1982) can be determined by using different substrates. Down to a depth of 20 cm there was a change in the preference of substrate decomposition from short-chained to long-chained ones. This qualitative shift may indicate that there existed a transitional zone between a zymogenic and an autochthonous microflora in the soil. An increasing chain length slowed down the absolute time for enzymatic hydrolysis (Fig. 5.3.1.6). However, the percentages of the data obtained for the profile and compared to the litter values, demonstrate a better hydrolysis of longer-chain substrates with depth, below 20 cm depth.

The "biochemical oxygen consumption" (*BOC*) represents another measure for microbial activities and for the availability of organic substances in soil. The *BOC* was determined by incubating a soil sample in oxygen-saturated water and by measuring the O_2 consumption. The profile of this oxygen consumption, measured after 5 days of incubation, corresponded well to the profiles obtained for enzymatic parameters (especially of dehydrogenases) and also for those of the counts of microorganisms. These measurements indicated that the oxygen consumption decreased rapidly with depth. As accompanying investigations have shown, the oxygen consumption could have been limited by the availability of micro- and macronutrients (Strasburger 1978) such as nitrogen and phosphorous.

Microbiology of the Unsaturated and Saturated Zone down to a Depth of 10 m

Profiles of DOC, C_{org} and C/N were derived from core samples collected at the same site in June 1986 (Sects. 2.6 and 4.2.3, this Vol.). The horizons with higher values are of special interest: a layer between 2 and 3 m, characterized by a substantial increase in pH value, a further horizon ranging from 4 to 5 m which includes the capillary zone, and the upper ground water horizons. Several soil samples taken from different locations within the test field revealed further specific properties of

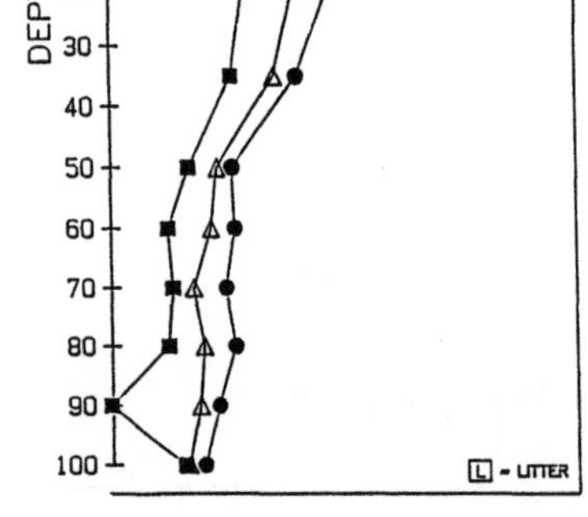

Fig. 5.3.1.6. Turnover of different fluorescein derivatives in the upper 100 cm of the profile: FD acetate (●); FD propionate (Δ); FD butyrate (■); *mU* milli-units

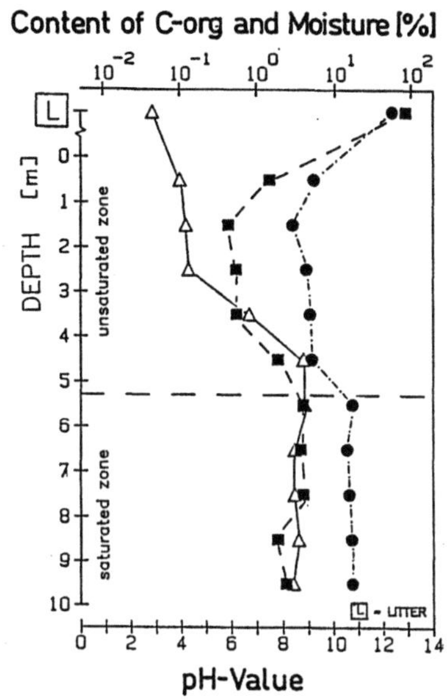

Fig. 5.3.1.7. Depth distribution of pH (Δ), organic carbon (■) and moisture content (●) down to 10 m

these zones with respect to microbial counts and activities. Figure 5.3.1.7 shows the depth profiles of pH value, C_{org} content and moisture content.

Distribution of Microorganisms down to a Depth of 10 m

Generally the numbers of bacterial colonies (aerobes and anaerobes) and the fungal CFU decreased constantly in the unsaturated zone with depth, followed by an increase in the ground water horizons (Fig. 5.3.1.8). A certain increase of heterotrophic bacterial numbers in the horizon was observed between 2 and 3 m depth, which presumably was the former fluctuation zone of the capillary fringe and the ground water table. Tests made with three different media (Table 5.3.1.1) at pH 7 and after incubation at 20 °C for 4 weeks did not reveal significant differences in bacterial numbers.

Distribution of Physiological Groups in the Upper 10 m of the Profile

The distribution profiles of different physiological groups (Figs. 5.3.1.9 and 5.3.1.10) refer to the nitrogen or sulfur cycle. Except for sulfur oxidizers, which were diminished in numbers especially in the unsaturated zone, the distribution profiles corresponded in many ways to those of the heterotrophic bacteria shown in Fig. 5.3.1.8. The numbers of bacteria with protein-ammonification potential and of nitrate-reducing microorganisms decreased in the unsaturated zone with depth, whereas their numbers increased substantially in the saturated zone, where a decrease in numbers of denitrifiers was observed. Here bacteria with protein-ammonification potential and nitrate reducers were detected only in some cases.

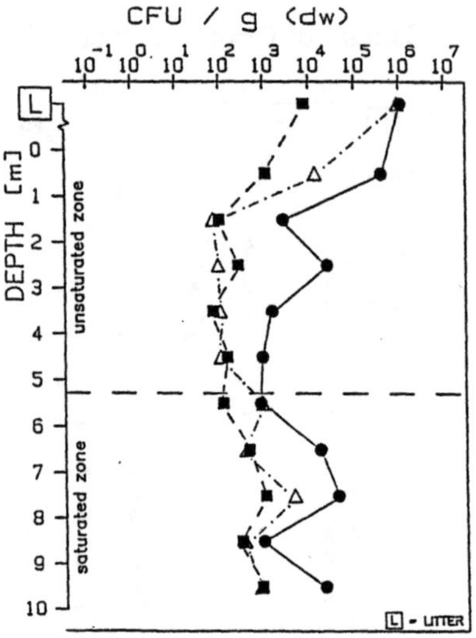

Fig. 5.3.1.8. Depth distribution of colony-forming units (*CFU*) of fungi (Δ) as well as aerobic (●) and anaerobic (■) heterotrophic bacteria down to 10 m

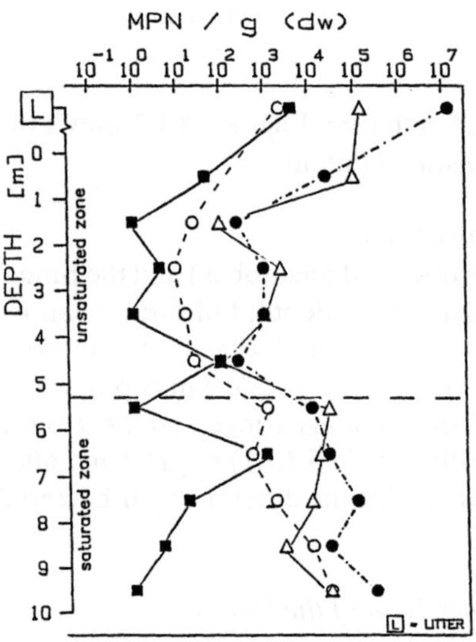

Fig. 5.3.1.9. Depth distribution of selected physiological groups of microorganisms down to 10 m: protein-ammonifiers (aerobic) (●); protein-ammonifiers (anaerobic) (○); denitrifiers (■); nitrate reducers (Δ); *MPN* most probable number

Distribution of Enzymatic Activities in the Upper 10 m of the Profile

In analogy to Fig. 5.3.1.5, the distribution of different enzymatic activities (catalase, BOC_{10}, acid phosphatases) is shown in Fig. 5.3.1.11. Here the profiles of the biochemical activities of saccharases, pectinases and xylanases (not shown) corresponded to those of the other enzymatic activities. With increasing depth there was an activity decrease of all enzymatic systems investigated. Apart from an activity increase of some enzymes in the horizon between 2 and 3 m depth, there was a significant increase in specific biochemical reactions in the ground water depending on depth. The profiles of bacterial and fungal colony numbers and CFU shown (Fig. 5.3.1.8) could be compared to the activity parameter "esterases" (Fig. 5.3.1.12): the

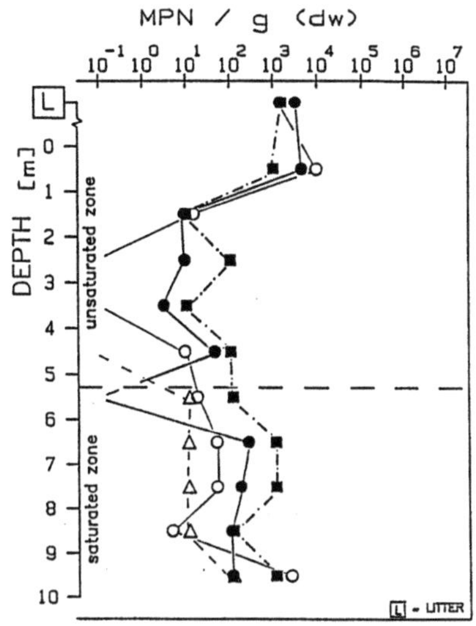

Fig. 5.3.1.10. Depth distribution of selected physiological groups of microorganisms representing the sulfur cycle, down to 10 m: $S_2O_3^{2-}$ oxidizers (■); S^0 oxidixers (Δ); SO_4^{2-} reducers (O); and SO_3^{2-} reducers (●); *MPN* most probable number

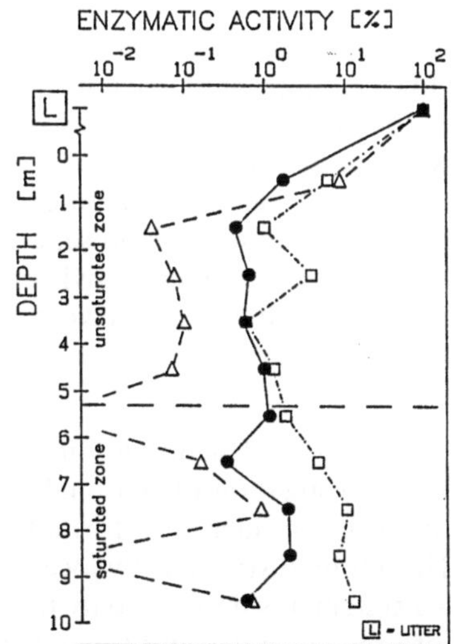

Fig. 5.3.1.11. Activities of calatase (●), acid phosphatase (□) and BOC_{10} (Δ) relative to those of the litter

correlation between activities and cell numbers shifted with increasing depth towards larger cell numbers, i.e. a larger number of microorganisms was responsible for only a slight increase in activity. In analogy to the transformed data of Fig. 5.3.1.6, the comparison between different substrates for esterases and the values measured in the litter revealed a shift in the enzymatic hydrolysis towards longer-chain substrates. The BOC_{10} (10 days of incubation) values (Fig. 5.3.1.11) confirmed the distribution of enzymatic activities within the profile.

Between 2 and 3 m, and also between 4 and 5 m, two layers with higher activities and higher numbers of microorganisms were observed in the unsaturated

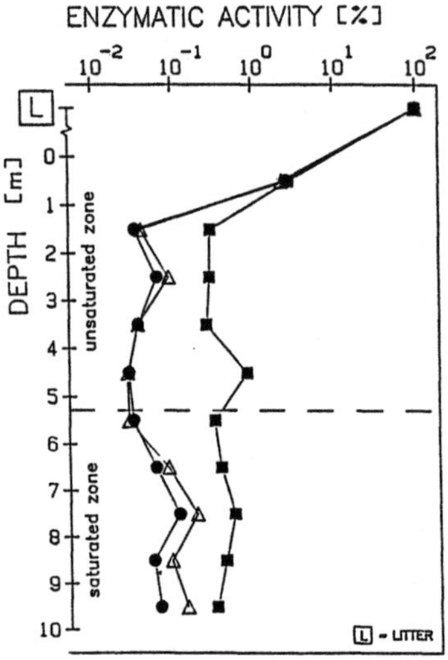

Fig. 5.3.1.12. Turnover of different fluorescein derivatives relative to those of the litter, down to depth of 10 m: FD acetate (●); FD propionate (Δ); FD butyrate (■)

zone characterized by a substantial increase in the respective values which may indicate different biochemical properties.

In the upper aquifer horizons higher polymeric organic substances may have been accumulated which were more resistant to decomposition. This accumulation resulted in increasing numbers of both bacterial and fungal CFU. Furthermore, oxygen may have been present in low concentrations, as indicated by fungal growth. Most of the enzymatic activities were also increased, but regarding their percental distribution there must have been a change in microbial populations or metabolic activities.

Examination of Seasonal Influences

The investigation of seasonal fluctuations in the individual horizons was complicated by the lack of information on the differences among the sampling sites. Therefore, distortions by local influences could not be excluded. Thus, the following statements describe only tendencies. The time-dependent CFU values for heterotrophic bacteria and fungi in the litter and bacterial counts down to 5 cm (Fig. 5.3.1.13) indicate that in analogy to the two maxima of bacteria observed in the litter in July 1986 and January 1987, the bacterial numbers in the layer from 0 to 5 cm twice reached a maximum, but some time later, presumably due to a vertical transmission of substrates. Furthermore, a maximum number of bacteria corresponded to a minimum number of fungi.

In Fig. 5.3.1.14a-c are shown the activity parameter "esterase", the BOC_5 the input of organic matter (leaf fall) into the system, and the values for CO_2 development in soil. The activities of esterases and *BOC* corresponded with the leaf fall and reached a local maximum especially in late autumn, whereas the values for CO_2 development in soil indicated a absolute maximum in July/August. A comparison with the seasonal numbers of bacteria demonstrates that there were hardly any

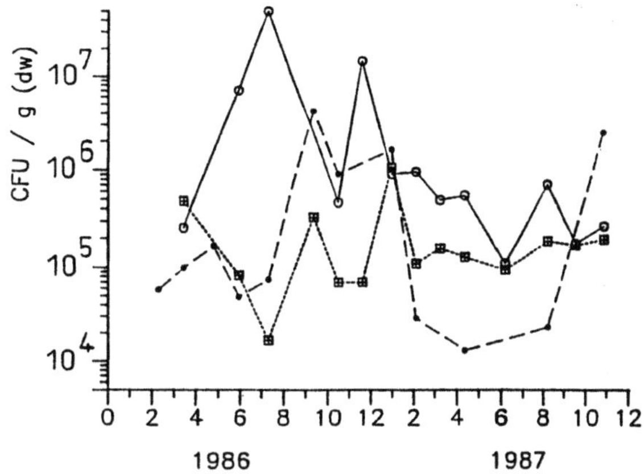

Fig. 5.3.1.13. Seasonal development of bacteria and fungi in the upper layers of the profile: bacteria in litter (○); fungi in litter (⊞); bacteria in 0–5 cm depth below the litter (●)

correlations with the development of activities. Only the numbers of fungal CFU corresponded approximately to the activity curves. For deeper horizons further sufficient data are not available.

The observed seasonal differences in the distribution of microorganisms and their activities probably resulted from multiple factors, from abiotic parameters as well as from various possible interactions of the participating organisms. Whereas the numbers of heterotrophic bacteria of the litter increased mainly in summer, the fungi (which are the first microorganisms to colonize the litter) gained numerically in importance during leaf fall in autumn and to some extent in spring. The formation of soil CO_2 reached a high peak during July and August, preferentially due to the seasonal respiration increase from roots in the transitional zone between the litter and the adjacent soil layer. Seasonal influences by the input of organic substance were observed in the time of leaf fall and the time when needles were shed, whereas data for the formation of necrotic root mass were lacking. The activity of esterases and *BOC* values did not correlate with data on CO_2 formation. This may indicate seasonal periods of increased microbial activities which were not directly correlated with the respective numbers of microorganisms.

5.3.1.5 Discussion

The forest ecosystem investigated, which at first sight seemed to be more or less homogenous, proved to be, from the ecological point of view, highly inhomogenous with respect to the soil microbiota present. On the level of microhabitats there may be totally different milieus adjacent to each other and consequently absolutely different conditions may prevail there for microbial growth. There are always correlations between the microorganisms and their microhabitats, which are influenced by external factors, such as temperature, the availability of free water, composition of the soil matrix, the nitrogen and CO_2 content, redox conditions, the nutrient supply and the presence of trace elements. An overlapping of interactions between microorganisms, the wickerwork of roots and small animals in soil is

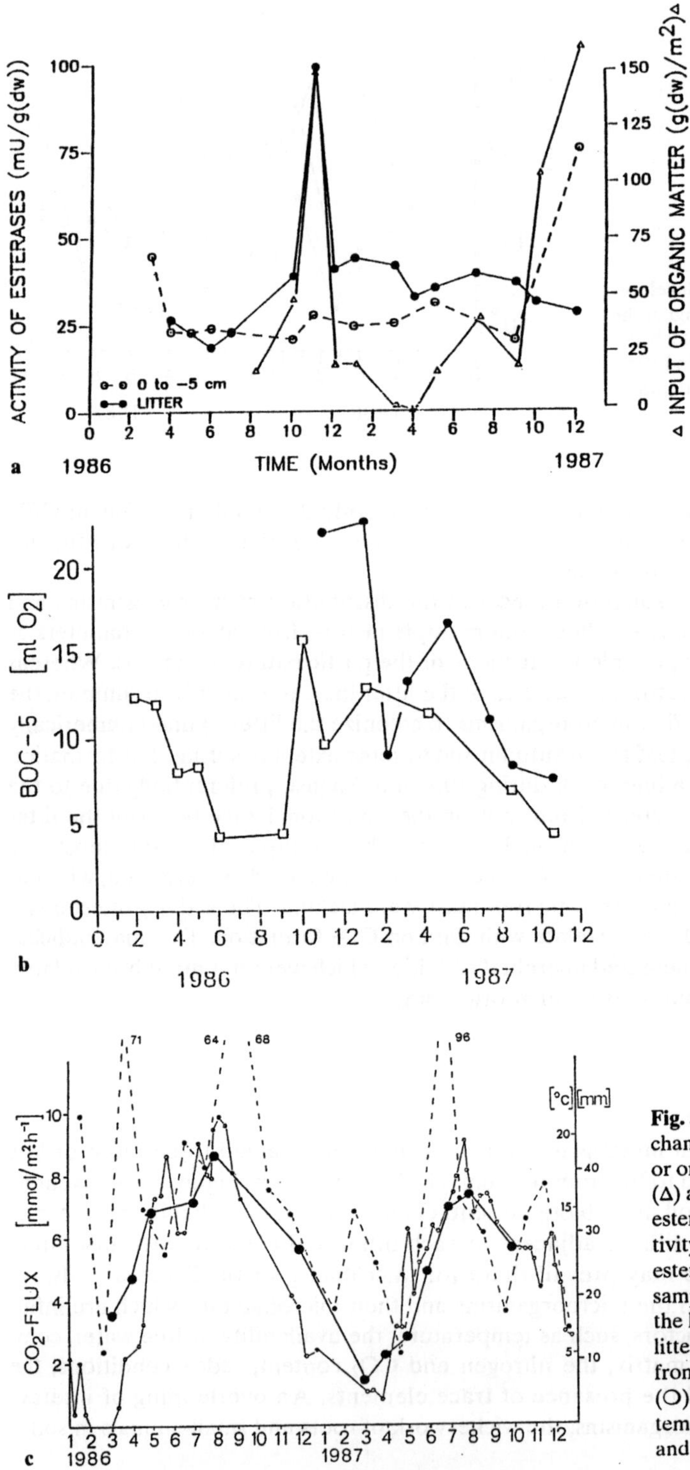

Fig. 5.3.1.14. **a** Seasonal changes of quantitative input or organic matter (leaf fall) (Δ) and activity of selected esterases; FD-A: esterase activity in 1 g litter (●), FD-B: esterase activity in 5-g sample from 0–5 cm below the litter (○); **b** *BOC-5* in 1 g litter (●), or 20 g sample from 0–5 cm below the litter (○). **c** CO_2 flux in soil (●), of temperature in the litter (○) and of rainfall (●)

also observed. Only a few specific methods allow an analysis of some of these phenomena.

The problems of methodology and interpretation have been discussed in Section 5.1.2 (see also Atlas and Bartha 1981 and Olsen and Bakken 1987). The choice of culture media and the conditions of the milieu have very selective functions.

A large number of bacteria which are unable to form colonies, but can be differentiated and detected by microscopy belong to the group of the so-called dwarf cells (Bakken and Olsen 1987). Their ecological significance is unclear, at least with respect to their measurable enzymatic activities. They may be considered as belonging to the autochthonous microflora in soil. Only very few of them are able to form colonies on agar. In comparison to the colony-forming heterotrophic "large" cells, the dwarf cells appear to represent a much smaller part of the microbial biomass with only very few physiological activities and an extremely low energy demand (Bakken and Olsen 1987).

When investigating the CFU numbers of fungi, no distinction was made between spores and fragments of hyphae. Therefore, activity measurements do not give informations on their participation in mass transformation. Activity measurements, such as that of "esterases" which comprise the activities of different enzymes, or other enzymatic transformations, also measurements of *BOC*, or measurements of CO_2 in situ do not differentiate between the participating organisms. For this reason, the total soil enzymatic activities measured cannot be deduced specifically from the necrotic mass of roots, the lysates of fungi and/or bacteria, the protozoa or even of small animals which were present in the soil. However, the measured enzymatic activities could, in part at least, be attributed to exoenzymes. A specific enzymatic activity depends not only on the species investigated but also on the site. In some cases, adsorption by soil particle surfaces and the resulting change of the molecular conformation may be very important. Also the inactivation and decomposition (turnover) of the exoenzymes, the availability of water (as reaction medium) and the local parameters of the micromilieu play a major role. Thus, the activity data of esterases measured in the site with the help of model substrates made it possible to draw specific conclusions only within the demonstrated limits. So it is reasonable to present activity data as relative activity distribution with depth. Only by additional activity measurements should it be possible to give more detailed information on substrate turnover rates and kinetics within a given in situ subsurface ecosystem.

5.3.2 Heterotrophic Bacterial Communities in the Bocholt Aquifer System

J. Kölbel-Boelke and A. Nehrkorn

5.3.2.1 Background

Until now, there is no investigation method known by which complete information on the "reality" of a microbial community in its natural environment can be obtained. Methods for isolating bacteia from their environment and their characterization in vitro are in principle selective, depending on isolation media and other growth conditions.

One possible way of describing bacterial communities became important in the last few years: the community analysis which rests on numerical taxonomy. This method has been employed in studies on various aquatic biotops, such as fresh water (Holder-Franklin 1981; Rosswall and Persson 1982; Witzel et al. 1982), estuaries and brackish water (Bölter 1977; Mallory et al. 1977), sea water (Austin et al. 1979), or drinking water (Dott et al. 1982).

This method was used to characterize heterotrophic bacterial communities of the Bocholt ground water environment (Kölbel-Boelke et al. 1988a,b). Our interest was focussed especially on the following questions: what would be the composition of bacterial communities in the ground water environment? Are there differences between those bacterial communities living attached to sediment particle surfaces and those that live free in the ground water? Are there community differences at different profiles and/or in different depths or layers of the same profile? Which physiological/biochemical capabilities can be observed with the bacteria communities of the ground water environment?

Answers were found to some of these questions, but many questions remained, and many more arose during this work.

5.3.2.2 Materials and Methods

Sediment and water samples were collected from the first stratum (up to 49 m deep) of the Bocholt aquifer system. The hydrogeochemical conditions of this sytem as well as the well positions are described in Section 4.1.1. The methods of sample collection and processing are described in Kölbel-Boelke et al. (1988b). For each sample, viable cell numbers were determined as CFU on a modified medium of Wolters and Schwartz (1956): medium P (0.1% Witte peptone; 0.01% glucose; 0.01% $K_2HPO_4 \cdot H_2O$; $FeSO_4$ $7H_2O$; agar; tap water; pH 7.2). This medium allowed a very efficient isolation of heterotrophic and oligotrophic bacteria (Nehrkorn 1968). Total cell numbers were determined by using the amido black 10B stain (Kölbel-Boelke et al. 1988b). Table 5.3.2.1 summarizes origin, kind and numbers of samples that were examined. Samples collected from different depths of the same bore hole and at the same sampling date were given similar sample numbers

Table 5.3.2.1. List of water and sediment samples for the determination of bacterial cell numbers and for the isolation of pure cultures

Sample	Borehole	Kind of sample water (W) sediment (S)	No. of samples examined (from different depths)		No. of isolates
			for viable cell counts	for isolations	
1–01	BOC-PES 5	W	10	5	246
2–02	BOC 2	S	7	6	268
1–03	BOC 1	W	17	6	352
2–04	BOC 4	S	6	6	341
2–05	BOC 5	S	18	18	717
2–06	BOC 7	S	21	7	266
1–08	BOC 3	W	5	4	181
1–09	BOC 5	W	17	8	341

which in the first position indicated the kind of sample (1 = water, 2 = sediment samples) but were processed separately. A more detailed list has been published by Kölbel-Boelke et al. (1988a).

About 40 to 60 colonies per sample were picked randomly from plates prepared for viable cell counts. Bacterial strains were characterized with regard to 155 different features: 40 characters described colony morphology, 37 cell morphology and 78 were physiological or biochemical characters (Kölbel-Boelke et al. 1988b). In order to test physiological and biochemical capabilities of these ground water bacteria, the media compositions, incubation conditions and test reactions were adapted to ground water conditions. Moreover, the tests were adapted to the microtitre plate method (Dott and Thofern 1980), a miniaturized test system which is suitable to characterize a large number of strains within acceptable time. All tests were described in detail by Kölbel-Boelke et al. (1988b). Characters of all strains were evaluated by similarity analysis, using computer programmes (Krambeck and Witzel 1983). The similarity values were calculated by the simple matching coefficient S_{SM} (Sokal and Michener 1958). Strains were grouped by complete linkage clustering (Sneath and Sokal 1973). Division into clusters and subclusters was carried out with similarity values of 80% and higher. Similarities between clusters were plotted as dendrograms.

The in vitro activities of single strains or communities (Dott and Trampisch 1983) were calculated from 58 biochemical tests that concerned mainly the degradation of carbohydrates, alcohols, organic acids, amino acids, nucleic acids, proteins, lipids, as well as some enzymatic properties (Kölbel-Boelke et al. 1988b).

5.3.2.3 Results

Cell counts

There were significant differences in the viable cell counts between water and sediment samples. In the former, viable cell counts (CFU) decreased rapidly from 0 to 5 m below surface. Below 5 m these numbers usually amounted to less than 100 CFU/ml, in many cases even to less than 50 CFU/ml. Sediment samples yielded mostly 10- to 100-fold higher viable cell numbers (10^2 to 10^4 CFU/g d.wt.); this varied strongly from sample to sample, probably due to the existence of different microhabitats. Particles rich in organic carbon (e.g. pieces of wood) were much more colonized than mineral particles. Vertical profiles of viable cell counts have been published by Kölbel-Boelke et al. (1988a). The profiles of viable cell numbers in water resembled those obtained for the concentrations of oxygen, nitrate and organic carbon (Kölbel-Boelke 1987).

Total cell counts usually yielded cell numbers two to three orders of magnitude higher than viable cell counts (Kölbel-Boelke et al. 1988a).

Comparison of Bacterial Communities from Different Boreholes

Studies of the morphotype diversity of bacterial isolates revealed great quantitative and qualitative differences in the community structures of different boreholes (Table 5.3.2.2). Even communities observed in boreholes about 5 m apart were structured very differently (Table 5.3.2.2: samples 2–02 and 2–04; 1–03 and 1–08). In biochemical degradation tests, each community showed its own preferences for

Table 5.3.2.2. Morphological composition of water and sediment communities from different boreholes in the Bocholt test field

Borehole (well) Sample	Water communities				Sediment communities			
	BOC-PES 5 1–01	BOC 1 1–03	BOC 3 1–08	BOC 5 1–09	BOC 2 2–02	BOC 4 2–04	BOC 5 2–05	BOC 7 2–06
% Gram-positive or -variable isolates	11	57	8	6	40	11	35	43
% Gram-negative isolates	89	43	92	94	60	89	65	57
Gram-positive or -variable isolates								
% rods	1	1	0	0	1	0	5	0
% coccoid rods/cocci	0	47	0	1	3	2	2	13
% coryneforms/pleomorphic cells	9	10	8	5	35	8	27	29
% branched filaments	1	0	0	0	1	1	1	1
Gram-negative isolates								
% straight rods	81	35	76	75	42	77	32	41
% curved rods	1	1	10	15	1	1	1	2
% unbranched filaments	1	5	0	0	12	7	3	0
% coccoid rods/cocci	4	1	2	1	1	1	1	11
% cells of variable forms	2	1	7	3	4	3	28	3
% isolates with flagellar movement	69	33	58	87	28	77	36	43
% isolates with gliding movement	0	5	0	0	11	7	3	0
% pigmented isolates	6	20	34	26	47	43	33	18
% isolates with stalks	0	4	3	6	1	1	2	0

specific substrates or substrate groups. Bacteria of sample 1–01 degraded best some mono- and disaccharides, whereas those of 2–06 preferred macromolecular substrates with the exception of polysaccharides (Kölbel-Boelke et al. 1988a).

Clustering of bacterial strains from the same borehole occasionally yielded for each community a matrix that could be divided into many clusters and subclusters.

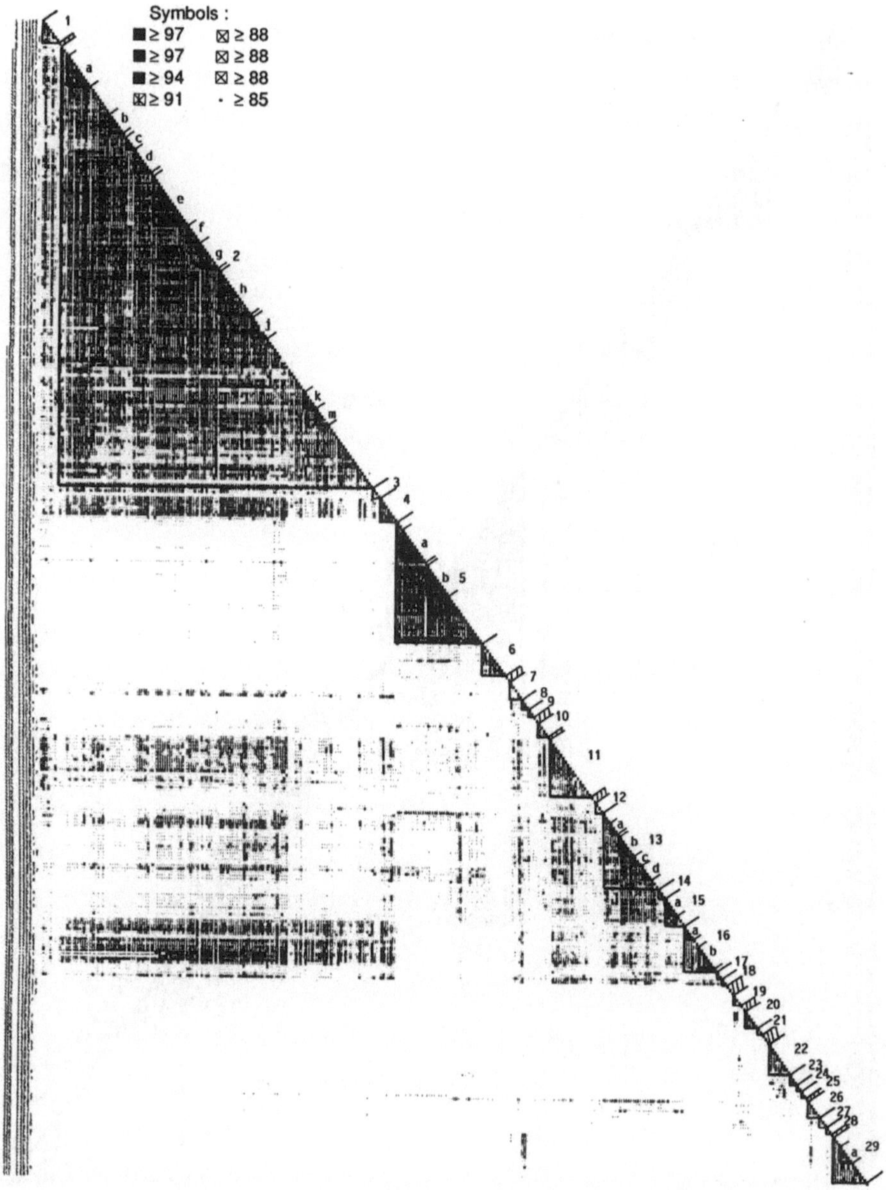

Fig. 5.3.2.1. Cluster matrix of water community 1–03 (well BOC1) with clusters and subclusters

As an example, the cluster matrices of isolates from samples 1–03 and 2–04 are shown in Figs. 5.3.2.1 and 5.3.2.2. Matrices of all other communities have been presented by Kölbel-Boelke (1987). The presence of such a great number of clusters and subclusters demonstrated the pronounced diversity of these communities. The similarities between all clusters of a single matrix were plotted as dendrograms

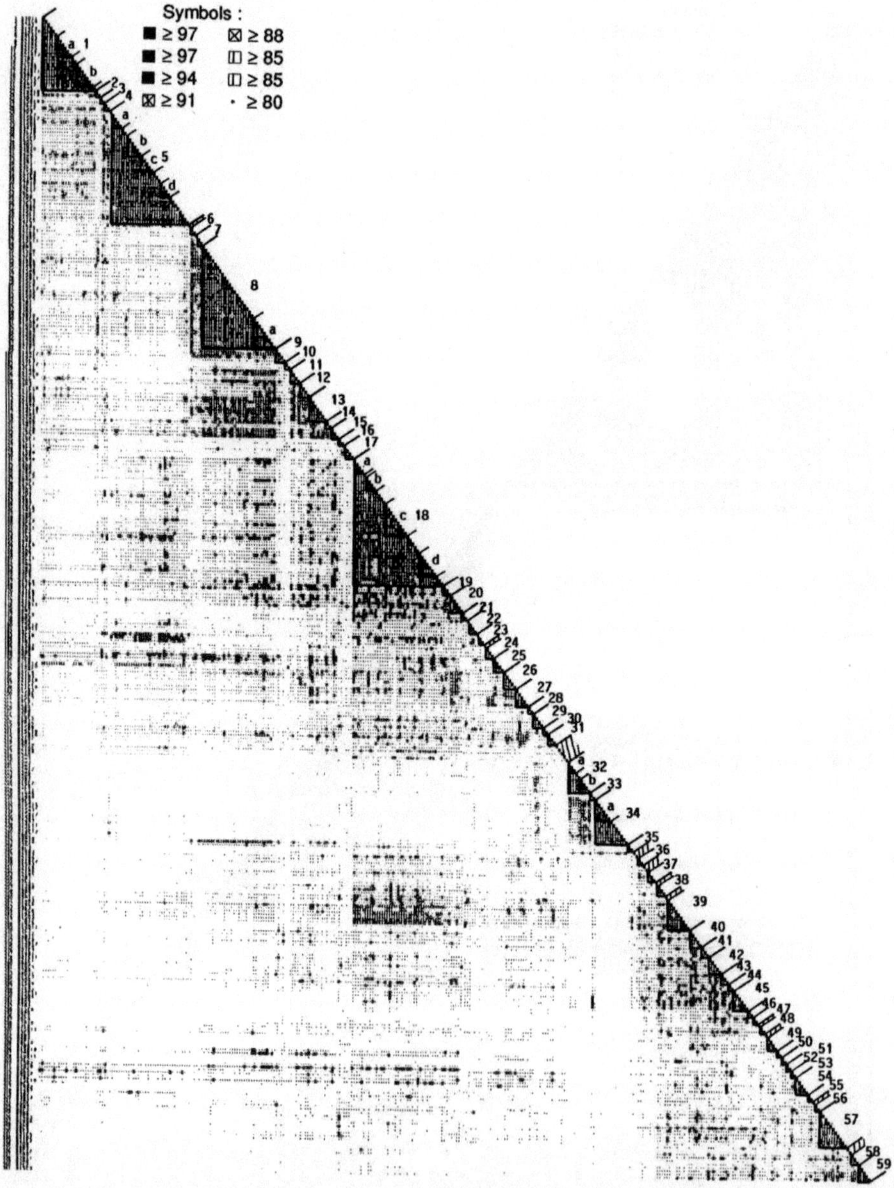

Fig. 5.3.2.2. Cluster matrix of sediment community 2–04 (well BOC4) with clusters and subclusters

(Figs. 5.3.2.3 and 5.3.2.4). Dendrograms of the other communities have been presented by Kölbel-Boelke (1987).

Bacterial communities isolated from different boreholes were clustered one with another to obtain information about their distribution in the various boreholes Figure 5.3.2.5 presents results of such pair clusterings between communities from different boreholes. Community designations (sample numbers) were arranged in this figure corresponding to the positioning of the boreholes in the test field. Communities processing similar strains (similarity ≥ 94%) were connected with lines if the strains were taxonomically similar, or with dotted lines if the taxonomic similarity was uncertain. Figure 5.3.2.5 demonstrates the general lack of similarity between communities from different boreholes in the same area or only restricted similarities. Even neighbouring communities, as in the case of those from samples 2–02 and 2–04 (or 1–03 and 1–08), were very different. This was also true for the comparison of a sediment (2–05) with a water community (1–09) which both had been isolated from borehole BOC5, but with a time delay of 8 months. Figure 5.3.2.5 shows that the direction of ground water flow did not influence or create a higher similarity of community compositions.

Altogether, about 2700 bacterial strains from seven different boreholes were characterized. Of these isolates 72% strained Gram-negatively and 28% were Gram-positive or Gram-variable (Table 5.3.2.2). A major group of isolates (52%) were Gram-negative, straight rods, and 17% were coryneform bacteria. It was not possible to identify all clusters on the genus or even up to the species level. Clusters that could be identified on the genus level very often belonged to *Pseudomonas (P. fluorescens, P. cichorii, P. putida, P. cepacia, P. maltophilia, P. putrefaciens)* and "*Arthrobacter/Corynebacterium*" (not distinguishable with our methods). Additionally, strains of the genera *Cytophaga (C. fermentans, C. johnsonae)*, *Flexibacter (F. succinans)*, *Flavobacterium* (*F. multivorum*), *Acinetobacter* (*A. calcoaceticus*), *Aeromonas* (*A. hydrophila*), *Alcaligenes* (*A. denitrificans*), *Caulobacter* (*C. bacteroides*), *Citrobacter* (*C. freundii*), *Enterobacter* (*E. agglomerans, E. cloacae*), *Chromobacterium* (*C. lividum, C. violaceum*), *Paracoccus, Xanthomonas, Arthrobacter, Bacillus, Micrococcus* (*M. agilis, M. roseus*), *Staphylococcus*, and *Streptomyces* were isolated. Strains of *Citrobacter* and *Enterobacter* occurred only in sediment samples, whereas strains of *Alcaligenes* were obtained only from ground water samples (Kölbel-Boelke 1987).

Comparison of Bacterial Communities from Different Depths of the Same Borehole

Bacterial communities isolated from ground water or sediment samples collected from different depths of the same boreholes as well as communities isolated from different sediment materials (e.g. sand, wood, peat, clay) obtained from the same sediment core, differed distinctly in their morphotype composition. But these showed in most cases very similar preferences for substrates that were degraded with highest frequencies (Kölbel-Boelke 1988). A dependence on aquifer depth could not be observed for the proportion of Gram-positive or Gram-negative isolates, for the occurrence of motile or pigmented bacteria, or for the calculated in vitro activities (Kölbel-Boelke et al. 1988a).

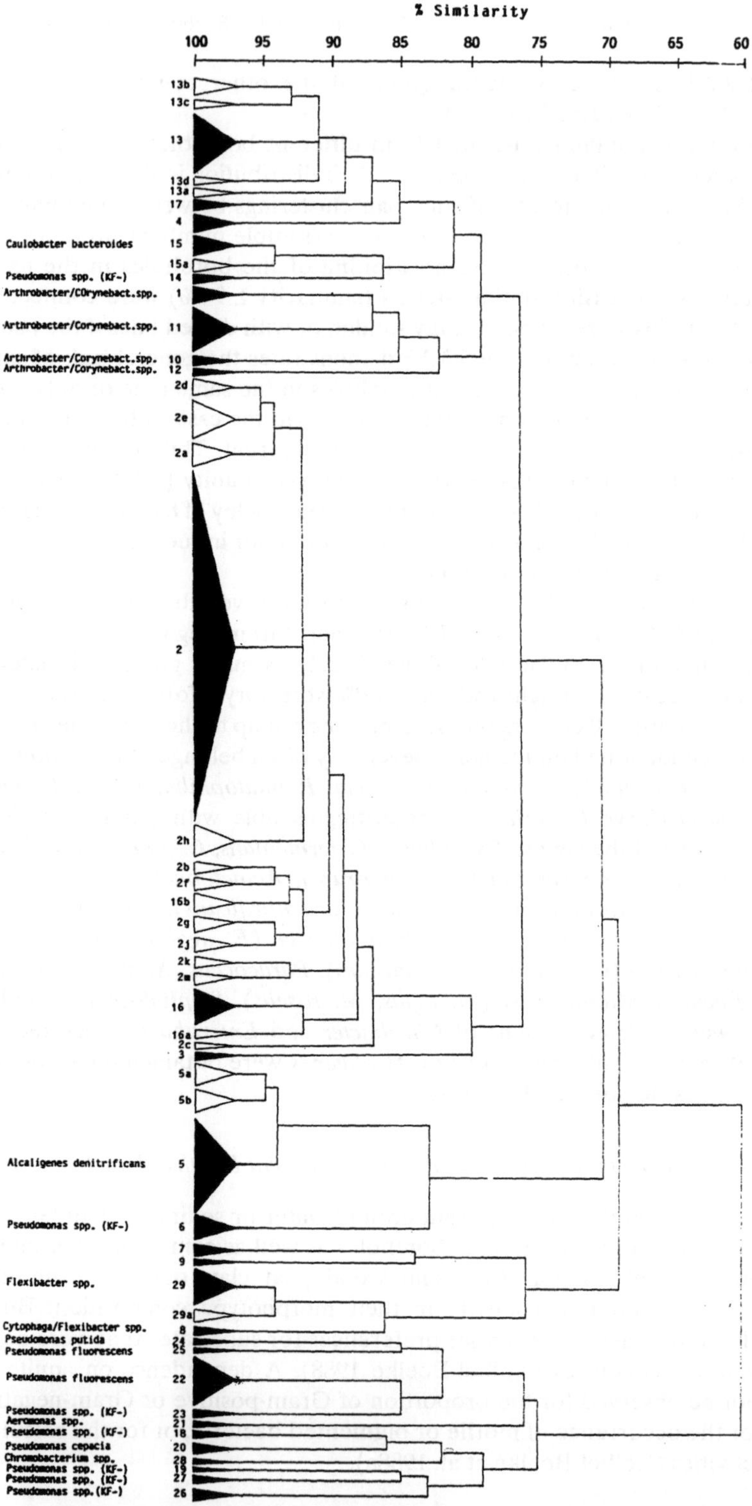
% Similarity
100
95
90
85
80
75
70
65
60
13b
13c
13
13d
13a
17
4
Caulobacter bacteroides
15
15a
Pseudomonas spp. (KF-)
14
Arthrobacter/COrynebact.spp.
1
Arthrobacter/Corynebact.spp.
11
Arthrobacter/Corynebact.spp.
10
Arthrobacter/Corynebact.spp.
12
2d
2e
2a
2
2h
2b
2f
16b
2g
2j
2k
2m
16
16a
2c
3
5a
5b
Alcaligenes denitrificans
5
Pseudomonas spp. (KF-)
6
7
9
Flexibacter spp.
29
29a
Cytophaga/Flexibacter spp.
8
Pseudomonas putida
24
Pseudomonas fluorescens
25
Pseudomonas fluorescens
22
Pseudomonas spp. (KF-)
23
Aeromonas spp.
21
Pseudomonas spp. (KF-)
18
Pseudomonas cepacia
20
Chromobacterium spp.
28
Pseudomonas spp. (KF-)
19
Pseudomonas spp. (KF-)
27
Pseudomonas spp.(KF-)
26

As cluster analysis demonstrated, at each borehole bacteria occurred regularly whose presence was restricted to only one or two of the samples examined. There were also, otherwise, bacteria which could be isolated from several, or even in some cases, all samples. Diversity of communities of the same borehole varied from one depth to another but did not decrease with increasing depth (Kölbel-Boelke 1987).

Comparison of Bacterial Communities from Water with Those from Sediment

With respect to morphotype composition, there existed distinct differences between ground water and sediment communities (Table 5.3.2.2; Kölbel-Boelke and Hirsch 1989). Most ground water communities contained low proportions of Gram-positive isolates (< 11%), whereas in sediment communities this percentage was usually three- to sevenfold higher (35–43%). This was confirmed by comparing the community structure of sediment with that of ground water samples from the corresponding depths of one well (BOC5): The community of 2–05 contained 43% Gram-positive bacteria but that of 1–09 only 5% (Kölbel-Boelke et al. 1988a). Ground water communities were characterized by higher proportions of Gram-negative straight or curved rods, whereas in sediment communities more *Arthrobacter*-like organisms, gliding filaments of the *Cytophaga/Flexibacter* group and Gram-negative bacteria of variable cell forms were observed (Table 5.3.2.2). Moreover, ground water samples contained more motile isolates than sediment samples which in contrary had a higher proportion of pigmented bacteria (Table 5.3.2.2).

The calculations of in vitro activities (Kölbel-Boelke et al. 1988a) yielded higher values for Gram-positive sediment bacteria than for Gram-positive water bacteria. Similar results were obtained when comparing Gram-negative sediment bacteria with Gram-negative ground water bacteria (Kölbel-Boelke et al. 1988a). Sediment and ground water bacteria in most cases degraded the same substrates with highest frequencies, but in sediment communities the numbers of positive isolates usually were higher. Distinct differences were observed between water and sediment bacteria for such substrate groups which generally were degraded less well (Kölbel-Boelke et al. 1988a; Kölbel-Boelke and Hirsch 1989).

A comparison of cluster matrices showed that sediment as well as water communities were composed very diversely. Sediment communities (Fig. 5.3.2.2) mostly contained more and smaller clusters than ground water communities (Fig. 5.3.2.1), which, in contrast, were often characterized by one to two larger clusters and a higher number of subclusters. The results of pair clusterings of communities from different boreholes (Fig. 5.3.2.5) demonstrated that among sediment communities as well as among ground water communities no higher similarities were observed than between sediment and ground water communities.

Fig. 5.3.2.3. Dendrogram and taxonomic characterization of clusters within water community 1–03. Remarks: Cluster 13: *Pseudomonas* spp. (KF-) and *Caulobacter* spp.; cluster 4: as C2; cluster 2: Gram +/ +- immobile white coccoides/cocci; cluster 16 and 3: as C2

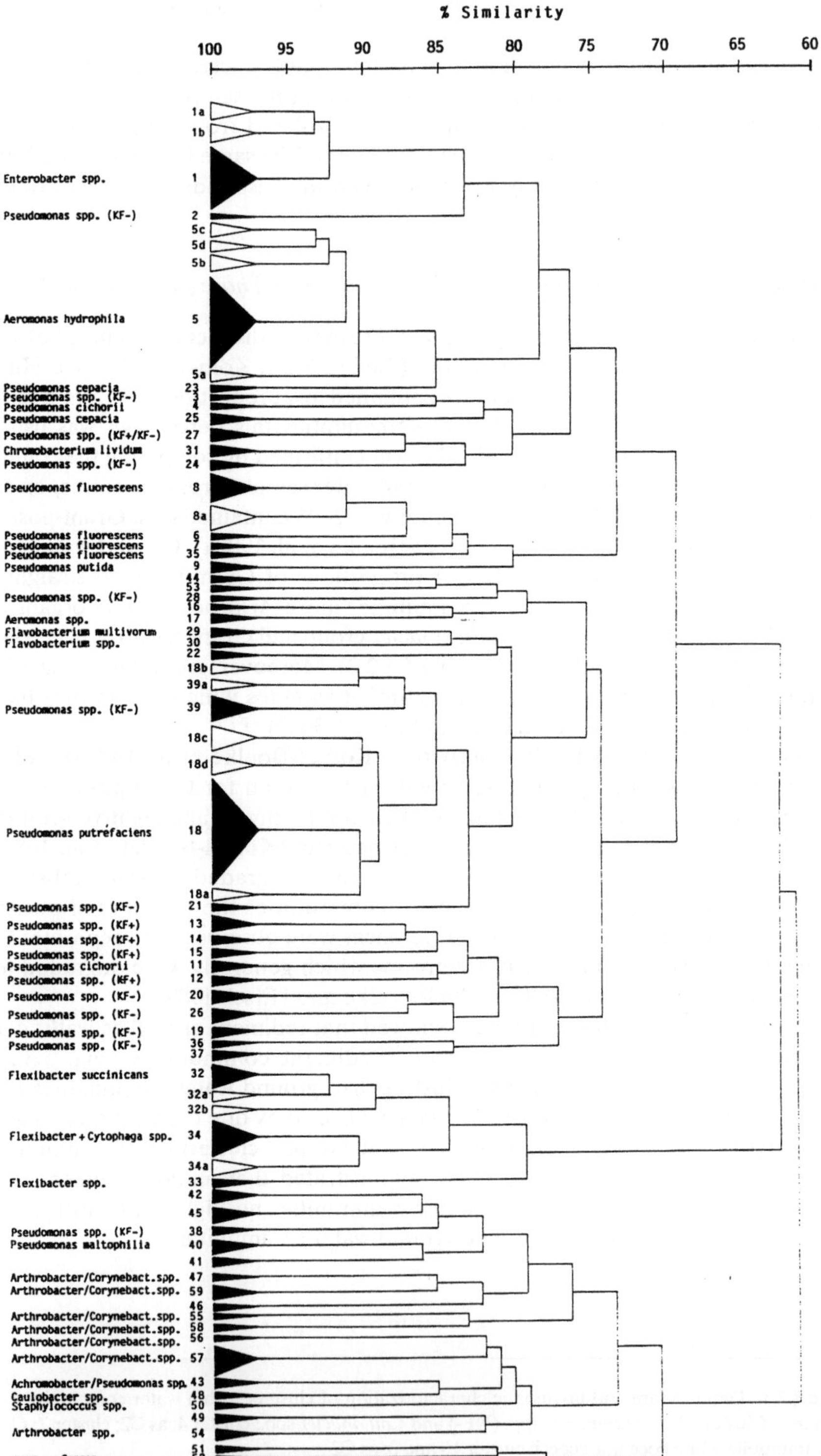
% Similarity
100 95 90 85 80 75 70 65 60
1a
1b
Enterobacter spp. 1
Pseudomonas spp. (KF-) 2
5c
5d
5b
Aeromonas hydrophila 5
5a
Pseudomonas cepacia 23
Pseudomonas spp. (KF-) 3
Pseudomonas cichorii 4
Pseudomonas cepacia 25
Pseudomonas spp. (KF+/KF-) 27
Chromobacterium lividum 31
Pseudomonas spp. (KF-) 24
Pseudomonas fluorescens 8
8a
Pseudomonas fluorescens 6
Pseudomonas fluorescens 7
Pseudomonas fluorescens 35
Pseudomonas putida 9
44
53
Pseudomonas spp. (KF-) 28
16
Aeromonas spp. 17
Flavobacterium multivorum 29
Flavobacterium spp. 30
22
18b
39a
Pseudomonas spp. (KF-) 39
18c
18d
Pseudomonas putrefaciens 18
18a
Pseudomonas spp. (KF-) 21
Pseudomonas spp. (KF+) 13
Pseudomonas spp. (KF+) 14
Pseudomonas spp. (KF+) 15
Pseudomonas cichorii 11
Pseudomonas spp. (KF+) 12
Pseudomonas spp. (KF-) 20
Pseudomonas spp. (KF-) 26
Pseudomonas spp. (KF-) 19
Pseudomonas spp. (KF-) 36
37
Flexibacter succinicans 32
32a
32b
Flexibacter + Cytophaga spp. 34
34a
Flexibacter spp. 33
42
45
Pseudomonas spp. (KF-) 38
Pseudomonas maltophilia 40
41
Arthrobacter/Corynebact.spp. 47
Arthrobacter/Corynebact.spp. 59
46
Arthrobacter/Corynebact.spp. 55
Arthrobacter/Corynebact.spp. 58
Arthrobacter/Corynebact.spp. 56
Arthrobacter/Corynebact.spp. 57
Achromobacter/Pseudomonas spp. 43
Caulobacter spp. 48
Staphylococcus spp. 50
49
Arthrobacter spp. 54
51
gram +- Coccen 52
gram + bzw. gram - Coccen 10

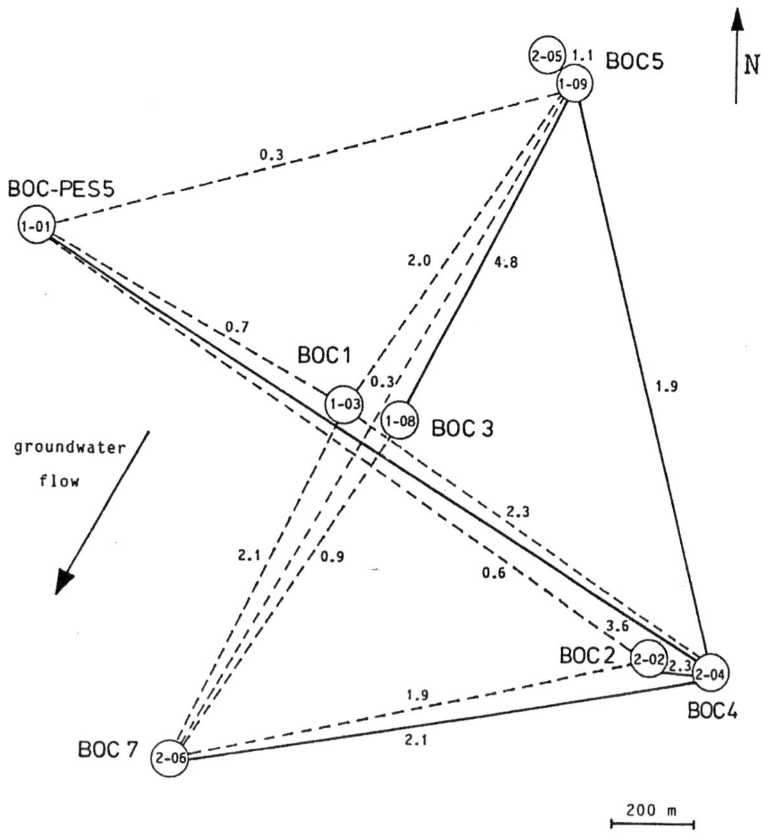

Fig. 5.3.2.5. Percentage proportions of very similar strains (similarity ≥ 94%) between communities from different boreholes

In Vitro Activities

All bacteria were tested for their ability to degrade 58 substrates of different chemical classes. As has been pointed out before, the communities from the different boreholes were composed of many different bacterial types whose "total activities" were calculated to range from 0 to 60% or 75% positive tests. However, each community showed its own activity pattern (Kölbel-Boelke et al. 1988a).

All communities possessed great biochemical versatility. Each of the 58 substrates could be degraded by at least one member of each community. The most frequent capability observed was the formation of extracellular phosphatase which was produced by 64% of all isolates. The most favourable substrates of the aquifer bacteria were macromolecular substances, with the exception of polysaccharides

◄

Fig. 5.3.2.4. Dendrogram and taxonomic characterization of clusters within sediment community 2–04

Table 5.3.2.3. Average percentage frequency of positive test results in the degradation of different groups of substrates

Borehole (well) Sample	BOC-PES 5 1–01	BOC 1 1–03	BOC 3 1–08	BOC 5 1–09	BOC 2 2–02	BOC 4 2–04	BOC 5 2–05	BOC 7 2–06
Monosaccharides (ARA, FRU, GAL, GLU, MNE, XYL)	78	30	21	56	55	70	29	31
Disaccharides (CEL, LAC, MAL, SAC)	64	23	8	22	44	63	23	23
Trisaccharides (MEZ, RAF)	33	28	3	1	24	26	7	13
Organic acids (ACE, CIT, FOR, GLA, LAT, MLO, NAT, OXA, PRO, SUC)	44	10	19	43	13	45	23	33
Amino acids (ADH, LDC, ODC, PA)	30	6	5	19	10	26	8	9
Polysaccharides (AMD, CHI, GLY, INU, XYA, ZEL)	10	12	12	7	18	24	15	9
Lipids (LEC, T20, T40, T60, T80)	29	28	53	62	43	59	57	69
Proteins (CAS, GEL)	68	19	19	52	54	69	28	57

Explanations: ARA = arabinose, FRU = fructose, GAL = galactose, GLU = glucose, MNE = mannose, XYL = xylose, CEL = cellobiose, LAC = lactose, MAL = maltose, SAC = sucrose, MEZ = melizitose, RAF = raffinose, ACE = acetate, CIT = citrate, FOR = formate, GLA = glutamate, LAT = lactate, MLO = malonate, NAT = tartrate, OXA = oxalate, PRO = propionate, SUC = succinate, ADH = arginine, LDC = lysine, ODC = ornithine, PA = phenylalanine, AMD = starch, CHI = chitin, GLY = glycogen, INU = inulin, XYA = xylan, ZEL = cellulose, LEC = lecithin, T20 = Tween 20, T40 = Tween 40, T60 = Tween 60, T80 = Tween 80, CAS = casein, GEL = gelatin.

and monosaccharides: Tween 40 (degradation by 64% of the isolates), Tween 60 (60%), DNA (56%), Tween 20 (51%), lecithin (49%), galactose (49%), mannose (48%), glucose (47%), gelatin (47%), arabinose (45%), and xylose (45%) (Kölbel-Boelke 1987). A very frequent character was the ability to cleave o-nitrophenyl-β-galactoside (ONPG; 57%) whereas only 27% of isolates degraded lactose. Table 5.3.2.3 lists the average frequencies of positive tests for degrading specific substrate groups; the listing separates communities from different boreholes. In comparing the results for different classes of carbohydrates, monosaccharides were more often degraded than disaccharides and least degraded were the trisaccharides. Low activity values were also calculated for polysaccharides, but within this group there existed great differences between better degradable substrates such as starch, chitin or glycogen, and those that were utilized only rarely, such as xylan and cellulose. Organic acids could be utilized best as sole carbon source mostly by those communities which also degraded monosaccharides with high frequency (Kölbel-Boelke 1987).

5.3.2.4 Discussion

Cell Counts

Viable cell counts in water samples agreed well with those reported by Wolters and Schwartz (1956) or Stetzenbach et al. (1986). In sediment samples, the viable cell numbers varied irregularly from depth to depth, as has been reported earlier by Hoos and Schweisfurth (1982). These differences could probably be due to the heterogeneous mineral composition of the aquifer environment and thus to the presence of many different microhabitats. The highest viable cell counts in the Bocholt sediment samples were in most cases about one order of magnitude lower than those obtained for other test fields that had been examined before (Hoos and Schweisfurth 1982; Wilson et al. 1983; Balkwill and Ghiorse 1985).

Amido black direct counts, in most cases, yielded cell numbers two to three orders of magnitude higher than viable cell counts on P-agar. Similar results were reported by Ghiorse and Balkwill (1983); Wilson et al. (1983), and Buchanan-Mappin et al. (1986).

Diversity of Bacterial Communities

All eight communities contained a great number of different bacterial morpho- and physiotypes. Similar results were reported by Hirsch and Rades-Rohkohl (1983a,b) and in Section 5.3.3. This agrees well with investigations of Jannasch (1978), who found that oligotrophic water environments are colonized by bacterial communities of great diversity.

Morphological and physiological characterization of all communities indicated strong differences clearly demonstrable by pair clusterings. These differences in communities from different boreholes (or from different depths of the same borehole) were probably due to differences in the prevailing hydrogeochemical conditions in the holes or layers. The presence of many microhabitats in these sites does not permit better resolution of these conditions.

Water and sediment communities showed differences in their morphotype composition (Kölbel-Boelke et al. 1988a; Kölbel-Boelke and Hirsch 1989). With regards to the Gram-stain, community structures of most water communities agreed well with those examined by Stetzenbach et al. (1986), who also observed high proportions of Gram-negatives. Isolated communities of sediment samples have not been studied so far by other authors. Community studies by TEM of thin sectioned bacteria that had been released from sediment particles (Ghiorse and Balkwill 1983; Wilson et al. 1983; Balkwill and Ghiorse 1985) indicated a predominance of Gram-positive bacteria (85–90%, Balkwill and Ghiorse 1985). Communities from the Bocholt aquifer system contained lower proportions of Gram-positive bacteria, which possibly could be explained by the special isolation conditions employed. These may tend to preferentially support growth of Gram-negative bacteria. Another explanation could be the assumption that only a fraction of the bacteria present could be released from the sediment particles (Balkwill et al. 1977). It might, however, also be possible that the communities in the Bocholt aquifer system actually do contain lower numbers of Gram-positives than the shallow aquifers investigated by Balkwill and Ghiorse (1985). Additionally, these authors found a decreasing diversity with increasing depth in the aquifer investigated which is in strong contradiction to our observations indicating no dependence of community diversity from aquifer depth. Results similar to ours were reported by Balkwill (1989) for deep aquifers (down to 265 m).

In Vitro Activities

The determination of biochemical capabilities primarily served to characterize and compare isolates or communities by numerical methods. But from these calculated in vitro activities it could not be predicted which substrates would be degraded in nature and which activity rates occurred in the natural habitat, since our tests were only done with pure cultures and under conditions which were certainly not representative for those in nature. Nevertheless, a great versatility of biochemical properties could be demonstrated. The generally high frequency of extracellular phosphatase production as well as the degradation of macromolecular substrates (lipids, proteins and DNA) or monosaccharides indicated that these activities might be important in the natural environment. In this case, of course, the assumption is that these substrates are present and the conditions amenable to their utilization.

5.3.3 The Natural Microflora of the Segeberger Forst Aquifer System

P. Hirsch and E. Rades-Rohkohl

5.3.3.1 Research Goals

Studies on the microbiology of Segeberger Forst ground water were initiated to describe the natural microflora of this unpolluted site (Fig. 1.1; Table 1.1). Thus, the early work concerned the methods of sampling uncontaminated water and sediment, the sample processing, and the microbial diversity and cell numbers. Of special interest were micro-environments of ground water organisms and the interactions of these with each other and with solid surfaces. An unexpected result of these studies was the large diversity of microbial morphology and physiology. Consequently, the distribution was also investigated, and attention was paid to possible signs of in situ activity.

All this work was preceded by studies on the subsurface spreading of faecal indicator bacteria (*E. coli*) and the interactions of the natural microflora with such *E. coli* cells (Hirsch and Rades-Rohkohl 1983b; Matthess et al. 1985).

5.3.3.2 Materials and Methods

Sample Collection and Processing

The collection of water samples from the Segeberger Forst (SEG) test area was described previously (Hirsch and Rades-Rohkohl 1983a,b). Water samples from

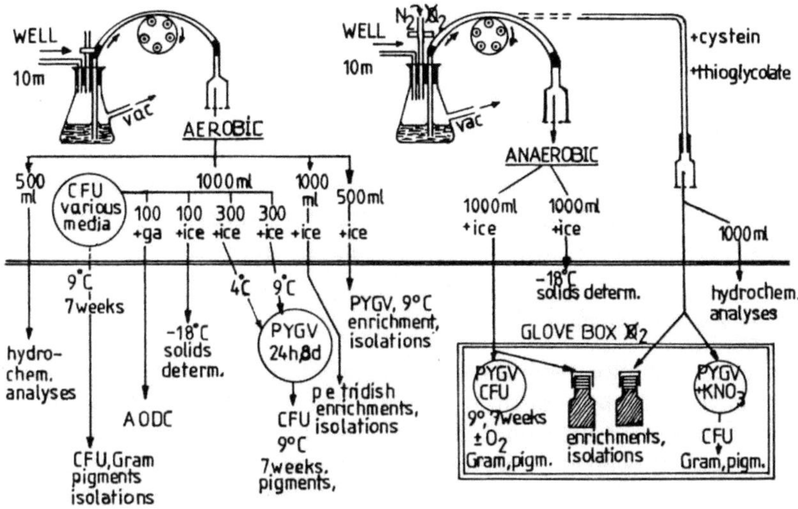

Fig. 5.3.3.1. A typical protocol for the processing of ground water samples. All work indicated *below the dividing line* was carried out in the laboratory

PVC sampling tube SEG8 (established in 1984) were drawn through a stainless steel tap at the end of the tube after treatment with 70% ethanol and flaming (Hirsch and Rades-Rohkohl 1988b). Sediment sampling was described in the same publication. Sample processing occurred at the sampling site (Fig. 5.3.3.1). Transport to the laboratory was over ice and lasted no longer than 1.5 h.

Microscopy

For LPM, agar-coated slides (Pfennig and Wagener 1986) and a Zeiss Photomicroscope II were used. Electron microscopic preparations were negatively stained with 1% phosphotungstic acid, or they were carbon-shadowed from an angle of 35°. Sediment samples were also studied employing a technique whereby a polymer film is lifted off from the sample by flotation (Waid 1973). For SEM of the inner PVC tube surface a section was fixed in glutaraldehyde (final concentration 3%), dehydrated in ethanol, and critical point dried with Freon (Kölbel-Boelke et al. 1988b).

Cell Counting Procedures

Total cell counts (AODC) and viable cell counts (CFU) were obtained as described previously (Zimmermann et al. 1978; Ghiorse and Balkwill 1982; Hirsch and Rades-Rohkohl 1988a). Colony counts on solid media were made weekly for up to 6 weeks to allow for slow growth of Gram-positive bacteria. When cell counts were determined for special physiological groups by employing selective media, the physiological activity to be tested was evaluated separately from mere growth on the particular medium.

Culture Media

Most media used were described in previous publications: PYGV, NB, YM, 337-C, 337G+NO_3^- (Hirsch and Rades-Rohkohl 1983a); 66a, MEA, 258 (Hirsch and Rades-Rohkohl 1983b); FEP, MOB, MRA, SO (Hirsch and Rades-Rohkohl 1988a). Medium P was used as by Kölbel-Boelke et al. (1988a); $PKNO_3$ was PYGV with 1% (w/v) KNO_3. Medium NA was Bacto Nutrient Agar. A lactate medium for sulfate reducing bacteria "LM" and mineral salts base "B" were suggested by Gerhardt et al. (1981). All cultures were grown at 9 °C, unless stated otherwise.

Enrichments and Isolations

Subsamples or dilutions hereof were inoculated into liquid media or onto agar plates and scanned frequently for up to 1 year for growth of different colony- or morphotypes; long incubation times favoured the development of protozoan blooms. The petri dish enrichment technique, especially helpful for the growth of attaching bacteria, was described by Hirsch et al. (1977). For isolations from plates, single colonies were patched onto master plates using sterile tooth picks. All different colonies were isolated.

Physiology

Tests for catalase, oxidase, starch hydrolysis, Voges-Proskauer or methyl red reactions and indol formation followed Skerman (1967). Nitrate reduction was tested with medium $PKNO_3$, pH 6.2–6.5. DNase activity was observed with Difco DNase Test Agar. The utilization of carbon sources (0.1%, w/v) was followed with mineral salts medium B. Control cultures without added C sources helped to detect possible oligotrophic growth. The utilization of organic acids (1%, w/v) was studied with Koser medium; growth with or without various nitrogen sources was determined according to Kölbel-Boelke et al. (1988a). Manganese oxidation was detected by the dark brown colouration of colonies growing on MOB agar. Manganese reduction was evident from the halo formation around the colonies on MRA. Iron oxide precipitation was tested with FEP agar and was recognized by the brown discolouration of the colonies and by the Berlin blue reaction. Resistance to antibiotics was studied with medium PYGV to which the antibiotics were added at concentrations of 10 or 50 µg/ml. For testing resistance against lysozyme, the enzyme was added at a concentration of 0.001% (w/v). Cleavage of the protocatechuic acid aromatic ring was followed according to Stanier et al. (1966) as modified by Gerhardt et al. (1981). Possible stimulation of pure cultures by 10% (w/v) addition of a humic preparation was studied in medium B. For tests of ecophysiological properties the isolates were grown in PYGV. They were subjected to temperatures of 4, 9, 22, 25, 30, 37 or 43 °C, or to NaCl concentrations of 1, 5, 10% (w/v), as well as to a pH range of 5–9.

Most of the physiological tests were performed in sterile microtiter plates. For these, liquid inocula were kept frozen at –20 °C. When needed, they were precultured for 3 weeks and from these precultures master inoculation plates were filled with active suspensions. Such master plates could then be frozen at –70 °C (at first) and then stored at –20 °C; they could serve as an inoculum for 22 experimental microtitre plates. After 4 weeks of incubation in a moist chamber the plates were evaluated by two persons. For each experiment three parallel plates were prepared.

5.3.3.3 Results

Sample Collection and Processing

Collecting uncontaminated samples was initially a major problem, as the Segeberger Forst wells had been established for geological studies rather than for sterile work by microbiologists. The methods finally adopted for collecting ground water were described previously (Hirsch and Rades-Rohkohl 1983a,b; 1988a; Sect. 5.1.2, this Vol.). Experience had shown that the wells had to be prepumped before the actual sampling in order to remove the well water microflora. Prepumping reduced the total and viable cell counts in the water samples approx. 500-fold to a constant level (Hirsch and Rades-Rohkohl 1988a).

Water samples were obtained by applying a vacuum to a PVC tube by using rotary tube pumps, or by pushing the water up with immersible pumps. Strong suction resulted in turbid samples; viable cell counts were then generally higher (Hirsch and Rades-Rohkohl 1988a). Filtration (0.45 μm pore width) reduced these cell counts. Sample storage for extended periods of time before processing resulted in microbial diversity changes as well as in altered viable counts (Hirsch and Rades-Rohkohl 1988a). Therefore, future water samples were processed immediately after collection, i.e. right on the site of the well.

Sediment samples were more difficult to obtain. Early drilling procedures resulted in heating of the sediment core, and only the careful application of a device which made it possible to collect the sediment core after preboring in a sterile plastic tube gave satisfactory results. A list of all samples investigated is given in Table 5.3.3.1. A total of 25 free water samples and three interstitial water samples as well as eight sediment samples were studied. Additionally, the inner surface of the PVC sampling tube SEG8 was investigated. A typical protocol for the processing of a water sample is shown in Fig. 5.3.3.1.

Morphological Diversity in Ground Water and Sediment Samples

Using phase contrast and electron microscopy, 107 morphologically distinct microorganisms (morphotypes, MT) were discovered in the subsurface samples (Sect. 5.1.1). Of these, 80 MT were prokaryotes and 27 eukaryotes (protozoa, lower fungi, one alga). The eukaryotes were generally found in aerobic water samples; several occurred also in aerobic enrichments made from sediment samples and after prolonged incubation. Amoebae and flagellates were most common among them. There were 19 bacteria with extracellular polymers (capsules), and 17 bacterial MT were motile. Some 20 bacteria were normally attached to surfaces, and 16 formed aggregates. Cyanobacteria were not found. The list and MT drawings are shown in Section 5.1.1.

Electron microscopy of stored water samples (without any additions) revealed an unusually large number of bacterial MT with surface appendages such as pili, fimbriae, spines or spires.

Table 5.3.3.1. List of samples investigated

Sample time (m/a)	No. of samples	Sample types[a]	Sample origin	Well numbers or drilling sites SEG	Depth (m)	Remarks
6/1981	3	W	Stainless steel well: gauge pipe	4,7,10	10	Preflow 10 a.15 1
6/1981	11	W	Multilevel PVC tube wells	5,6,9	5.65–19.65	Preflow 10–12 1
4/1982	2	S	Drilling cores	2	31.5–31.8	Above and below capillary zone
6/1983	2	W	Stainless steel well	7	10	
9/1983	2	S	Sterile sediment exposed 12 weeks	4	10/20	Colonization expt.
2/1984	2	S	Sterile sediment exposed 12 weeks	4	10/20	Colonization extp.
2/1984	2	P	Sterile sediment exposed 12 weeks	4	10/20	Colonization expt.
8/1984	1	S	Drilling core (PVC lined Stainless steel auger pipes)		6	
9/1984	1	S	Drilling core (“ ”)	8	10	New well established
9/1984	1	P	Drilling core (“ ”)	8	10	New well established
11/1984	1	W	PVC sampling tube	8	10	
8/1985	3	W	PVC sampling tube	8	10	Preflow 0.1.2 and 19 1
8/1985	1	W	Stainless steel well	7	10	Preflow 39 800 1
10/1985	2	W	PVC sampling tube	8	10	Samples turbid/clear
5/1986	2	W	PVC sampling tube	8	10	Aerobic/anaerobic
5/1986	1	T	PVC sampling tube: inner wall	8	10	For scanning EM

[a]W - water; S - sediment; P - pore water; T - tube wall.

Some of these appendages were flexible, others quite rigid (Hirsch and Rades-Rohkohl 1983a,b). Many of the MT found in subsurface samples were morphologically so characteristic that they could be affiliated with much certainty to known genera: *Gallionella, Nevskia, Leptospira, Seliberia, Planctomyces, Blastobacter, Pirellula, Prosthecobacter, Caulobacter, Thiodendron, Stella, Prosthecomicrobium, Hyphomicrobium, Hyphomonas and Nocardia (Amycolata).* A large proportion (25–30%) of the MT could be classified as oligotrophic, i.e. they grew and multiplied in extremely dilute media (Hirsch 1986; Hirsch and Rades-Rohkohl 1988b). Such media were PYGV or PYGV/100, the latter with approx. 4 mg/l organic carbon (Staley 1968).

The majority of MT were discovered in various types of enrichment cultures which had been incubated up to 1 year. Gram-postive bacteria often appeared later than the Gram-negative ones. When pure cultures were studied for possible changes of morphology under laboratory conditions, it was observed that those morphological features by which they had been classified initially in the natural samples or enrichments were stable in laboratory culture over several transfers.

SEG-sediment isolates (thought to be normally living in a biofilm, i.e. attached) were compared with isolates from interstitial water that ran off during sample retreaval or with those of the free ground water (after prepumping) (Kölbel-Boelke and Hirsch 1989). There were distinct differences: rod-shaped cells were more prevalent in free ground water, but sediment contained more attaching bacteria such as *Caulobacter* spp. More numerous in the sediment were also filamentous fungi, minute bacteria (0.2–0.4 µm diameter), pigmented organisms and Gram-positive forms. Free water organisms were more often motile and often contained polymer. Strains from interstitial water were mostly intermediate in their properties between the sediment and free water isolates.

Of special interest was the microflora living on the inner walls of the ground water wells. While forms that adhered to stainless steel well walls could not be studied with the techniques available, cell-counting data obtained after various periods of prepumping indicated a richer microflora (by number and MT) in the undisturbed water of both the stainless steel wells and those that had PVC tubes. The PVC plastic sampling tubes SEG8 were eventually pulled up and sections of the inner tube wall were cut off, fixed, and viewed with the scanning electron microscope after critical point drying (Figs. 5.3.3.2–5.3.3.4). Here a biofilm microflora was discovered which was rich in morphologically unusual forms, such as *Planctomyces, Caulobacter, Seliberia, Hyphomicrobium and Prosthecobacter* spp. Rod-shaped cells (*Seliberia*?) were found to stand upright, with one cell pole attached by a holdfast polymer. From previous experience (P. Hirsch unpubl.) it must be assumed that these forms could not be dislodged from their attachment sites by the pumping procedure, but all of these normally multiply through formation of motile "swarmer" cells, which then could also appear in the free ground water.

Determination of Total and Viable Cell Counts

Total cell counts were determined after acridine orange staining. The few data from water samples ranged from 12×10^3 to 71×10^3 cells/ml. Viable cell counts (CFU/ml) on PYGV were remarkably similar for aerobes in different depths and different wells (Table 5.3.3.2). They ranged from $0.3–10.8 \times 10^3$, with an average of about

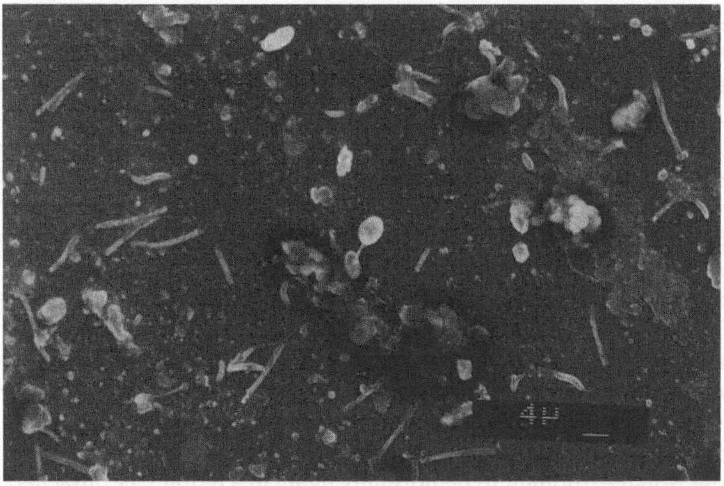

Fig. 5.3.3.2. SEM graphs of the inner surface of PVC sampling tube SEG8 from 10 m depth. Patchy growth of *Planctomyces* sp. (*centre*), *Caulobacter, Hyphomicrobium, Seliberia* spp.

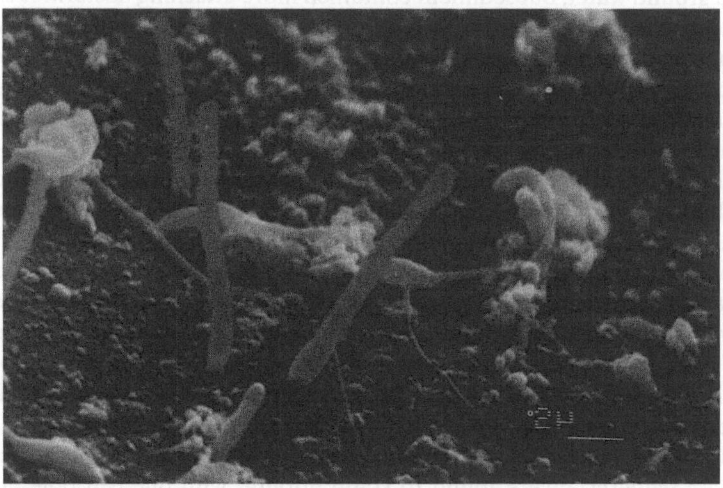

Fig. 5.3.3.3. SEM graphs of the inner surface of PVC sampling tube SEG8 from 10 m depth. Higher magnification of an area with *Seliberia* sp. standing upright

4.4×10^3 CFU/ml. Viable counts of anaerobes were determined only twice; in these cases they were found to be 0.4×10^3 or 4.5×10^3 CFU/ml, respectively (Table 5.3.3.2).

The use of selective media for counting cells of different physiological groups gave variable results: in most cases there were more manganese-reducing bacteria than oxidizing ones (Table 5.3.3.3). Iron oxide-precipitating activity was rare, and the numbers of sulfur-oxidizing bacteria were remarkably high. Anaerobic sulfate

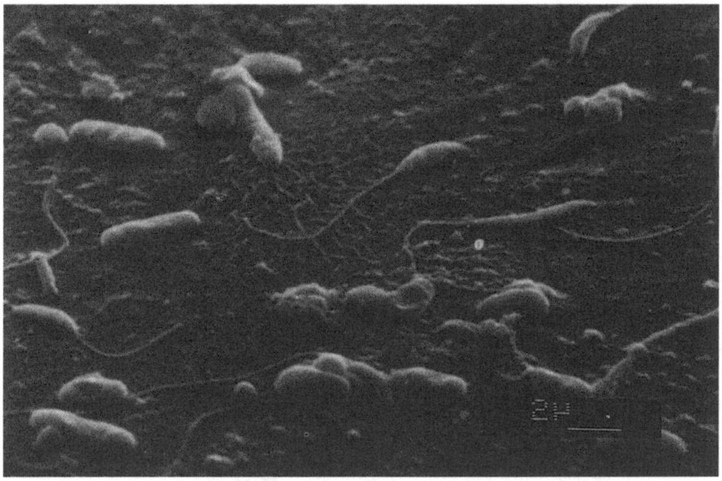

Fig. 5.3.3.4. SEM graphs of the inner surface of PVC sampling tube SEG8 from 10 m depth. High magnification of an area with *Caulobacter* spp. growing completely attached to the surface

Table 5.3.3.2. Average viable counts (one to three determinations) in ground water samples from the Segeberger Forst 1981–1986. All plates were inoculated at the site, five parallels per dilution. Medium PYGV; evaluation after 41–52 d. (Data from Hirsch and Rades-Rohkohl 1988a)

Depth (m)	Type of well[a]	Preflow discarded (1)	Total cell counts (AODC × 10^3)	Viable cell counts + (CFU × 10^3 ml^{-1})	
				Incub. aerobic	Incub. anaerobic
5.65	MTW:PVC	12	n.d.[b]	9.5 ± 9.3	n.d.
8.65	"	12	n.d.	1.5–3.4 ± 1.0	n.d.
9.80	GP:PVC	10	n.d.	2.7 ± 0.2	n.d.
10.00	"	15	12.0	4.3 ± 0.7	0.4 ± 0.05
10.00	GP:SS	30	n.d.	0.3 ± 0.02	4.5 ± 1.8
10.00	MP:SS	39 800	71.0	3.6 ± 1.2	n.d.
10.65	MTW:PVC	12	n.d.	1.7 ± 0.3	n.d.
12.65	"	12	n.d.	3.5 ± 1.5	n.d.
16.65	"	12	n.d.	8.9 ± 2.8	n.d.
19.65	"	12	n.d.	10.8 ± 1.6	n.d.

[a]MTW – multilevel tube well; GP – gauge pipe; SS – stainless steel well; MP – main pipe.
[b]not determined.

reducers were not found in our samples. The highest aerobic viable cell counts were always obtained with medium PYGV.

With sediment samples there existed the problem that adhering microorganisms could not be dislodged ("extracted") quantitatively (Balkwill et al. 1977). Viable cell counts were determined with the assumption that the extraction procedures employed would always release approximately the same proportion and types of sediment bacteria. Thus, total sediment cell counts were determined to be 1.9 × 10^6 cells/g d. wt. (6 m depth), or 1.02 × 10^6/g d. wt. (10 m depth). Two other samples

Table 5.3.3.3. Total and viable cell counts of various physiological types in SEG ground water samples from 10 m depth. Incubation at 9 °C in the dark

Sampling time (m/y)	Type of well	Total cell counts[a] (AODC; × 10^2 ml^{-1})	Aerobic viable cell counts: CFU × 10^2 ml^{-1} (water) or CFU × 10^2 g^{-1} dry sediment						Anaerobic CFU × 10^2 ml^{-1} (water) or CFU × 10^2 g^{-1} dry weight of sediment			
			Micro-organisms (PYGV)	Fungi (MEA)[b]	Mn-oxidizing (MOB)[b]	Mn-reducing (MRA)[b]	Fe-precipitating (FEP)[b]	S0 oxidizing (SO)[b]	Micro-organisms (PYGV)	NO_3-reducing ($PKNO_3$)	N_2-forming ($PKNO_3$)	SO_4-reducg. (LM)[b]
11/1984	PVC	n.d.[c]	8000[d]	6.0	n.d.	79.0	n.d.	4100	5800	n.d.	n.d.	n.d.
8/1985	PVC	n.d.	120	10.0	7.0	70.0	30.0	110	n.d.	n.d.	n.d.	n.d.
8/1985	SS[e]	710	36	5.0	1.3	23.0	2.0	16	n.d	n.d.	n.d.	n.d
5/1986	PVC	n.d.	400	0.6	46.0	71.0	n.d.	53	n.d.	n.d.	n.d.	n.d.
5/1986	PVC	120	43	0.2	4.0	1.8	n.d.	n.d.	4	0.1	0.1	n.d.
9/1986	Sediment	10200[f]	120	3.2	1.1	16.8	17.0	n.d.	100	n.d.	n.d.	n.d.

[a]Acridine orange direct counts;
[b]MEA - malt extract agar. MOB - medium of Mulder and van Veen (1963). MRA - medium of Schweisfurth (1968); FEP - medium of Wenzel and Schweisfurth (1979); SO - medium of Wieringa (1966); $PKNO_3$ - medium PYGV + 0.1% KNO_3; LM - lactate medium of Gerhardt et al. (1981).
[c]Not determined.
[d]Well only 2 months old.
[e]Stainless steel well.
[f]AODC × 10^2 g^{-1} dry sediment.

from the Radesforder Berg site (in the vicinity of the Segeberger Forst test site) from 31.5 to 31.8 m depth had 5.5×10^7 and 7.6×10^7 total cells/g d. wt. The viable counts per g dry sediment on PYGV medium were two orders of magnitude lower than the AODC (Hirsch and Rades-Rohkohl 1988a; Table 5.3.3.3).

Viable cell counts on selective media were generally of similar magnitude as those on PYGV.

When Hirsch and Rades-Rohkohl (1988a) compared CFU/ml on two similar media (PYGV and P), the data were nearly identical. However, microscopy and staining procedures revealed the fact that the two media had supported growth of quite different microorganisms. On PYGV agar, 86.8% of the colonies were rods and related MT, but spirilla were completely missing. On P-medium, on the other hand, the rods comprised only 57.7% of the colonies and 29.2% were spirilla, reflecting slightly more peptone and a different kind of peptone (Kölbel-Boelke et al. 1988a). This example shows that similar viable counts on certain media may arise from very dissimilar populations. Other factors that affected viable cell numbers were sample turbidity, the storage time before processing and the incubation time of the agar plates (Hirsch and Rades-Rohkohl 1988a).

Distribution of Ground Water Microorganisms

Studies of Kölbel-Boelke et al. (1988b; Sect. 5.3.2, this Vol.) have shown that in the Bocholt aquifer system microbial communities were quite different even in adjacent wells and in different depths, an indication for inhomogeneities of the subsurface, supply of oxygen or organic carbon etc. Therefore, sediments of the Radesforder Berg site (SEG2) and water from wells near Haltern (Westfalia) or Munich (Bavaria) were also studied for MT diversity (Table 5.3.3.4). Sediment and water samples at the Segeberger Forst test site contained a total of 93 different MT (25 eukaryotes and 68 bacteria); most common were straight or curved rods and *Caulobacter* spp. Sediment from the Radesforder Berg (SEG2) had the same major MT, but only 59 different ones. Again, less MT (44) were found in Munich water samples, but the major MT were also rods and caulobacters. The Haltern water samples were different: the low number of MT (28 only) and the complete lack of caulobacters indicated that these wells may have been under stress conditions. Also, the Segeberger Forst wells contained more yeasts and filamentous fungi compared to the other sites. These data indicate major differences among ground water wells of these locations.

Likewise, a comparison was made among four wells of the Segeberger Forst test site (Table 5.3.3.5). While the numbers of prokaryote MT vary little, eukaryotes were more distributed in PVC wells than in stainless steel ones. Differences between well types were also observed in the following experiment. Water samples were collected from PVC sampling tube SEG8 (10 m) after various amounts of preflow had been discarded. At the same time, water samples were also taken from stainless steel well SEG7 (10 m) after prepumping off and discarding 39 800 l. All samples were then plated on PYGV medium and colony pigmentation and Gram reaction compared of randomly selected 95–280 colonies. While the Gram reaction was nearly identical in all cases (12–13% Gram-positive and the rest negative), there were differences in pigmentation. Without or with only 1.2 l preflow, the PVC

Table 5.3.3.4. Morphotype distribution in ground water and sediments of the Segeberger Forst (SEG), Radesforder Berg (SEG 2), of wells near Haltern, Westphalia (HA) and of wells near Munich (MUC)

Morphotype description[a]	MT. no.	Numbers of different MT present in samples from			+
		SEG 4–11	SEG 2	HA	MUC
Cocci (various shapes)	1– 6	5	5	4	2
Rods (various shapes)	7– 22	16	15	6	7
Filaments	23– 24	2	1	1	2
Pointed rods	25	1	1	1	0
Curved rods a. spirilla	26– 31	6	5	5	4
Gallionella spp.	32	1	0	0	0
Nevskia ramosa	33	1	0	0	0
Coryneforms	34	1	1	0	1
Pointed short rods	35– 37	2	1	1	1
Spirochetes	38– 40	3	2	0	0
Spore formers	41– 45	3	2	0	4
Seliberia spp.	46– 47	2	1	0	0
Attached rods	48	0	1	0	0
Planctomyces spp.	49– 52	3	3	0	2
Budding cocci	53– 54	1	0	0	0
Budding rods	55– 58	1	4	2	1
Rosette forming rods	59	1	1	0	1
Pirellula spp.	60– 61	2	1	1	0
Prosthecobacter spp.	62– 64	3	1	1	3
Caulobacter spp.	65– 71	6	5	0	6
Thiodendron spp.	72	1	0	0	1
Stella spp.	73	1	1	0	0
Prosthecomicrobium spp.	74	1	1	0	0
Hyphomicrobium spp.	75	1	1	1	1
Hyphomonas spp.	76	1	0	1	0
Actinomycetes	77– 80	3	2	1	1
Amoebae	81– 83	3	1	0	1
Flagellates, colorless	84– 89	6	0	0	0
Choanoflagellate	90	1	0	0	1
Flagellates, colorless	91– 93	2	1	0	1
Cysts of algae	94– 95	2	1	0	0
Ciliates	96	1	0	0	0
Yeasts	97– 100	4	0	1	1
Filamentous fungi	101– 105	5	1	2	2
Heliozoon	106	0	0	0	1
Green alga	107	1	0	0	0
Total no. of MT		93	59	28	44

sampling tube SEG8 had 66–67% pigmented colonies, 15–29% of which were red. With increasing preflow the numbers of pigmented colonies dropped to 40%, but 89% of these were red. The water from stainless steel well SEG7 had, after 39 800 l preflow, 46% pigmented colonies, but none of these was red. It can be concluded that the stagnant water within the PVC sampling tube SEG8 contained a higher proportion of attached and red bacteria as compared to the stainless steel well.

Table 5.3.3.5. Comparison of the microbiota of different wells in the Segeberger Forst test area. (Data from Hirsch and Rades-Rohkohl 1988b)

Well no.	Well type	Depth sampled (m)	No. of morphotypes		Viable cell counts (CFU ml^{-1})
			Prokaryotic	Eukaryotic	
SEG4	PCV tube	10	56	11	4.3×10^3
SEG6	PVC tube well	5.65–19.65	46	18	$1.5–10.8 \times 10^3$
SEG5	Stainless steel well	6	40	6	3.6×10^3
SEG7	Stainless steel well	10;20	45	7	$0.26–6 \times 10^5$

Prepumping removed predominantly yellow, purple or brownish pigmented organisms. The stagnant water in PVC sampling tube SEG8 was also found to contain, after approx. 2 years of operation, a substantially higher concentration of DOC (82.0 mg/l) and only less than 0.1 mg/l NO_3^-. The corresponding values for prepumped wells were in the ranges of 5–18.4 mg/l DOC and 9.5–22.0 mg/l of nitrate. Such observations reflect strong differences of environmental conditions within these two well types, and, consequently, differences between the microbial communities living there. Although all the well water microorganisms must have come from the ground water, the surface structure and possible charge of the plastic well tube may have allowed for better growth of selected organisms.

Variations in the diversity of MT were even more evident when different depths of two wells were compared (Table 5.3.3.6). Some MT (numbers 10, 46, 47, and 97–100; Sect. 5.1.1, this Vol.) were found in all depths, but others occurred only in one or a few depths. Such great differences can only be explained by different sediment structure and composition. Riemer (1983) and Schröter (1983) described subsurface profiles of the Segeberger Forst test site and indicated the presence of clay lenses and coarser gravel within the finer grained sediment.

On a still smaller scale, the distribution of microbial MT was studied within individual samples of one well using electron microscopy. For this investigation the sample preparation followed the protocol suggested by Waid (1973). A thin polymer layer dislodged from the surface of sediment grains was transferred to an electron microscope grid and the preparation shadowed at a low angle. Several rod-shaped cells were seen to carry flagella; most bacteria occurred in isolated groups and/or clumps with wide void areas in between. The bacterial assemblages consisted of mixed MT living in close togetherness. Such patchy growth pattern may also explain the general lack of nutrients within this subsurface environment: only organisms living within one patch can exchange nutrients; the total patch community is dependent on the rare supply of humic substances arriving from above. Figure 5.3.3.5 shows a composite picture of sediment bacteria growing in patches.

A similar distribution pattern was also observed when PVC sampling tube SEG8 was investigated by scanning electron microscopy. Here the colonization was much denser (as has been pointed out above), which probably reflects the higher nutrient concentration within the PVC tube (Figs. 5.3.3.2–5.3.3.4).

Another important question concerns possible differences between the microflora which adheres to sediment particle surfaces (organic or inorganic) and that which lives within free ground water. Kölbel-Boelke and Hirsch (1989) have ad-

Table 5.3.2.6. Depth distribution of selected pure culture MT isolated from water samples of two Segeberger Forst multi-level wells. (Data partially from Hirsch and Rades-Rohkohl 1983b)

Well no.	Depth sampled (m)	Morphotypes of pure cultures observed[a]															Total no. of MT in aerobic enrichments
		A[b]	B	C	D	E	F	G	H	I	K	L	M	N	O	P	
SEG5	5.65	+	+	•	•	+	•	•	+	•	•	•	•	•	+	+	21
SEG5	8.65	•	+	+	•	+	+	+	+	+	•	+	+	•	+	•	21
SEG5	10.65	•	+	+	+	+	+	•	•	•	+	+	+	•	+	•	13
SEG5	12.65	•	+	•	•	•	+	•	•	•	+	•	•	•	+	+	16
SEG5	16.65	•	+	•	+	+	+	•	+	•	•	+	•	•	+	+	9
SEG5	19.65	•	+	•	•	•	+	•	•	•	•	+	•	•	+	•	13
SEG6	8.65	•	+	•	•	•	+	•	+	•	•	+	•	•	+	•	21
SEG6	10.65	•	+	•	•	+	+	•	+	•	•	•	•	•	+	+	22
SEG6	12.65	•	+	•	•	•	+	+	+	•	•	+	•	+	+	•	15

[a]For MT numbers see Section 5.1.1 (this Vol.).

[b]A - cocci (MT 2,3,4,5); B - rods (MT 10,46,47); C - rods, granulated (MT 11–13, 15); D - rods, pointed (MT 25); E - rods; light refractile (MT 13,14); F - rods with capsules (MT 10); G - coryneforms (MT 34); H- short rods (MT 7,8,35); I - rods, minute (MT 9); K - filamentous rods (MT 23); L - rods, vibrioid (MT 27–30, 33); M - *Caulobacter* spp. (MT 62–70); N - *Prosthecomicrobium* spp. (MT 74); O - yeasts (MT 97–100); P - filamentous fungi (MT 103–105).

dressed this problem in a comparative study of the Segeberger Forst and Bocholt aquifer systems. Morphological as well as physiological properties of both communities were studied. These data became even more meaningful since different sectors of the microbial communities were investigated: while in the Bocholt case all colonies growing on a plate were subcultured, isolated and tested, the Segeberger Forst study concentrated on only all different ones. While the numbers of different MT were practically alike, the MT composition was quite different in the case of the Segeberger Forst aquifer system. Sediment isolates were more often smaller, pigmented, Gram-positive and *Caulobacter*-shaped, while free aquifer water isolates were usually rod-shaped, motile and more often produced polymer. Physiologically, the free water isolates were more often capable of Mn oxidation or -reduction, of Fe(III) oxide precipitation, or of growth without added C sources. Water strains were more resistant to antibiotics.

Enrichments, Isolations and Culture Collection

A variety of different enrichment techniques were employed for the description of the ground water microflora (Hirsch and Rades-Rohkohl 1983a). Subsamples were either inoculated directly onto agar or into liquid media; inoculation occurred on the sampling site. Some of the samples were simply stored without additions. Finally, the Petri dish method (Hirsch et al. 1977) was employed successfully. This latter method yielded especially many attaching bacteria.

A total of 14 media were tested to find the largest number of morphologically different organisms (Hirsch and Rades-Rohkohl 1983a,b). For this, samples from Haltern wells (different depths) were plated and all different colonies and MT

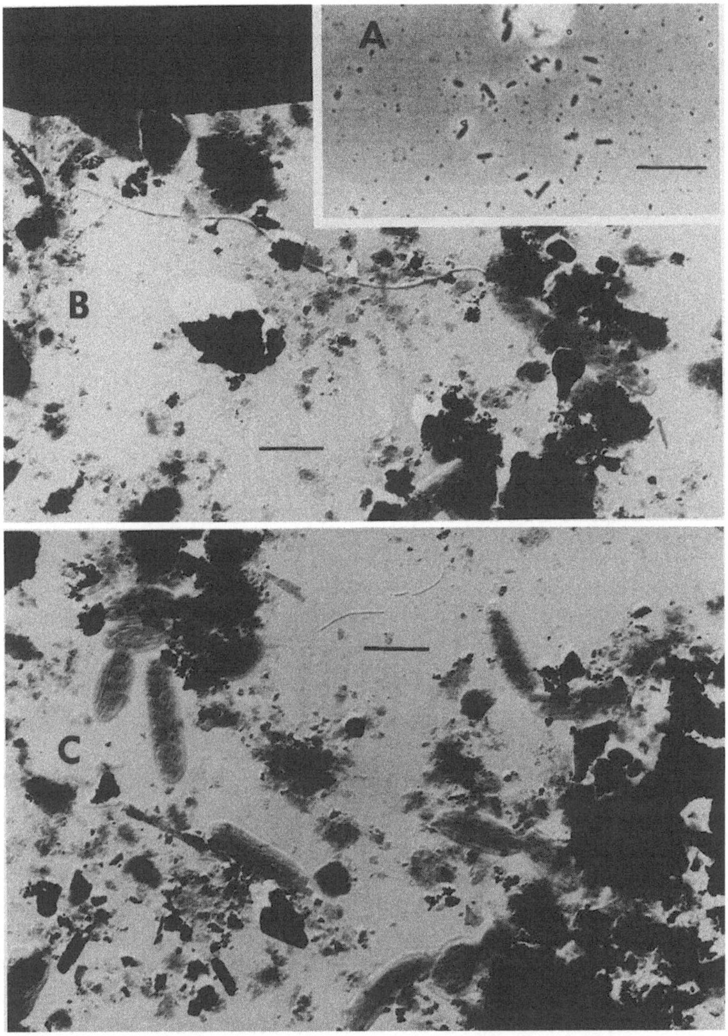

Fig. 5.3.3.5. Patchy growth of bacteria which were lifted off with a polymer layer from a sediment sample that had been prepared according to Waid (1973). **A** Phase contrast micrograph; *bar* 7 μm, **B** and **C** transmission electron micrographs with C/Pt shadow; *bar* 1 μm

isolated. Thus, the number of pure cultures obtained reflected the efficiency of a particular medium. Data of one such experiment are shown in Table 5.3.3.7. PYGV with low concentration of organic C was by far superior to the other media. It was also employed in dilutions (1:5, 1:10, 1:100). The comparison between PYGV and P medium has already been mentioned. In viable cell counting both media gave similar results (3.6×10^3 and 3.3×10^3 CFU/ml), but the MT composition was different (Hirsch and Rades-Rohkohl 1988a). Further enrichments were all carried out with PYGV.

Table 5.3.3.7. Efficiency of media for the growth, counting and isolation of aerobic ground water microorganisms. The water samples came from two wells near Haltern (Westphalia). Incubation: at 22 °C

Well no.	Depth (m)	No. of isolates obtained from medium[a]							
		PYGV	YM	NA	66a	MEA	258	337-C	337G,NO_2^-
V9	6.5	2	–	–	–	–	–	–	–
	7.7	14	12	6	6	2	7	3	30
	10.0	60	2	5	7	3	36	7	5
	15.0	39	19	–	16	8	19	5	6
	19.0	20	4	20	4	6	40	2	2
V11	5.5	1	–	3	3	–	3	3	–
	6.8	8	8	1	8	14	7	5	8
	8.4	48	6	3	1	2	7	3	7
	11.6	0	–	2	3	–	5	0	0
	13.6	13	4	6	9	20	11	11	8
	17.0	32	4	9	2	10	4	12	5
V9+V11	Total	238	59	55	59	65	139	50	71

[a]All different colonies and morphotypes were isolated.
For the description of media see Hirsch and Rades-Rohkohl (1983a, 1988b).

Incubation temperatures were likewise varied. As the ambient subsurface temperature was 9–10 °C, all microorganisms grew well at this temperature and usually less well at lower or higher ones (Fig. 5.3.3.6).

In aerobic liquid petridish enrichments, frequent observation revealed a sequence of bacterial blooms. An initial mass development of vibrios (3–8 d) was followed by blooms of rods, then *Caulobacter* spp., *Planctomyces* spp. and finally caulobacters again. After 3–4 months, predators appeared such as *Bdellovibrio* spp., spirochetes and protozoa. In contrast to water enrichments, those which had been inoculated with sediment samples contained large blooms of stalked and appendaged bacteria already within 2–3 weeks after inoculation. Anaerobic enrichments had their highest MT diversity (often pleomorphic and clostridial forms) after 3–5 weeks, depending on the medium. Here again: PYGV was optimal for the highest diversity. A constant composition of the aerobic community was usually reached after approx. 4 months.

Isolations were carried out by plating repeatedly all colonies of interest. In this fashion a culture collection of 523 strains was assembled; Table 5.3.3.8 indicates their numbers and origin. Most cultures were prokaryotic (457); there were 66 eukaryotes (27 yeasts, 38 filamentous fungi, 1 green alga). All isolates tested have so far survived either lyophilization or frozen storage at –20 °C.

Properties of the Ground Water and Sediment Isolates

Identification tests were carried out with those strains that exhibited reasonable growth in an acceptable time period (Tables 5.3.3.9–5.3.3.13, Fig. 5.3.3.6). The distribution of Gram reactions among the isolates reflected the fact that most of

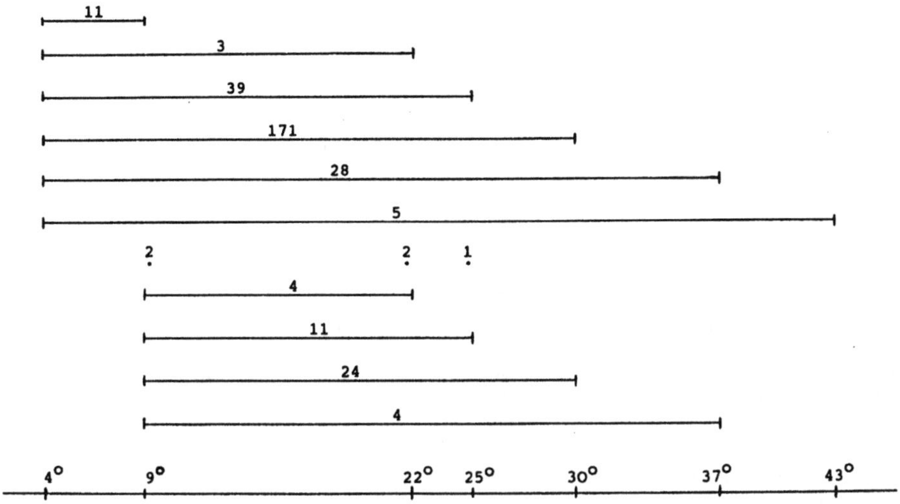

Fig. 5.3.3.6. Temperature tolerance ranges for growth of 305 aerobic subsurface isolates. The *numbers* show the strains positive for the particular range

Table 5.3.3.8. Culture collection of aerobic ground water microorganisms from the Segeberger Forst sites: origin

Total number of aerobic ground water and sediment strains: 523

1. Segeberger Forst test area: 387 strains
   - 1.1 from water samples: 328 strains
     - 1.1.1 multilevel tube well SEG5 (5.65–19.65 m): 106 strains
     - 1.1.2 multilevel tube well SEG6 (8.65–12.65 m): 52 strains
     - 1.1.3 multilevel tube well SEG9 (10–12.65 m): 25 strains
     - 1.1.4 stainless steel well SEG7 (10 m): 17 strains
     - 1.1.5 stainless steel well SEG10 (10 m): 9 strains
     - 1.1.6 stainless steel well SEG4 (10 m): 12 strains
     - 1.1.7 stainless steel well SEG7 (10 m; 39 800 l prepumped) 47 strains
     - 1.1.8 PVC sampling tube SEG8 (10 m): 60 strains
   - 1.2 from interstitial water: 26 strains
   - 1.3 from sediment: 33 strains
2. Radesforder Berg SEG3: 136 strains
   - 2.1 from sediment 31.5 m (capillary zone): 80 strains
   - 2.2 from sediment 31.8 m (ground water zone): 41 strains
   - 2.3 from sediment 31.9 m (Fe-horizon): 15 strains

them came from ground water. Corresponding observations were reported by other investigators (Ghiorse and Balkwill 1982, Kölbel-Boelke and Hirsch 1989). The high percentage of pigmented forms was surprising in view of their underground origin. Also widely distributed was the ability to precipitate ferric iron oxides, to reduce Mn^{4+}, and to reduce nitrate to nitrite (Table 5.3.3.9).

Table 5.3.3.9. Some physiological and enzymatic activities of aerobic ground water and sediment isolates

Activity/test	No. of strains			Percent positive
	Total	Negative	Positive	
Oxidase	403	168	235	58.3
Catalase	357	24	333	93.3
Nitrate reduction to NO_2^-	174	100	74	42.5
Nitrate reduction to N_2	116	100	16	13.8
DNase (extracellular)	281	194	87	31.0
Sialidase (extracellular)	202	194	8	4.0
Mn^{2+} oxidation	329	295	34	10.3
Mn^{4+} reduction	236	135	101	42.8
Fe^{3+} precipitation	313	123	190	60.7
Starch hydrolysis	266	204	62	23.3
Indol formation	143	73	70	49.0
Voges-Proskauer reaction[a]	172	49	123	71.5
Protocatechuic acid: degradation				
in meta-position	113	93	10	8.8
in ortho-position	141	94	47	33.3
in meta- *and* ortho-position	141	140	1	0.7
Humic preparation[b]: stimulation				
in liquid medium	153	108[c]	45	29.4
in agar medium	154	149	5	3.2

[a]Data combined from experiments at 9, 20 and 30 °C.
[b]Preparation from A_h horizon of Segeberger Forst test area, kindly provided by H. Klamberg, Marburg.
[c]Data combined: 57 strains without growth, 46 strains growing but not stimulated, 4 strains with variable results, and one strain which was inhibited by the A_h humic preparation.

The majority of all strains (85%) was capable of growing in the pH range of 5–8; only very few isolates had a more narrow range (Table 5.3.3.13). The temperature spectra (Fig. 5.3.3.6) showed 67% of the strains to be eurytherm, i.e. to grow between +4 and +30–43 °C, while only 15 isolates (5%) could be considered as psychrophiles.

The low concentration of organic C in SEG ground water samples may have been the reason for the large proportion (42.1%) of oligocarbophilic strains among the isolates (Table 5.3.3.10). Lack of nutrients in the Segeberger Forst subsurface environment was also indicated by the frequency of oligonitrophilic isolates (118 strains). The main C- and energy sources for the Segeberger Forst microorganisms may be organic acids, as indicated by the high percentage of isolates which could utilize these (Table 5.3.3.10). Only glycolate and tartrate were rarely utilized. Sugars and sugar alcohols also served as C sources, but less often under aerobic than anaerobic conditions (Table 5.3.3.11). This effect was especially evident with fructose. The concentrations of sugar compounds in these tests were quite important for the outcome, as shown in Table 5.3.3.11 for glucose, where the largest concentration supported growth of the highest number of isolates.

A HA preparation that was extracted from the A_h horizon of the SEG test area actually stimulated growth of 45 strains when tested in liquid culture and when compared to growth without this preparation. But only five strains were stimulated

Table 5.3.3.10. Carbon and nitrogen utilization by selected isolates

C or N source	No. of strains			Percent positive
	Total	Negative	Positive	
Growth without C source added	283	164	119	42.1
Growth w.pepton, yeast extract and glucose (medium B)	283	20	263	92.1
Growth with casein	283	52	231	81.6
Growth with organic acids (0.1%)				
Formate	31	9(4)[a]	22	71.0
Acetate	62	36(1)	26	42.0
Glycolate	54	48(4)	6	11.1
Propionate	50	8(1)	42	84.0
Lactate	86	27	59	68.6
Pyruvate	67	9	58	86.6
Malate	68	17(2)	51	75.0
Fumarate	51	6(2)	45	88.2
Succinate	71	9(2)	62	87.3
Glutamate	91	50(1)	41	45.1
Malonate	66	11(1)	55	83.3
Tartrate	32	28(2)	4	12.5
Citrate	70	14(1)	56	80.0
Citrate (0.2%)	90	18(4)	72	80.0
Growth without N source added	299	181	118	39.5
Growth with NH_4^+ as sole N source	299	97	202	67.6

[a]The negative data include (in brackets) the number of strains which were inhibited by the addition of the organic acid.

on the same medium solidified with agar (Table 5.3.3.9). The reasons for this effect are unknown. A number of isolates were also capable of cleaving protocatechuic acid after induction with p-OH-benzoate (Table 5.3.3.9). In most cases degradation of the aromatic ring occurred in ortho position; one strain could cleave in ortho and meta position.

The Voges-Proskauer test was carried out at different temperatures. Although most isolates originated from 9 °C enrichments, their activity in this test could be almost doubled by raising the temperature to 30 °C (Hirsch and Rades-Rohkohl 1988a). A slight increase of activity was also noted in the case of the methyl red test. This demonstrates the difficulty in comparing ground water isolates with other (mesophilic) strains under "standard" test conditions.

The tests for antibiotic resistance were carried out with antibiotic concentrations of 10 and 50 µg/ml (Table 5.3.3.12). Yeasts showed generally high resistance to most compounds, except for tyrocidin. Gram-negatives were least resistant to rifampicin and streptomycin, while Gram-positive isolates were most inhibited by tyrocidin, rifampicin and streptomycin.

As an example for possible interactions of the ground water isolates with polluting faecal bacteria, the addition of *Escherichia coli* K12 to pure cultures was tested (Table 5.3.3.13). About 20% of the strains demonstrated inhibitory (17.8%) or even stimulatory activity (2.8%) in such two-membered cultures.

Table 5.3.3.11. Acid formation from sugars and sugar alcohols

Carbon source	Concentration (%)	O_2	No. of strains			Percent positive
			Total	Negative	Positive[a]	
Glucose	0.025	+	167	131(63)[b]	36	21.6
Glucose	0.05	+	151	107(103)	44	29.1
Glucose	0.10	+	148	44(15)	104	70.3
Glucose	0.50	+	148	38(19)	110	74.3
Galactose	0.10	+	142	58(21)	84	59.2
Mannose	0.10	+	163	89(69)	74	45.4
Fructose	0.10	+	168	166(163)	2	1.2
Ribose	0.10	+	152	115(96)	37	24.3
Maltose	0.10	+	196	55(45)	141	71.9
Saccharose	0.10	+	183	177(101)	6	3.3
Mannitol	0.10	+	152	65(10)	87	57.2
Ribitol	0.10	+	151	108(11)	43	28.5
Glucose	0.025	–	165	52(49)	103	62.4
Glucose	0.05	–	236	9(9)	227	96.2
Glucose	0.10	–	180	22(16)	158	87.8
Glucose	0.50	–	179	20(18)	159	88.8
Galactose	0.10	–	167	25(25)	142	85.0
Mannose	0.10	–	156	80(80)	76	48.7
Fructose	0.10	–	147	126(126)	21	14.3
Ribose	0.10	–	153	108(107)	45	29.4
Maltose	0.10	–	161	86(81)	68	42.2
Saccharose	0.10	–	160	87(87)	73	45.6

[a]Medium CV with 0.8% bromothymolblue; positive - acid formation and growth, negative - growth, or not, and no acid formation.
[b]The negative data include (in brackets) the number of strains which were inhibited by the addition of the sugar or polyol.

Table 5.3.3.12. Resistance of 120 isolates against antibiotics. Concentrations in $\mu g\ ml^{-1}$; microtiter plate experiments

Antibiotic and target[a]		Percent of strains with resistance					
		Gram-positives[b]		Gram-negatives[c]		Yeasts[d]	
	(Concentrations)	10 µg	50 µg	10µg	50 µg	10 µg	50 µg
Bacitracin	C	61.9	61.9	88.8	80.9	100.0	100.0
Penicillin G	C	42.9	19.0	59.6	38.2	90.0	90.0
Nalidixic acid	D	95.2	95.2	90.9	95.5	100.0	100.0
Novobiocin	M	33.3	33.3	48.3	34.8	90.0	90.0
Rifampicin	R	9.5	4.8	11.2	2.2	90.0	70.0
Tyrocidin	PE	4.8	4.8	18.0	14.6	0	0
Trimethoprim	F	90.5	71.4	80.9	67.4	100.0	100.0
Erythromycin	P,S	19.0	9.5	67.4	38.2	90.0	90.0
Chloramphenicol	P	28.5	9.5	64.0	30.3	80.0	80.0
Streptomycin	P	9.5	4.8	11.2	4.5	70.0	60.0
Tetracyclin	P	14.3	4.8	20.2	7.9	90.0	80.0

[a]C - cell wall synthesis; D - DNA synthesis; M - Mg-chelation; R - RNA synthesis; PE - permeability, cation detergent; F - folic acid biosynthesis; P - protein synthesis; S - sporulation of actinomycetes.
[b]21 strains. [c]89 strains. [d]10 strains.

Table 5.3.3.13. Some further properties of the subsurface isolates

Property	No. of strains			Percent positive
	Total	Negative	Positive	
Gram staining reaction	360	321	39	10.8
Motility	439	173	266	60.6
Pigmentation[a]	523	417	106	20.3
Growth in presence of lysozyme (10 μg ml^{-1})	299	109	190	63.5
Growth between pH 5 and 8	143	21	122	85.3
Growth between pH 5 and 7	143	139	4	2.8
Growth between pH 6 and 8	143	135	8	5.6
Growth between pH 7 and 8	143	137	6	4.2
Growth only at pH 7	143	140	3	2.1
Interactions with *E. coli* K12 (ATCC 11229)	358	62(286)[b]	10	2.8

[a]Clear, white, or cream coloured colonies were considered to be non-pigmented.
[b]Strains which were indifferent are given in brackets.

Relationships to Submerged Surfaces

Differences between sediment organisms which are normally thought to adhere to solid surfaces and those that live within free water have been pointed out (Kölbel-Boelke and Hirsch 1989). It can be assumed that surface polymers confer the ability to adhere to the surface. Among the Segeberger Forst morphotypes there were approx. 50 MT which produced at least one type of surface polymer, and 24 MT which normally exhibited attachment to surfaces (Table 5.3.3.14). Many of these could be identified to the genus level, as has been pointed out before. Attachment

Table 5.3.3.14. Morphotypes observed in Segeberger Forst samples which produce surface polymers that may result in adhesion to ground water substrata

Polymer types	Prokaryotes		Eukaryotes	
	MT numbers	Total	MT numbers	Total
Capsules	3,4,5,7,13,14,16 17,18,20,28,31,34, 56	14	84,94,97,98,99 106,107	7
Stalks (excreted)	32,33,49,50,51,52	6		0
Sheaths	22	1		0
Fimbriae, rigid	2,35,50,60,68,75,76	7	90	1
Fimbriae, flexible	10,29,52	3		0
Fimbriae, polar	21	1		
Holdfasts	46–52,57,59,60,62–72	21	90	1
Flagella (prokar.)	27,28,32,33,47,52,65-69,74,75	13		0
Flagella (eukar.)		0	84–87,89–93,96	10
Total MT		47		16

of these bacteria to other microorganisms has often been observed, and it was also pointed out (Hirsch and Rades-Rohkohl 1983b) that the formation of such small aggregates could effectively slow down the transportation of these organisms through the subsurface pore system. Apart from such polymer-mediated attachment, ground water microorganisms were often covered with clay particles. This was especially found with sediment samples which were treated according to Waid (1973). Size and shape of these clay particles suggest a montmorillonite type of clay. Certainly, such surface coatings would also reduce motility (Fig. 5.3.3.6).

5.3.3.8 Attempts to Understand in Situ Activity

Still largely unsolved is the question of activity of the subsurface organisms. There is no doubt that there occur, within ground water and sediment, viable microorganisms in large numbers and with great diversity. However, all of these could conceivably survive there in a dormant state and initiate growth or other activities only when brought up to the surface and inoculated into culture media. Ghiorse and Balkwill (1982) pointed out that the presence of dividing cells in preparations could indicate in situ activity. Also the presence of intact flagella (Fig. 5.3.3.6) in such preparations has been connected with in situ activity. But intact flagella do not necessarily indicate motility. Experiments where sediment or water samples were brought up and then enriched with ^{14}C-labelled substrates to study their uptake are not conclusive for the same reasons. The only convincing evidence would be to demonstrate in situ growth and/or multiplication within the undisturbed environment. Such experiments have been tried by Marxsen (1982), and, in the case of the SEG aquifer, by Hirsch and Rades-Rohkohl (1990): washed and sterilized sediment was exposed for 12 weeks in a stainless steel well in 10 and 20 m depth. In one experiment the sediment was baked twice at 600 °C prior to exposure to remove the organic carbon. These experimental sediments were colonized completely to reach 10^5 viable cells/g d.wt., a level which surpasses even that of the normal natural sediment samples. The ground water must have supplied the necessary carbon and energy for this biomass to be formed, although the colonization could also have been facilitated by frequent entering of motile microorganisms into the experimental chambers.

5.3.3.4 Discussion

The data presented here show that the Segeberger Forst aquifer system contains a large variety of pro- and eukaryotic microorganisms. This high diversity could be recognized from the many different MT and from the high viable cell counts obtained with selective media. A highly diverse community normally indicates stability of a climax situation and rich interactions among the components of the ecosystem. It could thus be assumed that the Segeberger Forst subsurface environment was not polluted.

For optimal sample-collecting procedures it became necessary to culture the microorganisms immediately after the samples were obtained. It was also learnt that ground water, interstitial water and well water were quite different with respect to chemistry and the biota. Therefore, it became necessary to prepump the wells before sampling the "free ground water". The well water certainly must have been colonized from the aquifer; it only concentrated certain parts of the ground water community. This in turn could have increased the concentration of dissolved organic carbon, as has been observed with PVC sampling tube SEG8.

Observation of the organisms' morphology could also reveal their nutritional status. Most pure culture organisms were in vitro not larger than in the original samples, and the cells were not smaller than most common bacteria. This shows that the SEG isolates were not starvation-resistant "minicells", as have been described for unsaturated soil samples or ocean water. But the frequent presence of storage granules (poly-β-hydroxybutyric acid) and the formation, in natural samples, of extremely large prosthecae by *Caulobacter* and *Hyphomicrobium* spp. do indicate lack of nutrients in the sediment. Finally, the frequent ability to grow oligocarbophilically further stresses the paucity of organic carbon in the ground water environment of these organisms.

The cell counts determined here agreed well with those obtained by other authors who used similar methods and media (Ghiorse and Balkwill 1982). Although medium PYGV appeared to be optimal for the conditions employed here, the difference between viable and total cell counts (two to three orders of magnitude) could be due to the presence of highly specific microorganisms that require growth conditions which were not offered here.

The distribution of ground water organisms was found to be patchy. Just how patchy it can be deduced from the following calculations. Dr. D. Schenk (Kiel) kindly determined the inner surface of a 10 m sediment sample in which the total cell number was 10 cells/g d.wt.: the inner surface was found to be 27 m^2/g d.wt. It can be calculated, then, that the 10^6 cells would be spread over 2.7×10^{13} μm^2. Assuming a homogeneous distribution, there would have to be one cell per every 2.7×10^7 μm^2. However, with a patchy distribution and large voids between organisms it certainly would be difficult to find any cells using SEM or TEM. Indeed, there were only very few patches that could be found.

When testing growth of the 523 isolates with various C sources, it was found that organic acids were suitable C sources for many strains. Glucose was utilized by many strains aerobically, but other sugars were better substrates under anaerobic conditions. This observation, together with the high percentage of nitrate reducers (56.3%) and Mn reducers (42.8%) indicate the presence, within the sediment, of many microaerophilic or even anaerobic microhabitats. This was supported by the determinations of the average oxygen concentration in 10–20 m depth which was found in ground water to be 3–5 mg/l.

In unpolluted ground water environments, the HS and related aromatic compounds are probably the major sources of carbon, nitrogen and energy. But only a small part of the isolates was capable of growing with the humic preparation or protocatechuic acid. The other strains would then have to live at the expense of these aromatic and humic degraders or their excretions such as organic acids or polymeric substances. In this connection it is noteworthy that 31% of the isolates produced extracellular DNase, at least under laboratory conditions. Of course, the real in situ enzymatic activities are still unknown, and the sediment exposure experiments carried out in this study showed colonization activity only at the end, i.e. after 12 weeks, when the sediments could be retrieved. Further work needs to be done to obtain reasonable data on in situ growth rates.

5.3.4 Microbiological Observations of the Unsaturated Zone of a Quaternary Gravel Profile

I. Alexander, G. Freitag, J. Grossmann, B. Merkel, P. Udluft and I. Ullsperger

5.3.4.1 Background

The knowledge of microbiological activity in the unsaturated zone has up to now been quite poor. One reason is the difficulty in obtaining subsoil samples under sterile conditions from depths of more than 1 m. Sediment and pore water samples were collected from different depths down to 9 m of the Quaternary gravel profile in the southern part of the Munich gravel plain at the accessible research shaft in Munich-Großhadern (MUC3; Merkel et al. 1982; Ullsperger et al. 1988a); this allowed the examination of samples from the same location at different times.

5.3.4.2 Materials and Methods

Sampling Methods

The wall of the accessible research shaft was perforated at different depths by holes with a diameter of 3.5 cm. The samples were taken using a sterilized auger at a distance of about 20 to 80 cm from the shaft's wall. The holes were closed by rubber plugs after each sampling procedure. Additionally, several samples were taken simultaneously for chemical analysis and moisture control. Pore water was also sampled from different depths by means of ceramic suction cups. All samples were processed for microbiological investigations within less than 24 h. In seven depths the ground temperature and the CO_2 content of the ground air were continuously determined with a non-dispersive infrared sensor.

Bacterial Counts and Determination Methods

The samples were mixed with 250 m of sterile 0.85% NaCl-solution and stirred for 10 min. After sedimentation, 0.1 or 1.0 ml of the suspension was plated according to Koch's pour plate method. Bacterial counts were obtained by using "DEV" agar (1% peptone and 1% meat extract); counts of hydro-carbon decomposing bacteria were taken from TGE agar; both media are described in the German Drinking Water Decree (Trinkw V 1986). The bacterial populations were determined after incubation at 20 ± 2 °C and 36 ± 1 °C. Some of the samples were also investigated for nitrate- and sulfate-reducing bacteria as well as for *Escherichia coli* and coliform bacteria. Nitrate- and sulfate-reducing bacteria were determined using special media (Merck, Difco,: Sulfate Reducer API Agar and Nitrate Broth). Sulfate-reducing bacteria were detected from the reducing of sulfate to iron sulfide and were counted by the number of black spots occurring on the media. The number of nitrate-reducing bacteria was determined by nitrite tests in different dilutions of water (titre determination).

The bacterial population numbers were related to the fraction < 2 mm per g of dry sample to obtain relative numbers which are based on the surface size of a sample.

Pore equilibrium solutions were obtained by mixing the sediment samples with 250 ml of distilled water and waiting for 24 h.

The esterase activity in these equilibrium solutions and in pore water collected with suction cups was measured with fluorescence spectrometry according to Holzapfel-Pschorn et al. (1985). This method gives a rough estimation of the biological activity and is related mostly to the decay of organic matter.

Reproducibility and Error Estimation

The sampling and preparation procedures were standardized in order to obtain reproducible results. Most important were an exact timing and careful sterilization of all tools used. In some cases two or three parallel samples were taken to test reproducibility and to obtain error estimations. The variation of the bacterial population size was less than one order of magnitude. These results are consistent with the assumption that bacteria in water and rock material are inhomogeneously distributed and form clusters (Sects. 5.0, 5.1.2 and 5.3.2).

The determination of esterase activity in pore water samples was affected by the duration of sampling, the cup material and the pretreatment procedure of the sampling system. Comparisons were made between Al_2O_3 cups and plastic cups, and it was shown that the esterase activity was significantly increased in samples sucked through plastic cups. This might have been due to bacterial colonization of the unsterilized Al_2O_3 cups and more bacterial growth on the plastic cups. Another study was done with ceramic cups which were disinfected with formaldehyde (1%) and rinsed with sterilized distilled water. Such pretreated cups showed a significant decrease of esterase activity. In consequence, only disinfected ceramic cups were used to sample pore water for the determination of esterase activity.

5.3.4.3 Results and Discussion

Although Werner (1987) found only about 0.1 to 1% of the total bacterial population to be detectable by the methods described above, they were used in this preliminary study because they are standard methods and were easy to handle. The results were comparable to each other because the same methods were used in all cases.

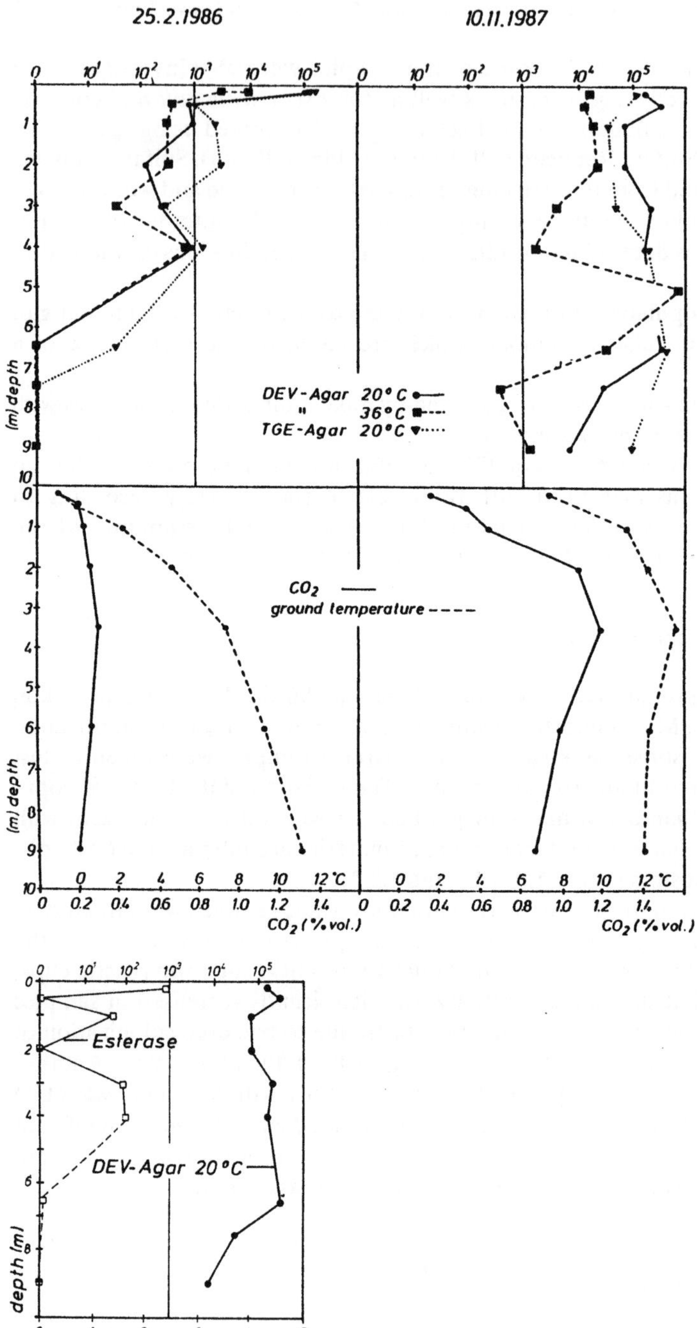

Fig. 5.3.4.1. Bacterial viable cell counts/g dry soil on DEV agar at the accessible research shaft. Incubation at 20 ± 2 °C and 36 ± 1 °C. *TGE-agar/20* °C indicates numbers of hydrocarbon decomposing bacteria. For comparison, ground temperature and CO_2 partial pressure in ground air are also given. Note differences between the two dates

Figure 5.3.4.1 displays bacterial viable counts in different depths of the accessible research shaft and at different times. Generally, the results in Fig. 5.3.4.1 prove that sampling with a sterilized auger was not linked to a significant contamination of the natural environment. The first sampling campaign in the research shaft had occurred in February 1986 (air temperature –7.6 °C) when the soil was frozen down to a depth of 50 cm. However, between 6 m and 9 m the ground temperature was 10 °C. Accordingly, relatively small numbers of bacteria were found (between 0 and 10^5/g). The second sampling was done in November 1987 with soil temperatures between 7.4 and 12 °C, and the third one occurred in March 1988. The year 1988 was characterized by a rather warm winter and due to this the numbers of bacterial viable counts in March 1988 were high. In all cases a good correlation between the bacterial cell counts and the CO_2 content of the ground air was found. The nutrient supply, especially the dissolved organic carbon content (DOC), was

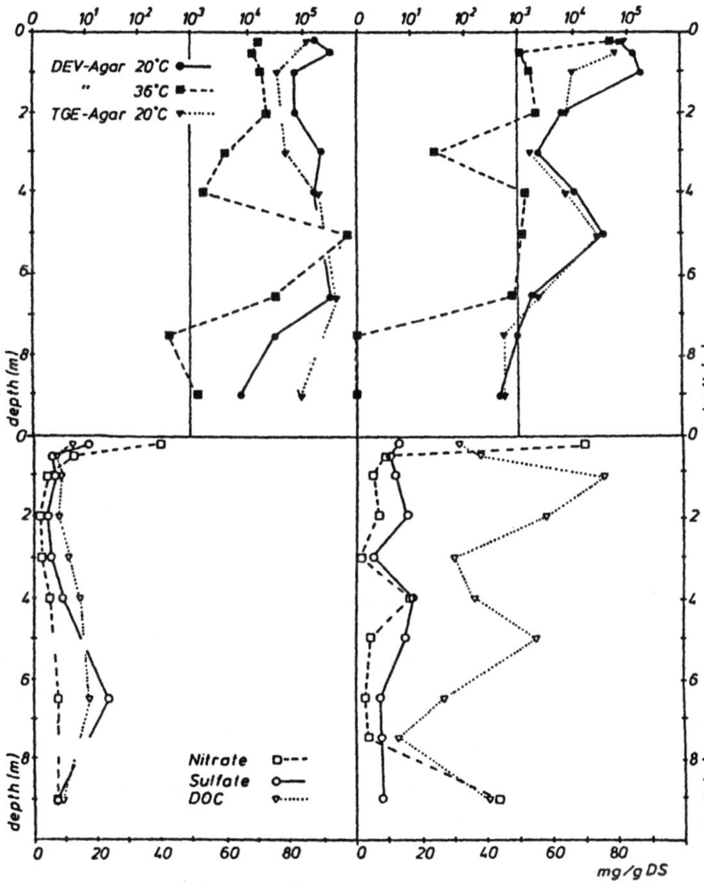

Fig. 5.3.4.2. Bacterial viable cell counts/g at the accessible research shaft after incubation at 20 ± 2 °C and 36 ± 1 °C. Hydrocarbon decomposing bacterial (TEC agar) and cells growing on DEV-agar are compared on two dates to DOC, nitrate and sulfate

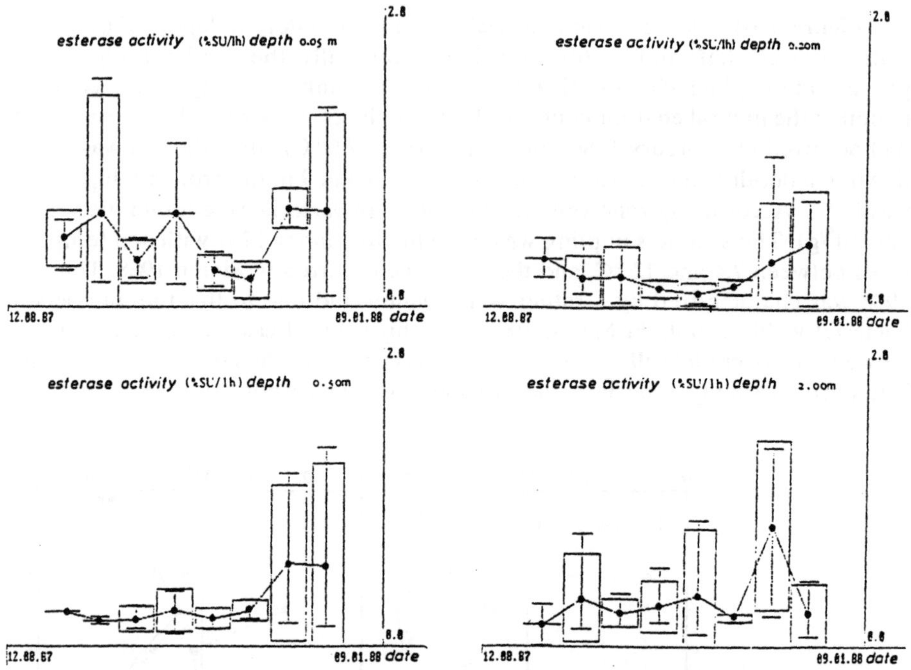

Fig. 5.3.4.3. Esterase activity in pore water of four depths. (*% SU/lh*) relative substrate consumption per litre and hour. The *line* displays the mean values at times of two weeks. Extreme values are marked by the *bars* and the standard deviation as *boxes*, respectively

much higher in March 1988 than in November 1987, when DOC was broken down nearly completely, but the bacterial cell counts were high throughout the whole profile (Fig. 5.3.4.2). Five months later the profile was characterized by a significant decrease of the bacterial numbers and by an increased nutrient supply. The correlation between DOC and bacterial numbers on DEV or TGE-Agar at 20 °C was statistically significant.

In general, there was no significant relationship between esterase activity and bacterial population size. However, a direct correlation between esterase activity and bacterial numbers would not be necessary because the enzymatic activity here was a general sum parameter linked to the decay of organic material. However, only some types of bacteria form esterases. Considering that only 0.1 to 1% of the total bacterial population might have been detected with the methods used, one could assume that most of these bacteria did not form esterases at all.

The esterase activity was determined continuously in Munich-Großhadern during summer and autumn 1987 at four different depths (Fig. 5.3.4.3). The results are characterized by pronounced variations of esterase activity but no significant seasonal variations. The recorded variations may have been related to the high inhomogeneity of the soil, or could have been due to the sampling and measuring procedures. On the contrary, increased microbiology activities should in any case occur during summer, even taking into account a moderate winter.

6 Hydrogeochemical and Geochemical-Hydraulic Models and Model Concepts

6.0 Introduction

H.D. Schulz

In the priority programme all research groups were encouraged to develop and use model concepts and computer models to evaluate the measured data. Notwithstanding the fact that models cover a very wide range of sophistication, computer models are extremely helpful for the description of the complex geochemical reaction systems, especially in combination with the transport processes in the unsaturated or saturated zones. Thus, the different sections in this chapter cover models that combine geochemical black boxes with different flow systems, models that use geochemical measurements for the calibration of flow systems, models that combine geochemical reaction systems with flow systems and models which try to achieve a better understanding of calculated saturated indices in seepage and ground water.

For most natural waters the calcite-carbonate system controls the usually high concentrations of dissolved calcium and hydrogen carbonate as well as the pH value and a variety of other dissolved components. At the beginning of the priority programme the dissolution and precipitation processes of carbonates for most researchers seemed to be important as well, and simple enough to achieve a quantitative understanding of the system including the reaction kinetics. The different laboratory studies and field measurements on carbonate systems, presented in Chapter 3 of this volume, are evaluated in the Sections 6.1.1, 6.1.2 and 6.2.2 of Chapter 6, which deal with the modelling of carbonate systems.

In Section 6.1.1 Dreybrodt applies his theory of calcite dissolution kinetics (Sect. 3.1.1) with a realistic flow system through a fracture aquifer in carbonate rocks. The result is a deterministic model that predicts the development of a natural karst system in a geological time scale. For this combination it is necessary to apply the theory of calcite dissolution kinetics which could only be tested in short-time laboratory experiments on a process in a time which is orders of magnitudes longer. The realistic results for the time necessary for the development for a karst system are a reliable proof for the validity of the kinetic theory and show that for the first time a geological/geochemical process is quantitatively understood.

The Sections 6.1.2 (Eichinger, Merkel and Udluft) and 6.2.1 (Schulz) refer to sections in Chapter 3 describing the laboratory experiments and field measurements on the calcite-carbonate system in the unsaturated zone. Besides different locations of the field studies, the differences between both sections lie especially in the regarded time scale. Whereas in Section 6.2.2 the calcite weathering is described by Schulz as a long-term process in its dependency on seepage velocity, calcite dissolution, mass balance and cation exchange reactions, in Section 6.2.2 Eichinger, Merkel and Udluft simulated the annual variations of dissolved carbonate in seepage water as dependent on climatic variations, seepage velocity and CO_2 produced by microbiological activities. The comparison between the measured concentrations of dissolved carbonate and calcium with the simulated concentrations of these ions shows that the entire process of the calcite-carbonate system in the unsaturated zone is well understood in its major and minor controls.

The release rate of feldspars measured in three different laboratory experimental setups (Sect. 4.3.2, this Vol.) are discussed in Section 6.1.3 (Dahmke, Matthess, Petersen and Schenk) with respect to the influence of grain sizes, flow rates, premineralization and pH changes. These evaluations are connected with the laboratory observations on the influence of organic ligands (Sect. 2.9, this Vol.), and are helpful for the discussion of the field observations (Sect. 4.3.2, this Vol.).

In Section 6.2.5 (Strebel, Böttcher and Duynisveld), a model is presented which attaches more importance to the flow system and the mass balance of components, while treating geochemical reactions only as source or sink terms in the calculations. The model uses in a one-dimensional tube or in a two-dimensional vertical cross-section for the simulation of the physical transport in ground water. A mass balance for chloride as a conservative tracer is used as a test for the model. In comparison with the nitrate and sulfate concentrations, it is necessary to incorporate sink or source terms in the calculations which are used to quantify the microbiological reactions in ground water for these constituents.

In Section 6.2.2 (Bugner and Mull) the development of sulfate concentrations in ground water, including microbiological decomposition, was simulated for an unconfined aquifer of gravel and sands. In order to identify a vertical cross-section through the aquifer, first a three-dimensional ground water flow model was calculated. For the unidirectional flow in the two-dimensional mass transport model with curvilinear elements the velocity data of the three-dimensional ground water flow model were used. After a calibration of the model using environmental isotopes, the sulfate reduction was incorporated by using a time-dependent exponential reaction kinetics comparable to that of a radioactive decay. Obviously, in this model the physical transport is much more sophisticated than geochemical reactions, since the major aim of the authors was to find a physical transport model which allows the incorporation of even more complex reactions. Thus, this model shows how, in future work, complex flow situations could be combined with different multicomponent geochemical reactions.

The models of the Sections 6.2.3 (Dörr, Schlosser, Stute and Sonntag) and 6.2.4 (Forster, Loosli and Weise) deal with the description of ground water flow systems using different isotope measurements. In these models geochemical reactions are not regarded and importance is attached only to the determination of residence times of ground water in the aquifer and to the calibration of ground water transport models. Both sections show quite clearly to what extent isotope techniques today are useful and necessary for this purpose.

Using the data of three test areas in Section 6.3 (Kölling and Schulz), mineral saturations were calculated with different geochemical equilibrium models like

WATEQ2 or PHREEQE. This study tries to find a better understanding of mineral saturation indices. Especially new is the quantitative description of kinetic steady-state concentrations which opens a possibility to understanding how saturation indices for different minerals with one or more common ion can be interpreted. Sensitivity analyses show to what extent the calculated saturation indices are reliable depending on the unavoidable uncertainties of concentration measurements and of thermodynamical data. For two test areas the limiting phases for the ion content of ground water are derived as an interpretation of geochemical model calculations.

These sections give examples for some hydrogeochemical models that are in use. Since many basic hydrogeochemical reactions are not yet understood, different researchers are still attempting in very different ways to model the complex total system.

6.1 Hydrogeochemical Models and Concepts

6.1.1 Development of Secondary Permeability of a Fracture Aquifer in Carbonate Rocks: a Model

W. Dreybrodt

6.1.1.1 Background

Primary permeability of carbonate rocks results from an interconnected network of small fissures and cracks comprising joints, bedding plane partings and possibly also faults. Initial apertures are about 1×10^{-2} cm (Motyka and Wilk 1984). Once by some geological event an input-output configuration for surface water is given, which supplies a hydraulic gradient, driving surface water through the net of fractures to an output, solutional attack of the CO_2-containing water starts to widen the fractures and secondary permeability originates. Figure 6.1.1.1 shows some of the many small fractures constituting a percolating pathway from a point input to a point output. By solutional widening a conduit arises which is also shown. This

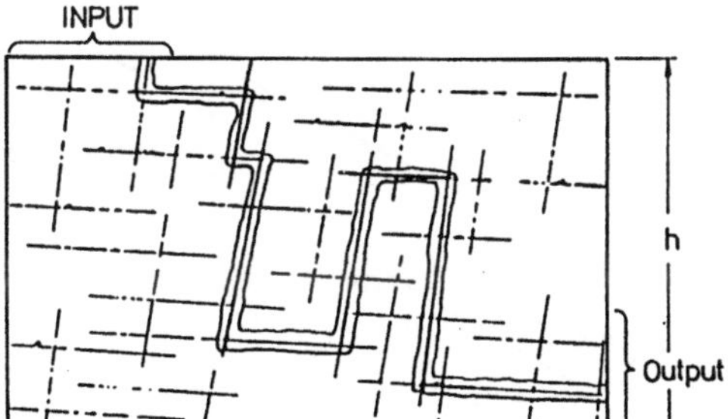

Fig. 6.1.1.1. Percolating net of fractures in the initial state of a fracture aquifer. Along one on the pathways a later growing conduit is shown

process, generally known as karstification, is intimately connected to the chemical kinetics of the system H_2O-CO_2-$CaCO_3$ (Dreybrodt 1988; Sect. 3.1.1). If dissolution proceeds very fast the calcite aggressive H_2O-CO_2 solution very quickly attains equilibrium and thus cannot penetrate deep into the rock without losing its solutional power. Therefore only surface denudation should be expected. On the other hand, if dissolution rates were extremely small, water can penetrate deeply into the rock but solutional widening is then extremely slow and it may take geologically unrealistic times until an aquifer typical for carbonate rocks develops. It is known (Bakalowicz 1982; Mylroie and Carew 1986) that minimum times of about 10 000 years are required until such an aquifer develops to maturity. It is of interest to model the development of carbonate aquifers since new land use programmes of limestone terrains provide human impact onto these systems and a deeper understanding of their development may prevent damage.

6.1.1.2 The Model

The basic question to answer is: provided the hydraulic gradient between input and output stays independent of time, what is the time dependence of the flow rate Q(t) of water driven through the pathway? How does this rate depend on geometrical parameters such as the length l of the pathway, its average aperture, a, and the magnitude of the hydraulic head, h, between input and output? And furthermore, how does this system depend on chemical parameters such as saturation concentration c_{eq} and chemical rate laws describing dissolution. To give answers to these questions the pathway is simplified to an initially uniform fracture of rectangular or ellipsoidal shape with an initial aperture a_o and breadth b_o. By solutional activity these initial parameters will not stay constant along the length of the fracture but will become functions of x, the coordinate along the length. Figure 6.1.1.2 shows a part of the fracture, between x and x+dx. The concentration of dissolved calcite, given as $c = (Ca^{2+})$ at the inflow x is c(x) and at x+dx it is c(x+dx) = c + dc. The flow velocity is v(x). Finally, dissolution proceeds by a rate law F(c). Mass balance now requires

$$F(c)P(x)dx = v(x)A(x)dc, \qquad (6.1.1.1)$$

stating simply that the amount of calcite dissolved from the wall with perimeter $P(x)$ is equal to the amount of calcite transported away by flow through the cross-

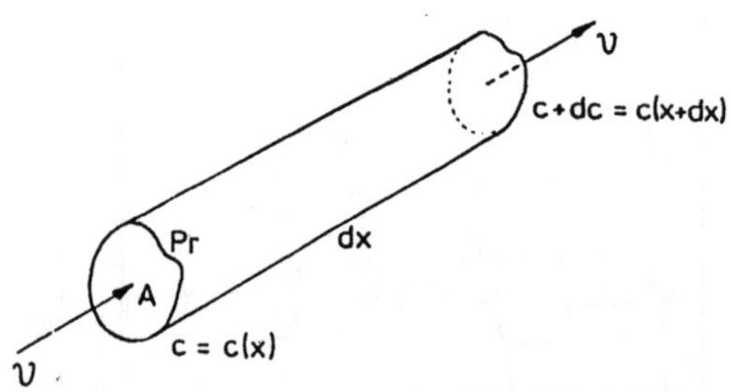

Fig. 6.1.1.2. Schematic model of a flow conduit at position x

sectional area $A(x)$. Now from the equation of continuity a constant flow rate results along the conduit

$$v(x)\,A(x) = Q. \tag{6.1.2.2}$$

Integrating Eq. (6.1.1.1) yields

$$Q \int_{c_o}^{c} \frac{dc}{F(c)} = \int_{o}^{x} P(x)\,dx. \tag{6.1.1.3}$$

Thus, one obtains the concentration c as a function of x along the pathway. To calculate Q one uses the equation of Hagen-Poiseuille for laminar flow (Beek and Mutzall 1975). From this a flow resistance R_f can be defined

$$R_f = \frac{12\,\eta}{\rho w^g} \int_0^1 \frac{dx}{a^3(x,t)b(x,t)M_o(x,t)}, \tag{6.1.1.4}$$

with the dynamic viscosity ρ, density of water ρ_w, and the gravitational acceleration g. M_o is a geometrical factor depending on the ratio *b/a*. It is given by Beek and Mutzall (1975: Fig. II 12) for rectangular and elliptical cross-sections with values between 1 and 0.3. The flow rate Q then results as

$$Q = \frac{h}{R} = \frac{I \cdot L_f}{R}; I = \frac{h}{L_f} \tag{6.1.1.5}$$

h is the hydraulic head and I the hydraulic gradient along the flowpath L_f.

To calculate $Q(t)$ as a function of time, one starts with the initially uniform fracture from which $R_o = R(0)$ can be easily calculated. Then with this value Eq. (6.1.1.3) is integrated numerically and the dissolution rates are obtained as a function $F(x)$. Using this function, one obtains for the fracture aperture $a(x,t)$ and the fracture breadth $b(x,t)$ at latter time t

$$a(x,t+\Delta t) = a(x,t) + \bar{F}(x)\,\Delta t \tag{6.1.1.6}$$

$$b(x,t+\Delta t) = b(x,t) + \bar{F}(x)\,\Delta t.$$

γ is a factor converting dissolution rates from $mol \cdot cm^{-2}/s$ into retreat of bedrock in cm/a and Δt is a suitable time increment. Thus, by iteration the time dependence of the flow rate and of the aperture width along the fracture can be calculated.

6.1.1.3 Dissolution Kinetics of $CaCO_3$

Dissolution rates of $CaCO_3$ by a calcite aggressive H_2O-CO_2 solution flowing laminarly under conditions of a closed system in a completely water filled fracture of given aperture *a* are given by Dreybrodt (Sect. 3.1.1, this Vol.)

$$F^{(1)}(c) = \alpha_k(x) \cdot (c_{eq} - c);\ \alpha_k(x) = \alpha_{ko} \cdot \left(1 + \frac{\alpha_{ko}\,a(x)}{6D_M}\right)^{-1}. \tag{6.1.1.7}$$

The kinetic constant α_{ko} depends on the initial CO_2 content of the solution and on the aperture a. It is almost constant for 5×10^{-3} cm $\leq a \leq 1$ cm (Buhmann and Dreybrodt 1985, Sect. 3.1.1, this Vol.) The effective constant α_k considers the influence of mass transport by diffusion for large apertures b. D_M is the coefficient of diffusion and its value is 1×10^{-5} cm²/s. Depending on temperature and initial CO_2-concentration of the solution prior to dissolution of calcite, values of α_{ko} are in the range of 2.5×10^{-5} cm/s. For pure calcite free of magnesium, this rate law holds up to saturation (Dreybrodt and Svensson 1989); for natural marble and icelandspar containing Mg, however, close to saturation a change in the chemical kinetics at the surface occurs and dissolution rates follow a higher order rate law (Plummer and Wigley 1976).

$$F^{(4)}(c) = \beta \cdot (c_{eq} - c)^4;\ \beta = 1.2 \cdot 10^{13}\ \mathrm{cm}^{10}/(\mathrm{mol}^3 \cdot \mathrm{s}). \quad (6.1.1.8)$$

Therefore in this calculation the first-order kinetics of Eq. (6.1.1.7) for concentrations $c \leq 0.9\ c_{eq}$ are applied and fourth-order kinetics for $c > 0.9\ c_{eq}$ are used. This implies an abrupt drop of dissolution rates by two orders of magnitude at $c = 0.9\ c_{eq}$, which in nature will be smoothed. However, it is convenient for the interpretation of the results.

6.1.1.4 Results

Figure 6.1.1.3 shows the profile of a fracture with $l = 20000$ cm, $b_o = 100$ cm, $a_o = 0.01$ cm, subject to a hydraulic gradient of $I = 0.1$, as it develops in time. Note that the scale of a is logarithmic.

In the early stage of evolution, i.e. for times below 2000 years, $c = 0.9\ c_{eq}$ is attained so quickly that practically fourth-order kinetics determine dissolution along the entire length of the fracture, which increases slowly in width a. As a

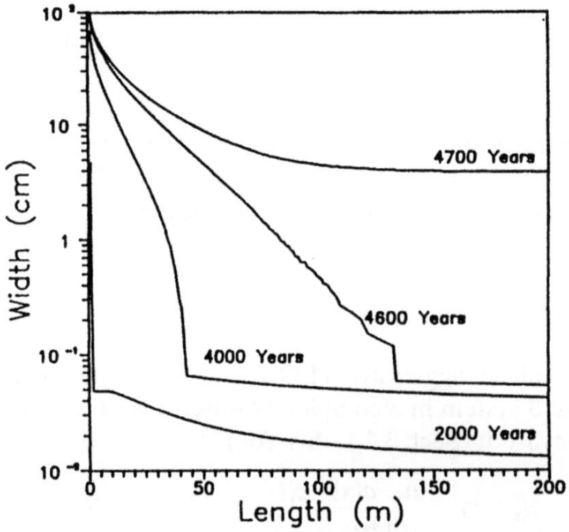

Fig. 6.1.1.3. Enlargement of a fracture with initial widths $a_o = 10^{-2}$ cm. The *numbers on the curves* gives the time in years after onset of water flow. Note that the widths $a(x,t)$ are plotted logarithmically. Parameters are $l = 200$ m, $a_o = 10^{-2}$ cm, $b_o = 100$ cm, $I = 0.1$. $\alpha_{ko} = 2.5 \times 10^{-5}$ cm/s, $\beta = 1.2 \times 10^{13}$ cm^{10}/(mol^{13}.s), $c_{eq} = 2 \times 10^{-6}$ mol/cm

consequence $Q(t)$ increases in time also, as shown by Fig. 6.1.1.4. The penetration length L_p, i.e. the length water can travel until under the regime of first-order kinetics the concentration has reached $c = 0.9\ c_{eq}$ is given (Dreybrodt 1988) by

$$L_p(t) = 1.15 \cdot \frac{Q(t)}{\alpha_{ko}\, b} = 9.4 \cdot 10^3 \cdot \frac{a_o^3 \cdot I}{\alpha_{ko}}\ [\mathrm{cm}]\,. \tag{6.1.1.9}$$

Thus, $L_p(t)$ increases in time and after about 2000 years a region where high dissolution rates determine the further widening of the aperture, starts to penetrate into the fracture, as can be seen by the two profiles which have developed after 4000 and 4900 years. The region of fourth-order dissolution correspondingly becomes smaller, but dissolution rates are now more evenly distributed along its length. Finally, the region of first-order dissolution reaches the output of the fracture and dissolution now quickly widens the end of the funnel with a dramatic increase in flow rate (Fig. 6.1.1.4). This event is called break-through and the corresponding time is the break-through time T. After break-through has occurred, dissolution proceeds with first-order kinetics exclusively and solutional widening is in the order of several 10^{-3} cm/a and fairly uniform over the entire length. If finally flow becomes turbulent at conduit widths of several mm due to the influence of eddy diffusion, a further increase in solutional widening occurs up to about 10^{-2} cm/a (Buhmann and Dreybrodt 1985b). This behaviour of time development of the profile is valid as long as $L_p(0) << 1$. For the initial aperture from Eqs. (6.1.1.9), (6.1.1.4) and (6.1.1.5) one obtains, using $\alpha_{ko} = 2.5 \times 10^{-5}$ cm/s:

$$L_p = 3.76 \times 10^8\ a_o^3 \cdot I\ (\mathrm{cm}). \tag{6.1.1.10}$$

This means for this case $L_p(0) = 37.6$ cm. Note that if one assumes first-order kinetics to prevail up to $c = c_{eq}$ the dissolution rates drop by a factor of 10 after each further travel over the length L_p, such that after 376 cm dissolution rates have dropped to 10^{-10} of the initial rate. Therefore, if first-order kinetics predominated up

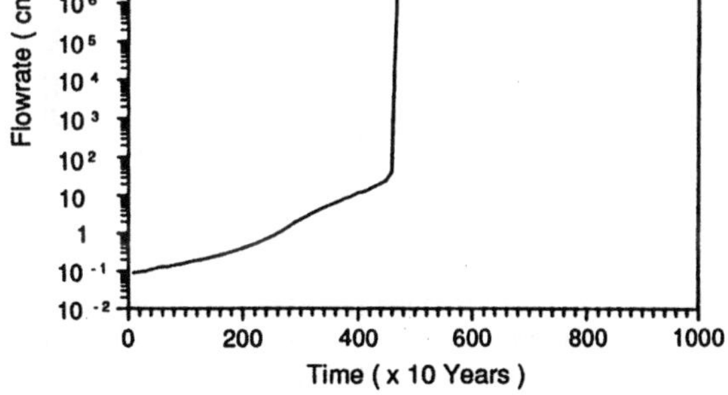

Fig. 6.1.1.4. Flow rate for the fracture in Fig. 6.1.1.3 as a function of time. Break-through time T is defined by the abrupt rise in flow rate

to saturation, secondary permeability in spatially extended fracture aquifers would not develop at all. This was first recognized by White (1977), who claimed a region of inhibited dissolution near equilibrium, to account for the development of secondary permeability of fracture aquifers in limestone rocks.

To further elucidate the behaviour of this model the break-through times were calculated as functions of the two sets of parameters which determine the problem.

The first set are the geometrical parameters a_o, L_f, and I. b_o is left constant b_o = 100 cm in all computer runs, discussed here.

Figure 6.1.1.5a shows the dependence of break-through time on the hydraulic gradient for several initial apertures a for L_f = 200 m. The chemical parameters are $\alpha_{ko} = 2.5 \times 10^{-5}$ cm/s, $\beta = 1.2 \times 10^{13}$ cm^{10}/(mol^3.s) and c_{eq} typically for limestone terrains is $c_{eq} = 2 \times 10^{-6}$ mol/cm^3. The results show that break-through times depend heavily on hydraulic gradient and initial apertures. Note that break-through times are plotted logarithmically. Figure 6.1.1.5b shows break-through times as function of fracture lengths for several initial apertures. These examples show that development of permeability depends heavily on the geometrical parameters.

The set of chemical parameters comprises α_{ko}, β and c_{eq}. The two kinetic constants α_{ko} and β are assumed to depend on the chemical composition and lithology of the rock. Their variation thus reflects the influence of different kinds of rock. c_{eq} is the saturation concentration and depends on the CO_2 concentration of the inflowing water.

Figure 6.1.1.6 shows the dependence of break-through times (now plotted linearly) on the first-order kinetic constant for various lengths with parameters c_{eq} = 2×10^{-6} mol/cm^3, $\beta = 1.2 \times 10^{13}$ cm^{10}/(mol$^3 \cdot$ s), $a_o = 10^{-2}$ cm, b_o = 100 cm, I = 0.1. For long (L_f > 200 m) fractures, although α_{ko} changes by two orders of magnitude, the variation in T is only by a factor between 2 and 3. The increase in T with increasing

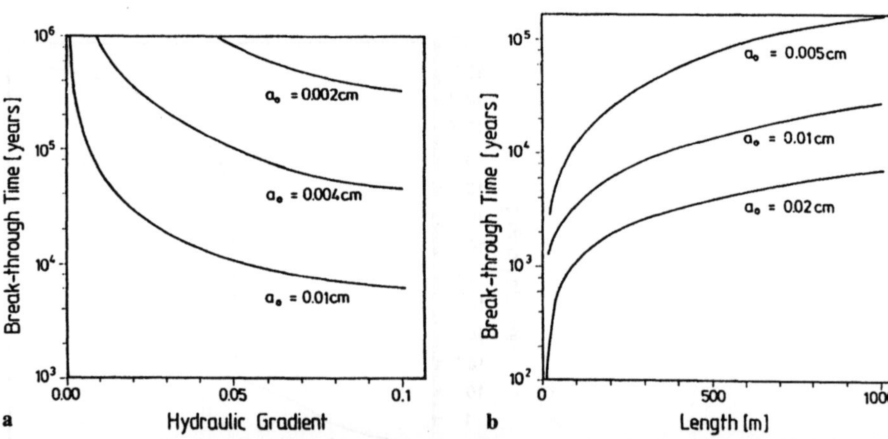

Fig. 6.1.1.5. Break-through times (**a**) in dependence on hydraulic gradient for different width of the initial aperture a_o (*numbers on the curves*) and (**b**) in dependence on lengths of a fracture for different widths of initial apertures. Other parameters as in Fig. 6.1.1.3

α_{ko} results from the fact that the corresponding first order penetration length [Eq. (6.1.1.10)] decreases. At short fracture lengths ($L_f < 50$ m) there is a significant variation in break-through times.

Figure 6.1.1.7 shows the variation of T with β for fractures of various lengths. It is interesting to note the minimum in the region of the experimental value of β, indicated by an arrow. The variation of break-through time is small for lengths below 200 m. With increasing lengths, the minimum becomes more pronounced and variations of T within a range of β of two orders of magnitude amount to a factor of about 3.

Finally Fig. 6.1.1.8 shows the dependence of T on c_{eq}. The saturation value of calcite depends on the CO_2 concentration of ground water, which is very much influenced by climate (Ford and Drake 1980, Drake and Ford 1981).

There is only a slight variation in break-through time for $c_{eq} > 2.5 \times 10^{-6}$ mol/cm^3. In the region between 2.5 and 1.5 × 10^{-6} mol/cm^3 a variation by a factor of 3 is

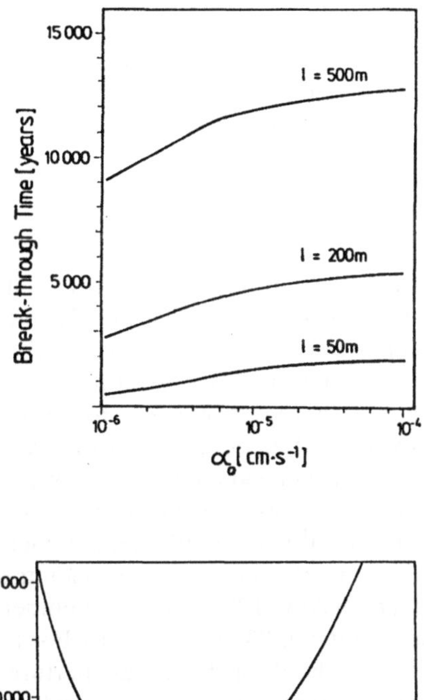

Fig. 6.1.1.6. Break-through times as function of kinetic constant α_{ko} for various lengths. Other parameters as in Fig. 6.1.1.3

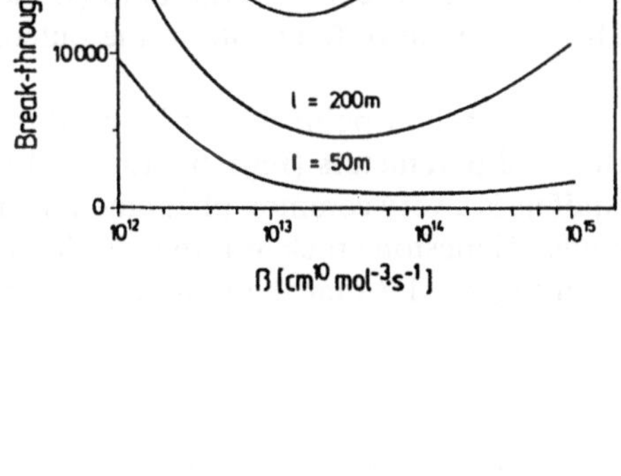

Fig. 6.1.1.7. Break-through times as function of kinetic constant β for various lengths. Other parameters as in Fig. 6.1.1.3.

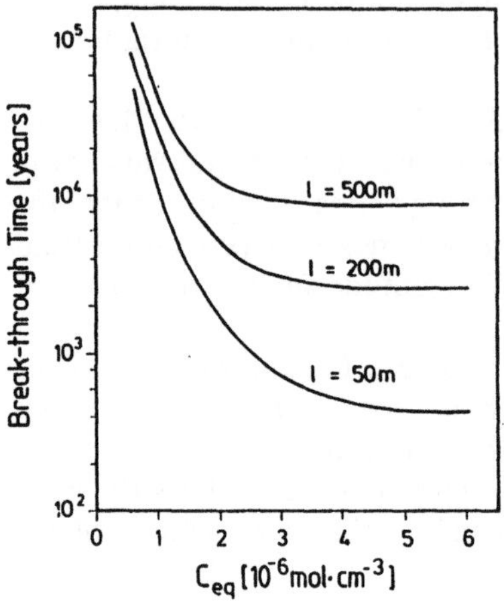

Fig. 6.1.1.8. Break-through times as function of equilibrium concentration c_{eq} for various lengths. Other parameters as in Fig. 6.1.1.3

observed. This is the region of karst waters in temperate and also tropical climates (Smith and Atkinson 1976). Below 1.5×10^{-6} mol/cm^3 a dramatic increase in break-through times is observed.

6.1.1.5 Discussion

Two important features are revealed by the data from Figs. 6.1.1.5 to 6.1.1.8. First, break-through times are mainly influenced by the set of geometrical parameters. This means that the geological setting of the net of primary fissures to a large extent determines the future evolution of the system. From the data in Fig. 6.1.1.5a,b, one may draw the conclusion that flow conduits in limestone areas in the length of several 100 m, developing from fractures of 0.01 cm aperture and hydraulic gradient $I > 0.05$ need about 10 000 years for their initiation until break-through. Then they grow with growth rates of about 10^{-2} cm/a (Dreybrodt 1988), which takes a further 10^4 years until channel conduits of 1 m widths have developed. There is almost no chance that large conduits develop at fractures of 0.004 cm or at hydraulic heads below 0.03. This last number is in agreement with a minimum hydraulic gradient of 0.03 postulated by Ford (1980). From our data one may conclude that the threshold for the initial aperture is about 0.01 cm. This is the average value of joint widths as observed by Motyka and Wilk (1984).

Minimum times of several 10 000 years for the development of large karst conduits are observed in the field (Bakalowicz 1982; Mylroie and Carew 1986). The agreement of field data to the outcome of our model allows confidence in its quality.

The second point is that break-through times are quite independent on the chemical parameters (Figs. 6.1.1.6–6.1.1.8). This shows that the evolution of karst aquifers is a very common phenomenon and occurs in the whole variety of different kinds of limestone rock as is the case in nature. It is especially interesting to observe from Fig. 6.1.1.7 that the minimum in the curve for 500 m length sets a range of values β for karstifiable rock in the region from $5 \times 10^{12} < \beta < 5 \times 10^{14}$. It is most interesting that the experimental value of β observed on natural icelandspar is in the center of this region.

Fig. 6.1.1.8 shows the intensity of karst development in the realm of natural karst waters, the hardness of which ranges from about an average of 1.7×10^{-6} mol/cm^3 in tropical climate to an average of 2×10^{-6} mol/cm^3 in temperate climate, although with much larger variations. Variation of break-through time in this region is small. There is a threshold, however, for water with hardness below 1×10^{-6} mol/cm^3. Such waters are encountered in areas without vegetation, i.e. bare limestone rock. In this case break-through times are larger by about a factor of 10 in the order of 10^5 years. This renders karst initiation unlikely on these areas. One therefore may conclude that initiation of karstification is only likely in vegetated areas with high soil CO_2. It should be noted, however, that once break-through is achieved, further development into a conduit aquifer with rates of retreat of bedrock in the order of several 10^{-2} cm/a proceeds also if surfaces are bare rock.

So far cases have been discussed where the penetration length in the initial fracture is short compared to its length, L_f. If this is not the case and $L_p(0) \approx 1$ then widening of the fracture is effected by first-order dissolution exclusively. Rates of fracture widening are then about 5×10^{-2} cm/a everywhere along the fracture. Thus, within a short time of about 50 years small channels with widths of a few mm may develop.

In case of a vadose cave situated 10 m below ground, hydraulic gradients up to 0.5 may be feasible. With an initial fracture system with widths in the order of about 0.01 cm and values of α_{ko} of about 1×10^{-5} cm/s, penetration lengths in the order of 10 m result [Eqs. (6.1.1.9) and (6.1.1.10)]. It should be noted that now L_p depends largely on α_{ko}. But even in the case, when $L_p ¢ 0.2 \cdot l$, break-through occurs in times of several tens of years. This is in agreement with observations of Gunn and Gagen (1987), who observed sinkholes in limestone quarries abandoned 50 years ago. Therefore in such situations formation of sinkholes in the short time of several tens of years seems possible. This may be of high importance for land use projects, where by lowering of the ground water table or by construction of canals and dams such situations might occur.

6.1.1.6 Conclusion

The proposed model describes the initial phase of the evolution of a fractured limestone mass to a karst aquifer by solutional widening by calcite aggressive water, percolating through pathways from a water input to an exsurgence. Large-scale aquifers can only develop by a concerted action of two regimes of dissolution kinetics. Far from equilibrium, fast first-order dissolution rates are dominant, such that ground water quickly approaches equilibrium with respect to calcite. Close to equilibrium, slow fourth-order kinetics enable slow dissolution over distances in the order of several 100 m. This enlarges fissures at a flow path but increases the flow rate through the rock, such that the realm of first-order dissolution propagates through the entire length of the fissure until it reaches the outlet. At that moment there is a dramatic increase in flow rate through the fracture and the evolution of the system proceeds quickly to mature conduit aquifers. The model estimates time and length scales of the evolution of aquifers in fractured limestone in accordance to field observations.

6.1.2 Some Aspects of Modelling the Carbon System in the Unsaturated Zone

B. Merkel, L. Eichinger and P. Udluft

6.1.2.1 Basic Annotations

Pore water movement, heterogeneous chemical reactions and solute transport through the unsaturated zone were studied in a field experiment carried out in the Munich Gravel Plain (Fig. 1.1, Table 1.1), which is characterized by carbonate gravels covered by a thin layer of organic material. Apart from meteorological data, the hydraulic heads, ground temperatures, carbon dioxide content of ground air as well as concentrations of main solutes and isotope contents were recorded for a 5 years' period. These data were used to validate a numerical model simulating reactions and processes within unsaturated carbonate gravels (Sect. 3.2.2, this Vol.; Merkel 1983).

The unsaturated zone is a three-phase system of liquid, gas and solid material with physical and chemical interactions at all phase boundaries. Modelling the carbon system in the unsaturated zone, the following variables have to be taken into account as boundary and/or initial conditions: quantity and quality of rain, evaporation as function of air temperature and humidity, transpiration of vegetation, microbial CO_2 production, chemical composition of seepage water, soluble minerals, CEC of the unsaturated zone solid phases, degassing of CO_2 from ground water and pore size distribution.

The parameters to be simulated are unsaturated water flow, heat flow, decay of organic matter, production of CO_2, transport of solutes concerning dispersion and diffusion in the liquid phase, interaction at the water/ground air boundary, chemical reactions at the water-rock interface (dissolution and precipitation, cation exchange), chemical reactions within the seepage water, and gas transport within the unsaturated zone.

Multi-phase flow and transport of colloids might be of interest, but were not taken into account in this study. Handling all these parameters in one model would require a rather sophisticated framework. It is more convenient to break down the problem into several modules, which can be developed and tested separately, if they are independent of each other. Pore water movement and transport of solutes, for example, can only be modelled separately if the water is slightly mineralized and thus the flow equation is not dependent on the density of the water. In the particular case the problem was solved in three steps: (1) computation of flow velocities and moisture content, (2) of ground air balance, and (3) of solute transport and reactions. The applicability of the model described is highly dependent on the mineralogical characteristics and the soil physics of the unsaturated zone (Merkel 1983).

6.1.2.2 Flow Velocity and Water Content

Ground water flow is determined by the coefficient of hydraulic conductivity which is basically a function of water density and water viscosity. The coefficient of hydraulic conductivity within the unsaturated zone, k_u, however, is furthermore dependent on the moisture content. The relationship between the hydraulic conductivity and the capillary pressure head within sandy gravels may be expressed by:

$$\log k_u = A + B \cdot \log p_c, \qquad (6.1.2.1)$$

with the hydraulic conductivity k_u (m/s), the capillary pressure head p_c (100 Pa), the regression coefficients A (4.0 to 5.0) and B (4.08), for values of p_c from −0.1 to −80 kPa. B was determined by non-linear regression analysis using the field data of a 5 years' period. However, it was not possible to find a value for A which fits the recorded tensiometer data for all depths and – due to hysteresis – even for two- and threefold tensiometer recordings in one depth (Ullsperger et al. 1988b). Therefore it was necessary to determine an individual value A for each tensiometer. The problem was solved by an iterative procedure using the total ground water recharge balance as boundary condition.

The conventional method to obtain flow velocities for the transport model is the direct calculation from recorded suction heads in different depths and interpolation between these points (Fig. 6.1.2.1). However, a most popular model, the Fickian-based convection equation, is not appropriate to solve the time-dependent unsaturated flow in a sandy gravel. A more deterministic approach is based on the Richards equation, which can be written in one dimensional form as

$$c(p_c)\,\frac{\delta H}{\delta t} = \frac{\delta}{\delta z}\left(k_u\,\frac{\delta p_h}{\delta z}\right) \qquad (6.1.2.2)$$

with the hydraulic head $p_{h(z,t)}$ (cm), the capillary pressure head $p_{c(z,t)}$ (cm), the hydraulic conductivity k_u (cm/h), the volumetric pore water content w (cm^3/cm^3), the specific moisture capacity $c(p_c)$ ($=\delta w/\delta p_c$), the vertical space coordinate z (cm) and the time t (h). The hydraulic head p_h in the unsaturated zone is the sum of the gravitational head and the capillary pressure head. Equation (6.1.2.2) can be solved by analytic solutions or by finite difference or finite element methods. Using finite differences, Vauclin et al. (1979) showed that weighting with geometric averages gives a more accurate solution than harmonic or arithmetic mean values.

A numerical model of the unsaturated flow in sandy gravels based on Eq. (6.1.2.2) could only be calibrated and verified with regard to monitored suction heads if the recorded pressure heads were standardized to remove the spatial variability and the inhomogeneities of the porous media (Merkel et al. 1983). The extreme non-linearity in the relationships between hydraulic conductivity and cap-

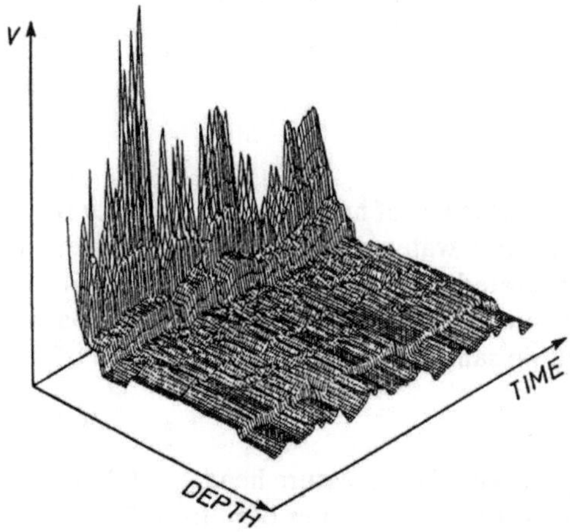

Fig. 6.1.2.1. Calculated flow velocities versus depth and time using Eq. (6.1.2.1) and tensiometer readings at seven depths during 1 year. Non-linear interpolation was done in between the measured levels

illary pressure (p_c) and specific moisture content and capillary pressure required an additional iterative circle instead of a straightforward computation as well as a small discretization in time and space. Thus, the numerical effort was considerable, but still stability problems remained for long-term simulation runs. In addition, the determination of ^3H and $\delta^{18}O$ isotope contents brought up severe interpretation problems. The ^3H content of ground air vapour was measured in different depths in 1981/83 at the forest area Munich-Harlaching (MUC2). A maximum was detected in 15 m depth. If this tritium peak is attributed to the atmospheric tritium peak of 1975, an average seepage velocity of 2 m/a results for the forest site (MUC2). The input function of $\delta^{18}O$ is well known; the difference between summer and winter is about 6 to 10‰. This input function could be traced even in 9 m depth very well and shows a time lag of about 6 months. Figure 6.1.2.2 displays the measured chronological sequence for $\delta^{18}O$ in two depths at the grass-covered site in Munich-Grosshadern (MUC1). In contrast to the seepage velocity calculated from ^3H contents (2 m/a), a mean seepage velocity of more than 20 m/a can be approximated from the $\delta^{18}O$ contents. This discrepancy cannot be explained, even by taking into account a higher ground water recharge at the grass covered site (about 360 mm/a) compared to that at the forest site (about 260 mm/a).

However, these results support the assumption that both seepage water and solutes are transported by a moderate, slow flow in micropores, and by a fast, perhaps turbulent flow through macropores. As a consequence, the numerical simulation of pore water movement in sandy gravels requires a "two-region" respectively "multi-region" model, which includes the problem of quick flow through macropores. Figure 6.1.2.2 shows the measured and calculated $\delta^{18}O$ contents in two depths of the unsaturated zone. The two-box model used calculates the unsaturated flow by solving the one-dimensional Richards equation twice for two independent pore systems. The transport modelling in both pore systems is based on a particle tracking method (Merkel et al. 1988), a diffusive exchange term between both pore systems and a final smoothing procedure. The good accordance between recorded

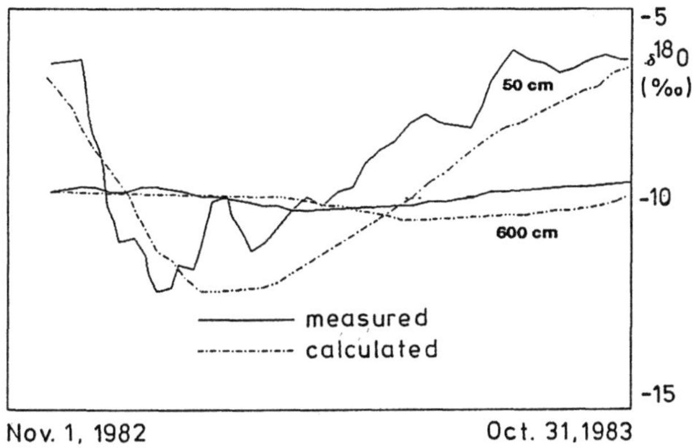

Fig. 6.1.2.2. Chronological sequence of measured and calculated ^{18}O contents at depths of 50 and 600 cm

and simulated data indicates the necessity of using a two- or multi-region flow model.

6.1.2.3 Simulation of CO_2 Ground Air Budget

The CO_2 concentration of ground air is mostly affected by the decay of organic matter, by diffusion, degassing out of and dissolution of CO_2 in pore water, by sorption of gas by soil and rock particles. The CO_2 concentration of the ground air is the main boundary condition for dissolution and precipitation of calcite, dolomite, and other carbonate minerals. The gas transport phenomena are mostly governed by diffusion, convection is only of minor importance.

Decay of Organic Matter

The decay of organic matter may be written as

$$R\text{-}CH_2 \longrightarrow CO_2 + H_2O + R \tag{6.1.2.3}$$

and is mainly controlled by temperature, pore water content, bacteria counts and quantity of organic matter. All these parameters are time-dependent and linked to sophisticated boundary conditions. For some of these parameters only very poor data are available.

Ground temperatures in different depths are easy to monitor and to simulate with a numerical model if the upper (e.g. subsurface or air temperature) and lower boundary conditions (e.g. ground water temperature) are known and the temperature transfer due to water transport is neglected. In this case the temperature distribution can be described with Fick's diffusion equation. A rough estimation of ground temperatures in all depths is provided by an empirical Eq. (6.1.2.4), which was compiled by Ullsperger et al. (1988b) from a 5 years' monitoring period in southern Bavaria.

$$t = 10.12 + (12.44 \cdot e^{(-0.232 z)} \cdot \sin(\frac{n}{8.28} + 3.133 - 0.2089 z) \qquad (6.1.2.4)$$

with the ground temperature t, the week n in the hydrological year and the depth z (m). Equation (6.1.2.4) is valid for Quaternary gravels, depths between 0 and 10 m and comparable meteorological boundary conditions (annual mean temperature 10 °C, annual precipitation 900 mm). Equation (6.1.2.4) has the advantage that recorded data are not required as boundary conditions. The measured temperature data in different depths (curve a) and the data from the numerical simulation (curve b) in Fig. 6.1.2.3 fit better than those of curve c, which was calculated with Eq. (6.1.2.4). Curve b was calculated with a finite difference approach, using the recorded air temperature as upper boundary, a diffusion coefficient of 50 cm/s and a constant lower boundary of 10 °C in a depth of 12 m. These results indicate that although seepage velocities are considerable, the impact of pore water movement on the heat transfer can be neglected. The empirical function which needs no temperature data provides a rough approximation, which might be sufficient as boundary condition for a thermodynamic submodel.

Pressure transducer tensiometers were used in the field study to read the suction head threefold in seven depths every second hour. The water content was calculated from the suction head with the help of Eq. (6.1.2.5) (Ullsperger et al. 1988b)

$$w = A + B \cdot \log p_c \qquad (6.1.2.5)$$

with the water content w (% vol.), the capillary pressure head p_c (100 Pa), the empirical constants A (15.0 to 16.3) and B (3.7 to 6.6). Equation (6.1.2.5) is valid for $-100 > p_c > -20000$ Pa and sandy gravels with a silt fraction between 0 and 0.1%.

Figure 6.1.2.4 displays rainfall distribution, ground temperature and CO_2 concentration of ground air in 0.2 and 0.5 m depth. The CO_2 content significantly in-

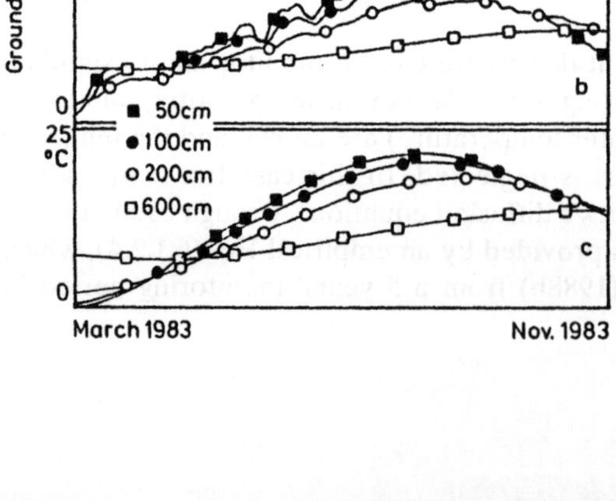

Fig. 6.1.2.3. Measured (**a**) and calculated ground temperatures (**b**, **c**) at different depths

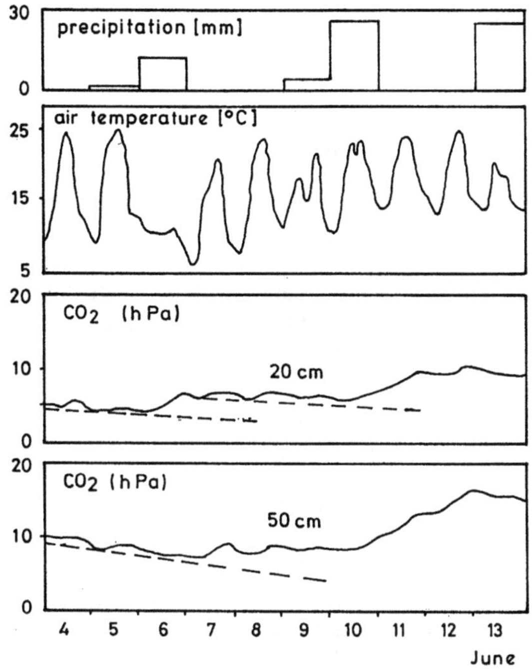

Fig. 6.1.2.4. Precipitation, temperature, and measured CO_2 concentration at depths of 20 and 50 cm

creases due to an increased CO_2 production from microbial activity. However, the decay reaction was stimulated only during summer by increasing moisture content, whereas during winter heavy rainfall events were correlated with decreased CO_2 contents due to CO_2 dissolution in the seepage water.

The numbers of bacteria may have an effect on the decay of organic matter as well. There is only little published quantitative information available up to now. Own investigations in the unsaturated zone showed a distinct decrease of the number of microorganisms with depth (Ullsperger et al. 1988a). Taking into account the dissolution of CO_2 in seepage water as well as CO_2 diffusion towards the atmosphere, the time-dependent CO_2 production was calculated from a 5 years' recording period of ground air CO_2. These data, DOC content, temperature, moisture content and a rough approximation of the bacteria counts in different depths were used to fit the coefficients of the empirical function:

$$Q = A \cdot w \cdot c \cdot \log(n) \cdot e^{(B.t)}, \qquad (6.1.2.6)$$

with the CO_2 production Q ($cm^3/(h.cm^2)$), the DOC in seepage water c (mg/l), bacteria counts n (CFU/ml), the temperature t (°C), the moisture content w (cm^3/cm^3), and the regression coefficients A (20 to 100) and B (≈0.1).

Diffusion

Diffusion is controlled by Fick's law, which can be written as

$$\frac{\delta c}{\delta t} = D_{eff} \cdot \frac{\delta^2 c}{\delta z^2} + c_{so} - c_{si} \ , \qquad (6.1.2.7)$$

with the concentration c, the time t (s), specific diffusion coefficient D_{eff} (cm^2/s), the distance z (cm) and the CO_2 source c_{so} and sink c_{si}. The temperature-dependent CO_2 diffusion coefficient $D_{M,CO2}$ in air is well known. Both tortuosity due to porosity and the varying water content affect the diffusion coefficient significantly. An empirical relation between capillary pressure head and the specific diffusion coefficient D_{eff} is given by Merkel (1983):

$$D_{eff} = \frac{0.02}{1 + e^{(-1.842 \cdot \ln p_c + 8.519)}} \ , \qquad (6.1.2.8)$$

with p_c (in 100 pa, $p_c < 0$). c_{so} and c_{si} in Eq. (6.1.2.7) are source and sink terms such as degassing, dissolution of CO_2 by seepage water or sorption effects. Replacing p_c with w from Eq. (6.1.2.1) gives the relationship between w and D_{eff}. Figure (6.1.2.5) shows monitored (a) and calculated (b and c) CO_2 concentrations versus time for 50, 200 and 600 cm depth. Curves b – unlike curves c – neglect the relation between w and D_{eff}. Curves d and e display the direct comparison of recorded and modelled data in 50 and 20 cm depth. Thus, Fig. 6.1.2.5 indicates that the dependence of D_{eff} on w has to be taken into account.

Calculated and measured data do not match well in 6 m depth for both models, indicating that at least in greater depths besides diffusion controlled CO_2 variations additional source and/or sink terms have to be taken into account. Source terms can be seen in microbial activity or degassing of CO_2 from seepage water. Since no evidence was found for rapid changes in the microbiological activity in greater depths, degassing from seepage water is probably responsible for some changes of CO_2 partial pressure. This is understandable in the case of a fast seepage water movement which does not provide enough time to equilibrate seepage with respect to ground air CO_2. Apparently dissolution of CO_2 has a much faster kinetic than degassing from water. Thus, seepage water, which has been equilibrated with high CO_2 concentrations in the upper soil layers, might be supersaturated for some time, if it is moved downwards considerably fast. The fact, that variation of ground air CO_2 in greater depths cannot be explained only be diffusion, again supports the assumption of a fast macropore based flow component.

Dissolution of CO_2 in Pore Water

The transfer of gas at the interface water/gas is described by Eq. (6.1.2.9) and by Henry's law (Stumm and Morgan 1981):

$$CO_2(g) \rightleftharpoons CO_{2(aq)}. \qquad (6.1.2.9)$$

Dissociation of $CO_{2(aq)}$ is described by:

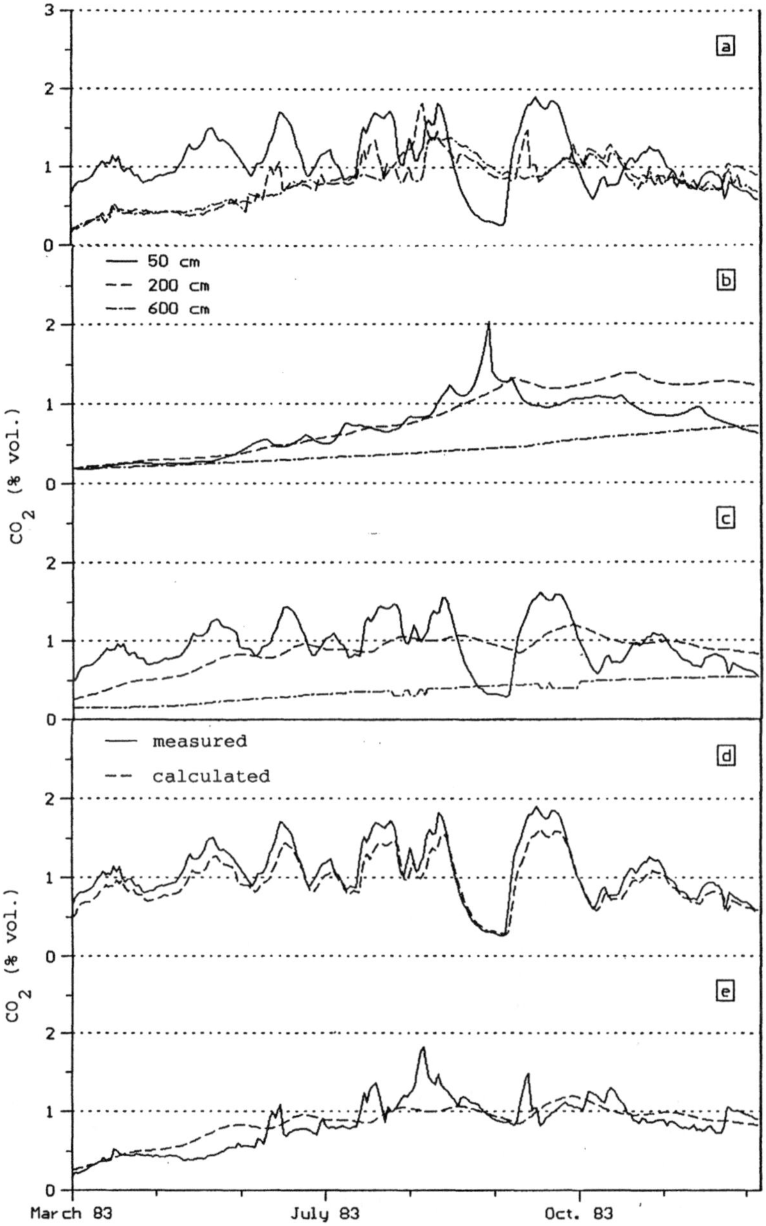

Fig. 6.1.2.5. Monitored CO_2 concentrations (Pa) at depths of 50, 200 and 600 cm. (**a**). Calculated CO_2 contents without (**b**) and with (**c**) taking into account the w-D_{eff} relationship. **d** Recorded and calculated p_{CO_2} at 50 cm depth taking into account the w-D_{eff} relationship. **e** Recorded and calculated p_{CO_2} at 200 cm depth taking into account the w-D_{eff} relationship. The model still works, however, some monitored peaks are not shaped by the model

$$CO_{2(aq)} + H_2O \rightleftharpoons HCO_3^- + H^+ \quad (6.1.2.10)$$

$$HCO_3^- \rightleftharpoons CO_3^{2-} + H^+. \quad (6.1.2.11)$$

The protons formed in reactions (6.1.2.10) and (6.1.2.11) will cause a carbonate dissolution:

$$CaCO_{3(s)} + H^+ \rightleftharpoons Ca^{2+} + HCO_3^-. \quad (6.1.2.12)$$

Equations (6.1.2.10)–(6.1.2.12) yield further CO_2 dissolution in the seepage water until equilibria are reached. The kinetics of the total process is controlled either by diffusion [Eq. (6.1.2.7)] or Eqs. (6.1.2.10) to (6.1.2.12). If the thickness of the water film is very small or the bulk phase is turbulently mixed, the chemical reactions [Eqs. (6.1.2.10) to (6.1.2.12)] will control the system (Merkel 1983; Buhmann and Dreybrodt 1985a). While the dissolution rate of calcite is well known and easy to calculate (Plummer et al. 1978), Eq. (6.1.2.7) deals with an unknown length where diffusion takes place. In a natural system of differently spaced pores, it is rather difficult to define an appropriate value for this length. Basically the problem was solved in the particular case assuming a dual pore system with macro- and micropores. Equilibria were assumed in the macropores with regard to an open system. Disequilibria with respect to the ground air CO_2 were supposed for the micropore system because of the slow kinetic in the micropore system due to diffusion controlled CO_2 exchange. A rather constant hydrogen carbonate content in the micropores may result.

Another significant reaction in a soil or subsoil system might be sorption and desorption of CO_2 by clay minerals and silt surfaces. However, there are only poor data available and therefore this reaction was not taken into account at all.

6.1.2.4 Treatment of Chemical Reactions

The activities of HCO_3^-, Ca^{2+}, Mg^{2+}, protons and other species may be calculated according to mass balance equations. This, however, requires to solve a set of equations for each grid point and time step of a numerical model. The empirical function (6.1.2.13) was compiled from equilibria calculations using the computer programme PHREEQE (Parkhurst et al. 1980) and a data set with varying temperatures and CO_2 partial pressures. The use of the regression equation requires less computer time and gives still satisfactory precision in comparison with the numerical solution of the mass balance equations. The maximum deviation from the correct solution is less than 0.25 mmol/l within the definite range.

$$(HCO_3^-) = 0.1809 + 0.9999.p' - 0.01425.t, \quad (6.1.2.13)$$

with $p' = 1.074 + 436.7p - 17030p^2 + 2.767 \times 10^5 p^3$, $p = p_{CO_2}$ (100 kPa), the temperature t (°C and the hydrogen carbonate content (HCO_3^-) (mmol/l). Equation (6.1.2.13) is valid within the following ranges: t between 1 and 25 °C, ionic strength < 10 mmol/l, and p_{CO_2} between 0.03 and 3 kPa.

Figure 6.1.2.6a displays measured (dashed lines) and calculated (solid lines) hydrogen carbonate concentrations using the regression function [Eq. (6.1.2.13)], the recorded CO_2 partial pressures and the ground temperatures. However,

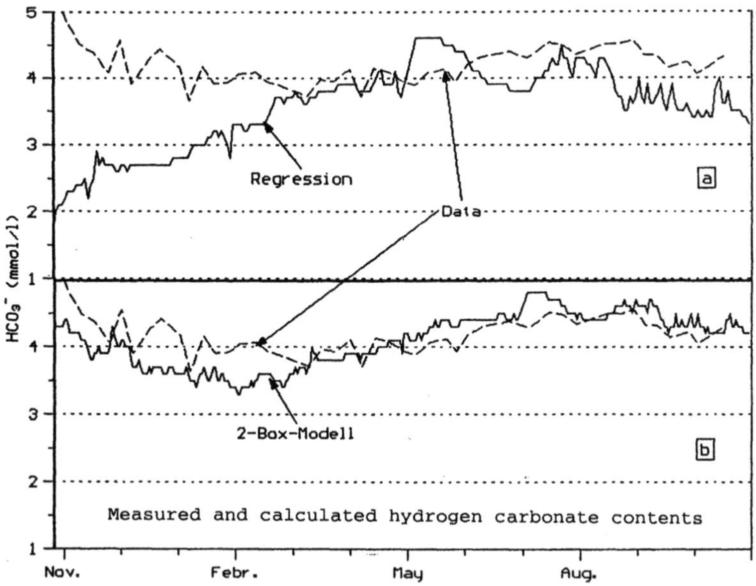

Fig. 6.1.2.6. Measured (**a**) and calculated (**b**) hydrogen carbonate concentrations versus time. Model **a** was calculated from recorded p_{CO_2} and temperature, assuming equilibrium with ground air. Model **b** was based on a dual pore system, assuming equilibrium between ground air and seepage water in macropores and a constant content of 5 mmol/l hydrogen carbonate in micropores

measured and calculated hydrogen carbonate concentrations do not match well; a thermodynamic approach brought similar results. This is noteworthy inasmuch as field and laboratory experiments have shown (Sect. 3.1.2) that the kinetics of carbonate dissolution can be assumed as quite fast. Equilibrium between ground air and pore water does not face the facts at all times, but lysimeter experiments showed carbonate equilibria after 1 to 3 days at most and time series analysis of ground air CO_2 versus specific electrical conductivity from field experiment confirmed these results with time lags from 7 to 20 days. Thus, the carbonate dissolution rates cannot be responsible for the observed phenomena.

On the other hand, it is to be considered that the water samples were extracted by suction cups. Even if by this procedure water chemistry is not changed significantly (Grossmann et al. 1988), water is sucked both from macro- and micropores. Taking into account the previously discussed dual pore model, the result could be improved. In Fig. 6.1.2.6b it is assumed that the seepage water sampled was a 1:1 mixture from macro- and micropores. The hydrogen carbonate concentration in the micropores was assumed to be 5 mmol/l and the macropore concentrations were calculated from Eq. (6.1.2.13). The rather good fit between measured and calculated data again indicates the importance of micro- and macropores for hydrochemical processes.

6.1.2.4 Conclusions

Numerical modelling was performed for the evaluation of field measurements in the unsaturated zone of a carbonate gravel terrain. Major results are:

- Mean flow velocities of 20 m/a were calculated from tritium contents of ground air vapour whereas flow rates of 2 m/a result from $\delta^{18}O$ data in seepage water.
- In spite of considerably fast carbonate dissolution kinetics, measured carbonate species in seepage water did not match the calculated ones; when equilibria with the ground air CO_2 partial pressure are assumed.
- The measured and calculated ground air CO_2 variations in different depths, taking into account the impact of moisture changes on the diffusion coefficient, fit well in the upper few metres of the unsaturated zone.

These observations could be explained with the help of a dual pore model and the following assumptions. Water in macropores is for most of the time characterized by carbonate equilibria with regard to ground air CO_2. Disequilibria with respect to ground air CO_2 are assumed for the micropore system due to slower diffusion controlled kinetics. The two-box model solves the one-dimensional Richards equation for two independent pore systems, the transport model for both pore systems is based on a particle tracking method and the chemical composition of the seepage water is calculated by a thermodynamic approach. The good fit between measured and simulated data supports the concept of a two-region flow and transport model. Concerning pollutant transport this phenomenon gains a special significance with regard to gravel terrains.

6.1.3 Methodical Concepts in Silicate/Water Interaction – a Comparison of Results

A. Dahmke, G. Matthess, A. Petersen and D. Schenk

6.1.3.1 Aim

This section aims to describe the influence of the selected experimental approach on the resulting dissolution rates. Special emphasis is given to the influence of the pretreatment of the samples, to the effect of abrupt pH changes, to the premineralization of solvents and – in open systems – to the flow velocity of water. The influence of complex forming ligands on the precipitation of secondary phases is described in Section 2.9, this Volume.

The following evaluation is restricted to those results of the laboratory experiments described in Section 4.3.1, which deal with the influence of methods on the determined release rates (Table 6.1.3.1).

Table 6.1.3.1. Dissolution rates for the experimental variations (calculated for Si release)

pH value	Grain size (mm)	Si release rate [mol/(cm^2 × s)] Anorthite	Albite	Orthoclase
Fixed pH value				
4	0.063–0.1	4.0×10^{-15} $\pm 1.6 \times 10^{-16}$	3.0×10^{-15} $\pm 2.8 \times 10^{-16}$	2.3×10^{-15} $\pm 1.8 \times 10^{-16}$
6	0.063–0.1	2.7×10^{-15} $\pm 3.3 \times 10^{-16}$	3.0×10^{-15} $\pm 2.8 \times 10^{-16}$	2.1×10^{-15} $\pm 3.1 \times 10^{-16}$
8	0.063–0.1	2.3×10^{-15} $\pm 2.2 \times 10^{-16}$	3.1×10^{-15} $\pm 4.3 \times 10^{-16}$	2.0×10^{-15} $\pm 4.2 \times 10^{-16}$
10	0.063–0.1	5.4×10^{-16} $\pm 1.2 \times 10^{-16}$	2.2×10^{-16} $\pm 2.4 \times 10^{-16}$	1.1×10^{-15} $\pm 3.4 \times 10^{-16}$
Free pH value				
7.5–8.5	< 0.063	1.6×10^{-15} $\pm 1.4 \times 10^{-16}$	7.9×10^{-16} $\pm 7.1 \times 10^{-17}$	1.3×10^{-15} $\pm 7.5 \times 10^{-17}$
7.5–8.5	0.063–0.1	2.2×10^{-15} $\pm 1.7 \times 10^{-16}$	1.2×10^{-15} $\pm 1.0 \times 10^{-16}$	1.5×10^{-15} $\pm 6.9 \times 10^{-17}$
7.5–8.5	0.2–0.4	9.3×10^{-16} $\pm 1.8 \times 10^{-16}$	4.2×10^{-16} $\pm 1.3 \times 10^{-16}$	1.2×10^{-15} $\pm 9.9 \times 10^{-17}$
7.5–8.5	0.8–2.0	5.5×10^{-15} $\pm 6.1 \times 10^{-16}$	4.3×10^{-15} $\pm 6.1 \times 10^{-16}$	4.0×10^{-15} $\pm 5.3 \times 10^{-16}$

A comparison of the Si release rates of the three feldspar dissolution experimental setups (Table 6.1.3.1) indicates that the highest dissolution rates are found in the batch experiments as closed systems for solid and fluid phases. The rates derived from fluidized bed reactor experiments (FBR) and from column experiments are usually lower, the lowest being in general those from the column experiments.

These findings may be due to the mechanical stress of the minerals' grain surfaces during the shaking process in batch experiments, which may cause abrasion and the production of reactive microcrystals. Furthermore, the shaking process favours the contact between almost the whole feldspar mineral surface and the aquatic phase, whereas especially at the column experiments the feldspar surface is partly blocked for the dissolution process by grain-grain contacts.

Depending on the flow rate, leaching rates between 3.5×10^{-16} and 1.5×10^{-15} mol/(cm^2.s) Si were measured in fluidized bed reactor experiments. Column experiments show Si release rates of the same magnitude (4.6×10^{-16}-1.3×10^{-15} mol/ (cm^2.s) Si (Dahmke 1988). These results are consistent with published data. The Si release rates (about 1.4×10^{-16} mol/(cm^2.s) derived from flow experiments by Knauss and Wolery (1986) are lower than those derived from batch experiments of Busenberg and Clemency (1976) and Holdren and Berner (1979) (e.g. albite 4.5 ×

10^{-16} mol/(cm^2.s) and 4.8 × 10^{-16} mol/(cm^2.s). Flow experiments of Holdren and Speyer (1985) and Chou and Wollast (1985) gave values of 5.8 × 10^{-16} mol/(cm^2.s) and 4.2 × 10^{-16} mol/(cm^2.s), respectively.

6.1.3.2 Influence of Grain Size on Leaching Rates

The influence of the grain size fractions on the leaching rates of the rock-water suspensions was deduced from the accumulation of Si in solution vs. time of the batch experiments.

The dissolved amount of feldspar was calculated by dividing the released Si amount by the average molar fraction of the feldspar-rich rock, neglecting the small amounts of impurities like micas and quartz. Holdren and Speyer (1985) postulated a theoretical relationship between the surface area in m^2.g^{-1} and the bulk reaction rate in mol/(g.s) at the final stage of the experiments:

$$R = (A_{min})^m \tag{6.1.3.1}$$

or transformed to

$$\ln(R) = m \ln (A_{min}), \tag{6.1.3.2}$$

with the dissolution rate R, the surface area A_{min} and an empirical exponent m specific for the relation between both variables. However, the postulated linearity between ln (R) and ln (A_{min}) implicating a constant m could neither be confirmed with the coarse (0.3–0.6 mm) nor with the fine grain size fractions (<0.036 mm) (Fig. 6.1.3.1). This is consistent with the experimental results of Holdren and Speyer (1985), who explained these findings by a constant density of lattice defects (unbalanced charge points) independent from an increase of specific surface. These authors assumed that in a certain range of grain sizes (10–600 μm) the distances between lattice defects are greater than or equal to the grain diameter, so that further crushing increases the specific surface, but does not increase the number of lattice defects.

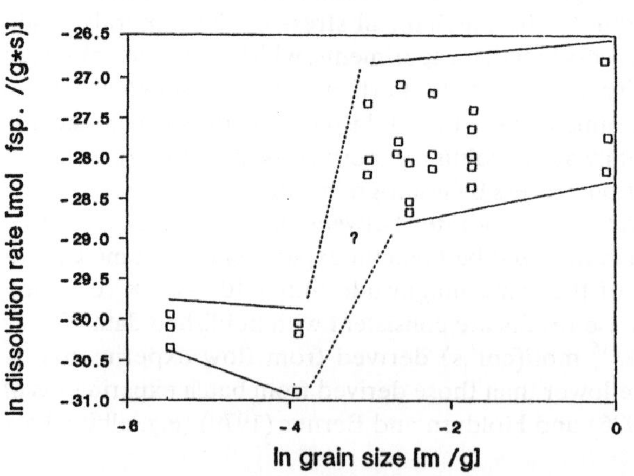

Fig. 6.1.3.1. Grain size vs. dissolution rate in logarithmic scale

The presented results indicate that there are no direct relationships between bulk release rates and specific mineral surface. Thus, the transfer of laboratory-measured release rates related to specific surfaces to natural weathering profiles with unknown surface properties implies high insecurities.

6.1.3.3 Influence of Flow Rates

The influence of contact time on the dissolution rates were investigated in the FBR and in the column experiments. Both types of experiments indicate, that at low residence time (55.4 h; equivalent to high flow rates) the concentration of the respective solutes in the outflow are the same as at high residence times (101.7 h; equivalent to low flow rates).

In the column experiments with a given rock material, about the same concentration in the solution is observed after an 83-cm flow length independently of the flow velocity (Fig. 6.1.3.2). This can be explained by an increase of dissolution rates with increasing flow velocities (decreasing residence times).

Consistent with these observations, the fluidized bed reactor experiments indicate a linear increase of the dissolution rates with increasing flow velocities (Fig. 6.1.3.3).

The slope of the relationship between flow rates (ml/s) and Si release [mol/(cm^2/s)] was found to be 10^{-13} mol/(cm^2.ml).

The observed relationship between dissolution rates and flow rates can be explained by the assumption that the dissolution process at the flow rates of the

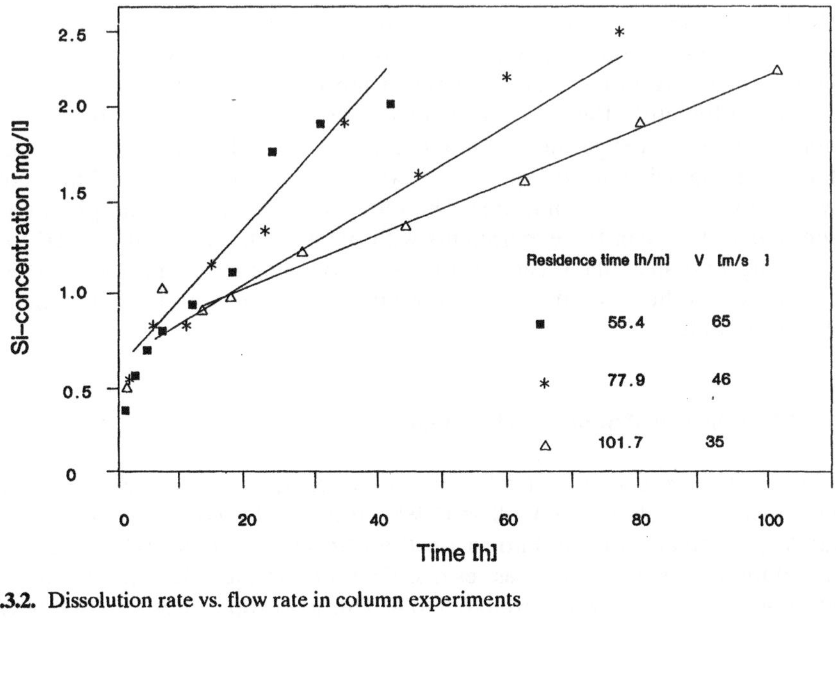

Fig. 6.1.3.2. Dissolution rate vs. flow rate in column experiments

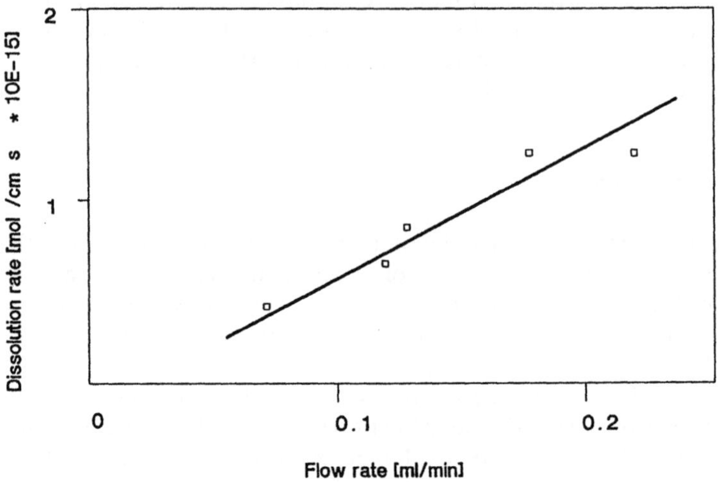

Fig. 6.1.3.3. Dissolution rate vs. flow rate in fluidized bed reactor experiments

FBR are influenced by partial equilibria of undefined solid phases or surface properties.

6.1.3.4 Influence of Premineralization

Both the initial and the final stage of albite dissolution in batch experiments with a CsCl premineralization result in a decrease of the Si dissolution rate and in a smaller final Si concentration compared with experiments with deionized water. The difference in the resulting ionic strength of the respective solutions is too small to explain sufficiently the differences of the dissolution rates up to one order of magnitude. But in experiments with deionized water, H^+ is the only reacting ion at the beginning, whereas in experiments with a premineralization Cs^+ may compete with the protons for sorbing sites. Therefore the experiments with pH 6 are more sensitive to Cs^+ than the experiments with pH 4.5, where the ratio of H^+ to Cs^+ is much higher than at pH 6. The sorption of Cs^+ on the mineral surfaces also results in an increase of the amounts of Na^+ in solution compared with the experiments with deionized water.

6.1.3.5 Influence of Sudden pH Changes

Whereas the dependency of feldspar dissolution rates on pH have recently been investigated by Chou and Wollast (1985), Blum and Lasaga (1987) and Petersen (1988), relatively little is known on the effects of sudden pH changes on the dissolution rates of feldspars as results. Therefore in the FBR experiments the pH value was abruptly lowered from 6.5 to 4. This pH drop causes as a first reaction a

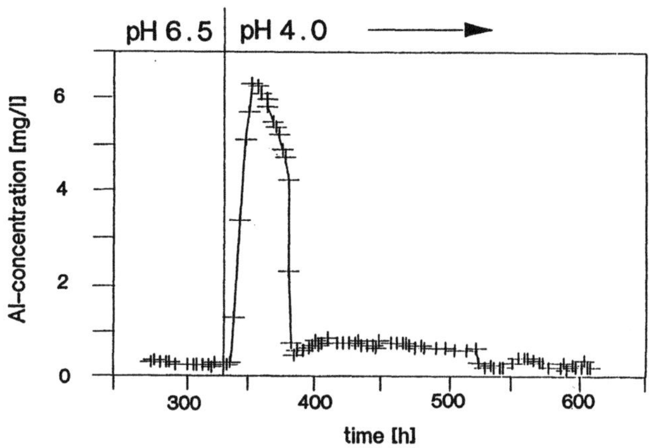

Fig. 6.1.3.4. Effects of sudden pH changes on Al release in fluidized bed reactor experiments

strong increase of the concentrations of all solutes (Fig. 6.1.3.4). Later, the solute concentrations decrease appreciably to a somewhat higher concentration than before the pH change.

This observation can be explained consistently with Holdren and Speyer (1985) by the assumption that the short initial increase of concentration after an abrupt pH change is caused by the fast dissolution of newly formed solid secondary phases.

6.2 Combination of Transport and Geochemical Reactions

6.2.1 Water Movement and Geochemical Reactions in the Unsaturated Zone of Sands with Low Calcite Content

H.D. Schulz

6.2.1.1 Model Concept for the Combination of Physical Transport and Geochemical Reactions

In order to model the complex combination of physical transport and different geochemical reactions in seepage water or ground water, it is not possible to use a general analytic solution of the problem. A problem like this can be solved by using a model concept similar to that of Schulz and Reardon (1983). In this case the total flow path through the unsaturated zone is divided into discrete mixing/reaction cells (Fig. 6.2.1.1). In this model the continuum of time is divided into different time steps. In the first time step the concentrations of meteoric water are given into the first cell. In all cells for each time step, different chemical reactions and/or equilibria can be simulated (Sect. 6.2.1.2). In the second time step the concentrations of all

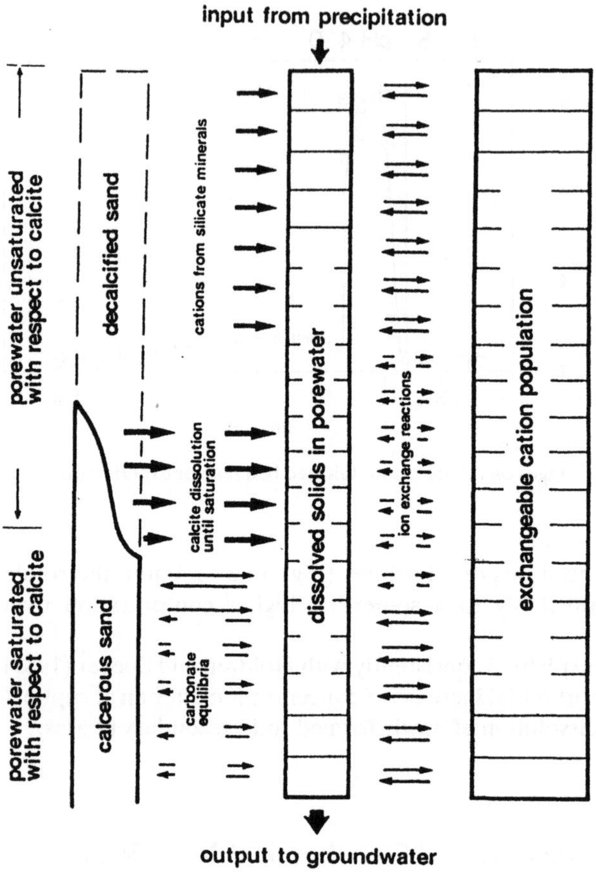

Fig. 6.2.1.1. Structure of the model for the combination of physical transport with geochemical reactions in the unsaturated zone of sands with low calcite contents

cells are moved by one position down. The concentrations of the precipitation are fed again into the first cell, the content of the last cell is regarded as output of the unsaturated zone into the ground water and it is taken out of the calculation.

If this type of model is used to simulate the reactions within the unsaturated zone, it is necessary to have a detailed knowledge about physical transport parameters such as seepage velocity, porosity of the material and dispersion. Furthermore, all geochemical reactions with quantitative importance have to be incorporated into the model. In this case three groups of geochemical reactions were regarded as being of quantitative importance. The first group of reactions is given by the cation exchange equilibria for the major cations calcium, magnesium, sodium and potassium. The second group of reactions combines equilibria necessary for the calculation of the calcite dissolution processes. The third group of reactions takes into account the processes of silicate weathering.

Of course, there are many more reactions and distinguishable geochemical subsystems in seepage water but it is known (Schulz 1977) that the three groups of reactions mentioned above control up to 80 or 90% of the geochemical composition of the seepage water.

The model was tested in the unsaturated zone of fluvioglacial sediments in northern Germany (Segeberger Forst). This study area with its coarse to medium sands with a low calcite content is described in detail by Schulz (1977) and Schröter (1983, 1984) (see also Sect. 3.1.2 this Vol.).

6.2.1.2 Geochemical Reactions

In this section the three geochemical subsystems of the model are discussed in more detail. This includes the question as to which reactions take place, and especially which input data and boundary conditions are necessary for the simulation.

Cation Exchange Equilibria

For all calculations in which the concentrations of calcium, magnesium, sodium and potassium in solution are concerned it is always necessary to take into account also the amount of these four cations on exchange sites. The very low CEC of this sandy material (1–2 meq/100 g) contains usually between one and two orders of magnitude more of these cations than the solution itself. Since cation exchange reactions usually reach equilibrium very fast (less than 1 h) the entire amount of the four cations has to be taken into account.

Schulz and Reardon (1983) describe a mathematical concept for the calculation of cation exchange reactions together with a non-iterative procedure for its solution. This concept assumes a constant CEC throughout the column of cells, which is not influenced by changing pH values, and cation exchange equilibria were installed for all cells in each time step. This approach proved to be suitable for most cation exchange reactions in the zone of weathering.

Calcite Dissolution in a System Open for CO_2

The calcite dissolution in the unsaturated zone is regarded as a process under conditions open for CO_2. This includes the assumption that the CO_2 diffusion is fast enough to establish a constant partial pressure of CO_2 within the unsaturated zone. Our CO_2 measurements at different depths and at different times of the year (Fig. 6.2.1.2) show that this assumption can hold true for the sandy material of the Segeberger Forst. In such a situation the equilibrium concentration of calcium in solution is only a function of p_{CO_2} if pH is carbonate-controlled.

Different groups of this DFG priority programme showed that calcite saturation usually is reached quite fast. Even for the sandy material of the Segeberger Forst with low calcite content it takes only between a few days and 1 or 2 weeks until calcium concentrations in solution are reached which are already very close to saturation. From this fact it can be deduced that it is not necessary to take into account the calcite dissolution kinetics for the simulation of the unsaturated zone with its very low seepage velocities and very long contact times between solution and solid phases.

Thus, in each time step of the model, the calcium concentration of calcite saturation was calculated from p_{CO_2}. The amount of calcite which was needed for

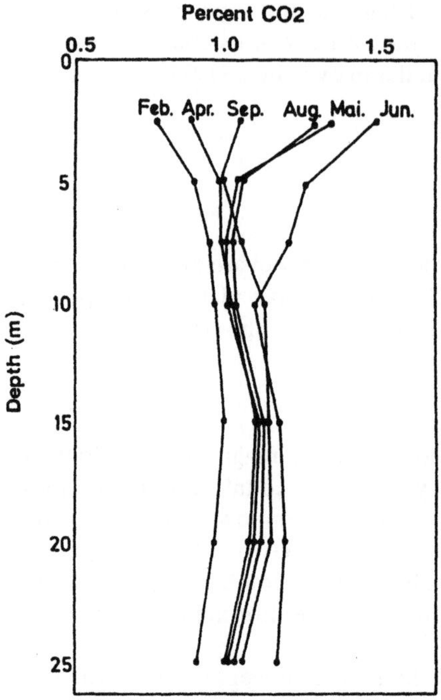

Fig. 6.2.1.2. Depth profiles of the CO_2 content of the ground air at different times of the year in the Segeberger Forst aquifer system

this procedure was subtracted from a given calcite content of each cell. If a cell after many time steps had no calcite content left, calcite equilibrium was no longer calculated for such a cell. Thus, after many time steps the upper cells of the column became calcite-free and a calcite leaching zone moved slowly down through the profile.

Simulation of Silicate Weathering

The very complex reactions of silicate weathering are not sufficiently understood to model them is as much detail as was done for the cation exchange equilibria and the calcite dissolution processes. Nevertheless, it was necessary to take them into account as a source for magnesium and potassium because these ions are necessarily included in the cation exchange reactions.

Ohse et al. (1983) described profiles through this dandy material with calculated undersaturations for different silicate minerals. From these profiles it can easily be deduced that a quantitatively important silicate weathering takes place only in the uppermost metres of the unsaturated zone of this study area. Thus, without considering details of silicate weathering, potassium and magnesium were added into the upper cells of the model at points where undersaturation was indicated (Ohse et al. 1983). The total amount of the two cations added in each time step is given by the difference between the concentrations in the precipitation and in ground water. With more knowledge, following the concepts in Section 6.1.3 on silicate weathering, it will be possible to extend the model onto these reactions.

6.2.1.3 Input Data and Model Results

The most important input value for the transport is the flow velocity. For this field site the seepage velocity is very well-known from other investigations (Eichinger and Schulz 1984). Figure 6.2.1.3 shows a profile of chloride concentrations in the pore water of the unsaturated zone of this study area. Figure 6.2.1.4 shows similar profiles for the concentrations of chloride and sodium in the pore water of another location of the same study area. In both figures the concentration profiles are compared with the total amount of annual precipitation. It is quite obvious in both figures that a high precipitation always coincides with low concentrations of these ions and vice versa. Thus, it is possible to document a 13 years' age profile for the pore water. From this a very reliable seepage velocity of about 0.6 m/a can be easily deduced.

Also from the Figs. 6.2.1.3 and 6.2.1.4 it follows that the dispersion is not very important in this situation. Otherwise it would not have been possible to distinguish different annual layers (Eichinger and Schulz 1984) of seepage water up to a depth of nearly 15 m below the surface. For this model a dispersion-coefficient of 5×10^{-10} m^2/s was assumed.

The chemical input concentrations into the first cell are given by the very low concentrations in the precipitation. Input data for the geochemical reactions of the

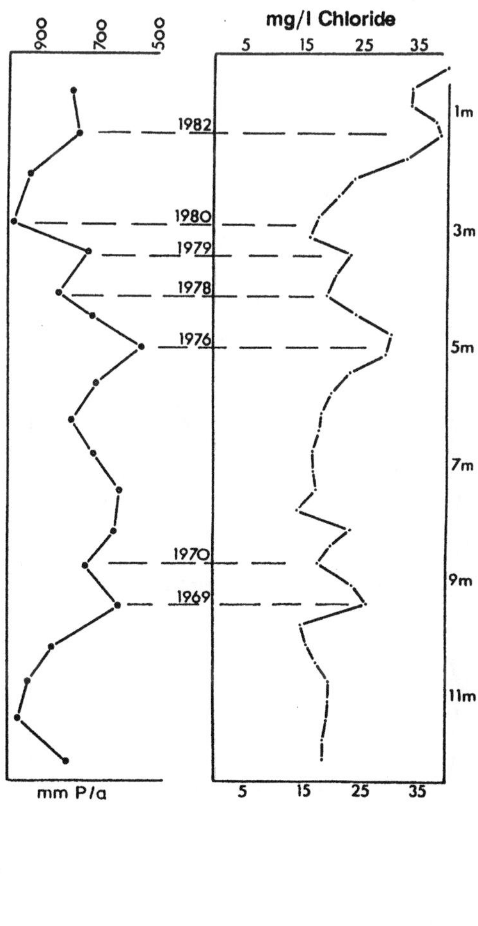

Fig. 6.2.1.3. Depth profiles of chloride concentrations in pore water (**right**) compared with values for the annual precipitation (**left**). The coincidence between both curves is used for the dating of the pore water

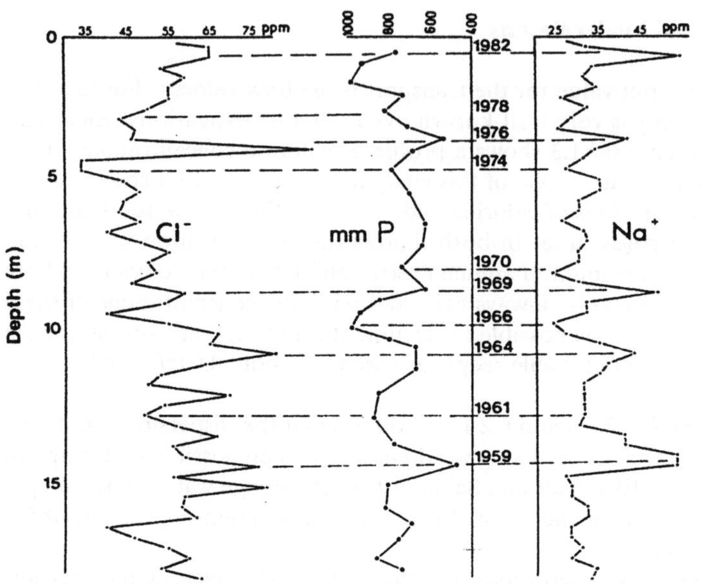

Fig. 6.2.1.4. Depth distribution of chloride and sodium contents of another location in the Segeberger Forst area. It allows the dating of pore water up to 13 years back to a depth of 15 m below surface

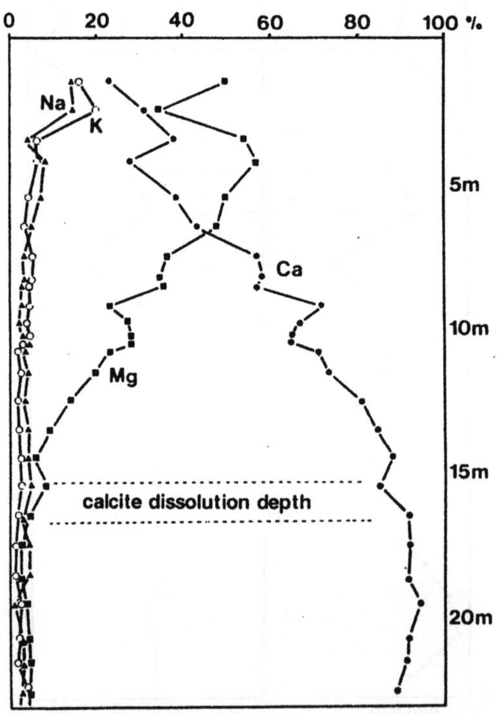

Fig. 6.2.1.5. Profile with measured percentages of exchangeable cation populations at the field site in the Segeberger Forst aquifer system

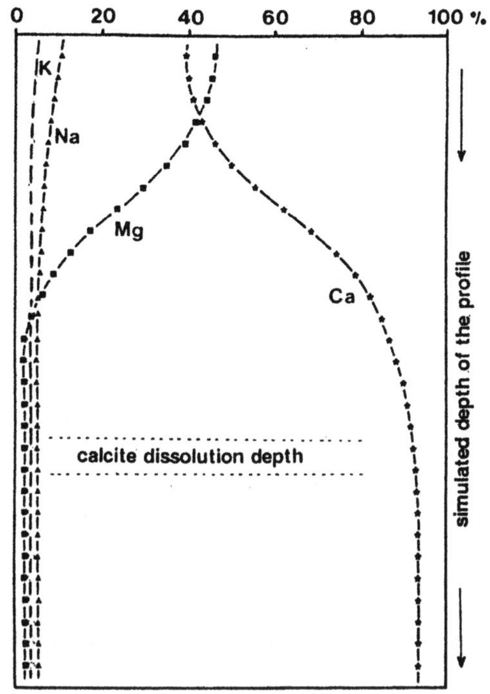

Fig. 6.2.1.6. Profile with percentages of exchangeable cation populations as a result of model calculations. Comparison with Fig. 6.2.1.5 shows that the reactions and the quantities of the model are in good agreement with the natural processes

total system of cells are only the selectivity coefficients of the cation exchange equilibria, the total CEC and the partial pressure of CO_2 of the profile. All values for this site are described in detail by Schulz and Baumann (1985).

Figure 6.2.1.5 shows a profile through the unsaturated zone with measured percentages of exchangeable cation population. This distribution of percentages for the different cations in different depths of the profile is quite typical for the study area. Figure 6.2.1.6 shows the same profile as a result of the simulation. The quite good agreement between both figures indicates that the processes and quantities obviously were correctly simulated.

6.2.2 Physical and Biochemical Processes Affecting Mass Transport in the Bocholt Aquifer System

C. Bugner and R. Mull

6.2.2.1 Background

Besides convection and dispersion, biochemical processes occurring during the transport of solutes through porous media can be a dominant mechanism affecting spatial and temporal concentration in the Bocholt aquifer system (Sect. 4.1.1 this Vol.). Vertical concentration profiles evaluated in multi-level wells (ML well) are also influenced by the input concentration at the ground water surface which may vary in space and time.

Previous studies mainly evaluate contaminant transport from point sources, such as sanitary landfills or areally limited test sites, considering microbial substrate utilization according to Michaelis-Menten equations (Sykes et al. 1982) as well as first-order kinetic reactions (Jones et al. 1987). Molz et al. (1986) extend an one-dimensional transport degradation model to include transport and utilization of both, the organic substrate and the necessary electron acceptor for biodegradation, as well as the growth dynamics of the biomass.

The microbial decomposition of organic matter with the coincident reduction of sulfate and the subsequent formation of pyrite is studied by Berner (1985). According to Leuchs (1988) and Leuchs and Obermann (Sect. 4.1.1, this Vol.), sulfate reduction occurs within small lenses of peat and wood embedded in the aquifer offering a high organic carbon content to enable biodegradation.

This study emphasizes the development of a model to simulate the transport degradation processes in porous media with respect to sulfate. It considers diffusive mass input at the ground water surface, and a macroscopic first-order reaction kinetic rate for the simulation of biodegradation. To evaluate the impact of convection and dispersion, the transport model is previously calibrated using environmental isotopes with known decay rates.

6.2.2.2 Mass Transport Scenario

The generic system in which the processes are studied consists of a 30- to 50-m-thick, unconfined aquifer of Tertiary and Quaternary gravel and sand underlain by impervious clay layers. The upper surface of the domain is a shallow water table which receives an areal recharge flux according to the land use (i.e. forest, arable land, grassland). The aquifer is bounded to the north and the south by two rivers. To the west it extends beyond the downstream discharge boundary, on the upstream side to the east a constant recharge is assumed, being equivalent to the amount of recharge in the remaining catchment.

The type of source is diffusive, since the mass input occurs simultaneously with the natural ground water recharge. The source function may vary over time and space.

Vertical concentration profiles monitored in four multi-level wells, which are situated along a flow path, are used for comparison between measured and simulated data.

The procedure and the set of differential equations are outlined in Fig. 6.2.2.1.

SET-UP OF 3-D GROUND WATER FLOW MODEL

$$S \cdot \frac{\partial h}{\partial t} = \frac{\partial}{\partial x_{i*}}\left(k_{fi*j*}\frac{\partial h}{\partial x_{j*}}\right) + q \qquad i^* j^* = 1,2,3$$

DETERMINATION OF INTERSTITIAL VELOCITY

$$v_{i*} = -\frac{k_{fi*j*}}{\theta_m} \cdot \frac{\partial h}{\partial x_{j*}}$$

ESTIMATION OF MEAN GROUND WATER AGE

$$t = \int \frac{\partial x_{j*}}{v_{i*}}$$

DETERMINATION OF MASS INPUT AREA AT GROUND WATER SURFACE

EVALUATION OF THE VERTICAL CROSS-SECTION

SET-UP OF 2-D MASS TRANSPORT MODEL

$$\alpha \cdot \frac{\partial c}{\partial t} = \frac{\partial}{\partial s_{i*}}\left(\alpha \cdot D_{i*j*}\frac{\partial c}{\partial s_{j*}}\right) - \frac{\partial}{\partial s_{j*}}(\alpha \cdot v_{i*} \cdot c) + \frac{q}{\theta_m} c^1 \qquad i^*, j^* = 1,2$$

CALIBRATION OF MASS TRANSPORT MODEL
USING ENVIRONMENTAL ISOTOPES

$$\alpha \cdot \frac{\partial c}{\partial t} = \frac{\theta_m}{\theta}\frac{\partial}{\partial s_{i*}}\left(\alpha \cdot D_{i*j*}\frac{\partial c}{\partial s_{j*}}\right) - \frac{\theta_m}{\theta}\frac{\partial}{\partial s_{j*}}(\alpha \cdot v_{i*} \cdot c) - \alpha \cdot \lambda \cdot c$$

ESTIMATION OF MACROSCOPIC DECOMPOSITION RATE

$$\frac{\partial c}{\partial t} = -\lambda(t) \cdot c$$

IMPLEMENTATION OF DECOMPOSITION RATE INTO
CALIBRATED MASS TRANSPORT MODEL

SIMULATION OF VERTICAL CONCENTRATION PROFILE

Fig. 6.2.2.1. Outline of proceeding

6.2.2.3 Ground Water Flow and Mass Transport Model

The hydrological and geological situation require the application of a three-dimensional ground water flow model, being the prerequisite to identify a vertical cross-section extending from the multi-level well BOC6 towards BOC7 according to the flowpaths in their vicinity. The numerical scheme and the setup of the model is shown in Fig. 6.2.2.2.

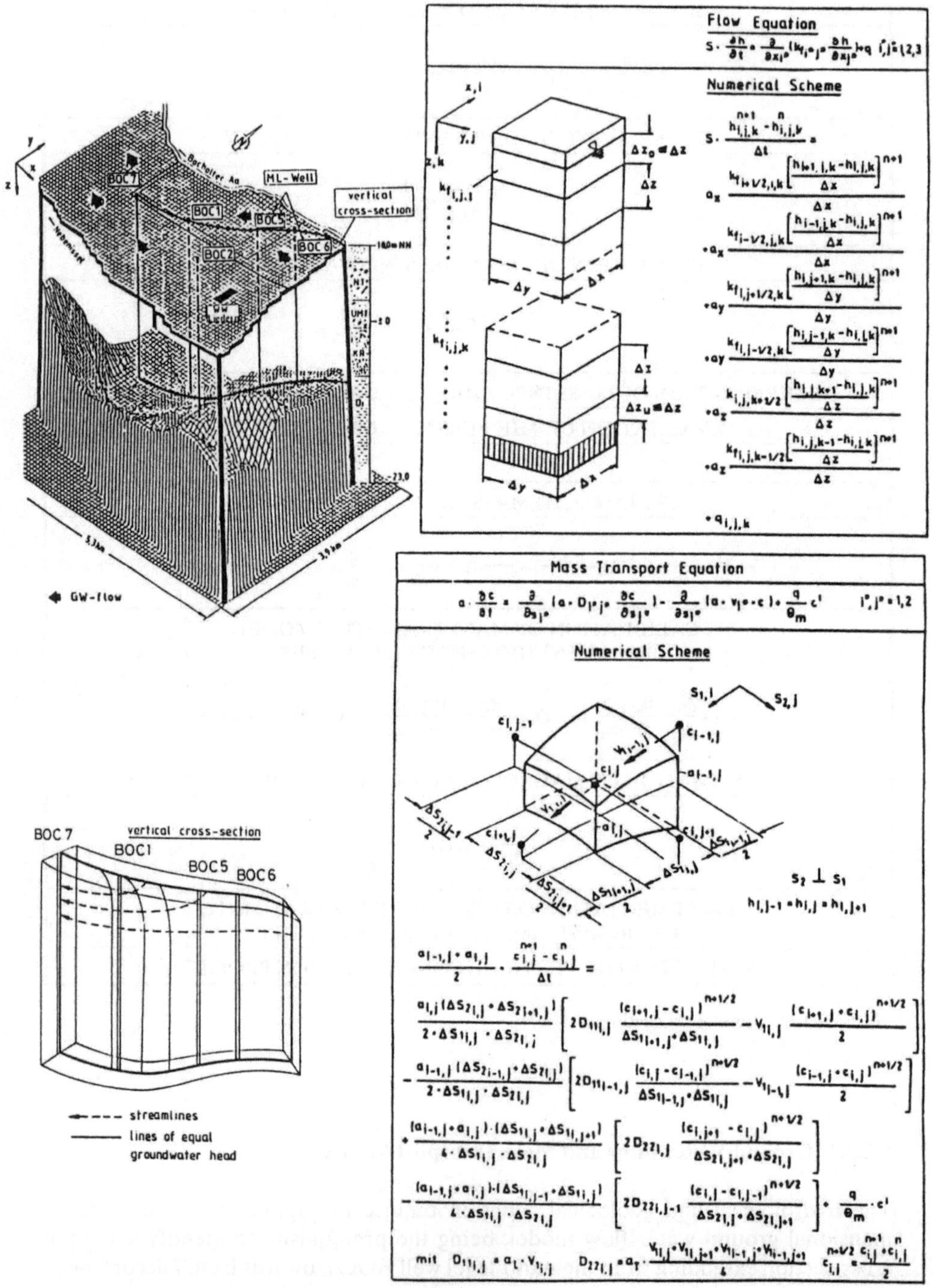

Fig. 6.2.2.2. Setup of the model and numerical scheme

The governing equations of ground water flow and mass transport are numerically solved by the method of finite differences. For the ground water flow model, the grid consists of rectangular prisms of equal size, the height of the upper and lower element is deformed accordingly to the position of the free water table and to the contours of the impervious layer, respectively.

The curvilinear elements for the two-dimensional mass transport model are formed by the streamlines and the lines of equal hydraulic heads both being determined applying an extended version of a bilinear interpolation scheme to describe flowpaths (Prickett et al. 1981) using velocity data of the ground water flow model. Because of the unidirectional flow this method reduces the number of components of the dispersion tensor and the velocity vector to be considered.

6.2.2.4 Calibration of the Mass Transport Model Using Environmental Isotopes

Considering a time interval of about 30 years since the agricultural use of fertilizers increased, and, consequently, the biochemical processes have been enforced by the solutes being recharged, it is necessary to determine first the processes of propagation being influenced by convection, and by longitudinal and transversal dispersion.

The concentration of tritium vs. depth at the multi-level wells as well as isolated measurements of krypton-85 and argon-39 present an useful data-set to calibrate the model, especially since their input functions over time have been carefully reported during the last 30 years, and their radioactive decay rates are known. Additionally, it can be assumed that their inputs are distributed homogeneously over the investigated area (Loosli 1979; Salvamoser 1982; Forster 1984).

The governing equation of mass transport including decay and the concept of mobile/immobile water zones within the pore space are shown in Fig. 6.2.2.1. Due to the relatively complete mixing of solutes into the zones of immobile water within each retention time considered, solute transport can be modelled using the effective velocity and dispersion multiplied by the ratio of effective porosity to total porosity (Goltz and Roberts 1986).

The source functions of argon-39, krypton-85 and tritium represent three different types of input functions: a constant input for argon-39 (Loosli 1979), an input which increases regularly with time for krypton-85, i.e. a quasi-step-input (Salvamoser 1982), and a sequence of peak-inputs over time for tritium (Weiss et al. 1979). The measured concentrations of the inert gases argon-39 and krypton-85 can be used uncorrected as input due to their ubiquitous distribution as a consequence of both the uniform production in the atmosphere and the permanent release by nuclear power plants. The tritium values need to be corrected for losses due to the seasonal variation of ground water recharge.

Various simulations, considering several combinations of dispersivities, give the best fit of data of the concentration of tritium using a longitudinal and transversal dispersivity of 10 m and 4 cm, respectively, and the mean annual tritum values being reduced by a factor of 0.8 (Fig. 6.2.2.3). Due to the unidirectional flow along the streamlines numerical dispersion can be neglected in the transversal direction.

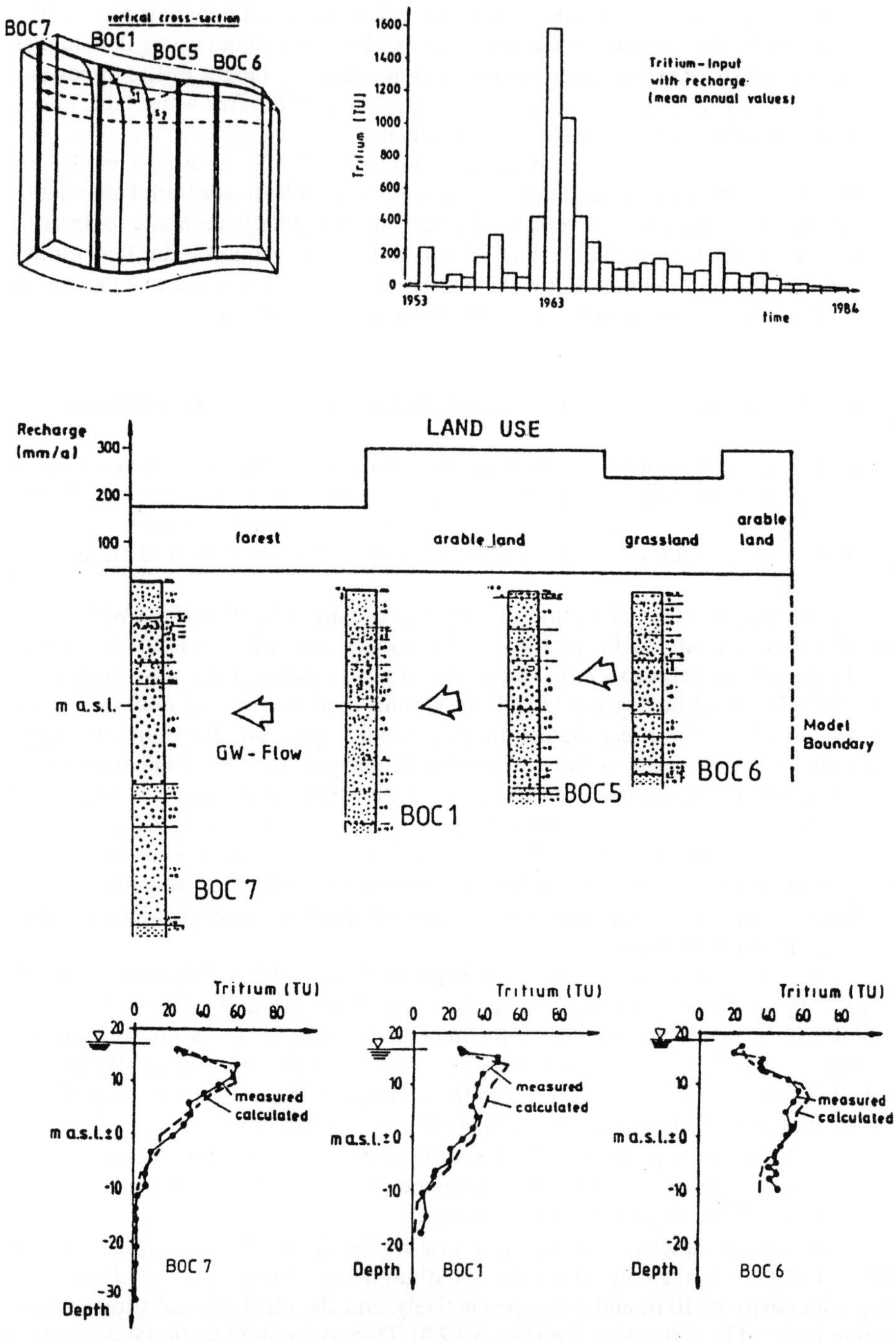

Fig. 6.2.2.3. Simulation of vertical concentration profiles of tritium

The calculated break-through curves at different depths and the vertical concentration profiles of the three environmental isotopes at the furthermost downstream multi-level well BOC7 are shown in Fig. 6.2.2.4. They show the smoothing effect of dispersion during mass transport and the importance to consider a vertical dispersivity for any interpretation of vertical concentration profiles.

6.2.2.5 Simulation of the Biochemically Affected Distribution of Sulfate

According to Leuchs and Obermann (Sect. 4.1.1, this Vol.), the observed decrease of sulfate vs. depth of the aquifer is partially subject to biodegradation which is enabled by the organic material within lenses of peat and wood embedded in the aquifer. Besides the physical processes of mass transport being described previously, the vertical distribution of sulfate is influenced, in addition, by the kinetics of biodegradation.

Determination of a Macroscopic Decomposition Rate

Taking the mean ground water ages deduced from the distribution of argon-39, krypton-85 and tritium, verified with the mass transport model, a comparison of the input concentrations at the ground water surface with the concentration at the relevant depth at the observation well and the elapsed time of transport gives a first estimate of the decomposition rate. A conservative tracer functions as a correction factor to cope with the influence of dispersion. An evaluation of all available data results in a time dependent decomposition rate shown in Fig. 6.2.2.5. This decomposition rate is then incorporated into the mass transport model.

Simulation of the Vertical Distribution of Sulfate

Instead of the radioactive decay rates, the mass transport model for sulfate is run using first-order decomposition rates within the predetermined range of 0.02 to 0.08 (a^{-1}). The setup of the model includes the same scenario as described in Sections 6.2.2.1 and 6.2.2.3. Due to ignorance of the type of input functions of sulfate, they are assumed to be constant or linearly increasing over the time interval of simulation, their respective maximum values are determined according to the areal distribution of the concentration of sulfate at the ground water surface (Fig. 6.2.2.6).

The comparison of the simulated vertical distribution of sulfate at the multilevel well BOC1 regarding the different decomposition rates and the two types of input functions verify the best fit if the rate constant keeps within the lower range of 0.02 a^{-1} and the input function is constant over time (Fig. 6.2.2.7).

These results are used for a further simulation of the distribution of sulfate over the whole vertical cross-section shown in Fig. 6.2.2.8. The comparison between the measured data and the calculated values at the four multi-level wells demonstrates an acceptable agreement.

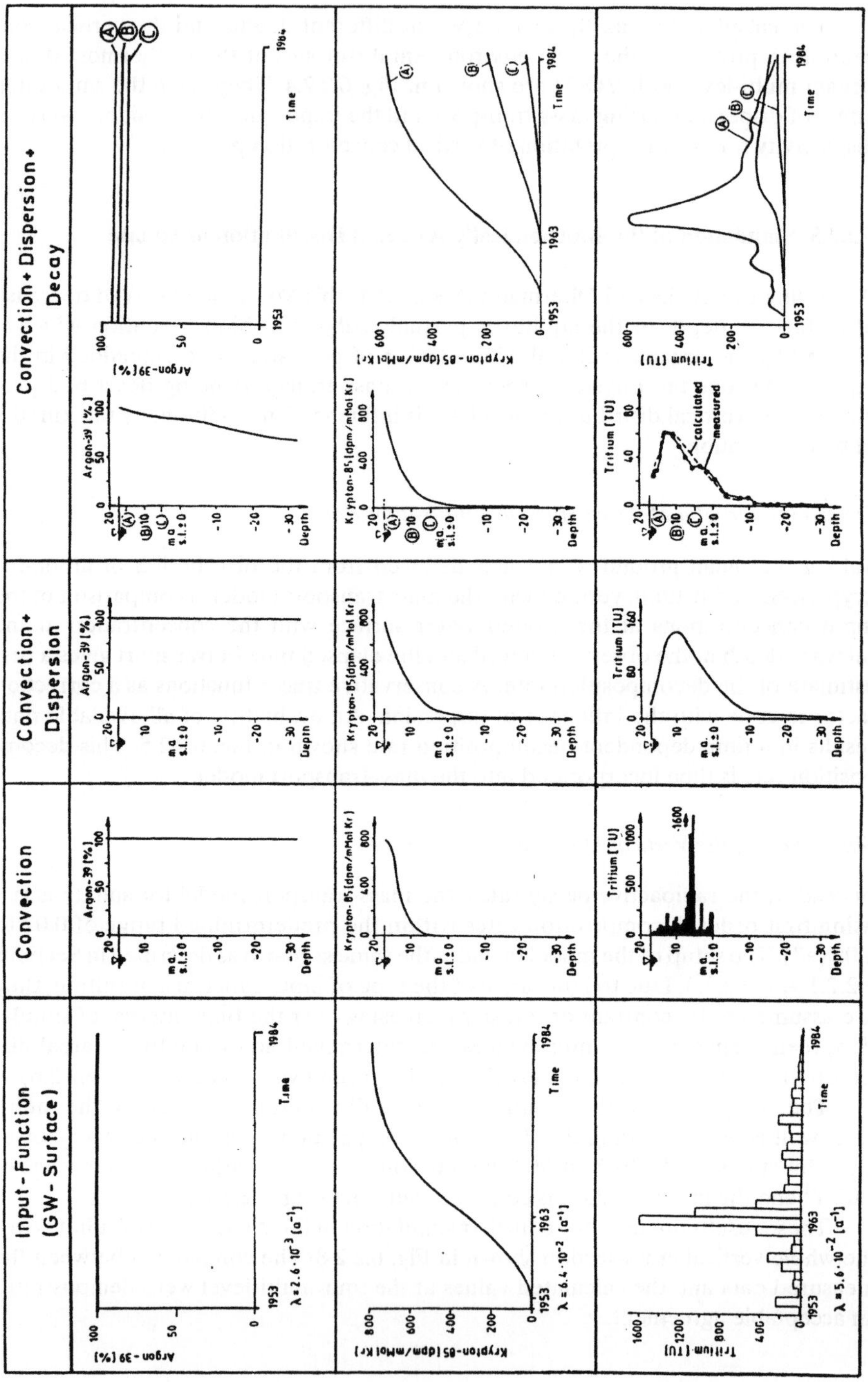
Input-Function (GW-Surface)
Convection
Convection+ Dispersion
Convection+ Dispersion+ Decay
Argon-39 [%]
Krypton-85 [dpm/mMol Kr]
Tritium [TU]
Time
Depth
m a. s.l.±0
1953
1963
1984
λ = 2.6 · 10⁻³ [a⁻¹]
λ = 6.4 · 10⁻² [a⁻¹]
λ = 5.6 · 10⁻² [a⁻¹]
calculated
measured

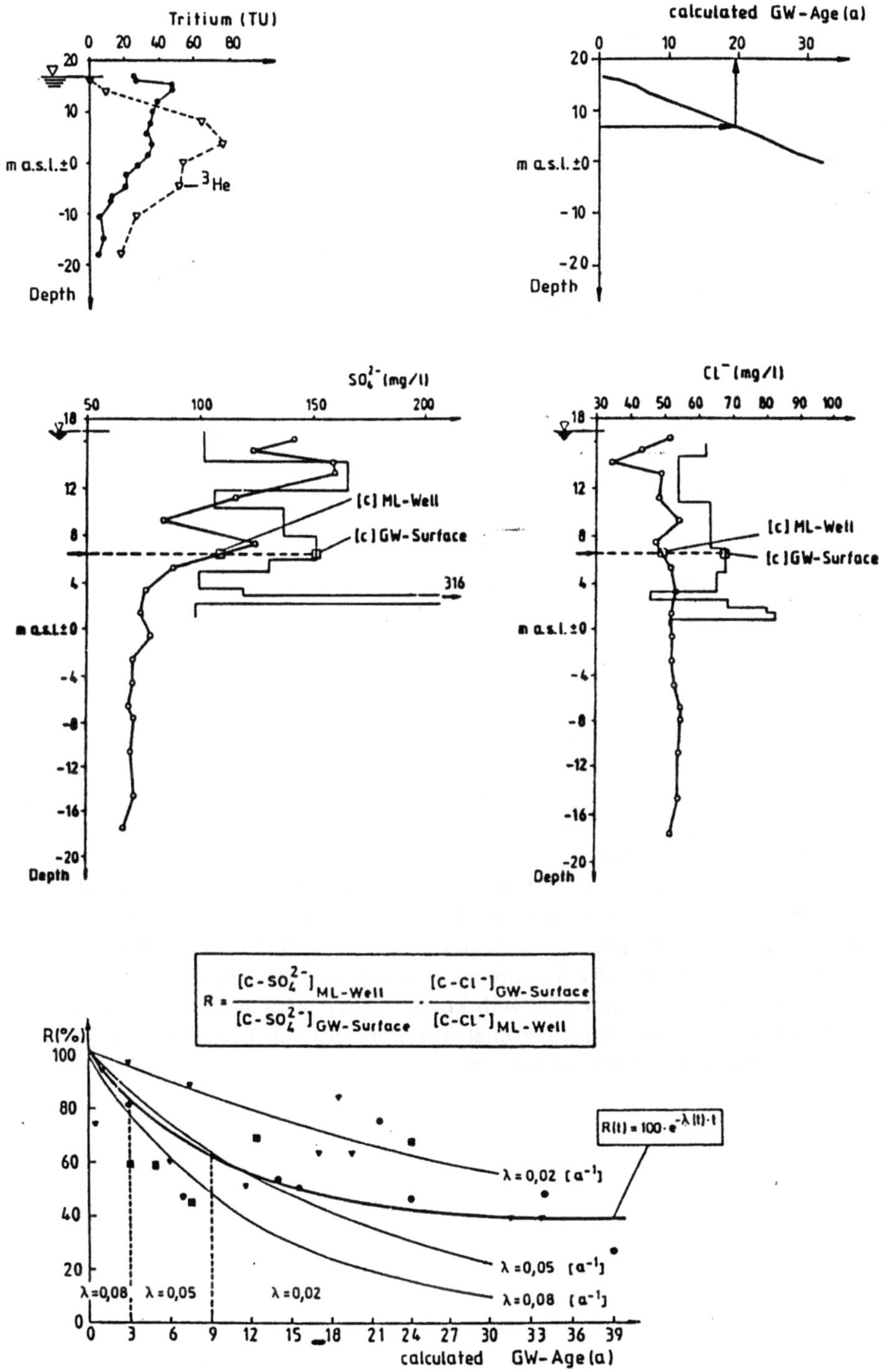

Fig. 6.2.2.5. Determination of a macroscopic decomposition rate

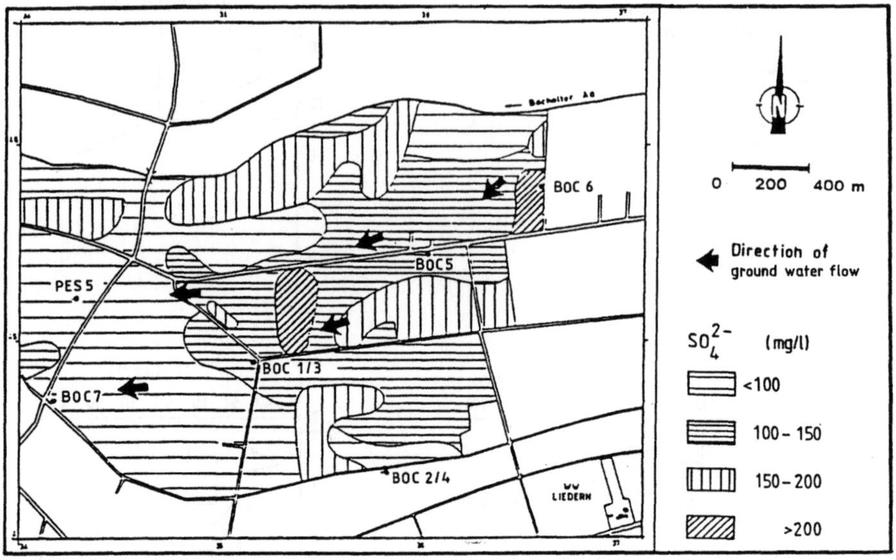

Fig. 6.2.2.6. Areal distribution of sulfate at the ground water surface

6.2.2.6 Conclusions

The presented simulations of mass transport in an aquifer with a numerical flow and mass transport model clarify the effects of convection, dispersion, decay, and decomposition as well as the type of input function on the vertical distribution of solutes being diffusively recharged at the top of the aquifer.

Especially, it is shown that even small transversal dispersivities such as the determined value of 4 cm, which results in a transversal dispersion coefficient being within the range of a diffusion coefficient, cause the occurrence of solute in depths of the aquifer where they are not expected considering the mean ground water age.

The analysis of the physical processes using environmental isotopes such as tritium, krypton-85 and argon-39 with known decay rates and input-functions enables the determination of the microbial decomposition rate of sulfate to be 0.02 (a^{-1}) for the investigated Bocholt aquifer system.

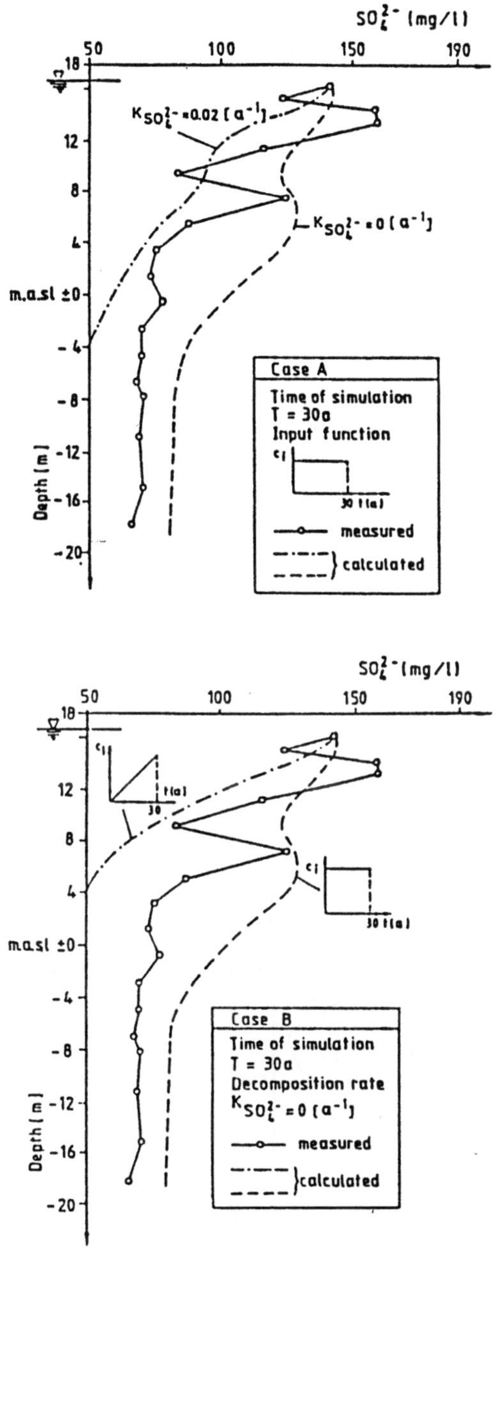

Fig. 6.2.2.7. Simulation of the vertical distribution of sulfate using different input functions

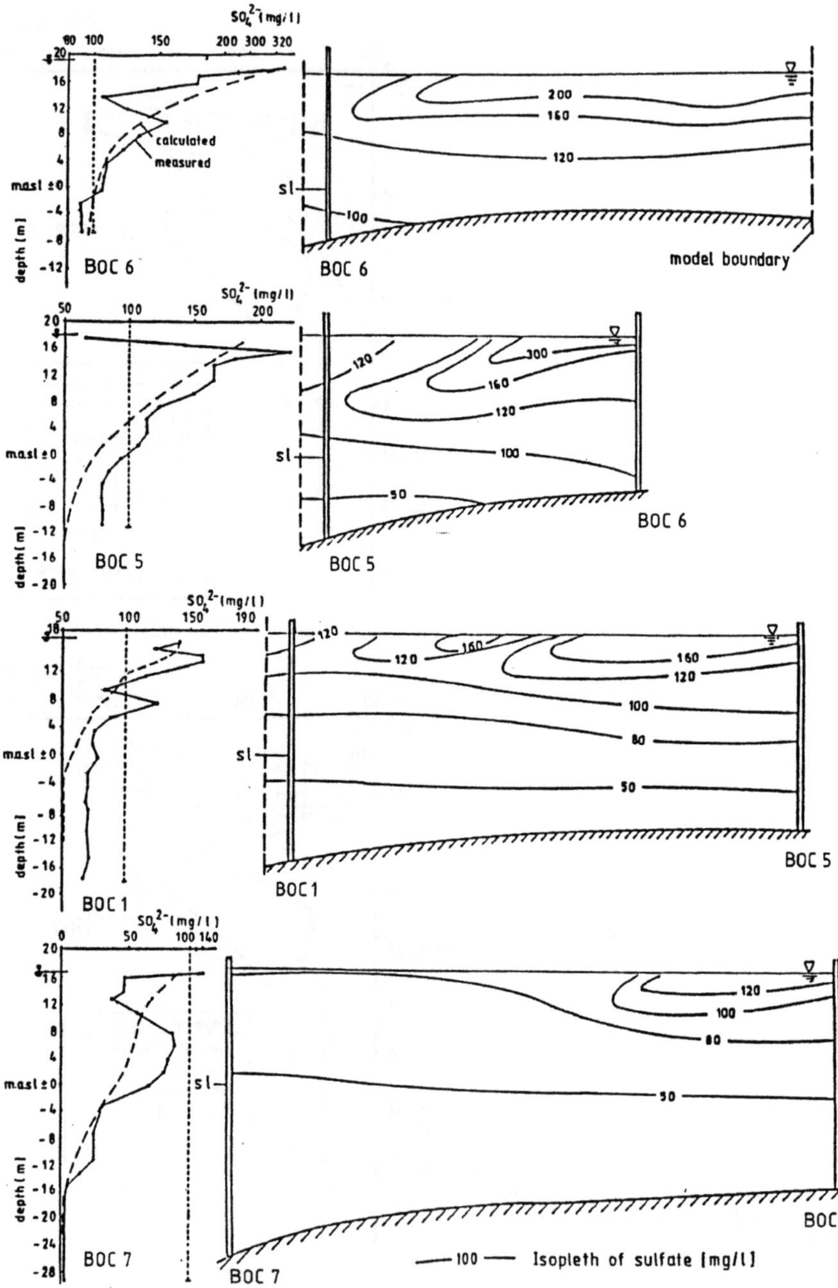

Fig. 6.2.2.8. Simulation of the distribution of sulfate within the vertical cross-section

6.2.3 Tritium and ^{3}He Measurements as Calibration Data for Ground Water Transport Models

H. Dörr, P. Schlosser, M. Stute and C. Sonntag

6.2.3.1 Background

Ground water transport modelling needs reasonable numbers for the hydraulic conductivity and for the coefficients of transversal and longitudinal hydrodynamic dispersion for the entire ground water body. In general, however, geohydraulic data are obtained only at few locations in an investigated area. Therefore, a ground water transport model has to be calibrated, i.e. modelled tracer distributions have to be adjusted to measured tracer distributions by matching geohydraulic parameters. Suitable tracers to calibrate ground water transport models in a shallow aquifer are tritium and tritiogenic ^{3}He.

The radioactive hydrogen isotope tritium (^{3}H) is produced mainly by reactions of protons and neutrons with oxygen and nitrogen in the upper atmosphere. After oxidation to HTO it takes part in the hydrological cycle. Tritium decays (β^-) with a half-life of $T_{1/2}$ = 12.43 a (Unterweger et al. 1980) to the stable helium isotope ^{3}He. The natural tritium level, a steady-state activity between tritium production and radioactive decay, is about 4–6 TR (1 TR: tritium ratio is the synonym to TU and means a ^{3}H/H-ratio of 10^{-18} (Taylor and Roether 1982)). Between 1953 and 1963 the tritium level in Central European precipitation increased to about 2000 TR due to stratospheric nuclear weapon tests (Weiss et al. 1979). After the test-stop agreement in 1963, the tritium level decreased to the present level of about 40 TR. Compared to the bomb tritium component the delivery of tritium from industrial production and nuclear power plants is low, although it will presumably be the dominant tritium source in future times.

The significant tritium peak in precipitation in the early 1960's is a transient tracer signal which can be used in studies of the water movement in the unsaturated zone (Sonntag 1986) and in shallow ground water (Münnich 1963; Matthess et al. 1968). However, the tritium peak in shallow ground water decreases with time due to radioactive decay and hydrodynamic dispersion. On the other hand, the amount of tritiogenic ^{3}He in ground water, i.e. ^{3}He produced by radioactive decay of tritium, increases with time, providing an alternative tracer signal. It can be considered as a conservative tracer for water movement. Diffusional loss of ^{3}He across the ground water table is of minor importance (Schlosser et al. 1988). While tritium in ground water decreases by radioactive decay

$$^3H(t) = {}^3H(t=0) \cdot e^{-t/T} \tag{6.2.3.1}$$

its decay nuclide ^{3}He increases according to

$$^3He(t) = {}^3H(t=0) \cdot (1-e^{-t/T}). \tag{6.2.3.2}$$

^{3}H (t = 0) stands for the tritium content, which the ground water has had at the time of infiltration, and T = 17.93 years is the mean life of tritium. The ground water age t, i.e. the time since the water is isolated from the unsaturated soil zone, results from Eqs. (6.2.5.1) and (6.2.5.2) to be:

$$t = T \cdot \ln\left({}^3He/{}^3H + 1\right) \qquad (6.2.3.3)$$

This $^3H/^3He$ ground water age is independent of the initial tritium content and, thus, independent of the bomb tritium input function.

In the present study the assumptions and restrictions for the use of ^3H-, ^3He-, $^3H/^3He$ ground water age profiles in the shallow Bocholt ground water system are discussed as calibration data for a ground water transport model (Sect. 6.2.2, this Vol.). Coefficients of transversal hydrodynamic dispersion are derived from the vertical distribution of ^3H and ^3He in ground water.

Ground water samples were obtained from various multi-level wells yielding representative water samples from the individual ground water horizons (Sect. 4.1.1, this Vol.). Tritium concentrations were measured by low-level β counting, ^3He by mass spectrometry (Schlosser et al. 1988).

6.2.3.2 Geohydraulic Conditions in the Investigated Area

The shallow Bocholt aquifer system is formed by rather high permeable River Rhine sediments of about 50 m total thickness above impermeable clay. Different grain size and hydraulic conductivity suggest subdividing the aquifer into three parts: the upper part of about 10 m thickness and the lower one (20 m thick) mainly consist of fine to medium sand. In the intermediate part (about 20 m thick) coarse sand and gravel prevail. Table 6.2.3.1 shows reasonable numbers for the hydraulic conductivity and the mean ground water flow velocity as well as estimates of the ground water age for these aquifer sections (Sect. 3.2.3, this Vol.).

The hydraulic conductivity data were derived from grain size distribution. The mean ground water flow velocity was calculated from Darcy's equation with a hydraulic gradient of 4.5×10^{-4}. The estimate of ground water age is based on the assumption of a logarithmic age distribution (Vogel 1970), using a mean annual ground water infiltration rate of 200 mm, and a mean total porosity of 0.35. From these data it is expected that most of the bomb tritium is contained in the upper part of the ground water body, whereas the lower one should be tritium-free.

Table 6.2.3.1. Hydraulic conductivity k_f, mean ground water flow velocity $\bar{v}$, and estimated ground water age t for the shallow Bocholt aquifer system

	k_f (m/a)	$\bar{v}$ (m/a)	t (a)
Upper section (0–10 m)	$3 \cdot 10^2 - 3 \cdot 10^3$	0.5–5	0–20
Middle section (10–30 m)	$3 \cdot 10^4$	50	20–100
Lower section (> 30 m)	$2.5 \cdot 10^3 - 2 \cdot 10^4$	3–25	>100

6.2.3.3 Bomb Tritium Input to the Shallow Bocholt Ground Water System

For modelling the vertical distribution of bomb tritium in ground water, the tritium input function has to be established from tritium data of precipitation over the last three decades. Without any tritium data of local precipitation, the bomb tritium data of annual mean precipitation from other locations in Central Europe had to be used for the period between 1953 and 1976 (Weiss et al. 1979). For the period between 1976 and 1985 unpublished tritium data of Hof, FRG, were used. These annual mean tritium concentrations of precipitation were reduced by 42%, which accounts for the seasonal variation of tritium in precipitation, and for the fact that ground water recharge predominantly occurs between October and March. The tritium input function for Bocholt is shown in Fig. 6.2.3.1. With reasonable numbers for the mean annual ground water recharge of 200 mm, a total porosity of 0.35 and a mean residence time of water in the unsaturated zone of 2 years (causing a 10% loss of tritium due to radioactive decay) this input function gives about 2500 TR·m (TR·m: TR·meters of water) for the (^{3}H + ^{3}He) inventory of the ground water body. The tritium inventory alone is obtained by a radioactive decay correction of each individual annual tritium input to the ground water body with respect to the date of sampling (1985) to be about 905 TR·m at this time. The estimate of the tritium input is subject to various sources of error (e.g. variability of tritium concentration in precipitation for different locations, variability of the ground water infiltration rate with space and time). The tritium input function used has an accuracy in the order of ± 20% (Atakan et al. 1974a).

6.2.3.4 Vertical Distribution of Tritium in the Shallow Bocholt Ground Water System

Figure 6.2.3.2 shows the vertical distribution of tritium in the multi-level wells BOC1, BOC4, and BOC7. The profiles show high tritium contents in the upper section of the ground water body. The intermediate section is characterized by a

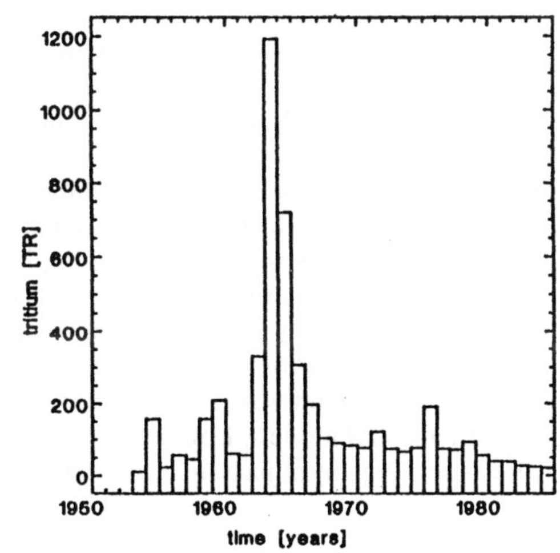

Fig. 6.2.3.1. Tritium content in precipitation at Bocholt from 1953 to 1985

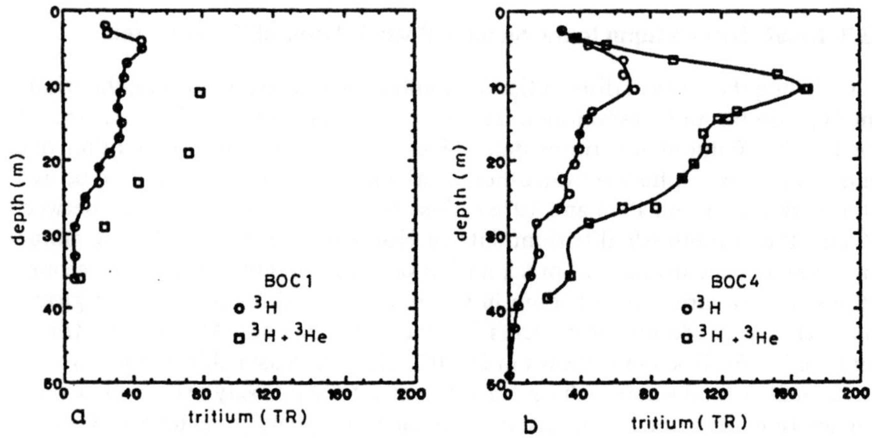

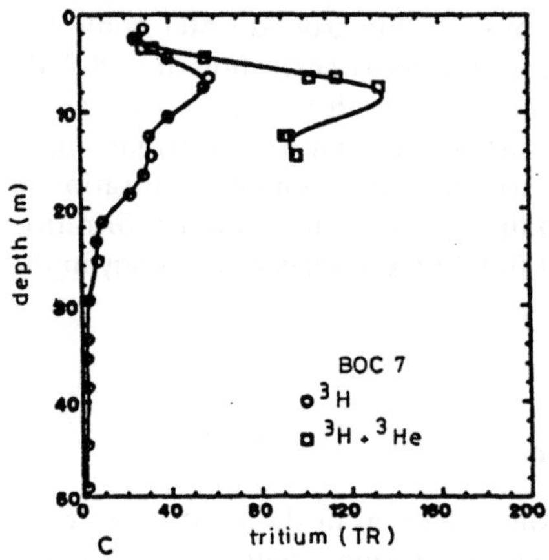

Fig. 6.2.3.2. Tritium- and ($^3H + ^3He$) profiles from multi-level-wells BOCl (**a**), BOC4 (**b**) and BOC7 (**c**)

continous tritium decrease down to 30 m below ground surface, where tritium in ground water is below the detection limit. Tritium inventories at the multi-level wells BOC1 (823 TR·m) and BOC7 (760 TR·m) are comparable with the estimate of 905 ± 181 TR·m, mentioned above. BOC4, however, shows a significantly higher tritium inventory (1305 TR·m), which is obviously due to comparatively high tritium contents at depths greater than 10 m below ground surface. This excess tritium in the intermediate aquifer section decreases in flow direction from BOC4 over BOC1 to BOC7. A plausible explanation of this surplus could be that a higher ground water flow velocity in the intermediate aquifer section leads to ground water at the well locations which is too young in comparison to the ground water ages expected from the logarithmic age profile. This shows that the tritium profiles cannot be modelled with only one dispersion coefficient for the entire ground water body. Due to the stratification of the aquifer the tritium distribution with depth is a result of a complex ground water flow pattern, which needs to be modelled with at least two-dimensional models. Longitudinal and transversal dispersion coefficients have to be determined by adjusting the modelled tritium distribution to measured profiles.

6.2.3.5 ^{3}H/^{3}He Dating of the Shallow Bocholt Aquifer System

^{3}He measurements give additional information on ground water age. Figure 6.2.3.2 also shows the (^{3}H + ^{3}He) profiles at BOC1, BOC4 and BOC7. Information on the ground water age can be derived in two different ways. First, the (^{3}H + ^{3}He) peak dates ground water which has been formed in the years 1962 to 1965. The (^{3}H + ^{3}He) peak is found in all profiles in 10 m depth, i.e. the water in this depth is about 20 years old (tritium peak in precipitation: 1963; 2 years movement in the unsaturated zone; sampling date: 1985). This absolute age can be compared with the formally calculated ^{3}H/^{3}He age using Eq. (6.2.3.3). The ^{3}H/^{3}He age (Fig. 6.2.3.3) shows a nearly linear increase from close to zero at the ground water surface to about 17 years at the (^{3}H + ^{3}He) peak. The ^{3}H/^{3}He ages below this peak remain constant, which is due to dispersive downward mixing of bomb tritium and its decay nuclide ^{3}He. Since ground water below the peak stems from infiltration water of the early bomb test period and from prebomb time, even relatively small admixtures of

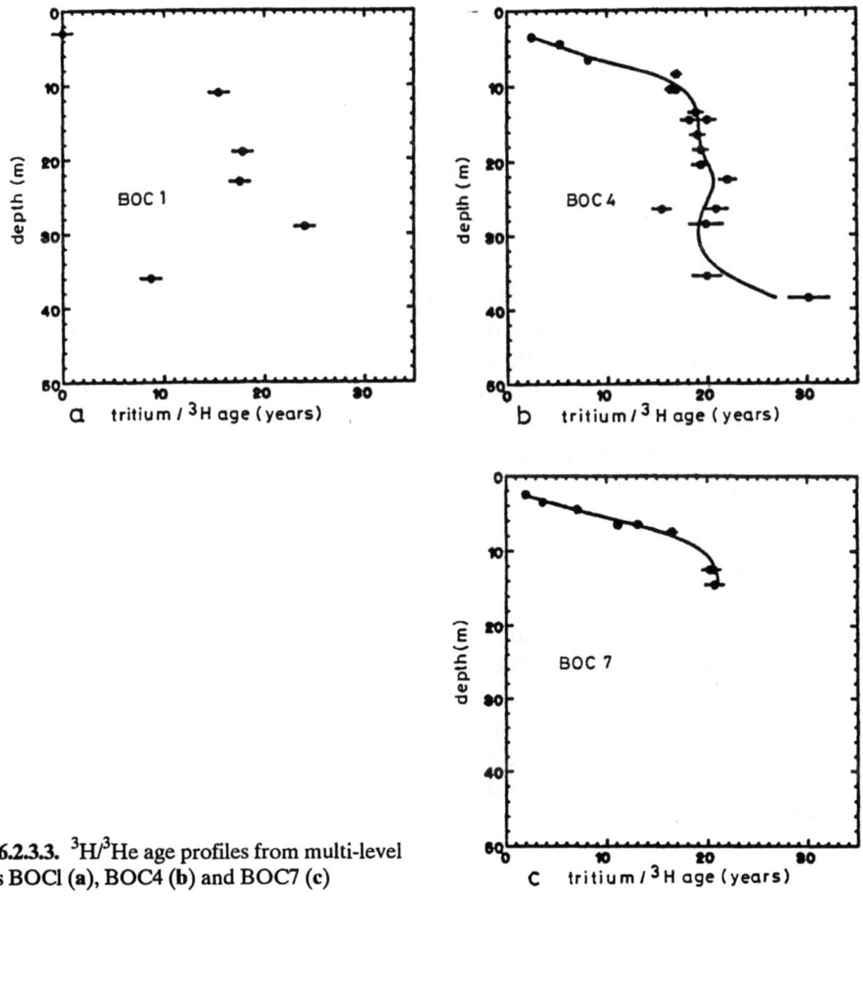

Fig. 6.2.3.3. ^{3}H/^{3}He age profiles from multi-level wells BOCl (**a**), BOC4 (**b**) and BOC7 (**c**)

ground water from the main bomb test time will govern ^{3}H- and ^{3}He- level there and, thus, limit the $^{3}H/^{3}He$ ground water age to about 17 years.

The difference between the age determined from the position of the ($^{3}H + ^{3}He$) peak (20 a) and the $^{3}H/^{3}He$ age of 17 years can be explained by dispersive mixing of water of different tritium concentration, which is associated with a non-linear behaviour of the apparent tritium/^{3}He ages. Moreover, incomplete ^{3}He confinement comes into play. The diffusive loss of tritiogenic ^{3}He across the ground water table was estimated by comparison of the inventories of tritium and of ($^{3}H + ^{3}He$) in the ground water with the time-integrated bomb tritium input. This gives a ^{3}He loss of about 20% (Schlosser et al. 1988). The ^{3}He confinement is a function of the vertical flow velocity and of the transversal dispersion coefficient. With transversal dispersion coefficients of 0.5–1.5 m^2/a a simple, one-dimensional box model yields vertical flow velocities of 0.15–0.25 m/a, below which the ^{3}He loss is increasing rapidly to high values (> 50%) (Schlosser et al. 1989). A ^{3}He confinement of about 80%, however, is good enough to use ($^{3}H + ^{3}He$) as a conservative tracer. In this case the vertical profiles of the tritium/^{3}He ground water age can be used for an independent calibration in ground water transport modelling (Sect. 6.2.2, this Vol.).

6.2.3.6 Estimation of Transversal Dispersion Coefficients

Although the measured vertical distribution of tritium and ^{3}He strongly indicates that ground water flow in the Bocholt aquifer system is at least a two-dimensional problem, nevertheless, the transversal dispersion coefficient was estimated from the observed ($^{3}H + ^{3}He$) profiles. Assuming a δ-function like tritium input to the ground water 20 years ago, the coefficient of longitudinal hydrodynamic dispersion D can be determined from $\sigma^2 = 2 \cdot D \cdot t$, where σ is the width of the Gaussian peak observed at time t after injection of the δ-peak. The coefficient of transversal hydrodynamic dispersion D_t is estimated from the width of the predominant ($^{3}H + ^{3}He$) peak in the upper aquifer section to be about 1.6 m^2/a at BOC1, 0.6 m^2/a at BOC4, and 0.5 m^2/a at BOC7. The dispersivity $\alpha_x = D_t/\bar{v}$, where $\bar{v}$ is the average ground water flow velocity, is 0.1–0.3 m if a ground water flow velocity of 5 m/a is assumed. These estimated dispersion coefficients can serve as a "first guess" in calibrating a ground water transport model.

6.2.3.7 Conclusions

Tritium and ^{3}He prove to be valuable tracers for ground water transport modelling in a shallow aquifer system. The suitability of tritium alone will decrease in the future due to radioactive decay. Additionally, ^{3}He measurements provide tritium/^{3}He ground water ages to serve as model-independent calibration data. Estimated dispersion coefficients from the vertical ($^{3}H + ^{3}He$) distribution give reasonable values, which have to be modified by adjusting modelled tritium and ^{3}He distributions to measured profiles.

6.2.4 ^{39}Ar-, ^{85}Kr-, ^{3}He- and ^{3}H Isotope Dating of Ground Water in the Bocholt and Segeberger Forst Aquifer Systems

M. Forster, H. Loosli and S. Weise

6.2.4.1 Methodical Background

Investigating radioactive ^{39}Ar- and ^{85}Kr contents in ground water permits the determination of mean residence times (MRT). This is generally done by comparing the measured isotope content of the ground water with the isotope contents of the atmosphere which serve as the "input signal" of the aquifer (e.g. Sects. 6.2.2 and 6.2.3, this Vol.). The use of the noble gas isotopes ^{39}Ar and ^{85}Kr for the purpose of dating ground waters offers the advantage of being independent of geochemical interactions between ground water and the aquifer itself, which affect the ^{14}C dating method (e.g. Eichinger 1987 and Dörr et al. 1987). In addition, the ^{39}Ar-dating method allows dating of ground waters with MRT in the range from 100 to 1000 years and is based on an input signal which is constant and uniform in time.

The ^{85}Kr-dating method also shows the advantage of a noble gas isotope method; its input signal, which has been rising since the late 1950's and its shorter half-life result in a dating range between around 2 and 30 years. When dating with ^{85}Kr, the time dependence of the input content of ^{85}Kr must be included.

In connection with "classic" ^{3}H- and ^{14}C isotope methods, the ^{39}Ar-, ^{85}Kr-, and ^{3}He methods offer valuable information when investigating ground water systems.

6.2.4.1 Production and Atmospheric Content of ^{39}Ar and ^{85}Kr

Following the reaction ^{40}Ar(n,2n)^{39}Ar, the cosmic radiation is the main source of the atmospheric ^{39}Ar level. Due to an equilibrium between n-induced production and radioactive decay of ^{39}Ar ($T_{1/2}$ = 269 years) the atmospheric ^{39}Ar content is 1.78 ± 0.07 mBq/l of Ar. This concentration is defined to be 100% mod. of ^{39}Ar. The atmospheric ^{39}Ar content even remained constant throughout the nuclear test series between 1952 and 1962 (Loosli 1979, 1983).

The sources of atmospheric ^{85}Kr are nuclear power and reprocessing plants. Due to the increasing use of nuclear industry, the atmospheric ^{85}Kr content rised since the early 1950's from negligible values (≤ 33 Bq ^{85}Kr/l Kr) to around 55 dpm ^{85}Kr/ml Kr in 1987. ^{85}Kr- and ^{39}Ar input functions are given in Fig. 6.2.4.1 as activities per liter Kr or per liter of Ar and can be compared with the input function for the ^{3}H- and ^{14}C isotope methods (Sect. 6.2.2, this Vol.).

Due to similar half-lives of ^{3}H (12.43 a) and ^{85}Kr (10.76 a) and due to different input functions, ^{85}Kr data are well suited to be compared with ^{3}H data. In contrast to ^{85}Kr, ^{3}H data can be ambiguous due to the ^{3}H input function.

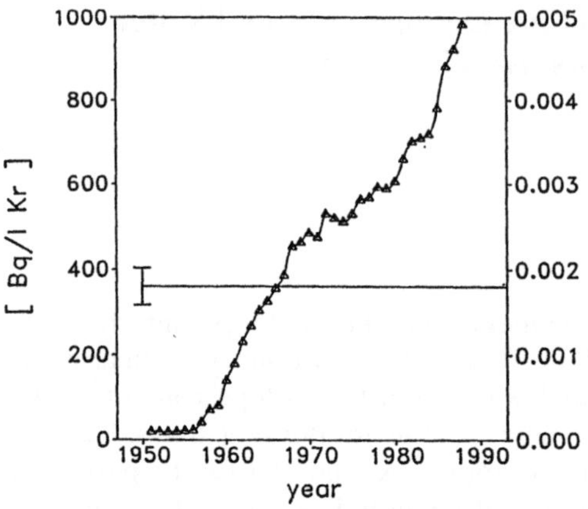

Fig. 6.2.4.1. Atmospheric levels of noble gas isotopes ^{39}Ar and ^{85}Kr. They represent the input functions when studying ground water output systems to determine ground water MRT (^{85}Kr values from 1980/87: pers. commun. from Weiss; earlier values: Salvamoser 1982)

6.2.4.2 Gas Sampling Procedure, Kr- and Ar Sample Preparation and ^{39}Ar- and ^{85}Kr Measuring Technique

As a result of the very low specific activity of ^{39}Ar in Ar and of the low Kr content dissolved in ground water, the ground water sample size is 10 m^3 for one Ar sample and 0.5 to 1.5 m^3 for one Kr sample. Therefore, the gases dissolved in the ground water are extracted already during the sampling procedure. This is done by spraying the water sample into an evacuated system, where the water is degassed and the components gas and water are pumped off continuously. The next preparation steps are done in the laboratory, where 2–3 l of Ar and around 0.8 ml of Kr are separated from the 200 l gas mixture. The noble gas fraction is purified by means of several, large-volume reaction ovens, followed by a special gas chromatographic system (H_2 carrier gas) which gives final purification and noble gas separation. After H_2 is removed from the Ar-H_2 or Kr-H_2 mixture, the Ar- or Kr-sample is filled into proportional counting tubes with selected low counting background. Counting the ^{39}Ar or the ^{85}Kr activity of the sample then gives the ^{39}Ar or ^{85}Kr activity of the known amount of noble gas filling (Forster and Maier 1987).

Due to the efforts necessary for one ^{39}Ar result (excluding a 3 weeks' measuring time, the laboratory sample preparation needs 1 week), until now both noble gas isotope methods have been only used in ground water systems, where detailed hydrochemical and ^{3}H-, ^{14}C-, ^{2}H-, and ^{18}O isotope investigations have been performed before.

6.2.4.3 Isotope Hydrological Situation in the Bocholt and the Segeberger Forst Aquifer Systems

Ground water flows in the Bocholt aquifer with velocities between 20 and 30 m/a (Leuchs 1988; Sect. 4.1.1, this Vol.). From modelling ground water flow, ground water MRT near 200 years seemed to be probable in aquifer depths of 20–50 m. Due to layers with relative low permeabilities, “ages” could be even higher and “classical" estimates from Leuchs (1988) gave MRT of 100 to 500 years for those ground waters which are found in greater depths in the Bocholt aquifer. As an extension to ^{3}He/^{3}H depth profiles in the ground water (which indicated MRT of 20 years for depths of 10 to 30 m), some indication for deeper ground waters showing higher MRT could be taken from ^{14}C- and ^{4}He excess data (Sonntag 1986 pers. commun.; Dörr et al. 1987).

The aquifer in Segeberger Forst consists of sandy and gravelly layers in which hydraulic conductivities between 1.3×10^{-4} and 1.5×10^{-3} m/s, flow velocities of <1 m/d and longitudinal dispersivities of 0.05–1.0 m have been determined (Schröter 1983, 1984). ^{3}H isotope contents in these ground waters have been investigated by Eichinger (pers. commun.) and indicated ground water model ages near or above 4 years for ground waters below 10 m. On this basis it had been of interest to perform combined ^{85}Kr- and ^{3}He/^{3}H determinations in order to substantiate these MRT and also to study ground water flow and mixing processes. Due to the 5 years' time interval since Eichinger's ^{3}H isotope investigations, a remeasurement of the ^{3}H depth profile appeared also to be promising.

6.2.4.4 Isotope Results in Bocholt Ground Waters

Samples for ^{39}Ar, ^{3}H and ^{14}C isotope measurements were collected in Bocholt ground waters in 1986, the results are listed in Table 6.2.4.1. To assess possible subsurface production of ^{39}Ar (Loosli 1983; Forster 1984; Forster and Maier 1987), U, Th and K contents of aquifer material have also been measured. Low contents were determined in the samples of the depth profile.

The comparison of ^{3}H and ^{14}C contents in ground water from wells BOC1 and BOC4 sampled in 1985 and 1986 show no change in the ground water situation within the given standard deviation. Therefore it is assumed that ^{3}H/^{3}He derived MRT as given in Dörr et al. (1987) are still valid in this ground water system.

To estimate MRT in the ground water system, a similar depth distribution is assumed to be valid in wells BOC1 and BOC4. All isotope results and especially the

Table 6.2.4.1. Isotope contents of ground waters from wells BOC1 and BOC4, ^{14}C- and ^{3}H contents are remeasured for comparison with former ^{14}C- and ^{3}H results. (Dörr et al. 1987). Given "errors" are the 2σ standard deviations

	Depth (m)	^{3}H (1986) (TU)	^{3}H (1985)[a] (TU)	^{14}C (1986) (pmc)	^{14}C (1985)[a] (pmc)	^{39}Ar (1986) (%mod.)
BOC1	13 + 15	34.2 ± 3.0	30	–		–
	25 + 26	14.6 ± 2.1	13	–		–
	36	8.7 ± 1.6	5	58 ± 5	60	84 ± 12
BOC4	8.5 + 10.5	70.0 ± 5.0	62–71	–		–
	29.5 + 32.5	12.2 ± 1.9	16–18	–		–
	48	≤1	≤0.23	43 ± 3	44	56 ± 10

[a]From Dörr et al. pers. commun.

^{39}Ar content indicate that the ground water from well BOC4 in 48 m has a higher MRT than the water from well BOC1 in 36 m. This is a general confirmation of the $^{3}H/^{3}He$-based depth distribution stated from Dörr et al. (1987). When comparing the ^{3}H contents in ground waters from both wells, the ground water system can be approximated to be a mixing system. In a first approach a two-component mixing is considered.

Following Dörr et al. (1987), a MRT of 20 years can be assumed for ground waters from well BOC1 at depths from 10 to 30 m. Taking into account the radioactive decay, an ^{39}Ar activity of 94%-mod. can be calculated for this water. This is used as the activity of the younger component of water which is mixing with an older component to yield the isotopic composition of the water found in well BOC1 in a depth of 36 m (Fig. 6.2.4.2). For the older component, a ^{3}H activity below the ^{3}H detection limit is assumed which is in agreement with the ^{3}H content measured in the deepest ground water in BOC4. These assumptions for the two mixing components in BOC1 and the isotope contents measured in ground water at a depth of 36 m yield the conclusion, that 70% of an old component with an ^{39}Ar activity of 80%-mod. is admixed. Its ^{39}Ar-MRT (depending on the standard deviation) should be between 20 and 130 years.

For the water at BOC4 a similar mixing model is assumed (Fig. 6.2.4.2): at most 3% of a BOC1 water found in 36 m is admixed as the "young" component to a component with old water to yield the isotope contents measured in ground water from BOC4 at a depth of 48 m. The ^{39}Ar-MRT, depending on the percentage of admixed components and on the flow model assumed, then varies between 150 and 300 years.

Another approach would be to assume vertical advection and diffusion in addition to a horizontal ground water flow. This would explain the depth-dependent decrease of ^{3}H activities in the two profiles. Five to ten TU of ^{3}H activity would be

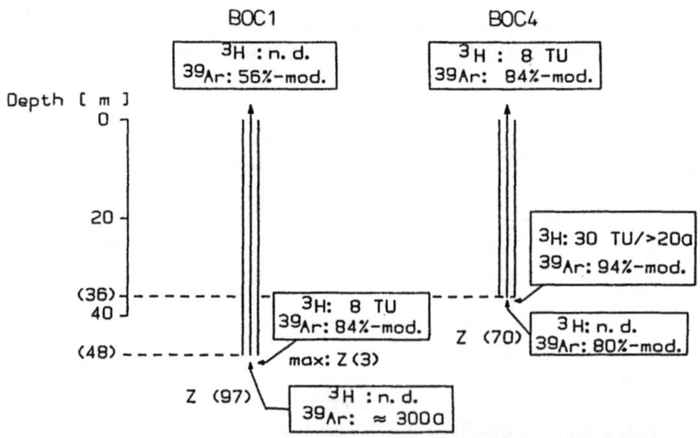

Fig. 6.2.4.2. Schematic diagram of the ground water mixing system in ground waters from wells BOC1 and BOC4. Mixing components are indicated as boxes, where also ^{3}H- and ^{39}Ar isotope contents are given. Admixed amounts of single components are indicated with *Z*, their weight percentage is assigned in *brackets*

allowed to reach a depth of 36 m, whereas in 48 m less than 1% of recent water can be found. A method suited to control this assumption would be the measurement of the ^{85}Kr profiles. From the measured ^{39}Ar activity it must be concluded that most deep water inflow is horizontal; its ^{39}Ar-based MRT would not change considerably that conclusion, which is considered in the mixing model mentioned above. These two models are apparently both possible and a decision which one is closer to reality could only be made if ^{85}Kr- and more ^{39}Ar results would be available. Due to tight tubes of the multi-level wells, the sampling of ground water from different aquifer depths for ^{39}Ar- and ^{85}Kr measurements was not possible.

Summarizing ^{39}Ar- and ^{3}H contents measured in ground waters from wells BOC1 and BOC4 in the Bocholt aquifer system it can be stated:

- Deep-lying ground waters of the well BOC1 (36 m) show a model-dependent MRT between 20 and 130 years. The depth profile of ^{3}H and ^{3}He results (Dörr et al. 1987) seems to allow the conclusion that deep ground waters are influenced by ground waters with MRT of about 100 years rather than close to 20 years.
- The MRT in the deepest ground waters of well BOC4 are between 150 and 300 years. Allowing mixing processes of ground waters with different, model-dependent MRT, the ^{39}Ar data support those "ages" which were obtained from hydrodynamic calculations (Leuchs 1988) and also confirm "age" estimations which were deduced from ^{14}C data and ^{4}He excess considerations (Dörr et al. 1987).

Together with ^{39}Ar determinations in Bocholt ground waters also ^{37}Ar contents have been measured. ^{37}Ar has a half-life of 35 days and can be produced in the underground by the reaction ^{40}Ca(n,α) ^{37}Ar in aquifers, where α-decaying U and Th can supply the necessary neutrons via (α,n)-reactions in the aquifer material. The ^{37}Ar/Ar measurements showed unexpectedly high values of about 3.3 mBq ^{37}Ar/l. Although U- and Th contents are not elevated in the aquifer material and no extreme ^{222}Rn contents have been measured in the ground water (Dörr pers. commun.), the measured ^{37}Ar content must be due to in situ ^{37}Ar production. Because ^{39}Ar can be produced in situ following ^{39}K(n,p)^{39}Ar one cannot exclude that a portion of ^{39}Ar measured in the Bocholt ground waters is due to subsurface production. Correction of the measured ^{39}Ar contents for this effect would raise MRT. Studies dealing with effects of enhanced ^{37}Ar escape from place of production through recoil and diffusion and with the effect of neutron-induced subsurface production of ^{37}Ar and ^{39}Ar are in progress.

6.2.4.5 Isotope Results in the Segeberger Forst Ground Waters

In ground waters from the sandy and gravelly Segeberger Forst aquifer no ^{39}Ar samples were taken because lower MRT were expected for those ground waters which are not covered by the ^{39}Ar dating range.

Samples for isotope measurements (2 m^{3} water per ^{85}Kr sample) were collected in well SEG5 (^{3}H depth profile), in well SEG7 at depths of 10 and 32 m and in well SEG4 at 20 m, each sampled for ^{3}H-, ^{3}He-, ^{85}Kr-, ^{222}Rn-, and ^{224,228}Ra contents.

With the exception of the ^{222}Rn- and $^{224,228}Ra$ contents, which showed no elevated values, isotope results analyzed in ground waters from the Segeberger Forst aquifer are given in Table 6.2.4.2. The nature of the sampling points (multi-level wells and fully screened wells) leads to the assumption that a certain vertical mixing of water from different depths cannot be excluded.

Because the tubes of multi-level well SEG6 were clogged, the remeasurement of the ^{3}H depth profile in this aquifer was done in well SEG5. The plausible assumption must be made that the depth profiles of the isotopes in the ground water are the same and have undergone the same evolution in time, since both wells lie only 4 m apart from each other. Figure 6.2.4.3 gives ^{3}H depth profiles of ^{3}H series

Table 6.2.4.2. Results of ^{3}H, ^{3}He, and ^{85}Kr isotope measurements in wells SEG4, SEG5, and SEG7 in Segeberger Forst ground waters

Well/depth (m)	Sampling date	^{3}H content (TU)	^{3}He content (TU)	^{85}Kr content (Bq ^{85}Kr/l Kr)
SEG4/20	14.04.88	32.5 ± 2.2	15 ± 6	400 ± 22
SEG 5/5.5	12.04.88	29.0 ± 2.0		
7.5		32.7 ± 2.4		
8.5		40.1 ± 2.7		
9.5		29.3 ± 2.0		
10.5		25.1 ± 1.7		
12.5		25.6 ± 1.8		
14.5		30.5 ± 2.1		
15.5		25.5 ± 1.9		
16.5		34.8 ± 2.4		
17.5		33.3 ± 2.2		
19.5		36.5 ± 2.5		
20.5		36.6 ± 2.5		
22.5		42.6 ± 2.9		
SEG7/10	13.04.88	31.1 ± 1.5	14 ± 2	770 ± 70
SEG7/32	13.04.88	32.6 ± 1.6	20 ± 2	420 ± 20

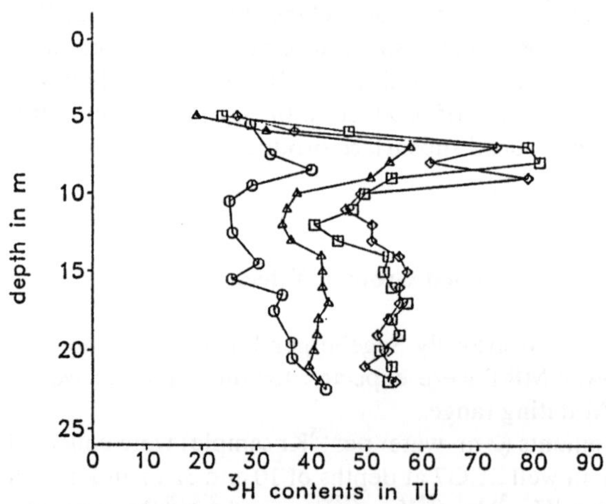

Fig. 6.2.4.3. ^{3}H depth profiles measured in 1983–1984 and 5 years later. □ and ◊ indicate the 1983 and 1984 ^{3}H measurements (Eichinger pers. commun.),○ mark the ^{3}H profile in ground water from well SEG5, measured in 1988; Δ, being the mean of 1983/1984 ^{3}H contents, is corrected with respect to ^{3}H decay for a sampling and measurement in 1988 assuming piston flow conditions (no mixing of ground waters with different

in 1983–1984 (Eichinger pers. commun.) and in 1988. Because ^{3}H contents are nearly constant below a depth of 10 m, they indicate ^{3}H-based ground water MRT at or above 4 years. For a comparison with the ^{3}H profile measured in well SEG5 in 1988 and for a discussion of the new results, the values of the ^{3}H profile measured in 1983–1984 were corrected for ^{3}H decay until 1988. From Fig. 6.2.4.3 it can be stated that both curves show the same ^{3}H peak followed by a narrow "minimum", and that from there the difference in the ^{3}H contents between both curves decreases with depth and vanishes at about 20 m.

Several conclusions can be drawn from this:

- No remarkable vertical flow processes of the ground water take place. Ground water flow is nearly horizontal; this can be derived from the depth of the ^{3}H peak, which remained quite constant within 5 years.
- No measurable ground water mixing processes take place in deepest aquifer layers. The ^{3}H contents found in well SEG5 in depths around 20 m support the assumption that the only reason for the depletion of ^{3}H is its decay. This can be concluded from coincident ^{3}H contents measured in 1988 with these 1983–1984 data, which are corrected only for ^{3}H decay.
- In general, the ^{3}H contents measured in 1988 are lower than the 1983–1984 values corrected for decay. This shows, that "ageing" or "admixing" of ground waters takes place in depths until around 20 m, which is more than can be described with the piston flow model. Admixture of ground water with lower ^{3}H contents is probable. A numerical description of isotope transport due to convection and dispersion verifies their distribution within the aquifer and even considers inhomogeneities of the aquifer. These calculations could represent ground water flow and fit the measured isotope contents in detail (Bugner pers. commun.).

^{85}Kr contents measured in samples taken from wells SEG7 and SEG4 are shown as bars in Fig. 6.2.4.4. For comparison, model-dependent ^{85}Kr values are also given which were calculated from ^{85}Kr input data (Fig. 6.2.4.1). The curves given in Fig. 6.2.4.4 imply the following assumptions:

1. The ground water flow system can be approximated with piston flow conditions and the ^{85}Kr input into the ground water system is undelayed.
2. Same as (1), yet the ground water flow system is described assuming a mixture of ground waters with an exponential age distribution.
3. As an example, the effect on ^{85}Kr data when the ^{85}Kr signal needs time to reach the ground water table is shown in curve c). There, the expected ^{85}Kr content in the ground water is calculated for the case that the atmospheric ^{85}Kr input into the ground water shows a 3 years' delay due to passing the soil before reaching the ground water table. As a first approximation, this situation is described with a ^{85}Kr input until 1985 and its correction to a sampling in 1988 only for ^{85}Kr decay. The ground water flow is described with an exponential mixing model (same as 2).

From Fig. 6.2.4.4 and following assumption (1), the ^{85}Kr content measured in ground water from well SEG7 at 10 m would show a MRT below 2 years (region I). Assuming exponential-mixing model flow, the MRT would be only slightly higher.

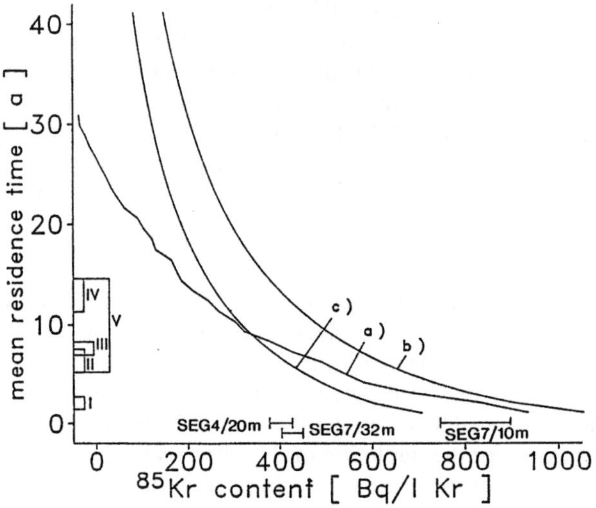

Fig. 6.2.4.4. Model-dependent ^{85}Kr contents and corresponding MRT of ground water. Measured ^{85}Kr contents are given as *bars* to be compared with ^{85}Kr-based model ages, which are given as ranges *I* to *V* and depend on the assumptions (1)-(3)

With respect to the comparatively high standard deviation of this ^{85}Kr determination, a MRT of up to 3 years can be derived, which does not fit with the hydrogeologic concept.

^{85}Kr contents in ground waters from wells SEG7 (32 m) and SEG4 (20 m) show MRT of 7 to 8 years following piston flow conditions, and 11 to 14 years when exponential mixing flow conditions are considered (region III curve a or IV curve b).

It can be seen from the ^{85}Kr content measured in the 10-m ground water sample from well SEG7 (Fig. 6.2.4.4) that the MRT of this ground water equals about that time which is assumed to be necessary for the ^{85}Kr input signal to pass the unsaturated zone. This effect tends to enlarge the model-dependent total "spread" in ground water model ages (II, III, and IV) and can overlay differences induced from considered, different ground water flow models (V).

Accepting all models and without further isotope measurements, ground water MRT for SEG7/32 m and SEG4/20 m can only be given as a range between 5 (region II, curve c) and 14 years. But, as is shown in the following, a few ^{3}He results will add valuable information.

^{3}He samples were measured in addition to ^{85}Kr. The ^{3}He results from Table 6.2.4.2 are shown in Fig. 6.2.4.5, where the ^{3}He contents, corrected to the He solubility and "excess air" contents, are given as TU (Sect. 3.2.3, this Vol.). The solid line indicates calculated values for a sampling in 1988 under the assumption of exponential flow conditions in this aquifer (Fig. 6.2.4.2).

Dependent on the flow model assumed, ground water MRT can be derived from this figure to be 4 to 5 years for well SEG7/10 m and 4 to 6 years for SEG4/20 m. When comparing with the ^{3}He-based MRT of ground water from well SEG7/32 m, which is 5 to 6 years, lower MRT of ground water must be assumed than were deduced from the ^{85}Kr measurements. This is understood to be an indication for the assumption (3), and ^{85}Kr-based model ages in the range II and III (Fig. 6.2.4.4).

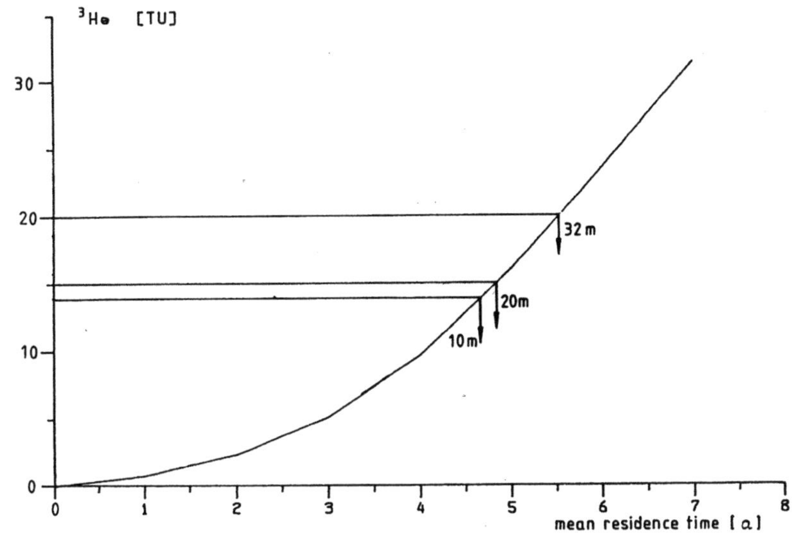

Fig. 6.2.4.5. ^{3}He contents given as TU equivalents and corresponding MRT of ground water. Exponential flow conditions are assumed. The diagram is valid for sampling in 1988 and based on "whole year ground water recharge". (Weiss 1986)

A study for quantifying a possible ^{3}He ageing effect during soil passage is in progress.

As a summary, noble gas isotope- and ^{3}H measurements in ground waters from Segeberger Forst indicate that:

- the ground waters from wells SEG4 and SEG7, below 10 m, show increasing age with increasing aquifer depth. This is reflected in the measured ^{85}Kr, ^{3}He, and ^{3}H contents.
- ground water flow is mainly horizontal and must be described with flow models which take into account effects of ground water mixing. The most simple model is exponential mixing. In contrast to this, ^{3}H decay seems to be the only sink for ^{3}H in the deepest parts of wells SEG5 and SEG6 (sampled in 1983–1984 and 1988).
- the ground waters seem to have MRT which are below 5 years at a depth of about 10 m (based on ^{85}Kr, ^{3}H, and ^{3}He contents). MRT increase again to more than 5 years when ground waters from depths below 10 m are sampled. This would fit the measured flow velocity v_a of 0.25 m/d at maximum in 7 to 12 m depths and between 0.01 and 0.1 m/d below 12 m (Schröter 1983, 1984).
- ground water dating with ^{85}Kr can be affected when ageing of the ^{85}Kr input signal is taken into account before it reaches the ground water table. This effect can cover differences which are induced when assuming different ground water flow models. This seems to be probable in the Segeberger Forst aquifer system, where no contradicting ^{85}Kr-, ^{3}He-, and ^{3}H-based ages are found, if, as an example, an ageing of the ^{85}Kr input is assumed.

6.2.5 Modelling of Mass Balance and of Microbial Transformations in the Fuhrberger Feld Sandy Aquifer

O. Strebel, J. Böttcher and W.H.M. Duynisveld

6.2.5.1 Concept of Model Approaches, Methods of Investigation

The quality of ground water extracted by a production well is the result of the solute input into ground water in the recharge area by non-point sources as well as of complex interactions within the aquifer. This is particularly the case if microbial transformations within the aquifer are involved. Thus, it is difficult to understand quantitatively the reason for increasing concentrations of particular solutes in the pumped ground water or to predict the future trend of these solute concentrations. The input-output balance of solutes in the ground water for a hydraulically defined recharge area (mass balance model) can be used to quantify the complex interactions including microbial transformations (e.g. identify sinks or sources within the defined ground water system) at a resolution in time in the order of several decades. Deterministic simulation models allow insight more detailed in time and space and describe quantitatively the processes of solute transport and transformation in the ground water. The results presented here are based on intensive investigations on solute input into the ground water (Strebel et al. 1984; Böttcher and Strebel 1985b, 1987, 1988a,b,c, 1989, 1990; Strebel and Böttcher 1989) and on solute concentrations of the ground water as a function of depth (Böttcher et al. 1985), carried out in the recharge area of the production well 1/Fuhrberg waterworks (Sect. 4.1.3.1, this Vol.).

Figure 6.2.5.1 shows the ground water recharge area of about 1000 ha of production well 1 defined according to detailed piezometric level contour maps. Detailed information on soils (mainly sandy podzol and gleyic-podzol soils) and on land use (arable land, coniferous forest, grassland) of this area were available. The mean long term ground water recharge per year, which depends on soil, land use and depth of ground water table, was determined by using a calibrated and validated soil water simulation model (Wessolek et al. 1985; Renger et al. 1986). To determine long term mean soil- and land use-specific solute concentrations of the ground water recharge, a hand auger slitprobe technique was applied to sample the upper decimetre of ground water (Strebel et al. 1984). At 43 sites with arable land (different crops), coniferous forest and grassland, samples of the uppermost ground water were taken at six sampling dates during the years 1983–1986 along eight sampling strips (black bars in Fig. 6.2.5.1). From these measurements mean solute concentrations representative for different land use units were calculated (Strebel and Böttcher 1989). Multiplying the yearly ground water recharge with its mean solute concentration gives the yearly solute input into the ground water (in kg/(ha•a)). Data from monitoring seven multi-level wells (see Fig. 6.2.5.1) allowed estimations of changes over the years of the solute concentration in the aquifer; to calculate solute influx over the southern boundary, data from the southern multi-level well were used. Pumping rates and solute concentrations of pumped ground water at production well 1 have been known since 1959.

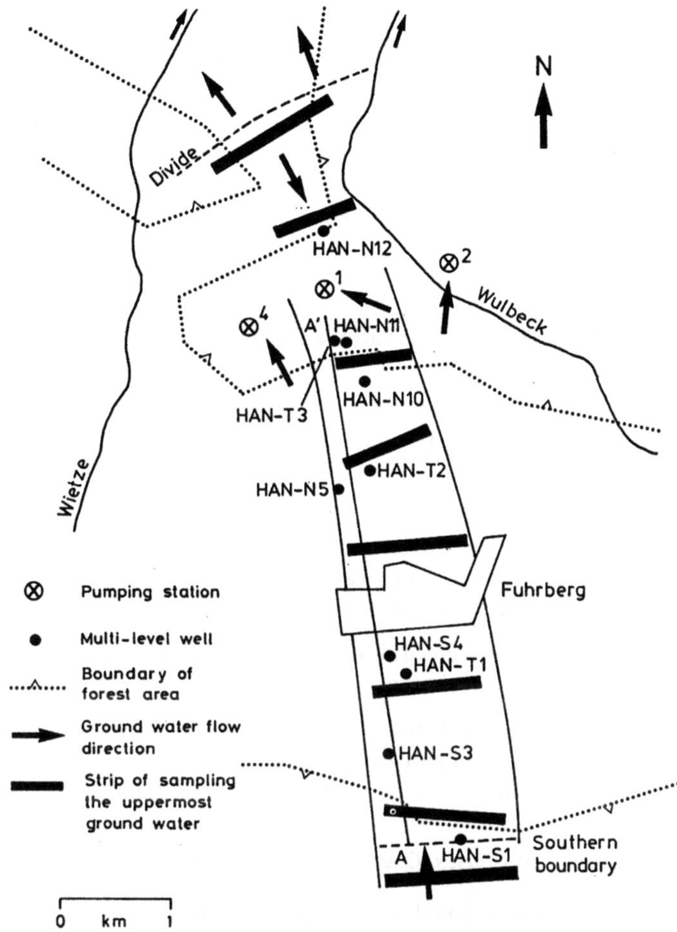

Fig. 6.2.5.1. Recharge area of production well 1/Fuhrberg waterworks, A-A' denotes the position of the cross-sections shown in Figs. 4.1.3.1, 6.2.5.3 and 6.2.5.4. (After Strebel et al. 1985)

6.2.5.2 Mass Balance Model

Methodical Approach

For a hydraulically defined ground water recharge area, without additional ground water inflow into the system and an aquifer consisting of non-reactive sediments, the following general mass balance equation for a specific solute holds with respect to recharge area and time period of balance:

$$I_g \pm S_g = O_g, \qquad \text{(kg)} \tag{6.2.5.1}$$

with I_g = solute input into ground water, O_g = solute output with pumped ground water, S_g = storage changes of solute in the aquifer system.

Provided reliable data for the terms in Eq. (6.2.5.1) are available, a balance deficit would indicate a source term and balance surplus a sink term, due to solute transformations in the ground water.

Furthermore, with reliable data of soil- and landuse-specific ground water recharge and solute input into ground water for a hydraulically defined ground water system, a mean solute concentration of the ground water and of the pumped well water, respectively, can be assessed, using:

$$\bar{c} = \frac{I_t}{R_t} \quad (\text{mg/l}) \qquad (6.2.5.2)$$

or for a recharge area with various soil and land-use units:

$$\bar{c} = \frac{\sum_{n=1}^{i} I_n}{\sum_{n=1}^{i} R_n} \quad (\text{mg/l}) \qquad (6.2.5.3)$$

with $\bar{c}$ = mean solute concentration of ground water or of pumped well water (mg/l), I_t = solute input into ground water per total recharge area (mg), R_t = ground water recharge per total recharge area (l),

I_n = solute input into ground water } per total area of soil and land-use unit n
R_n = ground water recharge.

The Eqs. (6.2.5.2) and (6.2.5.3) assume steady-state conditions in time with respect to solute input and ground water recharge (long-term mean values are used), solute transformations in the aquifer are not included for the time being.

Identification of Causes, Quantification of Transformations

Table 6.2.5.1 gives some land use-specific data of mean solute concentrations of the yearly ground water recharge in our recharge area. The balances are calculated for an area of about 1000 ha according to Eq. (6.2.5.1) and also include solute inflow over the southern boundary (Fig. 6.2.5.1). The results for chloride, nitrate and sulfate are listed in Table 6.2.5.2. The chloride balance is used as a test, source or sink terms can be excluded for chloride as a conservative solute. It gives a satisfying balance between input and output considering the unavoidable imperfections. The nitrate balance, however, is characterized by a significant loss in the aquifer, since there is a low output in spite of a considerable input. The sulfate balance indicates significant sulfate gains.

The reason for this imbalance of nitrate and sulfate is bacterial denitrification in ground water (Sect. 4.1.3.3, this Vol.). The balance deficit of nitrate is nearly covered if the total nitrate input with the ground water recharge (2937 t in Table 6.2.5.2) is denitrified; thereby, according to the stoichiometry of Eq. (4.1.3.1) 3231 t of sulfate are formed. However, in the sulfate balance, even if the total surplus is diminished by the amount due to sulfate formation, an unexplained sulfate surplus of 4594 t remains. In this context it must be considered that in the recharge area of well 1 240 ha (= about 25% of the area defined in Fig. 6.2.5.1) of

Table 6.2.5.1. Land use-specific mean concentrations of the yearly ground water recharge of chloride, nitrate and sulfate in the recharge area of production well 1. (After Strebel and Böttcher 1989)

Land use	Concentration mg/l		
	Chloride	Nitrate	Sulfate
Coniferous forest	18	3	107
Alder tree forest	13	33	100
Grassland	10	25	70
Arable land	24	131	64

Table 6.2.5.2. Input-output balance of chloride, nitrate and sulfate of the ground water of the recharge area of production well I/Fuhrberg waterworks in 1965–1982 (in metric tons). (After Strebel et al. 1985)

		Chloride	Nitrate	Sulfate
(a)	Solute input: Ground water recharge	787	2937	2146
	Ground water inflow southern boundary	2398	48	3357
	Sum input	3185	2985	5503
(b)	Change of solute content of the aquifer over balance period	≈ 0	≈ 0	≈ 572
(c)	Solute output: Ground water extracted at well 1	3266	≈ 80	12756
(d)	Balance: (a)-(b)-(c) =	*–81*	*2905*	*–7825*
(e)	Denitrification of nitrate input with ground water recharge		–2937	3231
(f)	Solute input from grassland areas converted into arable land		≈ 4000	≈ 700
(g)	Denitrification of nitrate input (f)		≈ –4000	≈ 4400
(h)	Final balance: (a+f)+(e+g)–(b)–(c) =	*–81*	*–32*	*506*

permanent grassland has been converted into arable land in the last three decades. The position of these plots and the date of ploughing were detected by comparing successive editions of topographic maps. The organic nitrogen content in the A-horizon on permanent grassland is distinctly higher than on arable soils. A detailed study on these sandy soils (Strebel et al. 1988) showed a decrease of the organic nitrogen content of about 10 t N/ha on permanent grassland to about 5 t N/ha on arable land; a new arable land-specific steady state of organic nitrogen content is reached in about 4 years after ploughing. About 75–80% of the organic nitrogen thus mineralized can be assumed to cause a very high nitrate input into ground water for a limited time period in the areas with grassland ploughing. Therefore, the conversion of 240 ha permanent grassland into arable land induces an additional nitrate input pulse into ground water of about 4000 t. Denitrified quantitatively, this nitrate input is equivalent to 4400 t of sulfate. If this amount is included in the sulfate balance (Table 6.2.5.2), a nearly equalized balance results. The remaining sulfate surplus can be considered as indication of some sulfate losses. However, the sulfate reduction proceeds only at a slow rate (half-life about 76–100 years, see Sect. 4.1.3.4, this Vol.; Böttcher et al. 1989). Therefore, sulfate losses by sulfate reduction are of only minor importance and not included in the balance.

Prediction of Solute Concentration of the Pumped Groundwater of Well 1

Figure 6.2.5.2 shows the measured concentrations of sulfate and nitrate in the pumped ground water of well 1 as a function of time. The sulfate concentration stepwise increases between 1967 and 1981, caused by two main periods of grassland ploughing, after which follows a continuous moderate decline. Model calculations, using Eq. (6.2.5.3) enlarged to include solute inflow over the southern boundary (Fig. 6.2.5.1) and assuming complete denitrification of the nitrate input into the ground water, indicate that the sulfate should further decrease to about 120 mg/l in the long term (Strebel and Böttcher 1985) (Fig. 6.2.5.2). Further model calculations with Eq. (6.2.5.3) indicate that, under the condition of present land use distribution in the total recharge area but without denitrification losses in the ground water, concentrations of about 66 mg sulfate/l and 58 mg nitrate/l should be expected in the long run. For further examples see Böttcher and Strebel (1985a) and Strebel and Böttcher (1987).

6.2.5.3 Kinetics of Denitrification and Sulfate Reduction in Ground Water

Modelling of transformations in deterministic simulation models requires data on the kinetics of the reactions involved under the actual conditions in the aquifer. The kinetics can be derived provided the vertical cross-sectional plane flow net (Fig. 4.1.3.1), solute concentrations as a function of ground water depth (Fig. 4.1.3.2) at various positions along the flow direction of ground water and lines of equal ground water ages (Fig. 6.2.5.3) are known. The calculated ground water ages were controlled by tritium measurements in the years 1983/84 (Böttcher et al. 1985). The tritium concentration as a function of depth (Fig. 6.2.5.4) shows a maximum in the

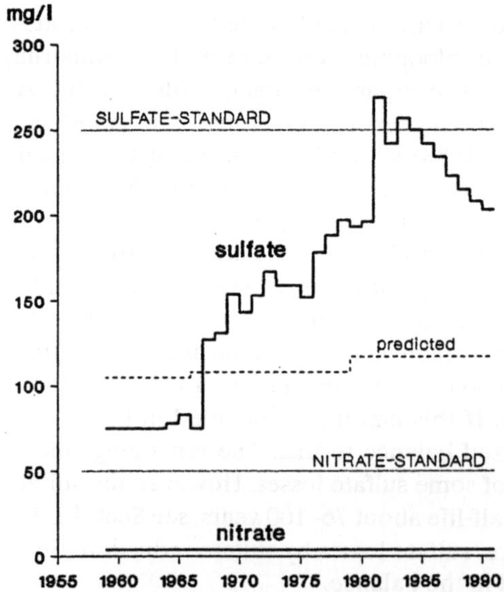

Fig. 6.2.5.2. Measured concentrations of nitrate and sulfate of the extracted ground water at well 1 as a function of time: *solid lines.* Predicted sulfate concentration without additional nitrate input due to ploughing of permanent grassland (mineralisation of organic soil N): *broken line.* (After Kölle et al. 1985; Strebel and Böttcher 1985)

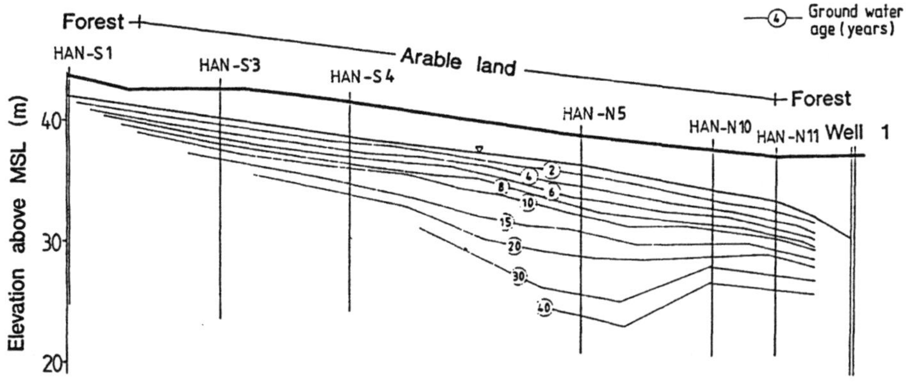

Fig. 6.2.5.3. Calculated ground water isochrones of the investigated vertical cross-section. (Böttcher et al. 1985)

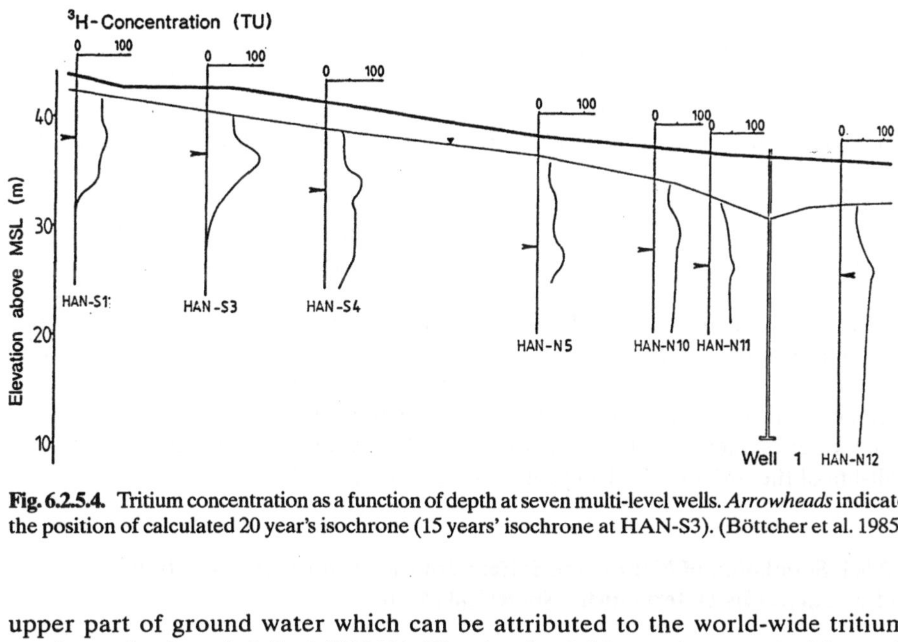

Fig. 6.2.5.4. Tritium concentration as a function of depth at seven multi-level wells. *Arrowheads* indicate the position of calculated 20 year's isochrone (15 years' isochrone at HAN-S3). (Böttcher et al. 1985)

upper part of ground water which can be attributed to the world-wide tritium maximum in the rainfall of 1963/64. The depth of this maximum coincides well with the depth of the calculated 20 year's isochrone (at multi-level well HAN-S3, where tritium was already measured in 1978, with the 15 years' isochrone). Thus, the calculated cross-sectional flow net and the calculated ground water isochrones are correct descriptions of the hydraulic conditions.

A thorough analysis of the depth functions of nitrate in ground water, based on the cross-sectional flow net, clearly confirmed denitrification at several multi-level wells (Sect. 4.1.3.3). The observed decrease of nitrate concentration due to denitrification can be attached to certain ground water ages using the calculated

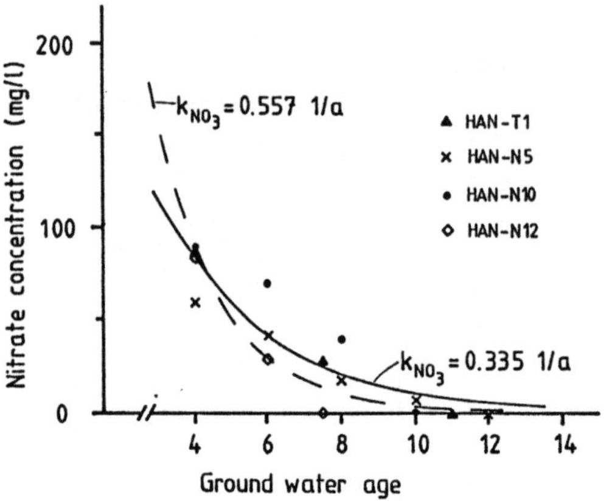

Fig. 6.2.5.5. Relation between nitrate concentration and age of ground water in the denitrification zone of the aquifer. *Lines* describe first-order reactions with different values of reaction constant *k*. (Böttcher et al. 1989)

ground water isochrones (Fig. 6.2.5.5). This procedure involves some imperfections, due to the combined effect of longitudinal dispersion and of variability in time of nitrate input concentrations at the recharge area of specific flow lines. However, the influence of the very weak transverse dispersion in vertical direction (Frind et al. 1990) may be neglected (nitrate profiles show steep vertical concentration gradients). If first-order reaction kinetics is adjusted to the experimental data, then rate constants k_{NO_3} between 0.335 and 0.557 a^{-1} are determined, the half-life ranges between 2.1 and 1.2 years (Böttcher et al. 1989).

Due to less clear data of the ground water age for the greater depths of the sulfate reduction zone (Sect. 4.1.3.4, this Vol.) and problems regarding the original sulfate concentration, the relation between sulfate concentration and ground water age is somewhat uncertain. Therefore, it is more difficult to apply the previous methodical approach for the determination of the kinetics of sulfate reduction. Nevertheless, the evaluation of field data supported by sulfate isotope analysis (Strebel et al. 1990) indicates that sulfate reduction can also be described by first-order kinetics. However, this reaction proceeds at a rate significantly slower, with a half-life of the order of 76–100 years (Böttcher et al. 1989).

6.2.5.4 Simulation of Nitrate and Sulfate Transport and Transformation in the Aquifer by Deterministic Numerical Models

The numerical models should describe quantitatively the coupled transport of nitrate and sulfate, with bacterial denitrification, production of sulfate and depletion of reserves of reduced sulfur compounds in the upper denitrification zone (Sect. 4.1.3.3), as well as with bacterial desulfurication and production of sulfides in the lower sulfate reduction zone (Sect. 4.1.3.4) of the ground water. The field data of solute input into the ground water, of solute concentrations as a function of ground water depth, of kinetics of the transformation reactions and of hydraulic conditions are the essential prerequisites for development, calibration and validation of the models.

Within the funding of this DFG Priority Programme only a rather short time was available for modelling work. The intensive model studies started were in full progress only later. Therefore, this section is confined to a short description of the methodical approach used. Depending on the main objective, three lines of modelling are followed:

Stream Tube Model

The 1D simulation considers the transport and transformation in specific stream tubes along ground water flow lines. For the numerical solution of the convection-dispersion equation a finite-difference method is used; longitudinal dispersion is included, but not the very weak transverse dispersion in vertical direction. Objective of the stream tube model is to test the coupling of transformation kinetics (nitrate reduction/depletion of reserves of reduced sulfur compounds/sulfate formation, sulfate reduction/sulfide formation), the overall depletion of reduced sulfur reserves and the solute concentration at the stream tube outlets as a function of time (e.g. "break through" of nitrate) for developing a long term prognosis. Results are published in Böttcher et al. (1989, 1990b).

Vertical Cross-Sectional Plane Model

The objective of this 2D simulation model is to obtain a better insight into the depth-specific dynamic processes of nitrate and sulfate migration and transformation in the aquifer (e.g. extent of zones of denitrification, of desulfurication, and of sulfide depletion). The boundary between denitrification and desulfurication zone varies as a result of the spatial differentiation of the nitrate input, and it migrates slowly downstream as the reserves of reduced sulfur compounds become gradually depleted. The model uses a finite-element solution which is formulated in principal direction coordinates and combined with a new time integration scheme. Results obtained to date indicate that the major characteristics of the measured solute distributions in the ground water can be well matched by the model (Frind et al. 1989, 1990, 1991).

2D Areal Plane Model

The objective of this 2D-horizontal model is to analyze and predict changes of solute concentrations in the extracted ground water of well 1. For numerical solution a finite-element method is used. First results confirm that the major stepwise increases in sulfate concentration of well 1 (Fig. 6.2.5.2) are mainly due to the effects of ploughing of permanent grassland in the past during two consecutive main periods of grassland conversion (Duynisveld et al. 1989).

6.3 Description of Geochemical Environments with Thermodynamic Equilibrium Models

M. Kölling and H.D. Schulz

6.3.1 Approach

This section approaches mineral/water equilibria from the hydrochemical side, keeping in mind that in the unsaturated zone the aqueous phase may be less than 1% wt. of the system. The evaluation is based on data from three sites (Table 1.1, Fig. 1.1): Tunisian playa waters (Kössl 1988; Sect. 3.3.1, this Vol.), ground water samples of the Seffent springs catchment area (Sect. 3.2.5, this Vol.) and seepage water samples from the Bocholt aquifer system (Leuchs 1988; Sect. 4.1.1, this Vol.).

6.3.2 Geochemical Models and Limiting Reactions

Thermodynamic equilibrium models like WATEQ2 (Ball et al. 1979) or PHREEQE (Parkhurst et al. 1980) mainly deal with complex equilibria and mineral equilibria. In these programmes, the species distribution is calculated on the base of thermodynamic constants for about 500 important inorganic complexes. Saturation indices (SI) for about 300 minerals are calculated which are defined as logarithms of the ratio of ion activity product (IAP) vs. the thermodynamic solubility product of the solid species K_s ($SI=\log(IAP/K_s)$). Thermodynamic equilibrium models can be linked to hydraulic models for the simulation of the interactive transport of substances in aqueous systems (Schulz and Baumann 1985).

Assuming ideal dissolution/precipitation behaviour (instant establishment of mineral equilibria) of all minerals in contact with a solution, calculations with a geochemical model like WATEQ2 would yield SI of 0 for a maximum of n-1 minerals in a solution with n ions (according to the Gibbs phase rule) and negative SI for all the other minerals that can be formed theoretically from these ions. The primary minerals, when brought into contact with the solution, dissolve until their equilibrium ion activity is reached, or – if the primary mineral is not the thermodynamically most stable phase – dissolve completely at length and the stable phase (indicated by a greater saturation index) is precipitated till equilibrium is established.

However, SI of natural waters in addition to undersaturated minerals usually are positive (supersaturation) or negative for many phases, but close to equilibrium (SI=0) for very few minerals. This may result from erroneous calculations or from the oversimplifications in the basic assumptions. The effects of the neglection of kinetic and transport phenomena will be discussed below.

6.3.2.1 Kinetics

The time necessary for a solution to equilibrate with a mineral may vary from seconds [e.g. amorphous $Fe(OH)_3$] to years (e.g. hematite) and depends besides on temperature on the mineral surface area and the grade of super-/undersaturation (Stumm and Morgan 1981).

Indicators for concentration control caused by reaction kinetics are saturation values approaching zero along a profile (Table 6.3.1). Fig. 6.3.1 shows the fast

Table 6.3.1. Water analyses along a profile in the Seffent springs catchment area (30.03.84)

		AAC1	AAC2	AAC3	AAC4	AAC6	AAC7
T	(°C)	5.0	9.8	10.1	9.9	10.1	10.1
pH		7.39	6.85	7.55	7.29	7.12	7.18
E_H	(mV)	192	172	258.	141	260.	275.
O_2	(mg/l)	10.5	7.5	7.1	4.5	5.1	5.0
Ca^{2+}	(mg/l)	50.	140.	88.	145.	137.	149.
Mg^{2+}	(mg/l)	9.2	9.6	3.2	8.7	8.8	6.9
Na^+	(mg/l)	9.3	10.6	15.5	6.6	5.3	8.8
K^+	(mg/l)	16.1	2.2	9.2	1.8	1.4	1.8
NH_4^+	(mg/l)	0.85	0.26	0.27	0.34	0.28	0.29
Fe^{2+3+}	(mg/l)	0.14	0.15	<0.01	0.02	<0.01	<0.01
Mn^{2+}	(mg/l)	0.13	0.06	<0.01	0.07	<0.01	<0.01
Al	(mg/l)	0.063	0.042	0.003	0.012	0.009	0.008
Si	(mg/l)	4.8	4.0	6.9	10.5	11.5	11.7
HCO_3^-	(mg/l)	75.0	232.	215.	343.	329.	282.
SO_4^{2-}	(mg/l)	56.0	168.	40.	81.	70.	121.
Cl^-	(mg/l)	19.0	30.	31.	27.	23.	30.
NO_3^-	(mg/l)	65.0	5.6	15.	11.	19.	27.
NO_2^-	(mg/l)	0.21	0.11	0.03	0.38	0.03	0.02
F^-	(mg/l)	0.09	0.19	0.57	0.08	0.09	0.08

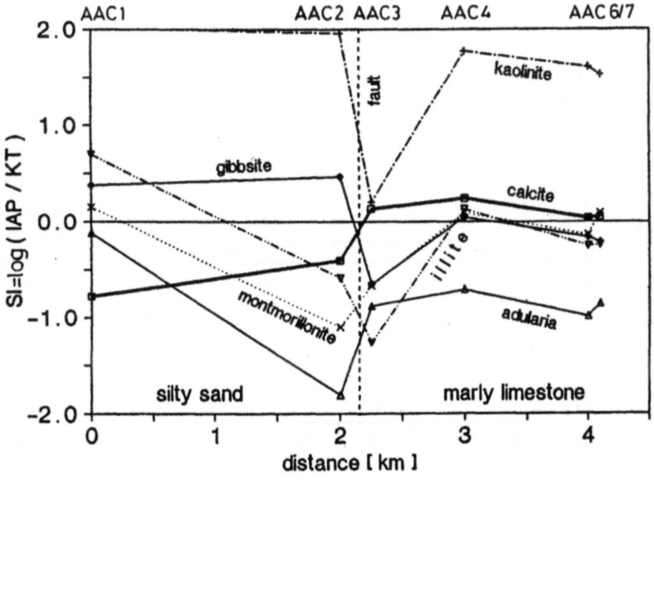

Fig. 6.3.1. SI values for different minerals along a profile in the Seffent springs catchment area (30.03.84)

equilibration of the water with respect to calcite as the aquifer mineralogy changes from silty fine sand to marly limestone at a fault between AAC2 and AAC3 in the Seffent springs catchment area. Saturation indices for calcite change from undersaturation (SI= –0.4) in the sandy aquifer to a slight supersaturation within a 200 m distance between the fault and well AAC3 (SI= +0.2). The calculated supersaturation may be caused by either analytical errors or by incongruent dissolution of the natural carbonates (Fig. 6.3.2). The SI development of the phyllosilicates along the flow path (Fig. 6.3.1) indicates significantly slower reactions. The first SI values close to 0 were calculated for samples from well AAC4 in a 1500 m distance downstream from the fault. Here a kinetic equilibrium (i.e. geochemical steady state condition; Paces 1978) between adularia, gibbsite/montmorillonite and kaolinite as described below (Fig. 6.3.3) may be assumed.

Assuming a simple equilibration process (congruent dissolution) of a 1:1 mineral the reaction kinetics may be described by the reaction rates of one ion (e.g. Ca reaction rates for calcite equilibration)

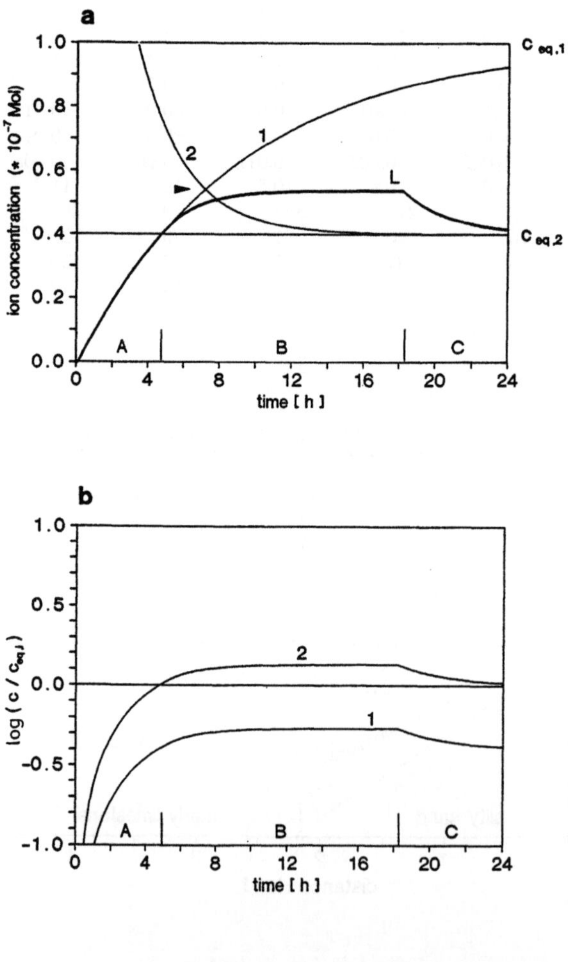

Fig. 6.3.2. KINEQ calculation. Kinetic equilibrium between two 1:1 phases with a common ion. **a** Ion concentration of the solution for congruent dissolution of the primary mineral (*1*), precipitation of the secondary mineral (*2*) and resulting ion concentration of the solution for both reactions occurring simultaneously (*L*). **b** Corresponding saturation values

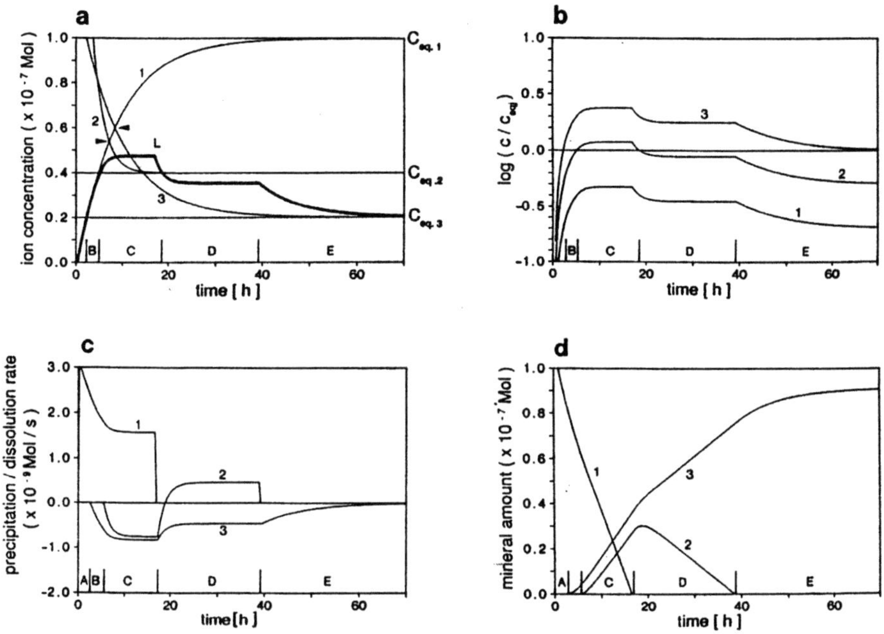

Fig. 6.3.3. KINEQ calculation. **a** Ion concentration of solution for dissolution of primary phase (*1*), precipitation of intermediate phase (*2*) and stable secondary phase (*3*) and resulting ion concentration (*L*) when all reactions may occur simultaneously. **b** Corresponding SI values. **c** Corresponding dissolution (positive) and precipitation rates (negative). **d** Corresponding mineral amounts

$$\frac{dc}{dt} = k \cdot (c_{eq.} - c) \tag{6.3.1}$$

and the corresponding development of the concentration of this ion in the solution by

$$c = c_{eq.} \cdot [1 - \exp(-k \cdot t)] + c_o \cdot \exp(-k \cdot t), \tag{6.3.2}$$

with dc/dt: reaction rate, k: kinetic factor of the reaction, c_{eq}: equilibrium ion concentration for a certain mineral, c: actual ion concentration of the solution, c_o: initial ion concentration of the solution.

In natural systems incongruent dissolution occurs more often than simple equilibration of a single solid phase with the solution. In this case all mineral reactions with their specific kinetic factors k_i and specific equilibrium concentrations of the common ion $c_{eq.i}$ are to be considered.

W. Dreybrodt (pers. commun. 1989) contributed the following mathematical solution of this problem:

$$\frac{dc}{dt} = \Sigma\, [k_i \cdot (c_{eq.i} - c)] \tag{6.3.3}$$

or

$$\frac{dc}{dt} = -c \cdot \Sigma k_i + \Sigma (k_i \cdot c_{eq.i}) \tag{6.3.4}$$

divided by Σk_i

$$\frac{dc}{\Sigma k_i \cdot dt} = -c + \frac{\Sigma (k_i \cdot c_{eq.i})}{\Sigma k_i} \tag{6.3.5}$$

This expression is simplified by introducing c′ defined by Eq. (6.3.6)

$$c' = -c + \frac{\Sigma (k_i \cdot c_{eq.i})}{\Sigma k_i} \tag{6.3.6}$$

and since

$$\frac{\Sigma(k_i \cdot c_{eq.i})}{\Sigma k_i} = \text{const.} \tag{6.3.7}$$

by Eq. (6.3.8)

$$\frac{dc}{dt} + -\frac{dc'}{dt} \tag{6.3.8}$$

If Eqs. (6.3.6) and (6.3.8) are substituted in (6.3.5), then

$$-\frac{dc'}{dt} \cdot \frac{1}{\Sigma k_i} = c' , \tag{6.3.9}$$

which is the derivation of an exponential function of the form

$$c' = b \cdot \exp(-\Sigma k_i \cdot t). \tag{6.3.10}$$

If this expression is substituted for c′ in (6.3.6), it follows

$$c = -b \cdot \exp(-\Sigma k_i \cdot t) + \frac{\Sigma(k_i \cdot c_{eq.i})}{\Sigma k_i} \tag{6.3.11}$$

and setting t = 0 b is found as

$$b = \frac{\Sigma(k_i \cdot c_{eq.i})}{\Sigma k_i} - c_o \tag{6.3.12}$$

hence the ion concentrations of a solution with i phases competing for one ion is

$$c = \frac{\Sigma(k_i \cdot c_{eq.i})}{\Sigma k_i} \cdot [1 - \exp(-\Sigma k_i \cdot t)] + c_o \cdot \exp(-\Sigma k_i \cdot t) \tag{6.3.13}$$

Consequently the steady-state ion concentration for a given set of minerals reacting simultaneously is given by

$$c_{eq.tot} = \frac{\Sigma(k_i \cdot c_{eq.i})}{\Sigma k_i} \tag{6.3.14}$$

which is a mean of the equilibrium values for the single minerals weighted by the kinetic factor.

The time constant $\tau = 1/\Sigma\, k_i$, after which 63% saturation is reached since c (assuming $c_o = 0$)

$$c(\tau) = \frac{\Sigma(k_i \cdot c_{eq.i})}{\Sigma k_i} \cdot [1 - \exp(-\Sigma k_i \cdot \tau)] = \frac{\Sigma(k_i \cdot c_{eq.i})}{\Sigma k_i} \cdot [1 - \exp(-1)] \qquad (6.3.15)$$

becomes obviously smaller, the more minerals compete for or contribute to one ion since $\Sigma\, k_i$ increases with every mineral participating the kinetic equilibrium.

The examples in Figs. 6.3.2 and 6.3.3 were calculated with the iterative model KINEQ. The iterative algorithm instead of the mathematical solution given above provides the possibility of adjusting different k values for every time step. This is important since in natural systems k is an intrinsic value depending on surface areas. However, in the simulations presented here the kinetic factors k were kept constant. The model KINEQ allows for two or more minerals to dissolve or precipitate according to their kinetic characteristics on the base of Eq. (6.3.1). The values for k and $c_{eq.}$ in the examples below are chosen arbitrarily. Real (intrinsic) kinetic factors k as defined here have been determined, e.g. by Schulz (1988) for the overall dissolution reaction of different carbonates.

Figure 6.3.2 presents calculated results of a simulation, where a limited amount of mineral 1 with an equilibrium ion concentration $c_{eq.1} = 1 \times 10^{-7}$ mol reacts with a solution with $c_o = 0$ (e.g. distilled water) and dissolves with a rate constant $k_1 = 3 \times 10^{-5}\ s^{-1}$ to attain equilibrium (Fig. 6.3.2a, curve 1). If a thermodynamically more stable secondary phase 2 with a lower equilibrium ion concentration ($c_{eq.2} = 4 \times 10^{-8}$ mol) and faster kinetics ($k_2 = 1 \times 10^{-4}\ s^{-1}$) is considered, it will start to form from mineral 1 as the solution exceeds $c_{eq.2}$ (Fig. 6.3.2a, region B). The solution then attains a kinetic equilibrium with an ion concentration that is between $c_{eq.1}$ and $c_{eq.2}$ (Fig. 6.3.2a, curve L), being closer to the equilibrium concentration of the reaction with faster kinetics.

After the primary mineral is used up, the solution will equilibrate with mineral 2 (Fig. 6.3.2a, region C). In this sample, constant kinetic factors are assumed. In natural systems the intrinsic kinetic factor of mineral 1 would decrease, as its amount decreases, since k is a function of the surface area. The decrease in the ion concentration of the solution will therefore be much smoother than shown in Fig. 6.3.2a, and it will start already in region B. Fresh precipitates of mineral 2 blocking the surface of mineral 1 would have the same effect (decrease of k_1 with decreasing of reactive surface area).

Figure 6.3.2b shows the corresponding saturation values $\log(c/c_{eq.i})$ which are comparable to the saturation indices calculated by the thermodynamical models. At the partial or kinetic equilibrium (Fig. 6.3.2, region B), which is the usual state of many natural systems, the permanently undersaturated mineral 1 is transformed to the thermodynamic stable mineral 2.

The results of a simulation for a system with three minerals are shown in Fig. 6.3.3 and Table 6.3.2. In Fig. 6.3.3a the ion concentration curve (L) for a dissolving primary mineral (1) and two secondary minerals (2 and 3) is plotted, where the less stable phase (2) has faster kinetics (here: $k_2=1 \times 10^{-4}s^{-1}$) than the primary mineral and the stable mineral (here: $k_1=k_3=3 \times 10^{-5}s^{-1}$). In a first step (region A) only dissolution of mineral 1 occurs, whereas minerals 2 and 3 have not yet formed. In step 2 (region B) small amounts of mineral 3 will precipitate, whereas in step 3 (region C) both minerals 2 and 3 precipitate. The overall ion concentration is lower than the partial equilibrium ion concentrations (arrows), if only precipitation of minerals 2 or 3 was assumed. After the primary mineral is completely dissolved, mineral 2 is being transformed to mineral 3 (region D). When phase 2 is resorbed completely, the ion concentration of the solution (curve L) approaches the equilibrium value for the stable secondary mineral 3 (region E).

The corresponding saturation values for the three minerals (Fig. 6.3.3b) indicate that in region C and D mineral 2 is the limiting phase for the ion concentration of the solution with saturation values close to zero, whereas in region E mineral 3 will limit the dissolved ion content. The dissolution rates (Fig. 6.3.3c) of mineral 1 are kept constantly high as mineral 2 precipitates due to the limited ion concentration, maintaining a constant undersaturation of the primary mineral. Although phase 2 is controlling the ion content of the solution in regions C and D, its amount (Fig. 6.3.3d) is smaller than the amounts of the primary and the stable secondary phases (1 and 3). This difference becomes even greater, when real initial amounts are chosen, since in this calculation a very small initial amount of mineral 1 (1.3×10^{-7} mol) was assumed, to show all reaction features until equilibration of the stable phase within a short time span (250 reaction steps of 1000 s each), because at considerably longer time steps the model will start to oscillate.

Partial equilibria as presented in Figs. 6.3.2 and 6.3.3 have been described by Lasaga (1981a,b) between nepheline and gibbsite or by Paces (1978) and Dahmke (1988) between feldspar, amorphous $Al(OH)_3$, gibbsite and kaolinite (cf. Sect. 4.3.1, this Vol.). Thorstenson and Plummer (1977) showed that a calculated slight supersaturation of pure calcite (Fig. 6.3.1) may result from the dissolution of natural Mg-bearing calcite and precipitation of the thermodynamically stable low Mg calcite.

Table 6.3.2. Saturation indices of three water samples with respect to different Si, Al, and Fe minerals, that may result from kinetic equilibria. Values in brackets are theoretical for [Al] = 0.01 mg/l (no Al data for the Tunisian samples)

	Aachen	Bocholt	Tunisia
Well	AAC4	BOC1/3 21 m	TN26
Date	30.03.84	29.08.84	04.05.84
SiO_2am.	−0.23	−0.65	−0.74
Quartz	+0.82	+0.39	+0.26
Adularia	−0.71	−0.99	(−0.89)
Gibbsite	−0.04	+0.92	(+0.19)
Kaolinite	+1.77	+2.89	(+1.30)
$Fe(OH)_3$am.	−0.59	+0.24	−0.09
Goethite	+3.23	+4.39	+4.18
Hematite	+26.00	+28.10	+28.30

For most samples SI values indicate that $SiO_{2am.}$ is slightly undersaturated and quartz is slightly supersaturated due to a partial equilibrium between the two phases (Table 6.3.2, Fig. 6.3.2; Ohse et al. 1984).

Partial equilibria between feldspar as the primary Al phase, gibbsite as a secondary phase with fast kinetics and intermediate solubility and kaolinite as the stable phase may be assumed for many samples like those from Bocholt and Aachen (Table 6.3.2, Fig. 6.3.1). For the sample from Tunisia (gypsum water with moderate ionic strength $I = 0.05$), where Al was not analyzed, a similar constellation of saturation indices is calculated assuming an Al concentration $[Al] = 0.01$ mg/l. The total amounts of the intermediate secondary phase (Fig. 6.3.3d) and the change in the total amounts of the primary phase may be below the detection limit.

Extremely fast kinetics of amorphous $Fe(OH)_3$ compared to those of goethite and hematite may explain why $Fe(OH)_{3am.}$ has saturation indices very close to 0 in many samples, whereas the water is highly supersaturated with respect to the thermodynamic stable phases goethite and especially hematite (cf. Scheffer and Schachtschabel 1982). In the solid phase of the sample from Bocholt no amorphous $Fe(OH)_3$ but significant amounts of Ca-siderite (SI=+1.4) as a metastable phase with respect to goethite and hematite were detected (Leuchs 1988; Sect. 4.1.1, this Vol.).

6.3.2.2 Surface Blocking

Secondary minerals tend to precipitate close to the place where water chemistry (e.g. increase of calcium activity and pH at calcite grains) changes towards their supersaturation. This process causes a decrease of the reactive surface area and therefore a decrease in dissolution rates shifting the ion activity to the equilibrium value for the secondary phase. The overall effect is an increase in time necessary to establish a constant ion activity. Well-known effects are inhibition of calcite dissolution by apatite precipitates (Schulz 1988) and coatings of amorphous $Fe(OH)_3$ on calcite particles. In batch experiments with feldspars and aqueous solutions of organic ligands, Petersen (1988) determined a decrease of Al release rates despite constantly negative saturation indices of feldspar, which he explained by surface blocking (Sect. 2.8, this Vol.).

6.3.2.3 Anthropogenic and Microbial Influences

The nitrate and chloride content of natural waters are not controlled by mineral reactions, but by the anthropogenic input as fertilizer salts etc. The nitrate, ammonium and methane content of natural water is mainly caused by microbial conversion of organic substance (Sect. 5.2, this Vol.). The amount of these ions in solution is only indirectly limited by mineral reactions through the presence of redox sensitive minerals (manganese, iron, sulfate and sulfide minerals) in the solid phase. For example, the release rate of sulfate by pyrite weathering is controlled by the activity of *Thiobacillus ferrooxidans* (Schwartz 1958; Toran 1987; Kölling 1990).

Table 6.3.3. Saturation indices for calcite and (F,CO_3) apatite in the Seffent springs catchment area

	AAC1	AAC2	AAC3	AAC4	AAC6	AAC7
Calcite	–0.78	–0.40	0.14	0.24	0.04	0.05
(F,CO_3) apatite	–0.01	–0.12	5.38	2.86	1.51	1.99

6.3.3 Calculation Errors

6.3.3.1 Number of Parameters Analyzed

In PO_4-bearing waters – especially in anthropogenically influenced waters – fluor apatite may be the limiting phase for the calcium activity in solution (Schulz 1988). If phosphate or fluoride data are missing, the model calculations yield a constant undersaturation for calcite without any solid Ca phase in equilibrium, which is not understandable in the presence of calcite, since this mineral is known to have fast reaction kinetics. In the decalcified sandy aquifer of the Seffent springs catchment area (AAC1, AAC2) the Ca contents are obviously limited by apatite solubility. After entering the carbonatic sediments, Ca concentration is controlled by calcite dissolution, causing a high supersaturation of apatite, which decreases along the pathway (Table 6.3.3). This indicates apatite precipitation with a much lower rate for the (F, CO_3)-apatite phase considered here (Fig. 6.3.2) compared to that of calcite.

For some minerals with low solubility products the equilibrium concentrations of the dissolved components are below detection limit. If instead of the real – but unknown – ion concentrations the concentrations of the detection limit are used arbitrarily, the calculations may yield a positive saturation index (indicating supersaturation). If data for important ligands like fluoride, which significantly affect the distribution of dissolved cation species, are missing, calculated activities of many free cations, and consequently many saturation indices, are too high. In this case sensitivity analyses may be used to check whether phases close to equilibrium are present at reasonable concentrations of the non-measured ion.

6.3.3.2 Analytical Errors

Due to the mineral stoichiometry, analytical errors cause errors of different level in the calculated saturation indices. An Al concentration which is one order of magnitude greater than the equilibrium value produces a SI of +1 for amorphous $Al(OH)_3$, but a SI of +4 for $Al_4(OH)_{10}SO_4$. Therefore a SI of +1 for a complex mineral may be considered as an equilibrium value within the analytical error, whereas calcite certainly is supersaturated at a calculated SI of +0.5 (Sect. 6.3.4, this Vol.).

6.3.3.3 Thermodynamic Constants and Data Base

Thermodynamic constants are based on experimental data. Values from different authors rarely are exactly equal depending on the experimental equipment, the method of measurement and evaluation and the mineral samples used. Slight differences in mineral composition or grade of crystallinity may cause differences of solubility data in orders of magnitude (Stumm and Morgan 1981). Values for the temperature dependence of solubility constants differ also. Since the programme output lists, besides saturation indices, the ion activity products and thermodynamic constants separately, effects of different log K_s values can be checked easily.

Solubility constants cannot be determined by equilibration experiments for different smectites (May et al. 1986), since the composition of the solution is controlled by the formation of amorphous $Al(OH)_3$ and gibbsite rather than by simple smectite dissolution (Fig. 6.3.3). Similar results have been presented by other authors, e.g. Churchman and Jackson (1976) proposed secondary Si- and Al phases from montmorillonite equilibration studies.

Most equilibrium programmes take into account a great variety of inorganic mineral and complex equilibria, whereas organic ligands are considered only in a few programmes (e.g. GEOCHEM, Sposito and Mattigod 1980).

As mentioned in Section 2 (this Vol.), the complexation capacity of organic molecules is based on different functional groups. The complexation capacity may be described by a (conditional) function of the metal ion activity rather than by a thermodynamical constant (Klamberg et al. 1989). If the dependency between the detectable concentration of an organic molecule with a defined molecular weight and the activity of its functional groups is known, mass spectra of organic molecules may be converted into activities of functional groups, that can be treated as several single ions (ligands) in the computer model. This concept may lead to a more universal and for modelling purposes more practicable approximation of the organic complexation capacity.

6.3.4 Sensitivity Analyses

Sensitivity analyses were carried out by using WATEQ2 as a subroutine in computer programmes that scan values for parameters with great detection uncertainties, like the redox potential (Kölling 1986) or the Al concentrations, over a reasonable range.

The slope of the curves between Al contents and the saturation indices with respect to different Al minerals in a water from the Seffent spring catchment area is the same for all normally mineralized waters (Fig. 6.3.4), whereas their relative position depends upon the pH and the activity of the other mineral components such as Si, K, Mg, and Na.

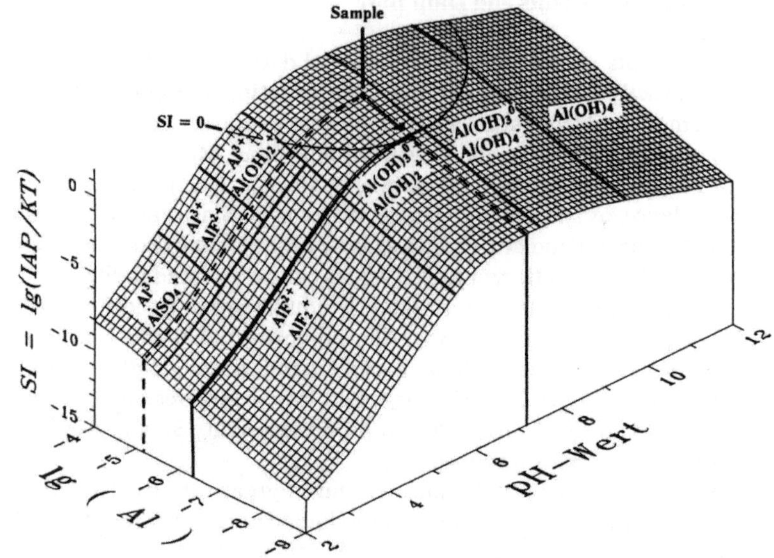

Fig. 6.3.4. SI sensitivity of different Al phases to changes in Al activity. Original data (*broken line*) from Seffent springs catchment area, well AAC7, 31.05.83

Plotting of SI vs. two parameters, the following problems may be solved:

- Is a water in equilibrium with a mineral within the analytical error?
- How may the water composition change during equilibration of a (sub) system?
- Which are the dominant soluble species of a given ion under different conditions? (The change in dominant soluble species is indicated by buckles in the SI surface.)

The gibbsite curve in Fig. 6.3.4 is a cross section through the SI surface in Fig. 6.3.5 at the measured pH of 7.21. Assuming a gibbsite equilibrium, the marked line (SI=0) represents possible changes of Al with pH or vice versa. This plot also visualizes the relatively high solubility of Al at low and high pH values by lowered SI. Since the saturation state of many minerals depend on more than three parameters, it is more useful to construct 2D graphs for all parameters separately (Fig. 6.3.4). The sensitivity analyses show that calculations with even little deviations from the analyzed concentration yield equilibrium values for possible limiting solid phases which are not found by calculations with the original data. This holds especially for the sensitivity of the SI to the E_H value charged with an analytical uncertainty of at least ± 50 mV (Kölling 1986). Figure 6.3.6a shows the E_H sensitivity of the saturation state of some iron minerals for analytical data from the Bocholt aquifer system (Well BOC1/3, 21 m). Due to the strong sensitivity of ferric species to changes in the E_H value in the lower E_H range (Fig. 6.3.6b), the E_H influence on the SI may be as great as 5 SI units per 100 mV (Na-jarosite) for minerals containing ferric iron. Therefore, a standard deviation of ±50 mV from the

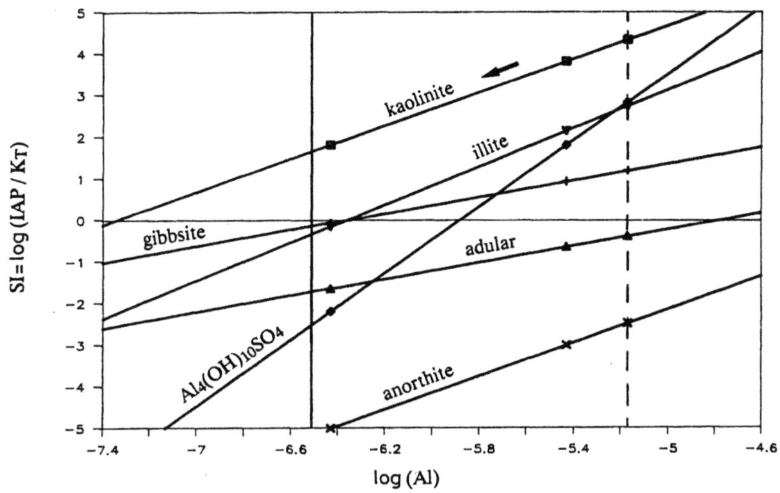

Fig. 6.3.5. Sensitivity of gibbsite saturation indices to changes in pH and Al activity. The dominant Al species under different pH-Al conditions are noted on the surface. Same original data as in Fig. 6.3.4

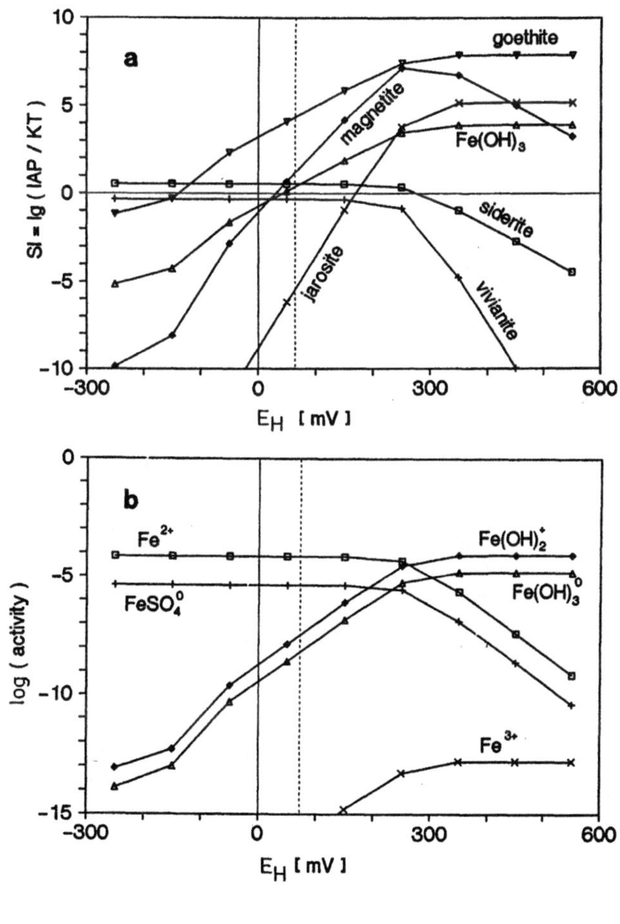

Fig. 6.3.6. **a** SI sensitivity to changes in E_H for different iron minerals. **b** Change of iron species distribution with E_H. Original data from test area Bocholt, well BOC1/3, 21 m. Original E_H = 67 mV

measured E_H of 67 mV means a deviation in the SI of ± 1.8 for magnetite and ±0.9 for amorphous $Fe(OH)_3$, whereas the SI of solid phases containing ferrous iron (siderite, vivianite) is independent of E_H over a wide range. By EDAX analyses, Leuchs (1988) found fresh precipitates of Ca siderite (SI=1.4), which is more stable than the pure siderite considered in Fig. 6.3.6a down to (very theoretical) Ca activities of (Ca)=2.5×10^{-8} mg/l, of samples from this location. However, the model calculations indicate by saturation indices close to zero (SI=0.24) amorphous $Fe(OH)_3$ to be the limiting phase for the iron content of the water, which may be transformed to the more stable Ca siderite in a second reaction step.

6.3.5 Limiting Phases

The procedures described above make it possible in many cases to define safe or possible limiting mineral reactions. Table 6.3.4 shows limiting phases for different ions from model calculations with data from the Seffent springs catchment area (Sect. 3.2.5, this Vol.).

The evaluation of data from Tunesian playa waters (Kössl 1988, Sect. 3.3.1, this Vol.) allowed the comparison of saturation state data with the mineral association found as macrocrystals in the playa sediments. Despite all analytical and calculation difficulties due to ionic strengths up to 4, the SI data and formation of many carbonates (calcite, aragonite, dolomite, magnesite) and sulfates (gypsum,

Table 6.3.4. Limiting phases for the ion content of a water sample from the Seffent springs catchment area. (Well AAC4, 30.03.84)

Parameter	Limiting phase	Formula	Saturation index
pH	Calcite	$CaCO_3$	0.242
Ca	Calcite	$CaCO_3$	0.242
Mg	Illite	$K_{0.6}Mg_{0.25}Al_{2.3}Si_{3.5}O_{10}(OH)_2$	0.133
Na	–		
	Montmorillonite?	$(H,Na,K)_{04}Mg_{0.5}Fe_{0.3}Al_{1.5}Si_{3.8}(OH)_2$	0.09
K	Illite	$K_{0.6}Mg_{0.25}Al_{2.3}Si_{3.5}O_{10}(OH)_2$	0.133
	Montmorillonite?	$(H,Na,K)_{0.4}Mg_{0.5}Fe_{0.3}Al_{1.5}Si_{3.8}(OH)_2$	0.09
Fe	Montmorillonite?	$(H,Na,K)_{0.4}Mg_{0.5}Fe_{0.3}Al_{1.5}Si_{3.8}(OH)_2$	0.09
Al	Illite	$K_{0.6}Mg_{0.25}Al_{2.3}Si_{3.5}O_{10}(OH)_2$	0.133
	Gibbsite	$Al(OH)_3$	0.048
	Montmorillonite?	$(H,Na,K)_{0.4}Mg_{0.5}Fe_{0.3}Al_{1.5}Si_{3.8}(OH)_2$	0.09
Si	Quartz	SiO_2	0.823
	$SiO_{2(am.)}$	SiO2	–0.234
	Illite	$K_{0.6}Mg_{0.25}Al_{2.3}Si_{3.5}O_{10}(OH)_2$	0.133
CO_3	Calcite	$CaCO_3$	0.242
Cl	–	(Input)	
SO_4	–	(Input/microbial/redox)	
NO_3	–	(Input/microbial redox)	
NO_2	–	(Microbial/redox)	
NH_4	–	(Microbial/redox)	

Table 6.3.5. Phases controlling the ion content of a water sample from the Bocholt aquifer system (Well BOC1/3, 21 m)

Para-meter	Limiting phase	Formula	Saturation index
pH	Calcite	$CaCO_3$	0.098
Ca	Calcite	$CaCO_3$	0.098
Mg	Montmorillonite	$(H,Na,K)_{0.4}Mg_{0.5}Fe_{0.3}Al_{1.5}Si_{3.8}(OH)_2$	0.958
Na	(Input)		
	Montmorillonite	$(H,Na,K)_{0.4}Mg_{0.5}Fe_{0.3}Al_{1.5}Si_{3.8}(OH)_2$	0.958
K	Montmorillonite	$(H,Na,K)_{0.4}Mg_{0.5}Fe_{0.3}Al_{1.5}Si_{3.8}(OH)_2$	0.958
	Adularia	$KAlSi_3O_8$	−0.995
Fe	Siderite	$FeCO_3$	0.537
	$Fe(OH)_3$am.?	$Fe(OH)_3$	0.365
	FeS am.	FeS	0.471
Al	Gibbsite	$Al(OH)_3$	0.921
	Montmorillonite	$(H,Na,K)_{0.4}Mg_{0.5}Fe_{0.3}Al_{1.5}Si_{3.8}(OH)_2$	0.958
	Adularia	$KAlSi_3O_8$	−0.995
Si	Quartz	SiO_2	0.387
	$SiO_{2(am.)}$	SiO_2	−0.653
CO_3	Calcite	$CaCO_3$	0.098
S	FeS am.	FeS	0.471
PO_4	Vivianite	$Fe_3(PO_4)_28H_2O$	−0.323
Cl	–	(Input)	
SO_4	–	(Input/microbial/redox)	
NO_3	–	(Input/microbial/redox)	
NO_2	–	(Microbial/redox)	
NH_4	–	(Microbial/redox)	

anhydrite, mirabilite, thenardite, jarosite, epsomite) were in good agreement. Saturation indices for jarosite, which was found only in a small region fo the investigated site area, were close to equilibrium or positive only in samples from exactly this region. Chloride minerals were significantly undersaturated even in these highly mineralized waters, indicating that contrary to sulfates, the precipitation of chlorides needs further concentration by evaporation.

The evaluation of data from the Bocholt aquifer system indicates a partial equilibrium between adularia as a primary phase and montmorillonite/gibbsite as secondary phases. Kaolinite as the stable secondary phase has a SI of 2.89 in this analysis (Table 6.3.5).

6.3.6 Conclusions

- Model calculations with hydrogeochemical data along profiles allow the study of saturation state changes, from which information about on-going reactions in an aquifer can be derived.
- The solid phase mainly controlling the activity of a certain dissolved ion approaches a saturation index close to zero.

- Sensitivity analyses should be carried out for parameters with great uncertainties and important non-analyzed parameters to check the reliability of calculated saturation states.
- Reaction kinetics may be more important than the saturation state of a specific mineral.
- The total amounts of limiting phases are often – due to both fast precipitation and dissolution kinetics – below the detection limits, whereas significant amounts of the primary (undersaturated) and stable or metastable (supersaturated) phases are found in the solid phase.
- Indicators for a kinetic or partial equilibrium are saturation values of mineral couples approaching a constantly slight super- and undersaturation, respectively.

7 Perspectives and Needs for Future Work

F. Frimmel, P. Hirsch, G. Matthess, H.D. Schulz and E. Usdowski

The results of the priority programme presented here indicate that the very complex interactions between subsurface solid and aqueous phases in the presence of organic compounds and mediated by microbial activities can only be successfully approached by interdisciplinary groups. The newly developed techniques of sample collection, measurements and evaluation will also be useful for the investigation of other natural and contaminated subsurface systems. These techniques, as well as the scientific results reported here, provide presently the background for the current priority programme Schadstoffe im Grundwasser (Detrimental substances in ground water) of the Deutsche Forschungsgemeinschaft.

The present analytical methods yield excellent results in general. However, further development of analytical methods for the study of subsurface organic substances, especially for humic substances, is necessary for the understanding of microbial activities and geochemical processes. Oxidation and reduction reactions, and sorption-desorption phenomena, or dissolution and precipitation influence the persistence, mobility, and transport of substances in the unsaturated and saturated zones. A better understanding of sorption and microbial degradation of organic and inorganic pollutants in subsoils and aquifers is necessary. Pollutant binding properties of humic substances – especially in aquifers with low content of organic carbon – need to be investigated to possibly protect drinking water supplies from future pollution. For this purpose, knowledge of the molecular structure of the organic substances and their physico-chemical properties has to be improved. In this vein, another area of great interest should be the study of extracellular microbial enzyme sorption and the enzymes' longevity in the ground water environment.

To obtain reliable measurements is very difficult and time-consuming. The following aspects need consideration:

- The sample collection, especially in the unsaturated zone, as well as the sample treatment, may introduce errors. These difficulties have been overcome, partially, and standard methods have been suggested in order to gather comparable data. Nevertheless, further work is required in this respect.
- All drilling methods are connected with possible contamination of the ground water and the rocks adjacent to the walls and bottom of the drillhole. This holds also for the coring techniques, where the installation and removal of the coring apparatus causes considerable pressure fluctuations in the drilling fluid.
- Methods should be used which conserve the natural configurations of the organic substances as much as possible.
- Investigating identical samples with different methods and samples of varying origin with the same method will allow the examination and confirmation of results and the extent of their interpretation.

Keeping in mind the fundamental role of organic material as reservoir for biological activity, as mediator for numerous physical and chemical processes and as indicator for the environmental status quo, it is clear that reliable and powerful analytical tools are needed to obtain all the information necessary for a better understanding of this group of substances. The limited applicability and interpretation of in situ measurements and the necessity of isolation and concentration procedures for the use of sophisticated analytical methods has lead to the conclusion that a detailed and comprehensive chemical structure of the organic acids and humic material in the water-unsaturated zone cannot be achieved at present. Therefore the functionality like aromaticity, hydrophilicity, molecular size etc. of these substances has come more into the focus. It is obvious that operational definitions cannot be avoided. However, they have to be used in a precise, clear and consistent way to serve scientific progress. As an example, the isolation and preparation of reference and standard material has brought a wide comparability of methods and has lead to the development and validation of numerous experimental approaches. As pitfalls, denaturing reactions and artefact formation occur and have to be kept in mind when interpreting the results.

One of the most attractive goals for future work has to be seen in the development of methods sensitive enough for direct application to original samples. Especially in aquatic systems, hydrophilic high molecular organic acids are difficult to characterize and have been neglected. Liquid chromatography in combination with powerful detection systems seems to be promising for these investigations.

In the past, many characterization data have been expressed as average numbers. Because of the heterogeneity of substances in natural systems, distribution functions will give a more precise picture of reality than average numbers. It is an experimental challenge for future work and it will be a difficult task to include these functions into our model concepts.

Even though there are many results on the interaction of metals with organic substances, relatively little is known on the specific bond formation. This also applies to the function of hetero-atoms like nitrogen, phosphorus and sulfur. For a more advanced understanding of the biochemical reactions in the underground, these problems have to be tackled. In addition to the inorganics, organic micropollutants and especially xenobiotics in their interaction with humic substances have to be investigated. Degradation, transformation and translocation are only a few aspects of growing importance due to the awareness that ground water pollution has become a severe danger. It is also necessary to investigate these processes not only under aerobic but also under anaerobic conditions, for from the findings in the field of waste water treatment it can be extrapolated that different end products and metabolites have to be expected.

Putting all details together might end up with an integrated view of the ground water system in which the pool of organics and their specific character and structure reflect the environmental strain which is not only below but also above the surface. Multidisciplinary work and an intense exchange of ideas amongst the researchers is the most urgent need for this.

Field measurements, laboratory experiments and models which describe the $CaCO_3$-H_2O system have shown that this system is understood to some extent. Thus, computer models have led to a number of realistic simulations of natural processes. This is due to the fact that first-order kinetics may be applied to dissolution reactions. However, this is true only as long as the solutions are distinctly undersaturated with respect to the solid. Close to saturation, complex and higher-order kinetics appear to govern various dissolution processes which are not yet really understood. This situation becomes even more complicated if magnesium-bearing calcite is involved.

The production of the dissolved carbonate species is controlled not only by inorganic chemical reactions, but also by biological activities which usually change the composition of the stable carbon isotopes. This fact impedes the evaluation of the mass transfer out of the unsaturated zone and the determination of carbon-14 ages. Nevertheless, carbon isotopes may be a useful and powerful tool for the understanding of subsurface geochemical processes. Their measurement should be combined with that of other isotopes, such as ^{2}H, ^{3}H, ^{18}O, ^{3}He, ^{39}Ar, and ^{85}Kr, to allow for a check up of the validity of the derived model concept.

If carbonate rock dissolution is enhanced by sulfuric acid from atmospheric pollution, model considerations with respect to the nature of a hydrogeochemical system become more complicated, especially if the transition of a hydrogeochemical system from "open" to "closed" conditions is examined. Using the stable isotopes of sulfur allows for the necessary distinction between the quantities of sulfate derived from pollution and sulfate which originates from the dissolution of calcium sulfate minerals.

The weathering processes of silicate minerals obviously are controlled at very low concentrations by complex dissolution and precipitation reactions. The understanding of these reactions under laboratory conditions is advanced enough to build up realistic reaction models. The evaluation of weathering rates from natural systems still includes some uncertainties, especially in the estimation of the true reaction surfaces. This evidence holds even more in the case of the unsaturated zone. Further work will have to be done in order to reach a better understanding of silicate weathering in natural subsurface systems. This includes both field investigations and laboratory experiments with natural sediments. The application of the activated-surface complex theory seems to allow the incorporation of experimental data into realistic reaction/transport models of the silicate weathering processes.

Future work on carbonate and silicate systems should be pursued in the same direction as before, in principle, but focussed on a single representative area for field work to which laboratory experiments may be easily adapted and for which a multitude of parameters must be measured simultaneously.

Dissolution of carbonate rocks eventually leads to an increase of the fissures and voids which results in an increase of hydraulic conductivity and – on a macroscopic scale – to karstification. However, the insoluble components of limestones and dolomites may clog the subsurface flow path, thus impeding further dissolution processes. Similar effects of geochemical processes, both with respect to an increase of permeability by dissolution and to a decrease by clogging due to secondary precipitates, have been reported from silicate systems as well. Further work should investigate the quantitative role of geochemical processes in the development of the hydraulic conductivity in both carbonate and silicate rocks.

In ground water, microbiological redox reactions between organic material and oxygen, nitrate or sulfate are still described mainly by empirical functions. The reason for this is that such reactions where bacterial enzymes are involved still suffer from a lack of data for microbial in situ activity, although microbial activity data from pure laboratory studies have been published.

As microorganisms (in the ground water environment: mainly bacteria) respond directly to "local" conditions in the bacterial size range, their microhabitat and not the macroenvironment determines their present in situ activity. A bacterial microcolony or even a single cell may be attached to a sediment particle and be more or less in contact with the passing pore or ground water. But only a few micrometers away the local redox potential, pH, availability of nutrients or gases could be dramatically different and could be changing at different rates, due to substrate exhaustion or production by other bacteria. Further research in ground water microbiology should be focussed on the measurement and description of such microhabitats and on their stability. Techniques for the in situ determination of microbial activities would have to be developed. "In situ" in this context does not mean "within a sediment or water sample brought up to the surface", but it describes the activities that occur really down within the possibly undisturbed subsurface environment. For this, much cooperation between microbiologists and geochemists will be necessary.

In the context of the development of in situ bioremediation techniques bacteria will eventually be introduced into the ground water environment to clean up polluted sites. The fate of such organisms, possibly alien to the natural microbial population and certainly interacting with these, will have to be studied so that it could be predicted. Furthermore, in a geoecological approach, a chemical or even physical anthropogenic influence on the natural (native) bacterial population could lead to enhanced or impeded in situ activity of these organisms. In this connection, maximally tolerable concentrations of such pollutants or physical impacts will have to be determined.

Finally, much of the work presented in this Volume has been carried out on fairly shallow aquifers. Information on deeper aquifers is accumulating, and there certainly are differences within the depths of study sites. Eukaryotic microorganisms (i.e. fungi, yeasts, protozoa) have only been found in the upper 10 m of the subsurface environment, while new and hitherto unknown organisms could be expected to show up in much greater and consequently warmer depths.

Depending on the scientific background of the individual author, there are very remarkable differences between the models concerned with the combinations of transport and geochemical reactions. Usually the more elaborate transport models of two or even three dimensions, which also take into account hydrodynamical dispersion, regard only relatively trivial geochemical reactions. On the other hand, models with more detailed geochemical reactions usually use only one-dimensional transport models. A future cooperation between modellers of these two types should lead to generally better results. For the general problem of combining realistic physical transport with complex geochemical and microbiological reactions, the transport models are presently far ahead of the models of geochemical reactions and those again are more advanced than models which deal with a general description of in situ microbiological activities. Thus, besides the necessary cooperation, geochemists and especially microbiologists will have to improve their knowledge and their quantitative descriptions of the respective natural systems.

References

Aagaard P, Helgeson HC (1982) Thermodynamic and kinetic constraints on the reaction rates among minerals and aqueous solutions. I. Theoretical considerations. Am J Sci 282:237–285

*Abbt-Braun G (1987) Untersuchungen zur Struktur isolierter Huminstoffe; allgemeine Charakterisierung und massenspektrometrische Bestimmungen. PhD Thesis, Univ München

Abbt-Braun G, Frimmel FH (1990a) Alkylierungsreaktionen als Schlüssel für die Charakterisierung isolierter aquatischer Huminstoffe. Vom Wasser 74:307–324

Abbt-Braun G, Frimmel FH (1990b) Restmetallgehalte isolierter Huminstoffe aus Erde, Moor und Deponie. Acta Hydrochim et hydrobiologica 18:649–656

*Abbt-Braun G, Frimmel FH, Schulten HR (1989) Structural investigations of aquatic humic substances by pyrolysis field ionization mass spectrometry and pyrolysis gaschromatography/mass spectrometry. Water Res 12:1579–1591

*Abbt-Braun G, Frimmel FH, Schulten HR (1990) Strukturelle Charakterisierung isolierter aquatischer Huminstoffe – Anwendbarkeit, Grenzen und Vergleich ausgewählter Methoden. Vom Wasser 74:325–338

Abdul AS, Gibson TL, Rai DN (1990) Use of humic acid solution to remove organic contaminants from hydrogeologic systems. Environ Sci Technol 24:328–333

Aiken GR (1985) Isolation and concentration techniques for aquatic humic substances. In: Aiken GR, McKnight DM, Wershaw RL, MacCarthy P (eds) Humic substances in soil, sediment, and water. John Wiley, New York, pp 363–386

Aiken GR (1988) A critical evaluation of the use of macroporous resins for the isolation of aquatic humic substances. In: Frimmel FH, Christman RF (eds) Humic substances and their role in the environment. John Wiley, New York, pp 15–28

Akopodje EG (1985) The occurrence of bassanite in some Australian arid-zone soils. Chem Geol 47:361–364

*Al-Azawi A, Ebhardt G (1983) Redoxverhältnisse des Grundwassers in der Altneckaraue (Bergstraße, Nördlicher Oberrheingraben). Z Dtsch Geol Ges 134:833–843

*Al-Azawi A, Ebhardt G (1984) Bodengas-Eintrag in das Grundwasser im nördlichen Oberrheingraben (W. Deutschland). In: Udluft P, Merkel B, Prösl KH (eds) Recent investigations in the zone of aeration. Proc Internat Symp Munich, West Germany, Okt. 1984, Dept Hydrogeol Hydrochem Techn Univ Munich, pp 131–141

Albers JH (1976) Feinstratigraphie, Faziesanalyse und Zyklen des Untercampans (Vaalser Grünsand = Hervien) von Aachen und dem niederländisch-belgischen Limburg. Geol Jahrb A34:3–68

Albertsen M (1977) Labor- und Felduntersuchungen zum Gasaustausch zwischen Grundwasser und Atmosphäre über natürlichen und verunreinigten Grundwässern. PhD Thesis, Univ Kiel

Albertsen M (1979) Kohlendioxidhaushalt in der Gasphase der ungesättigten Bodenzone, dargestellt am Beispiel eines Podsols. Z Pflanzenernähr Bodenkd 142:39–56

Albertsen M, Matthess G (1980) Verhalten relevanter Radioisotope in drei typischen Sandböden des Norddeutschen Flachlandes. Meyniana Kiel 32:7–72

Albertsen M, Matthess G, Pekdeger A, Schulz HD (1980) Quantifizierung von Verwitterungsvorgängen. Geol Rdsch 69/2:532–545

Albrechtsen HJ (1988) Groundwater microbiology: a literature review. Lossepladsprojekted report P 6-1 ISBN 87-503-7118-5, 76 pp

---

*Denotes work carried out as part of the priority programme.

Aleem MIH, Hoch GE, Varner JE (1965) Water as the source of oxidant and reductant in bacterial chemosynthesis. Proc NAS Biochem 54:869–873

Al-Hashimi WS, Hemingway JE (1973) Recent dedolomitization and the origin of the rusty crusts of Northumberland. J Sediment Petrol 43:82–91

*Amberger A, Schmidt HL (1987) Natürliche Isotopengehalte von Nitrat als Indikatoren für dessen Herkunft. Geochim Cosmochim Acta 51:2699–2705

Andersson KK, Hooper AB (1983) O_2 and H_2O are each of the source of one O in NO_2-produced from NH_3 by *Nitrosomonas*: 15N-NMR evidence. FEBS Lett 164:236–240

Andres E (1984) Kohlenstoff-Umsatz und –Bilanz des Bodens eines Buchenwald-Ökosystems auf Kalkstein. Gött Bodenkd Ber 80

Ardakani MS, Stevenson FJ (1972) A modified ion-exchange technique for the determination of stability constants of metal-soil organic matter complexes. Soil Sci Soc Am Proc 36:884–890

Ares J (1986) Identification of aluminium species in acid forest soil solution on the basis of Al:F reaction kinetics: reaction paths in solutions. Soil Sci 141:339–407

Atakan Y (1972) Bomben-Tritium-Hydrologie in einem Grundwasserkörper mit freier Grundwasseroberfläche. PhD Thesis, Univ Heidelberg

Atakan Y, Roether W, Münnich KO, Matthess G (1974a) The Sandhausen shallow-groundwater tritium experiment. In: Isotope techniques in groundwater hydrology, vol 1. IAEA-SM-182/2, Vienna, pp 21–43

Atakan Y, Roether W, Matthess G, Münnich KO (1974b) Felduntersuchungen von Fließvorgängen in einem Porengrundwasserleiter mittels Farbstoffindikatoren. Gas- Wasserfach 115:159–164

Atkinson B, Fowler HW (1974) The significance of microbial film in fermenters. In: Ghose TK, Fiechter A, Blakebrought N (eds) Advances in biochemical engineering. Springer, Berlin Heidelberg, New York, pp 221–277

Atlas MR, Bartha R (1981) Microbial ecology. Addison-Wesley, Reading, pp 94–95

Austin B, Garges S, Conrad B, Harding EE, Colwell RR, Simudu U, Taga N (1979) Comparative study of aerobic heterotrophic bacterial flora of Chesapeake Bay and Tokyo Bay. Appl Environ Microbiol 37:704–714

Backes CA, Tipping E (1987) An evaluation of the use of cation-exchange resin for the determination of organically-complexed Al in natural acid waters. Int J Environ Anal Chem 30:135–143

Bailey A (1976) Effects of temperature on the reaction of silicates with the aqueous solutions in the low temperature range. Proc Inst Symp on Rock–Water–Interaction, Int Assoc Geochim Cosmochim, Prague pp 375–380

Bakalowicz M (1982) La genèse de l'aquifer karstique vue par un geochimiste. Reun Monogr Sobre Karst-Larra 82:159–174

Bakken R, Olsen RA (1987) The relationship between cell size and viability of soil bacteria. Microb Ecol 13:103–114

Balkwill DL (1989) Numbers, diversity, and morphological characteristics of aerobic, chemoheterotrophic bacteria in deep subsurface sediments from a site in South Carolina. Geomicrobiol J 7 (1–2):33–52

Balkwill DL, Ghiorse WC (1985) Characterization of subsurface bacteria associated with two shallow aquifers in Oklahoma. Appl Environ Microbiol 50:580–588

Balkwill DL, Rucinsky TE, Casida LE Jr (1977) Release of microorganisms from soil with respect to transmission electron microscopy viewing and plate counts. Antonie Leeuwenhoek J Microbiol 43:73–81

Ball JW, Jenne EA, Nordstrom DK (1979) WATEQ2 – A computerized chemical model for trace and major element speciation and mineral equilibria of natural waters. In: Jenne EA (ed) Chemical modeling in aqueous systems. Am Chem Soc Symp, Series 93, Washington DC, pp 815–835

Barbee GC, Brown KW (1986) Comparison between suction and free-drainage soil solution samplers. Soil Sci 141:149–154

Bard AJ, Faulkner LR (1980) Electrochemical methods, fundamentals, and applications. John Wiley, New York

Bargon E (1969) Bodenkte Hessen 1: 25 000, Bl. 6217 Zwingenberg a.d. Bergstr. m. Erl. HLfB, Wiesbaden

Barnett PJ (1978) Quaternary geology of the Simcoe Area, southern Ontario; Ontario Div. Mines, GR162, 74, accompanied by mpas 2369, 2370, and 2371, scale 1:50 000

Bauer (1964) Kalkabtragungsmessungen in den österreichischen Kalkhochalpen. Erdkunde 18:95–102

*Bauhus W (1988) Hydrochemische und isotopenphysikalische Untersuchungen an Grundwässern tieferer Aquifere im Raum Hamburg. Münster Forsch Geol Paläontol 67:157 pp
*Bauhus W, Loehnert EP (1987) Groundwater management of the Hamburg deep aquifer system (FRG) using environmental isotopes. In: Int Symp Groundwater Monitoring and Management, Proc Complex II. Int Assoc Sci Hydrol, Dresden, 15 pp
*Baumann J (1986) Lösungskinetik von Kalzit in porösen Gesteinen, PhD Thesis, Ber Rep Geol-Paläontol Inst Univ Kiel 15:1–132
*Baumann J, Schulz HD (1984) Lösungskinetik von Kalzit im Sickerwasserbereich von Sanden mit geringem Kalkgehalt. Proc Int RIZA–Symp 2:547–557
*Baumann J, Buhmann D, Dreybrodt W, Schulz HD (1985) Calcite dissolution kinetics in porous media. Chem Geol 53:219–228
Baumgarten HJ, Levetzow R (1969) Untersuchungen zur hygienischen Beschaffenheit von im Handel befindlicher Speisegelatine. Arch Lebensmittelhygiene 20:38–42
Bear J (1972) Dynamics of fluids in porous media. American Elsevier, New York, 764 pp
Beck T (1968) Mikrobiologie des Bodens. Bayrischer Landwirtschaftsverlag, München
*Becker R (1987) Untersuchung zum Sorptionsverhalten von Boden-Huminsäuren gegenüber Metall-Ionen mit der statischen und dynamischen Methode. PhD Thesis, Univ Marburg
Beek WJ, Mutzall KMK (1975) Transport phenomena. John Wiley, New York
Beek CGEM van, Kooij D van der (1982) Sulfate-reducing bacteria in ground water from clogging and nonclogging shallow wells in the Netherlands river region. Ground Water 20:298–302
Beeman RE, Suflita JM (1987) Microbial ecology of a shallow unconfined groundwater aquifer polluted by municipal landfill leachate. Microb Ecol 14:39–54
Beese F, Homeyer B, Meyer B (1975) Der Wasser- und Bioelementhaushalt von Ackerparabraunerden aus Löß in den nördlichen Randgebieten der mitteldeutschen Schwelle (Süd-Niedersachsen). Gött Bodenkd Ber 33:102–132
*Behrens H (1984) Chemie des Radioiods in Oberflächengewässern und Böden unter dem Einfluß von biogeochemischen Vorgängen. In: Proc Tagung Strahlung und Radionuklide in der Umwelt. Bonn, pp 21–24
*Behrens H (1985) Speciation of radioiodine in aquatic and terrestrial systems under the influence of biogeochemical processes. In: Bulman RA, Cooper JR (eds) Speciation of fission and activation products in the environment. Elsevier, London pp 223–230
Beijerinck MW (1913) Oxidation des Manganbicarbonates durch Bakterien und Schimmelpilze. Folia Microbiol 2:123–134
Bellamy LJ (1975) The infrared spectra of complex molecules, Vol. 1. Chapman and Hall, London
Bellamy LJ (1980) The infrared spectra of complex molecules, Vol. 2. Chapman and Hall, London
Beloin RM, Sinclair JL, Ghiorse WC (1988) Distribution and activity of microorganisms in subsurface sediments of a pristine study site in Oklahoma. Microbiol Ecol 16:85–97
Benade W (1954) Mikrobiologie der Moore. Länderverl Wien, 128 pp
Bengtsson G (1989) Growth and metabolic flexibility in groundwater bacteria. Microb Ecol 18:235–248
Bernard AJ (1973) Metallogenic processes of intra-karstic sedimentation. In: Amstutz GC, Bernard AJ (eds) Ores in sediments, VIII. Int Sed Congress Heidelberg 1971, Springer, Berlin Heidelberg New York, pp 43–58
Berner H (1985) Sulphate reduction, organic matter decomposition and pyrite formation. Philos Trans R Soc Lond 315/A:25–38
Berner RA (1963) Electrode studies of hydrogen sulfide in marine sediments. Geochim Cosmochim Acta 27:563–575
Berner RA (1964) Distribution and diagenesis of sulphur in some sediments from Gulf of California. Mar Geol 1:117–140
Berner RA (1970) Sedimentary pyrite formation. Am J Sci 268:1–23
Berner RA (1978) Rate control of mineral dissolution under earth surface conditions. Am J Sci 278:1235–1252
Berner RA (1980) Early diagenesis: a theoretical approach. Princeton University Press, Princeton, N.J.
Berner RA (1981) Kinetics of weathering and diagenesis. In: Lasaga AC, Kirkpatrick RJ (eds) Kinetics of geochemical processes. Rev Mineral, 8:111–134, Min Soc Am, Washington, DC
Berner RA (1985) Sulphate reduction, organic matter decomposition and pyrite formation. Philos Trans R Soc Lond A315:25–38

Berner RA, Holdren GR (1977) Mechanism of feldspar weathering: some observational evidence. Geology 5:369–372
Berner RA, Morse JW (1974) Dissolution kinetics of calcium carbonate in seawater. IV: Theory of calcite dissolution. Am J Sci 274:108–134
Bewley RJF, Stotzky G (1983) Effects of cadmium and simulated acid rain on ammonification and nitrification in soil. Arch Environ Contam Toxicol 12:285–291
Black AP, Christman RF (1963) Characteristics of colored surface water. J Am Water Works Assoc 55:753–770
Blum A, Lasaga A (1987) Role of speciation in the low temperature dissolution of minerals. Nature 331/4:431–433
Bodes-Fischer H (1985) Zum Sorptionsverhalten von Huminstoffen gegenüber Metall-Ionen. PhD thesis, Univ Marburg
Bogdanova VM (1965) Distribution of microflora in rocks and waters of tertiary deposits of the northern part of Caspian lowland (furmanova area). Microbiology (translation) 34:300–303
Bögli A (1978) Karsthydrographie und physische Speläologie. Springer, Berlin Heidelberg New York
Bölter M (1977) Numerical taxonomy and character analysis of saprophytic bacteria isolated from the Kiel Fjord and the Kiel Bight. In: Rheinheimer G (ed) Microbial ecology of a brackish water environment. Springer, Berlin Heidelberg New York
Bonde GJ (1985) Bacteriological characteristics of groundwater resources. Aqua 1:21–26
Bone TL, Balkwill DL (1986) Improved flotation technique for microscopy of in situ soil and sediment microorganisms. Appl Environ Microbiol 41:462–468
Bone TL, Balkwill DL (1988) Morphological and cultural comparison of microorganisms in surface soil and subsurface sediments at a pristine study site in Oklahoma. Microb Ecol 16:49–64
*Böttcher J, Strebel O (1985a) Mittlere Nitratkonzentration des Grundwassers in Sandgebieten in Abhängigkeit von der Bodennutzungsverteilung. Wasser Boden 37:383–387
*Böttcher J, Strebel O (1985b) Räumliche Variabilität der Stoffanlieferung an das Grundwasser bei Gley-Podsolen aus Fein- bis Mittelsand unter Acker. Mitt Dtsch Bodenkd Ges 43:199–204
*Böttcher J, Strebel O (1985c) Redoxpotential und E_g/pH-Diagramme von Stoffumsetzungen in reduzierendem Grundwasser (Beispiel Fuhrberger Feld). Geol Jahrb Reihe C/40:3–34
*Böttcher J, Strebel O (1987) Deterministische und stochastische Einflüsse auf die Variabilität der Stoffanlieferung an das Grundwasser bei Sandböden. Mit Dtsch Bodenkd Ges 55:121–126
*Böttcher J, Strebel O (1988a) Spatial variability of groundwater solute concentrations at the water table under arable land and coniferous forest. Part 1: Methods for quantifying spatial variability (geostatistics, time series analysis, Fourier transform smoothing). Z Pflanzenernähr Bodenkd 151:185–190
*Böttcher J, Strebel O (1988b) Spatial variability of groundwater solute concentrations at the water table under arable land and coniferous forest. Part 2: Field data for arable land and statistical analysis. Z Pflanzenernähr Bodenkd 151:191–195
*Böttcher J, Strebel O (1988c) Spatial variability of groundwater solute concentrations at the water table under arable land and coniferous forest. Part 3: Field data for a coniferous forest and statistical analysis. Z Pflanzenernähr Bodenkd 151:197–203
*Böttcher J, Strebel O (1988d) Ermittlung des Redoxpotentials (E_g-Wert) an Böden und Grundwässern aus dem zeitlichen Verlauf der Elektrodendepolarisation. Z Pflanzenernähr Bodenkd 151:363–368
*Böttcher J, Strebel O (1988e) Ermittlung des Redoxpotentials aus dem zeitlichen Verlauf der Elektrodendepolarisation. Mitt Dtsch Bodenkd Ges 56:271–274
*Böttcher J, Strebel O (1989) Einfluß der Bodennutzung auf die Stoffkonzentration der Grundwasserneubildung bei Sandböden. Kali-Briefe (Büntehof) 19(9):629–648
Böttcher J, Strebel O (1990) Quantification of deterministic and stochastic variability components of solute concentration at the groundwater table in sandy soils. In: Roth K, Flühler H, Jury WA Parker JC (eds): Field-scale water and solute flux in soils, Birkhäuser, Basel, pp 129–140
*Böttcher J, Strebel O, Duynisveld WHM (1985) Vertikale Stoffkonzentrationsprofile im Grundwasser eines Lockergesteins-Aquifers und deren Interpretation (Beispiel Fuhrberger Feld). Z Dtsch Geol Ges 136:543–552
Böttcher J, Strebel O, Duynisveld WHM (1989) Kinetik und Modellierung gekoppelter Stoffumsetzungen im Grundwasser eines Lockergesteins-Aquifers. Geol Jahrb Reihe C 51:3–40
*Böttcher J, Strebel O, Voerkelius S, Schmidt HL (1990a) Using isotope fractionation of nitrate nitrogen and nitrate-oxygen for evaluation of microbial denitrification in a sandy aquifer. Hydrol 114:413–424

Böttcher J, Strebel O, Duynisveld WHM (1990b) Microbial denitrification in the groundwater of a sandy aquifer: Kinetics and stream-tube model. Mitt Dtsch Bodenkd Ges 60:265–270
Bowen HJM (1979) Environmental chemistry of the elements. Academic Press, London, 26 pp
Bracewell JM, Haider K, Larter SR, Schulten HR (1989) Thermal degradation relevant to structural studies of humic substances. In: Hayes MHB, Mac Carthy P, Malcolm RL, Swift RS (eds) Humic substances II: In search of structure. John Wiley, Chichester, pp 181–222
Braitsch O (1971) Salt deposits. Their origin and composition. Springer, Berlin Heidelberg New York
Brassard P, Kramer JR, Collins PV (1990) Binding site analysis using linear programming. Environ Sci Technol 24:195–201
Breddin H, Brühl, Dieler H (1960) Das Blatt Aachen-Nordwest der praktisch-geologischen Grundkarte 1:5000 des Aachener Stadtgebietes. Geol Mitt 1:251–428
Breemen N van (1973) Acid sulphate soils Vol I, Wageningen, pp 66–130
Breen A, Stahl DA, Flesher B, Sayler G (1989) Characterization of *Pseudomonas geomorphus*: a novel groundwater bacterium. Microb Ecol 18:221–233
Brend'amour H-J (1986) Untersuchung von Stoff-Flüssen in der wasser-ungesättigten und – gesättigten Zone in quartären Sedimenten (Segeberger Forst, Schleswig-Holstein). PhD Thesis, Ber Rep Geol Paläontol Inst Univ Kiel 17:196 pp
Broecker W, Olson EA (1961) Lamont radiocarbon measurements VIII. Radiocarbon 3:176
Bromfield SM (1956) Oxidation of manganese by soil microorganisms. Aust J Biol Sci 9:238–252
Buchanan-Mappin JM, Wallis PM, Buchanan AG (1986) Enumeration and identification of heterotrophic bacteria in groundwater and in a mountain stream. Can J Microbiol 32:93–98
Buddrus J, Burba P, Herzog H, Lambert J (1989) Quantitation of partial structures of aquatic humic substances by one-dimensional and two-dimensional solution carbon-13 nuclear magnetic resonance spectroscopy. Analyt Chem 61(6):628–631
Buffle J (1988) Complexation reactions in aquatic systems. Ellis Horwood, Chichester
Buhmann D (1984) Ein neues kinetisches Modell zur Berechnung der Lössung und Ausscheidung von Kalk bei Verkarstungsprozessen und seine experimentelle Bestätigung. PhD Thesis, Univ Bremen
Buhmann D, Dreybrodt W (1985a) The kinetics of calcite dissolution and precipitation in geologically relevant situations of karst areas: 1. Open system. Chem Geol 48:189–211
Buhmann D, Dreybrodt W (1985b) The kinetics of calcite dissolution and precipitation in geologically relevant situations of karst areas: 2. Closed system. Chem Geol 53:109–124
Buhmann D, Dreybrodt W (1987) Calcite dissolution kinetics in the system H_2O–CO_2–$CaCO_3$ with participation of foreign ions. Chem Geol 64:89–102
Bull HB (1951) Physical biochemistry, 2nd edn. John Wiley, New York
Bulman RA (1985) The speciation of iodine in the environment. In: Bulman RA, Cooper JR (eds) Speciation of fission and activation products in the environment. Elsevier, London pp 213–222
Burba P, Willmer PG (1988) An ion-exchange study on the mobility of heavy metals bound to aquatic humic substances. Proc 4th Int Meet 1988, Huelva, Spain, Humic Substances Society (IHSS)
Busenberg E, Clemency CH V (1976) The dissolution kinetics of feldspars at 25 °C and 1 atm CO_2–partial pressure. Geochim Cosmochim Acta 40:41–49
Busenberg E, Plummer LN (1985) Kinetic and thermodynamic factors controlling the distribution of SO_4^{2-} and Na^+ in calcites and selected aragonites. Geochim Cosmochim Acta 49:713–725
Busson G (1967) Le Mesozoique saharien, 2 ème Partie: L'Extreme – Sud Tunisien. Ed Centre Nat Rech Sci, Paris
Büttner G, Lamersdorf N (1985) Exkursionsführer Sommer 1985. Ber Forschungszentr Waldökosysteme/Waldsterben 331–357
Buurman P, Van Reeuwijk LP (1984) Proto-imogolite and the process of podzol formation: a critical note. Soil Sci 35:447–452
Campbell PGC, Bisson M, Bougie R, Tessier A, Villeneuve JP (1983) Speciation of aluminium in acidic fresh waters. Anal Chem 55:2246–2252
Carpenter AB (1980) The chemistry of dolomite formation. I: The stability of dolomite. In: Zenger DH (Ed) Concepts and models of dolomitization. Soc Econ Paleontol and Mineral Spec Publ 28:111–121
Carslaw HS, Jäger JC (1959) Conduction of heat in solids, 2nd edn. Clarendon, Oxford
Ceratzki W (1971) Eine Saugvorrichtung für kapillar gebundenes Bodenwasser. Landbau Völkenrode 21:13–14
*Cevrim M (1984) Die Zink–Blei–Vererzungen des Aladag–Gebietes, Türkei, mit Betrachtungen zur Paläokarstentwicklung. PhD Thesis, TH Aachen, 184 pp

*Cevrim M, Echle W (1988) Zink–Minerale der Pyroaurit-Sjögrenit und Berthierin-Gruppe als Produkte kinetischer Umwandlungsprozesse im Karstmilieu. Fortschr Mineral 66 Beih 1:20
*Cevrim M, Echle W, Friedrich G (1986) Paleokarst related Zn-Pb mineralization of Aladag Mountains. Bull Geol Soc Turk 29/C:27–41 (in Turkish)
Champ DR, Gulens J, Jackson RE (1979) Oxidation-reduction sequences in groundwater flow systems. Can J Earth Sci 16:12–23
Chiou CT, Malcolm RL, Brinton TJ, Kile DE (1986) Water solubility enhancement of some organic pollutants and pesticides by dissolved humic and fulvic acids. Environ Sci Technol 20:502–508
Chou L, Wollast R (1984) Study of the weathering of albite at room temperature and pressure with a fluidized bed reactor. Geochim Cosmochim Acta 48:2205–2217
Chou L, Wollast R (1985) Steady-state kinetics and dissolution mechanisms of albite. Am J Sci 285:963–993
Christman RF, Gjessing ET (1983) Aquatic and terrestrial humic material. Ann Arbor Science, Ann Arbor, Michigan
Christman RF, Norwood DL, Seo Y, Frimmel FH (1989) Oxidative degradation of humic substances from freshwater environments. In: Hayes MHB, Mac Carthy P, Malcolm RL, Swift RS (eds) Humic substances II. In search of structure. John Wiley, Chichester, pp 33–68
Christophersen N, Seip HM, Njos A (1980) Simulation of flow patterns and ion–exchange in soil percolation experiments. Water Air Soil Pollut 14:159–170
Churchman GJ, Jackson ML (1976) Reaction of montmorillonite with aqueous solutions: solute activity control by a secondary phase. Geochim Cosmochim Acta 40:1251–1259
Cicerone RJ (1981) Halogens in the atmosphere. Rev Geophys Space Phys 19:123–139
Civijic J (1918) Hydrographie souterraine et évolution morphologique du karst. Trav Inst Geogr Alp Univ Grenoble 6:4
Claypool GE, Kaplan IR (1974) The origin and distribution of methane in marine sediments. In: Kaplan IR (ed) Natural gases in marine sediments. Plenum, New York, pp 99–139
Correns CW (1940) Die chemische Verwitterung der Silikate. Naturwissenschaften 28:369–376
Correns CW, von Engelhardt W (1938) Neue Untersuchungen über die Verwitterung des Kalifeldspates. Chem Erde 12:1–22
Cronan CS, Schofield CL (1979) Aluminium leaching response to acid precipitation: effects on high-elevation watersheds in the northeast. Science 204:304–306
Crutzen PJ (1983) Atmospheric Interactions–Homogeneous gas reactions of C, N and S containing compounds. In: Bolin B, Cook RB (eds) The major biogeochemical cycles and their interactions. John Wiley, New York, pp 67–112
Cullimore DR, McCann AE (1977) The identification, cultivation and control of iron bacteria in ground water. In: Skinner E (ed) Aquatic microbiology. Soc Appl Bacteriol Symp Ser No 6 Academic Press, London, pp 219–261
D'Ans J (1933) Die Lösungsgleichgewichte der Systeme der Salze ozeanischer Salzablagerungen. Kali-Forsch.-Anstalt, Berlin
*Dahmke A (1988) Lösungskinetik von Feldspat-reichen Gesteinen und deren Bezug zu Verwitterung und Porenwasser-Chemie natürlicher Sander-Sedimente. PhD Thesis, Ber Rep Geol–Paläontol Inst Univ Kiel 20:168
*Dahmke A, Matthess G, Pekdeger A, Schenk D, Schulz HD (1986) Near-surface geochemical processes in Quaternary sediments. J Geol Soc Lond 143:667–672
*Dahmke A, Matthess G, Petersen A, Schenk D (1989) Gleichgewichts- und Ungleichgewichts beziehungen zwischen Porenlösungen und festen Substanzen der Verwitterungszone quartärer Sedimente. Ber Rep Geol Paläontol Inst Univ Kiel 25:51
David MB, Driscoll CT (1984) Aluminium speciation and equilibria in soil solutions of an Adirondack Mountain forest soil. Geoderma 33:297–309
Davies ND, De Wiest RJM (1967) Hydrogeology. John Wiley, New York - London - Sydney, 463 pp
Davis GH, Newton KG (1969) Numerical taxonomy of some named coryneform bacteria. J Gen Microbiol 56:195–214
Dawson HJ, Ugolini FC, Hrutfiord BF, Zachara J (1978) Role of soluble organics in the soil processes of a podzol Central Cascades, Washington. Soil Sci 126:290–296
De Haan H (1974) Effect of a fulvic acid fraction on the growth of a *Pseudomonas* from Tjeukemeer (The Netherlands). Freshwater Biol 4:301–310
De Haan H, De Boer T (1978) A study of the possible interactions between fulvic acids, amino acids, and carbohydrates from Tjeukemeer, based on gel filtration at pH 7.0. Water Res 12:1035–1040

Deines P, Langmuir D, Harmon RS (1974) Stable carbon isotope ratios and the existence of a gas phase in the evolution of carbonate ground waters. Geochim Cosmochim Acta 38:1147–1164

De Nobili M, Gjessing ET, Sequi P (1989) Sizes and Shapes of Humic Substances by Gel Chromatography. In: Hayes MHB, MacCarthy P, Malcolm RL, Swift RS (eds) Humic Substances II. In search of structure. John Wiley, Chichester, pp 561–591

DEV, Deutsche Einheitsverfahren zur Wasser-, Abwasser- und Schlammuntersuchung (1971–1988). (Ed) Fachgruppe Wasserchemie GDCh, 3rd ed, parts 1–20, Verlag Chemie, Weinheim/Bergstr

Dibble WE Jr, Tiller WA (1981) Non-equilibrium water/rock interaction. I. Model for interface-controlled reactions. Geochim Cosmochim Acta 45:79–92

*Dietze G (1985) Bindungsformen und Gleichgewichte von Aluminium in Sickerwasser saurer Böden. Ber Forschungszentr Waldökosysteme/Waldsterben 16:1–118

*Dietze G (1986) Tiefenverteilung der Al-Bindungsformen bzw. Gleichgewichte im Sickerwasser ausgewählter Standorte beim Transport durch die ungesättigte Zone. Ber Forschungszentr Waldökosysteme/Waldsterben B/3:1–24

DiSpirito AA, Hooper AB (1986) Oxygen exchange between nitrate molecules during nitrite oxidation. J Biol Chem 261:10534–10537

Dockins WS, Olson GJ, McFeters GA, Turbak SC (1980) Dissimilatory bacterial sulfate reduction in Montana groundwaters. Geomicrobiol J 2:83–98

Domergue C, Dumon E, de Lapparent AF, Possel P (1952) Monographies Regionales l'ème Serie, Tunisie – No. 7 (Sud et Extrême-Sud Tunis). XIX Congr Geol Int, Tunis

Domke K (1972) Fraktionierung "freier" und "gebundener" Braunhuminsäuren und Grauhuminsäuren mittels Gelfitration, Elektrophorese bzw. organischer Lösungsmittel. Arch Acker-Pflanzenbau Bodenkd 16:3–11

Donnan FG (1911) Theorie der Membrangleichgewichte und Membranpotentiale bei Vorhandensein von nicht dialysierenden Elektrolyten – Ein Beitrag zur physikalisch–chemischen Physiologie. Z Elektrochem 17:572–581

Dörr H (1984) Die Untersuchung des Gas– und Wasserhaushalts in der ungesättigten Bodenzone mit Hilfe von ^{222}Rn und CO_2. PhD Thesis, Univ Heidelberg

Dörr H (1986) Der Abbau der organischen Substanz im Boden. Heidelb Geowiss Abh 6:71–80

Dörr H, Münnich KO (1980) ^{14}C and ^{13}C in soil CO_2. Radiocarbon 22/3:909–918

*Dörr H, Münnich KO (1986) Annual variations of the ^{14}C-content of soil CO_2. Radiocarbon 28/2A:338–345

*Dörr H, Münnich KO (1987) Annual variation in soil respiration in selected areas of the temperate zone. Tellus 39/B:114–121

Dörr H, Münnich KO (1990) ^{222}Rn flux and soil air concentration profiles in West Germany. Soil ^{222}Rn as tracer for gas gransport in the unsaturated soil zone. Tellus 42B:20–28

*Dörr H, Sonntag Ch, Regenberg W (1987) A field study on the initial ^{14}C content as a limiting factor in ^{14}C groundwater dating. In: Isotope techniques in water resources development. IAEA SM-299/038, Vienna, 1987, pp 73–86

Dott W, Thofern E (1980) Qualitative und quantitative Bestimmung von Bakterienpopulationen aus aquatischen Biotopen. 2. Mitteilung: Anwendung miniaturisierter Testsysteme zur Identifizierung und Biotypisierung von Bakterien unter Verwendung der Vielpunktbeimpfungsmethode. Zbl Bakt Hyg, 1 Abt Orig B 170:99–107

Dott W, Trampisch HJ (1983) Qualitative und quantitative Bestimmung von Bakterienpopulationen aus aquastischen Biotopen. 5.Mitteilung: Vergleichende Untersuchungen an zwei Schnellsandfiltern. Zbl Bakt Hyg, 1 Abt Orig B 177:141–155

Dott W, Tuschewitzki GJ, Thofern E (1982) Methylotrophe Bakterien im Trinkwasserbereich. 2. Mitteilung: Biochemisch/physiologische Charakterisierung der Isolate. Zbl Bakt Hyg, 1 Abt Orig B 176:189–201

Drake JJ, Ford DC (1981) Karst solution: a global model for groundwater solute concentrations. Trans Jpn Geomorphol Un 2:223–230

Drews G (1974) Mikrobiologisches Praktikum, 2nd edn. Springer, Berlin Heidelberg New York

Drews G (1976) Mikrobiologisches Praktikum, 3rd edn. Springer, Berlin Heidelberg New York

Dreybrodt W (1988) Processes in karst systems: physics, chemistry and geology. Springer Series in Physical Environments, vol. 4. Springer, Berlin Heidelberg New York Tokyo

Dreybrodt W, Svensson U (1989) The effect of $MgCO_3$ to the solubility of calcite: enhanced aggressiveness? In: Kosa A (ed) Proc 10th Int Congr on Speleol, Budapest, pp 177–178

Driscoll CT (1984) A procedure for the fractionation of aqueous aluminum in dilute acidic waters. Int J Environ Anal Chem 16:267–283

Droubi A, Cheverry C, Fritz B, Tardy Y (1976) Géochimie des eaux des sels dans les sols des polders du lac Tschad: application d'un modèle thermodynamique de simulation de l'evaporation. Chem Geol 17:165–177

Dubach P, Metha NC (1963) The chemistry of soil humic substances. Soils Fert 26:293–300

Dubinina GA (1978) Mechanism of the oxidation of divalent iron and manganese by iron bacteria growing at a neutral pH of the medium. Mikrobiologiya 47:471–478 (translation)

Duynisveld WHM, Strebel O, Böttcher J, Kinzelbach W (1989) Longterm prognosis of the groundwater quality as influenced by landuse and landuse changes in the Fuhrberger Feld using numerical solute transport models. In: Kobus HE, Kinzelbach W (eds) Contaminant transport in groundwater. Balkema, Rotterdam, pp 89–96

DVWK Fachausschuß Grundwasserbiologie (1988) Bedeutung biologischer Vorgänge für die Beschaffenheit des Grundwassers. Schriftenr DVWK 80:322

Dzombak DA, Fish W, Morel FMM (1986) Metal-humate interactions. 1. Discrete ligand and continuous distribution models. Environ Sci Technol 20:669–675

Edmunds WM (1977) Groundwater geochemistry – controls and process. In: Papers and proceedings, groundwater quality, measurement prediction and protection. Water Research Centre, Medmenham, England, pp 115–147

Edmunds WM, Miles DL, Cook JM (1984) A comparative study of sequential redox processes in three British aquifers. IAHS Publ 150:55–70

Ehrlich HL (1976) Manganese as energy source for bacteria. In: Nriagu JD (ed) Environmental biogeochemistry, vol 2. Ann Arbor Science, Michigan, pp 633–644

Ehrlich HL (1978) Inorganic energy sources for chemolithotrophic and mixotrophic bacteria. Geomicrobiol J 1/1:65–83

Ehrlich GG, Godsy EM, Goerlitz DF, Hult MF (1983) Microbial ecology of a creosote–contaminated aquifer at St. Louis Park, Minnesota. (In: Nash III CH Underkofler LA (eds) Dev Ind Microbiol 24:235–245) Microb Ecol 11:205–219

*Eichinger L (1987) Isotopic and geochemical studies investigating the genesis of carbon isotope content in young groundwaters. In: Proc Int Symp Use of Isotope Techniques in Water Resources Development. IAEA, Vienna, pp 87–134

Eichinger L, Merkel B (1982) Untersuchungen zum Kohlenstoffhaushalt der ungesättigten und gesättigten Zone in einem Karbonatschotter – Methoden und erste Ergebnisse. In: Beitr über hydrol Tracemethoden u ihre Anwendungen. Ges f Strahlen- und Umweltforschung, München, GSF-Bericht R 290:113–132

*Eichinger L, Merkel B (1984) Grundluftuntersuchungen – ein Vergleich zwischen einem Wald– und einem Freiflächenstandort der Münchener Schotterebene. In: Wald and Wasser, Prozesse im Wasser und Stoffkreisläufe von Waldgebieten, Proc National park verwaltung Bayrischer Wald vol II, pp 669–721

*Eichinger L, Schulz HD (1984) Bestimmung der Sickerrate in einem Sander Norddeutschlands durch Messungen der Umweltisotopen und Chlorid–Konzentrationen des Sickerwassers. In: Udluft P, Merkel B, Prösl KH (eds) Proc RIZA-Symp, Dept Hydrogeology Hydrochemistry, Techn. Univ. München, Vol II, pp 547–558

*Eichinger L, Merkel B, Nemeth G, Salvamoser J, Stichler W (1984) Seepage velocity determination in unsaturated quaternary gravel. A comparison of various methods. In: Udluft P, Merkel B, Prösl KH (eds) Proc RIZA–Symposium, vol I, Dept Hydrogeology Hydrochemistry Techn Univ München, pp 303–313

Ephraim J, Marinsky JA (1986) A unified physicochemical description of the protonation and metal ion complexation equilibria of natural organic acids (humic and fulvic acids). 3. Influence of polyelectrolyte properties and functional heterogeneity on the copper ion binding equilibria in an Armadale horizons Bh fulvic acid sample. Environ Sci Technol 20:367–376

Ephraim J, Alegret S, Mathuthu A, Bicking M, Malcolm RL, Marinsky JA (1986) A unified physicochemical description of the protonation and metal ion complexation equilibria of natural organic acids (humic and fulvic acids). 2. Influence of polyelectrolyte properties and functional group heterogeneity on the protonation equilibria of fulvic acid. Environ Sci Technol 20:354–366

Esser N (1980) Bombentritium–Zeitverhalten seit 1963 im Abfluß mitteleuropäischer Flüsse und kleinerer hydrologischer Systeme, PhD Thesis, Univ Heidelberg

Farkasdi G, Golwer A, Knoll K-H, Matthess G, Schneider W (1969) Mikrobiologische und hygienische Untersuchungen von Grundwasserverunreinigungen im Unterstrom von Abfallplätzen. Städtehygiene 20:25–31

Feldberg SW (1969) Digital simulation: solving electrochemical diffusion-kinetic problems. Electroanal Chem 3:199–296

Figura P, McDuffle B (1980) Determination of labilities of soluble trace metal species in aqueous environmental samples by anodic stripping voltametry and chalex column and batch methods. Anal Chem 52:1433–1439

*Finger W (1989) Isolierung und Charakterisierung von Boden-Fulvinsäuren und Untersuchung ihrer Komplexbildung mit Metall-Ionen. PhD Thesis, Univ Marburg

*Finger W, Post B, Klamberg J (1990) Wechselwirkungen zwischen Boden–Huminstoffen und Metall-Ionen. I. Isolierung und Charakterisierung von Boden–Huminstoffen. Z Pflanzenernähr Bodenkd, 153:305–312

Fish W, Dzombak DA, Morel FMM (1986) Metal–humate interactions. 2. Application and comparison of models. Environ Sci Tecnol 20:676–683

Flaig W (1964) Chemische Untersuchungen an Huminstoffen. Z Chem 4:253–265

Flaig W, Beutelspacher H, Rietz E (1975) Chemical composition and physical properties of humic substances. In: Giesking JE (ed) Soil components, vol 1. Organic compounds. Springer, Berlin Heidelberg New York

Flipse WJ Jr, Katz BG, Lindner JB, Markel R (1984) Sources of nitrate in a sewered housing development, central Long Island, New York. Ground Water 22:418–426

Florence TM (1986) Electrochemical approaches to trace element speciation in water. A review. Analyst 111:489–505

Fontes JC, Garnier JM (1979) Determination of the initial ^{14}C activity of the total dissolved carbon: a review of the existing models and a new approach. Water Resour Res 15/2:399–413

Ford DC (1980) Threshold and limit effects in karst hydrology. In: Coates DR, Vitek ID (eds) Threshold in geomorphology. Allen and Unwin, London, pp 345–362

Ford DC, Drake JJ (1980) Spatial and temporal variations in karst solution rates: the structure of variability. In: Thorn CD (ed) Space and time in geomorphology. Allen and Unwin, London, pp 147–170

Forster M (1984) C-14- und Ar-39-Gehalte in einem Grundwasserleiter des Saarlandes. Vergleich zweier Grundwasserdatierungsmethoden. PhD Thesis, Univ München

Forster M, Maier P (1987) The Munich facility for argon and krypton preparation and low level measurement of 39-Ar and 85-Kr in the Hydrosphere. In: Low level measurements and their application to environmental radioactivity. World Scientific Publ, Singapore

Frank JR (1981) Dedolomitization in the Taum Sauk Limestone (Upper Cambrian), SE Missouri. J Sediment Petrol 51:7–18

Frimmel FH (1977) Komplexierung durch Huminstoffe eines Braunwassersees. Vom Wasser 49:1–10

Frimmel FH (1985) Metal pollutants and humic substances. Water Supply 3:49–60

*Frimmel FH, Bauer H (1987) Influence of photochemical reactions on the optical properties of aquatic humic substances gained from fall leaves. Sci Total Environ 62:139–148

Frimmel FH, Christman RF (1988) Humic substances and their role in the environment. John Wiley, Chichester

Frimmel FH, Geywitz J (1983) Zur koordinativen Bindung von Metallionen an Gewässerhuminstoffe. Fresenius Z Anal Chem 316:582–588

*Frimmel FH, Hopp W (1986) Stability spectra for the description of copper-humic complexes. A fluorescence quench study. Fresenius Z Anal Chem 325:68–72

Frimmel FH, Niedermann H (1980) Komplexierung von Metallionen durch Gewässerhuminstoffe. I. Ein Braunwassersee als Huminstofflieferant. Z Wasser Abwasserforsch 13/4:119–124

*Frimmel FH, Trampisch H (1984) Fractionation and characterization of humic substances by HPLC and different detection systems. In: Hayes MHB, Swift RS (eds) Volunteered Papers 2nd Int Conf Int Humic Substances Society

Frimmel FH, Immerz A, Niedermann H (1983) Heavy metal interaction with aquatic humus. Int J Environ Anal Chem 14:105–115

Frind EO, Duynisveld WHM, Strebel O, Böttcher J (1989) Simulation of nitrate and sulfate transport and transformation in the Fuhrberger Feld aquifer, Hannover, Germany. In: Kobus HE, Kinzelbach W (eds) Contaminant transport in groundwater. Balkema, Rotterdam, pp 97–104

Frind EO, Duynisveld WHM, Strebel O, Böttcher J (1990) Modelling of multi-component transport with microbial transformation in groundwater: the Fuhrberg case. Water Resour Res 26:1707–1719

Frind EO, Strebel O, Duynisveld WHM, Böttcher J (1991) Transport and transformation of nitrate in groundwater: The Fuhrberger Feld case. In: Schadstofftransport in Grund- und Oberflächengewässern. 21 Int Wasserbau-Seminar 1990/91, Techn Hochschule Aachen (in press)

*Fritsch P (1990) Stoffumsetzungen und Stofftransport in tonigen Deckschichten-Untersuchungen zur Genese des reduzierten Grundwassers in der Altneckaraue (Oberrheingraben). PhD Thesis TH Darmstadt

*Fritsch P, Ebhardt G (1988) Stoffumsetzungen in heterogenen Porensystemen toniger Auen-Deckschichten der Altneckaraue/Oberrheingraben. Z Dtsch Geol Ges 139:407–414

Froehlich PN, Klinkhammer GP, Bender ML, Luedtke NA, Heath GR, Cullen D, Dauphin P, Hammond D, Hartmann B, Maynard V (1979) Early oxidation of organic matter in pelagic sediments of the eastern equatorial Atlantic: suboxic diagenesis. Geochim Cosmochim Acta 43:1075–1090

Fuchs F (1982) Die Verwendung der Gelfiltrations-Chromatographie zur Untersuchung höhermolekularer organischer Wasserinhaltsstoffe. Veröff Lehrstuhl Wasserchemie und DVGW-Forschungsstelle Engler-Bunte-Institut Universität Karlsruhe 19:54–95

Fuchs F (1985) Gelchromatographische Trennung von organischen Wasserinhaltsstoffen, I, II. Vom Wasser 64:129–144, 65:93–106

Fuchs F (1986) Gelchromatographische Trennung von organischen Wasserinhaltsstoffen, III. Vom Wasser 66:127–136

Fuchs F, Raué B (1981) Charakterisierung hochmolekularer Wasserinhaltsstoffe durch Fraktionierung mittels Gelfiltration. Vom Wasser 57:95–106

Fuchs O (1980) Physikalische Grundlagen und Eigenschaften der Lösungen von nieder- und hochmolekularen Verbindungen. In: Gnamm H, Fuchs O: Lösungsmittel und Weichmachungsmittel. Vol 1. 8th edn. Wissenschaftliche Verlagsgesellschaft, Stuttgart, pp 1185

Fuchsman CH (1984) The humic acid problem and the prospects of peat utilization. In: Fuchsman CH, Spigarelli SA (eds) Proc Int Symp on Peat utilization. Bemidji 1983, pp 477–486

Füchtbauer H (1950) Die nichtkarbonatischen Bestandteile des Göttinger Muschelkalkes mit besonderer Berücksichtigung der Mineralneubildungen. Heidelb Beitr Mineral Petrol 2:235–254

Füchtbauer H (1962) Fazies, Porosität und Gasinhalt der Karbonatgesteine des norddeutschen Zechsteins. Z Dtsch Geol Ges 114:484–531

Füchtbauer H, Goldschmidt H (1965) Beziehungen zwischen Ca-Gehalt und Bildungsbedingungen der Dolomite. Geol Rundsch 55:29–40

Fuksa J (1981) Microorganisms in oil-polluted groundwater environment. Verh Int Verein Limnol 21:1376–1380

Furrer G, Stumm W (1986) The coordination chemistry of weathering: I. Dissolution kinetics of δ-Al_2O_3 and BeO. Geochim Cosmochim Acta 50:1847–1860

Gadel F, Bruchet A (1987) Application of pyrolysis-gaschromatography-mass spectrometry to the characterization of humic substances resulting from decay of aquatic plants in sediments and waters. Water Res 21:1195–1206

Gallikowski CA, Hirsch P (1989) Microorganisms of some uncontaminated soils (regoliths) of the McMurdo Dry Valleys (Ross Desert), South Victoria Land, Antarctica. Polar Biol (submitted)

Gamble DS, Underdown AW, Langford CH (1980) Copper (II) titration of fulvic acid ligand sites with theoretical, potentiometric, and spectrophotometric analysis. Anal Chem 52:1901–1908

Gardener WS, Landrum PF (1983) Characterization of ambient levels of ultraviolet-absorbing dissolved humic materials in natural waters by aqueous liquid chromatography. In: Christman RF, Gjessing ET (eds) Aquatic and terrestrial humic materials. Ann Arbor Science, Ann Arbor, Michigan

Garrels RM, Christ CL (1965) Solutions, minerals and equilibria. Harper and Row, Evanston

Garrels RM, MacKenzie (1971) Evolution of sedimentary rocks. Norton, New York, 397 p

Gehlen M, Trampisch HJ, Dott W (1985) Physiological characterization of heterotrophic communities from selected aquatic environments. Microb Ecol 11:205–219

Gerhardt P, Murray RGE, Costilow RN, Nester EW, Wood WA, Krieg NR, Phillips GB (eds) (1981) Manual of methods for general bacteriology. American Society for Microbiology, Washington DC

Ghiorse WC, Balkwill DL (1982) Microbial characterization of subsurface environments. In: Ward CH, Giger W (eds) Proc. 1st Int Conf on Groundwater Quality Research. Wiley Interscience New York, pp 387–401

Ghiorse WC, Balkwill DL (1983) Enumeration and morphological characterization of bacteria indigenous to subsurface environments. Dev Ind Microbiol 24:213–224
Ghiorse WC, Balkwill DL (1985) Microbial characterization of subsurface environments. In: Ward CH, Giger W, McCarty PL (eds) Ground water quality. John Wiley, New York, pp 536–556
Ghiorse WC, Wilson JT (1988) Microbial ecology of the terrestrial subsurface. Adv Appl Microbiol 33:107–172
Ghosh K, Schnitzer M (1980) Macromolecular structures of humic substances. Soil Sci 129/5:266–276
Gjessing ET (1976) Physical and chemical characteristics of aquatic humus. Ann Arbor Science, Ann Arbor, Michigan
Glasstone S, Laidler K, Eyering H (1941) The theory of rate processes. McGraw-Hill, New York
Glückauf E (1955) Theory of chromatography. Part 9. The "theoretical plate" concept in column separation. Trans Faraday Soc 51:34–43
Gmelin L (1973) Handbuch der Anorganischen Chemie, 8th edn, part C3: Kohlenstoff. Verlag Chemie, Weinheim/Bergstr
*Goebelbecker J (1985) Sedimentpetrographische und hydrogeochemische Untersuchungen an den dolomitischen Carbonatgesteinen des Djebel Dahar, Südtunesien. PhD Thesis, Univ Karlsruhe
Goldsmith JR, Heard HC (1961) Subsolidus phase relations in the system $CaCO_3$–$MgCO_3$. J Geol 69:45–74
Goltz M, Roberts PV (1986) Three-dimensional solutions for transport in an infinite medium with mobile and immobile zones. Water Resour Res 22/7:1139–1148
Golwer A, Matthess G, Schneider W (1970) Selbstreinigungsvorgänge im aeroben und anaeroben Grundwasserbereich. Vom Wasser 36:64–92
Golwer A, Knoll K-H, Matthess G, Schneider W, Wallhäusser KH (1972) Mikroorganismen im Unterstrom eines Abfallplatzes. Gesundheits–Ingenieur 93:142–152
Golwer A, Knoll K-H, Matthess G, Schneider W, Wallhäusser KH (1976) Belastung und Verunreinigung des Grundwassers durch feste Abfallstoffe. Abh Hess Landesamt Bodenforsch 73:131
Gonzales B, Merino A, Almeida M, Vicna R (1986) Comparative growth of natural bacterial isolates on various lignin-related compounds. Appl Environ Microbiol 52:1428–1432
Goossens DA, Philippaerts JG, Gijbels R, Pijpers AP, van Tendeloo SR, Althaus E (1989) A SIMS, XPS, SEM, TEM and FTIR study of feldspar surfaces after reacting with acid solution. In: Miles DL (ed) Proc Water-Rock Interaction WRI-6. Balkema, Rotterdam, pp 271–274
Gottfreund J, Schwisfurth R (1983) Mikrobiologische Oxidation und Reduktion von Manganspecies. Fresenius Z Anal Chem 316:634–638
Gottfreund J, Schmitt G, Schweisfurth R (1983) Chemische und mikrobiologische Untersuchungen an Komplexverbindungen des Mangans. Mitt Dtsch Bodenkd Ges 38:325–330
Gottfreund E, Gerber I, Schweisfurth R (1984) Quantitative Verteilung verschiedener physiologischer Bakteriengruppen im Lockergestein zweier Grundwasserleiter bis 90 m Tiefe. Mitt Dtsch Bodenkd Ges 38:522–530
Gottfreund J, Schmitt G, Schweisfurth R (1985a) Wertigkeitswechsel von Manganspecies durch Bakterien in Nährlösungen. Landwirtsch Forsch 38:80–86
Gottfreund E, Gottfreund J, Gerber I, Schmitt G, Schweisfurth R (1985b) Occurrence and activities of bacteria in the unsaturated and saturated underground in relation to the removal of iron and manganese. Water supply 3 Berlin "A":109–115
Gottfreund E, Gottfreund J, Schweisfurth (1985c) Mikrobiologische Grundlagen der subterrestrischen Enteisenung und Entmanganung. Forum Städte–Hyg 38:178–183
Gounot AM, DiRuggiero J, Haroux C (1987) Bacterial manganese transformations in groundwaters. In: Giovannozzi-Sermanni G, Nannipieri P (eds) Current perspectives in environmental biogeochemistry: CNR-IPRA, Rome pp 371–832
Graentzel A (1980) Dünnfilm-UV-Meßeinrichtung. Manual, Physikalische Werkstätte, Karlsruhe
Greenland DJ, Hayes MHB (eds) (1978) The chemistry of soil constituents. John Wiley, New York
*Grossmann J (1988) Physikalische und chemische Prozesse bei der Probenahme von Sickerwasser mittles Saugsonden. PhD Thesis, TU München, 147 pp
*Grossmann J, Freitag G, Merkel B (1985) Eignung von Nylon- und Polyvinylidenfluorid membranfiltern als Materialien zum Bau von Saugkerzen. Z Wasser- Abwasserforsch 18:187–190
*Grossmann J, Quentin KE, Udluft P (1987) Sickerwassergewinnung mittels Saugkerzen – eine Literaturstudie. Z Pflanzenernähr Bodenkd 150:258–261
*Grossmann J, Merkel B, Udluft P (1988) Calcite-carbon equilibrium in soil water samples. Z Wasser-AbwasserForsch 21:177–181

Gunn J, Gagen P (1987) Limestone quarrying and sinkhole development in the English Peak District. In: Beck BF, Wilson WL (eds) Karst hydrology: engineering and environmental applications. Balkema, Rotterdam

*Hack A, Leuchs W, Obermann P (1984) Der Salzsprung im Grundwasser. Geowiss unserer Zeit 6:194–200

Haider K, Schulten HR (1985) Pyrolysis field ionization mass spectrometry of lignin, soil humic compounds and whole soil. J Anal Appl Pyrolysis 8:317–331

Haines Bl, Waide JB, Todd RL (1982) Soil solution nutrient concentrations sampled with tension and zero-tension lysimeters: Report of discrepancies. Soil Sci Soc Am J 46:658–661

Hansen EA, Harris AR (1975) Validity of soil-water samples collected with porous ceramic cups. Soil Sci Soc Am Proc 39:528–536

Hantush MS (1956) Analysis of data from pumping tests in leaky aquifers. Trans Am Geophys Union 37:702–714

Hantush MS (1964) Hydraulic of wells. In: Advances of hydroscience, Vol. 1. Academic Press, New York, pp 281–432

Harper AM, Duewer DL, Kowalski BR, Fasching JL (1977) In: Kowalski BR (ed) Chemometrics and application. ACS Symp Ser 52:14–51

Hashimoto J, Jackson ML (1960) Rapid dissolution of allophane and kaolinite-halloysite after dehydration. 7th Natl Conf Clays Clay Miner 3:102–113

Hässelbarth U, Lüdemann D (1967) Die biologische Verockerung von Brunnen durch Massenentwicklung von Eisen- und Manganbakterien. Bohrtechnik-Brunnenbau-Rohrleitungsbau 10/11, 12 pp

Hauhs M, Dise N (1989) Depletion of exchangeable base cations in an acid forest soil at Lange Bramke, West Germany. Geoderma (in press)

Hayes MHB (1985) Extraction of humic substances from soil. In: Aiken GR, McKnight DM, Wershaw RL, MacCarthy P (eds) Humic substances in soil, sediment and water. John Wiley, New York

Hayes MHB, MacCarthy P, Malcolm RL, Swift RS (1989) Humic substances II. In search of structure. John Wiley Chichester

Heinrichs H, König N, Schultz R (1985) Atom-Absorptions- und Emissionsspektroskpische Bestimmungsmethoden für Haupt- und Spurenelemente in Probelösungen aus Waldkökosystemuntersuchungen. Ber Forschungszentr Waldökosysteme/Waldsterben 8:1–92

Helgeson HC (1971) Kinetics of mass transfer among silicates and aqueous solutions. Geochim Cosmochim Acta 35:421–469

Helgeson HC, Murphy WM, Aagaard P (1984) Thermodynamic and kinetic constraints on reaction rates among minerals and aqueous solutions. II. Rate constants, effective surface area, and the hydrolysis of feldspar. Geochim Cosmochim Acta 48:2405–2432

Hem JD (1973) Chemistry of manganese in natural waters. US Geol Surv Water-Supply Pap 1667, Washington DC, pp 64+42+18+28

Hem JD (1985) Study and interpretation of the chemical characteristics of natural water, 3rd edn. US Geol Surv, Water-Supply Pap 2254, Washington DC, 263 pp

*Hempfling R, Schulten HR (1988) Characterization and dynamics of organic compounds in forest humus studied by pyrolysis-gas chromatography/electron ionization mass spectrometry and pyrolysis-(high resolution) field ionization mass spectrometry. J Anal Appl Pyrolysis 13:319–325

Hempfling R, Schulten HR (1990a) Direct chemical characterization of dissolved organic matter in water by pyrolysis-field ionization mass spectrometry. Int J Environ Anal Chem 43:55–62

Hempfling R, Schulten HR (1990b) Chemical characterization of the organic matter in forest soils by Curie point pyrolysis GC/MS and pyrolysis field ionization mass spectrometry. Org Geochem 15:131–145

Hempfling R, Ziegler F, Zech W, Schulten HR (1987) Litter decomposition and humification in acidic forest soils studied by chemical degradation, IR and NMR spectroscopy and pyrolysis-field ionization mass spectrometry. Z Pflanzenernähr Bodenkd 150:179–186

Hempfling R, Zech W, Schulten HR (1988) Chemical composition of the organic matter in forest soils II: Moder profile. Soil Sci 146:262–276

Herman JS (1982) The dissolution kinetics of calcite, dolomite and dolomite rocks in carbon dioxide water system. PhD Thesis, Pennsylvania State Univ

Herman JS, White WB (1985) Dissolution kinetics of dolomite: effects of lithology and fluid flow velocity. Geochim Cosmochim Acta 49:2017–2026

Hernandez HA (1981) The analysis of nitrogen-bearing additives in polyolefins. Antek Appl New 1/1:4–5

Hessen DO (1985) The relation between bacterial carbon and dissolved humic compounds in oligotrophic lakes. FEMS Microbiol Ecol 31:215–223

Heumann KG (1988) Isotope dilution mass spectrometry. In: Adams F, Gijbels R, van Grieken R (eds) Inorganic mass spectrometry. Wiley, New York, pp 301–376

*Heumann KG, Seewald H (1985) Gleichzeitige Bestimmung von Chlorid, Bromid- und Iodidspuren in Wässem durch massenspektrometrische Isotopenverdünnungsanalyse. Fresenius Z Anal Chem 320:493–497

Heumann KG, Schindlmeier W, Zeininger H, Schmidt M (1985) Application of an economical and small thermal ionization mass spectrometer for accurate anion trace analyses. Fresenius Z Anal Chem 320:457–462

Hilden HD, Suchan KH (1974) Neue Untersuchung über Verbreitung, Mächtigkeit und Grundwasserführung der Halterner Sandfazies. Fortschr Geol Rheinl Westfalen 20:79–90

Hirsch P (1972) Neue Methoden zur Beobachtung und Isolierung ungewöhnlicher oder wenig bekannter Wasserbakterien. Z Allg Microbiol 12:203–218

Hirsch P (1986) Microbial life at extremely low nutrient levels. Adv Space Res 6:287–298

Hirsch P, Müller M (1986) Methods and sources for the enrichment and isolation of budding, nonprosthecate bacteria from freshwater. Microb Ecol 12:331–341

Hirsch P, Pankratz SH (1970) Study of bacterial populations in natural environments by use of submerged electron microscope grids. Z Allg Microbiol 10:589–605

Hirsch P, Rades-Rohkohl E (1983a) Microbial diversity in a groundwater aquifer in northern Germany. Dev Ind Microbiol 24:183–200

Hirsch P, Rades-Rohkohl E (1983b) Die Zusammensetzung der natürlichen Grundwasser-Mikroflora und Untersuchungen über ihre Wechselbeziehungen mit Fäkalbakterien. DVWG Schriften Wasser 35:59–80 (Eschborn)

*Hirsch P, Rades-Rohkohl E (1988a) Some special problems in the determination of viable counts of groundwater microorganisms. Microb Ecol 16:99–113

*Hirsch P, Rades-Rohkohl E (1988b) Die Viefalt mikrobieller Morphotypen im Grundwasserbereich des Segeberger Forstes. Z Dtsch Geol Ges 139:343–353

Hirsch P, Rades-Rohkohl E (1990) Microbial colonization of aquifer sediment exposed in a groundwater well in Northern Germany. Appl Environ Microbiol 56:2963–2966

Hirsch P, Müller M, Schlesner H (1977) New aquatic budding and prosthecate bacteria and their taxonomic position. Soc Appl Bacteriol Symp Ser No 6. Academic Press, London pp 107–133

Hodges SC (1987) Aluminium speciation: a comparison of five methods. Soil Sci Soc Am J 51:57–64

*Hoffmann M (1986) Hydrogeochemische Vorgänge im Aquifersystem der Dorstener Kreidemulde unter Berücksichtigung des Leakage-Phänomens. PhD Thesis, RWTH Aachen

*Hoffmann M, Langguth HR, Larue J (1987) Das Leakage-Phänomen in der Dorstener Mulde, beobachtet mit Hilfe von Multi-level-Pegeln. Bull Cent Hydrogeol, Univ Neuchatel 7:49–63

Holder-Franklin MA (1981) Methods in studying population shifts in aquatic bacteria in response to environmental change. In: Carlisle DB (ed) Environment Canada, Scientific Series 124, Inland Waters Directorate, Ottawa, Canada

Holdren GR, Adams JE (1982) Parabolic dissolution kinetics of silicate minerals: an artifact of non-equilibrium precipitation processes? Geology 10:186–190

Holdren GR Jr, Berner RA (1979) Mechanism of feldspar weathering. I. Experimental studies. Geochim Cosmochim Acta 43:1161–1171

Holdren GR Jr, Speyer PM (1985) Stoichiometry of alkali feldspar dissolution at room temperature and various pH values. Am J Sci 285:994–1026

Hollocher TC (1984) Source of the oxygen atoms in nitrate in the oxidation of nitrate by *Nitrobacter agilis* and evidence against P-O-N-anhydride mechanism in oxidative phosphorylation. Arch Biochem Biophys 233:721–727

Hollocher TC, Tate ME, Nicholas DJD (1981) Oxidation of ammonia by *Nitrosomonas europaea.* J Biol Chem 256:10834–10836

Holthusen H (1981) Lösungs-, Transport- und Immobilisationsprozesse durch das Sickerwasser in der ungesättigten Bodenzone – Genese der Beschaffenheit des oberflächennahen Grundwassers. PhD Thesis, Univ Kiel

Holzapfel-Pschorn A, Obst U, Haberer K (1985) Fluoreszenzspektroskopie zum empfindlichen Nachweis von Enzymaktivitäten in der Trinkwasseraufbereitung. Gewässerschutz Wasser Abwasser 79:352–366

Hoos E, Schweisfurth R (1982) Untersuchungen über die Verteilung von Bakterien von 10–90 Meter unter Bodenoberkante. Vom Wasser 58:103–112

*Hoppe J, Merkel B, Eichinger L, Hoffmann M (1987) Nachweis von Carbonatausfällung im Boden mit Hilfe von Isotopenuntersuchungen. Z Wasser-Abwasserforsch 20:16–21

*Hötzl H, Istrate G, Kössl H, Smykatz-Kloss W, Wohnlich S (1984) Hydrogeochemische und mineralogische Untersuchungen in einer südtunesischen Sabkha. In: Udluft P, Merkel B, Prösl K-H (eds) Proc RIZA-Symp Dept Hydrogeology Hydrochemistry Techn Univ München, pp 109–118

Huffman EWD Jr, Stuber HA (1985) Analytical methodology for elemental analysis of humic substances. In: Aiken GR, McKnight DM, Wershaw RL, MacCarthy P (eds) Humic substances in soil, sediment and water. John Wiley, New York

Hund W (1988) Hydrogeologische Untersuchung eines Umweltschadens in einem pleistozänen Grundwasserleiter in Hamburg-Rissen. PhD Thesis, Univ Münster

Imreh L (1965) Les minéralisations plumbo-zinciferes de la province metallifère du Zamanti (secteur nord). MTA Bull 65:92–118

Ingerson E, Pearson F (1964) Estimation of age and rate of motion of groundwater by the ^{14}C-method. In: Miyake (ed) Recent research in the fields of hydrosphere, atmosphere and nuclear geochemistry. Tokyo, pp 263–283

International Humic Substances Society, In: Spiteller M (1985) Beiträge zur Struktur und Dynamik von Huminstoffen. Gött Bodenkd Ber 84

Irving HM, Williams RJP (1948) Order of stability of metal complexes. Nature 162:746–747

Irwin WJ (1982) Analytical pyrolysis: a comprehensive guide. Marcel Dekker, New York

Ittekkot V, Spitzy A, Lammerz U (1982) Data on dissolved carbohydrates and amino acids in world rivers: a documentation by Mitt Geol–Paläontol Inst Univ Hamburg. SCOPE/UNEP Spec Vol 52:575–584

Ivshina JB, Oborin AA, Nesterenko OA, Kasumova SA (1981) *Rhodococcus* bacteria in ground water from oil field in the perm cisural region. Microbiology 50:531–538 (translation)

Iwanoff A (1981) Zur Profildifferenzierung einer saalezeitlichen Moräne in Hamburg-Rissen. Verh Naturwiss Ver Hamburg 24:113–122

Jackson RE, Inch KJ (1980) Hydrogeochemical processes affecting the migration of radionuclides in a fluvial sand aquifer at the Chalk River Nuclear Laboratories. NHRI, Paper No 7, Sci Ser No 104

Jackson TA (1974) Humic matter in natural waters and sediments. Soil Sci 119/1:56–63

*Jacob H (1987) Wasserhaushalt und Stofftransport in der ungesättigten Bodenzone eines Waldökosystems: Untersuchungen in der Schwetzinger Hardt. PhD Thesis, Univ Heidelberg

*Jacob H (1990) Chemischer Stoffeintrag ins Grundwasser im Staatswald Schwetzinger Hardt. Heidelb Geowiss Abh (in press)

James BR, Riha SJ (1984) Soluble aluminium in acidified organic horizons of forest soils. Can J Soil Sci 64:637–646

Jannasch HW (1978) Microbial ecology of aquatic low nutrient habitats. In: Shilo M (ed) Strategies of microbial life in extreme environments. Verlag Chemie, Weinheim, pp 243–260

Jennings JN (1985) Karst geomorphology. Blackwell, Oxford

Jones RL, Hornsby AG, Rao PSC, Anderson MP (1987) Movement and degradation of aldicarb residues in the saturated zone under citrus groves on the Florida ridge. J Cont Hydr 1:265–285

Jordan H (1976) Geologische Karte von Niedersachsen 1:25 000 Erläuterungen zu Blatt Osterode Nr. 4227. Niedersächsisches Landesamt für Bodenforschung, Hannover

Jörgensen BB (1977) Bacterial sulfate reduction within reduced microniches of oxidized marine sediments. Mar Biol 41:7–17

Jung WK, Schweisfurth R (1979) Manganoxidierende Bakterien. III. Wachstum und Manganoxidation bei Pseudomonas manganoxidans. Schweiz Z Allg Mikrobiol 16:587–597

Jungk A (1964) Über die selektive Kationensorption von Torf. Z Pflanzenernähr Düng Bodenkd 104:206–215

Kalterherberg J (1982) Ingenieurgeologische Untersuchungen bei der Gründung von Erweiterungsbauten der RWTH Aachen in Seffent-Melaten, NW von Aachen. Mitt Ing Hydrogeol Aachen 12:13–43

Kanazawa S, Filip Z (1986) Distribution of microorganisms, total biomass and enzyme activities in different particles of brown soil. Microb Ecol 12:205–215

Karamanos RE, Rennie DA (1980) Changes in Natural ^{15}N abundance associated with pedogenic processes in soil. I. Changes associated with saline seeps. Can J Soil Sci 60:337–344

*Katlein R (1986) Bestimmung von Komplexkapazitäten mit der Mangandioxid-Methode. PhD Thesis, Univ Marburg

Kay JM, Nedderman RM (1985) Fluid mechanics and transfer processes. Cambridge University Press, Cambridge

Keller C (1986) Radioiod in Oberflächengewässern und in der Atmosphäre. GIT Suppl Umwelt 1:47–50

Kepkay PE, Nealson KH (1987) Growth of a manganese oxidizing *Pseudomonas* sp. in continuous culture. Arch Microbiol 148:63–67

Kern MD (1960) The hydration of carbon dioxide. J Chem Educ 37:14–23

Kerndorff H, Schnitzer M (1980) Sorption of metals on humic acids. Geochim Cosmochim Acta 44:1701–1708

Khairy AH, Ziechmann W (1981) Die Veränderung von Huminstoffen in alkalischer Lösung. Z Pflanzenernähr Bodenkd 144:407–422

Kinniburgh DG, Miles DL (1983) Extraction and chemical analysis of interstitial water from rocks and soils. Environ Sci Technol 17:362–368

*Kirchhoff A (1985) Hydrogeochemische und isotopenchemische Untersuchungen an karbonathaltigen Grundwässern im Göttinger Raum. Z Dtsch Geol Ges 136:575–583

*Klamberg H, Matthess G, Pekdeger A (1989) Organo-metal complexes as mobility-determining factors of inorganic toxic elements in porous media. In: Bar-Yosef B, Barrow NJ, Goldshmid J (eds) Inorganic contaminants in the vadose zone. Springer, Berlin Heidelberg New York Tokyo, pp 3–17 (Ecological Studies 74)

Knapp G (1978) Erläuterungen zur Geologischen Karte der nördlischen Eifel 1:100 000. Mit einem Beitrag von Hans Hager, 2nd edn. Geologisches Landesamt Nordrhein-Westfalen, Krefeld

Knauss KG, Wolery TG (1986) Dependence of albite dissolution kinetics on pH and time at 25°C. Geochim Cosmochim Acta 50:2481–2497

Kneißel M (1985) Fraktionierte Extraktion und oxidativer Abbau einer isolierten Bachsediment-Huminsäure. PhD Thesis, Univ Marburg

*Kölbel-Boelke J (1987) Diversität und in vitro–Aktivitäten heterotropher Bakteriengemeinschaften in einem pleistozänen, sandig–kiesigen Grundwasserleiter. PhD Thesis Univ Bremen

*Kölbel-Boelke J (1988) Diversität heterotropher Bakteriengemeinschaften in einem pleistozänen, sandig–kiesigen Grundwasserleiter bei Bocholt/Westfalen. Z Dtsch Geol Ges 139:355–363

*Kölbel-Boelke J, Hirsch P (1989) Comparative physiology of bio-film and suspended organisms in the groundwater environment. In: Charaklis WG, Wilderer PA (eds) Structure and function of biofilms. Dahlem Konferenzen. John Wiley, Chichester, pp 221–238

*Kölbel-Boelke J, Anders EM, Nehrkorn A (1988a) Microbial communities in the saturated groundwater environment. II Diversity of bacterial communities in a pleistocene sand aquifer and their in vitro activities. Microb Ecol 16:31–48

*Kölbel-Boelke J, Tienken B, Nehrkorn A (1988b) Microbial communities in the saturated groundwater environment. I Methods of isolation and characterization of heterotrophic bacteria. Microb Ecol 16:17–29

Kölle W (1982) Redoxprozesse im Grundwasserleiter. Veröff Bereich u Lehrst Wasserchemie u DVGW-Forschungsstelle am Engler-Bunte-Inst/Univ Karlsruhe 20:301–315

*Kölle W, Werner P, Strebel O, Böttcher J (1983) Denitrifikation in einem reduzierenden Grundwasserleiter. Vom Wasser 61:125–147

*Kölle W, Strebel O, Böttcher J (1985) Formation of sulfate by microbial denitrification in a reducing aquifer. Water Supply 3, Berlin "B":35–40

*Kölle W, Strebel O, Böttcher J (1987) Reduced sulfur compounds in sandy aquifers and their interaction with groundwater: Proc Int Symp Groundw Monitoring and Mangm Dresden. Complex 1/11:1–14

*Kölle W, Strebel O, Böttcher J (1990) Reduced sulfur compounds in sandy aquifers and their interactions with groundwater: Intern Assoc Hydrological Sci Publ 173:23–30

Kölling M (1986) Vergleich verschiedener Methoden zur Bestimmung des Redoxpotentials natürlicher Wässer. Meyniana 38:1–19

Kölling M (1990) Modellierung geochemischer Prozesse im Sickerwasser und Grundwasser. Beispiel: Die Pyritverwitterung und das Problem saurer Grubenwässer. PhD Thesis, Berichte, Fachbereich Geowissenschaften, Univ Bremen 8:135 pp

König N, Baccini P, Ulrich B (1986) Der Einfluß der natürlichen organischen Substanzen auf die Metallverteilung zwischen Boden und Bodenlösung in einem sauren Waldboden. Z Pflanzenernähr Bodenkd 149:68–82

Kononova MM (1966) Soil organic matter. Pergamon, New York

Korotkov MG, Frolov VP (1982) Study of the composition of fractions of organic matter in typical chernozem by the method of sequential extraction by organic solvents and water. Izv Timiryazev S-kh Akad: 67–73

Korzh VD (1984) Ocean as a source of atmospheric iodine. Atmos Environ 18:2707–2710

Kosa A (1980) Gypsum karst discovered in Libya. Karszt es Barlang 1980–I:23

Kosa A, Smykatz-Kloss W (1978) Solution phenomena in the dolomites of North Tripolitania. Karszt es Barlang 1978–I/II:43–48 (in Hungarian)

*Kössl H (1988) Mineralogische und hydrogeochemische Untersuchungen im Gipskarstgebiet von Foum Tatahouine (SE-Tunesien) und im südwestdeutschen Gipskeuper (Obersontheim, Raum Schwäbisch-Hall). PhD Thesis, Univ Karlsruhe

Kramarenko LE (1962) Bacterial biocenoses in underground waters of some mineral fields and their geological importance. Microbiology (translation) 31:566–571 = Microbiologya 31:694–701 (in Russian)

Krambeck HJ, Witzel KP (1983) Classification of aquatric bacterial strains: an example of numerical taxonomy in limnology. EDV Med Biol 14:45–49

Kühn H, Weller H (1977) 6-jährige Untersuchungen über Schwefelzufuhr durch Niederschläge und Schwefelverluste durch Auswaschung (in Lysimetern). Z Pflanzenernähr Bodenkd 140:431–440

Kumada K, Kawamura Y (1986) On the fractionation of humic acids by a fractional precipitation technique. Soil Sci Plant Nutr 14:198–200

Kunz B, Stumm W (1984) Kinetics of nucleation and growth of calcium carbonate. Vom Wasser 63:279–293

Kupfahl HG, Meisl L, Kümmerle E (1972) Blatt 6217 Zwingenberg a.d. Bergstraße. Erläuterung der Geologischen Karte von Hessen 1:25 000, Wiesbaden, 267 pp

Kuznetsova ZI (1957) Izucheniye Kolichestvennogo Soderzhaniya Bakteriy v Podzemnykh Vodakh Neftyanykh Mestorozhdeniy (A study of the quantitative content of bacteria in the ground waters of oil deposits). Mikrobiologiia 26

Kuznetsova ZI (1961) Raspredeliniye i Ekologiya mikroorganizmov v Glubokikh Podzemnykh Vodakh Nekotorykh Territoriy SSSR. Sb. "Geologischeskaya Deyatel'nost' Mikroorganizmov" (The distribution and ecology of the microorganisms in the deep ground waters of certain territories of the USSR, in the symposium volume. The geological activity of microorganisms). Tr Inta Mikrobiol AN SSSR 9

Kyuma K (1966) A fractional precipitation technique applied to soil humic substances. Soil Sci Plant Nutr 10:128–137

Ladd TJ, Ventullo RM, Wallis PM, Costerton JW (1982) Heterotrophic activity and biodegradation of labile and refractory compounds by ground water and stream microbial populations. Appl Environ Microbiol 44:321–329

Lagache M (1965) Contribution a l'étude de l'alteration des feldspaths, dans l'eau entre 100 et 200 °C, sous diverses pression de CO_2 et application a la synthèse des minéraux argileux. Bull Soc Fr Miner Crist 88:223–253

Langguth HR, Treskatis CH (1987) Umgekehrte Wasserspiegelreaktionen bei hydrologischen Pumpversuchen – der Nachweis von "NOORDBERGUM-Effekt" und einer entsprechenden Druckabnahme ("RHADER-Effekt") beim Wiederansteig in halbgespannten Aquiferen. bbr 10:369–376

Langguth HR, Voigt R (1980) Hydrogeologische Methoden. Springer, Berlin Heidelberg New York

Langmuir I (1916) The constitution and fundamental properties of solids and liquids. J Am Chem Soc 38:2221–2295

Lasaga AC (1981a) Rate laws of chemical reactions. In: Lasaga AC, Kirkpatrick RJ (eds) Kinetics of geochemical processes. Mineral Soc Am Rev in Mineral 8, Chelsea, Michigan, pp 1–68

Lasaga ACA (1981b) Transition state theory. In: Lasaga AC, Kirkpatrick RJ (eds) Kinetics of geochemical processes. Mineral Soc Am Rev in Mineral 8, Chelsea, Michigan, pp 135–169

Lauritzen SE (1986) Hydraulics and dissolution kinetics of a phreatic conduit. Proc 9th Int Congr Speleol 1:20–22

La Zerte BD (1984) Forms of aqueous aluminum in acidified catchments of Central Ontario: a methodological analysis. Can J Fish Aquat Sci 41:766–776

La Zerte BD (1985) Correction. Can J Fish 42:400

Leenheer JA (1985) Fractionation techniques for aquatic humic substances. In: Aiken GR, McKnight DM, Wershaw RL, MacCarthy P (eds) Humic substances in soil, sediment and water. John Wiley, New York

*Leuchs W (1985) Mikrobiell beeinflußte Stoffverschiebungen zwischen Grundwasser und Grundwasserleiter in einem pleistozänen Aquifer. Z Dtsch Geol Ges 136:553–561

*Leuchs W (1988) Vorkommen, Abfolge und Auswirkungen anoxischer Redoxreaktionen in einem pleitozänen Porengrundwasserleiter. Besond Mitt Dtsch Gewässerkd Jahrb 52:1–106

Liao W, Christman RF, Johnson JD, Millington DS, Hass JR (1982) Structural characterization of aquatic humic material. Environ Sci Technol 16:403–410

Lillich W, Kuckelkorn KF, Hofmann W (1973) Untersuchungen zum Grundwasserhaushalt im repräsentativen Lockergesteinsgebiet Fuhrberger Feld bei Hannover – Bilanzjahre 1967 und 1968. Beih Geol Jahrb 107:67

Lindquist B (1932) The beach forests of Sweden. In: Ruebel E (ed) Die Buchenwälder Europas. Veröff Geobot Inst Buebel, Zürich

Lippmann F (1973) Sedimentary carbonate minerals. Springer, Berlin Heidelberg New York

Littman S (1985) Zur chemisch-analytischen Untersuchung von Eiprodukten und Beurteilung ihres hygienischen Zustands an Hand des Gehaltes an verderbspezifischen organischen Säuren. Dtsch Lebensm Rudsch 81/11:345–350

Loehnert E (1970) Grundwasserchemismus und Kationenaustausch im norddeutschen Flachland. Z Dtsch Geol Ges, Sonderh Hydrogeol Hydrogeochem: 139–159

Loehnert EP (1985) The impact of groundwater and the role of hydrogeology on a city's growth – case study of Hamburg, FRG. In: Mem 18th Congr Int Assoc Hydrogeologists, Part 2. Cambridge, pp 178–186

Loehnert EP (1987) Tiefe Grundwässer in Raum Hamburg. DVWK Schriften 81:184–193, Parey, Hamburg

Loehnert EP, Paluska A (1977) Groundwater salinisation in the Elbe valley (City of Hamburg). In: Proc 5th Salt Water Instruction Meeting. Medmenham, pp 98–107

Loehnert EP, Sonntag C (1981) Grundwasser-Versalzung im Raum Hamburg im Licht neuer Isotopendaten. Z Dtsch Geol Ges 132:559–574

*Loehnert EP, Bauhus W, Sonntag C (1986) Mechanism of groundwater salinization in the Hamburg region, FR Germany – facts and new findings. In: Proc 9th Salt Water Intrusion Meeting. Delft, pp 613–628

Loosli H (1979) Eine Altersbestimmungsmethode mit ^{39}Ar. Habilitation Thesis, Univ Bern

Loosli HH (1983) A dating method with ^{39}Ar. Earth Planet Sci Lett 63:51

Lorenz I (1986) Untersuchung von Metall-Huminstoff-Komplexen mit der Gelchromatographie. PhD Thesis, Univ Marburg

Luce RW, Bartlett RW, Parks GA (1972) Dissolution kinetics of magnesium silicates. Geochim Cosmochim Acta 36:35–50

Lüpkes G (1974) Drei neue stygobionte Peritrichen aus dem interstitiellen Grundwasser. Arch Hydrobiol 73:394–402

MacCarthy P, Rice IA (1985) Spectroscopic methods (other than NMR) for determining functionality in humic substances. In: Aiken GR, McKnight DM, Wershaw RL, MacCarthy P (eds) Humic substances in soil, sediment and water. John Wiley, New York, pp 527–561

Magaritz M, Goldenberg L, Kafri U, Arad A (1980) Dolomite formation in the seawater-freshwater interface. Nature 287:622–624

Malcolm RL (1989) Application of solid-state ^{13}C-NMR spectroscopy to geochemical studies of humic substances. In: Hayes MHB, MacCarthy P, Malcolm RL, Swift RS (eds) Humic substances II. In search of structure. John Wiley, Chichester, pp 339–372

Mallory LM, Austin B, Colwell RR (1977) Numerical taxonomy and ecology of oligotrophic bacteria isolated from estuarine environment. Can J Microbiol 23:733–750

Mantoura RFC, Riley JP (1975) The analytical concentration of humic substances from natural waters. Anal Chim Acta 76:97–106

Marinsky JA, Ephraim J (1986) A unified physicochemical description of the protonation and metal ion complexation equilibria of natural organic acids (humic and fulvic acids). 1. Analysis of the influence of polyelectrolyte properties on protonation equilibria in ionic media: fundamental concepts. Environ Sci Technol 20:349–354

Mariotti A, Létolle R (1977) Application de l'étude isotopique de l'azote en hydrologie et en hydrogéologie – analyse des résultats obtenus sur un exemple précis: le bassin de Mélarches (Seine-et-Marne, France). J Hydrol 33:157–172
Martin F, Saiz-Jimenez C, Cert A (1977) Py-GC-MS of Soil humic fractions I. The low building point compounds. J Soil Sci Am 41:1114–1118
Martin JP, Filip Z, Haider K (1976) Effect of montmorillonite and humate on growth and metabolic activity of some actinomycetes. Soil Biol Biochem 8:409–413
Marxsen J (1981) Fluoreszenzmikroskopische Untersuchung der Bakterienflora und Bestimmung ihrer heterotrophen Aktivität in organisch belastetem und unbelastetem Grundwasser sandig kiesiger Ablagerungen. Int J Speleol 11:173–201
Marxsen J (1982) Ein neues Verfahren zur Untersuchung der bakteriellen Besiedlung grundwasserführender sandiger Sedimente. Arch Hydrobiol 95:221–233
Marxsen J (1988) Investigations into the number of respiring bacteria in groundwater from sandy and gravelly deposits. Microb Ecol 16:65–72
Mast MA, Drever JI (1987) The effect of oxalate on the dissolution rate of oligoclase and tremolite. Geochim Cosmochim Acta 51:2559–2568
Matthess G (1973) Die Beschaffenheit des Grundwassers. Borntraeger, Berlin Stuttgart, pp 324
Matthess G (1982) The properties of groundwater. John Wiley, New York, p 406
Matthess G (1984) The role of natural organics on water interaction with soil and rock. IAHS Publ 150:11–20
Matthess G, Ubell K (1983) Allgemeine Hydrogeologie – Grundwasserhaushalt. Borntraeger, Berlin, Stuttgart 438 pp
Matthess G, Thilo L, Roether W, Münnich KO (1968) Tritiumgehalte im Wasser tieferer Grundwasserstockwerke. Wasser/Abwasser 108:353–355
Matthess G, Alexander J, Althaus H, Dizer H, Filip Z, Havemeister G, Hirsch P, Jung KD, Kaddu-Mulindwa D, Käss W, Lopez J, Milde G, Naffer A, Pekdeger A, Rades-Rohkohl E, Riemer R, Ritter R, Sacré C, Schroeter J, Seidel K, Seiler KP (1985) Lebensdauer von Bakterien und Viren in Grundwasserleitern. Ber UBA 2, E Schmidt, Berlin, 108 pp
Matzner E (1988) Der Stoffumsatz zweier Waldökosysteme im Solling. Ber Forschungszentr Waldökosysteme/Waldsterben, Reihe A, 40:1–217
Mauersberger K (1981) Measurement of heavy ozone in the stratosphere. Geophys Res Lett 8:935–937
May HM, Kinniburgh DG, Helmke PA, Jackson ML (1986) Aqueous dissolution, solubilities and thermodynamic stabilities of common aluminosilicate clay minerals: kaolinite and smectites. Geochim Cosmochim Acta 50:1667–1677
Mayer R (1971) Bioelement-Transport im Niederschlagswasser und in der Bodenlösung eines Waldökosystems. Gött Bodenkd Ber 19:1–119
Mayer R (1981) Natürliche und anthropogene Komponenten des Schwermetallhaushalts von Waldökosystemen. Gött Bodenkd Ber 70:1–292
Mehra OP, Jackson ML (1960) Iron oxide removal from soils and clays by a dithionite-citrate system buffered with sodium bicarbonate. Proc 7th Nat Conf Clays Clay Miner 5:317–327
Mehta NC, Dubach P, Deuel H (1963) Untersuchungen über die Molekulargewichtsverteilung von Huminstoffen durch Gelfiltration an Sephadex. Z Pflanzenernähr Düng Bodenkd 102:128–137
Meiners HG (1985) Hydrogeologische Untersuchungen der langfristigen Grundwassergewinnungsmöglichkeiten im Bereich der Halterner Sande mit Hilfe eines mathematischen Grundwassermodells. Forschungsber Land Nordrhein-Westfalen, 93 pp
Meiwes KJ, Hauhs M, Gerke H, Asche N, Matzner E, Lamersdorf N (1984) Die Erfassung des Stoffkreislausfs in Waldökosystemen. Ber Forschungszentr Waldökosysteme/Waldsterben 7:68–142
Merkel B (1983) Untersuchungen zum Kohlenstoffhaushalt im gesättigten und ungesättigten Bereich quartärer Carbonatschotter. Hydrochem Hydrogeol Mitt München 6:219
*Merkel B, Prömper R (1984) Eine Kunststoff-Saugkerze in Sandwichbauweise zur Gewinnung von Sickerwasserproben. Z Wasser Abwasserforsch 17:204–205
Merkel B, Nemeth G, Udluft P, Grimmeisen W (1982) Hydrogeologische und hydrochemische Untersuchungen in der ungesättigten Zone eines Kiesgrundwasserleiters. Teil 1: Entwicklung und Erstellung eines begehbaren Probenahmeschachtes zur Boden-, Wasser- und Luftuntersuchung. Z Wasser Abwasserforsch 15:191–195

Merkel B, Schmitz G, Haverkamp R, Czirwitzky HJ (1983) Temperature affected pollutant transport phenomena in unsaturated soil. In: 21st Int Assoc Hydraulic Res Congress, Melbourne, Aug 1985, pp 187–193
*Merkel B, Grossmann J, Faust A (1988) Simulation of groundwater transport taking into account thermodynamical reactions. Proc VII Int Conf MIT, USA, June 1988. Dev Water Sci 36/2:281–286
Meuzelaar HCL, Haverkamp J, Hileman FD (eds) (1982) Pyrolysis mass spectrometry of recent and fossil biomaterials: compendium and atlas. Elsevier, Amsterdam
Meyer JL, Edwards RT, Risley R (1977) Bacterial growth on dissolved organic carbon from a blackwater river. Microbiol Ecol 13/1:13–29
*Michaelis J (1987) Hydrogeochemische Modellbetrachtungen zur Kalkauflösung und Entwicklung karbonatischer Grundwässer im Göttinger Wald unter spezieller Berücksichtigung der stabilen Isotope des Kohlenstoffs und Schwefels. PhD Thesis, Univ Göttingen
Michaelis J, Usdowski E, Menschel G (1984) Model considerations for the evolution of carbonate groundwaters in the Schwäbische Alb (Federal Republic of Germany). Isotope Geosci 2:197–204
Mikita A, Steelink C, Wershaw RL (1981) Carbon-13 enriched NMR method for the determination of hydroxyl functionality of humic substances. Anal Chem 53:1715–1717
Moaledj K (1984) Über schnelle miniaturisierte Bestimmungsverfahren für Bakterienpopulationen in aquatischen Ökosystemen. Arch Hydrobiol 11:99–121
Molin G, Nilsson I (1985) Degradation of phenol by *Pseudomonas putida* ATCC 11172 in continuous culture at different ratios of biofilm surface to culture volume. Appl Environ Microbiol 49:402–407
Molz FJ, Widdowson MA, Benefield LD (1986) Simulation of microbial growth dynamics coupled to nutrient and oxygen transport in porous media. Water Resour Res 22/8:1207–1216
Mook WG (1976) The dissolution-exchange model for dating ground-water with ^{14}C, interpretation and environmental isotope and hydrochemical data in groundwater hydrology. IAEA, Vienna 1976:213–225
Mook WG, Bommerson JC, Staverman WH (1974) Carbon isotope fractionation between dissolved bicarbonate and gaseous dioxide. Earth Planet Sci Lett 22:169–176
Morse JW (1974) Dissolution kinetics of calcium carbonate in sea water. V. Effects of natural inhibitors and the position of the chemical lysocline. Am J Sci 274:638–647
*Moschik A (1990) Hydrogeochemische Vorgänge in organischen Ablagerungen unterschiedlicher Genese aus der wassergesättigten Zone, die vom Wasser durchströmt werden. PhD Thesis Univ Berlin
Möse JR, Brantner H (1966) Mikrobiologische Studien an manganoxidierenden Bakterien. Zentralbl Bakt II. Abt 120:480–495
Motyka I, Wilk Z (1984) Hydraulic structure of karst-fissures Triassic rocks in the vicinity of Olkusz (Poland). Kras i Speleol 14/5:11–24
Mucci A, Morse JW (1984) The solubility of calcite in seawater solutions of various magnesium concentration, $I_t = 0.697$ m and 25 °C and one atmosphere total pressure. Geochim Cosmochim Acta 48:815–822
Mueller H, Bandaranayake W (1982) An automated method for the determination of dissolved organic carbon in seawater using continuous thin-film UV oxidation. Mar Chem 12:59–68
Mulder EG, van Veen WL (1963) Investigations on the *Sphaerotilus – Leptothrix* group. Antonie Leeuwenhoek J Microbiol 29:121–153
Müller G (1965) Bodenbiologie. Fischer, Stuttgart
Munn D, Jackson DE (1980) Dedolomitization of lower Carboniferous dolostone in the Wirksworth area, Derbyshire, England. Geol Mag 117:607–612
Münnich KO (1963) Atombomben Tritium als Indikator in der Hydrologie. Phys Blätter 9:418–421
Münnich KO (1983) Moisture movement in the unsaturated zone. In: IAEA (ed) Guidebook on nuclear techniques in hydrology. Tech Rep Ser No 91, IAEA, Vienna, pp 203–222
Murray JW, Dillard JG, Giovanoli R, Moers H, Stumm W (1985) Oxidation of Mn(II): Initial mineralogy, oxidation state and ageing. Geochim Cosmochim Acta 49:463–470
Mylroie JE, Carew JL (1986) Minimum duration for speleogenesis. 9th Congr Int Espeleol Barcelona 1:249–251
Nash JC (1979) Compact numerical methods for computers: linear algebra and function minimization. Adam Hilger, Bristol
Nehrkorn A (1968) Grundwasserkeimzahlen bei Anwendung nährstoffarmer Substrate. Gesundheits-Ingenieur 89:88

Neitzke M (1984) Einfluß von Aluminium auf die Keimung und Keimlingsentwicklung der Buche unter Berücksichtigung von Wechselwirkungen mit Calcium und Stickstofform. PhD Thesis, Univ Münster

Nestaas IS, Terjesen SG (1969) The inhibiting effect of scandium ions upon the dissolution of calcium carbonate. Acta Chem Scand 23:2519–2531

Neuman SP, Witherspoon PA (1969) Theory of flow in a confined two-aquifer system. Water Resour Res 5/4:803–816

Nielsen H, Rambow D (1969) S-Isotopenuntersuchungen an Sulfaten hessischer Mineralwässer. Notizbl Hess L Amt Bodenforsch 97:352–366

Nielsen H, Ricke W (1964) Schwefel-Isotopen-Verhältnisse von Evaporiten aus Deutschland, ein Beitrag zur Kenntnis von ^{34}S im Meerwasser-Sulfat. Geochim Cosmochim Acta 28:577–591

Nilsson SI, Bergkvist B (1983) Aluminum chemistry and acidification processes in a shallow podzol on the Swedish westcoast. Water Air Soil Pollut 20:311–329

Nye PH, Tinker PB (1977) Solute movement in the soil-root system. Blackwell, Oxford

Obermann P (1976) Möglichkeiten der Anwendung des Doppelpackers im Beobachtungsbrunnen bei der Grundwassererkundung. – Bohrtechnik-Brunnenbau – Rohrleitungsbau 27:93–96

Obermann P (1982) Hydrochemische/hydromechanische Untersuchungen zum Stoffgehalt von Grundwasser bei landwirtschaftlicher Nutzung. Besond Mitt Dtsch Gewässerkd Jahrb 42:217 pp

Ochiai M, Nakajima T (1985) Decomposition of organic matter extracted from sessile microbes. Verh Int Verein Limnol 22:2744–2748

Ohse W (1983) Lösungs- und Fällungserscheinungen im System oberflächennahes unterirdisches Wasser/gesteinsbildende Minerale – Eine Untersuchung auf der Grundlage der chemischen Gleichgewichts-Thermodynamik. PhD Thesis, Univ Kiel

Ohse W, Matthess G, Pekdeger A (1983) Gleichgewichts- und Ungleichgewichtsbeziehungen zwischen Porenwässern und Sedimentgesteinen im Verwitterungsbereich. Z Dtsch Geol Ges 134:345–361

Ohse W, Matthess G, Pekdeger A, Schulz HD (1984) Interaction water-silicate minerals in the unsaturated zone controlled by thermodynamic equilibria. IAHS Publ 150:31–40

Olleros T (1983) Kinetischer Isotopeneffekt der Arginase- und Nitratreduktase-Reaktion; ein Beitrag zur Aufklärung der entsprechenden Reaktionsmechanismen. PhD Thesis, TU München

Olsen RA, Bakken R (1987) Viability of soil bacteria: optimization of plate count technique and comparison between total counts and plate counts within different size groups. Microb Ecol 13:59–74

Olson GJ, Dockins WS, McFeters GA, Iverson WP (1981) Sulfate-reducing and methanogenic bacteria from deep aquifers in Montana. Geomicrobiol J 2:327–340

Otsuki A, Hanya T (1966) Fractional precipitation of humic acids from recent sediment by use of N,N-dimethylformamide and its infra-red absorption spectra. Nature 212:1462–1463

Paces T (1973) Steady-state kinetics and equilibrium between ground water and granitic rocks. Geochim Cosmochim Acta 37:2641–2663

Paces T (1978) Reversible control of aqueous aluminum and silica during the irreversible evolution of natural waters. Geochim Cosmochim Acta 42:1487–1493

Pagenkopf GK (1978) Introduction to natural water chemistry. Environ Sci Technol 3:272 pp

Palciauskas VV, Domenico PA (1976) Solution chemistry, mass transfer, and the approach to chemical equilibrium in porous carbonate rocks and sediments. Geol Soc Am Bull 87:207–214

Parker JM, James RC (1985) Autochthonous bacteria in the chalk and their influence on groundwater quality in East Anglia. J Appl Bacteriol Symp Suppl: 15–25

Parkhurst DL, Thorstenson DC, Plummer LN (1980) PHREEQE – a computer programme for geochemical calculations. USGS Water Resour Inv 80/96:210

Pauli FW (1975) Heavy metal humates and their behavior against hydrogen sulfide. Soil Sci 119/1:98–105

Pearson FJ Jr, Hanshaw BB (1970) Sources of dissolved carbonate species in groundwater and their effects on carbon-14 dating. Isotope Hydrology 1970. IAEA, Vienna pp 271–285

Pearson FJ, Donald JR, Fischer W, Plummer LN (1978) Correction of ground-water chemistry and carbon isotopic composition for effects of CO2 outgassing. Geochim Cosmochim Acta 42:1799–1807

Pekdeger A (1977) Labor- und Felduntersuchungen zur Genese der Sicker- und Grundwasserbeschaffenheit. PhD Thesis, Univ Kiel

Perdue EN (1985) Acidic functional groups of humic substances. In: Aiken GR, McKnight DM, Wershaw RL, MacCarthy P (eds) Humic substances in soil, sediment, and water. John Wiley, New York

Perdue EM, Gjessing ET (1990) Organic acids in aquatic ecosystems. John Wiley, Chichester

Perdue EM, Lytle CR (1983) A critical examination of metal-ligand complexation models: application to defined multiligand mixtures. In: Christman RF, Gjessing ET (eds) Aquatic and terrestrial humic materials. Ann Arbor Science, Ann Arbor, Michigan, pp 295–313

Perthuisot JP (1976) Une sebkha sulfatée sodique en pays sedimentaires. La sebkha Oum el Krialate (Sudtunisien). Geol Mediterr 3:265–274

*Petersen A (1988) Laboruntersuchungen zum Einfluß organischer Komplexbildner auf die Kinetik der Feldspatverwitterung. PhD Thesis, Ber Rep Geol-Paläontol Inst Univ Kiel 21:201

Petrovic R (1976) Rate control in feldspar dissolution: II. The protective effect of precipitates. Geochim Cosmochim Acta 40/2:1509–1521

Petrovic R, Berner RA, Goldhaber MB (1976) Rate control in dissolution of alkali feldspar. I. Studies of residual feldspar grains by X-ray photoelectron spectroscopy. Geochim Cosmochim Acta 40:1675–1686

Pfennig N, Wagener S (1986) An improved method of preparing wet mounts for photomicrographs of microorganisms. J Microbiol Meth 4:303–306

Pickens JF, Cherry JA, Gilham RW, Merrit WF (1977a) Field studies of dispersion in a shallow sandy aquifer. In: Proc Invitational Well Testing Symp, 19.–21.10.1977, pp 55–62

Pickens JF, Cherry JA, Grisak GE (1977b) A multilevel device for groundwater sampling and piezometric modelling. Groundwater, Worthington, Ohio, 16/5:322–327

Pierce RH, Felbeck GT (1975) A comparison of three methods of extracting organic matter from soils and marine sediments. In: Povoledo D, Golterman HL (eds) Humic substances, their structure and function in the biosphere. Proc Int Meeting at Nieuwersluis 1972, Pudoc, Wageningen, pp 493–526

Pleskow YV, Filinovskii VY (1975) The rotating disc electrode. Studies in Soviet Science. Consultants Bureau, New York

Plummer LN, Wigley TLM (1976) The dissolution of calcite in CO_2-saturated solutions at 25 °C and 1 atmosphere total pressure. Geochim Cosmochim Acta 40:191–202

Plummer LN, Jones BF, Truesdell AH (1976) WATEQF – a Fortran-4 version of WATEQ: a computer program for calculating chemical equilibrium of natural waters. US Geol Surv Water Resour Invest 76–13:61

Plummer LN, Wigley TML, Parkhurst DL (1978) The kinetics of calcite dissolution in CO_2-water system at 5 ° to 60 °C and 0.0 to 1.0 atm CO_2. Am J Sci 278:179–216

Plummer LN, Parkhurst DL, Wigley TLM (1979) Critical review of calcite dissolution and precipitation. Am Chem Soc Symp Ser 93:537–573

*Post B (1987) Chromatographische und spektroskopische Charakterisierung von Boden-Huminstoffen nach sukzessiver Lösungsmittel-Extraktion. PhD Thesis, Univ Marburg

*Post B, Hempfling R, Klamberg H, Schulten H-R (1988) Zur Charakterisierung von Boden-Huminstoffen. Fresenius Z Anal Chem 331:273–281

Powels AD, Tom A, Eijkel GB, Boon JJ (1987) Characterization of beach wood lignins and its holocellulose and xylan fractions by pyrolysis-gaschromatography/mass spectrometry. J Anal Appl Pyrolysis 11:417–436

*Prenzel J (1986) A numerical scheme for the calculation of coupled chemical equilibria. Geoderma 38:31–40

Preuss A, Thier HP (1982) Quantitative Analyse natürlicher Dickungsmittel durch Methanolyse und Capillargaschromatographie. Lebensm Unters Forsch 175:93–100

Prickett TA, Naymik TG, Lonnquist CG (1981) A "random-walk" solute transport model for selected groundwater quality evaluations. Dept Ener Nat Res I11 (USA) Bull 65:57–61

Puchelt H (1967) Schwefelisotopenuntersuchungen an österreichischen und deutschen Schwefelquellen. Arch Phys Ther 19/3:357–365

Puchelt H, Nielsen H (1967) Untersuchungen über die Verteilung der Schwefelisotope in den Grubenwässern des Ruhrreviers. Glückauf-Forschungshefte Essen 6:303–310

Raben G (1985) Exkursionsführer Sommer 1985. Ber Forschungszentr Waldökosysteme/Waldsterben 111–114

Raspor B, Nürnberg HW, Valenta P (1981) Voltammetric studies of the Zn(II) chelates with NTA and EDTA and the kinetics of their formation in Lake Ontario. Water Limnol Oceanogr 26:54–66

Rauch HW, White WB (1977) Dissolution kinetics of carbonate rocks, 1. Effects of lithology on dissolution rate. Water Resour Res 13/2:381–394

Reardon EJ, Fritz P (1978) Computer modelling of groundwater ^{13}C and ^{14}C isotope compositions. J Hydrol 36:201–224

Reardon EJ, Allison GB, Fritz P (1979a) Recharge in northern clime calcarous sandy soils: soil water chemical and carbon-14 evolution. Geochim Cosmochim Acta 44:1723–1735
Reardon EJ, Allison GB, Fritz P (1979b) Seasonal chemical and isotopic variations of soil CO_2 at Trout Creek. Ontario. J Hydrol Amsterdam 43:355–371
Reardon EJ, Mozeto AA, Fritz P (1980) Recharge in northern clime calcarous sandy soils: soil water chemical and carbon-14 evolution. Geochim Cosmochim Acta 44:1723–1735
Reddy MM (1977) Crystallization of calcium carbonate in the presence of trace concentrations of phosphorous-containing anions. J Crystal Growth 41:287–295
Reeder RJ (1981) Electron optical investigation of sedimentary dolomites. Contrib Mineral Petrol 76:148–157
*Regenberg W (1987) Hydrogeologie und Hydrochemie im Bereich der Schwetzinger Hardt. PhD Thesis, Univ Heidelberg
*Reifenhäuser C (1989) Charakterisierung und Quantifizierung von Iodspezies in Wässern durch radionalytische und massenspektrometrische Methoden. PhD Thesis, Univ Regensburg
*Reifenhäuser C, Heumann KG (1990) Development of a definitive method for iodine speciation in aquatic systems. Fresenius J Anal Chem 336:559–563
Renger M, Strebel O, Giesel W (1974) Beurteilung bodenkundlicher, kulturtechnischer und hydrologischer Fragen mit Hilfe von klimatischer Wasserbilanz und bodenphysikalischen Kennwerten. Z Kulturtech Flurbereinig Berlin Heidelberg 15:148–166, 206–221, 263–271, 353–366
Renger M, Strebel O, Wessolek G, Duynisveld WHM (1986) Evapotranspiration and groundwater recharge – a case study for different climate, crop patterns, soil properties and groundwater depth conditions. Z Pflanzenernähr Bodenkd 149:371–381
Rengpipat S, Landworthy TA, Zeikus JG (1988) *Halobacteroides acetoethylicus* sp. nov., a new obligately anaerobic halophile isolated from deep subsurface hypersaline environments. Syst Appl Microbiol 11:28–35
Riemer R (1983) Untersuchungen zum Transportverhalten von Bakterien in sandigen und kiesigen Filtermedien. PhD Thesis Univ Kiel, pp 1–161
Rodrigues JD (1983) The Nordbergum effect and characterization of aquitards at the Rio Major mining project, vol 21/2. Ground Water. Worthington, Ohio, pp 200–207
Rössner B, Schwedt G (1984) Analysenverfahren zur Bestimmung anorganischer Anionen – Vergleich von photometrischen Verfahren zur Nitrat – und Phosphatbestimmung. Vom Wasser 62:115–124
Rosswall T, Persson IB (1982) Functional description of bacterial populations from seven Swedish lakes. Limnologica (Berlin) 14:1–16
Saar RA, Weber JH (1980) Comparison of spectrofluorometry and ion-selective electrode potentiometry for determination of complexes between fluvid acid and heavy-metal ions. Anal Chem 52:2095–2100
Saar RA, Weber JH (1982) Fulvic acid: modifier of metal-ion chemistry. Environ Sci Technol 16/9:510A–516A
Saiz-Jimenez C, De Leeuw JW (1986) Chemical characterization of soil organic matter fractions by analytical pyrolysis-gas-chromatography-mass spectrometry. J Anal Appl Pyrolysis 9:99–119
Saiz-Jimenez C, Boon JJ, Hedges J, Hessels JKC, De Leeuw JW (1987) Chemical characterization of recent and buried woods by analytical pyrolysis. Comparison of pyrolysis data with ^{13}C-NMR and wet chemical data. J Anal Appl Pyrolysis 11:437–450
Saleh FY, Ong WA, Chang DY (1989) Structural features of aquatic fulvic acids. Analytical and preparative reversed phase high-performance liquid chromatography separation with photodiode array detection. Anal Chem 61:2792–2800
Salfeld JC (1964) Fraktionierung eines Huminstoffpräparats mit wasserhaltigen organischen Lösungsmitteln. Landbauforsch Völkenrode 14:131–136
Salvamoser J (1982) ^{85}Kr-Grundwasser-Meßmethodik, Modellüberlegungen und Anwendung auf natürliche Grundwassersysteme. PhD Thesis, Univ München
*Salvamoser J (1986) Quantitative separation of admixed young groundwater and surface water with the H-3 Kr-85 method. Int Assoc Hydrol Sci Publ 156:355–363
*Schaefer KW, Usdowski E (1987) Modelle der Kalkauflösung und $^{13}C/^{12}C$-Entwicklung von Karbonatgrundwässern. Z Wasser-Abwasser-Forsch 20:69–81
Scheffer F, Ulrich B (1960) Humus und Humusdüngung, Vol 1. Enke, Stuttgart
Scheffer F, Schachtschabel P (1982) Lehrbuch der Bodenkunde. 11th edn. Enke Stuttgart

Scheffer F, Ziechmann W, Pawelke G (1960) Über die schonende Gewinnung natürlicher Huminstoffe mit Hilfe milder organischer Lösungsmittel. Z Pflanzenernähr Düng Bodenkd 90:56–69

*Schenk D, Matthess G, Dahmke A, Petersen A (1987) Zur Untersuchung der Silikat-Lösungskinetik unter Grundwasser- und Bodenbedingungen- Ein Methodenkonzept-. Meyniana 39:23–39

Schenk D, Petersen A, Matthess G (1989) Acceleration and retardation of silicase weathering due to organic substances. In: Miles DL (ed) Water-rock interaction WR1–6. Proc Rotterdam/Brookfield 1989, pp 605–607

Schlegel HG (1985) Allgemeine Mikrobiologie, 6th edn with contrib of K Schmidt. Thieme, Stuttgart

Schlichting E, Bljme HP (1966) Bodenkundliches Praktikum. Parey, Hamburg, 209 pp

*Schlosser P, Stute M, Dörr H, Sonntag Ch, Münnich KO (1988) Tritium/^{3}He dating of shallow groundwater. Earth Planet Sci Lett 89:353–362

Schlosser P, Stute M, Sonntag C, Münnich KO (1989) Tritiogenic ^{3}He in shallow groundwater. Earth Planet Sci Lett 94:245–256

Schnitzer M (1973) Chemical, spectroscopic and thermal methods for the classification and characterization of humic substances. In: Povoledo D, Golterman HL (eds) Humic substances, their structure and function in the biosphere. Centre of Agriculture Publishing and Documentation, Wageningen, pp 293–308

Schnitzer M (1975) Chemical, spectroscopic and thermal methods for the classification and characterization of humic substances. In: Povoledo D, Golterman HL (eds) Humic substances, their structure and function in the biosphere. Proceedings of an international meeting held at Nieuwersluis 1972, Pudoc, Wageningen, pp 293–310

Schnitzer M, Khan SH (1972) Humic substances in the environment. Marcel Dekker, New York

Schnitzer M, Preston CM (1986) Analysis of humic acids by solution and solid-state carbon-13 NMR. Soil Sci Soc Am J 50:326–332

Schnitzer M, Skinner SIM (1968) Alkali versus acid extraction of soil organic matter. Soil Sci 105:392–396

Schnürer J, Rosswall T (1982) Fluorescein diacetate hydrolysis as a measure of total microbial activity in soil and litter. Appl Environ Microbiol 43:1256–1261

Schoeller H (1962) Les eaux souterraines. Masson, Paris, 642 pp

Schofield CL (1984) Aqueous aluminum survey. ALBIOS 1984, Ann Rep 208, EPRI Proj RP 2365–01

Scholtis A (1986) Transport von anorganischen und organischen Substanzen durch die wasserungesättigte Zone unter Einwirkung unterschiedlicher Modell-Niederschlagswässer. PhD Thesis, Ber Rep Geol-Paläontol Inst Univ Kiel 13:154

Schröter J (1983) Der Einfluß von Textur- und Struktureigenschaften poröser Medien auf die Dispersivität. PhD thesis, Univ Kiel

Schröter J (1984) Mikro- und Makrodispersivität poröser Grundwasserleiter. Meyniana Kiel 36 pp 1–34

Schubert J (1948) The use of ion exchangers for the determination of physical-chemical properties of substances, particularly radiotracers in solution. J Phys Coll Chem 52:340–356

Schulten H-R (1984) Relevance of analytical pyrolysis to biomass conversion. J Anal Appl Pyrolysis 6:251–272

*Schulten H-R (1986) Pyrolyse-Massenspektrometrie zur Charakterisierung von Bodenhuminstoffen und Böden. Mitt Dtsch Bodenkd Ges 45:141–146

*Schulten H-R (1987) Pyrolysis and soft ionization mass spectrometry of aquatic/terrestrial humic substances and soils. J Anal Appl Pyrolysis 12:149–186

Schulten H-R, Beckey HD, Boerboom AJH, Meuzelaar HLC (1973a) Pyrolysis field desorption mass spectrometry of deoxyribonucleic acid. Anal Chem 45:2358–2362

Schulten H-R, Beckey HD, Meuzelaar HLC, Boerboom AJH (1973b) High resolution field ionization mass spectrometry of bacterial pyrolysis products. Anal Chem 45:191–195

*Schulten H-R, Abbt-Braun G, Frimmel FH (1987a) Time resolved pyrolysis field ionization mass spectrometry of humic material isolated from freshwater. Environ Sci Technol 21:349–357

Schulten H-R, Simmleit N, Müller R (1987b) High-temperature, high-sensitivity pyrolysis field ionization mass spectrometry. Anal Chem 59:2903–2908

Schulten H-R, Hempfling R, Zech W (1988) Discriminating horizons in a moder profile by field ionization mass spectrometry and pattern recognition. Geoderma 41:211–222

Schulten H-R, Hempfling R, Haider K, Gröblinghoff FF, Lüdemann H-D, Fründ R (1990) Characterization of cultivation effects on soil organic matter. Z Pflanzenernhähr Bodenkd 153:97–105

Schulz HD (1970) Chemische Vorgänge beim Übergang vom Sickerwasser zum Grundwasser. PhD Thesis, TH Aachen

Schulz HD (1972) Grundwasserneubildung berechnet aus der Chlorid-Bilanz. Geol Mitt Aachen 12:53–60
Schulz HD (1977) Die Grundwasserbeschaffenheit der Geest Schleswig-Holsteins. Bes Mitt Dtsch Gewässerkd Jahrb 40:1–41
*Schulz HD (1988) Labormessung der Sättigungslänge als Maß für die Lösungskinetik von Karbonaten im Grundwasser. Geochim Cosmochim Acta 52/11:2651–2657
Schulz HD, Albertsen M (1981) Bodenfeuchte und Grundwasserneubildung – Statistische Auswertung und Anwendung eines Analogmodells. Meyniana 33:23–39
*Schulz HD, Baumann J (1985) Modellvorstellung zum Vorgang der Kalklösung in natürlichen Sanden. Z Dtsch Geol Ges 136:407–416
Schulz HD, Reardon EJ (1983) A combined mixing cell/analytical model to describe two-dimensional reactive solute transport for unidirectional groundwater flow. Water Resour Res 19:493–502
Schwartz W (1958) Die Bakterien des Schwefelkreislaufes und ihre Lebensbedingungen. Freiberger Forsch-H C44:1–13
Schwarzenbach G (1956) Organische Komplexbildner. In: XV Int Kong pure and applied Chemistry Lissabon, 1956, Birkhäuser, Basel
Schweisfurth R (1968) Untersuchungen über manganoxidierende und -reduzierende Mikroorganismen. Mitt Int Ver Limnol 14:179–186
Schweisfurth R, Schauermann HH (1984) Eisen- und Manganentfernung im Untergrund. DVGW-Schriftenr Wasser 105:241–271
Schwertmann U (1964) Differenzierung der Eisenoxide des Bodens durch Extraktion mit Ammoniumoxalat-Lösung. Z Pflanzenernahr 105:194–202
Seip HM, Müller L, Naas A (1984) Aluminum speciation: comparison of the spectrophotometric analytical methods and observed concentrations in some acidic aquatic systems in southern Norway. Water Air Soil Pollut 23:81–95
Selenka F, Hack A, Heuser H-P (1986) Bakteriologische Vorgänge bei der unterirdischen Aufbereitung von nitratreichem Grundwasser bei Zufuhr von aufbereitetem Abwasser oder Erdgas. DVGW – Schriftenr Wasser 106:117–130
Senesi N, Sipos S (1985) Molecular weight distribution, analytical and spectroscopic characterization of humic fractions sequentially isolated by organic solvents from a brown coal humic acid. Org Geochem 8:157–162
Senesi N, Griffith SM, Schnitzer M, Townsend MG (1977) Binding of Fe^{3+} by humic materials. Geochim Cosmochim Acta 41:969–976
Senesi N, Testini C, Polemio M (1983) Chemical and spectroscopic characterization of soil organic matter fractions isolated by sequential extraction procedure. J Soil Sci 34:801–813
Shimp R, Pfänder FK (1985) Influence of naturally occurring humic acids on biodegradation of monosubstituted phenols by aquatic bacteria. Appl Environ Microbiol 49:402–407
Shotyk W, Sposito G (1988) Fluorescence quenching and aluminium complexation by a chestnut leaflitter extract. Soil Sci Soc Am J 52:1293–1297
Shuman MS, Cromer JJL (1979) Copper association with aquatic fulvic and humic acids. Estimation of conditional formation constants with a titrimetric anodic stripping voltammetry procedure. Environ Sci Technol 13:543–545
Shuman MS, Collins BJ, Fitzgerald PJ, Olson DL (1983) Distribution of stability constants and dissociation rate constants among binding sites on eustuarine copper-organic complexes: rotated disk electrode studies and an affinity spectrum analysis of ion selective electrode and photometric data. In: Christman RF, Gjessing ET (eds) Aquatic and terrestrial humic materials. Ann Arbor Science, Ann Arbor Michigan, pp 349–370
Simmleit N, Schulten H-R (1989) Analytical pyrolysis and environmental research. J Anal Appl Pyrolysis 15:3–28
Sinclair JL, Ghiorse WL (1989) Distribution of aerobic bacteria, protozoa, algae and fungi in deep subsurface sediments. Geomicrobiol J 7:15–31
Sjöberg EL (1978) Kinetics and mechanism of calcite dissolution in aqueous solutions at low temperatures. Stockholm Contrib Geol 32:1–92
Skerman VBD (1967) A guide to the identification of the genera of bacteria, 2nd edn. Williams and Wilkins, Baltimore
Skougstad MW, Scarboro GF (1968) Water sample filtration unit. Environ Sci Technol 2:298–301
Sládecková A (1972) Periphyton as indicator of the ground water quality. Verh Int Verein Limnol 18:1011–1017

Slavek J, Wold J, Pickering WF (1982) Selective extraction of metal ions associated with humic acids. Talanta 29:743–749

Smeerdijk DG van, Boon JJ (1987) Characterization of subfossil *Sphagnum* leaves rootlets of Ericaceae and their peat by pyrolysis-high-resolution gas-chromatography-mass spectrometry. J Anal Appl Pyrolysis 11:377–402

Smith DJ, Atkinson TC (1976) Process, landforms and climate in limestone regions. In: Derbyshire E (ed) Geomorphology and climate. Wiley, New York, pp 367–409

Smykatz-Kloss W (1974) Differential thermal analysis. Application and results in mineralogy. Springer, Berlin Heidelberg New York (Minerals and Rocks, vol 11)

Smykatz-Kloss W (1980) Mg-deficient dolomite from the Djebel Neusa, Tripolitania. Proc 6th Int Conf Thermal Anal Bayreuth, pp 301–306

*Smykatz-Kloss W, Istrate G, Hötzl H, Kössl H, Wohnlich S (1985) Vorkommen und Entstehung von Bassanit, $CaSO_4.1/2H_2O$, im Gipskarstgebiet von Foum Tatahouine, Südtunesien. Chemie der Erde 44:67–77

Sneath PHA, Sokal RR (1973) Numerical taxonomy. The principles and practice of numerical classifications. Freeman, San Francisco

Snelting H (1979) Mini-screens sampling system. Q Rep Nat Inst Water Supply/Netherlands 16:2

Sokal RR, Michener CJ (1958) A statistical method for evaluating systematic relationships. Kansas Univ Sci Bull 38:1409–1438

Sonntag C (1986) The role of tracer data for modelling soil-water flow in the unsaturated zone. In: Mathematical models for interpretation of tracer data in groundwater hydrology. IAEA-TEC-DOC-381. IAEA, Vienna, pp 117–137

Spies RH (1984) Charakterisierung der Sorptionseigenschaften eines Sediments des Roten Wassers durch Trennstufenzahl, Sättigungskapazität und Sorptionskonstante. PhD Thesis, Univ Marburg

Spiteller S (1981) Kapillar-GC-MS von Huminsäureabbauprodukten eines Podsols. Z Pflanzenernähr Bodenkd 144:472–485

Sposito G, Mattigod SV (1980) GEOCHEM: A computer program for the calculation of chemical equilibria in soil solutions and other natural systems. UCR, Dept Soil Environ Sci (University California), Riverside, CA

Staley JT (1968) *Prosthecomicrobium* and *Ancalomicrobium*: new prosthecate freshwater bacteria. J Bacteriol 95:1921–1942

Stainer RY O, Palleroni NJ, Doudoroff M (1966) The aerobic pseudomonads: a taxonomic study. J Gen Microbiol 43:159–271

Stedman DH (1981) Analysis by chemoluminescence. Antek Application News 1/1:1–3

Steelink C (1985) Implications of elemental characteristics of humic substances. In: Aiken GR, McKnight DM, Wershaw RL, MacCarthy P (eds). Humic substances in soil, sediment and water. John Wiley, New York, pp 457–476

Steelink C, Mikita MA, Thorn KA (1983) Magnetic resonance studies of humates and related model compounds. In: Christman RF, Gjessing ET (eds) Aquatic and terrestrial humic materials. Ann Arbor Science, Ann Arbor, pp 83–106

Steelink C, Wershaw RL, Thorn KA, Wilson MA (1989) Application of liquid-state NMR spectroscopy to humic substances. In: Hayes MHB, MacCarthy P, Malcolm RL, Swift RS (eds) Humic substances II. In search of structure. John Wiley, Chichester, pp 281–308

Sterritt RM, Lester JN (1984) Comparison of methods for the determination of conditional stability constants of heavy metal-fulvic acid complexes. Water Res 18:1149–1153

Stetzenbach LD, Kelley LM, Sinclair NA (1986) Isolation, identification and growth of well-water bacteria. Ground Water 24:6–10

Stevenson FJ (1982) Humus chemistry: Genesis, composition, reactions. John Wiley, New York

Stevenson FJ, Butler JHA (1969) Chemistry of humic acids and related pigments. In: Englinton G, Murphy MTJ (eds) Organic geochemistry. Springer, Berlin, pp 534–557

Stevenson FJ, Goh KM (1971) Infrared spectra of humic acids and related substances. Geochim Cosmochim Acta 35:471–483

Stobbe PC, Wright JR (1959) Modern concepts of the genesis of podzols. Soil Sci Soc Proc 23:161–164

Strasburger E (1978) Lehrbuch der Botanik, 31st edn. G. Fischer, Stuttgart, pp 331

Straub H (1988) Probenahme von CKW-haltigen Sickerwässern mittels Saugkerzen. Z Wasser-Abwasser-Forsch 21:155–157

*Strebel O, Böttcher J (1985) Einfluß von Bodennutzung und Bodennutzungsänderungen auf die Stoffbilanz eines reduzierenden Aquifers im Einzugsgebiet eines Förderbrunnens. Wasser Boden 37:111–114

*Strebel O, Böttcher J (1987) Ermittlung von Ursachen und Prognosen der mittleren Nitratkonzentration im Grundwasser von Wassereinzugsgebieten. Mitt Dtsch Bodenkd Ges 53:305–309

*Strebel O, Böttcher J (1989) Solute input into groundwater from sandy soils under arable land and coniferous forest: determination of area-representative mean values of concentration. Agric Water Manage 15 (3):265–278

*Strebel O, Böttcher J, Renger M (1984) Einfluß von Boden und Bodennutzung auf die Stoffanlieferung an das Grundwasser. In: Udluft P, Merkel B, Prösl K-H (eds) Proc RIZA-Symp, Dept Hydrogeology Hydrochemistry Techn Univ München, pp 663–669

*Strebel O, Böttcher J, Kölle W (1985) Stoffbilanzen im Grundwasser eines Einzugsgebietes als Hilfsmittel bei Klärung und Prognose von Grundwasserqualitätsproblemen (Beispiel Fuhrberger Feld). Z Dtsch Geol Ges 136:533–541

*Strebel O, Böttcher J, Eberle M, Aldag R (1988) Quantitative und qualitative Veränderungen im A-Horizont von Sandböden nach Umwandlung von Dauergrünland in Ackerland. Z Pflanzenernähr Bodenkd 151:341–347

*Strebel O, Böttcher J, Fritz P (1990) Use of isotope fractionation of sulfate-sulfur and sulfate-oxygen to assess bacterial desulfurication in a sandy aquifer. J Hydrol 121:155–172

Stumm W, Baccini P (1978) Man-made chemical perturbation of lakes. In: Lerman A (ed) Lakes: chemistry, geology, physics. Springer, Berlin Heidelberg New York, pp 91–126

Stumm W, Morgan JJ (1970) Aquatic chemistry. Wiley Interscience, New York, 583 pp

Stumm W, Morgan JJ (1981) Aquatic chemistry 2nd edn. Wiley Interscience, New York, 780 pp

Stumm W, Furrer G, Wieland E, Zinder B (1985) The effects of complex-forming ligands on the dissolution of oxides and alumosilicates. In: Drever JI (ed) The chemistry of weathering. -NATO ASI Series C: Mathematical and physical sciences. 149:55–74

Suarez DL (1986) A soil water extractor that minimizes CO_2 degassing and pH errors. Water Resour Res 22:876–880

Susic M, Boto KG (1989) High-performance liquid chromatographic determination of humic acid in environmental samples at the nanogram level using fluorescence detection. J Chromatogr 482:175–187

Swett K (1968) Authigenic feldspars and cherts resulting from dolomitization of illitic limestones: a hypothesis. J Sediment Petrol 38:128–135

Swift RS (1985) Fractionation of soil humic substances. In: Aiken GR, McKnight DM, Wershaw RL, MacCarthy P (eds). Humic substances in soil, sediment and water. John Wiley, New York

Sykes JF, Soyapak S, Farquhar GJ (1982) Modelling of leachate organic migration and attenuation in groundwaters below sanitary landfills. Water Resour Res 18:135–145

*Sztuka F (1985) Zur Genese oberflächennaher Wässer im Hamburger Raum – Untersuchungen in der ungesättigten Zone am Beispiel ausgewählter Profiltypen. PhD Thesis, Univ Hamburg

Talsma T, Hallam PM, Mansell RS (1979) Evaluation of porous cup soil–water extractors: physical factors. Aust J Soil Res 17:417–422

Tamers MA (1967) Surface-water infiltration and groundwater movement in arid zones of Venezuela. Isotopes in hydrology, IAEA, Vienna pp 339–351

Tamers MA, Scharpenseel HW (1970) Sequential sampling of radiocarbon in groundwater. Isotopes in hydrology. IAEA, Vienna, pp 241–256

Taylor CB, Roether W (1982) A uniform scale for reporting low-level tritium measurements in water. Int J Appl Radiat Isot 33:377–382

Tekeli O (1980) Structural evolution of the Aladag Mountains in the Taurus Belt. Bull Geol Soc Turk 23:11–14

Thauer RK, Jungermann K, Decker K (1977) Energy conservation in chemotrophic anaerobic bacteria. Bacteriol Rev 41:100–180

*Thielen H, Bodes H, Lorenz I, Klamberg H (1984) Wechselwirkungen zwischen Huminstoffen und Metall-Ionen in wäßrigen Systemen. In: Udluft P, Merkel B, Prösl K-H (eds) Proc RIZA-Symp Dept Hydrogeology Hydrochemistry Techn Univ München 2:579–589

Thoma G, Esser N, Sonntag C, Weiss W, Rudolph J, Leveque P (1979) New technique of in-situ soil-moisture sampling for environmental isotope analysis applied at Pilat sand dune near Bordeaux. HETP modelling of bomb tritium propagation in the unsaturated and saturated zones. In: Isotope hydrology 1978, vol 2. IAEA, Vienna, pp 753–766

Thorn PM, Ventullo RM (1988) Measurement of bacterial growth rates in subsurface sediments using the incorporation of tritiated thymidine into DNA. Microb Ecol 16:3–16

Thorstenson DC, Plummer LN (1977) Equilibrium criteria for two - component solids reacting with fixed composition in an aqueous phase - example: the magnesian calcites. Am J Sci 277:1203–1223

Thurman EM (1986) Organic geochemistry of natural waters. Martinus Nijhoff/Dr W Junk, Dordrecht

Thurman EM, Malcolm RL (1981) Preparative isolation of aquatic humic substances. Environ Sci Technol 15:463–466

Tikhomirov FA, Kasparov SV, Prister BS, Kurbatov VM (1978) Study of the interaction of iodine-131 with humic acids by gel chromatography. Pochvovedeniye 12:52–59

Toran L (1987) Sulfate contamination in groundwater from a carbonate-hosted mine. J Contaminant Hydrol 2:1–29

Tranvic LJ, Höfle MG (1987) Bacterial growth in mixed cultures on dissolved organic carbon from humic and clear waters. Appl Environ Microbiol 53:482–488

Trinkwasserverordnung (1986) Verordnung über Trinkwasser und über Wasser für Lebensmittelbetriebe (Trinkwasserverordnung - TrinwV) vom 22. Mai 1986. BGBL 7 vom 28. Mai 1986, Bonn, Teil I:760–773

Trudgill ST, High CJ, Hanna FK (1981) Improvements to the micro-erosion meter. Br Cave Res Group Tech Bull 29:3–17

Truesdell AH, Jones BF (1974) WATEQ, a computer program for calculating chemical equilibria of natural waters. J Res US Geol Surv 2/2:233–248

Trypton/Glukose/Fleischextrakt-Agar (TGE) (1971) Stand methods, 19th edn. Am Publ Health Assoc New York

Tuschall JR, Brezonik PL (1981) Evaluation of the copper anodic stripping voltammetry complexometric titration for complexing capacities and conditional stability constants. Anal Chem 53:1986–1989

*Ullsperger I, Alexander I, Freitag G, Grossmann J, Merkel B (1988a) Mikrobiologische Untersuchungen in ungesättigten quartären Kiesen und ihrer Bodenauflage. Z Dtsch Geol Ges 139:365–375

*Ullsperger I, Merkel B, Quentin KE, Udluft P (1988b) Hydrogeologische und hydrochemische Untersuchungen in der ungesättigten Zone eines Kiesgrundwasserleiters. Teil II: Ergebnisse langjähriger bodenhydrologischer Messungen. Z Wasser-Abwasser-Forsch 21:56–66

Ulrich B (1981a) Ökologische Gruppierung von Böden nach ihrem chemischen Bodenzustand. Z Pflanzenernähr Bodenkd 144:289–305

Ulrich B (1981b) Theoretische Betrachtung des Ionenkreislaufs in Waldösystemen. Z Pflanzenernähr Bodenkd 144:647–659

Ulrich B (1984) Deposition von Säure und Schwermetallen aus Luftverunreinigungen und ihre Auswirkungen in Waldökosystemen. In: Merian E (ed): Metalle in der Umwelt. Verl Chemie, Weinheim, pp 163–170

Ulrich B (1985) Natürliche und anthropogene Komponenten der Bodenversauerung. Mitt Dtsch Bodenkd Ges 1–31

Ulrich B (1989) Effects of acid deposition on forest ecosystems in Europe. In: Adriano DC, Johnson AH (eds) Acidic precipitation, vol 2. Biological and ecological effects. Springer, Berlin Heidelberg New York, pp 189–212

Ulrich B, Khanna PK (1971) Methodische Untersuchungen über Kationengehalt der Bodenlösung und Schofield'sche Potentiale. Gött Bodenkund Ber 19:121–148

Ulrich B, Matzner E (1983) Abiotische Folgewirkung der weiträumigen Ausbreitung von Luftverunreinigungen. Luftreinhaltung Forschungsber 1983 UBA 10402615:1–119

Unterweger MP, Coursey BM, Schima FJ, Mann WB (1980) Preparation and calibration of the 1987 NBS tritiated-water standards. Int J Appl Radiat Isot 31:611–614

Usdowski E (1982) Reactions and equilibria in the systems CO_2-H_2O and $CaCO_3$-CO_2-H_2O. A review. Neues Jahrb Mineral Abh 144:148–171

Usdowski E, Hoefs J, Menschel G (1979) Relationship between ^{13}C and ^{18}O fractionation and changes in major element composition in a recent calcite depositing spring. A model of geochemical variations with inorganic $CaCO_3$ precipitation. Earth Planet Sci Lett 42:267–276

Usdowski HE (1967) Die Genese von Dolomit in Sedimenten. Springer, Berlin Heidelberg New York

Valeton I, Khoo F (1980) Petrography and geochemistry of the basal tills of the Saale glaciation in north-western Germany. Meded Rijks Geol Dienst 34:30–44

Vartiainen T, Liimatainen A, Kauranen P (1987) The use of TSK size exclusion columns in determination of the quality and quantity of humus in raw waters and drinking waters. Sci Total Environ 62:75–84

Vauclin M, Haverkamp R, Vachaud G (1979) Etude de la resolution numérique de l'équation d'eau en milieu non saturé. Univ Sci Med Grenoble Rapp HL–78–1

Ventullo RM, Larson RJ (1985) Metabolic diversity and activity of heterotrophic bacteria in ground water. Environ Toxicol Chem 4:759–771

Ventullo RM, Ladd TI, Costerton JW (1983) Distribution and activity of the free-living and attached bacteria in ground water. Proceedings, Environmental Chemistry Division of the American Chemical Society, Washington, DC. August 1983, pp 282–285

Verruijt A (1969) Elastic storage of aquifers. In: De Wiest RJM (ed) Flow through porous media. Academic Press, New York, pp 331–376

Voerkelius S (1989) Isotopendiskriminierungen bei Nitrifikation und Denitrifikation, Grundlagen und Auswirkungen in der Zuordnung von NO_3^- und N_2O. PhD Thesis, TU München

Vogel JC (1970) ^{14}C dating of groundwater. In: Isotope hydrology 1970. IAEA, Vienna, pp 225–240

Vogel JC, Grootes PM, Mook WG (1970) Isotope fractionation between gaseous and dissolved carbon dioxide. Z Phys 230:225–238

Völtz H (1979) Das System hydrogeologischer und hydrochemischer Karten im Norden und Nordwesten der Niederrheinischen Bucht. Mitt Ing Hydrogeol 10:74 pp

Vortisch W (1982) Clay mineralogical studies of some tills in northern Germany. Geol Palaeontol 15:167–192

Wahlen M, Yoshinari T (1985) Oxygen isotope ratios in N_2O from different environments. Nature 313:780–782

Waid JS (1973) A method to study microorganisms on surface films from soil particles with the aid of the transmission electron microscope. Bull Ecol Res Comm (Stockholm) 17:103–108

Wallis PM, Ladd TI (1983) Organic biogeochemistry of ground water at a mountain coal mine. Geomicrobiol J 3:49–78

Watson SW, Novitsky TJ, Quinby HL, Valois FW (1977) Determination of bacterial number and biomass in the marine environment. Appl Environ Microbiol 33:940–946

Weber JH (1988) Binding and transport of metals by humic materials. In: Frimmel FH, Christman RF (eds) Humic substances and their role in the environment. John Wiley, Chichester

Wedepohl KH (1969) Handbook of geochemistry, Vol. 1. Springer, Berlin Heidelberg New York, 442 pp

Weis M, Abbt-Braun G, Frimmel FH (1989) Humic-like substances from landfill leachates – characterization and comparison with terrestrial and aquatic humic substances. Sci Total Environ 81/82:343–352

Weise S (1986) Heliumisotopen-Gehalte im Grundwasser – Messung und Interpretation. PhD Thesis, Univ München

Weiss W, Bullacher J, Roether W (1979) Evidence of pulsed discharges of tritium from nuclear energy installations in central European precipitation. In: Behaviour of tritium in the environment. IAEA-SM-232/18, Vienna, pp 17–30

Welch MJ, Lifton JF, Seck JA (1969) Tracer studies with radioactive oxygen-15. Exchange between carbon dioxide and water. J Phys Chem 73:3351–3356

Wenzel A, Schweisfurth R (1979) Quantitative Bestimmung von eisenpräzipitierenden Bakterien aus Boden und Wasser unter Verwendung verschiedener Fe(II) und Fe(III) – Komplexe im Medium. Mitt Dtsch Bodenkd Ges 29:493–506

Werner P (1987) Die Bedeutung biologischer Prozesse bei der Wasseraufbereitung durch Filtration. Wasser-Kalender 1987:92–116

Wershaw RL (1985) Application of NMR spectroscopy for determining functionality in humic substances. In: Aiken GR, McKnight DM, Wershaw RL, MacCarthy P (eds) Humic substances in soil, sediment and water. Geochemistry, isolation, and characterization. John Wiley, New York, pp 561–582

Wershaw RL, Bohner GE (1969) Pyrolysis of humic and fulvic acids. Geochim cosmochim Acta 33:757–762

Wessolek G, Renger M, Strebel O, Sponagel H (1985) Einfluß des Grundwasserflurabstandes auf die Grundwasserneubildung unter Acker, Grünland und Nadelwald. Z Kulturtechn Flurbereinig 26:130–137

Westrich JT, Berner RA (1984) The role of sedimentary organic matter in bacterial sulphate reduction: the G-model tested. Limnol Oceanogr 29:236–249

Weyandt R (1990) Verteilung von Mikroorganismen und bodenenzymatischen Aktivitäten in der ungesättigten und gesättigten Bodenzone eines Waldsystems. PhD Thesis, Univ Saarland, Homburg (in press)

*Weyandt R, Schweisfurth R (1989) Verteilung von Mikroorganismen und bodenenzymatischen Aktivitäten in einem Tiefenprofil vom Oberboden bis in die gesättigte Zone. Z Dtsch Geol Ges 139:331–334

White DC, Fredrickson HL, Gehron MJ, Smith GA, Martz RF (1983) The groundwater aquifer microbiota: biomass, community structure, and nutritional status. Dev Ind Microbiol 24:189–199

White WB (1977) Conceptual models for carbonate aquifers: revisited. In: Dilamarter RR, Csallany SC (eds) Hydrologic problems in karst regions. Eastern Kentucky University, Bowling Green, pp 176–187

Whitehead DC, Tinsley J (1964) Extraction of soil organic matter with dimethylformanide. Soil Sci 97:34–42

Wieringa KT (1966) Solid media with elemental sulphur for detection of sulphur-oxidizing microbes. Antonie Leeuwenhoek J Microbiol 32:183–186

Wigley TML (1975) Carbon 14 dating of groundwater from closed and open system. Water Resour Res 11:324–328

Wigley TML (1977) WASPEC – a computer program for determining the equilibrium specification of aqueous solutions. Br Geomorphol Res Group Tech bull Norwich, 20:48

Wigley TML, Plummer LN (1976) Mixing of carbonate waters. Geochim Cosmochim Acta 40:989–995

Wigley TML, Plummer LN, Pearson FJ (1978) Mass transfer and carbon isotope evolution in natural water systems. Geochim Cosmochim Acta 42:1117–1139

Wilson JT, McNabb JF, Balkwill DL, Ghiorse WC (1983) Enumeration and characterization of bacteria indigenous to a shallow watertable aquifer. Ground Water 21:134–142

Wilson MA (1981) Application of nuclear magnetic resonance spectroscopy to the study of the structure of soil organic matter. J Soil Sci 32:167–186

Wilson MA (1989) Solid-state nuclear magnetic resonance spectroscopy of humic substances: basic concepts and techniques. In: Hayes MHB, MacCarthy P, Malcolm RL, Swift RS (eds). Humic substances II. In search of structure. John Wiley, Chichester, pp 309–338

Wilson MA, Barren PF, Gillans AH (1981) The structure of freshwater humic substances as revealed by 13C-NMR spectroscopy. Geochim Cosmochim Acta 45:1743–1750

Wilson MA, Philip RP, Gillam AH, Gilbert TD, Tate KR (1983) Comparison of the structure of humic substances from aquatic and terrestrial sources by pyrolysis-gas chromatography-mass spectrometry. Geochim Cosmochim Acta 47:497–502

Wilson MA, Vasallo AM, Perdue EM, Reuter JH (1987) Compositional and solid-state NMR study of humic and fulvic acid fractions of soil organic matter. Anal Chem 59:551–558

Wippern J (1964) Die Bauxite des Taurus und ihre tektonische Stellung (in Turkish). MTA Bull 58:47–70

Witzel KP, Overbeck HJ, Moaledj K (1982) Microbial communities in Lake Plußsee – an analysis with numerical taxonomy of isolates. Arch Hydrobiol 94:38–52

Wollast R (1967) Kinetics of the alteration of K-feldspar in buffered solutions at low temperatures. Geochim Cosmochim Acta 31:635–648

Wolters N, Schwartz W (1956) Untersuchungen über Vorkommen und Verhalten von Mikroorganismen in reinen Grundwässern. Arch Hydrobiol 51/4:500–541

Yoshida N (1984) Nitrogen isotope studies on the geochemical cycle of nitrous oxide. PhD Thesis, Department of Chemistry, Tokyo Institute of Technology

Yoshinari T, Wahlen M (1985) Oxygen isotope ratios in N_2O from nitrification at a wastewater treatment facility. Nature 317:349–350

Ziechmann W (1980) Huminstoffe. Verlag Chemie, Weinheim

Zimmermann R, Iturriaga R, Becker-Birch J (1978) Simultaneous determination of the total number of aquatic bacteria and the number thereof involved in respiration. Appl Environ Microbiol 36:926–935

ZoBell CE (1943) The effect of solid surfaces upon bacterial activity. J Bacteriol 46:39–56

Subject Index

D

Q

R

MIX
Papier aus verantwortungsvollen Quellen
Paper from responsible sources
FSC® C105338

If you have any concerns about our products, you can contact us on
ProductSafety@springernature.com

In case Publisher is established outside the EU, the EU authorized representative is:
Springer Nature Customer Service Center GmbH
Europaplatz 3, 69115 Heidelberg, Germany

Printed by Libri Plureos GmbH
in Hamburg, Germany